COASTAL ENVIRONMENTS AND GLOBAL CHANGE

Coastal Environments and Global Change

Edited by

Gerd Masselink

School of Marine Science and Engineering
Plymouth University
Plymouth
UK

and

Roland Gehrels

Environment Department
University of York
York
UK

This work is a co-publication between the American Geophysical
Union and Wiley

This edition first published 2014 © 2014 by John Wiley & Sons, Ltd

Registered Office
John Wiley & Sons, Ltd, The Atrium, Southern Gate, Chichester, West Sussex, PO19 8SQ, UK

Editorial Offices
9600 Garsington Road, Oxford, OX4 2DQ, UK
The Atrium, Southern Gate, Chichester, West Sussex, PO19 8SQ, UK
111 River Street, Hoboken, NJ 07030–5774, USA

For details of our global editorial offices, for customer services and for information about how to apply for permission to reuse the copyright material in this book please see our website at www.wiley.com/wiley-blackwell.

Library of Congress Cataloging-in-Publication Data

Coastal environments and global change / edited by Gerhard Masselink & Roland Gehrels.
      pages   cm
   Includes bibliographical references and index.
   ISBN 978-0-470-65660-0 (cloth) – ISBN 978-0-470-65659-4 (pbk.)   1. Coast changes.   2. Coastal ecology.   3. Environmental degradation.   4. Global warming.   5. Coastal zone management.   6. Water levels.   I. Masselink, Gerhard.   II. Gehrels, W. Roland.
   QC903.C585 2014
   551.45′7–dc23
                                  2013046580
A catalogue record for this book is available from the British Library.

Wiley also publishes its books in a variety of electronic formats. Some content that appears in print may not be available in electronic books.

Cover image: Cusps on the beach at Man O'War Cove, near Lulworth in Dorset, southern England. Photo: Roland Gehrels.
Cover design by Design Deluxe

Set in 9/11.5pt Trump Mediaeval by SPi Publisher Services, Pondicherry, India
Printed in Singapore by Ho Printing Singapore Pte Ltd

1   2014

Contents

Contributors

WILLIAM P. ANDERSON, JR. *Department of Geology, Appalachian State University, Boone, NC, USA*

EDWARD J. ANTHONY *Aix Marseille Université, Institut Universitaire de France, Europôle Méditerranéen de l'Arbois, Aix en Provence Cedex, France*

DANIEL C. CONLEY *School Marine Science and Engineering, University of Plymouth, Plymouth, UK*

DUNCAN FITZGERALD *Department of Earth and Environment, Boston University, Boston, MA, USA*

ROLAND GEHRELS *Environment Department, University of York, York, UK*

IOANNIS GEORGIOU *Department of Earth and Environmental Sciences, University of New Orleans, New Orleans, LA, USA*

SYTZE VAN HETEREN *Geological Survey of the Netherlands, Utrecht, The Netherlands*

PAUL KENCH *School of Environment, The University of Auckland, Auckland, New Zealand*

AART KROON *Center for Permafrost (CENPERM), Department of Geosciences and Natural Resource Management, University of Copenhagen, Copenhagen, Denmark*

CATHERINE E. LOVELOCK *The School of Biological Sciences, The University of Queensland, St Lucia, QLD, Australia*

GERD MASSELINK *School of Marine Science and Engineering, Plymouth University, Plymouth, UK*

GLENN A. MILNE *Department of Earth Sciences, University of Ottawa, Ottawa, Canada*

MICHAEL MINER *Marine Minerals Program, Bureau of Ocean Energy Management, Gulf of Mexico Region, New Orleans, LA, USA*

ROBERT J. NICHOLLS *Faculty of Engineering and the Environment and Tyndall for Climate Change Research, University of Southampton, Southampton, UK*

KARL F. NORDSTROM *Institute of Marine and Coastal Sciences, Rutgers – the State University of New Jersey, New Brunswick, NJ, USA*

ROSHANKA RANASINGHE *Department of Water Science Engineering, UNESCO-IHE, Delft, The Netherlands*

KERRYLEE ROGERS *School of Earth and Environmental Sciences, University of Wollongong, Wollongong, NSW, Australia*

GERBEN RUESSINK *Department of Physical Geography, Faculty of Geosciences, Institute for Marine and Atmospheric Research Utrecht, Utrecht University, Utrecht, The Netherlands*

WAYNE STEPHENSON *Department of Geography, University of Otago, Dunedin, New Zealand*

MARCEL J.F. STIVE *Faculty of Civil Engineering and Geosciences, Delft University of Technology, Delft, The Netherlands*

ADAM D. SWITZER *Earth Observatory of Singapore, Nanyang Technological University, Nanyang Avenue, Singapore*

RICHARD S.J. TOL *School of Economics, University of Sussex, Brighton, UK*

COLIN D. WOODROFFE *School of Earth and Environmental Sciences, University of Wollongong, Wollongong, NSW, Australia*

About the Companion Website

This book is accompanied by a companion website:

www.wiley.com/go/masselink/coastal

The website includes:

- Powerpoints of all figures from the book for downloading
- PDFs of tables from the book

1 Introduction to Coastal Environments and Global Change

GERD MASSELINK[1] AND ROLAND GEHRELS[2]

[1] School of Marine Science and Engineering, Plymouth University, Plymouth, UK
[2] Environment Department, University of York, York, UK

1.1 Setting the scene

1.1.1 What is the coastal zone?

At the outset of this book, it is important to articulate clearly what we mean by 'coast', because the term means different things to different people. For most holidaymakers, the coast is synonymous with the beach. For birdwatchers, the coast generally refers to the intertidal zone; while for cartographers, the coast is simply a line on the map separating the land from the sea. Coastal scientists and managers tend to take a broader view.

According to our perspective, the coast represents that region of the Earth's surface that has been affected by coastal processes, i.e. waves and tides, during the Quaternary geological period (the last 2.6 M years). The coastal zone thus defined includes the coastal plain, the contemporary estuarine, dune and beach area, the shoreface (the underwater part of the beach), and part of the continental shelf and, in areas of isostatic or tectonic uplift, fossil raised shorelines (Fig. 1.1). At a first glance, it seems rather arbitrary and perhaps odd to take such a long-term view of the timescale involved with coastal processes and geomorphology. However, as we will see later (Chapter 2), the Quaternary was a period characterized by significant changes in sea level. In the past, eustatic, or global, sea level has been considerably lower than at present (>100 m) during cold glacial periods, but also somewhat higher (up to 10 m) during some of the warm interglacial periods. This implies that coastal sediments and landforms have the potential to extend considerably beyond the zone of contemporary coastal processes. In areas of former glaciations, where isostatic processes have caused crustal uplift, fossil coastal landforms can be found far above the present shoreline (Fig. 1.2a). Similarly, in tectonically active coastal areas, fossil shorelines can also be significantly displaced (Fig. 1.2b). In a lateral sense our definition means that the coastal zone can span hundreds of kilometres, especially

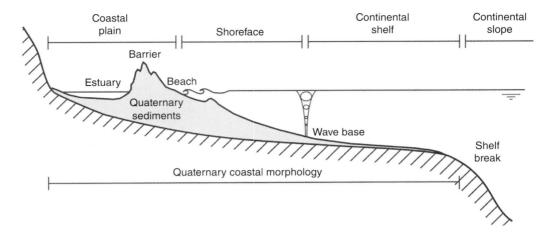

Fig. 1.1 Spatial extent of the coastal zone, including the coastal plain, shoreface and continental shelf. Note that the widths of these zones are globally highly variable. (Source: Masselink et al. 2011. Reproduced with permission of Hodder & Stoughton Ltd.)

Fig. 1.2 (a) Postglacial raised beaches at Porsangerfjord, Finnmark, Norway; (b) fossil coastal notch in Barbados formed in the last interglacial (c. 125,000 years ago) and raised above sea level by tectonic processes; and (c) view from Prawle Point (south Devon, UK) looking east, showing an apron of periglacial solifluction deposits emplaced on a raised shore platform presumed to date to the last interglacial. The fossil interglacial sea cliff is also visible. (Source: Photographs by Roland Gehrels.)

(a)

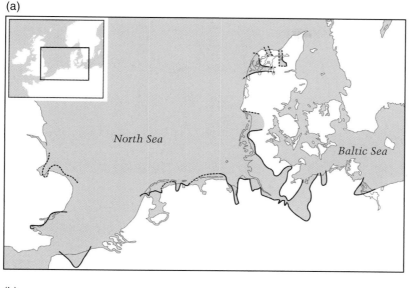

(b)

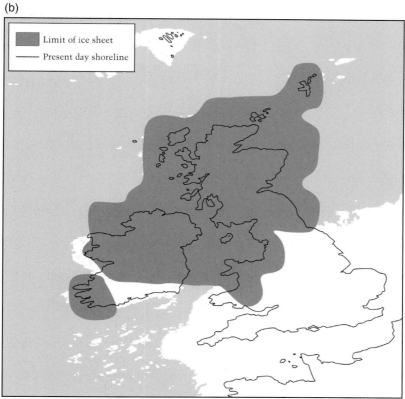

Fig. 1.3 (a) Coastline around the North Sea during the last interglacial, around 125,000 years ago (Source: Adapted from Streif 2004. Reproduced with permission of Elsevier); and (b) land area (in white) around the British Isles during the Late Glacial Maximum, around 20,000 years ago (Source: Adapted from Brooks et al. 2011).

in areas with broad continental shelves and shallow seas. For example, Fig. 1.3a shows the position of the coastline in northwest Europe during the last interglacial when sea level was several metres higher than today. During the Last Glacial Maximum the shoreline was close to the present-day continental shelf edge (Fig. 1.3b). Because coastal evolution is cumulative, i.e. the contemporary coastal landscape is partly a product of coastal processes and landforms in the past (Cowell and Thom, 1994), we need to take this long-term perspective.

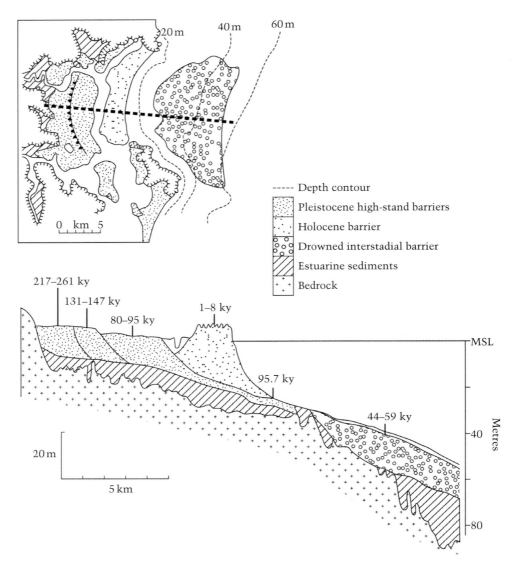

Fig. 1.4 Coastal morphology of the Tuncurry embayment, New South Wales, Australia, showing the presence of five barrier systems: the contemporary barrier, a drowned barrier on the inner shelf, and three high-stand barriers. Each of these barriers is of a different age and formed at a different relative sea level. MSL, mean sea level. (Source: Adapted from Roy et al. 1994. Reproduced with permission from Cambridge University Press and Masselink et al. 2011.)

Figure 1.4 shows an interpretive map and cross-section of the Tuncurry embayment in New South Wales, Australia. Here, research has demonstrated the presence of at least five coastal barrier systems of various ages (see Chapter 8), each of which is associated with a different sea level (Roy et al., 1994). In addition to the contemporary barrier system, there are three so-called highstand barriers to the landward (ages c. 240ky, 140ky and 90ky BP) and one drowned barrier system to the seaward on the continental shelf (age c. 50ky BP). To understand fully the dynamics of the present barrier system, in addition to contemporary coastal processes and sea level, the evolution and configuration of

these older barriers also have to be taken into account. For example, the drowned barrier system can supply (and probably has supplied) sediment to the contemporary barrier, whereas the highstand barriers have provided the substrate on which the present-day barrier has developed.

Figure 1.2c shows a scenic view from Prawle Point in Devon, UK. At this location, periglacial solifluction deposits (locally known as 'head') were emplaced during the last glacial period on a raised shore platform that formed during the preceding interglacial when sea level was several metres higher than present. The 'head' is an important sediment source for contemporary beaches, while rocky shore platforms are re-occupied during consecutive interglacial highstands. So here also, present-day coastal geomorphology is significantly affected by past coastal processes and landforms. In fact, erosional coastal features, especially when carved into resistant rocks, are often polygenetic (i.e. the product of more than one sea level) and rocky coast morphology can rarely be explained solely in terms of contemporary processes and sea level (Trenhaile, 2010).

1.1.2 Coastal zone and society

The coastal zone, representing the interface between the land and the sea, is of interest to a range of coastal scientists, including geographers, geologists, oceanographers and engineers. Societal concern and interest are, however, concentrated on that area in which human activities are interlinked with both the land and the sea. This area of overlap is referred to as the 'coastal resource system' and is of great societal importance, often serving as the source or backbone of the economy of coastal nations. The most obvious use of the coastal zone is providing living space, and the coast is clearly a preferred site for urbanization. For example, 23% of the global population currently live within 100 km of the coast and less than 100 m above sea level. Population density in coastal areas is three times larger than average, and projected population growth rates in the coastal zone are the highest in the world (Small and Nicholls, 2003). In addition, 21 of the 33 megacities (cities with more than eight million people; the projected top five for 2015 are Tokyo, Mumbai, Lagos, Dhaka and Karachi) can be considered coastal cities (Martinez et al., 2007). It is worth pointing out, however, that the dynamic definition of the coastal zone at the start of this section (based on sediments, sea-level history and coastal processes) is different from the static definition generally used by planners and demographers, based on some arbitrary distance from the coastline and/or elevation above sea level.

Human occupation is, however, but one of many uses of the coastal resource system and an extraordinarily wide range of resources and activities essential to our society take place in the coastal zone, including navigation and communication, living marine resources, mineral and energy resources, tourism and recreation, coastal infrastructure development, waste disposal and pollution, coastal environmental quality protection, beach and shoreline management, military activities and research (Cicin-Sain and Knecht, 1998). Unfortunately, there can be fierce competition for coastal resources by various users (or stakeholders) and these may result in conflicts, and possible severe disruption, or even destruction, of the functional integrity of the coastal resource system. Such conflicts are especially prevalent in the case of incompatible uses of the coastal zone (e.g. land reclamation versus nature conservation; coastal protection versus tourism; waste disposal versus fisheries).

The dramatic growth in coastal population and uses has placed increased pressure on the coastal resource system and has led, in many cases, to severely damaged coastal ecosystems and depleted resources. In addition, overdevelopment of the coast in terms of urbanization and infrastructure has significantly increased our vulnerability to coastal erosion and flooding, whilst at the same time the increased reliance on hard coastal engineering structures for coastal protection has reduced our resilience. To make matters worse, global climate change resulting in a rise in sea level and potentially an increase in storminess (or at least a change in wave climate) will provide additional pressure on the coastal zone. An integrated approach is required for the management of activities and conflicts in the coastal zone (Integrated Coastal Zone Management, ICZM; see section 7.4 and Chapter 17), but what is also essential, is a thorough understanding of the key processes driving and controlling coastal environments.

1.1.3 Scope of this book and chapter outline

The focus of this book, therefore, is to provide a description of the various coastal environments, including their functioning and governing processes, and also to evaluate how they might be affected by global change and how coastal management may assist in dealing with coastal problems arising from climate change. To provide the theoretical framework and the scope of this book, this chapter will first discuss the dominant paradigm for coastal research ('morphodynamics'). This is followed by a summary of the dominant elements of climate change relevant to the coastal zone and finally a description of the various approaches used for modelling coastal change.

1.2 Coastal morphodynamics

1.2.1 Research paradigm

In science, the term 'paradigm' refers to the 'set of practices that defines a scientific discipline at any particular period of time' (Kuhn, 1996). It relates to the overall research approach adhered to by the majority of the researchers in a certain scientific discipline and encompasses a large number of elements, including methods of observation and analysis, the types of questions asked and the topics studied, the theoretical framework of the discipline, and even mundane issues such as the key scientific journal(s) of the discipline. In the vernacular, it can simply be translated as the most common way to study a subject or, even, the way a subject should be studied ('exemplar'). As a

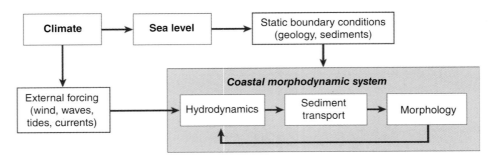

Fig. 1.5 Conceptual diagram illustrating the morphodynamic approach, showing the coastal morphodynamic systems and the environmental boundary conditions (sea level, climate, external forcing and static boundary conditions). (Source: Masselink 2012. Reproduced with permission from Pearson Education Ltd.)

discipline evolves over time, it is imperative that our knowledge and understanding thereof increases, concurrent with an increased sophistication of the research tools and analysis methods. As this happens, the relevant questions and methods of addressing these are likely to change as well; in other words, the paradigm changes. Thomas Kuhn (1922–1996), a leading philosopher of science, argued that science progresses by means of abrupt paradigm shifts, generally initiated by key scientific discoveries and/or novel research tools shedding new light on hitherto unobservable phenomena.

The dominant paradigm in coastal research up to World War II was observation and classification of coastal landforms, mainly in the context of geology and sea-level change, with coastal scientists primarily being concerned with describing and mapping the coast. During the 1950s and 1960s, the emphasis changed from observation to explanation, and this required a better understanding of the actual processes involved in driving and controlling coastal landforms and evolution. This development occurred right across the disciplines of geomorphology and physical geography, and is referred to as the process revolution (Gregory, 2000). A key tool of this paradigm was conducting actual measurements of (coastal) processes, either in the laboratory or in the field, and formulating empirical models and theories to explain these observations. Coastal landforms were very much considered the mere product of the processes, but it quickly became apparent that not only is the morphology shaped by processes, but it also provides feedback to these processes. In other words, the geomorphology is an active player, rather than a passive responder to the forcing, and has some degree of control over its own development. This notion initiated a new paradigm, referred to as the 'morphodynamic approach', and this approach was eloquently and comprehensively introduced to coastal geomorphologists by Wright and Thom (1977) in a benchmark paper in *Progress in Physical Geography* (ironically, a journal now rarely used as an outlet for coastal research).

There have been subsequent developments in geomorphology and physical geography that have contributed to a refining of the morphodynamic paradigm, involving concepts such as chaos theory and non-linear dynamics (Richards, 2003). However, these are all directly reliant on the key notion of mutual feedback between process and form, and are therefore not fundamentally different from the morphodynamic approach. It has been argued that the most current paradigm involves interactions between physical and socio-economic systems, and has materialized in a new scientific field: Earth System Science. Others maintain that this is merely a rebranding of the old discipline of Geography (Pitman, 2005). We leave such musings behind and focus on what the morphodynamic paradigm represents.

1.2.2 Coastal morphodynamic systems

According to the coastal morphodynamic paradigm, conceptualized in Fig. 1.5, coastal systems (e.g. salt marsh, beach, tidal basin) comprise three linked elements (morphology, processes and sediment transport) that exhibit a certain degree of autonomy in their behaviour, but are ultimately driven and controlled by environmental factors (Wright and Thom, 1977). These environmental factors are referred to as 'boundary conditions', and include the solid boundary (geology and sediments; Chapter 3), climate (section 1.3) and external forcing (wind, waves, storms, tides and tsunami; Chapters 4 and 5), with sea level (Chapter 2) serving as a meta-control by determining where coastal processes operate. When contemporary coastal systems and processes are considered, human activity should also be taken into account. In fact, along many of our coastlines human activities, such as beach nourishment, construction of coastal defences, dredging and land reclamation, are more important in driving and controlling coastal dynamics than the natural boundary conditions and can therefore not be ignored

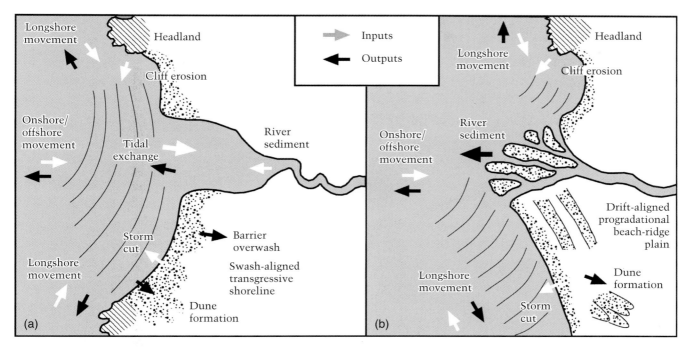

Fig. 1.6 Sediment budgets on: (a) estuarine; and (b) deltaic coasts. (Source: Masselink et al. 2011. Reproduced with permission of Hodder & Stoughton Ltd and adapted from Carter and Woodroffe 1994 with permission from Cambridge University Press.)

(Chapter 17). Moreover, through climate change, humans are altering the boundary conditions themselves (sea-level rise and changes to the wave climate).

Unless long-term coastal change (centuries to millennia) is considered, the boundary conditions can be viewed as given and constant, although it should be borne in mind that external forcing is stochastic (random), and the dynamics of coastal systems arise from the interactions between the three linked elements:

(1) Processes: This component includes all processes occurring in coastal environments that generate and affect the movement of sediment, resulting ultimately in morphological change. The most important of these are hydrodynamic (waves, tides and currents) and aerodynamic (wind) processes. Along rocky coasts, weathering is an additional process that contributes significantly to sediment transport, either directly through solution of minerals, or indirectly by weakening the rock surface to facilitate mobilization by hydrodynamic processes (Chapter 15). In addition, biological, biophysical and biochemical processes are important in salt marsh (Chapter 10), mangrove (Chapter 11) and coral reef (Chapter 16) environments. River outflow processes are important in deltas (Chapter 13).

(2) Sediment transport: A moving fluid imparts a stress on the bed, referred to as 'bed shear stress', and if the bed is mobile this may result in the entrainment and subsequent transport of sediment. The ensuing pattern of erosion and deposition can be assessed using the sediment budget

(Fig. 1.6). If the sediment balance is positive (i.e. more sediment is entering a coastal region than exiting), deposition will occur and the coastline may advance, while a negative sediment balance (i.e. more sediment is exiting a coastal region than entering) results in erosion and possibly coastline retreat. This makes quantifying the sediment budget a fundamental means for understanding coastal dynamics, as well as providing a tool for assessing and predicting future coastal change.

(3) Morphology: The three-dimensional surface of a landform or assemblage of landforms (e.g. coastal dunes, deltas, estuaries, beaches, coral reefs, shore platforms) is referred to as the morphology. Changes in the morphology are brought about by erosion and deposition, and are, in part, recorded in the stratigraphy (section 1.2.4).

It is worth emphasizing that the morphodynamic approach is scale-invariant, i.e. the approach can be applied regardless of the spatial scale of the coastal feature under investigation. For example, at the smallest scale, the approach can be applied to wave and tidal bed forms; at the largest scale, to tidal basins or entire delta systems. Importantly, the spatial and temporal scales of coastal morphodynamic systems are related (Fig. 1.7): the larger the spatial scale of the coastal system, the longer the timescale associated with the dominant process(es) and the associated coastal morphodynamics. The spatio-temporal relationship is, however, not linear: some coastal systems respond faster than one would expect on the basis of their

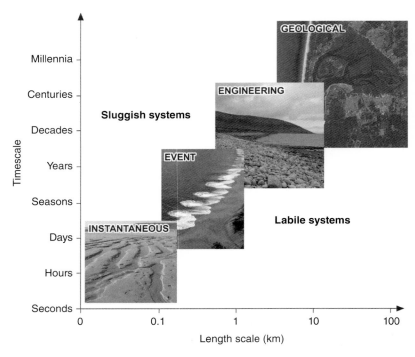

Fig. 1.7 Relationship between spatial and temporal scales of coastal systems. Sluggish and labile systems are those that respond relatively slow and fast, respectively. (Source: Adapted from Cowell and Thom 1994. Imagery © 2013 Terrametrics. Map data © 2013 Google.) For colour details, please see Plate 1.

size (labile systems; e.g. sandy barriers without dunes), whereas other coastal systems exhibit a relatively slow response (sluggish systems; e.g. rocky coasts). The timescale of the response of a coastal system also depends, of course, on the magnitude of the forcing, and the classic magnitude-frequency concept (Wolfman and Miller, 1960) is as relevant now as it was when it was introduced in geomorphology.

1.2.3   Morphodynamic feedback

A characteristic of coastal morphodynamic systems is the presence of strong links between form and process (Cowell and Thom, 1994). The coupling mechanism between processes and morphology is provided by sediment transport and is relatively easy to comprehend. There is, however, also a link between morphology and processes to complete the morphodynamic feedback loop.

As an example, under calm wave conditions sand is transported on a beach in the onshore direction resulting in beach accretion and the construction of a feature known as the 'berm' (Fig. 1.8). During berm construction, the seaward slope of the beach progressively steepens and the top of the berm increases in elevation relative to sea level through accretion; both morphological developments

Fig. 1.8 Photograph of a developing berm on a sandy beach. Berms are swash-formed features that usually develop as part of beach recovery following storm erosion. On tidal beaches they are found just above the high-tide level. This particular berm formed after a period of energetic waves and is well defined with a small runnel located to the landward. The photo was taken at high tide and the berm is still being overtopped by swash action and is therefore still being constructed. (Source: Photograph by Gerd Masselink.)

have profound effects on the wave-breaking processes and sediment transport (Masselink and Puleo, 2006). The steepening of the beach makes it increasingly difficult for the onshore-directed uprush flow to transport sediment up-slope, whilst at the same time the down-slope transport by the offshore-directed backwash flow is enhanced. Additionally, the wave breaker type may change from energetic plunging, which entrains large amounts of sediment that become advected into the uprush promoting onshore transport, to surging, which is less favourable to the uprush. The increased elevation of the berm reduces the frequency of waves reaching the top of the berm, leading to a progressive reduction in the vertical accretion rate. At some stage during beach steepening, the hydrodynamic conditions may be sufficiently altered to stop further onshore sediment transport, and berm construction will cease.

The berm development discussed above is a relatively simple example of morphodynamic feedback and other examples of feedback between morphology and processes include estuarine infilling and tidal currents, foredune development and aerodynamics, delta lobe growth and hydraulic gradient, salt marsh accretion and tidal inundation frequency, mangrove establishment and sedimentation processes, and coral reef development and wave attenuation. In all cases, due to the close coupling between process and form, cause and effect are not readily apparent. This gives rise to the 'chicken-and-egg' nature of coastal morphodynamics whereby it is often not clear whether the morphology is the result of the hydrodynamic processes, or vice versa. In a developing morphodynamic system, process and form co-evolve, and this is one of the key factors that make it so difficult to predict reliably long-term coastal development: small errors in predicting either the morphological change or the hydrodynamic processes end up magnifying dramatically over time. In more technical parlance, coastal morphodynamic systems are therefore also described as 'complex' or 'non-linear'.

The feedback between morphology and processes is fundamental to coastal morphodynamics, and can be negative or positive.

• **Negative feedback** is a damping mechanism that acts to oppose changes in morphology and is a stabilizing process, eventually resulting in equilibrium. An example of negative feedback is the berm development discussed previously. However, morphological adjustment involves a redistribution of sediment and this requires a finite amount of time. The time it takes to attain equilibrium defines the relaxation time and is a measure of the morphological inertia within the system (de Boer, 1992). The relaxation time depends on the volume of sediment involved in the morphologic adjustment (i.e. the spatial scale of the landform) and the energy level of the forcing that controls the sediment transport rate. For large coastal landforms, the relaxation time generally exceeds the time between changes in environmental conditions, and in these cases it is unlikely that equilibrium is ever reached.

• **Positive feedback** pushes a system away from equilibrium by modifying the morphology such that it is even less compatible with the processes to which it is exposed. A morphodynamic system driven by positive feedback seems to have a 'mind of its own' and exhibits self-forcing behaviour. An example of positive feedback is the infilling of deep estuaries by marine sediments due to asymmetry in the tidal flow. In a deep estuary, flood currents are stronger than ebb currents and this tidal asymmetry results in a net influx of sediment and infilling of the estuary. As the estuary is being infilled, the tidal asymmetry increases even more as friction and shoaling effects are enhanced by the reduced water depths (Friedrichs and Aubrey, 1988). In turn, the increase in tidal asymmetry speeds up the rate of estuarine infilling. This constitutes positive feedback between the estuarine morphology and the tidal processes, resulting in rapid infilling of the estuary. Eventually, intertidal salt marshes and tidal flats start developing in the estuary and this marks a reversal in feedback. As the intertidal areas become more extensive, the flood asymmetry of the tide progressively decreases so that the estuarine morphology approaches steady state as sediment imports during flood and exports during ebb equilibrate.

One of the most powerful and exciting explanations for coastal features to have emerged from the last two decades of coastal morphodynamic research is the notion of self-organization, or emergence, which refers to the development of morphological features with a specific shape and/or spacing that has arisen from the mutual interactions between form and process. In other words, the template for the morphology is not directly related to that of a specific hydrodynamic phenomenon, but has emerged from the morphodynamic interactions (i.e. feedback). The notion of self-organization has now become well established in a wide range of disciplines (Gallagher and Appenzeller, 1999), including geomorphology (Murray et al., 2009), and a range of coastal features are now interpreted as being self-organizing features, including rhythmic features such as wave ripples, beach cusps, bar morphology and cuspate shoreline features (Coco and Murray, 2007; Fig. 1.9). One of the main challenges of research into self-organization has been to identify the dominant length scales of the rhythmic shoreline features and, in addition to empirical techniques, the dominant tool has been the application of numerical modelling (section 1.4). An example of the application of a numerical model to explain cuspate features in coastal lagoons is discussed in Box 1.1.

Fig. 1.9 Flying spit in the Sea of Azov, Ukraine. The formation of these features has intrigued coastal scientists for decades, but numerical modelling by Ashton and Murray (2006a, b), based on the relation between the longshore sediment transport rate and the deep-water wave angle (see Box 1.1), seems to have provided a satisfactory explanation for their formation. (Source: Image © 2013 Terrametrics. Map data © 2013 Google.)

CONCEPTS BOX 1.1   Self-organization of elongate water bodies

The long axis of some elongate water bodies (e.g. coastal lagoons) exhibit wave-formed features, such as sandy spits and capes, and in some instances a series of almost-circular lakes outline a larger basin, suggesting that opposing cuspate shoreline features have joined, segmenting the lake along its long axis (Fig. 1.10). Zenkovich (1967) suggested a qualitative model whereby the formation of cuspate forms and the eventual segmentation of elongate water bodies could be attributable to waves generated by winds blowing across the long fetch parallel to the main axis, arriving with crests at angles greater than 45° relative to the long coastlines. Recent numerical studies (Ashton and Murray, 2006a, b) have investigated how such high-angle waves lead to the initial formation and subsequent self-organization of cuspate features, and further work by Ashton et al. (2009) has suggested how the growth of cuspate shoreline features in elongate water bodies may eventually lead to a segmentation of the water body into smaller, round water bodies (Fig. 1.10).

The physical basis of the models of Ashton and co-workers is shown in Fig. 1.11. It is based on the notion that the rate of longshore sediment transport is maximized when the angle between the crests of deep-water waves and the shoreline is approximately 45°;

thus, for both smaller and larger wave angles, long-shore transport rates decrease away from this 'flux-maximizing' angle. When the angle between the deep-water wave crests and the shoreline is greater than 45°, the sediment flux along the convex-seaward crest of a perturbation decreases in the flux direction, because the angle between the waves and the local shoreline is increasing, moving progressively farther away from the flux-maximizing angle (Fig. 1.11a). The resulting sediment accumulation at this location will lead to a growth of the perturbation (positive feedback). When deep-water waves approach from smaller angles, the sediment flux along the convex-seaward crest of the perturbation increases in the flux direction, because the angle between the waves and the local shoreline is moving progressively closer to the flux-maximizing angle (Fig. 1.11b). This results in erosion at this location, leading to a smoothing out of the perturbation and a straightening of the coastline (negative feedback).

According to the model of Ashton et al. (2009), a large number of small cuspate features initially develop in elongate water bodies, but, as the morphology evolves and feedback between the different cuspate forms start to become significant, the number of cuspate features decreases, while their size increases (left panels of the

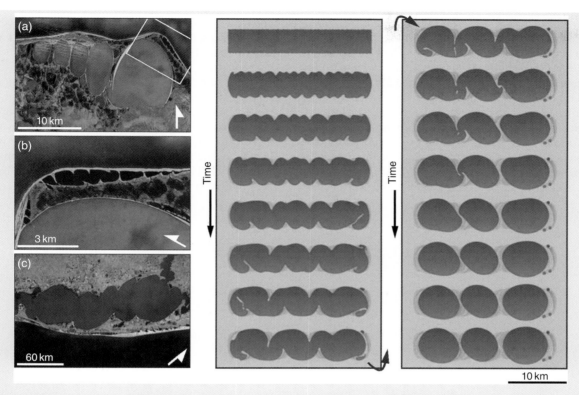

Fig. 1.10   Natural examples of enclosed water bodies with cuspate features and segmented water bodies. (a) Laguna Val'karkynmangkak, Russia; (b) inset of (a); and (c) Lagoa Dos Patos, Brazil. The results of a numerical simulation of the formation of cuspate features and segmented water bodies are shown in the right panels. (Source: Ashton et al. 2009. Reproduced with permission of the Geological Society of America.) For colour details, please see Plate 2.

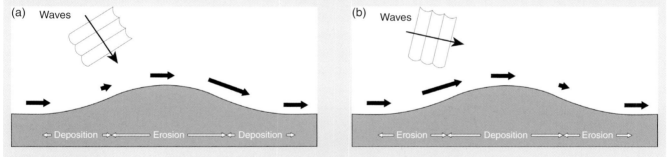

Fig. 1.11   Schematic illustrations of shoreline change caused by high-angle and low-angle waves. When the angle between the deep-water wave crests and the shoreline is less than the flux-maximizing angle (<45°) shoreline perturbations will be smoothed out, whereas for angles greater than the flux-maximizing angle (>45°) the perturbations will grow. (Source: Adapted from Coco and Murray 2007. Reproduced with permission of Elsevier.)

model simulation in Fig. 1.10). Such increase in spacing and amplitude is a well-known phenomenon of numerical self-organization models. Once the cuspate features extend significantly offshore (approximately half-way across the water body), opposing cuspate features start affecting each other by providing shelter from waves that propagate along the long-axis of the water bodies. A new dynamic emerges and opposing cuspate features that are initially offset, grow together, eventually merging and segmenting the water body into smaller, round water bodies (right panels of the model simulation in Fig. 1.10). Conditions conducive to the development of segmented elongate water bodies are relatively shallow water bodies with an energetic wind regime and non-cohesive (e.g. sand or gravel) shores. Inhibiting factors are shoreline vegetation (stabilizing the shoreline), significant tidal flows (which would become faster as the flow is constricted) and low sedimentation rates (cuspate spit growth may be too long to occur compared to other long-term environmental changes).

1.2.4 Coastal evolution and stratigraphy

As the coastal system evolves over time, its evolution is recorded in the sediments (clay, silt, sand and gravel) in the form of the stratigraphy. It is important to realize that stratigraphic sequences are a record of the depositional history and that erosional events are only represented by gaps or discontinuities in the stratigraphic record. Stratigraphy is the realm of geologists and sedimentologists, but, because it provides insights into the geomorphological evolution of coastal landforms as well as the history of the governing coastal processes, it is also of considerable interest to coastal scientists in general. The stratigraphy

can be particularly useful when dating has provided the age of certain coastal deposits; this information can then be used to quantify rates of accretion, and also to help with the reconstruction of the sea-level history.

As an example, Fig. 1.12 shows the stratigraphy of a salt marsh in southern New Zealand (Gehrels et al., 2008). Accumulations of salt-marsh sediment in this part of the southern hemisphere are generally very thin, because sea level during much of the middle and late Holocene was only slightly higher than at present and little accommodation space was available for salt-marsh deposits to fill. The intertidal sands that form the substrate of the salt marsh were

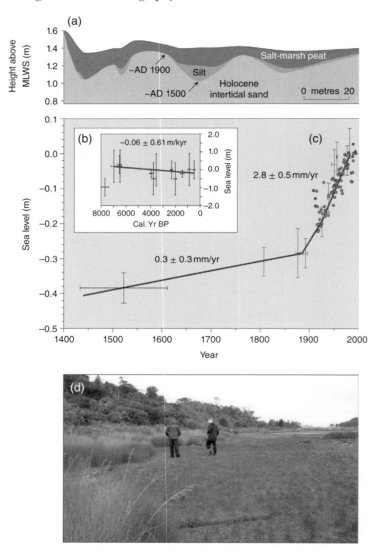

Fig. 1.12 Deriving sea-level history from salt-marsh stratigraphy. (a) Stratigraphy of a salt marsh in southern New Zealand. The marsh developed in the past half millennium on a substrate of late Holocene intertidal sands. The sands were deposited in the middle and late Holocene when sea level was slightly higher than present. MLWS, mean low water spring. (b) Since about AD 1900, accumulation has been very rapid as a consequence of the accommodation space provided by the sharp sea-level acceleration (c). The crosses in (b) and (c) represent dated samples of shells and plant material, respectively, which can be related to former sea levels (with vertical and age uncertainties). Different coloured dots in (c) represent annual measurements of sea level from two nearby tide gauges. Cal. Yr BP, calibrated years before present. (d) The photo shows an overview of the marsh, which can be found on the Catlins coast in southeastern New Zealand, near the village of Pounawea. (Source: Adapted from Gehrels et al. 2008. Reproduced with permission of John Wiley & Sons.) For colour details, please see Plate 3.

deposited during this time. By around 500 years ago, a salt-marsh environment had developed on the sands. The microfossils in the salt-marsh sediments show that the silts were deposited in an upper salt-marsh environment, close to the limit of the high spring tides. These microfossils are single-celled organisms (protists) called foraminifera. They are particularly useful in this context because they live in narrow vertical niches in the intertidal zone and can be precisely related to sea level. The transition therefore signifies that, following the deposition of the intertidal sands, the sea level must first have dropped to a low stand, before rising to a level that allowed upper salt-marsh grasses to colonize the sandy substrate. Thus, there is a significant time hiatus between the deposition of the intertidal sands and the formation of the salt marsh. Salt-marsh accretion was initially very slow; in 400 years the surface of the salt marsh only rose by about 10cm. Around AD 1900, however, a remarkable change occurred. This change is reflected in the sediments as a transition from silty to highly organic salt-marsh deposits. The microfossils show that the surface of the marsh remained close to the highest spring tide level, indicating that about 40cm of sea-level rise took place after c. AD 1500, but 30cm of this occurred in the last 100 years. The rising sea level has preserved the organic salt-marsh sediments very well, whereas the sediments that were deposited during the preceding centuries have lost their organic content due to frequent subaerial exposure.

This example clearly shows how: (1) sea-level change can control the stratigraphy and sediment types of the coastal zone; (2) sea-level rise provides the accommodation space in which sediments can accumulate; and (3) the sediments provide an archive from which sea-level changes can be reconstructed. A slowly rising sea level allowed salt marshes to colonize emerged tidal-flat deposits, first slowly, but in the last 100 years very rapidly. This rise is being recorded by various tide gauges (Hannah, 2004), but the sediments in the coastal system also bear witness to the sea-level acceleration. The rapid sea-level rise, which commenced around the beginning of the 20th century, appears to be a worldwide feature (Gehrels and Woodworth, 2013) and is due to climate change (Woodworth et al., 2009; Mitchum et al., 2010).

1.3 Climate change

1.3.1 Quaternary climate change

Throughout Earth's history, climate has always been changing, but at the onset of the Quaternary, about 2.6 million years ago (Gibbard et al., 2009), the closure of the Isthmus of Panama appears to have triggered a major change in the world's ocean circulation (Sarnthein et al., 2009). Since that event, the Earth has known over 50 glacial-interglacial cycles. The most complete record of these cycles is preserved in the marine sedimentary record. Analyses of oxygen isotopes in marine sediment cores have shown that the Quaternary contains 103 marine isotope stages (Raymo et al., 1989; Gibbard et al., 2009); the evenly numbered stages are cold (glacials and stadials), the odd-numbered stages are warm (interglacials and interstadials). This subdivision is a far cry from the 'classic' four glacial and interglacial periods that had been recognized in Europe and North America by the end of the 19th century. Climate change, glaciations and sea-level change are clearly the defining features of the Quaternary.

The most conspicuous consequence of Quaternary climate change that is relevant to the coastal zone is the growth and demise of ice sheets and the resulting changes in the level of the world's oceans, with amplitudes of up to 150m. Glaciations have also produced significant vertical changes in the level of the solid earth surface through the loading and unloading by ice and water. These isostatic changes affect both the land (glacio-isostasy) and the sea floor (hydro-isostasy) and they can produce vertical shifts of the coastal zone of up to 500m.

Sea-level changes during the past million years have been reproduced by model simulations (Fig. 1.13). Model results compare well with the longest Quaternary sea-level record hitherto obtained, that from the Red Sea, which spans 470,000 years (Fig. 1.13). The Red Sea is an evaporative basin, separated from the Arabian Sea by a shallow sill, which turns highly saline when sea level drops. Oxygen isotope ratios of seawater are sensitive to salinity changes. Because deep-sea foraminifera take up their oxygen from seawater, the oxygen isotopes in shells of foraminifera preserved in cores is a good measure of the level of the Red Sea and it allows the reconstruction of sea-level changes over several glacial-interglacial cycles. Both modelled and proxy records show that over millennial timescales, sea level behaves remarkably predictably, with lowstands during the coldest periods of 120 ± 10m below present sea level, and highstands during the peak of interglacials, to within 10m of present sea level. What is less certain, however, is the behaviour of sea level on centennial timescales, particularly during sea-level highstands. It has been suggested that during the last interglacial, when sea level was up to 9m higher than present (Kopp et al., 2009), sea-level fluctuations were very rapid, with rates of rise of, on average, 1.6m per century (Rohling et al., 2008). If correct, sea level during the last interglacial (marine isotope stage 5e) may be a reasonable analogue for future sea-level changes, when sea-level rise is predicted by some authors to be of similar magnitude (e.g. Vermeer and Rahmstorf, 2009). Marine isotope stage 11 may be the best analogue for future climate, because orbital (Milankovitch) forcing was broadly similar to present and near-future conditions. Perhaps reassuringly, sea-level behaviour during stage 11 was less erratic than during stage 5e (Rohling et al., 2010), but further research into sea-level changes during previous interglacials is needed, especially from a wider range of archives, to determine which sea-level behaviour is 'typical' for global conditions that are a few degrees warmer than the present.

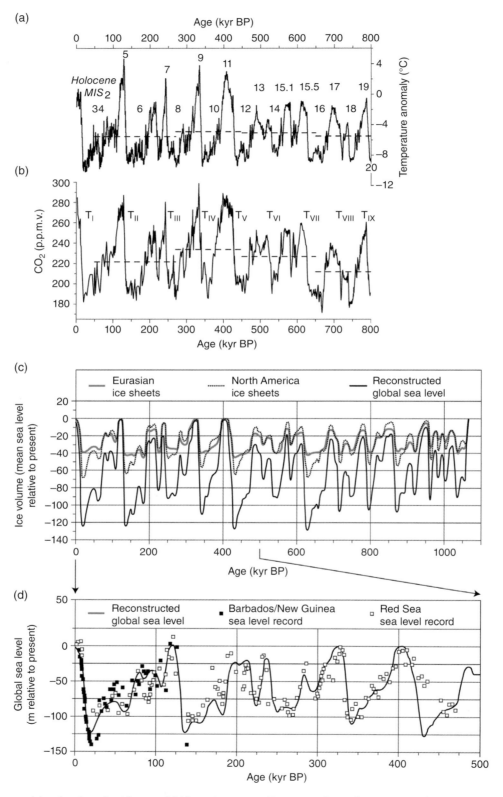

Fig. 1.13 Temperature (a) and carbon dioxide record (b) from Antarctica. (Source: Lüthi et al. 2008. Reproduced with permission of Nature Publishing Group.) (c) Modelled global sea-level record for the past 1M years, with contributions from Eurasian and North American ice sheets (Source: Bitanja et al. 2005). In (d), the model output is compared with the longest Quaternary sea-level record in the world, the record from the Red Sea (Source: Siddall et al. 2003), and with coral reef data from Barbados and New Guinea (Source: Lambneck and Chappell 2011).

1.3.2 Present and future climate change

Although sea-level change is an ultimate driver of coastal change (in the sense that it controls the position of the coast on the Earth's surface), it is not the only factor related to climatic change that affects the coast (Nicholls et al. 2007). Global warming raises the temperature of coastal waters, produces ocean acidification, changes storm patterns, and increases precipitation with major effects on coastal systems. Many climate-driven changes to our coasts are already underway (Table 1.1).

Temperature change and atmospheric greenhouse gas concentrations are intrinsically linked (Box 1.2). Future impacts of climate change depend to a large extent on the

Table 1.1 Climate drivers and their effects on coasts. (Source: Adapted from Nicholls et al. 2007.)

Climate driver	Trend	Effects
CO_2 concentration	Rising, 0.1 pH unit since 1750	Ocean acidification
Sea-surface temperature	Rising, 0.6°C since 1950	Circulation changes, sea-ice reduction, coral bleaching and mortality, species migration, algal blooms
Sea level	Rising, 1.7 ± 0.5 mm/yr since 1900	Flooding, erosion, saltwater intrusion, rising groundwater table and impeded drainage
Storm intensity	Rising	Erosion, saltwater intrusion, coastal flooding
Storm frequency, storm tracks, wave climate	Uncertain	Altered storm surges and storm waves
Run-off	Variable	Alterations in flood risk, water quality, fluvial sediment supply, circulation and nutrient supply

---

CONCEPTS BOX 1.2   Climate change and radiative forcing

The energy derived from the Sun controls the Earth's climate, but solar energy is reflected, absorbed and re-emitted by the Earth's surface and its atmosphere. The properties of the Earth's surface, through albedo effects, and the composition of the atmosphere, primarily through concentrations of greenhouse gases, play a critical role in regulating the Earth's temperature.

Climate change occurs because all three controlling mechanisms (the Sun's energy, the properties of the Earth surface, and the composition of the atmosphere) are subject to change on various timescales. The amount of solar energy that reaches the Earth varies with changes in the orbit of the Earth (e.g. Milankovitch cycles). The Earth's albedo (or its reflectivity) changes with the waxing and waning of ice sheets, which is an example of positive feedback in the climate system. Solar energy is reflected and absorbed in the atmosphere by dust particles and aerosols, which also have an effect on cloudiness, producing cooling through feedbacks. The most important greenhouse gases are water vapour, carbon dioxide (CO_2) and methane (CH_4), and their concentrations are subject to change through natural causes (e.g. volcanic gas emissions and exchange with the ocean) and through human emissions. Water vapour is the strongest greenhouse gas. Its concentration in the atmosphere depends on surface temperature and is therefore also prone to feedback.

The term 'radiative forcing' is used to describe how certain factors can alter the balance between incoming and outgoing energy. It is expressed in Watts per square metre (W/m^2) and is positive if a factor causes warming and negative if it causes cooling. The Intergovernmental Panel on Climate Change (IPCC) reports radiative forcing relative to a pre-industrial background at 1750. The total contribution of greenhouse gases to radiative forcing during a certain time period is determined by its change in concentration and by its strength, or effectiveness, in affecting the balance between incoming and outgoing energy. For example, in 2005 CO_2 had a radiative forcing of 1.49 to 1.83 W/m^2, whereas cloud albedo effects generated a cooling of -1.8 to -0.3 W/m^2. The net total contribution of anthropogenic factors was 0.6–2.4 W/m^2 (90% confidence range), mostly due to the emissions of greenhouse houses since the Industrial Revolution.

Some greenhouse gases (e.g. CO_2, CH_4 and nitrous oxide, NO_2) are stable and persist in the atmosphere for decades or longer. Changes in their concentrations over time have been accurately measured in gas bubbles preserved in ice cores (e.g. Fig. 1.13b) and, since the 1950s, by instruments. The concentration of atmospheric CO_2 has increased from 280 ppm (parts per million) in pre-industrial times to 400 ppm in 2013. As a consequence, the average global temperature during the 20th century increased by 0.74 ± 0.18 °C (Solomon et al., 2007), while sea level rose by 0.17 ± 0.03 m (Church and White, 2006). The IPCC states that sea level will continue to rise for centuries or millennia, even if radiative forcing were to be stabilized.

Table 1.2 Selected features of the Intergovernmental Panel on Climate Change (IPCC) emission scenarios from the Special Report on Emission Scenarios (SRES). Data are for the year 2100 and are from Nakićenović et al. (2000). Temperature forecasts are from Solomon et al. (2007). (Source: Data from Nakićenović et al. 2000.)

Family			A1		A2	B1	B2
Scenario group	1990	A1F1	A1B	A1T	A2	B1	B2
Population (billion)	5.3	7.1	7.1	7	15.1	7	10.4
World gross domestic product (GDP) (trillion 1990$US/yr)	21	525	529	550	243	328	235
CO$_2$ emissions from fossil fuels (GtC/yr)	6.0	30.3	13.1	4.3	28.9	5.2	13.8
Percentage of carbon-free energy usage	18	31	65	85	28	52	49
Range of projected temperature increase (°C)	0	2.4–6.4	1.7–4.4	1.4–3.8	2.0–5.4	1.1–2.9	1.4–3.8

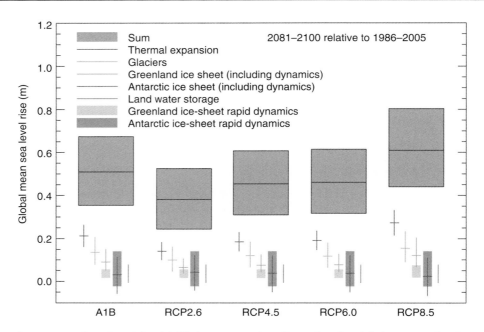

Fig. 1.14 Projections from process-based models with likely ranges and median values for global mean sea-level rise and its contributions in 2081–2100 relative to 1986–2005 for the four RCP scenarios and scenario SRES A1B used in the AR4. From Church et al. (2014). (Source: Meehl et al. 2007. Reproduced with permission of the IPCC.)

effects of human activities and the amount of carbon dioxide (CO$_2$) and other greenhouse gases that are likely to be emitted. In their assessment reports, the Intergovernmental Panel on Climate Change (IPCC) developed a range of emission scenarios that reflect a range of potential development projections of our planet. The set of scenarios that was used in the Fourth Assessment Report (AR4) was published in the Special Report on Emissions Scenarios (SRES) and are termed the SRES scenarios (Nakićenović et al., 2000). For these projections, the future state of the world was described in demographic, economic and political terms as 'storylines', resulting in four families of projections (Table 1.2). The A1 family has three 'marker scenarios'; the A2, B1 and B2 family have one each. The six marker scenarios are used by climate modellers to drive their climate models. For sea-level change

in the year 2100, this resulted in projections that range between 0.18 m for the lower end of the B1 scenario to 0.59 m for the upper end of the A1F1 scenario (excluding rapid ice-sheet dynamics). These changes were driven by projected temperature rises of 1.1 °C for the low end of the temperature range forecast under the B1 scenario, to 6.4 °C for the maximum rise considered possible under the A1F1 scenario. By the time of the publication of the Fifth Assessment Report in early 2014 modelling of ice-sheet dynamics had improved. The latest IPCC projection for the high-emission scenario is 0.52–0.98 m (relative to 1986–2005), producing rates of sea-level rise of up to 16 mm/yr (Church et al., 2014). The emission scenarios, except for A1B, have been replaced by so-called representative concentration pathways RCPs; (van Vuuren, 2011; see Chapter 2; Fig. 1.14).

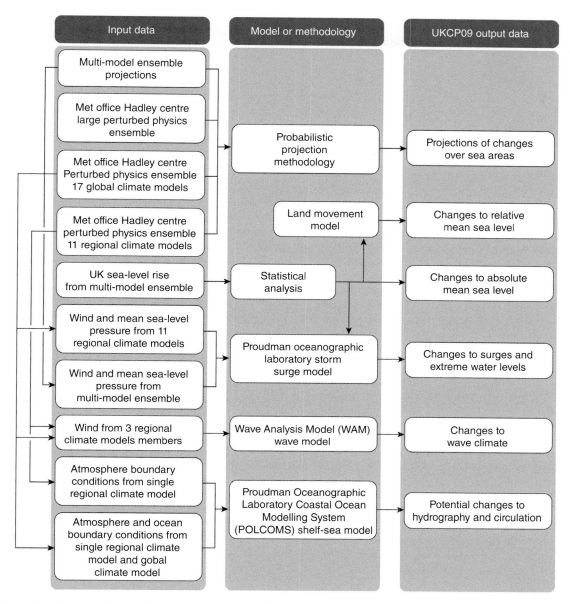

Fig. 1.15 Example of a regional approach to predicting sea-level changes. UKCP09, United Kingdom Climate Projections 2009. (Source: Adapted from Lowe et al. 2009. © UK Climate Projections, 2009.)

IPCC projections, including those for sea level, are global in scope. For coastal impact assessment they should ideally be downsized to a regional scale that is of practical use to coastal planners and managers (Gehrels and Long, 2008). This has been done in the UK, for example, where the UK Climate Impact Programme (UKCIP) has translated the SRES storylines into national and regional scenarios relevant to the UK economy. Moreover, the projections for sea-level change also include processes that act on a regional and local scale, including local land movement, tides, wind and wave climate (Fig. 1.15; Lowe et al., 2009).

Since the publication of the IPCC AR4 in 2007 there has been much debate about the accuracy of the sea-level predictions, mainly because they failed to include adequately dynamical glaciological processes such as ice-stream acceleration, basal lubrication and shelf breakup. The IPCC-estimated range of 0.09–0.17 m in the AR4 was almost certainly too low to account for these processes, but realistic modelling is notoriously difficult. Alternative semi-empirical projections that bypass the modelling difficulties include those based on the relationship between historical temperatures and sea level (Vermeer and Rahmstorf, 2009) and those based on palaeodata from the last interglacial (Rohling et al., 2008). The maximum sea-level predictions for the year 2100 based on these approaches are 1.6–1.9 m. In the UK, the H++ sea-level scenario (Lowe et al., 2009), which estimates that regional sea-level rise could be as high as 1.9 m by the year 2100, is partly based on the last interglacial analogue of Rohling et al. (2008). How accurate these projections are remains to be seen, but it is interesting to note that since IPCC predictions began, in 1990, global sea level has followed a path than overlaps with the upper range of their predictions (Rahmstorf et al., 2007).

1.4 Modelling coastal change

1.4.1 Need for adequate models

Climate represents a key environmental boundary condition for coastal systems. Climate change is therefore expected to have a major effect on coastal processes and coastal morphology. The two most important consequences of climate change are sea-level change (Chapter 2) and increased storminess (Chapters 4 and 5), both resulting in coastal erosion and flooding. There are, however, many other consequences, including the melting of permafrost cliffs and reduction in ice cover, resulting in increased erosion rates along cold coasts (Chapter 14), changes in precipitation affecting cliff instability and cliff erosion (Chapter 15), and the increase in sea-surface temperature causing coral bleaching (Chapter 16).

The ability to forecast confidently the consequences of climate change to the coastal zone is of paramount importance, not only for mitigating any adverse changes, but also to help reduce our vulnerability to environmental changes in the coastal zone through planning (Kay and Alder, 1999). This is not an easy task because predicting climate change effects to the coast comprises a number of linked steps: (1) consider appropriate greenhouse gas emission scenarios arising from our behaviour; (2) application of coupled ocean–atmosphere models to predict climate change; (3) evaluating the effect of climate change on sea level and wave climate; and (4) predicting the effect of the change in coastal drivers on nearshore sediment transport and morphological change. During each step, the feedback between drivers and responders needs to be considered, and with each step the amount of uncertainty in the predictions increases. We will focus here on the final step: models that link the coastal processes to geomorphology and evolution.

Any model is a representation, and therefore a simplification, of the real world, but the degree of abstraction (or its reverse: the level of complexity) varies hugely amongst models. In this section we consider, on a scale from simple to complex, different types of models: conceptual, empirical, behaviour-oriented, and process-based morphodynamic models. A special class of models to be discussed are physical models, which are scaled-down versions of the real world. The terminology used here is somewhat loose; for example, all models can be considered conceptual and the two most sophisticated models both have a strong empirical basis, but through the examples shown here the main characteristics of the different types of models will be made clear. For ease of comparison, the examples used to illustrate the different modelling approaches all pertain to the same coastal process: the response of barrier systems to storms and sea-level rise.

1.4.2 Conceptual models

Conceptual models provide a qualitative description, often in graphic form, of coastal systems and their main governing processes and functioning. They are generally developed by synthesizing generalities from a large number of field observations and are the result of inductive reasoning. They are often linked to classifications where they help identify different states (e.g. Australian beach state model; Wright and Short, 1984), and can also be used to predict qualitative changes in the environment by recognizing sequential stages of development (e.g. coral-reef island formation model; Kench et al., 2005). They are useful pedagogic tools when they can help bring across complex issues and enable case studies to be placed in a more general scientific framework (e.g. ternary delta model; Galloway, 1975). Conceptual models also help to identify key processes that can then be formalized in more sophisticated models to be used for predictions. They are, however, significantly oversimplifications of the real world and their practical use is generally limited to describing the current, but not the future, state of coastal systems. If used for prediction, conceptual models can at best predict the direction of change, but not the rate of change.

Figure 1.16 represents a conceptual model of the response of (gravel) barriers to increased wave conditions and raised water levels (Orford et al., 2003). According to this model, the critical factor in determining the response of barriers to increased hydrodynamic forcing is the difference in height between the elevation of the crest of the barrier and the wave run-up level, known as 'freeboard'. Positive freeboard occurs when the maximum run-up does not reach the barrier crest, and this will result in a relatively minor morphological change to the seaward face of the barrier. When the freeboard is zero or has a small negative value, the maximum run-up just reaches the crest of the barrier; this is referred to as 'overtopping' and causes

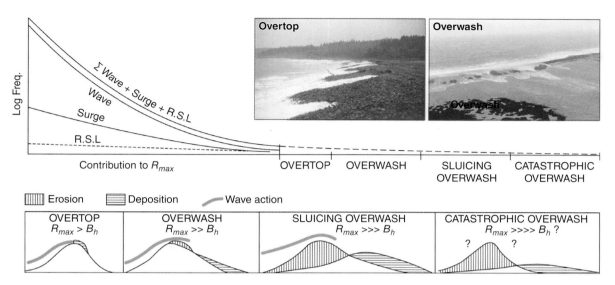

Fig. 1.16 Conceptual model showing the different stages of barrier response to increasing wave and water-level conditions. R_{max} and B_h refer to maximum run-up height and barrier height, respectively. For positive freeboard $R_{max} < B_h$, and for negative freeboard $R_{max} > B_h$. The maximum run-up height is the summation of storm surge and wave run-up, but for long timescales the relative sea-level rise also needs to be taken into consideration. The photographs represent overtopping of a Grand Desert gravel barrier in Nova Scotia, Canada and overwashing at Hurst Castle spit in Hampshire, UK (Source: Andrew Bradbury). RSL, relative sea level. (Source: Adapted from Orford et al. 2003. Reproduced with permission of American Society of Civil Engineers.)

sediment accretion on the top of the barrier, leading to an increase in the barrier crest elevation. A relatively large negative freeboard is accompanied by 'overwashing' of the barrier, with run-up events frequently extending across the crest of the barrier and running down the landward face of the barrier. This causes flattening of the barrier crest and deposition of sediment behind the barrier in the form of overwash. Increasing the negative freeboard even more may result in 'sluicing overwash' or even the wholesale destruction of the barrier feature. This conceptual model usefully illustrates the different stages of barrier response as a function of the intensity of the forcing, but does not indicate the exact thresholds for the morphological responses, nor the rate of morphological change.

1.4.3 Empirical models

In contrast to conceptual models, empirical models express relations between different elements of coastal systems in quantitative terms through the use of equations or parameters. Their formulation generally relies on a statistical analysis of numerical data collected in the field or the laboratory. Empirical models are particularly useful when they are combined with conceptual models, to provide the latter with a more quantitative foundation (e.g. the use of the dimensionless fall velocity and the relative tide range for classifying beaches; Masselink and

Short, 1993). When the empirical correlations are of a more generic form, for example the relationship between tidal prism and cross-sectional area of tidal inlets in estuarine environments (Townend, 2005), these equations can be used in more comprehensive models. The most widespread use of empirical models for predicting future coastal change is to quantify current change using statistical techniques (cliff recession, coastal retreat, salt-marsh accretion), and extrapolating this change into the future. By necessity, such models are site-specific.

When the interest is in understanding the functioning of coastal systems, empirical models are generally based on comparing or combining hydrodynamic and geomorphological parameters. In the case of barrier response to storms, the barrier breach model proposed by Bradbury (1998) is a good example. The model shown in Fig. 1.17 is based on the notion that the likelihood of a barrier breaching depends on the balance between the disturbing forces (parameterized by the wave steepness) and the resisting forces (parameterized by an inertia parameter based on barrier geometry and wave height). This notion is similar to that encapsulated by the conceptual model discussed earlier. However, the conceptual model has been taken one step further by parameterizing the disturbing and stabilizing forces, and using field observations to identify the thresholds between overwashing, overtopping and no change to the barrier crest.

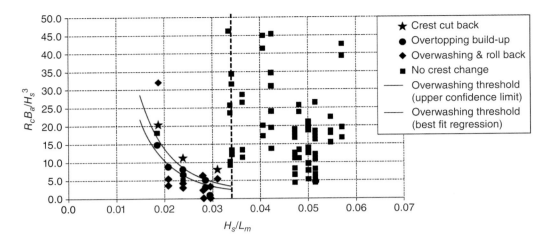

Fig. 1.17 Testing of the empirical model of Bradbury (1998) using Hurst Spit, UK. The model is based on barrier inertia parameter $R_c B_a / H_s^3$ and wave steepness H_s / L_m, where R_c is the barrier freeboard, B_a is the cross-sectional area of the barrier above still water level, H_s is the significant wave height, and L_m is the deep water wave steepness based on the mean wave period T_m ($L_m = g T_m^2 / 2\pi$, where g is gravity). The line represents the overwashing threshold, whereby conditions below the line predict barrier overwash. (Source: Data from Bradbury et al. 2005.)

1.4.4 Behaviour-oriented models

Behaviour-oriented models are realizations of coastal systems that attempt to reproduce the dominant behaviour without too much concern about the actual processes (e.g. estuarine equilibrium model of Townend and Pethick, 2002; rocky coast evolution model of Trenhaile, 2000). They tend to aggregate the complex processes into a number of simple parameterizations that can be used not only to conceptualize, but also quantify, coastal behaviour. Such models make quantitative predictions of coastal evolution and are therefore more sophisticated than conceptual models, and they may include parameterizations from empirical models. Behaviour-oriented models are often computationally efficient, because only the behaviour is modelled, rather than the detailed processes. They are therefore very useful for sensitivity analysis – for exploring 'what if' questions – thereby illuminating which aspects of the study are most in need of further study, and where more empirical data are most needed. They are most appropriate for systems that are not very well understood and/or that are very complex, because it is for these systems that sophisticated process-based models are not available or simply not good (enough). They are also useful when the input parameters are not very well constrained – there is no point in using a sophisticated model when the input data are not reliable (GIGO; garbage in, garbage out).

One of the most widely used behaviour-oriented models in coastal research, despite its shortcomings (Cooper and Pilkey, 2004), is the application of the Bruun rule to predict the effect of sea-level rise on barrier systems (Dean, 1991). According to this model, the underwater shoreface

profile is described by a simple exponential profile whose overall steepness is only a function of the sediment size, and its spatial extent is determined by the closure depth, itself a function of wave conditions and sediment size. Under conditions or rising or falling sea levels, it is assumed that the profile shape is maintained, but the profile is shifted up or down, respectively, while mass is being preserved. This simple model has been used as the basis for the Shoreface Translation Model (STM) model of Cowell et al. (1995), who added a number of capabilities to the model, including back-barrier sedimentation, longshore sediment transport and dune formation. Figure 1.18 illustrates output of the STM and shows that realistic behaviour of the response of barriers to sea-level rise can be reproduced.

1.4.5 Process-based morphodynamic models

Process-based morphodynamic models include all the relevant hydrodynamic processes and link these to the morphology through sediment transport. The sediment continuity equation is used to update the evolving morphology (Box 1.3). Feedback between morphology and hydrodynamics is accounted for and such models essentially cycle through the morphodynamic loop depicted in Fig. 1.5. The models therefore include both negative and positive feedback. These models can include elements of conceptual, empirical and behaviour-oriented models, but the key element is that process-based models attempt to account for the actual processes. Figure 1.20 shows the output produced by XBeach, a process-based morphodynamic model specifically designed to predict the response

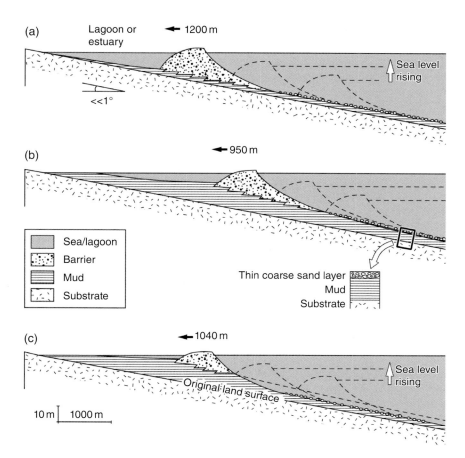

Fig. 1.18 Example of output of the Shoreline Translation Model (STM) of Cowell et al. (1995) showing the response of a barrier to sea-level rise. In Case (a), a small amount of mud deposition in the back barrier lagoon has little effect on barrier size or coastal behaviour, but with mud accumulating almost as fast as sea level is rising, as in Case (b), the barrier decreases in size and recession slows down. In Case (c), a deficit in the nearshore sediment budget, for example induced by a negative littoral drift differential, is superimposed on Case (b), resulting in a reduction in the size of the barrier and an increase in the recession rate. In addition to showing the different stages of barrier recession, the model also reproduces the stratigraphy, and in all cases lagoonal muds conformably blanket the substrate beneath and to the seaward of the barrier; after the barrier passes, wave reworking ensures that the bed is soon veneered by a coarse lag. (Source: Roy et al. 1994. Reproduced with permission of Cambridge University Press.)

of sandy barriers to hurricanes (Roelvink et al., 2009). The model includes all the key hydrodynamic processes, such as wave transformation (refraction, breaking and swash), nearshore currents and sediment transport, and is able to reproduce adequately barrier overwash and even breaching. Due to the nature of coastal morphodynamics, characterized by multiple positive and negative feedbacks, and the cumulative properties of coastal evolution, the most realistic, and therefore sophisticated, process-based models tend to work best at the shorter timescales. Over longer timescales, models' predictions rapidly diverge from observations or the model simply runs off the rails and crashes.

Whereas conceptual, empirical and behaviour-oriented models are mostly used in an explorative sense as research tools, the predictions generated by process-based numerical models are increasingly being used as a basis for policy decisions. This elevated status of this class of models, therefore, requires a lot more scrutiny with regards to the modelling results. Oreskes et al. (1994) warns of the dangers of overconfidence in modelling results due to an inappropriate consideration of what numerical models really are. It is general modelling practice to compare model predictions with observations, and when the former are consistent with the latter, usually after extensive model calibration (i.e. manipulation of the model input parameters to obtain a match between observation and model output), the model is considered verified (or validated). To claim that a model is verified is to say that its truth has been demonstrated, which implies its reliability as a basis for decision-making. Numerical models are, however, representations of the truth and can even be

METHODS BOX 1.3 Sediment continuity equation

Process-based morphodynamic models compute instantaneous flow velocities and sediment transport rates, and, because morphodynamic models take account of the morphodynamic feedback, the morphology is updated regularly to reflect the effect of the evolving morphology on the hydrodynamics. In the interest of computing efficiency, the time step for computing the hydrodynamics is generally smaller than that for updating the morphology. The ratio between the morphological step time and the hydrodynamic step time is generally $O(100–1000)$ and model output is very sensitive to this ratio.

The equation used for updating the morphology on the basis of the sediment transport rates is the sediment continuity equation, which in its differential formulation reads as:

$$\frac{dh}{dt} = (1-n)\left(\frac{dQ_x}{dx} + \frac{dQ_y}{dy}\right) + \frac{dV}{dt} \qquad (1.1)$$

where dh/dt is the change in bed elevation over time, $(1-n)$ is sediment porosity, dQ_x/dx is cross-shore gradient in the cross-shore sediment flux, dQ_y/dy is longshore gradient in the longshore sediment flux, and

dV/dt represents local sediment loss and gains, for example due to local sediment production, abrasion, nourishment and dredging. The sediment fluxes Q in the equation are volumetric fluxes (cubic metres per unit of time); if the fluxes are mass fluxes (kilograms per unit of time), then the sediment density will also need to be taken into account.

Process-based morphodynamic models generally operate on a rectangular grid; therefore flow velocities and sediment transport rates are computed numerically for each individual grid cell. Figure 1.19 illustrates how eqn. 1.1 is applied for a single grid cell. In the present example, the size of the grid cells are $\Delta x = 5\,m$ and $\Delta y = 10\,m$, and the morphological time step dt is 5 minutes. It is further assumed that: cross-shore transport Q_x into and out of the grid cell is $27\,m^3$ and $23\,m^3$, respectively; the longshore transport into and out of the grid cell is $124\,m^3$ and $123\,m^3$, respectively; $n = 0.4$; and $dV/dt = 0$. In this case, the total volume of sediment entering the grid cell over dt is $5\,m^3$ ($\Delta Q_x = 4\,m^3 + \Delta Q_y = 1\,m^3$). This gain in sediment is distributed over the $50\,m^2$ grid cell; therefore, the increase in bed elevation h is $0.1\,m$. Since the morphological time step was 5 minutes, the rate of bed-level change dh/dt is $0.02\,m$ per minute.

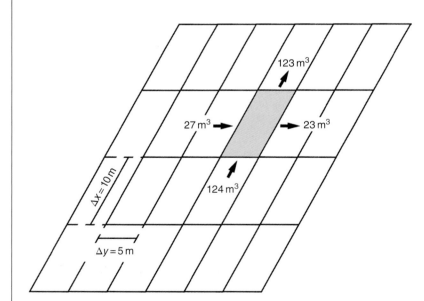

Fig. 1.19 Schematic illustrating the use of the sediment continuity equation for grid-based numerical models. See text for explanation.

considered a form of highly complex scientific hypotheses. The philosopher Karl Popper has taught us that one cannot prove a hypothesis, but one should attempt to refute it; therefore a model can never be verified. Anyone involved

with modelling is aware that good agreement between model predictions and observations can be attained in more than one way (by tuning different parameters), and such good agreement does not constitute confirmation

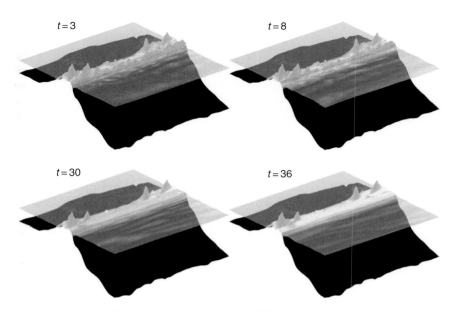

$t=3$ $t=8$

$t=30$ $t=36$

Fig. 1.20 Numerical simulation using the XBeach model of the impact of Hurricane Ivan (the largest of five hurricanes to strike to US coast in 2004) on a 2-km wide section of Santa Rosa Island, a narrow barrier island in Florida. The Gulf of Mexico is in the lower right and the back barrier bay in the upper left of all four panels. The simulation was run over 36 hours and the four panels represent the water-level variation and barrier morphology at different time steps. The maximum storm surge and significant wave height used for forcing the model were 1.75 m and 7 m, respectively, and were attained at $t = 18$ hours. Note the extensive overwashing at $t = 30$ hours and the complete destruction of the dunes in the central part of the modelled region. (Source: McCall et al. 2010. Reproduced with permission of Elsevier.) For colour details, please see Plate 4.

that the model is 'true'. At best, observations can support, or confirm, the model, and the greater the number and diversity of confirming observations, the more probable it is that the conceptualization embodied in the model is not flawed. But, whatever the level of sophistication and the agreement between predictions and observations, a model will always be a representation, and care should be taken in interpreting model results.

1.4.6 Physical models

Prior to the availability of computers, the best way to investigate the effect of coastal response to changing boundary conditions, other than making field observations, was through the use of physical modelling: the construction of physical models of the coastal system case in laboratories. Most commonly, the models are scaled-down versions of reality, but some large laboratory facilities enable the construction of models at the prototype scale (1 : 1). This methodology has been widely used in coastal engineering, for example, to investigate the effect of submerged breakwaters on shoreline stability (Ranasinghe and Turner, 2006), and still provides a good alternative to numerical process-based models.

An example of a physical model experiment designed to investigate the stability of gravel barriers is the BARDEX experiment held in the Delta Flume (length = 250 m; width = 5 m; depth = 7 m) in The Netherlands (Williams et al., 2012; Fig. 1.21). During this project, a fine-gravel barrier ($D_{50} = 10$ mm) was constructed in the wave flume, with a height of 4 m and a width at its base of 50 m. The barrier was heavily instrumented by a range of equipment, including bed-level sensors, current meters, water depth sensors, and was regularly profiled to measure its morphology. The barrier was constructed with a 'sea' and a 'lagoon' at either side, and by raising and lowering the water level in the lagoon (by 1 m relative to the sea level), the water table in the barrier was elevated and lowered in a controlled and reproducible manner relative to the mean sea level. At the end of the experiment, the barrier was subjected to increasingly elevated water-level conditions, eventually causing extensive overwashing of the barrier. Such measurements could not have been collected in a field setting, and the data will not only be used to increase our understanding of barrier processes, but also to help formulating and calibrating numerical models.

The main issue with physical models is scale (except for the prototype scale) and although scaling relationships are

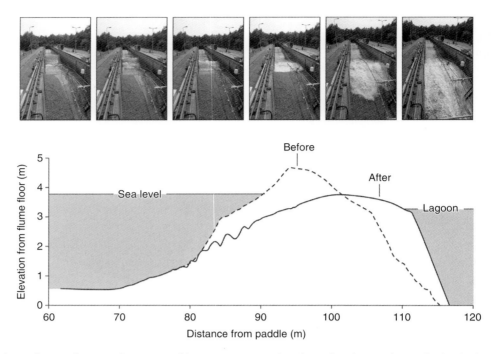

Fig. 1.21 Physical simulation of overwash on a gravel barrier constructed in the Delta Flume, The Netherlands, during the BARDEX experiment. (Source: Williams et al. 2012. Reproduced with permission of Elsevier.) The sequence of photos in the top panels represents the various stages of wave transformation of one of the overwashing waves. The bottom panel shows the morphological change after 2.5 hours of exposure to overwash, characterized by a lowering of the barrier crest by about 1 m and the development of an extensive washover deposit at the back of the barrier. The wave conditions during this test were characterized by a significant wave height of 0.8 m and a peak wave period of 8 s.

well established for interpreting the results of small-scale laboratory experiments, particularly in models that contain sediment, module scaling is never perfect. Nevertheless, physical models remain a valuable tool for coastal research.

1.5 Summary

This introductory chapter provides the theoretical framework and the scope of this book. The coast is defined as that region of the Earth's surface that has been affected by coastal processes during the Quaternary (the last 2.6M years). This definition is dynamic rather than static, and considers sediments, sea-level history and coastal processes. This book also follows the morphodynamic paradigm, according to which processes are linked with morphology through sediment transport. As the title suggests, an important focus of this book is on how global change affects coastal processes and landforms. The largest impact of future global change on coasts will be through an increase in the rate of global sea-level rise. A long-term perspective is crucial, as coastal evolution is recorded in coastal stratigraphy, which provides useful analogues for present-day and future coastal behaviour under various sea-level regimes. Models of coastal change include those

based on descriptions of processes (conceptual models), those that quantify processes (empirical models) and behaviour (behaviour-oriented models), those that include morphodynamic feedback (process-based morphodynamic models), and those that are a scaled-down version of the real world (physical models). These models are useful tools for studies of coastal impact assessment and are critical to explain and forecast coastal change.

• A long-term perspective on coastal evolution is useful because the present-day coast is a product of modern and past coastal processes and is often, at least partly, controlled by inherited landforms.

• According to the coastal morphodynamic paradigm, coastal systems comprise three linked elements (morphology, processes and sediment transport) that exhibit a certain degree of autonomy in their behaviour, but are ultimately driven and controlled by environmental factors.

• Coastal systems change through positive morphodynamic feedback, and stabilize through negative morphodynamic feedback. An important property of coastal systems is self-organization, or emergence, which refers to the development of morphological features with a specific shape and/or spacing that arises from the mutual interactions between form and process.

- During the 20th century, the global temperature increased by 0.7 °C and sea level rose by 0.2 m. Currently, sea level is rising at an accelerated rate and is likely to rise about 1 m in the next 100 years.
- Models can be used to forecast coastal change due to climate change. However, regardless of the level of model complexity, it should be borne in mind that any model is a representation, and therefore a simplification, of the real world, and model predictions should be interpreted with care.

Key publications

Church, J.A., Woodworth, P.L., Aarup, T. and Wilson, W.S. (Eds), 2010. *Understanding Sea-Level Rise and Variability*. Wiley-Blackwell, Chichester, UK.
[*State-of-the-art treatise on the current understanding of sea-level changes*]

Coco, G. and Murray, A.B., 2007. Patterns in the sand: from forcing templates to self-organisation. *Geomorphology*, **91**, 271–290.
[*Very accessible discussion of self-organization applied to a range of coastal phenomena*]

Cowell, P.J. and Thom, B.G., 1994. Morphodynamics of coastal evolution. In: R.W.G. Carter and C.D. Woodroffe (Eds), *Coastal Evolution*. Cambridge University Press, Cambridge, UK, pp. 33–86.
[*Excellent account of the morphodynamic paradigm*]

Masselink, G. and Hughes, M.H., 2003. *Introduction to Coastal Processes and Geomorphology*. Arnold, London, UK.
[*Introductory book providing good framework for the current book*]

Nicholls, R.J., Wong, P.P., Burkett, V.R., et al., 2007. Coastal systems and low-lying areas. In: M.L. Parry, O.F. Canziani, J.P. Palutikof, P.J. van der Linden and C.E. Hanson (Eds), *Climate Change 2007: Impacts, Adaptation and Vulnerability*. Contribution of Working Group II to the Fourth Assessment Report of the Cambridge University Press, Cambridge, UK, pp. 315–356.
[*Comprehensive and up-to-date account of the impact of climate change on coastal systems*]

References

Ashton, A.D. and Murray, A.B., 2006a. High-angle wave instability and emergent shoreline shapes, Part 1: modelling of sand waves, flying spits and capes. *Journal of Geophysical Research*, **111**, F04011, doi: 10.1029/2005JF000422.

Ashton, A.D. and Murray, A.B., 2006b. High-angle wave instability and emergent shoreline shapes, Part 2: wave climate analysis and comparisons to nature. *Journal of Geophysical Research*, **111**, F04012, doi: 10.1029/2005JF000423.

Ashton, A.D., Murray, A.B., Littlewood, R., Lewis, D.A. and Hong, P., 2009. Fetch-limited self-organisation of elongate sand bodies. *Geology*, **37**, 187–190, doi: 10.1130/G25299A.

Bitanja, R., van de Wal, R.S.W. and Oerlemans, J., 2005. Modelled atmospheric temperatures and global sea levels over the past million years. *Nature*, **437**, 125–128.

Bradbury, A.P., 1998. Predicting breaching of shingle barrier beaches – recent advances to aid beach management. *Proceedings 35th Annual MAFF Conference of River and Coastal Engineers*, Keele. DEFRA, London, 05.3.1–05.3.13.

Bradbury, A.P., Cope, S.N. and Prouty, D.B., 2006. Predicting the response of shingle barrier beaches under extreme wave and water level conditions in Southern England. *Coastal Dynamics 2005*, pp. 1–14, doi: 10.1061/40855(214)94.

Brooks, A.J., Bradley, S.L., Edwards, R.J. and Goodwyn, N., 2011. The palaeogeography of Northwest Europe during the last 20,000 years. *Journal of Maps*, **2011**, 573–587, doi: 10.4113/jom.2011.1160.

Carter, R.W.G. and Woodroffe, C.D. (Eds), 1994. *Coastal Evolution*. Cambridge University Press, Cambridge, UK.

Church, J.A. and White, N.J., 2006. A 20th century acceleration in global sea-level rise. *Geophysical Research Letters*, **33**, doi: 10.1029/2005GL024826.

Church, J.A., Clark, P.U., Cazenave, A. et al., 2014. Sea level change. In: T.F. Stocker, D. Qin, G.-K. Plattner et al. (Eds), *Climate Change 2013: The Physical Science Basis*. Contribution of Working Group I to the Fifth Assessment Report of the Intergovernmental Panel on Climate Change. Cambridge University Press, Cambridge, UK, and New York, NY, USA.

Cicin-Sain, B. and Knecht, R.W., 1998. *Integrated Coastal and Ocean Management*. Island Press, Washington, DC, USA.

Coco, G. and Murray, A.B., 2007. Patterns in the sand: from forcing templates to self-organisation. *Geomorphology*, **91**, 271–290.

Cooper, J.A.G. and Pilkey, O.H., 2004. Sea-level rise and shoreline retreat: time to abandon the Bruun Rule. *Global and Planetary Change*, **43**, 157–171.

Cowell, P.J. and Thom, B.G., 1994. Morphodynamics of coastal evolution. In: R.W.G. Carter and C.D. Woodroffe (Eds), *Coastal Evolution*. Cambridge University Press, Cambridge, UK, pp. 33–86.

Cowell, P.J., Roy, P.S. and Jones, R.A., 1995. Simulation of large-scale coastal change using a morphological behaviour model. *Marine Geology*, **126**, 45–61.

De Boer, D.H., 1992. Hierarchies and spatial scale in process geomorphology: a review. *Geomorphology*, **4**, 303–318.

Dean, R.G., 1991. Equilibrium beach profiles: characteristics and applications. *Journal of Coastal Research*, **7**, 53–84.

Friedrichs, C.T. and Aubrey, D.G., 1988. Non-linear tidal distortion in shallow well-mixed estuaries: a synthesis. *Estuarine, Coastal and Shelf Science*, **27**, 521–545.

Gallagher, R. and Appenzeller, T., 1999. Beyond reductionism. Introduction to Special Issue on Complex Systems. *Science*, **284**, 79.

Galloway, W.E., 1975. Process framework for describing the morphologic and stratigraphic evolution of deltaic depositional systems. In: M.L. Broussard (Ed.), *Deltas, Models for Exploration*. Houston Geological Society, Houston, TX, USA, pp. 87–98.

Gehrels, W.R. and Long, A.J., 2008. Sea level is not level: the case for a new approach to predicting UK sea-level rise. *Geography*, **93**, 11–16.

Gehrels, W.R. and Woodworth, P.L., 2013. When did modern rates of sea-level rise start? *Global and Planetary Change*, **100**, 263–277.

Gehrels, W.R., Hayward, B.W., Newnham, R.M. and Southall, K.E., 2008. A 20th century sea-level acceleration in New Zealand. *Geophysical Research Letters*, **35**, L02717, doi: 10.1029/2007GL032632.

Gibbard, P.L., Head, M.J., Walker, M.J.C. and the Subcommission on Quaternary Stratigraphy, 2009. Formal ratification of the Quaternary System/Period and the Pleistocene Series/Epoch with a base at 2.58 Ma. *Journal of Quaternary Science*, **25**, 96–102.

Gregory, K.J., 2000. *The Changing Nature of Physical Geography.* Arnold, London, UK.

Hannah, J., 2004. An updated analysis of long-term sea-level change in New Zealand. *Geophysical Research Letters*, **31**, L03307, doi: 10.1029/GRL2003GL019166.

Kay, R. and Alder, J., 1999. *Coastal Planning and Management.* Routledge, London, UK.

Kench, P.S., McLean, R.F. and Nicholls, S.L., 2005. A new model of reef-island evolution: Maldives, Indian Ocean. *Geology*, **33**, 145–148.

Kopp, R.E., Simons, F.J., Mitrovica, J.X., Maloof, A.C. and Oppenheimer, M., 2009. Probabilistic assessment of sea level during the last interglacial stage. *Nature*, **462**, 863–867.

Kuhn, T.S., 1996. *The Structure of Scientific Revolutions* (3rd edn). University of Chicago Press, Chicago, IL, USA.

Lambeck, K. and Chappell, J. 2001. Sea level change through the last glacial cycle. *Science*, **292**, 679–686.

Lowe, J.A., Howard, T.P., Pardaens, A., et al., 2009. *UK Climate Projections Science Report: Marine and Coastal Projections.* Met Office Hadley Centre, Exeter, UK.

Lüthi, D., Le Floch, M., Bereiter, B. et al., 2008. High-resolution carbon dioxide concentration record 650,000–800,000 years before present. *Nature*, **453**, 379–382.

Masselink, G., 2007. Coasts. In: J. Holden (Ed.), *An Introduction to Physical Geography and the Environment* (2nd edn). Pearson Education, Harlow, UK, pp. 467–507.

Masselink, G. and Hughes, M.H., 2003. *Introduction to Coastal Processes and Geomorphology.* Arnold, London, UK.

Masselink, G. and Puleo, J., 2006. Swash zone morphodynamics. *Continental Shelf Research*, **26**, 661–680, doi: 10.1016/j.csr.2006.01.015.

Masselink, G. and Short, A.D., 1993. The influence of tide range on beach morphodynamics: a conceptual model. *Journal of Coastal Research*, **9**, 785–800.

Martinez, M.L., Intralawan, A., Vazquez, G., Perez-Maqueo, O., Sutton, P. and Landgrave, R., 2007. The coasts of our world: ecological, economic and social importance. *Ecological Economics*, **63**, 254–272.

McCall, R.T., van Thiel de Vries, J.S.M., Plant, N.G., et al., 2010. Two-dimensional time dependent hurricane overwash and erosion modeling at Santa Rosa Island. *Coastal Engineering*, **57**, 668–683, doi: 10.1016/j.coastaleng.2010.02.006.

Meehl, G.A., Stocker, T.F., Collins, W.D., et al., 2007. Global climate projections. In: S. Solomon, D. Qin, M. Manning et al. (Eds), *Climate Change 2007: The Physical Science Basis.* Contribution of Working Group I to the Fourth Assessment Report of the Intergovernmental Panel on Climate Change. Cambridge University Press, Cambridge, UK, pp. 747–845.

Mitchum, G.T., Nerem, R.S., Merrifield, M.A. and Gehrels, W.R., 2010. Modern sea level change estimates. In: J. Church,

P. Woodworth, T. Aarup and S. Wilson (Eds), *Understanding Sea-level Rise and Variability*. Wiley-Blackwell, Chichester, UK, pp. 122–142.

Murray, A.B., Lazarus, E., Ashton, A., et al., 2009. Geomorphology, complexity, and the merging science of the Earth's surface. *Geomorphology*, **103**, 496–505.

Nakićenović, N., Davidson, O., Davis, G., et al., 2000. *Special Report on Emissions Scenarios.* A Special Report of Working Group III of the Intergovernmental Panel on Climate Change. Cambridge University Press, Cambridge, UK.

Nicholls, R.J., Wong, P.P., Burkett V.R., et al., 2007. Coastal systems and low-lying areas. In: M.L. Parry, O.F. Canziani, J.P. Palutikof, P.J. van der Linden and C.E. Hanson (Eds), *Climate Change 2007: Impacts, Adaptation and Vulnerability.* Contribution of Working Group II to the Fourth Assessment Report of the Intergovernmental Panel on Climate Change. Cambridge University Press, Cambridge, UK, pp. 315–356.

Orford, J.D., Jennings, S. and Pethick, J., 2003. Extreme storm effect on gravel-dominated barriers. In: R.A. Davis (Ed), *Proceedings of the International Conference on Coastal Sediments 2003.* CD-ROM published by World Scientific Publishing Corp. and East Meets West Productions, Corpus Christi, TX, USA.

Oreskes, N., Shraderfrechette, K. and Belitz, K., 1994. Verification, validation, and confirmation of numerical models in the Earth Sciences. *Science*, **263**, 641–646.

Pitman, A.J., 2005. On the role of geography in earth system science. *Geoforum*, **36**, 137–148.

Rahmstorf, S., Cazenave, A., Church, J.A., et al., 2007. Recent climate observations compared to projections. *Science*, **316**, 709.

Ranasinghe, R. and Turner, I.L., 2006. Shoreline response to submerged structures: a review. *Coastal Engineering*, **53**, 65–79.

Raymo, M.E., Ruddimann, W.F., Backman, J., Clement, B.M. and Martinson, D.G., 1989. Late Pliocene variation in Northern Hemisphere ice sheets and North Atlantic deep water circulation. *Paleoceanography*, **4**, 413–446.

Richards, K.S., 2003. Geography and the physical sciences tradition. In: S.L. Holloway, S.P. Rice and G. Valentine (Eds), *Key Concepts in Geography.* Sage Publications, London, UK, pp. 51–72.

Roelvink, D., Reniers, A., van Dongeren, A., van Thiel de Vries, J., McCall, R. and Lescinski, J., 2009. Modeling storm impacts on beaches, dunes and barrier islands. *Coastal Engineering*, **56**, 1133–1152, doi: 10.1016/j.coastaleng.2009.08.006.

Rohling, E.J., Grant, K., Hemleben, Ch., et al., 2008. High rates of sea-level rise during the last interglacial period. *Nature Geoscience*, **1**, 38–42.

Rohling, E.J., Braun. K., Grant, K., et al., 2010. Comparison between Holocene and Marine isotope stage-11 sea-level histories. *Earth and Planetary Science Letters*, **291**, 97–105.

Roy, P.S., Cowell, P.J., Ferland, M.A. and Thom, B.G., 1994. Wave-dominated coasts. In: R.W.G. Carter and C.D. Woodroffe (Eds), *Coastal Evolution: Late Quaternary Shoreline Morphodynamics*, Cambridge University Press, Cambridge, UK, pp. 121–186.

Sarnthein, M., Bartoli, G., Prange, M., et al., 2009. Mid-Pliocene shifts in ocean overturning circulation and the onset of Quaternary-style climates. *Climates of the Past*, **5**, 269–283.

Siddall, M., Rohling, E.J., Almogi-Labin, A., et al., 2003. Sea-level fluctuations during the last glacial cycle. *Nature*, **423**, 853–858.

Small, C. and Nicholls, R.J., 2003. A global analysis of human settlement in coastal zones. *Journal of Coastal Research*, **19**, 584–599.

Solomon, S., Qin, D., Manning, M. et al., 2007. Technical summary. In: S. Solomon, D. Qin, M. Manning et al. (Eds), *Climate Change 2007: The Physical Science Basis*. Contribution of Working Group I to the Fourth Assessment Report of the Intergovernmental Panel on Climate Change. Cambridge University Press, Cambridge, UK.

Streif, H., 2004. Sedimentary record of Pleistocene and Holocene marine inundations along the North Sea coast of Lower Saxony, Germany. *Quaternary International*, **112**, 3–28.

Townend, I. 2005. An examination of empirical stability relationships for UK estuaries. *Journal of Coastal Research*, **21**, 1042–1053.

Townend, I. and Pethick, J.S., 2002. Estuarine flooding and managed retreat. *Philosophical Transactions of the Royal Society of London A*, **360**, 1477–1495.

Trenhaile, A.S., 2000. Modelling the evolution of wave-cut shore platforms. *Marine Geology*, **166**, 163–178.

Trenhaile, A.S., 2010. The effect of Holocene changes in relative sea level on the morphology of rocky coasts. *Geomorphology*, **114**, 30–41.

van Vuuren, D.P., Edmonds, J., Kainuma, M., et al. 2011. The representative concentration pathways: an overview. *Climatic Change*, **109**, 5–31.

Vermeer, M. and Rahmstorf, S., 2009. Global sea level linked to global temperature. *Proceedings of the National Academy of Sciences*, doi: 10.1073/pnas.0907765106.

Williams, J., Buscombe, D., Masselink, G., Turner, I.L. and Swinkels, C., 2012. Barrier dynamics experiment (BARDEX): aims, design and procedures. *Coastal Engineering*, **63**, 3–12.

Wolfman, M.P. and Miller, J.P., 1960. Magnitude and frequency of forces in geomorphic processes. *Journal of Geology*, **68**, 17–26.

Woodworth, P.L., White, N.J., Jevrejeva, S., Holgate, S.J., Church, J.A. and Gehrels, W.R., 2009. Evidence for the accelerations of sea level on multi-decade and century timescales. *International Journal of Climatology*, **29**, 777–789, doi: 10.1002/joc.1771.

Wright, L.D. and Short, A.D., 1984. Morphodynamic variability of surf zones and beaches – a synthesis. *Marine Geology*, **56**, 93–118.

Wright D.L. and Thom, B.G., 1977. Coastal depositional landforms: a morphodynamic approach. *Progress in Physical Geography*, **1**, 412–459.

Zenkovich, V.P., 1967. *Processes of Coastal Development*. Oliver and Boyd, London, UK.

2   Sea Level

GLENN A. MILNE

Department of Earth Sciences, University of Ottawa, Ottawa, Canada

2.1   Introduction

2.1.1   What is sea level?

If asked the question in the title of this section, most people would reply, 'It is the level of the sea surface.' While this is not incorrect, a more specific definition is required so that this property of the Earth System can be measured and therefore studied in a quantitative fashion. To do this, a reference height or surface must be chosen so that the level, or height, of the sea surface can be measured relative to it.

The most common reference surface is the solid Earth (or sea floor); measurements of the height of the sea surface relative to the ocean floor represent what is known as 'relative sea level' (Fig. 2.1). Different methods for measuring relative sea level are outlined in section 2.1.3. Based on this definition, at any location in the ocean, a change in relative sea level can be produced by a vertical shift in either the sea surface or the sea floor, and so a measurement of relative sea-level change is ambiguous in the sense that it provides no information on the contribution to the change from each bounding surface. For example, at coastal locations, a local sea-level rise leading to a landward motion of the coastline, or marine transgression, could be driven by land subsidence and/or sea-surface rise. A retreat of the coast, known as a marine regression, could be driven by land uplift and/or sea-surface fall.

A second reference surface used in sea-level measurement is known as the 'reference ellipsoid' (Box 2.1), which is a surface of ellipsoidal geometry that best fits the mean (or time-averaged) level of the ocean surface. This conceptual surface is centred at the Earth's centre of mass, and so this reference can also be considered as a geocentric reference. Measurements of sea-surface height relative to the reference ellipsoid (or geocentre) represent what is known as 'geocentric' or 'absolute' sea level (Fig. 2.1). The second term is a little misleading, as the height measurement is still relative. However, compared to the reference surface used to define relative sea level, the reference ellipsoid is a more stable surface and so the measurements provide a more robust measure of height

Coastal Environments and Global Change, First Edition. Edited by Gerd Masselink and Roland Gehrels.
© 2014 John Wiley & Sons, Ltd. Published 2014 by John Wiley & Sons, Ltd. Companion Website: www.wiley.com/go/masselink/coastal

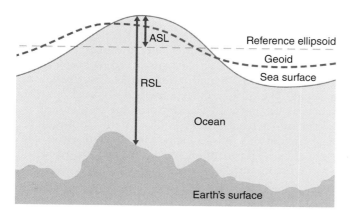

Fig. 2.1 Schematic diagram depicting relative sea level (RSL), defined as the height between the sea surface and the sea floor. This quantity is zero at coastlines. The geoid is an equipotential of the Earth's gravity field that approximates the mean position of the sea surface over time periods exceeding a few decades. Geocentric or absolute sea level (ASL) is the height of the sea surface with respect to the reference ellipsoid. The reference ellipsoid is a geometric surface that best fits the geoid. See Box 2.1 for more information on the geoid and reference ellipsoid. Note that displacements between the sea surface, geoid and reference ellipsoid have been exaggerated for the purpose of clarity.

shifts in the sea surface. Absolute sea level is measured using satellite altimetry (see section 2.1.3).

In the context of coastal evolution, relative sea level is the more relevant of the two sea-level definitions given above, as vertical motion of both the sea floor and the sea surface control the location of coastlines, and therefore where coastal processes will operate at a given time. For this reason, relative sea level will be the focus of this chapter and so, hereafter, the term 'sea level' will be used to mean relative sea level.

A key theme in this book is the influence of climate change on coastal evolution, and an important issue of concern in this regard is our ability to predict future changes in a warming climate. Measurements of sea-surface height play an important role in understanding recent climate-driven sea-level change and underpin our ability to predict future changes. Therefore, absolute sea-level changes have an important place in this chapter, albeit secondary to relative changes.

2.1.2 Processes affecting sea level

From the definitions of sea level given in section 2.1.1, it follows that any process that affects a height shift in either the sea surface or the sea floor at a given location will produce a change in relative sea level. Only those processes that influence the sea surface will significantly impact absolute sea level. There are a large number of processes

operating over a range of spatial and temporal scales that will deflect either, or both, the upper and lower bounding surfaces of the ocean (Fig. 2.3). The aim of this section is to provide an overview of these processes and to emphasize those that operate over century and longer timescales, as these have the greatest influence on coastal evolution during the Quaternary (in terms of defining the location of the coastal zone during this period).

Over relatively short timescales (seconds to days), vertical deflections of the ocean surface are driven mainly by interactions of the ocean with the atmosphere. The rapidity of these changes reflects the fact that the atmosphere and the ocean can flow relatively quickly (compared to other components of the climate system, such as ice and the solid Earth). Two short period changes that are not related to atmosphere–ocean interactions are tsunamis and tides; the former are most commonly driven by rapid vertical movement of the ocean floor due to earthquakes and the latter due to changes in gravity associated with changes in the position of the Earth relative to the Moon, Sun and other planets in the solar system (Pugh, 2004). Short-period changes play an important role in coastal evolution as they supply energy to the coastal zone that drives erosional and depositional processes (see Chapters 4 and 5). However, they are not relevant in this chapter where the focus is on sea-level changes as an external control on the location of the coastal zone over time periods of centuries to millions of years.

Over these longer time periods, vertical motion of the sea surface can occur through a variety of processes, including: (1) changes in the amount (mass) or volume of water in the oceans; (2) changes in gravity; and (3) changes in ocean-basin volume. The latter is largely due to vertical motion of the ocean floor and so will be discussed later in this section, after processes affecting this motion are discussed.

Changes in the amount or volume of water in the oceans are due to mass exchange between the ocean and other components of the Earth System as part of the hydrological cycle (e.g. between continental ice reservoirs and the ocean) or due to changes in water temperature and salinity causing expansion or contraction of the water column. (Note that these temperature and salinity changes lead to density changes, which affect the flow of ocean water and, therefore, changes in ocean surface height.) During the Quaternary, mass exchange was the more dominant of the two processes due to the large oscillations in continental ice volume associated with the so-called Milanković cycles. However, in the past century, ocean water volume changes are thought to be as important as land ice melt in driving sea-level change and may be the dominant process in the coming decades (see section 2.3.4). These climate-related processes will be discussed further in sections 2.2 and 2.3.

CONCEPTS BOX 2.1   The geoid and reference ellipsoid

The free surface of any fluid at rest will always lay perpendicular to the direction of external forces applied across it. If it did not, the material would flow until it did. This is because fluids cannot support a shear stress. For example, pour some water into a glass and tilt the glass – you will observe that the surface of the water, once it reaches equilibrium, remains horizontal (i.e. perpendicular to the local gravitational force). In the same way, the equilibrium ocean surface would lie perpendicular to the

gravity field at every location over the oceans and, by definition, would represent an equipotential of the Earth's gravity field. This hypothetical surface is known as the geoid. The adjective 'hypothetical' is used because the oceans are never in equilibrium, due to a variety of phenomena such as interactions with the atmosphere, ocean circulation, tides, changes in Earth rotation and earthquakes. If all perturbing forces could be 'switched off' then the ocean surface would become perfectly still and

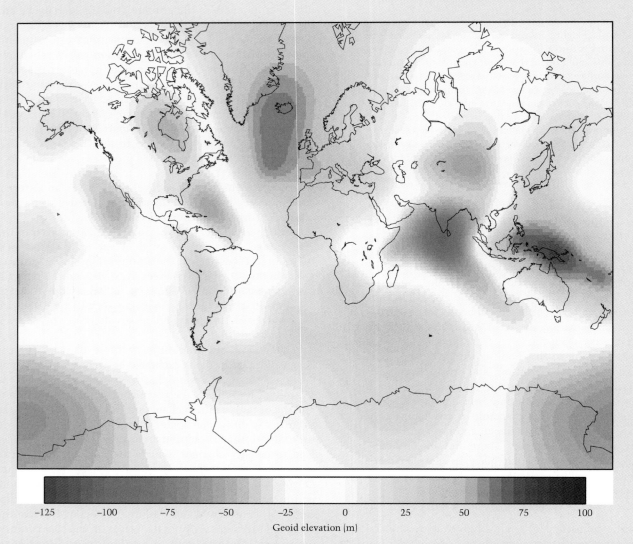

Geoid elevation (m)

Fig. 2.2   Map of geoid height relative to the reference ellipsoid. In general, these heights are a few 10s of metres, but they can reach ~100 m in some regions due to the existence of strong lateral variations in sub-surface rock density. The geoid model shown here is EGM2008 (Source: Pavlis et al. 2012), which is available via http://earth-info.nga.mil/GandG/wgs84/gravitymod/egm2008/index.html. For colour details, please see Plate 5.

lie on an equipotential of the gravity field. By making measurements of the Earth's gravity field, geodesists are able to arrive at estimates of the geoid like that shown in Fig. 2.2.

Inspection of Fig. 2.2 shows that the geoid has undulations of up to ~100 m in amplitude (e.g. there is a low in the geoid immediately south of India). These undulations reflect lateral variations in density structure within the Earth, as well as topography on the ocean floor. Of course, these density variations move in time on timescales of millions of years through flow within the mantle and movement of the tectonic plates, and so the geoid will also evolve.

The reference ellipsoid is a geometrical surface that best matches the shape of the geoid. Both the geoid and the reference ellipsoid are used as reference surfaces for geodetic purposes. Geodesy is a branch of geophysics aimed at better defining the shape of the Earth and changes in this shape with time. For more information on the geoid and the reference ellipsoid, the reader can refer to any general text on geophysics such as Fowler (2005) or more specialist books such as Lambeck (1988).

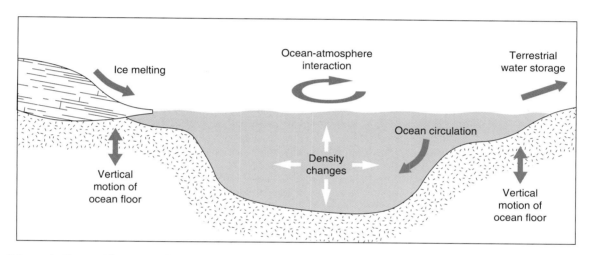

Fig. 2.3 Schematic diagram illustrating the processes that perturb the vertical position of the ocean surface and ocean floor and therefore affect sea level. Note that the ocean surface has undulations due to variations in gravity and forces caused by circulation in the atmosphere and ocean. (Source: Adapted from Milne et al. 2009. Reproduced with permission of Nature Publishing Group.)

Over time periods of centuries and longer, the mean height of the sea surface approximates the geoid (see Fig. 2.2 and Box 2.1) as the effects of shorter period changes average out. Since the height of the geoid is influenced by changes in gravity, it follows that any process influencing gravity will result in a height change of the sea surface. Any process that leads to a redistribution of mass either on the Earth's surface or in its interior will result in a perturbation to gravity. For example, the mass redistribution associated with Quaternary glaciations, as water is transferred between oceans and ice sheets, results in a significant vertical shift (several tens of metres) in the ocean surface that varies in amplitude around the globe. This will be discussed in section 2.2.3. Sediment redistribution can also lead to a sea-level signature; for example, the sediment build-up in major river deltas will amplify the local gravity field and lead to a rise in the adjacent ocean surface. Mass redistribution within the solid Earth is driven by either internal buoyancy forces associated with convection in the Earth's mantle or loading of the surface due to surface-mass redistribution (such as the two examples given earlier).

The internal mass redistribution of the Earth resulting from flow in the mantle and surface-mass loading leads to vertical motion of the ocean floor and, therefore, a change in sea level. The mantle convection process is responsible, for example, for uplift of the Huon Peninsula during the Quaternary resulting in a steady but large marine regression and the exposure of several coral terraces (Fig. 2.4; see Chapter 16). The largest shifts due to surface-mass redistribution during the late Quaternary are associated

Fig. 2.4 Coral terraces on the Huon Peninsula (New Guinea) that have been raised out of the ocean due to local uplift of the solid Earth associated with tectonic processes. (Source: Yokoyama and Esat 2011.)

with the growth and ablation of large ice sheets in areas such as Canada and Fennoscandia, with uplift of the Earth's surface of ~500m or more in parts of Canada following periods of deglaciation. The redistribution of sediment is also responsible for amplifying sea-level rise in the vicinity of large depocentres (such as the Mississippi Delta) due to the isostatic subsidence of the solid Earth in response to the sediment load. Areas like this are also affected by compaction of the sediments, which lowers the land surface and increases the rate of relative sea-level rise even further.

Another process that influences sea level through perturbing both the ocean surface and ocean floor is changes in Earth rotation. The processes leading to internal and external mass redistribution discussed earlier lead to a shift in the orientation of the Earth relative to its spin axis. This shift, known as true polar wander, produces a vertical deflection of both the ocean surface and the ocean floor (Milne and Mitrovica, 1998; Mound and Mitrovica, 1998). The influence of true polar wander on sea level will be discussed in section 2.2.3.

Thus far we have considered how vertical deflections of the ocean floor can produce local to regional changes in sea level. They can also influence global sea levels through their impact on ocean basin volume. If the mean height of the ocean floor were to increase over

time, the mean height of the ocean surface would also increase (assuming that the amount of water in the oceans remained constant). Of course, changes in the area of the oceans would also have to be taken into account in such an analysis. The influence of changing ocean basin geometry on sea-surface height will be discussed in section 2.2.3.

To conclude this sub-section, it is useful to define two commonly used terms in sea-level science. 'Mean sea level' refers to the mean, or time-averaged, sea-level change (relative or absolute) at a specific location. The period of averaging is usually a year or longer, and so the influence of short-period processes (e.g. atmospheric circulation, tides) are largely removed in calculating mean sea level at a given location. A second term that is widely used is 'eustasy' or 'eustatic sea level'. This refers to a component of sea-level change that is globally uniform. For example, changes in the mean height of the global ocean surface associated with changes in ocean basin volume due to the mantle convection processes are called 'tectono-eustasy'. Changes in the mean height of the ocean surface due to the melting of ice sheets and the consequent addition of water to the oceans are known as 'glacio-eustasy'. Note that, while both of these processes lead to a change in mean sea-surface height at the global scale, the actual sea-level changes (absolute and relative) that result from these processes are not globally uniform due to spatially variable changes in gravity and height of the sea floor that also occur. Therefore eustatic sea-level change is a concept that pertains to the global mean of the sea-level change associated with a given process. It is not a directly measureable quantity.

2.1.3 Observing sea level

A variety of different techniques have been used to observe changes in sea level. Most recently, sea levels have been monitored using satellite altimetry – a technique whereby the height between an orbiting satellite and a terrestrial surface (e.g. the ocean, an ice sheet) is determined via the travel time of an electromagnetic wave. This technique (described in more detail in section 2.3.2) provides near global determinations of changes in sea-surface height relative to a geocentric reference (e.g. the reference ellipsoid) and so provides a measure of absolute sea-level change. As stated earlier, and described in some detail in section 2.3, these data have revolutionized our understanding of sea-level changes on sub-annual to decadal timescales and play a central role in determining the influence of climate on sea-level changes over the past two decades.

A second approach to measuring sea level is through the use of an instrument known as a tide gauge. There are a handful of different types of tide gauges, all of which provide a measure of sea-surface height relative to a local land height reference (see section 2.3.2 for more details). In comparison to the satellite data, tide gauges give a measure of relative sea-level change at specific locations. As their name suggests, these instruments were originally developed and deployed to study ocean tides. However, tide-gauge data have been used to study a range of phenomena covering a broad spectrum of frequencies (e.g. from storms and tsunamis to tectonics and isostasy). An example of a typical tide-gauge record is shown in Fig. 2.5. This record illustrates the large range of frequencies that are characteristic of the processes affecting sea level (see Fig. 2.3). Figure 2.5 shows that, over decadal and longer timescales, sea level at Newlyn (in the southwest of England) is steadily rising. This rise is likely dominated by two main components: (1) increase in sea-surface height due to contemporary climate change (through land ice melting and/or ocean warming); and (2) steady subsidence of the sea floor due to the isostatic adjustment of this area in response to the melting of the British-Irish and Fennoscandian ice sheets (see section 2.3). One of the strengths of tide-gauge data is that the lengths of the time series at many locations are long enough to filter out the often large and dominant shorter-term variability associated with atmosphere–ocean interaction so that secular variations, such as those due to climate change, can be better isolated.

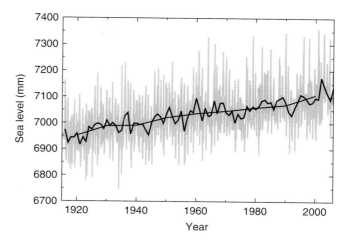

Fig. 2.5 Sea-level changes recorded at the tide gauge in Newlyn, UK. The grey line shows the time series of monthly mean values and the black lines show the time series of annual mean and decadal mean (smoothest line) values. Comparing the two indicates that there is high-amplitude variation at sub-annual periods. Note that the absolute values on the *y*-axis are not significant; these values are made arbitrarily large so that no negative values will be recorded. (Source: Woodworth et al. 2009. Reproduced with permission of John Wiley & Sons.)

Prior to the use of tide gauges, direct measurements of sea level do not exist. Pre-instrumental sea levels can be estimated, or reconstructed, using information from the geological record. More specifically, sea levels from the distant past are reconstructed by measuring the height (above or below present mean sea level) of morphological features or fossilized remnants of biological organisms that have a known height relationship to mean sea level. A large variety of such markers have been used, including erosional features such as shore platforms and notches (see Fig. 1.2), macrofossils such as corals and mangrove roots, and microfossils such as foraminifera and diatoms. The accuracy and precision of a given sea-level reconstruction depends on a number of factors, including the local tidal range and the accuracy of the height relationship between a given marker and palaeo mean sea level. More information on reconstructing ancient sea levels is given in section 2.2.2.

2.1.4 Chapter outline

The previous three sections have introduced the fundamental concepts and observational techniques required to discuss and understand sea-level changes at a variety of temporal and spatial scales. The remainder of this chapter builds upon this discussion, to consider observations and their interpretation through the Quaternary and up to the present day. Section 2.2 considers sea-level change during the Quaternary, with an emphasis on the late Quaternary, the most data-rich part of this period. Section 2.3 considers instrumented sea-level records with a focus on the 20th century (tide gauges) and the past few decades (satellite altimetry), and ends with a brief description of sea-level projections for the future.

2.2 Quaternary sea-level change

2.2.1 Introduction

The Quaternary is a period in the Earth's history characterized by the repeated growth and melting of large continental ice sheets in the northern and southern hemispheres. During this period, changes in land-ice volume were influenced by changes in the Earth's orbit around the Sun, as described in the Milanković theory (Ruddiman, 2008). Quaternary sea-level reconstructions (e.g. Fig. 1.13) demonstrate that fluctuations in land ice were the dominant driver of the global sea-level response. The repeated growth and demise of the ice sheets led to a considerable isostatic response of the solid Earth, which is an important component of the ice-driven signal. Other climate-related processes, such as changes in ocean density structure (due to temperature and salinity changes) and the resulting impact on ocean circulation, and changes in the configuration of secular wind patterns, also influenced sea levels during the Quaternary. However, the amplitude of these changes was relatively small compared to the signal (>100 m) associated with the ice sheets and so they will not be discussed in any detail.

Plate tectonics and mantle convection associated with the slow cooling of the Earth result in secular sea-level changes that affect global sea levels by many tens of metres over tens of millions of years (e.g. Miller et al., 2005; Moucha et al., 2008). The influence of this process on sea levels at the global scale during the Quaternary is relatively small compared to the ice-driven signal. However, in some tectonically active areas this process can lead to a large, background sea-level trend upon which the more rapid and repetitive ice-driven signal is superimposed. The raised coral terraces in the Huon Peninsula are a good example (see Fig. 2.4). As discussed later, the possible influence of tectonic processes on local and regional sea levels must be considered when interpreting observations from the Quaternary.

2.2.2 Sea-level observations

In this section, the observational techniques used to quantify sea-level changes during the Quaternary will be reviewed. The discussion shall begin with isotopic measurements of fossil organisms in deep-ocean sediments as these data provide a measure of changes in ice volume

during the Quaternary. However, these data do not provide a measure of relative sea-level changes at a given location. Because there was large spatial variability in sea-level change during the Quaternary, other techniques must be applied to quantify local changes in sea level. As introduced in section 2.1.3, these are based on the use of morphological and biological indicators that have a known height relationship to mean sea level and will be discussed towards the latter half of this sub-section.

The evaporation and transpiration process that takes water from the ocean and eventually precipitates it as snow onto the ice sheets leads to a depletion of lighter ^{16}O isotopes in the oceans and an enrichment of these in the continental ice sheets (see, for example, Ruddiman, 2008). Therefore, as more ice accumulates on the continents, the isotopically 'heavier' the oceans become (i.e. the mean $\delta^{18}O$ value will become more positive). It is the sensitivity of this quantity to land-ice volume that has enabled scientists to produce time series of changes in ocean mass – often referred to as glacio-eustatic sea level – throughout the Quaternary (see Waelbroeck et al., 2002, and references therein). It is possible to reconstruct the $\delta^{18}O$ of ancient seawater because certain micro-organisms that live in the ocean secrete calcareous shells that are in isotopic equilibrium with the water. Therefore, when these organisms die and sink to the ocean floor, they effectively record the $\delta^{18}O$ of ocean water at that place and time. The $\delta^{18}O$ of fossil shells obtained from marine sediment cores are measured to arrive at the type of plot shown in Fig. 2.6. It should be noted that the $\delta^{18}O$ of these shells also depends on the ambient ocean temperature and sometimes other quantities (e.g. salinity). Therefore, studies that attempt to constrain changes in ice volume tend to use species that live in the deep ocean (known as benthic or bottom dwelling species) where the changes in ocean water temperatures during the Quaternary were smaller than those in the shallow ocean.

The data shown in Fig. 2.6 represent, arguably, the foundation of Quaternary climate research. These data support the hypothesis put forward by Milanković that the global-scale waxing and waning of ice sheets is related to changes in the Earth-Sun orbit. A significant proportion of the variability in the $\delta^{18}O$ occurs at frequencies that match well with those known to exist in the Earth-Sun orbit. However, there are aspects of this correlation that remain poorly understood and these motivate an active component of contemporary research (see section 2.2.3). It is conventional to plot the $\delta^{18}O$ values in reverse order (see Fig. 2.6) so that peaks correspond to periods of relatively high eustatic sea level (low ice volume) and vice-versa. The scaling from $\delta^{18}O$ to sea level is approximately –0.1‰ to 10 m (glacio-eustatic) rise. The record shown in Fig. 2.6 depicts large fluctuations in glacio-eustatic sea level that exceed 100 m in magnitude during the late Quaternary.

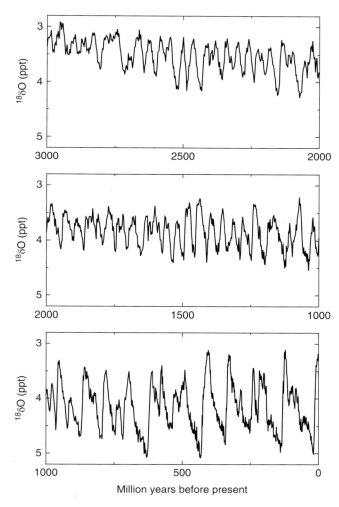

Fig. 2.6 A δ¹⁸O time series for the past three million years. Note that the values are plotted in reverse order so that the variation is proportional to ice volume (or glacio-eustatic sea-level change). Each frame shows the same *y*-axis range to enable direct comparison between each 1000-yr period. (Source: Data from Lisiecki and Raymo 2005.)

The δ¹⁸O data extracted from marine sediments provide a powerful constraint on changes in global ice volume during the Quaternary and earlier. However, local sea-level change can depart significantly from the glacio-eustatic signal. This is clearly demonstrated in reconstructions of past sea level using morphological and biological indicators from specific locations. A selection of reconstructions is shown in Fig. 2.7 that illustrate the large spatial variability in sea level during different periods following the last glacial maximum (LGM), which ended around 20 kyr BP.

In areas that are distant from major centres of glaciation (i.e. northern North America, northwest Eurasia,

Antarctica), the observations indicate a monotonic rise in local sea level (Fig. 2.7a) from the end of the most recent glacial maximum (~20 kyr BP) to the mid-Holocene (~7 kyr BP). This sea-level rise is dominated by the addition of glacial meltwater to the oceans, i.e. the glacio-eustatic signal. The sea-level rise shown in Fig. 2.7(a) contrasts markedly with the large sea-level fall reconstructed at sites once covered by ice (e.g. Fig. 2.7b). In these locations, the isostatic uplift of the solid Earth dominates the net sea-level change. The two data sets shown in Fig. 2.7(a) and Fig. 2.7(b) represent two end member sea-level responses. When the rate of global melting slowed down around 7 kyr BP (due to the disappearance of the large ice sheets in North America and Fennoscandia), a sea-level fall is recorded in many equatorial regions due to the influence of solid Earth isostatic effects (see next section). This is illustrated well in Fig. 2.7(c), which shows a reconstruction from Australia.

The reconstructions shown in Fig. 2.7 are comprised of a series of sea-level index points (SLIPs), which define the height of a past mean sea level (relative to the present mean sea level) at a given location. The two key components of a SLIP – height and time – have an associated uncertainty that depends on a number of factors; some of these are introduced below in a brief overview of the key elements involved in reconstructing a SLIP.

The geological indicators used to determine the change in relative sea level can be classed as either biological or morphological. Biological indicators are fossils of organisms that lived close to mean sea level. Common macrofossils used to reconstruct past sea levels include corals and mangroves; common microfossils include foraminifera, diatoms and pollen. Morphological indicators are erosional or depositional features that are formed in the coastal zone, such as shore platforms, tidal notches, raised beaches and marine deltas. When a sea-level indicator is identified and believed to be *in situ* (i.e. located in the place it was originally formed or deposited), then its height is measured relative to a local datum. Reducing this height measurement to a value for past mean sea level (with an associated uncertainty) is non-trivial. The general procedure used to perform this data reduction involves the use of a concept known as the indicative meaning of a given sea-level marker; this concept is illustrated in Fig. 2.8. In short, the indicative meaning of a given indicator is the height relationship of that marker to the tidal range.

The uncertainty or precision of a given SLIP is directly related to its indicative range. For example, many of the SLIPs shown in Fig. 2.7 are based on a particular species of coral (*Acropora palmata*), which, in today's environment, is found within a ~5 m vertical zone below low tide. Therefore, the uncertainty on these SLIPs is ±2.5 m. Some of the other SLIPs in Fig. 2.7 are based on another species of coral (*Porites asteroides*), which is commonly found up

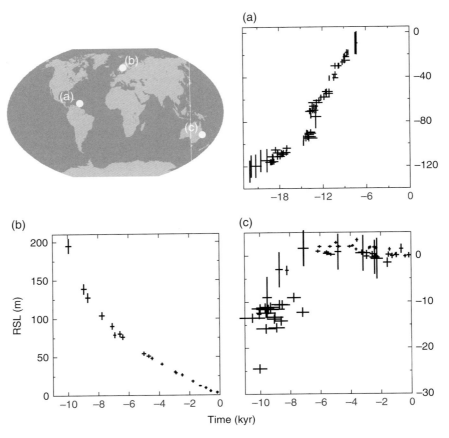

Fig. 2.7 Past sea level reconstructed from the geological record at three locations: (a) Barbados (Source: Milne et al. 2009. Based on data from Fairbanks 1989 and Bard et al. 1990. Reproduced with permission of Nature Publishing); (b) Angerman River, Sweden (Source: Milne et al. 2009. Based on data from Lambneck et al. 1998. Reproduced with permission of Nature Publishing); (c) Cleveland Bay, Australia (Source: Milne et al. 2009. Based on data from Woodroffe 2009. Reproduced with permission of Nature Publishing). These data show that sea-level changes during the past 10 to 20 millennia have exhibited large spatial variation due to the response of the solid Earth to the large-scale surface mass redistribution of the most recent deglaciation.

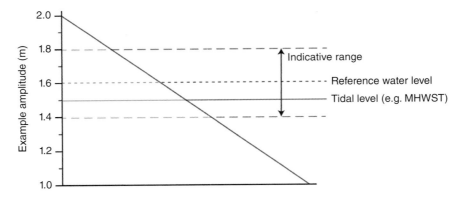

Fig. 2.8 Schematic diagram illustrating the concept of indicative meaning used to reconstruct past sea levels from the geological record. The indicative range is the vertical height over which the sea-level indicator existed while forming. The indicative meaning is the height of the middle of the indicative range relative to a known contemporary tidal reference height (mean high water spring tide (MHWST) is used as an example in the diagram). When the indicator is encountered in the fossil environment, the vertical difference between its height and the indicative meaning of its modern counterpart equals the change in the level of MHWST (all measured relative to modern MHWST). The uncertainty that is attached to the estimate of former sea level is determined by the indicative range but can also include surveying errors, sediment compaction, changes in tidal range, etc.). (Source: Edwards 2007. Reproduced with permission from Elsevier.)

to depths of ~20 m in the contemporary environment. Thus, the error bar on these SLIPs (e.g. the oldest points) is considerably larger at ±10 m. In Holocene sea-level studies using microfossils, the vertical distribution of the sea-level indicators in the contemporary environment is often quantified using regression models and the vertical error is statistically determined (e.g. Gehrels et al., 2008). An underlying assumption made in assigning a height uncertainty to a SLIP is that the indicative range of the contemporary species (or contemporary analogue) is representative of the species being used to reconstruct sea level in the past. A second assumption commonly adopted is that the tidal range at the time the marker was formed or deposited is the same as it is today. Some recent studies have shown that this is not the case in some areas, with changes exceeding several metres at certain times in the past (Neill et al., 2010). If the change in tidal range is known, a correction can be applied. Another issue that can impact the accuracy of a given SLIP is sediment compaction. SLIPs obtained from sediment cores might have been displaced downwards since the time of deposition due to compaction of the underlying sediment. This problem can be avoided by selecting only indicators obtained from sediments that rest on a hard substrate. When this is not practical, some procedures can be used to estimate the magnitude of the compaction (Bird et al., 2004).

There often exists evidence in the geological record that can provide useful information on the height of a past sea level, even though it is not possible to assign a robust indicative meaning. Such evidence includes, for example, shells that can live many tens of metres below mean sea level, therefore providing a height that contemporary relative sea level must have been above. This type of information is known as 'limiting' as it specifies a height that is an upper or lower limit on relative sea level at a given time. Examples of upper limiting markers include freshwater peat, tree stumps and archaeological finds (assuming they are *in situ*).

The chronological component of a SLIP is generally obtained by carbon-14 (^{14}C) dating of organic material. This is relatively straightforward for biological markers. When morphological indicators are used, applying this dating technique requires that organic material can be found, *in situ*, within the feature. If the dated material is not *in situ*, then the date obtained is only a minimum age. The ^{14}C dating technique is limited to the past ~50,000 years and so other techniques must be applied to reconstruct sea level at earlier times. Most SLIPs dating before 50,000 yr BP are from corals, and so the uranium-thorium dating method can be used. Indeed, the uranium-thorium method is used in preference to ^{14}C dating when possible due to larger uncertainties inherent in the latter technique (e.g. Bard et al., 1990). The ^{14}C technique breaks down for the past ~200 years, and so sea-level reconstructions for this recent period must rely on other methods, such as lead-210 (^{210}Pb), volcanic ash horizons, and even anthropogenic signals such as a commonly observed caesium spike in 1963–1964 due to nuclear bomb testing, and metal concentrations representing pollution events (Marshall et al., 2007).

The majority of existing SLIPs document sea-level changes during the Holocene period. There are considerably fewer data from the pre-Holocene due to the presence of ice in near-field areas and the logistical issues in coring offshore sediments deeper than ~10 m in intermediate- and far-field areas. Prior to the last glacial maximum, the quantity of data drops dramatically, with the majority obtained from coral terraces in tectonically uplifting regions such as Barbados and the Huon Peninsula.

In summary, measurements of δ^{18}O from fossil shells in ocean sediments provide a continuous record of land-ice volume change that spans the entire Quaternary. The total isotopic signal includes a significant temperature component and so isolating the ice volume information is not straightforward. These isotopic data do not contain information on the spatial variability of sea-level change and so are complemented by site-specific reconstructions of sea level using geological indicators that have a quantifiable relationship to mean sea level. These data demonstrate that the spatial variability in Quaternary sea levels has been large (in excess of 500 m) and is dominated by the isostatic response of the solid Earth to the mass exchange between ice sheets and the oceans.

2.2.3 *Interpretation of the observations*

In the case of δ^{18}O data, much of the interpretation has focused on the separation of the ice volume and ocean temperature signals, as well as the correlation of periodicities and amplitudes in the ice volume signal with changes in the Earth-Sun orbit. This sub-section will begin with a brief discussion of the latter. The remainder and bulk of this sub-section will focus on the interpretation of site-specific sea-level observations. These data contain information on the volume and spatial distribution of continental ice through time as well as physical properties of the solid Earth due to the large isostatic component in the observations. As a consequence, accurate interpretation of these data requires the use of geophysical models that simulate sea-level change associated with ice-ocean mass flux. The basic components of these models will be outlined before going on to discuss the interpretation of site-specific data.

The δ^{18}O time series shown in Fig. 2.6 displays a number of distinct features, some of which have a straightforward interpretation and others that do not. As mentioned earlier, interpretation of this record has focused on correlating the variations in δ^{18}O with changes in the

Earth-Sun orbit in order to test the idea that the growth and melting of large ice sheets is controlled by an astronomical forcing (as first suggested by James Croll and then pursued in more detail by Milutin Milanković). Many studies have demonstrated that the dominant frequencies in the $\delta^{18}O$ data relate to those found in the orbital changes (see Ruddiman, 2008, and references therein), providing clear evidence in support of at least a pacing from this mechanism. However, the orbital changes are consistent throughout the past few million years, whereas the changes in ice volume are not. Note, from Fig. 2.6, that the amplitude of the dominant oscillations in $\delta^{18}O$ increases at specific times – around 2.6 Myr BP and 1 Myr BP. The former is widely accepted to be the inception of large ice sheets in the northern hemisphere. This hypothesis is supported by a range of observational evidence in the geological record, such as the appearance of ice-rafted debris in north Atlantic sediments around this time. As for a causal mechanism, a variety of hypotheses have been put forward that continue to be tested (e.g. Ruddiman, 2008). These include, for example, closure of a narrow sea-way between the Pacific and Atlantic oceans (the Isthmus of Panama) and a subsequent shift in oceanic circulation and therefore climate in the north Atlantic; and a secular cooling driven by the drawdown of atmospheric carbon dioxide leading to summer temperatures that were not high enough to melt the snowfall from the previous winter.

The transition at ~800 kyr BP, known as the mid-Pleistocene transition, marks the onset of land ice volume fluctuations that are a factor of ~2 greater than before (see Fig. 2.6). There is also the emergence of greater power or energy in the frequency spectrum of the ice response at periods around ~100 kyr. These two changes are clearly related, as larger ice sheets take longer to grow and melt. The cause of the transition is not widely agreed upon. Some have suggested it is due to changes in the basal conditions of ice sheets as previous glaciations stripped the land surface of soil leaving a higher friction rock-ice interface, which permits the ice sheets to support greater shear stress at the base and therefore grow thicker (Clark et al., 2006). Another recent hypothesis calls upon the change from land- to marine-based ice in east Antarctica that acts to better synchronize the response of northern and southern hemisphere ice, resulting in a larger global volume change (Raymo et al., 2006). This topic continues to be an active area of research.

Interpreting site-specific sea-level data is also not straightforward, given the variety of mechanisms that can cause local sea level to change. These are commonly summarized in an equation of the form (e.g. Shennan, 2007):

$$S_{rec}(\theta,\phi,t) = S_{eus}(t) + S_{iso}(\theta,\phi,t) + S_{tect}(\theta,\phi,t) \\ + S_{other}(\theta,\phi,t) \tag{2.1}$$

which relates the observed sea-level change to processes that contribute to this change during the Quaternary. In this equation, the term $S_{rec}(\theta,\phi,t)$ represents the reconstruction of relative sea level at some past time (t) relative to the present value at latitude (θ) and longitude (ϕ). (Note that there will be an uncertainty associated with a sea-level reconstruction.) The first three terms on the right-hand side correspond to the processes that are thought to be the dominant contributors to sea-level changes during the Quaternary, from left-to-right: glacio-eustatic changes associated with variations in land ice volume; isostatic and gravitational changes in response to the ice-ocean mass exchange; and tectonic changes due to the response of the solid Earth and gravity field to mantle flow. The ice volume or eustatic term is given by:

$$\Delta S_{eus}(t) = -\frac{\rho_{ice}\Delta V_{ice}(t)}{\rho_{water}A_{ocean}} \tag{2.2}$$

in which ρ_{ice} and ρ_{water} are the densities of ice and water, $\Delta V_{ice}(t)$ is the volume of global grounded ice (relative to the present value), and A_{ocean} is the area of the ocean basins. Note that this expression of the glacio-eustatic sea-level change assumes that the ocean-basin area does not change with time. In reality, the ocean area would have changed considerably (5–10%) during a glacial cycle (due to sea-level regression/transgression and the advance/retreat of marine-based ice sheets). A more accurate version of eqn. 2.2 would include a time-varying ocean area; in this case, the equation takes the form of an integral over time (Lambeck et al., 1998).

The fourth term on the right-hand side, $S_{other}(\theta,\phi,t)$, is included for completeness and includes processes that most certainly affected sea levels during the Quaternary, but have not commonly been considered when interpreting observations under the assumption that their influence is relatively small. Examples of these processes include: compaction of the sediments that hosted the sea-level marker; changes in sea-surface height through ocean water temperature and salinity change; secular changes in prevailing wind patterns; sediment redistribution and the resulting isostatic and gravitational responses. Note that changes in tidal range through time are not explicitly included in eqn. 2.1 as these do not directly influence mean sea level; they do, however, influence the interpretation of a given SLIP when reconstructing ancient mean sea level (see previous section), and so will influence the value determined for $S_{rec}(\theta,\phi,t)$.

Accurate interpretation of the observed sea-level change at a given locality requires that each of the terms in eqn. 2.1 can be quantified. This is difficult to achieve and generally requires the use of models as well as observations of properties other than sea level. In most studies, the aim is to use the observations to arrive at an estimate of one of the first three components in the right-hand side of eqn. 2.1. For example, some researchers measure sea-level change to study tectonic processes, while others are more interested in climate change. The remainder of this sub-section will focus on the latter.

The first two terms on the right-hand side of eqn. 2.1 are the dominant sea-level components that relate to climate change during the Quaternary. Therefore, to accurately isolate the climate signal for interpretation (assuming processes grouped into the S_{other} term are negligible), the signal due to tectonic processes must be estimated and removed. The magnitude of the tectonic signal is generally small for the post-LGM period in areas that are tectonically stable. In areas near active plate margins, a significant correction could be required even for data of Holocene age. For data that are relatively old, dating to the early and mid-Quaternary, a significant tectonic signal will most likely be embedded in the observations, even when these are sourced from areas far removed from active plate boundaries (see, for example, Conrad and Husson, 2009).

Once the tectonic signal has been corrected for (or at least determined to be insignificant), the observations can be used to infer information on climate change. This is done through comparing output from a model that simulates the processes resulting in $S_{eust}(t)$ and $S_{iso}(\theta,\phi,t)$ to the corrected data. Models that simulate the sea-level response to mass exchange between land ice and oceans have been in existence since the 1970s (Farrell and Clark, 1976). The key components of these models, known as glacial isostatic adjustment (GIA) models, are outlined in Box 2.2.

Applications of GIA models have demonstrated that the complex spatial and temporal variations evident in the site-specific observations can be largely explained by isostatic and gravitational processes (e.g. Clark et al., 1978). As a result, observations of relative sea level have commonly been applied to constrain two key parameter sets in GIA models: ice-sheet histories and Earth viscosity structure (Milne and Shennan, 2007). The remainder of this sub-section will summarize the results of these types of model application, with an emphasis on constraining past ice-sheet histories.

It is useful to begin this discussion with an examination of some model output to illustrate the complexity of the sea-level response to changes in land ice. Figure 2.10 shows predictions of relative sea level – relative to the glacio-eustatic value – during the mid-Holocene, when glacio-eustatic sea level was several metres below the current value. The pattern shown in this figure reflects the processes at work (as described in Box 2.2). For example, in areas once covered by ice, such as Canada and northwest Eurasia, sea levels are higher than at present by tens of metres due, largely, to the active land uplift in these regions causing a sea-level fall (see Fig. 2.7). Peripheral to

these uplifting regions are areas of land subsidence leading to a sea-level rise and therefore sea levels that are significantly lower than the glacio-eustatic value. Both of these features in near- and intermediate-field areas are largely driven by the isostatic signal associated with the ice loading/unloading.

In far-field areas, there is a smaller, but significant, amplitude of spatial variability. In these areas, the influence of loading due to sea-level changes (known as 'hydro-isostasy') becomes more evident. For example, the thin region of high relative sea levels in shallow shelf areas (e.g. Indonesia) is due to this process. One consequence of this effect is the prevalence of a sea-level highstand in many equatorial areas (see Fig. 2.7). The influence of Earth rotation on sea level is also more apparent in far-field areas – the relatively high sea levels adjacent to southern South America are largely due to this effect.

As stated earlier, a common application of sea-level observations, near and far field, is the inference of Earth viscosity structure (see Milne and Shennan, 2007, and references therein). This property of the Earth's interior governs the rate and nature of planetary cooling and therefore tectonic evolution (e.g. Davies, 1999). Studies of this type date back to the classic analyses of Haskell and Vening-Meinesz in the 1920s and 1930s, who modelled the elevation of ancient shorelines in Fennoscandia (see Ekman, 1991, and references therein). The interpretation of near-field sea-level observations with the aim of inferring Earth viscosity has advanced significantly since these initial analyses, with recent studies using more sophisticated models, improved sea-level observations, and techniques to reduce the sensitivity of the viscosity inference to uncertainties in the ice history (Nakada and Lambeck, 1989; Mitrovica, 1996).

The use of sea-level observations to infer information of past ice evolution is a highly topical and important application of the data given the insights it provides on the response of ice sheets to climate change. The information obtained plays a central role in both constraining past changes in land ice on millennial to century timescales, as well as testing the accuracy of ice models that are employed to predict the future response of the present ice sheets to a specified climate scenario. The use of near-field records to infer the time distribution of ice extent also has a long history (see Milne and Shennan, 2007, and references therein). These typically focus on the deglacial history of the ice sheet following the LGM. In the same way that inferences of Earth viscosity are dependent to some degree on uncertainties in the ice model, inferences of ice-sheet changes are dependent on uncertainties in the Earth viscosity model. A key element in this application of sea-level data therefore involves reducing the sensitivity of the inference to unknowns in the Earth viscosity model. This is commonly done by carrying out a careful and

CONCEPTS BOX 2.2   Sea-level change and glacial isostatic adjustment (GIA) models

Models that compute sea-level change due to changes in land ice have been in development since the 1970s, beginning with the ground-breaking work of Farrell and Clark (1976), who presented a theory for accurately computing sea-level change due to the growth and melting of land ice. This theory can be expressed in the form of an equation known as the 'sea-level equation'. A concise form of this equation is:

$$S(\theta,\phi,t) = C(\theta,\phi)[G(\theta,\phi,t) - R(\theta,\phi,t) + H(t)] \qquad (2.3)$$

in which $S(\theta,\phi,t)$ is the change in sea level at a specific location with latitude (θ) and longitude (ϕ). The terms $G(\theta,\phi,t)$ and $R(\theta,\phi,t)$ represent the vertical deflections of the geoid and the Earth's solid surface, respectively, over the entire globe. $H(t)$ is a mass conservation term to ensure that the mass lost/gained by land ice is gained/lost by the oceans. This term includes two components, the eustatic change, as defined in eqn. 2.2, and a change known as syphoning, which accounts for changes in ocean basin volume over time as the ocean floor and ocean surface are perturbed through isostatic and gravitational changes (see Mitrovica and Milne, 2002). The term $C(\theta,\phi)$ is known as the ocean function and is defined as unity over ocean areas and zero over land areas. Note that this version of the sea-level equation does not account for changes in ocean area that occur during the glacial cycles. A newer version of this equation that does incorporate this effect is described in Mitrovica and Milne (2003). This paper, along with Kendall et al. (2005), is a good starting point for readers interested in learning more about the theory and current algorithms used to solve the sea-level equation.

The general elements of a glacial isostatic adjustment (GIA) model are shown in Fig. 2.9. An algorithm to solve the sea-level equation is embedded within the structure of these models. In essence, there are two key aspects: (1) a forcing, which comprises the surface-mass exchange between ice and oceans (of which the sea-level change is an important component) and the change in Earth rotation; and (2) the deformation response of the solid Earth to these forcing elements. The ice-history model can be developed in a number of ways, drawing on constraints from a variety of disciplines, including glacial geology, glaciology and GIA (see discussion in section 2.2.3). The Earth model used to compute the isostatic deformation in response to the forcings shown in Fig. 2.9 is spherical in geometry, with properties (density and rheology structure) that vary with depth only, or, in some more recent models (e.g. Wu et al., 2005), in three dimensions. The rheology adopted in most studies is that of a Maxwell visco-elastic body, which can exhibit an immediate elastic response as well as a lagged (linear) viscous response. Once the ice and Earth model components are defined, the other components shown in Fig. 2.9 can be computed (i.e. changes in Earth rotation and the sea-level response).

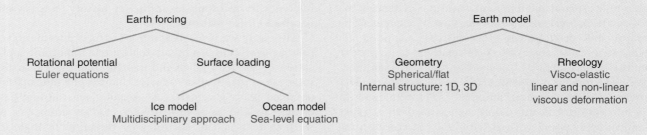

Fig. 2.9   Schematic illustration of the key elements comprising a model of the glacial isostatic adjustment (GIA) process. (Source: Adapted from Milne and Shennan 2007. Reproduced with permission of Elsevier.)

extensive forward modelling analysis to map out the sensitivity of model predictions to a subset of key parameters (e.g. Lambeck et al., 1998; Simpson et al., 2009). Recent studies have applied automated methods within a statistical framework to perform this task in a more robust manner (e.g. Tarasov et al., 2012).

Far-field relative sea-level data have most commonly been applied to constrain past changes in global land-ice volume. Given that isostatic and gravitational processes contribute significantly to sea-level changes even in the far field, it is necessary to remove these component signals in order to accurately determine the glacio-eustatic value. This procedure has been carried out most commonly for data at the LGM (e.g. Yokoyama et al., 2000) and during the mid-to-late Holocene (e.g. Nakada and Lambeck, 1989). The former provides a value that can be compared

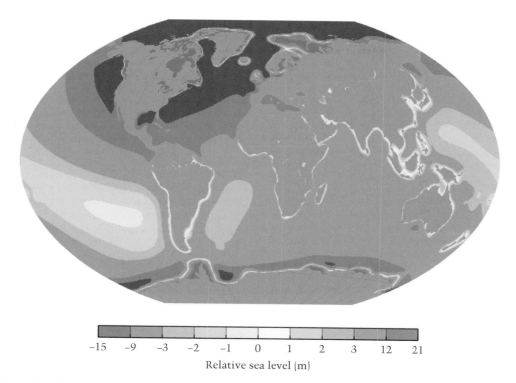

-15   -9   -3   -2   -1   0   1   2   3   12   21

Relative sea level (m)

Fig. 2.10   Relative sea level at 7 kyr BP from a glacial isostatic adjustment (GIA) model. Note that the model eustatic value is about –5 m at this time. For colour details, please see Plate 6.

to the $\delta^{18}O$ maxima at that time, and the latter provides an important background, or reference, rate of ice volume changes prior to the human-induced warming of our planet in the past century or so. This procedure has also been applied to observations of relative sea level during the last interglacial (Kopp et al., 2009). This period is of particular interest as it was a time when global temperatures were warmer than present and so serves as a partial analogue to gauge the response of the present ice sheets to projected warming in the coming centuries.

While far-field data are most commonly employed to infer changes in global ice volume, they also contain information on the spatial distribution of ice through time. These data are sensitive to changes in the volume partitioning of ice in different geographic regions rather than the more detailed regional changes in ice distribution. As a consequence, these data can provide relatively robust constraints on the large-scale evolution of global ice. One recent study made striking use of this sensitivity to place constraints on the ice sheets responsible for a rapid and large amplitude sea-level rise around 14 kyr BP (Clark et al., 2002). This rise of ~15 m in a few hundred years was first proposed from the interpretation of the Barbados data (see Fig. 2.7), which has since been confirmed with the addition of data from Indonesia

(Hanebuth et al., 2000) and Tahiti (Deschamps et al., 2012). The rapidity and magnitude of this reduction of ice volume (known as meltwater pulse 1A) leads to a distinct pattern of sea-level change that can be used to constrain or 'fingerprint' the melt source. The results of this particular study indicate that the existing data are most compatible with a large reduction in the volume of Antarctic ice at this time. Reconciling this scenario with geomorphological records of Antarctic ice retreat and models of Antarctic ice evolution is difficult and is the topic of ongoing research (Milne and Shennan, 2007).

In summary, the interpretation of sea-level data is made complex due to the number of processes that can influence sea level. The preceding discussion focuses on interpreting the climatic signal in sea-level data. To do this accurately requires that the signals due to tectonic, S_{tect}, and other, S_{other}, processes are either insignificant or removed. Once this is done, the data can be interpreted through the use of a model, in this case a model of glacial isostatic adjustment. Isolating the component signal of interest in the data can be difficult due to the sensitivity of predictions to both ice and Earth (viscosity) model parameters. As a consequence, great care is required to make robust inferences of either of these two parameter sets from relative sea-level observations.

2.3   Recent and future sea-level change

2.3.1   Introduction

The focus of this section is the past few centuries, and the 20th century in particular. As discussed in section 2.1.3, sea levels have been measured using instruments during this period: tide gauges provide data that span the 20th century (and longer at a handful of sites), and satellite altimetry data constrain near-global absolute sea-level changes since 1992. Sea-level changes during the past century or so are of particular interest from a climate perspective given the recent increase in global mean temperatures. As mentioned earlier, sea level is sensitive to climate through steric changes, changes in land ice, and changes in prevailing wind directions. The following section (2.3.2) will summarize some key results from the available data and section 2.3.3 will review the current interpretation of the data with a focus on the climate-induced signal. The final section (2.3.4) will briefly describe recent projections of future sea-level change and their uncertainties.

2.3.2   Sea-level observations

Tide gauges are devices that measure the height of the ocean surface relative to a local, land-based benchmark. Tide-gauge records provide a record of relative sea-level changes at distinct locations from as far back as a few centuries ago and up to the present day. There are a variety of different gauge designs and types of benchmark (see, for example, Pugh 2004). The global network of tide gauges has grown in number with time, as illustrated in Fig. 2.11. Stations in operation the longest (>60 years) tend to be found along the coasts of Europe, North America and Japan. The data from this network are available for downloading from the Permanent Service for Mean Sea Level (www.psmsl.org).

As evident from previous sections, many processes affect sea level and this fact is reflected in tide-gauge data from different locations. Figure 2.12 shows three tide-gauge records, each showing a signal that is dominated by a different process. The data from Stockholm show a steady fall of sea level over the past century of around 30 cm. This fall is dominated by isostatic uplift of the land in this region due to the melting of the large ice sheets that covered northwest Europe during the LGM. The data from Nezugaseki show a rapid sea-level rise of about 20 cm shortly after 1960. This distinct event was caused by rapid subsidence of this site during an earthquake. Finally, the data from Manila show rising sea level with a distinct acceleration around 1963 associated with increased subsidence due to groundwater mining. The results shown in Fig. 2.12 (and in Fig. 2.5) highlight the difficulty in interpreting tide-gauge records over long

(multi-decadal) timescales, as different processes can produce secular (long-term) sea-level trends.

A common application of tide-gauge data is to determine changes in global mean sea level during the 20th century. The main purpose in doing this has been to assess whether the recent climate warming is evident in the observed sea-level response. This has been done by comparing changes in rate (or accelerations and decelerations) during the tide-gauge monitoring period, as well as comparing the mean rate of rise during this recent period to estimates of this quantity during the late Holocene (from the geological record). As discussed earlier, changes in sea-surface height (absolute sea level) are more sensitive to climatic processes. Therefore, in many of these studies, a key element of the methodology adopted involved attempting to remove, or at least reduce, the signal associated with vertical land motion. For example, this can be done through a careful site selection process, using models to predict and remove this component signal, or, more recently, through the use of Global Positioning System technology to directly measure the land-motion component of the signal (e.g. Wöppelmann et al., 2009). Another potential problem in using tide-gauge data to arrive at a global mean sea-level change is the uneven spatial distribution of gauges – especially those that have been in place long enough to span most of the 20th century (see Fig. 2.11).

Figure 2.13 shows a recent estimate of global mean sea level since 1880 AD. Most estimates of this type indicate that the average rate of mean rise for this period has been within the range of 1–2 mm/yr, with estimates for the latter half of the 20th century giving values closer to 2 mm/yr (see, for example, Church et al., 2014). They also indicate a period of faster rise between, approximately, 1930 and 1960. An interpretation of this feature will be provided in the next sub-section.

Not surprisingly, there is significant spatial variability in sea-level change during the 20th century. This is illustrated in Fig. 2.12, in which the inter-site variability for these particular stations is dominated by local vertical land motion. There is also considerable variability within tide-gauge based reconstructions of change in different regions, as shown in Fig. 2.14, which indicates that processes affecting absolute sea level also add significantly to the spatial variability.

Satellite altimetry is a technique used to determine the height from an orbiting satellite to a chosen surface (e.g. the ocean) by measuring the time taken for a radar pulse to be emitted, reflected from the surface of interest, and then received back at the satellite. In order to determine the height of the ocean surface to a geodetic height datum (such as the reference ellipsoid) with sufficient accuracy, additional measurements are required. For

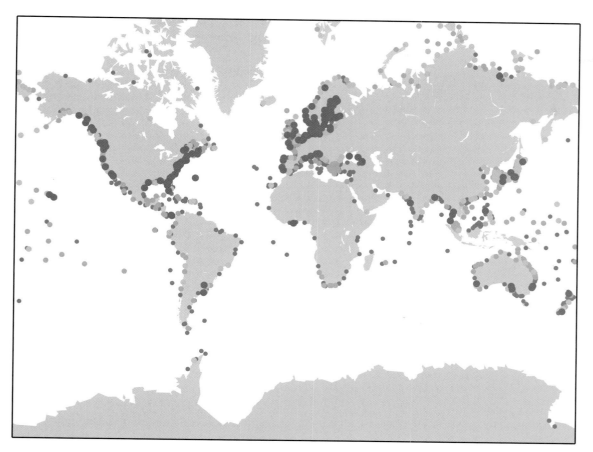

Fig. 2.11 Map showing the location and time span of data for the Revised Local Reference network of tide gauges. The different colours indicate the period over which a given gauge has been operational: purple, less than 20 years; green, 20–40 years; orange, 40–60 years; red, greater than 60 years. These gauges are regularly levelled relative to a local, land-based benchmark to check for vertical motion of the structure the tide gauge is mounted on (e.g. a pier). (Source: Data from Permanent Service for Mean Sea Level, from Holgate et al. 2013.) For colour details, please see Plate 7.

example, an instrument known as a gradiometer provides data used to determine the delay caused by water content in the atmosphere. Also, measurements of the position of the satellite are obtained using different techniques, such as laser ranging and the Global Positioning System, to ensure that the position of the satellite relative to the geocentric reference is known as accurately and precisely as possible. For more information on the technique of satellite altimetry, see Nerem and Mitchum (2000) and Pugh (2004).

Estimates of absolute sea level from various satellite missions (beginning with the launch of TOPEX/Poseidon in the early 1990s through to the Jason series) have revolutionized our understanding of sea-level change due to the almost complete cover of the ocean basins. As sea-surface height changes are much less sensitive to the influence of

vertical land motion compared to relative sea-level changes, the altimeter satellites provide data that include a more direct measure of climate-driven sea-level changes compared to tide-gauge data. One limitation of the altimetry data in studying the sea-level response to climate change is the relatively short monitoring period, although this is becoming less of an issue now that the time series (with TOPEX/Poseidon and Jason results spliced together) spans more than 20 years.

Figure 2.15 shows the global mean sea level curve obtained from altimeter observations for the period 1993 to the end of 2009. Note that a 60-day smoothing has been applied to these data and seasonal signals have been removed. The contribution from GIA, which causes a lowering of the global ocean surface at a rate of ~0.3 mm/yr, has also been removed from the data. The remaining

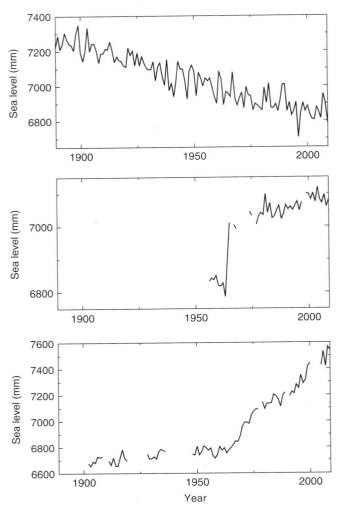

Fig. 2.12 Annual mean time series from three different tide gauges: Stockholm, Sweden (top); Nezugaseki, Japan (middle); and Manila, Philippines (bottom). Each frame covers the same time period. Note that the vertical axis range is different in each frame. See discussion of these data at www.psmsl.org/train_and_info/geo_signals/. (Source: Data from Permanent Service for Mean Sea Level, from Holgate et al. 2013.)

signal exhibits considerable interannual variability. Some of the largest fluctuations about the mean are associated with changes in wind patterns that are linked to the El Niño Southern Oscillation in the Pacific Ocean. For example, the large fluctuations in 1997–1998 and in recent years are thought to be due to this phenomenon. The mean rate of change over the 17-year period spanned in Fig. 2.15 is 3.2 ± 0.6 mm/yr (90% confidence interval). This is a significantly faster rise compared to that estimated for the 20th century from tide-gauge records.

Figure 2.16 shows the spatial variation in the rate of sea-level change for the same period covered in Fig. 2.15.

The spatial variation in the rates over this period is around 20 mm/yr, which is almost an order of magnitude larger than the global mean. There are large areas where the rates are close to 10 mm/yr (western Pacific) and others where there has been a marked sea-level fall over the same period (eastern Pacific). The large spatial variability in the Pacific is linked to the El Niño Southern Oscillation, which has a strong influence over decadal timescales.

2.3.3 Interpretation of the observations

For many of the same reasons discussed in section 2.2.3, interpreting instrumental records of recent sea-level change is not straightforward. A primary focus of much past and recent research has been to quantify the contribution of climate-related processes to the observed changes. This sub-section will summarize recent progress in this regard.

The global mean sea-level change inferred from the tide-gauge observations has received much attention in terms of understanding the contributions to the mean rate of rise during the 20th century. Most studies have focused their attention on the latter half of the 20th century, as observations of ocean temperature change are considerably more abundant (though still lacking in geographic and depth coverage) for this period. The most recent results conclude that summing the contributions due to ocean warming and land-ice melt leads to a value that overlaps with the observed value to within uncertainty (Church and White, 2010), with the thermosteric and land-ice contributions being of similar magnitude at about 50% of the total observed change. Changes in the amount of water stored on land through anthropogenic activity (e.g. dam building) have also been considered as a possible contributor to the observed signal. Recent studies indicate that the magnitude of this component is likely to be small, but fluctuations in land-water storage are a main contributor to interannual sea-level variability (Llovel et al., 2011).

It is of interest to compare the rate of global mean sea-level change during the 20th century to that of the preceding centuries and millennia. If the rate is significantly different, then one can argue that the cause of the change is the recent increase in global temperatures. Proxy records indicate that sea levels were relatively stable during the past few millennia, with long-term rates of change (after the influence of isostatic processes is removed) being effectively zero to within observational uncertainty (not more than a few tenths of a millimetre per year fall or rise over multi-century time periods). The mean rate during the 20th century is therefore anomalously high in this context. This interpretation is supported by site-specific, high-resolution records of sea-level change reconstructed from salt marshes

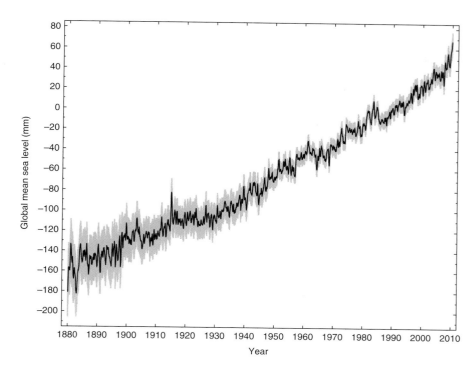

Fig. 2.13 A reconstruction of global mean sea level from tide-gauge data. The grey bar indicates the uncertainty. Note that the time series is zeroed at 1990. (Source: Data from Church and White 2011.)

and corals. Many of these records show a clear increase in the rate of change between the 19th and 20th centuries (e.g. Gehrels et al., 2008).

Reductions in the rate of global mean sea-level rise during the 20th century are known to be linked to volcanic activity (Church et al., 2005). Large volcanic eruptions release sulphur dioxide into the atmosphere, which scatters incoming radiation, leading to a rapid and large-scale cooling. There is a distinct lack of volcanic activity in the period 1930–60, and so this can at least partly explain why the rate of rise is relatively high in this middle part of the century (see Fig. 2.14).

The spatial variability in rates of sea-level change during the 20th century is to be expected given the processes (land ice melting and ocean warming) believed to dominate the signal. Both of these processes lead to a spatially variable sea-level response. For example, Fig. 2.17 shows the spatial variability in sea-level change estimated from changes in ocean temperature and salinity for the latter part of the 20th century, as well as the pattern of change associated with an assumed rate of melting from the ice sheets. Other processes also play a role in adding to this variability; for example, melting of mountain glaciers, vertical land motion due to tectonic processes as well as the ongoing isostatic adjustment to the last major deglaciation (~20–7 kyr BP), as well as long-term ocean dynamical changes (e.g. Miller and Douglas, 2007). Adding all of these processes together results in a complex pattern of sea-level changes during the 20th century that is reflected in the tide-gauge observations (see Fig. 2.14).

Compared to the tide-gauge data, interpreting the altimetry data is made easier by the dramatic improvement in the data control on changes in sea level as well as the processes that contribute to these changes over the past decade. Information on mass changes in the cryosphere (particularly the ice sheets) has improved through the application of satellites that measure the altitude of the ice surface via altimetry and the velocity field of the ice surface via the technique known as Interferometric Synthetic Aperture Radar (InSAR). A third technique, satellite gravimetry, has resulted in key information on mass changes in the large ice sheets, glaciers and ice caps through the Gravity and Climate Experiment (GRACE) satellite gravity mission, launched in 2002. The data obtained from this mission have also provided important constraints on mass changes over the oceans and land that, as discussed later, feed into current interpretations of contemporary sea-level change. One other observational initiative that plays a fundamental role in understanding recent sea-level change is the Argo ocean observing programme, in which a network of floats deployed in the ocean traverse the water column, making measurements of ocean water temperature, salinity and velocity.

Based on the ocean-altimeter measurements, the average rate of global mean sea-level rise was 3.2 ± 0.8 mm/yr for the period 1993–2011 (Leuliette and Willis, 2011). This rate is significantly greater than that inferred for the 20th century and suggests that the processes responsible for the 20th century global mean rise have accelerated considerably during this time. Interpretation of data resulting

46   GLENN A. MILNE

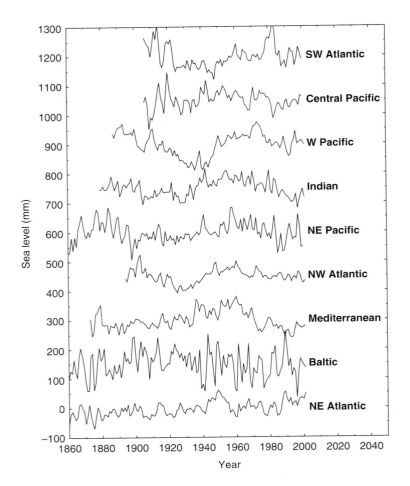

Fig. 2.14 Sea-level curves determined from tide-gauge data that represent different ocean basins. These results demonstrate that there is significant spatial variation even at this larger scale. This can be compared to the inter-site variability evident in Fig. 2.12. Note that each time series has been offset along the y-axis by an arbitrary amount so as to avoid overlap. (Source: Woodworth et al. 2009. Reproduced with permission of John Wiley & Sons.)

from the observation programmes outlined above support this hypothesis. Over the period spanned by ocean-altimeter observations, results for the first decade (1993–2003) show that the contributions from land-ice melt and ocean warming to the global mean rise were of similar magnitude (each being ~50% of the observed rise). From 2003 to 2008, the observations indicate that the thermosteric contribution has decreased and the land-ice contribution increased, such that the latter was responsible for ~75% of the total rise. Whether this change reflects shorter-term variability of the climate system rather than a longer-term trend will become evident as the monitoring period is extended.

One issue that has arisen in attempting to close the sea-level budget in the past few years, during which both GRACE and Argo data have been available, is the short time series involved and the difficulty in isolating a meaningful, secular, trend from the observations, given the dominance of interannual variability. The most recent papers indicate that the budget can be closed to within observational uncertainty (e.g. Cazenave et al., 2008; Leuliette and Willis, 2011).

Spatial variability in Fig. 2.16 correlates well with the pattern of sea-level change estimated from ocean-temperature measurements (e.g. Willis et al., 2011). If the current interpretation that land-ice melt has become dominant in recent years is correct, the spatial pattern associated with this process should become evident. However, at the time of writing, a robust ice-melt fingerprint has yet to be detected.

2.3.4 Estimating future sea levels

Models are required to produce a quantitative estimate of how sea level will change in the coming decades to centuries. To date, most studies have focused on projecting global mean sea level for the 21st century. Note that the term 'projection' has a different meaning to 'prediction': a prediction is a single estimate with a quantifiable uncertainty; a projection is an estimate that is based on a specific assumption related to a necessary model input – in the case of sea level, the amount of future greenhouse gas concentration in the atmosphere is assumed for a given projection. An accurate prediction of this quantity is not possible given

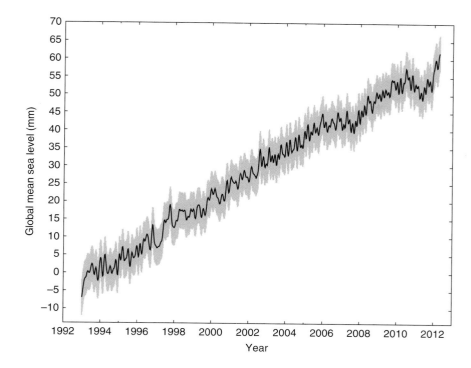

Fig. 2.15 A determination of global mean sea level from altimeter data. The data have been corrected for a number of effects (see text). The rate of rise is approximately double that determined for the 20th century from tide gauges (see Fig. 2.13). The grey bar indicates an adopted uncertainty of 5 mm. (Source: Data from AVISO: www.aviso.oceanobs.com)

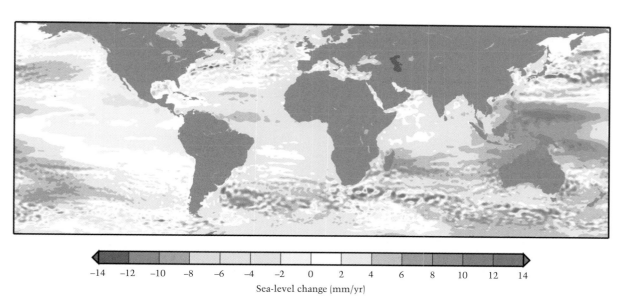

Fig. 2.16 Estimated trends in sea level for the period 1993 through to the end of 2009 from satellite altimeter data. The spatial variability in rate is a few centimetres and thus is an order of magnitude larger than the global mean rise (3.2 mm/yr) over the same period. Note that the altimeter satellite missions only measure ocean height between latitudes 66° north and south, hence the truncation of the map. (Source: Data from AVISO: www.aviso.oceanobs.com) For colour details, please see Plate 8.

that this will depend on a number of socio-economic, political and technological factors that are inherently unpredictable. Therefore, future changes in climate and sea level are projected for a handful of chosen greenhouse gas concentration scenarios, or so-called representative concentration pathways (van Vuuren et al., 2011). Global

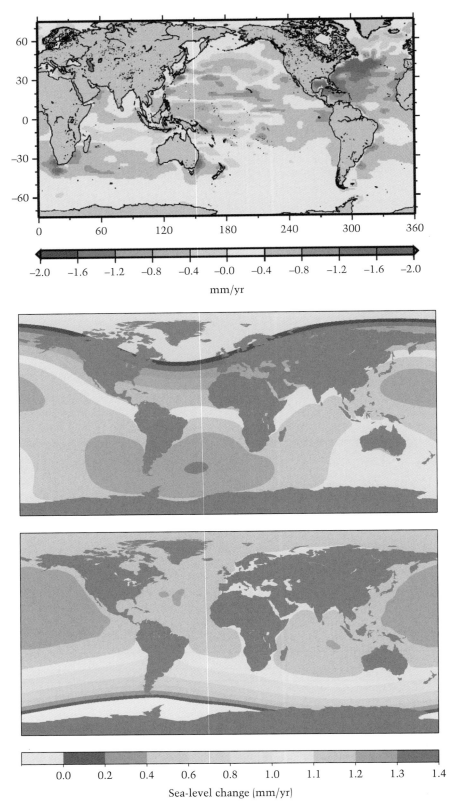

Fig. 2.17 Top frame shows an estimate of the rates of sea-level change for the period 1950–2003 due to changes in ocean temperature and salinity. (Source: Berge-Nguyen et al. 2008. Reproduced with permission from Elsevier.) The bottom two frames show the modelled relative sea-level response assuming a mass loss equivalent to 1 mm/yr of glacio-eustatic rise for the Greenland (middle frame) and West Antarctic (bottom frame) ice sheets. These model results include the influence of changes in gravity and Earth rotation. Note that the response of the solid Earth and gravity field result in a sea-level fall in the vicinity of the ablating ice sheet and an enhanced sea-level rise (relative to the glacio-eustatic value) in the far field. One consequence of this is that low-latitude regions will experience the greatest sea-level rise associated with melting of the polar ice sheets. (Source: Adapted from Milne et al. 2009. Reproduced with permission of Nature Publishing Group.) For colour details, please see Plate 9.

mean sea-level projections have been based on two classes of model: one that is relatively complex and simulates the relevant processes (e.g. changes in ocean temperature and the resulting sea-level response due to changes in ocean density structure and circulation); and one that does not attempt to simulate the processes but, instead, applies a semi-empirical relationship based on past observations of sea level and temperature or radiative forcing.

Assessment reports by the Intergovernmental Panel on Climate Change provide a good resource for the results obtained from process models; for example, Church et al. (2014) present a suite of results that place global mean sea level between a few decimetres to just under a metre by 2100. As noted by the authors of this report, these estimates are based on a certain statistical confidence level (67–100%), and so there is a 0–33% chance of global mean sea-level rise at 2100 lying outside this range. However, a large response of the Antarctic ice sheet is probably the only viable mechanism that could result in sea levels being significantly higher (several tens of centimetres) than the upper bound of just under 1 m. See Church et al. (2014) for more details on the various contributions to the total projected global mean sea-level rise.

A number of results from semi-empirical models have been published since 2007 (e.g. Rahmstorf, 2010). All of these apply a scaling relationship between temperature (or radiative forcing) and global mean sea level that is calibrated to match observations in the recent past (centuries to millennia). This scaling is then used to arrive at projections of global mean sea level for the coming century based on output from climate models forced by specific greenhouse gas scenarios. In comparing projections based on process models to those obtained from semi-empirical models, it is evident that the semi-empirical models consistently project a higher range of values. There are pros and cons associated with each type of model, and understanding the differences in model output is an active research topic.

Observations of past sea levels demonstrate that the changes over a range of timescales (10^6–10^0 yr) are spatially variable. Therefore, the changes projected in the future will also exhibit a significant degree of variability. Projections of spatial variability in sea level have been published (Slangen et al., 2011). It should be noted, however, that there remain large uncertainties in projecting future sea levels at regional scales (Church et al., 2014).

As mentioned earlier, the majority of projections focus on where sea level will be by the end of the 21st century. Of course, sea levels will continue to change beyond 2100 in response to past and contemporary forcings of the climate system and tectonic processes. The erosional and depositional processes that are active in coastal zones will also play a role in determining how local relative sea level will change in the future (e.g. Anderson et al., 2010). None of the models currently used to project future changes incorporate these processes.

2.4 Summary

Changes in sea level are an external control on the location of the zone in which coastal processes operate. Constraining past sea-level changes is therefore an integral part of understanding the evolution of the coastal zone as evident in the geological record. During the Quaternary, sea levels fluctuated with magnitudes in the order of 100 m with large spatial variation (100s of metres) in these changes due to the deformation of the solid Earth and changes in the gravity field associated with the mass flux between land ice and oceans as large ice sheets grew and melted. These processes were the major control on sea level and coastal evolution during the Quaternary. Whether the large amplitude and quasi-periodic changes in land ice that characterized the Quaternary will continue in the future remains to be seen. A growing body of evidence suggests that the recent warming trend will likely continue for at least several centuries as the system responds to an anthropogenic forcing. Sea levels will continue to change in the future and these changes are expected to be at an increasing rate of rise in most locations over the coming centuries due to the projected rise in global mean temperatures. The future response of the Earth's coastlines will reflect these rate increases as well as changes to other governing processes described in this book.

• Sea level is defined in two ways: (1) the height of the sea surface relative to the sea floor (known as relative sea level); and (2) the height of the sea surface relative to a geodetic reference such as the reference ellipsoid (known as absolute sea level).

• A variety of processes operating over a wide range of temporal and spatial scales influence sea level. As a consequence, observations of sea-level change are difficult to interpret.

• The ice-volume component of Quaternary sea-level changes has been estimated by measuring the isotopic signature of microfossils in ocean sediments. The resulting time series provide a continuous record of ice-volume changes during the Quaternary (and earlier). Comparing these data to known changes in the Earth-Sun orbit support the hypothesis that the ice sheets were responding to the climate changes associated with this forcing.

• Ancient sea level can be reconstructed using biological and morphological evidence found in the geological record. These site-specific reconstructions demonstrate that Quaternary sea-level change was dominated by the growth and melting of large continental ice sheets, and that a large component of the observed changes was due to the response of the solid Earth (deformation and change in rotation) and gravity field to the mass exchange between land ice and the oceans.

• Sea levels have been measured using tide gauges and satellites in the past few centuries and decades, respectively. These observations indicate that global mean sea level is rising at a few millimetres per year. This rate is an order of magnitude higher than that estimated from proxy data for the past few millennia, indicating that sea levels have responded to a warming of the climate in the past century or so.

- The rise in global mean sea level since the 1950s has been interpreted to be due to ocean warming and land ice melting (each contributing ~50% to the total). Both of these processes induce large spatial variability in the sea-level response.
- Projections of future global mean sea-level rise fall within the range of a few decimetres to a few metres in the coming century. There will be substantial spatial variability about the global mean and so an important element of contemporary research is aimed at improving the accuracy of regional and local sea-level projections.

Key publications

Church, J.A., Woodworth, P.L., Aarup, T. and Wilson, W.S. (Eds), 2010. *Understanding Sea-Level Rise and Variability*. Wiley-Blackwell, Chichester, UK.
[*State-of-the-art treatise on the current understanding of sea-level changes*]
Elias, S. (Ed.), 2007. *Encyclopedia of Quaternary Science*. Elsevier, London, UK.
[*Sections under 'Sea Level Studies' provide a comprehensive account on reconstructing Quaternary sea levels from the geological record*]
Oceanography, Special Issue on Sea Level, **24**(2), 2011.
[*Up-to-date and accessible volume on sea-level change on a range of timescales*]
Pugh, D., 2006. *Changing Sea Levels: Effects of Tides, Weather and Climate*. Cambridge University Press, Cambridge, UK.
[*Good reference text on sea-level changes due to a variety of phenomena*]

Acknowledgements

Anisa Ramia produced a number of the figures in this chapter.

References

Anderson, J.B., Rodriguez, A.B., Milliken, K.T., Simms, A., and Wallace, D., 2010. Is predicted coastal impact from accelerated sea-level rise underestimated? *Earth Observation Systems*, **91**, 205–206.

Bard, E., Hamelin, B., Fairbanks, R.G. and Zindler, A., 1990. Calibration of the 14C timescale over the past 30,000 years using mass spectrometric U–Th ages from Barbados corals. *Nature*, **345**, 405–410.

Berge-Nguyen, M., Cazenave, A., Lombard, A. et al., 2008. Reconstruction of past decades sea level using thermosteric sea level, tide gauge, satellite altimetry and ocean reanalysis data. *Global and Planetary Change*, **62**, 1–13.

Bird, M.I., Fifield, L.K. and Goh, B., 2004. Calculating sediment compaction for radiocarbon dating of intertidal sediments. *Radiocarbon*, **46**, 421–435.

Cazenave, A., Lombard, A. and Llovel, W., 2008. Present-day sea level rise: a synthesis. *Comptes Rendus Geoscience*, **340**, 761–770.

Church, J., White, N. and Arblaster, J., 2005. Significant decadal-scale impact of volcanic eruptions on sea level and ocean heat content. *Nature*, **438**, 74–77.

Church, J.A. and White, N.J., 2011. Sea level rise from the late 19th to the early 21st century. *Surveys in Geophysics*, **32**, 585–602.

Church, J.A., Clark, P.U., Cazenave, A. et al., 2014. Sea level change. In: T.F. Stocker, D. Qin, G.-K. Plattner et al. (Eds), *Climate Change 2013: The Physical Science Basis*. Contribution of Working Group I to the Fifth Assessment Report of the Intergovernmental Panel on Climate Change. Cambridge University Press, Cambridge, UK, and New York, NY, USA.

Clark, J.A., Farrell, W.E. and Peltier, W.R., 1978. Global changes in postglacial sea level: a numerical calculation. *Quaternary Research*, **9**, 265–287.

Clark, P.U., Mitrovica, J.X., Milne, G.A. and Tamisiea, M., 2002. Sea-level fingerprinting as a direct test for the source of global meltwater pulse 1A. *Science*, **295**, 2438–2441.

Clark, P.U., Archer, D., Pollard, D. et al. 2006. The Middle Pleistocene transition: characteristics, mechanisms, and implications for long-term changes in atmospheric pCO2. *Quaternary Science Reviews*, **25**, 3150–3184.

Conrad, C.P., and Husson, L., 2009. Influence of dynamic topography on sea level and its rate of change. *Lithosphere*, **1**, 110–120.

Davies, G.F., 1999. *Dynamic Earth: Plates, Plumes and Mantle Convection*. Cambridge University Press, Cambridge, UK.

Deschamps, P., Durand, N., Bard, E., Hamelin, B., Camoin, G., Thomas, A.L., Henderson, G.M., Okuno, J., Yokoyama, Y., 2012. Ice-sheet collapse and sea-level rise at the 1 Bølling warming, 14,600 years ago. *Nature*, **483**, 559–564.

Edwards, R.J., 2007. Low energy coasts sedimentary indicators. In: S. Elias (Ed.), *Encyclopedia of Quaternary Sciences*. Elsevier, London, UK, pp. 2994–3006.

Ekman M., 1991. A concise history of postglacial land uplift research (from its beginning to 1950). *Terra Nova*, **3**, 358–365.

Fairbanks, R.G., 1989. A 17,000-year glacio-eustatic sea level record: influence of glacial melting rates on the Younger Dryas event and deep-ocean circulation. *Nature*, **342**, 637–642.

Farrell, W.E. and Clark, J.A., 1976. On postglacial sea level. *Geophysical Journal of the Royal Astronomical Society*, **46**, 647–667.

Fowler, C.M.R., 2005. *The Solid Earth: An Introduction to Global Geophysics* (2nd edn). Cambridge University Press, Cambridge, UK.

Gehrels, W.R., Hayward, B.W., Newnham, R.M. and Southall, K.E., 2008. A 20th century sea-level acceleration in New Zealand. *Geophysical Research Letters*, **35**, L02717, doi: 10.1029/2007GL032632.

Hanebuth, T., Stattegger, K. and Grootes, P.M., 2000. Rapid flooding of the Sunda Shelf: a Late-Glacial sea-level record. *Science*, **288**, 1033–1035.

Holgate, S.J., Matthews, A., Woodworth, P.L., et al., 2013. New data systems and products at the Permanent Service for Mean Sea Level. *Journal of Coastal Research*, **29**(3), 493–504, doi: 10.2112/JCOASTRES-D-12-00175.1.

Kendall, R., Mitrovica, J.X., Milne and G.A., 2005. On post-glacial sea level – II. Numerical formulation and comparative results on spherically symmetric models. *Geophysical Journal International*, **161**, 679–706.

Kopp, R.E., Simons, F.J., Mitrovica, J.X., Maloof, A.C. and Oppenheimer, M., 2009. Probabilistic assessment of sea level during the last interglacial stage. *Nature*, **462**, 863–867.

Lambeck, K., 1988. *Geophysical Geodesy*. Oxford Science Publications. Clarendon Press, Oxford, UK.

Lambeck, K., Smither, C. and Johnston, P., 1998. Sea-level change, glacial rebound and mantle viscosity for northern Europe. *Geophysical Journal International*, **134**, 102–144.

Leuliette, E.W., and Willis, J.K., 2011. Balancing the sea level budget. *Oceanography*, **24**(2), 122–129.

Lisiecki, L., and Raymo, M., 2005. A Pliocene-Pleistocene stack of 57 globally distributed benthic δ18O records. *Paleoceanography*, **20**, PA1003, doi: 10.1029/2004PA001071.

Llovel, W., Becker, M., Cazenave, A. et al., 2011. Terrestrial waters and sea level variations on interannual time scale. *Global and Planetary Change*, **75**, 76–82.

Marshall, W.A., Gehrels, W.R., Garnett, M.H., Freeman, S.P.H.T., Maden, C. and Xu, S., 2007. The use of 'bomb spike' calibration and high-precision AMS ^{14}C analyses to date salt-marsh sediments deposited during the past three centuries. *Quaternary Research*, **68**, 325–337.

Miller, K.G., Kominz, M.A., Browning, J.V. et al. 2005. The Phanerozoic record of global sea-level change. *Science*, **310**, 1293–1298.

Miller, L., and Douglas, B.C., 2007. Gyre-scale atmospheric pressure variations and their relation to 19th and 20th century sea level rise. *Geophysical Research Letters*, **34**, L16602, doi: 10.1029/2007GL030862.

Milne, G.A. and Mitrovica, J.X., 1998. Postglacial sea-level change on a rotating Earth. *Geophysical Journal International*, **133**, 1–19.

Milne, G.A. and Shennan, I., 2007. Isostasy. In: S. Elias (Ed.), *Encyclopedia of Quaternary Sciences*. Elsevier, London, UK, pp. 3015–3023.

Milne, G.A., Gehrels, W.R., Hughes, C. and Tamisiea, M.E., 2009. Identifying the causes of sea-level change. *Nature Geoscience*, **2**, 471–478.

Mitrovica, J.X., 1996. Haskell (1935) revisited. *Journal of Geophysical Research*, **101**, 555–569.

Mitrovica, J.X. and Milne, G.A., 2002. On the origin of postglacial ocean syphoning. *Quaternary Science Reviews*, **21**, 2179–2190.

Mitrovica, J.X. and Milne, G.A., 2003. On post-glacial sea level – I. General theory. *Geophysical Journal International*, **154**, 253–267.

Moucha, R., Forte, A.M., Mitrovica, J.X. et al., 2008. Dynamic topography and long-term sea level variations: there is no such thing as a stable continental platform. *Earth and Planetary Science Letters*, **271**, 101–108.

Mound, J.E. and Mitrovica, J.X., 1998. True polar wander as a mechanism for second-order sea-level variations. *Science*, **279**, 534–537.

Nakada, M. and Lambeck, K., 1989. Late Pleistocene and Holocene sea-level change in the Australian region and mantle rheology. *Geophysical Journal International*, **96**, 497–517.

Neill, S.P., Scourse, J.D. and Uehara, K., 2010. Evolution of bed shear stress distribution over the northwest European shelf seas during the last 12,000 years. *Ocean Dynamics*, **60**, 1139–1156.

Nerem and Mitchum 2000. Observations of sea level change from satellite altimetry. In: B.C. Douglas, M.S. Kearney and S.P. Leatherman (Eds), *Sea Level Rise: History and Consequences*. Academic Press, San Diego, CA, USA, pp. 121–164.

Pavlis, N.K., Holmes, S.A., Kenyon, S.C. and Factor, J.K., 2012. The development and evaluation of the Earth Gravitational Model 2008 (EGM2008), *Journal of Geophysical Research: Solid Earth*, **117**, B04406, doi:10.1029/2011JB008916.

Pugh, D., 2004. *Changing Sea Levels: Effects of Tides, Weather and Climate*. Cambridge University Press, Cambridge, UK.

Rahmstorf, S., 2010. A new view on sea level rise. *Nature Climate Change*, **4**, 44–45.

Raymo, M., Lisiecki, L. and Nisancioglu, K., 2006. Plio-pleistocene ice volume, Antarctic climate, and the global δ18O record. *Science*, **313**, 492–495.

Ruddiman, W.F., 2007. *The Earth's Climate: Past and Future* (2nd edn). W.H. Freeman and Company, New York, USA.

Shennan I., 2007. Overview. In: S. Elias (Ed.), *Encyclopedia of Quaternary Sciences*. Elsevier, London, UK, pp. 3015–3023.

Simpson, M.J.R., Milne, G.A., Huybrechts, P. and Long, A.J., 2009. Calibrating a glaciological model of the Greenland ice sheet from the last glacial maximum to present-day using field observations of relative sea level and ice extent. *Quaternary Science Reviews*, **28**, 1630–1656.

Slangen, A.B.A., Katsman, C.A., van de Wal, R.S.W., Vermeersen, L.L.A. and Riva, R.E.M., 2011. Towards regional projections of twenty-first century sea-level change based on IPCC SRES scenarios. *Climate Dynamics*, doi: 10.1007/s00382-011-1057-6.

Tarasov, L., Dyke, A.S., Neal, R.M. and Peltier, W.R., 2012. A data-calibrated distribution of deglacial chronologies for the North American ice complex from glaciological modeling. *Earth and Planetary Science Letters*, **315–316**, 30–40.

van Vuuren, D.P., Edmonds, J., Kainuma, M. et al., 2011. The representative concentration pathways: an overview. *Climatic Change*, **109**, 5–31.

Waelbroeck, C., Labeyrie, L., Michel, E. et al., 2002. Sea level and deep water temperature changes derived from benthic foraminifera isotopic records. *Quaternary Science Reviews*, **21**, 295–305.

Willis, J.K., Chambers, D.P., Kuo, C.-Y. and Schum, C.K., 2011. Global sea level rise: recent progress and challenges for the decade to come. *Oceanography*, **23**(4), 26–35.

Woodroffe, S.A., 2009. Testing models of mid to late Holocene sea-level change, North Queensland, Australia. *Quaternary Science Reviews*, **28**, 2474–2488.

Woodworth, P.L., White, N.J., Jevrejeva, S., Holgate, S.J., Church, J.A. and Gehrels, W.R., 2009. Evidence for the accelerations of sea level on multi-decade and century timescales. *International Journal of Climatology*, **29**, 777–789, doi: 10.1002/joc.1771.

Wöppelmann, G., Letetrel, C., Santamaria, A. et al., 2009. Rates of sea-level change over the past century in a geocentric reference frame. *Geophysical Research Letters*, **36**, L12607, doi: 12610.11029/12009GL038720.

Wu, P., Wang, H. and Schotman, H., 2005. Postglacial induced surface motions, sea-levels and geoid rates on a spherical, self-gravitating laterally heterogeneous earth. *Journal of Geodynamics*, **39**, 127–142.

Yokoyama, Y., Lambeck, K., De Deckker, P., Johnston, P. and Fifield, L., 2000. Timing of the Last Glacial Maximum from observed sea level minima. *Nature*, **406**, 713–716.

3 Environmental Control: Geology and Sediments

EDWARD J. ANTHONY

Aix Marseille Université, Institut Universitaire de France, Europôle Méditerranéen de l'Arbois, Aix en Provence Cedex, France

3.1 Geology and sediments: setting boundary conditions for coasts

3.1.1 Coastal diversity: a heritage of geology and sediments

Coasts are extremely diverse in their morphology and sedimentary composition, and this diversity is largely explained by geology and by the sediments accumulating on coasts (Fig. 3.1). The geology determines the primary boundary condition within which the coast is formed and within which the processes of coastal evolution operate. The coastal boundary corresponds to the land-ocean interface, and involves a consideration of the overall tectonic framework and lithology. Embedded in these primary boundary conditions are sediment type and input. Three other important factors are the coastal space available for sediment to accumulate, called 'accommodation space', the topography, and the orientation of the coast, all of which reflect primary geological controls as well as the influence of sediment storage and redistribution in the coastal zone under the influence of coastal processes. Overall, therefore, coasts may range, in their morphology, from primary types hinged

Coastal Environments and Global Change, First Edition. Edited by Gerd Masselink and Roland Gehrels.
© 2014 John Wiley & Sons, Ltd. Published 2014 by John Wiley & Sons, Ltd. Companion Website: www.wiley.com/go/masselink/coastal

Fig. 3.1 Illustration of the influence of geology on coastal geomorphology. The photograph shows sand and gravel beaches (the gravel has accumulated in the breaker zone) between headlands at Miraflores, Lima, Peru. These embayed beaches are backed by cliffs cut into alluvial deposits comprising ancient raised beaches ('rasa') on the tectonically active coastal margin of South America (see also Fig. 3.4).

on the bedrock geology, to forms that reflect a large diversity of arrangement of sedimentary facies.

Although sediment type and input depend on geology, they are strongly influenced by climate. Climate directly affects the rate at which rocks are altered physically and chemically to eventually liberate sediments, and indirectly influences the transport of these sediments to the coast. Sediment supply to the coast is significantly influenced by climate change through modification of rates of landscape erosion, river flow, and storm intensity, all contributing to the release and transport of sediments. Sediment supply to coasts is also strongly affected by human activities, especially through agriculture and land use in river catchments, and through dams, thus mitigating the potential effects of a warmer climate on enhanced fluvial sediment discharge. This can, in turn, exacerbate coastal erosion induced by climate change and sea-level rise, especially in deltaic areas. Sediment supply along certain coasts has also been perturbed considerably by port and engineering structures, the latter generally aimed at

controlling erosion. On many developed coasts, cliffs have been stabilized and no longer provide sediments for the adjacent coasts, whereas many open beaches are subjected to restricted longshore drift due to the construction of groynes and breakwaters.

3.1.2 Spatial and temporal scales: from global tectonics to local geological controls

In order to understand the way geology and sediments form key environmental controls on coasts, an approach based on a consideration of the spatial and temporal scales within which these two controls operate is required. These scales range from global to local, with sediments derived directly or indirectly from the breakdown of rocks becoming more influential at smaller spatial scales. The primary geological control on coasts may be considered in terms of global tectonics, which are responsible for the formation and destruction of coasts, and for vertical and horizontal coastal movements at long (millions of years) timescales. At shorter

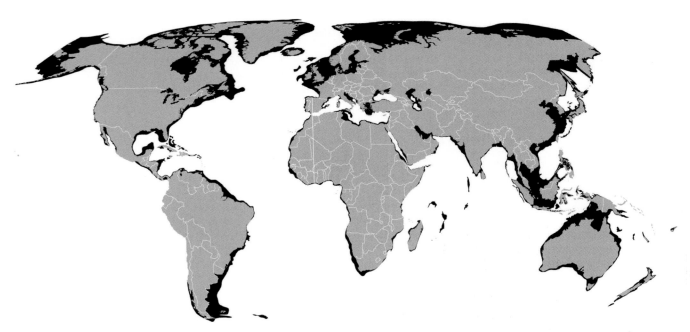

Fig. 3.2 World map showing variations in continental shelf width that are largely a product of the global tectonic regime and sediment trapping along plate margins.

timescales, the morphology of a coast is determined by both the prevailing regional/local tectonic pattern and by the rock characteristics or lithology, including sediments, and by the quantity of sediments available. The tectonic pattern is generally a heritage of the overall plate tectonic context but it should be noted that marked differences in tectonic uplift or subsidence patterns can occur over short coastal distances (sometimes <1 km), especially where still active faults or fracture zones affect the coast. Another aspect of regional tectonics important to the development of coasts is where isostatic rebound of the Earth's crust has occurred in response to both ice cap weight during the last glaciation and to water weight over the continental shelf following the postglacial rise in sea level. The regional/local tectonic pattern influences the position of the shore relative to sea level, and thus the boundaries within which processes that mould coasts operate, whereas the lithology determines the type of rock at the coast. Both of these parameters also strongly influence sediment supply to coasts.

The width of the continental shelf, and the gradient of the shoreface and the backshore zone are other extremely important criteria that reflect jointly geological and inherited morphological controls. Continental shelf width is strongly hinged on the age and level of tectonic activity of the continental margin. It plays a dominant role in determining the sediment accommodation space,

and whether a coast is dominated by waves or tides. The concept of accommodation space is disccussed in section 3.3.2. Waves are highest over steep, narrow continental shelves but can be considerably dissipated where the continental shelf is broad and shallow, especially near the coast, whereas tides can be strongly amplified over broad, shallow shelves (Fig. 3.2). Wide, shallow continental shelves are generally characterized by low-gradient shorefaces and backshore zones, whereas narrow continental shelves exhibit steep shorefaces and backshore zones. The gradient of the shoreface and backshore zone strongly influences the response of the coastal depositional environment to sea-level rise, for example, through affecting the rate of transgression and the provision of accommodation space.

3.2 Geology and coasts

3.2.1 The pervasive role of plate tectonics

Plate tectonics are at the origin of the world's coastlines and are also a determining factor in the regional tectonic pattern exhibited by a coast. The basic tectonic processes of plate breakup and accretion are the initiators of coasts and the oceans and seas they border. Collision, and to a certain extent subduction processes, is, on the other hand, responsible for the destruction of coasts. These are extremely complex processes that include

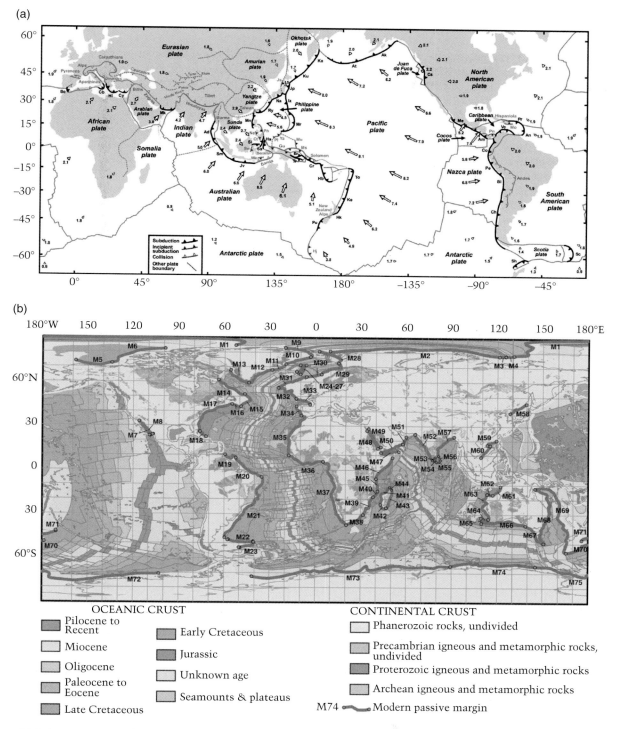

Fig. 3.3 Global tectonic maps showing: (a) active; and (b) passive plate margins. Each type of plate margin has a fundamental influence on coastal geomorphology. In (a), plates and all subduction zones on Earth (including incipient subduction zones), collision zones and the velocities of the major plates are shown (Source: Schellart and Rawlinson 2010. Reproduced with permission of Elsevier.). The subduction zones are: Ad, Andaman; Ak, Alaska; Am, Central America; An, Lesser Antilles; At, Aleutian; Be, Betic-Rif; Bl, Bolivia; Br, New Britain; Cb, Calabria; Ch, Chile; Co, Colombia; Cr, San Cristobal; Cs, Cascadia; Cy, Cyprus; Ha, Halmahera; Hb, New Hebrides; Hk, Hikurangi; Hl, Hellenic; Iz, Izu-Bonin; Jp, Japan; Jv, Java; Ka, Kamchatka; Ke, Kermadec; Ku, Kuril; Me, Mexico; Mk, Makran; Mn, Manila; Mr, Mariana; Na, Nankai; Pe, Peru; Pr, Puerto Rico; Pu, Puysegur; Ry, Ryukyu; Sa, Sangihe; Sc, Scotia; Sh, South Shetland; Sl, North Sulawesi; Sm, Sumatra; To, Tonga; Tr, Trobriand; Ve, Venezuela. Incipient subduction zones: Ct, Cotobato; Gu, New Guinea; Hj, Hjort; Js, Japan Sea; Mo, Muertos; Ms, Manus; Mu, Mussau; Ne, Negros; Ph, Philippine; Pl, Palau; Pn, Panama; Sw, West Sulawesi; We, Wetar; and Ya, Yap. The green circles in (b) divide margins into age sectors. (Source: Bradley 2008. Reproduced with permission of Elsevier.) For colour details, please see Plate 10.

crustal/lithosphere extension, subsidence due to sedimentation, subduction, and magmatic underplating. Well before the concept of plate tectonics appeared in the 1960s, Suess (1892, in Davies, 1980) distinguished between Pacific-type coasts, the structural trend of which is parallel to the coast, and Atlantic-type coasts with a discordant trend. Coasts located along convergent plate boundaries correspond largely to the Pacific-type coasts, whereas Atlantic-type coasts are embedded in plates of which the boundaries are accreting. Davies (1980) showed that the distinction between Pacific- and Atlantic-type coasts was still largely valid at the global scale.

The recognition of the important role of global tectonics is at the basis of a geophysical classification of coasts proposed by Inman and Nordstrom (1971). These authors highlighted the large-scale geomorphic and structural differences among various types of plate boundaries, each exhibiting a distinctive set of geological and morphological characteristics, and recognized three major types of coasts: (1) collision coasts; (2) trailing edge coasts; and (3) marginal sea coasts. Collision coasts, also termed leading edge coasts, are associated with coastal margins subject to convergence. This classification recognizes continental collision coasts, such as western North and South America, and island arc collision coasts, such as the coasts of island arcs flanking the Pacific (Japan, Philipines, Indonesia, New Guinea). Davies (1980) convincingly argued that the

important distinction is between collision coasts lying along a zone of convergence and all the others, which do not. The first group comprises plate-edge coasts, and the second group comprises plate-embedded coasts. More commonly today, the distinction employed in the literature between Atlantic or trailing edge and Pacific or collision-type coasts is between 'active' and 'passive' margins (Fig. 3.3). The characteristics of active and passive margin coasts are discussed next and their main differences in terms of age, relief, landforms, tectonics, sediment characteristics and drainage are summarized in Table 3.1.

3.2.1.1 Active margin coasts

Active margin coasts are relatively straight, steep and rocky, and flank high, tectonically mobile hinterlands exhibiting structural lineaments parallel to the shore. Convergence processes lead to compressive forces that result in the build-up of mountains such as the Rocky Mountains in North America. The compressive forces affecting these margins generate faulting, folding, block tilting, elastic rebound and gliding, which also serve to render the rock masses more fragile and susceptible to landslides and other mass movements that feed fluvial systems in sediment (e.g. Julian and Anthony, 1996). Although the vigorous relief associated with active margins is potentially favourable to abundant sediment supply to coasts, the high elevations generally block

Table 3.1 General characteristics of coasts on active and passive plate margins. (Source: Short 1999. Reproduced with permission of John Wiley & Sons.)

	Active margin	Passive margin
Age	Young (1 to 10s of millions of years)	Old (100s of millions of years)
Relief	Steep, mountainous	Low-gradient plains
Landforms	High mountains and volcanoes	Coastal aggradation plains
	Narrow continental shelf	Wide, low continental shelf
	Deep-sea trough	Continental slope
Tectonics	Active, earthquakes	Quiescent, stable
Weathering	Physical, mass movements	Chemical, fluvial, rivers
Drainage	Short, steep streams	Long, meandering rivers
Sediments		
Quantity	Low	High
Size	Fine to coarse	Fine
Sorting	Poor	Well
Colour	Dark	Light
Composition	Unstable minerals	Stable minerals
Coastal landforms	Rocky, few beaches	Extensive barriers and deltas
Wave attenuation	Low	Moderate to high
Tide range	Minimal amplification	Enhanced
Examples	West coast of the Americas	East coast of the Americas
	New Zealand	Southern Africa
	Iceland	North Alaska
	Japan	India

large rivers, which tend to be diverted towards the opposite plate-embedded passive margin coasts. Sediment input occurs through smaller rivers that actively downcut the steep relief, often initially guided by major faults. Their sediment is either lost directly offshore or trapped by strong wave action in small, narrow embayments. An important exception to this are the active margin coasts flanking island arcs in Asia, which receive some of the world's high sediment supply rivers emanating from the Himalayas (Changjiang or Yangtze and Huang He Rivers), Indonesia (Mahakam River) and Papua-New Guinea (Fly and Sepik Rivers).

High cliff shores on active margins are common, and may alternate with short stretches of embayed beaches built by high ocean waves. The vigorous and youthful morphology of such coasts is generally maintained by active uplift and by the formation of well-defined coastal terraces such as the 'rasa' of South America (Fig. 3.4). These geological conditions are coupled with steep, narrow wave-dominated continental shelves subject to little or no tidal amplification, with, therefore, relatively low tidal ranges. Neotectonic and Holocene deformations can have a marked effect on the shoreface, sometimes enhancing the capacity for onshore wave-dominated sediment transport from a progressively uplifted inner continental shelf, as in the case of the Kujukuri beach-ridge plain in eastern Japan (Tamura et al., 2007, see also section 3.3.8).

The tectonic regime prevailing along active margin coasts is responsible for significant volcanic and earthquake activity. Magmatic underplating of the subducted oceanic crust results in the formation of explosive volcanoes, as in the Cascade Range in Oregon and in the Andes. The earthquakes affecting these active margins can lead to abrupt vertical shoreline changes involving uplift, subsidence, or both. Earthquakes commonly generate tsunamis that lead to transient flooding, although the shoreline changes expected from such flooding are far less important than those to be expected from tsunamis affecting lower-elevation passive margin coasts (see next section). Tsunami waves impinging on active margins are likely to undergo reflection from steep coastal slopes (Dawson and Stewart, 2007) and be canalized where deep valleys cut into the coastal topography.

3.2.1.2 *Passive margin coasts*

Unlike active margins, which are relatively young when considered in terms of geological timescales, passive margins, also called rifted, trailing edge or divergent margins, have existed somewhere on Earth almost continually since 2740 Ma, and the present-day passive margins (see Fig. 3.3), which are not yet finished with their lifespans, have a mean age of 104 Ma, and a maximum age of 180 Ma

(Bradley, 2008). Passive margin coasts are associated with accreting plates. Inman and Nordstrom (1971) recognized three types based on their location relative to the plate: Amero-, Afro- and Neo-trailing edge coasts. Both Amero- and Afro-type coasts exhibit well-developed, mature margins with relatively limited tectonic activity. The major structural lineaments are generally orthogonal to the coast, as originally recognized by Suess (1892, in Davies, 1980). These lineaments are generally associated with transform fault fracture zones inherited from the opening of rifted margins.

Amero-, Afro- and Neo-trailing edge coasts represent significantly different boundary conditions for coastal evolution:

• **Amero-trailing edge coasts**, found along the eastern seaboard of both North and South America, exhibit wide, low plains and deltas, linked to broad continental shelves (see Fig. 3.2). Although these shelves may be wave-dominated, tidal amplification induced by the shelf morphology and bathymetry can be important, leading to locally tide-dominated conditions. These accretionary coasts are embedded in plates subject to collision or subduction at one of the two (continental) plate edges, the other plate edge being in a submarine zone of spreading. The collision or subduction zone exhibits high mountainous relief (e.g. Rocky Mountains in North America, Andes in South America), which provide abundant sediment. Major river flow and sediments are generally diverted towards the trailing edge coast, leading to the formation of large deltas and broad, sediment-rich continental shelves. Examples include the Amazon in South America, the Mississipi in North America, and the Indus and Ganges-Brahmaputra rivers on the Indian subcontinent.

• **Afro-trailing edge coasts** are associated with continental plate margins that are passive on both sides and subject to subsidence. These conditions generate low coastal relief and a relatively broad continental shelf. Fluvial sediment input is, however, less important than on Amero-trailing edge coasts because of the absence of high relief on the continental plate, and this difference in fluvial sediment supply constitutes a fundamental distinction between these two types of passive margin coasts (Davies, 1980). The coasts of Australia (Short and Woodroffe, 2009) and Africa are classic examples of relatively low fluvial sediment-supply coasts. Although these coasts are fed by rivers with large drainage basins, such as the Nile, the Congo and the Niger in Africa, and the Murray, Murrumbidgee and Darling in Australia, there is no preferential large-scale tectonic routing of these rivers, and since they are not associated with major mountain catchments, their sediment loads are generally low compared to those of Amero-trailing edge coasts. On these relatively ancient passive margins, deltas formed by the big rivers constitute

(a)

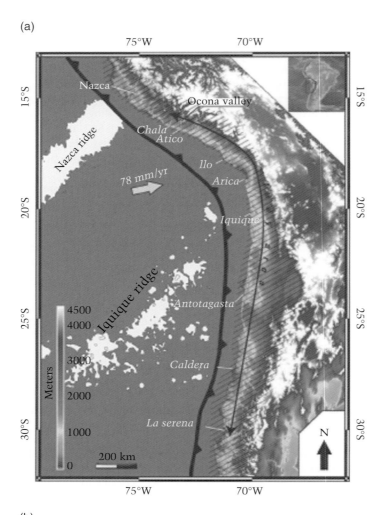

(b)

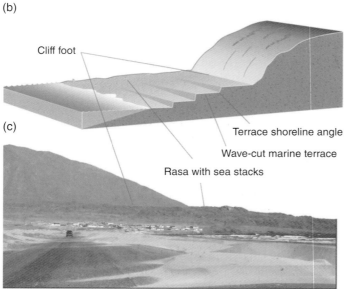

Cliff foot

(c)

Terrace shoreline angle

Wave-cut marine terrace

Rasa with sea stacks

Fig. 3.4 Effects of neotectonic uplift on an active margin coast. The map (a) shows the topography of the Central Andes, the rate of migration of the Pacific plate and the subduction zone associated with tectonic uplift of this active margin. Most of the Pacific coast of the Central Andes, between 15°S and 30°S, displays a wide (a couple of kilometres) planar feature, gently dipping seawards and backed by a cliff. It is now widely accepted that the inner edges of these terrace features, called 'rasa', correspond to sea-level highstands. The presence of these rasa argues for a recent and spatially continuous uplift of the margin over this 1500 km long coast. A rasa is a wave-cut surface limited at its continental edge by a cliff foot. (b) General sketch of a rasa, which locally can be occupied by marine terraces. (c) Example of a rasa and cliff at Tanaka, southern Peru (15.75°S); the cliff foot is at ~300 m above mean sea level (MSL). (Source: Regard et al. 2010. Reproduced with permission of Elsevier.) For colour details, please see Plate 11.

major local sediment depocentres that are generally subject to subsidence, as in the case of the and Niger river, whereas the Congo river-mouth, which loses sediment to a deep submarine canyon, has not developed into a delta. Because of their relatively low relief, passive margin coasts lying adjacent to tectonically active margins may be significantly affected by tsunamis caused by earthquakes. Tsunamis, such as that generated by the 24 December 2004 Sumatra-Andaman earthquake in the Indian Ocean, can result in extensive coastal erosion, sediment transport and deposition in a few minutes on Afro-trailing edge coasts.

• **Neo-trailing edge coasts**, such as those bordering the Arabian Peninsula, are associated with much more youthful coastal relief, and exhibit both volcanic and earthquake activity. The coastal topography is typically rugged, composed of cliffs backed by mountains and/or plateaus. These coasts are devoid of continental shelves because of their young age, which has not left enough time for both the development of important feeder drainage basins and the sediment supply needed for such shelves to develop through subsidence as sediment accumulates. As these coasts are formed from an initial rift valley, their geological fabric, generally composed of granite, is progressively modified by a host of geomorphic, biological, chemical and physical weathering, and hydrodynamic processes that go with ocean basin enlargement and coastal sediment accumulation.

3.2.2 The role of Quaternary ice sheets and isostatic rebound on high-latitude coasts

The influence of geology on coasts is also manifest in regional crustal movements that are adjustments to loading and unloading by ice (glacio-isostasy) in high-latitude areas covered by extensive Quaternary ice sheets (Fig. 3.5). The effects of differential crustal depression due to ice loading and isostatic rebound following ice melt and removal of the weight of ice are manifested on glaciated coasts by progressive uplift of Holocene shorelines. Raised series of beach ridges are particularly illustrative of this process. The rate of rebound following ice unloading is generally very rapid in the first few thousand years, slowing down considerably thereafter. Crustal loading and depression due to the weight of ice leads to the formation of a marginal bulge or uplifted area outside the margins of the ice sheet. On the east coast of North America, this bulge extends up to 800–1000 km south of the Laurentide Ice Sheet margin. Collapse of this bulge following ice melt and isostatic rebound of the depressed crust has induced subsidence of these marginal areas, resulting in the longshore variability of Pleistocene shorelines, as shown by work carried out by Scott et al. (2010), and this effect is deemed to be responsible for the ongoing postglacial sea-level rise along this part of the eastern North American coast

(Fig. 3.5). Similarly, the south coast of England is also still experiencing a relative rise in sea level due to the collapse of the forebulge. The southwest coast of England is subsiding rapidly (by over 1 mm/yr), because it is located in the forebulge zone of the former Scandinavian Ice Sheet and the forebulge zone of the smaller British-Irish Ice Sheet.

3.2.3 Water loading of continental shelves

In a manner similar to that of the isostatic adjustments associated with the loading and unloading of the crust by ice, loading and unloading of ocean basins following Quaternary changes in ocean volume deforms the crust, vertically affecting shorelines. Hydro-isostatic adjustments, are, however, less important than those due to ice, because the magnitude of sea-level change (<200 m) is much less than that of the thickness of ice sheets (>5000 m). Differential crustal movements along the east coast of Australia and west coast of Africa are caused by hydro-isostasy, whereas along broad continental shelves in forebulge zones, hydro-isostasy can exercabate coastal subsidence. For example, the Atlantic coast of Nova Scotia is sinking by more than 2 mm/yr, and c. 40% of this is due to water loading of the shelf (Gehrels et al., 2004).

3.2.4 Lithology, sediment texture and coasts

The composition of rocks is an important coastal environmental factor, both in terms of morphology and coastal sedimentary composition. Lithology depends on the geophysical setting and geological history of an area. Landform development depends on various rock characteristics such as chemical composition, jointing and tensile strength. These characteristics are, however, largely mediated by climate through weathering. As a result, the lithology of the hinterland may be more important than that of the immediate underlying coastal rocks in influencing the composition of coastal sediments (Woodroffe, 2003). Davies (1980) noted that the nature of coastal lithology changes rapidly and subtly along most coasts.

A discussion of coastal plan shape admittedly also throws in the old concept of 'capes and bays' geomorphology, where, on lithological grounds, hard-rock capes are considered as being resistant to long-term erosion compared to bays. This is not as simple as it sounds, however. Crenulated coasts composed of bays and headlands indeed generally reflect long-term longshore differences in rock resistance to erosion. Many headlands are commonly associated with rocks that are more resistant than those in the adjacent bays, whereas others develop on the basis of subtle differences in joint density, bedding thickness, and other structural influences (Trenhaile, 2002). Rock resistance, itself dependent on various factors such as rock strength, chemical composition, bedding thickness, joint density, and variations in rock strike and dip relative to

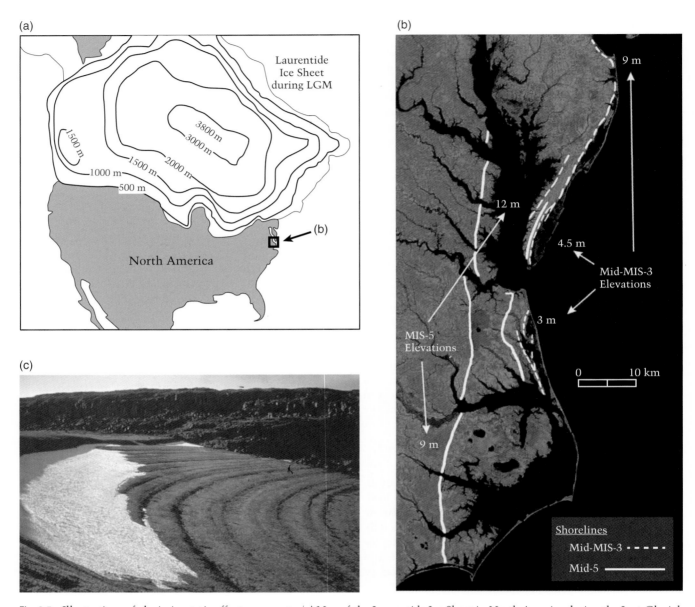

Fig. 3.5 Illustrations of glacio-isostatic effects on coasts. (a) Map of the Laurentide Ice Sheet in North America during the Last Glacial Maximum (LGM). (b) Aerial photograph of North Carolina, USA, showing approximate locations and alongshore variations in elevations of dated mid MIS-3 (Last Glacial) and MIS-5 (Last Interglacial) shorelines. The high elevations of the MIS-3 shorelines result from the effects of glacio-isostatic rebound of the Laurentide Ice Sheet. (Source: Scott et al. 2010. Reproduced with permission of Elsevier.) (c) Photo showing successive Holocene 'stair-case' raised beach ridges due to uplift resulting from glacio-isostatic rebound following ice removal at Kativik, Hudson Bay, Canada. (Source: Courtesy of Jean-Marie Dubois.)

the direction of incoming waves, is, however, only one of numerous interactive parameters that determine the actual plan shape of a rocky coast. Other factors include shoreline topography, wave characteristics, headland and bay spacing parameters, cliff characteristics, including sediment delivery and removal at the bases of cliffs, and the characteristics of beaches associated with rock shores (Trenhaile, 2002).

The influence of lithology on coastal embayments in New South Wales, Australia, has also been analysed by Bishop and Cowell (1997). They showed that on this tectonically stable coastline, which drains ancient river systems, embayment size (the length of the embayment from headland to headland at sea level) was related to the size of the river draining to the embayment (catchment area) rather than to coastal lithology, contrary to earlier assumptions. These authors highlighted the larger variability of embayment sizes and the greater abundance of 'under-sized' embayments in provinces exhibiting strong coast-parallel structures that are thought to facilitate the 'opening-up' of large catchments inland. They further suggested that the fundamental control of catchment area on embayment length reflected both the long-term (Cenozoic) development of drainage networks in the bedrock fluvial domain and their subsequent drowning in the Late Quaternary (postglacial) marine transgression. Shoreline configuration reflected, thus, the interplay of long-term fluvial influences, Quaternary sea-level fluctuations, and Late Quaternary sediment supply and accommodation volumes. Bishop and Cowell (1997) concluded that high sea levels along embayed bedrock coasts resulted in a more crenulate and compartmentalized coast with smaller embayments because the sea level penetrates up into the lower-order drainage network; conversely, longer embayments and more open, less compartmentalized coastlines were associated with the lower sea levels resulting from eustatic sea-level fall and/or uplift of the land. They reasoned that Late Quaternary sedimentation governed the extent to which palaeochannel drainage networks are buried, whereby the densities of palaeochannels, interfluves (and hence headlands) along the coastline were reduced. The effect of this deposition (coastal progradation) therefore corresponded to sea-level regression or to tectonic uplift of the coast and should tend to enhance littoral sediment transport in contemporary coastal processes.

Headland embayed beaches are a common feature of rock shorelines (see Fig. 3.1). The length and spacing of these beaches depends entirely on the pre-existing bedrock topography (Short and Masselink, 1999). The plan shape and orientation of both capes (such as cuspate forelands like Dungeness in the UK and the sedimentary capes of the northeastern USA) and large embayments filled with beach and dune sediments are generally determined by alongshore wave patterns and energy gradients.

The nature of the sediments reaching the coast is strongly influenced by climate, although local to regional lithological controls may be dominant, thus explaining sometimes sharp contrasting coastal sediment heterogeneity independent of the hydrodynamic conditions. Long-term weathering, especially of granite catchments over large shield areas such as West and Central Africa that have remained within the intertropical domain over more than 100 Ma, has generated very thick saprolite mantles. This has resulted in abundant clay stored in delta and ocean sediments, and extensive beach and beach-ridge plains comprising quartz sand. The prevalence of physical

weathering due to lower temperatures explains the more coarse sediment textures generally found along mid- to high-latitude coasts. The abundance of gravel barriers in such latitudinal settings (e.g. Orford et al., 2002) contrasts with the dearth of such gravel beaches and the commonality of quartz-dominated sandy beach deposits in the tropics.

Rock coasts are similarly more abundant in the mid to high latitudes. A widely stated figure is that 80% of the world's coast is rocky (Emery and Kuhn, 1982), although according to Naylor et al. (2010) little effort has ever been made to validate this claim. Rock shores occur on passive margins in the tropics where sediment supply has been limited or where mild tectonic activity has led to uplift. Where bedrock occurs at the coast, the shoreline morphology depends jointly on rock characteristics such as tensile strength, jointing, porosity, microporosity and dip, as well as on the weathering and hydrodynamic conditions. Where significant sediment input has occurred on the coast, the bedrock can be completely mantled, sometimes under several kilometres of sediment in basinal settings associated with passive margins, especially where deltas have been active over millions of years. Although rock shores are essentially erosional, they may be commonly associated with a suite of sedimentary facies generally deposited under high-energy conditions.

3.2.5 Other regional to local boundary conditions: coastal orientation and gradient

Both coastal orientation and gradient are influenced by the geological conditions and inherited morphology. Most of the world's coasts have a large-scale plan orientation hinged on the initial rift orientation for passive margins or on the structural lineaments created by convergence for active margins. Geological control on coastal orientation is likely to be dominant on bold cliff coasts composed of hard homogeneous rocks. Given the predominance of rock shores in the world, the orientation of a significant proportion of coasts is thus controlled by the geological fabric. This orientation, can, however, be locally independent of geological constraints where abundant sediment is reworked over time into coastal plan shapes that adjust to waves and currents.

On passive margin coasts cut by shore-parallel fracture zones, the differential horizontal movements associated with the transform faults in these fracture zones, some of which are still subject to (generally mild) seismic activity, have been responsible for important changes in both shoreline orientation and continental shelf width. This is the case in the West African coast from Sierra Leone to Guinea, where major continental shelf offsets with strong impacts on coastal morphology have been caused by the Sierra Leone and Guinea Fracture Zones (Fig. 3.6).

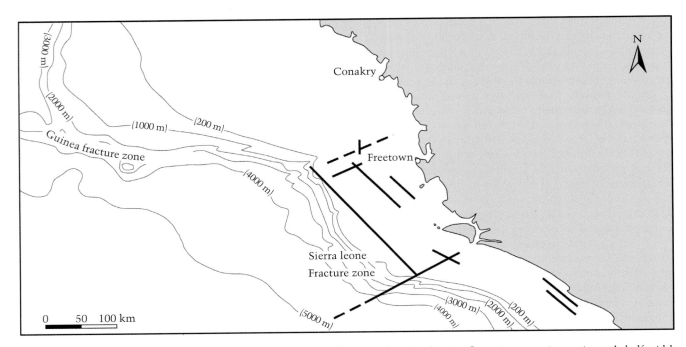

Fig. 3.6 Geological sketch of part of the West African continental margin showing the significant increases in continental shelf width related to offsetting by fracture zones. Fracture zones and their transform faults enable differential rates of ocean basin accretion associated with the Mid-Oceanic Ridge. These fracture zones show mild tectonic activity (occasional earthquakes) and their effect on the geomorphology of this coast has been fundamental. The narrow shelf south of the Sierra Leone Fracture Zone is a wave-dominated, low tide range coast with massive progradation (by several kilometres), shoreline advance and wave-formed sandy beach ridges. The broader shelf to the north shows transition between wave- and tide-domination up to the vicinity of Freetown, before becoming a tide-dominated shelf north of Freetown, associated with significant tide-range amplification and wave dissipation over the broad shallow shoreface. The coast in this tide-dominated zone comprises extensive open-coast mangroves and cheniers.

Another largely geologically mediated boundary condition is substrate gradient. It determines, in the case of barrier or reef shores, for instance, the degree of detachment or attachment relative to the mainland. The width of the backshore zone is also determined in part by the gradient of the substrate over which the shoreface translates. Steeper substrates offer less backshore accommodation space than do low-gradient substrates. Roy et al. (1994) have suggested that sediment-starved cliffed coasts with no backshore typically evolve in settings where the underlying substrate is steeper than the active or wave-affected shoreface (which is the inner part of the continental shelf connected to the shore), whereas barrier coasts evolve where the underlying substrate has a lower gradient than that of the active shoreface (Fig. 3.7). With the most favourable substrate gradient, levels of frictional dissipation of wave energy are such that there is sufficient energy to enable wave buildup of beaches under conditions of available sediment. Roy et al. (1994) predicted, for instance, sand barrier development on shoreface gradients between 0.05° and 0.8°, and optimum beach development on a gradient of 0.1°. Shorefaces with very weak gradients may be characterized by extreme levels of dissipation of

swell wave energy, often accompanied by tide-dominated conditions, as in the case of the shallow eastern English Channel and southern North Sea.

3.3   Sediments and coasts

Sediments are a fundamental component of coasts as they constitute the materials that are moulded by processes into a range of coastal lithosomes. Coastal sediments are also important archives of environmental change and Earth history, including the provision of absolute chronological frameworks through various dating methods of both organic and mineral sediments. Much of the world's coasts receive sediments composed of mud, sand and gravel. Most sediment is directly terrigenous, derived from continents as rocks are weathered and/or eroded. Sediments may also be derived from the shoreface, or be supplied by organic production. In both cases, much of the initial building blocks of these sediments are from terrestrial weathering and erosion. The sediments accumulating within the coastal boundary are organized by waves, currents and wind flow into various bodies or lithosomes

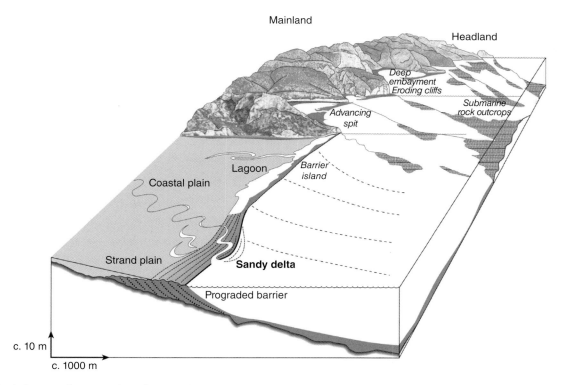

Mainland

Headland

Deep
embayment
Eroding cliffs

Submarine
rock outcrops

Advancing
spit

Lagoon

Barrier
island

Coastal plain

Strand plain

Sandy delta

Prograded barrier

c. 10 m

c. 1000 m

Fig. 3.7 Block diagram depicting the influence of variablility in coastal lithology, substrate gradient and sediment supply on the development of coastal barriers, with a change from a sediment-starved bedrock-dominated coast with bold headlands, deep embayments and offshore rock reefs, to a sediment-rich coastal plain comprising a range of wave-formed barrier deposits. The barrier range shown here includes fully attached bay and mainland beaches to partially attached spits and bay barriers (bb), detached barrier islands (bi) with tidal deltas (td), beach-ridge plains or strandplains (sp), and inner shelf sand bodies (ssb). (Source: Roy et al. 1994. Reproduced with permission of Cambridge University Press.)

associated with estuaries, deltas and open-ocean coasts, as well as more or less sheltered embayments. Depending on the hydrodynamic conditions, sediment type and abundance, these deposits range from estuarine channel-fill sand-bank systems, mudflats, sand flats and vegetated tidal wetlands along generally tide-dominated open or embayed estuarine and deltaic settings, to mud banks and beach, dune and barrier deposits in wave-dominated settings. Where sediments are not supplied to the coast, erosional shores commonly composed of cliffs and bluffs cut in hard or soft rock may be formed.

Sediments are described on the basis of their sedimentary or petrological properties. These include grain size and shape. Particle size distribution has been used for decades, both to describe the basic dimensions of sediments and to infer sources and processes of transport and deposition (Box 3.1). Process interpretations of grain sizes are still sources of debate (Flemming, 2007), although new computer technology and the use of increasingly more sophisticated statistical procedures and modelling techniques are creating new analytical perspectives. The extraction of environmental information from grain size is strongly conditioned by sampling strategy and data analysis methods. It is not clear, for instance, to what extent data yielded by the various types of devices in particle analysis and the various methods of analysis are actually comparable. Image segmentation techniques based on textures recorded by digital cameras are improving the remote-sensed identification and mapping of grain sizes, and their spatial and temporal variations (Buscombe et al., 2010).

Other less commonly used sediment properties include the packing of grains, their degree of imbrication or stacking, their porosity, and their shape. Particle shape has been much less exposed to the heated debates that have been waged on size. The early emphasis in the 1960s and 1970s on particle shape and surface characteristics, notably via scanning electron microscope analyses aimed at highlighting diagnostic environmental proxies, has now subsided, although this approach must not be discarded. Blott and Pye (2008) have insisted on the fundamental importance of particle shape and on the difficulties inherent in attempts at characterizing this parameter. Their review of particle shape also introduces new methods of characterization and classification, especially with regards to

CASE STUDY BOX 3.1   Sediment in the English Channel

The surficial seabed sediments in the English Channel (Fig. 3.8a) show eastward fining and large-scale coastal transport of fine sand from the Somme estuary to Belgium (Anthony and Héquette, 2007). Sand-rich Holocene to modern clastic deposits in the eastern English Channel and the southern North Sea coasts of France and Belgium occur extensively as nearshore sand bank, estuarine tidal flat, aeolian dune and beach sub-environments (Fig. 3.8b). Collection and analysis of a very large number (665) of sand samples from these

(a)

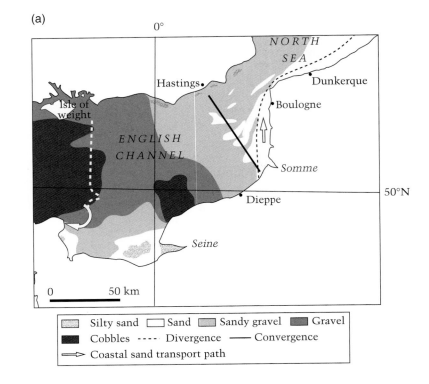

(b)

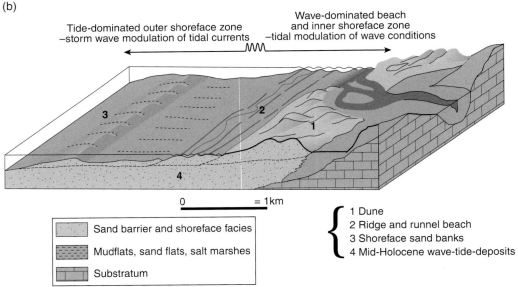

Fig. 3.8 (a) Surficial seabed sediments in the English Channel; (b) schematic Holocene stratigraphy of coastal clastic deposits in the eastern English Channel and the southern North Sea coasts of France and Belgium; and

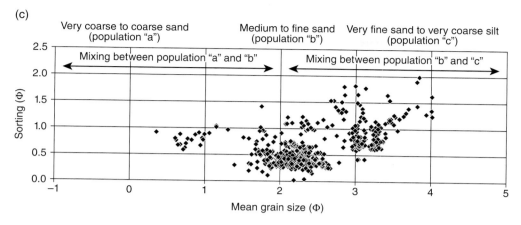

Fig. 3.8 (Continued) (c) sediment analysis of a large number of sand samples from these coastal deposits. (Source: Anthony and Héquette 2007. Reproduced with permission of Elsevier.)

deposits suggest the presence of three different populations (Fig. 3.8c): a largely dominant (83%) medium to fine quartz sand population ('b'), and finer-grained (14%) and coarser-grained (4%) populations ('c' and 'a', respectively). The distribution of these populations among the four sub-environments reflects tide- and storm-dominated sorting and transport processes, and a variable degree of mixing.

The different sediment populations are derived from a mixture of very fine-grained to very coarse-grained fluvial, outwash and paraglacial sediments deposited on the beds of the eastern English Channel and southern North Sea during the late Pleistocene lowstand. The nearshore sand bank environment, which also corresponds to the main offshore source area of the coastal deposits, exhibits population heterogeneity, reflecting the variability of hydrodynamic conditions and sediment sorting in this zone. The nearshore topography of tidal ridges, banks and troughs in these tidal seas leads to variable bed and tide- and storm-induced shear stress conditions. These conditions only allow for the mobilization and onshore transport of some of the finer fractions (populations 'b' and 'c'), leaving an offshore mixture of these finer populations with coarser, less mobile sediments (population 'a'). Once in the coastal zone, these two finer populations undergo further hydrodynamic sorting and segregation. Variably sorted, very fine sands to silts (population 'c') are trapped in the low-energy estuarine tidal flat sub-environment, whereas the highly homogeneous population 'b' is further sorted in aeolian dune and beach sub-environments. This sorting occurs via a coastal sand transport pathway linking the Somme estuary mouth to the southern North Sea bight, where tidal range and wave energy decrease relative to the English Channel. Since this sand transport pathway enables longshore transport of hydrodynamically sorted, medium to fine sand, derived directly from the immediate nearshore zone, it has further contributed to a net flux of this sand population from the eastern English Channel seabed to the southern North Sea.

roundness and circularity, considered as proxies for sphericity, and irregularity, which is relevant, for instance, in the characterization of irregular or branching sedimentary particles such as chert and coral.

Coastal sediments, mineral as well as organic, are archives of environmental conditions, both in their coastal context and as far as changes in distant source areas are concerned. Sediment provenance for sand-sized particles has long reposed on mineralogical, heavy mineral and magnetic mineral analyses. Notably, recent developments concern suspension-size sediments, with classical mineralogical and geochemical analyses being complemented by the analysis of radionuclide signatures (Caroll and Lerche, 2003). As archives of environmental conditions, sediments also provide materials for absolute chronological dating, through both upgraded non-radiactive and radioactive methods covering various time ranges. High-resolution chronological studies of Holocene deposits based on ^{14}C radiocarbon dating are now a routine part of many coastal studies, as is shorter-range dating using various methods, ranging from optically stimulated luminescene (OSL) techniques to radionuclide signatures, such as those of ^{7}Be, ^{137}Cs and ^{210}Pb. Radionuclide signatures can be used to determine mud history and dynamics in the vicinity of delta mouths, notably *in situ* mud residence times, accumulation rates, steady-state versus event-type sedimentation patterns, and deduction of mud migration rates. The reader is referred to relevant texts on the

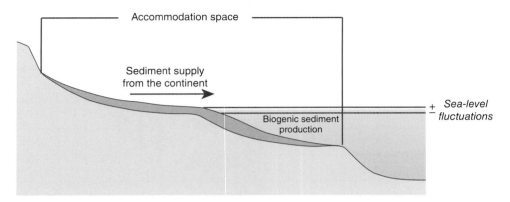

Fig. 3.9 A simple illustration of the concept of sediment accommodation space and its relationship to changes in sea level caused by eustasy or tectonics. The coastal sediment accommodation space can comprise both marine and non-marine space, and sediment supply from both the continent and the sea.

subject for more detailed presentations, notably Caroll and Lerche (2003) and Walker (2005).

3.3.1 Coastal sediment stacking over time: sequence stratigraphy and sea-level change

Vertical movements of the Earth and major changes in climate affect sea level. Sea-level changes control, in turn, the position of the shoreline and the locus of coastal sedimentation. Sea-level changes lead, over time, to the stranding of shorelines or significant bodies of coastal sediments offshore or inland. The coastal sedimentary record archives the stratigraphic sequences associated with shoreline trajectories, a field of study termed 'sequence stratigraphy'. Sequence stratigraphy has been considered by Catuneanu et al. (2009) as the most recent and revolutionary paradigm in the field of sedimentary geology, probably because it represents a significant modernization of geological thinking and methods of stratigraphic analysis. Catuneanu et al. (2009) have suggested that other, more conventional types of stratigraphy, such as biostratigraphy, lithostratigraphy, chemostratigraphy or magnetostratigraphy, are mostly concerned with data collection, whereas sequence stratigraphy has a clear interpretation component, which addresses issues such as: (1) the reconstruction of the environmental controls at the time of sedimentation, notably the issue of eustatic versus tectonic controls on sedimentation; and (2) predictions of sedimentary facies architecture in yet unexplored areas. Readers interested in sequence stratigraphy are further referred to Catuneanu (2006).

3.3.2 Sediment accommodation space

One important paradigm in sequence stratigraphy is the space that is made available within a basin for sediment to be deposited and the amount of sediment supplied.

Accommodation space, as far as the coast is concerned, refers to the space available for the deposition of sediment of marine or terrigenous origin. The concept is, therefore, a key one for coastal studies, and a useful reference text is that of Coe (2003). Accommodation space is chiefly governed by changes in relative sea-level (Fig. 3.9), but is regionally or locally modulated by the inherited morphology of the basin of deposition. Where there is no accommodation space available, sediments will be transported to an area of (positive) accommodation space where they can be deposited. Sediment by-passing occurs, thus, in areas of zero accommodation space. In an area where the amount of accommodation space is negative, sediment previously deposited will be eroded and transported to an area of positive accommodation space. This occurs because all sedimentary systems tend towards an equilibrium profile, where the available accommodation space is balanced by the amount of sediment supplied. This equilibrium can be offset by changes in either the rate of sediment supply or the rate of change of accommodation space, leading to changes in shoreline position either associated with regression (seaward shoreline translation) or transgression (landward shoreline shoreline).

3.3.3 Terrigenous sediment supply

Worldwide, rivers supply about 90% of coastal sediments (Woodroffe, 2003). The amount of sediment supplied annually by the world's rivers has been estimated at 10–20 billion metric tons, although there is considerable uncertainty concerning volumes, because of the effects of human intervention (Ericson et al., 2006; Syvitski et al., 2009). A useful notion in highlighting the joint influence of geology and sediment supply on coasts is to consider sediment input in terms of a 'source-to-sink' perspective (Fig. 3.10). The source-to-sink approach shows that terrigenous

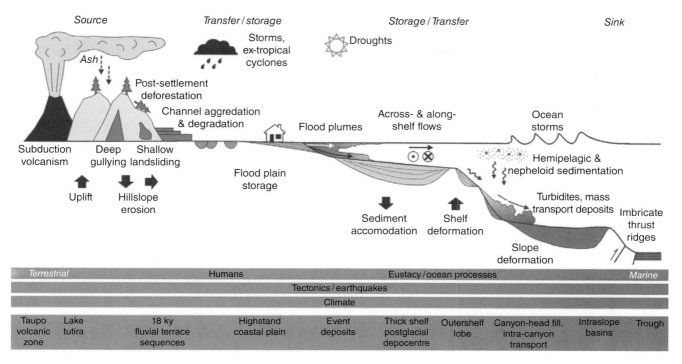

Fig. 3.10 Diagram illustrating the source-to-sink concept, using the Waipaoa Sedimentary System (WSS) in New Zealand as an example. (Source: Carter et al. 2010. Reproduced with permission of Elsevier.) For colour details, please see Plate 12.

sediment supply occurs at several temporal and spatial scales, animated by short-term events such as extreme floods, storms, earthquakes and volcanic eruptions, which may be superimposed on long-term tectonic and sea-level trends, depending on the geophysical and climatic context of both the sediment-supplying hinterland and the nature of the coast. As mentioned earlier, the largest supplies of terrigenous sediments are associated with inflowing rivers concerning Amero-trailing edge coasts and active margins facing island arcs in Asia. The Waipaoa Sedimentary System (WSS) of northeastern New Zealand (Box 3.2) is a good example of a mid-latitude, high sediment-input system (Carter et al., 2010). This sedimentary system extends across the convergence zone of the Australian and Pacific plates, resulting in significant influences of both tectonic and climatic drivers. Overall, the WSS is undergoing uplift, and this has been contributing to valley incision since the Last Glacial Maximum. Uplift is both gradual and intermittent, the latter relating to fault- and subduction-related earthquakes. Carter et al. (2010) have shown that such seismic activity destabilizes the landscape and, together with frequent storms, generates large amounts of sediment. As this sediment passes through the system, part is captured in traps formed by local subsidence of the coastal plain and the middle continental shelf where large thicknesses of Holocene and probably older sediment are sequestered (c. 200m under the coastal plain). Major

volcanic eruptions also affect the sediment flux, via direct input from ash-fall and by enhanced erosion following volcanic destruction of terrestrial vegetation. The source-to-sink concept is useful for understanding sediment budgets, cascades, transfer processes and their timescales, from upland catchments to the coast, especially as increased river damming and climate change in the future (see sections 3.4 and 3.5) will affect the sediment budgets of coasts, and as sea-level rise creates accommodation space for more sediment necessary to maintain shorelines in their present positions. Phillips and Slattery (2006) have suggested that sediment supply to the ocean by rivers crossing coastal plains needs to be interpreted with caution because of the influence, notably on sediment transport capacity, of changes in slope and accommodation space due to sea-level variations.

Steep coasts associated with active margins are also potentially subject to mass-wasting processes involving, notably, landslide activity and sediment evacuation downstream by steep-gradient streams. Sediment transfer from short source-to-sink segments, coupled with mountain hydrological regimes, regulate patterns of river channel aggradation and coastal sediment supply in such geomorphic settings, a fine example of which is the steep French Riviera margin in the Mediterranean (Anthony and Julian, 1999). In this coastal setting, sediment transfers from existing alpine mountain landslides, or from

CONCEPTS BOX 3.2   From source to sink

The source-to-sink concept is useful in terms of under-standing sediment budgets, sediment cascades, sediment transfer processes, and their timescales within the inter-linked system from the upland catchment to the coast (see Fig. 3.10). This is especially important as increased river damming and climate change in the future will affect the sediment budgets of coasts, and as sea-level rise creates accommodation space and the necessity for more coastal sediment to maintain shorelines in their present positions. A typical source-to-sink sedimentary system responds to drivers that operate on different spatial and temporal scales, at frequencies ranging from daily weather events to the 10^4–10^5 yr predictable rhythms of Milankovitch cycles, to the even less frequent events such as plate-driven seamount collisions with the conti-nental margin. Thus, both episodic events and slow behavioural shifts are important components of a source-to-sink system and these are recorded by the sedimentary archives found both in upland parts of the system and on the coast. On active margins, such as the Hikurangi mar-gin in New Zealand, the effects of tectonism are perva-sive, manifested as uplift that contributes onshore to valley incision and degradation that followed aggradation and valley fill during the Last Glacial Maximum (Carter et al., 2010). Such uplift thus reactivates valley systems, and part of the sediment supply generated by this regional uplift and incision is captured in local zones of subsid-ence that form sediment depocentres.

Tectonic change in base levels associated with fault- and subduction-related activity can have a significant impact on the system. Earthquakes associated with fault- and subduction-induced processes are another important component of the sediment supply process in such active margins. Compressive forces render fragile the upland terrain, which is further destabilized by seis-mic ground-shaking, generating landslides and other mass movements that liberate sediment for river trans-port to the coast. Sediment supply to the coast thus exhibits temporal and spatial variability, depending on earthquake intensity, debris availability, permanence of any landslide dams, and the intensity/frequency of heavy rainfalls and snow melts. Active plate settings, such as that of New Zealand, are also characterized by volcanic eruptions, which deliver sediment directly, in the case of the Waipaoa Sedimentary System (WSS), as airfall ash, while potentially enhancing sediment input to the coast through the destruction of terrestrial vegetation, the impact of which may last for a century or more.

On active margins, the cumulative effect of high mountain relief, climatic, tectonic and volcanic drivers, is to enhance fluvial discharge and its subsequent accu-mulation in coastal and shelf sinks. These conditions explain why collision and Amero-type coasts in embed-ded plates are associated with high-sediment discharge rivers debouching from the mountain catchments located at the opposite active margin. These drivers are far less active in source-to-sink sedimentary systems associated with passive margins. There may also be dif-ferences in weathering type between mid- to high-lati-tude systems, associated with the delivery of relatively coarse and poorly chemically weathered sediments, and tropical source-to-sink systems, where chemical weath-ering, sometimes over prolonged geological periods, leads to high river solute and fine-grained loads. Short source-to-sink systems, such as those flanking the Andes coastal margin and parts of the Mediterranean margin, may also be associated with the delivery of coarse sedi-ment to the coast, compared to longer systems. Finally, source-to-sink sediment supply can also be significantly impacted by human deforestation of the landscape and its conversion for agricultural use. As a consequence, in the WSS, for instance, hill-slope erosion has increased and the fluvial sediment discharge has considerably expanded. The exponential increase in dams worldwide has, however, drastically reduced sediment supply to coasts via deltas.

various minor mass-wasting processes, to stream chan-nels may result following bursts of heavy, concentrated rainfall, while earthquakes may also trigger slope insta-bilities that feed stream channels. High-magnitude flood-ing and massive sediment transport downstream are generally related to unpredictable, extreme rainfalls. Both mass movements and channel sediment storage pose serious hazards to downvalley settlements and infrastructure.

3.3.4 Sediment redistribution from river-mouth to coast

The dynamic processes prevailing at river-mouths are fundamental in the supply of sediments to the adjacent coasts. River-mouths occupy a transitional zone charac-terized by a change from fluvial to wave and tidal domi-nance, including salt-water influence, by marked changes in geomorphology and bathymetry, and by bidi-rectional sediment movement. These variations are particularly significant in open-mouthed estuaries and deltas (see Chapters 12 and 13). Hydrodynamic processes prevailing in the transitional fluvio-marine zone generally act to trap fluvial sediment in estuaries, and to limit its export to the sea. This occurs essentially through the large-scale effects of water mixing or salt-wedge development and landward-directed residual tidal flow. Strong deltaic flows that directly supply sediment to the coasts are thus important for the development of alluvial coasts.

Fine-grained sediment comprises varying proportions of mineral grains, polymineralic floccules and faecal pellets. Effective trapping depends on the relationship between sediment settling velocity and estuarine morphology and dynamics, with potential loss, through export to the sea of the lightest particles, and quasi-permanent deposition of rapidly settling particles. Grain size depends on numerous environmental factors, but an important consideration in estuarine dynamics and shoreface plume discharge is that of fundamental size changes brought about by processes of aggregation, which result in the binding together of small organic and inorganic particles by bacteria, other organisms, and organic detritus into porous aggregates or flocs. Biological mediation is an important part of the processes involved in organizing sediment recycling (Wolanski, 2007). The dynamically active process of flocculation alters the settling velocity of cohesive sediments, modifying vertical concentration gradients, and, consequently, deposition and accumulation.

In large river systems such as the Amazon and the Changjiang, or during episodic high river-discharge spates even in small rivers, the momentum of the freshwater outflow may preclude saltwater intrusion, and the complete sediment-laden freshwater column flows into the ocean. Through this process, mud transported by rivers may be delivered to adjacent shores, either from coherent mud banks associated with longshore mud streaming, as on the South American coast (Fig. 3.11) west of the Amazon, from loci of muddy inner-shelf sedimentation, as in the case of the Atchafalaya River on the Mississippi-influenced coast, or from alongshore dissemination of particulate mud, as on several of the Asian deltas such as the Mekong. Long-term deposition from these processes may lead to the formation of long stretches of muddy coasts associated with such deltas, and their distinct coastal geomorphologies (mangroves, cheniers, salt marshes and tidal flats).

Deltas are the major pathways of fluvial bedload sediment supply to coasts. Bedload supply is particularly important during storms and important river-flood events. Such bedload generally forms bars, which are the fundamental river-mouth sediment reservoirs from

Fig. 3.11 Photograph of a mud bank on the coast of French Guiana, South America, showing stages of colonization by mangroves, comprising young individual plants in the foreground and denser and more mature stands in the background. The 1500-km long mangrove coast of South America between the Amazon and the Orinoco river-mouths is one of the world's muddiest coasts. The Amazon is the world's largest river system, with a drainage basin of 6.1 million km^2. The mean annual water discharge has been estimated at about 180,000 m^3/s at Obidos, 900 km upstream of the mouth, leading to strong fluvial outflow onto the inner shelf. Part of the huge suspended sediment discharge of this river (about 750 million metric tons per year), composed essentially of clay-sized minerals derived from intense tropical weathering of rock debris from the Andes mountains, accumulates up to 150 km offshore, but feeds the formation of mud banks further inshore, which are conveyed alongshore from Brazil to Venezuela by waves and currents (Anthony et al., 2010). The mud banks, which may number more than 20 at any time, are each up to 5 m thick, 10–60 km long and 20–30 km wide, and each may contain one to several times the entire annual mud supply of the Amazon. As these banks migrate alongshore, they dissipate waves from the Atlantic, protecting the terrestrial shore. Temporary mud attachment to the shore leads to massive short-term (months to a years) coastal progradation (from hundreds of metres to over 1 km), followed by equally spectacular coastal erosion by waves a few years later as the bank continues its migration downdrift.

which coasts are fed in sandy and gravel sediments, especially under conditions of wave redistribution through longshore currents. These bar deposits are commonly sandy to gravelly, mainly because strong river flows and wave action in river-mouths inhibit mud deposition and/or resuspend any mud that may be deposited in shallow water.

3.3.5 Carbonate sediments

Carbonate sediments form a significant proportion of the Earth's sedimentary record, and biological constructions are, therefore, an important component of coasts (Spencer and Viles, 2002). Mountain-building processes resulting from the collision of plates, the margins of which are carbonate-rich, also mean that carbonate rocks are part of the terrigenous sediment supply to coasts following weathering and erosion. Carbonate platforms and their associated coral reefs are the largest coastal structures built solely by plants and animals (Hopley, 2005). Coral taphonomic groups and processes operate, under ambient wave and hydrodynamic conditions, and sediment supply and nutrient inputs, to produce distinct and preservable skeletons and/or traces, or to generate thick deposits sometimes exceeding several hundreds of metres, and several kilometres in the case of coral atolls. Although coral reefs are commonly associated with tropical shores, these forms constitute a continuum that goes well into the mid-latitudes, where low carbonate production end-members associated with restricted framework development and dominant siliciclastic sediment composition also occur. Coral growth is generated by coral-zooxanthellae symbiosis. Calcium carbonate secretion leads to the formation of a skeletal framework that is locally infilled by sedimentation and cementation. At the larger reef scale, detrital accumulations, which may be derived from coral rubble, may form while further reef sedimentation is assured by infill of backreef and lagoon environments. Such sediment supply is important as sediments provide a surface over which reefs can prograde. Infill sediment may range from terrigenous mud where nearby rivers debouch, to fine sand and gravel, often derived from reworking of the reefs. These coarser components may be siliciclastic sand in marginal reef environments such as cooler mid-latitude settings, or may be composed of reef rubble consisting essentially of bioclastic skeletal sands transported by waves and currents from reef organisms such as coral, algae, foraminifera and molluscs, especially in the case of reef islands. Even on coasts where coral reefs have not developed, carbonate sediments derived from broken-down shells are variably admixed with terrigenous sediments. Shells and skeletal remains are broken down and driven shoreward by wave action. Coral coasts and atolls are discussed further in Chapter 16.

Other sources of carbonate sediments are mangroves, salt marshes and seagrass colonies. The death of mangroves and salt marshes can result in the formation of significant levels of subsoil peat that contributes to coastal accretion, especially where terrigenous sediment inputs are low. Seagrass colonies, such as *Posidonia oceanica*, are important along many coasts, such as in the Mediterranean.

Apart from trapping terrigenous sediments, seagrass colonies also contribute to the dissipation of waves and tidal currents, thus encouraging coastal sedimentation. The destruction of seagrass colonies on many Mediterranean shores, through the construction of harbours, marinas and artificial beaches, has contributed to the erosion of adjacent beaches.

3.3.6 Sediment supply from soft cliffs

Cliff erosion accounts for only about 5% of sediments supplied to coasts (Woodroffe, 2003) and is, thus, significantly less important than river supply. Coastal cliffs are a prominent feature of both 'hard' rock and 'soft' rock coasts. The retreat of soft-rock cliffs may release sediment for *in situ* beach development over a shore platform or for transport alongshore onto beach sinks, whereas muds released by cliff erosion are generally deposited over calmer coastal and estuarine depositional environments such as mudflats and salt marshes. Cliff erosion along the Holderness coastline in England, for example, has released sand and gravel for the barrier spit of Spurn Head across the Humber estuary, while the muds have accumulated on salt marshes on the North Sea coast. Other examples include the eastern English Channel coasts of both England and France, where cliff erosion (Fig. 3.12) has sourced gravel beaches, including the remarkable gravel barriers of Dungeness in Kent and Cayeux in Picardy (see

Fig. 3.12 'Soft' rock (chalk) cliffs, Three Sisters, Sussex, England. The gravel beach fronting the cliffs is composed of flint clasts liberated by cliff erosion. In the foreground on the left, the lower beach comprises a sandy apron. Both gravel and sand overlie a chalk platform that crops out in the middle background of the photograph. (Source: Courtesy of Dr Uwe Dornbusch.)

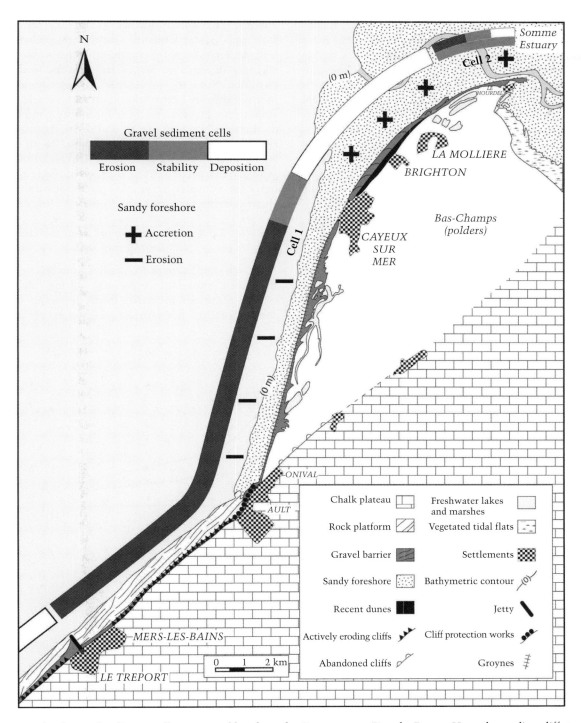

Fig. 3.13 Example of coastal sediment cells on a gravel beach on the Cayeux coast, Picardy, France. Here, the eroding cliffs to the south provide sediment for the gravel spit system. Sediment cells provide a conceptual (and shoreline management) tool for monitoring sediment transport continuities or discontinuities and sediment budgets along the coast. Each cell comprises an erosional, a transport, and a depositional sector, and is bound updrift and downdrift by cell boundaries that may be natural or artificial (engineering or port structures). (Source: Adapted from Anthony and Dolique 2001.)

section 3.3.7). Cliff retreat is commonly episodic, highly variable alongshore, and related to a large variety of mass movements and failure events, such as block falls, debris flows, mudflows, rotational slumps, and slides. These processes appear to be highly site-specific, sometimes with considerable intra-site variability, as a function of local rock parameters, cliff height, wave energy, groundwater conditions, the climate setting and beach characteristics. These processes are particularly important in episodic erosion events in areas of low wave energy.

3.3.7 Longshore sediment transport

Longshore transport enabling the redistribution of sediment alongshore may operate within the framework of one or several sediment cells (Fig. 3.13) with bounding limits to longshore drift (Carter, 1988). The sediment cell notion is important, both to coastal management issues and coastal geomorphic development, because of the relevance of cell boundaries to sediment flux continuity alongshore and the calculation of coastal sediment

budgets. The redistribution of sediments from cliff recession and coastal landslides, which may also be important local sources of sediment release to the coast, strongly depends on wave- and tide-induced longshore currents. Note that the same concept applies to sediment transport across the shoreface (see later), which sometimes involves a strong longshore component. The distinction between swash and drift-alignment, respectively designating shores associated with weak and strong rates of longshore drift (Davies, 1980), is now strongly entrenched in the literature, and is also a useful basis for considering process variations and long-term shore development patterns.

3.3.8 Sediment supply from the inner continental shelf

Although significant sand supplies from deltas may enrich the inner shoreface, especially during high-discharge spates, sediment inherited from the inner continental shelf (shoreface) from earlier sea-level stands may also be

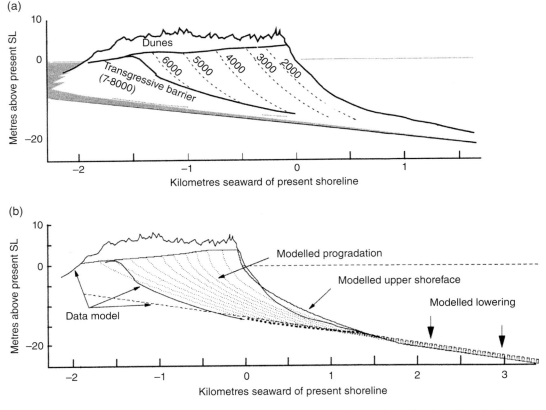

Fig. 3.14 An example of modelling of a prograded beach-ridge plain sourced by sediment from the shoreface, the Tuncurry barrier in southeast Australia. (a) Sketch of barrier morphology with isochrons in years BP. SL, sea level. (b) Shoreface translation model simulation output based on chronological and stratigraphic data. Progressive incorporation of shoreface sand onto the prograding beach-ridge plain is accompanied by shoreface lowering. (Source: Cowell et al. 2003b. Reproduced with permission of The Coastal Education & Research Foundation, Inc.)

(c)

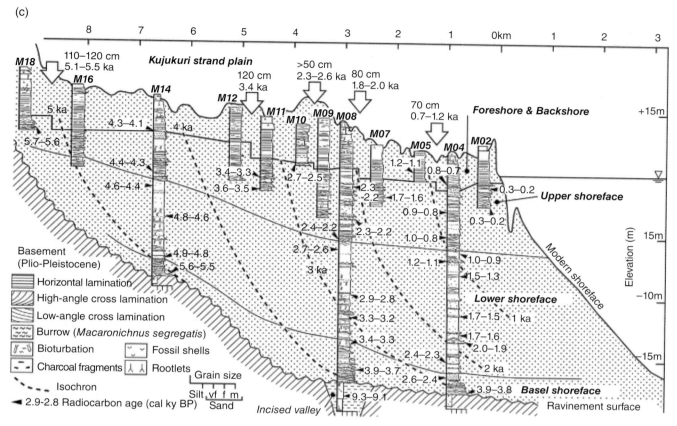

Fig. 3.14 (Continued) (c) Example of the complex beach-ridge stratigraphy of the Kujukuri strand plain, Japan, wherein the basal contact expresses a series of rapid sea-level falls. Isochrons based on radiocarbon ages of molluscan shells show that the beach has prograded seaward since 6 ka, forming regressive beach deposits overlying a ravinement surface that cuts the Plio-Pleistocene basement. The base of the foreshore facies steps down seaward, implying relative sea-level falls in response to tectonic uplift. (Source: (c) Tamura et al. 2007. Reproduced with permission of John Wiley & Sons.)

reworked by waves and currents to source beach and barrier coasts (Fig. 3.14). The shoreface has long been recognized as a fundamental component of long-term coastal development because of its central role in sand exchanges between the continental shelf and the coast. There has been a considerable refinement of these concepts of sediment exchange and it is now clear that over long timescales (centuries to millennia), lateral translation of the shoreface may determine the overall evolution of many of the world's wave-dominated coasts, especially (but not only) under conditions of rapid sea-level variations (Box 3.3). A useful and scientifically coherent approach in the analysis of the sediment links between the shelf and the coast is to consider this domain within the framework of an aggregated metamorphology, termed the 'coastal tract' by Cowell et al. (2003a), and characterized by long-term and large-scale exchanges that

are achieved through what these authors aptly refer to as 'sediment sharing' (Fig. 3.15).

3.3.9 Boulders on the shore: an enigmatic issue

Many rock shores comprise accumulations of gravel to boulder-sized deposits overlying bedrock (Fig. 3.16). Past studies of such rocky shoreline deposits have been largely descriptive, but these deposits have drawn attention over the last few years, mainly within the framework of the debate revolving around whether the megaclasts are transported and deposited by storms or tsunamis (Felton and Crook, 2003; Nott, 2004). Boulders on beaches can also accumulate simply from the *in situ* weathering of resistant bedrock. The role of ice in the transport and deposition of boulders may also be preponderant in cold-climate areas (e.g. Dionne, 2002).

CONCEPTS BOX 3.3   Sediment movement from inner shelf to coast

The large-scale metamorphology formed by what Cowell et al. (2003a) call the 'coastal sediment tract' (see Fig. 3.15) aggregates smaller-scale geomorphic domains, such as the outer and inner shoreface, the surf zone, the beach-face and the backshore, although these domains are often treated separately in site studies for convenience. Identifying sediment supply from the shelf to the beach-face is not an easy matter, and there are very few studies on this. Apart from direct measurements of sediment dynamics on the lower shoreface, methods of identifying such sediment supply include: (1) measurements of shoreface bathymetric changes over several decades; (2) measurement and numerical modelling of shoreface sediment transport; (3) behaviour modelling of the shoreface sediment-transport regime and of its profile; (4) radiometric dating of shoreline progradational features (such as beach ridges) at sites where sources of sediment supply other than the shoreface are limited, non-existent or cannot be estimated; (5) analyses of the sedimentology (grain sizes and bedforms) of the lower shoreface; and (6) analyses of regional sediment budgets. Each of these methods carries uncertainties and cannot be considered as conclusive. Sediment transport on the shoreface involves another poorly known aspect of fluid-bed interactions in regards to the spatial and temporal distributions of bed smoothness or roughness domains and their relationship to storm events, and to grain size and bedforms.

Under conditions of unlimited sediment supply from the shoreface (inherited, for instance, from earlier lower sea-level phases), a favourable shoreface gradient (see Fig. 3.7), and a simple configuration of a stable or slowly falling sea level, barriers may prograde through long-term sediment transfer from a shoreface in disequilibrium with high ambient wave energy. These can be ideal conditions for the generation of thick wave-formed coastal deposits (see Fig. 3.14). The transport mechanism is one of wave orbital asymmetry-induced transfer of sand shoreward from a shallow, more or less sand-rich shoreface in disequilibrium with the high wave energy conditions, to establish over time an inner shelf profile in equilibrium with the wave energy conditions. Supply from the shoreface is expected to diminish with time (over millennia), as the shoreface sand stocks are progressively exhausted through incorporation into the subaerial barrier, but this situation would depend on substrate potential to liberate sediment, as well as on sea level. Under a stable sea level, with time these shoreface sand sources may run out, as equilibrium between the wave forcing and the shoreface morphology is attained, leading to cessation of shoreline progradation, and to stationary barriers, such as some of the beach-ridge barriers in Australia and West Africa.

Attainment of equilibrium would depend on the wave climate. In certain high-energy swell wave settings of southeastern Australia, shoreface equilibrium and cessation of barrier progradation occurred relatively early during the Holocene, whereas similar equilibrium was attained at a much later time in the lower-energy swell setting of West Africa (Anthony, 2009). Overall, because of the moderate wave energy levels in West Africa, and probably off the coast of Brazil as well, the depths to which the shoreface has been planed to supply sand for barrier progradation are less than those of the higher wave energy coasts of southeastern Australia. In the latter setting, shoreface sand stocks were rapidly incorporated into prograding beach plains following the stabilization of sea level in the middle Holocene. This pattern of barrier progradation has been successfully modelled by Cowell et al. (2003b), using shoreface translation model simulation of the Holocene development of the Tuncurry barrier in Australia (see Fig. 1.4).

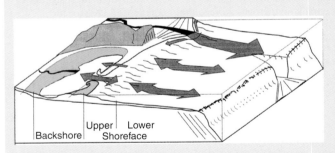

Fig. 3.15 Diagram illustrating the 'coastal sediment tract' across which occurs sediment sharing between the coast and the continental shelf at various timescales. (Source: Cowell et al. 2003b. Reproduced with permission of The Coastal Education & Research Foundation, Inc.)

Cliff-top storm deposits (CTSDs) are an increasingly documented category of heterogeneous deposits associated with cliff tops at elevations of up to 50 m above sea level in areas exposed to extreme storm waves, as in parts of the British Isles (e.g. Hansom and Hall, 2009) and Britanny, France (Suanez et al., 2009). Such deposits may range from boulders to sand. CTSDs are considered as being generated largely by the wave quarrying of blocks during major storms from the cliff top that are then transported by bores across the cliff-top platforms and ramps. Nott (2004) has reported that there are various locations in the Southern Ocean, North Atlantic Ocean

(a)

(b)

Fig. 3.16 (a) Megaclast deposit; and (b) boulder beach along the west coast of Ireland, County Clare. On several coasts of the world where such deposits occur, there is debate as to whether they were deposited by storm waves or by tsunamis. (Source: Photographs by Gerd Masselink.)

and the tropical to sub-tropical waters of the Pacific, Indian and Atlantic Oceans where intense tropical cyclones may lead to the emplacement of cliff-top deposits, the classic example being that of the Pacific island of Niue, where deposition of boulders by waves on cliff tops up to 30 m above sea level occurred in 1991 and 2004 during severe Tropical Cyclones Ofa and Heta, respectively.

3.4 Human impacts on sediment supply to coasts

Human impacts on coastal sediment supply date back to ancient times, essentially through modification of river catchments via agriculture and land use (Hoffmann et al., 2010), through river damming, and through the development of coastal settlements and activities. Disentangling the human contribution to the perturbation of coastal sediment budgets from that of natural parameters is, however, a difficult task. Changes in land use and agricultural practice have been shown to cause fluctuations in sediment supply from river catchments. Sediment sequestering in dams is a worldwide phenomenon and there is a likelihood that the goals of sustained economic development in many developing countries, where population increases are set to continue, will involve more important dam constructions in the future, thus significantly impacting many deltas (Fig. 3.17), and therefore, sediment supply to coasts. Syvitski et al. (2009) have suggested that humans have simultaneously increased the sediment transport by global rivers through soil erosion by 2.3 ± 0.6 billion metric tons per year, and yet reduced the flux of sediment reaching the world's coasts by 1.4 ± 0.3

billion metric tons per year because of retention within reservoirs. Over 100 billion metric tons of sediment are now sequestered in reservoirs constructed largely within the past 50 years, and African and Asian rivers carry greatly reduced sediment loads compared to Indonesian rivers, which deliver much more sediment to coastal areas.

Human engineering on shores dates back to ancient times, especially regarding ancient harbours in the Mediterranean (Marriner and Morhange, 2007). Large stretches of the coastal wetlands of northwestern Europe have been empoldered over the last nine centuries. Massive engineering interventions, with far-reaching consequences on coastal sediment transport and coastal stability, are, however, products of coastal urbanization and economic development over the last few decades. The construction of groynes, breakwaters and sea walls to contain erosion has, in many places, perturbed the longshore transport of sediments from source areas, such as river-mouths and cliffs, generating erosion downdrift. Cliff stabilization, generally aimed initially at protecting farmland and settlements, has, for example, been an important cause of beach erosion on the coasts of Normandy and Picardy, in France, leading coastal settlements into a spiral of construction of beach protection structures.

3.5 Climate change, geology and sediments

The direct significant impacts to be expected on coastal processes and sea level from climate change have been widely discussed in the literature. Geology at the human

Fig. 3.17 Global distribution of 40 deltas in the world, showing the potentially contributing drainage basin area of each delta (grey) and the large reservoirs (>0.5 km³ maximum capacity) in each basin. (Source: Ericson et al. 2006. Reproduced with permission of Elsevier.)

timescale is not affected by climate change, other than through sediment supply from catchments. The impact of climate change on geology and sediments associated with river catchments has been widely described in the literature, and numerous examples can be found in various earth and environmental science journals. Climate change is having a significant impact on mountain catchments (Beniston, 2003), which are the most important sources of coastal sediments at the global scale. Changes in temperature and rainfall regimes affect both glacier and torrential stream activity, and enhance the release of sediment from physically weathered mountain catchments, as well as from agricultural lands susceptible to increased erosion. The increases in flooding capacities of some Asian rivers, such as the Indus, are dramatically illustrated by the catastrophic northern hemisphere summer 2010 floods that affected Pakistan, India and China, fed in sediment by landslides following heavy monsoon rainfalls. This may by no means be globally true, as reduced stream discharges, such as in Mediterranean rivers, may lead to a reduction in sediment supply to the coast.

While the impact of climate change on sediment release from weathered catchments is a certainty, there has been little work on how coastal sediment budgets will be influenced by changes affecting fluvial systems upland, notable exceptions being the work of Ericson et al. (2006) and Syvitski et al. (2009) on some of the world's deltas. One problem stems from upland storage capacities in river floodplains and channel systems at spatial and temporal scales that may vary widely. Another problem related to this is that of sediment sequestering by dams, as stated earlier.

Coasts themselves will be directly affected in terms of sediment supply through enhanced erosion and redistribution of sediments within longshore cells or between the coastal and offshore zones. There have been numerous case studies of coastal erosion purportedly attributed in part or in whole to climate change, and a useful synthesis of the potential impact of climate change on coastal sediment budgets has been discussed by Ranasinghe and Stive (2009) for low coasts composed of clastic sediments. These authors examined, in particular, the potential for longshore and cross-shore redistributions of coastal sediment, and between coasts and inlets that will differentially impact shores in the future. On low coasts, the erosion to be expected from rising sea level associated with climate change will be countered by increasing human intervention, with impacts on the longshore redistribution of sediments.

Much less has been done on the impact of climate change on the erosion of rock coasts, which, it must be recalled, are the world's dominant coastal type and are traditionally considered as erosional coasts susceptible to release sediment. Naylor et al. (2010) noted the lack of consideration of rock coasts in scenarios of climate change induced coastal erosion, which, because of their slow rates of erosion are consequently considered as being relatively resilient to climate change. These authors have, however, pointed out the susceptibility of hard-rock cliffs to erode more actively due to climate change. Soft-rock cliffs (bluffs) generally erode actively, and thus contribute sediment to the coast. They are therefore likely to release sediment more actively under the impact of climate change (e.g. Dawson et al., 2009).

3.6 Summary

Geology is a fundamental control on coasts because it sets the primary boundary conditions within which coasts are formed and within which the processes of coastal change operate, with the overall tectonic framework and lithology playing significant roles. The coastal topography, orientation, and overall sediment accommodation space, are hinged on these primary boundary conditions. Geology also controls the type and rate of sediment supply to coasts, mediated by climate. Coasts may range, in their morphology, from primary types hinged on the bedrock geology, to accumulation forms reflecting a large diversity of arrangement of the sedimentary facies. Geology at the human timescale is not affected by climate change, other than through sediment supply from catchments.

• Tectonic setting controls patterns of landform development, as well as the size, relief and orientation of drainage basins, and affects sediment supply. It also imprints a primordial trend to coastal orientation and the development of the adjacent continental shelf. The shelf width and morphology affect, in turn, wave energy and tidal range.

• While global tectonics have a basic imprint on gross coastal characteristics, they must not mask, however, the importance of second-order aspects, such as lithology, local to regional tectonic constraints such as faulting, ice loading and sediment supply, in determining coastal morphology.

• A long-term perspective on coastal evolution is useful because the present-day coast is a product of modern and past coastal processes and is often, at least partly, controlled by inherited landforms.

• The influence of climate and sediment supply can become largely dominant over the geological heritage. Climate has a determining influence on both rock breakdown to liberate sediments and transport of these sediments to the coast. By affecting river flow and storm intensity, climate change will impact rates of landscape degradation that lead to the release of sediments for coasts.

• Overall, while large-scale modification of upland catchments that provide sediments for coastal landform development are deemed to have a significant impact on sediment supply to coasts, notably through reservoir-sediment retention that affects many of the world's deltas, there is a need for better understanding of the rates and scales of sediment transit from source to coastal sink – hence the utility of the sediment source-to-sink approach.

Key publications

Davies, J.L., 1980. *Geographical Variation in Coastal Development*. Longman, London, UK.
[*Provides a good analysis of the influence of geology and tectonics on coasts*]

Inman, D.L. and Nordstrom, C.E., 1971. On the tectonic and morphological classification of coasts. *Journal of Geology*, 79, 1–21.
[*Essential reading on the relationship between coasts and global tectonics*]

Masselink, G. and Hughes, M.H., 2003. *Introduction to Coastal Processes and Geomorphology*. Arnold, London, UK.
[*Introductory text with a useful presentation of the influence of geology on coasts*]

Woodroffe, C.D., 2003. *Coasts: Form, Process and Evolution*. Cambridge University Press, Cambridge, UK.
[*Provides a useful discussion of the effects of geology on coasts*]

References

Anthony, E.J., 2009. *Shore Processes and their Palaeoenvironmental Applications*. Developments in Marine Geology, Volume 4. Elsevier, Amsterdam, the Netherlands.

Anthony, E.J. and Dolique, F., 2001. Natural and human influences on the contemporary evolution of gravel shorelines between the Seine estuary and Belgium. In: J.R. Packham, R.E. Randall, R.S.K. Barnes and A. Neal (Eds), *The Ecology and Geomorphology of Coastal Shingle*. Westbury Academic and Scientific Publishers, Otley, UK, pp. 132–148.

Anthony, E.J. and Héquette, A., 2007. The grain size composition of coastal sand from the Somme estuary to Belgium: sediment sorting processes and mixing in a tide- and storm-dominated setting. *Sedimentary Geology*, **202**, 369–382.

Anthony, E.J. and Julian, M., 1999. Source-to-sink sediment transfers, environmental engineering and hazard mitigation in the steep Var river catchment, French Riviera, southeastern France. *Geomorphology*, **31**, 337–354.

Anthony, E.J., Gardel, A., Gratiot, N. et al., 2010. The Amazon-influenced muddy coast of South America: a review of mud-bank-shoreline interactions. *Earth-Science Reviews*, **103**, 99–121.

Beniston, M., 2003. Climate change in mountain regions: a review of possible impacts. *Climate Change*, **59**, 5–31.

Bishop, P. and Cowell, P., 1997. Lithological and drainage network determinants of the character of drowned, embayed coastlines. *The Journal of Geology*, **105**, 685–699.

Blott, S.J. and Pye, K., 2008. Particle shape: a review and new methods of characterization and classification. *Sedimentology*, **55**, 31–63.

Bradley, D.C., 2008. Passive margins through earth history. *Earth-Science Reviews*, **91**, 1–26.

Buscombe, D., Rubin, D.M. and Warrick, J.A., 2010. A universal approximation of grain size from images of noncohesive sediment. *Journal of Geophysical Research*, **115**, F02015.

Carroll, J. and Lerche, I., 2003. *Sedimentary Processes: Quantification using Radionuclides*. Elsevier, Amsterdam, the Netherlands.

Carter, L., Orpin, A.R. and Kuehl, S.A., 2010. From mountain source to ocean sink – the passage of sediment across an active margin, Waipaoa Sedimentary System, New Zealand. *Marine Geology*, **270**, 1–10.

Carter, R.W.G., 1988. *Coastal Environments*. Academic Press, London, UK.

Catuneanu, O., 2006. *Principles of Sequence Stratigraphy*. Elsevier, Amsterdam, the Netherlands.

Catuneanu, O., Abreu, V., Bhattacharya, J.P. et al., 2009. Towards the standardization of sequence stratigraphy. *Earth-Science Reviews*, **92**, 1–33.

Coe, A.L. (Ed.), 2003. *The Sedimentary Record of Sea-Level Change*. Cambridge University Press, Cambridge, UK.

Cowell, P.J., Stive, M.J.F., Niedoroda, A.W. et al., 2003a. The coastal-tract (Part 1): a conceptual approach to aggregated modeling of low-order coastal change. *Journal of Coastal Research*, **19**, 812–827.

Cowell, P.J., Stive, M.J.F., Niedoroda, A.W. et al., 2003b. The coastal tract (Part 2): applications of aggregated modeling of low-order coastal change. *Journal of Coastal Research*, **19**, 828–848.

Davies, J.L., 1980. *Geographical Variation in Coastal Development*. Longman, London, UK.

Dawson, A.G. and Stewart, I., 2007. Tsunami deposits in the geological record. *Sedimentary Geology*, **200**, 166–183.

Dawson, R.J., Dickson, M.E., Nicholls, R.J. et al., 2009. Integrated analysis of risks of coastal flooding and cliff erosion under scenarios of long term change. *Climate Change*, **95**, 249–288.

Dionne, J.C., 2002. The boulder barricade at Cap à la Baleine, north shore of Gaspé Peninsula (Québec): nature of boulders, origin, and significance. *Journal of Coastal Research*, **18**, 652–661.

Emery, K.O. and Kuhn, G.G., 1982. Sea cliffs: their processes, profiles, and classification. *Geological Society of America Bulletin*, **93**, 644–654.

Ericson, J.P., Vörösmarty, C.J., Dingman, S.L., Ward, L.G. and Meybeck, M., 2006. Effective sea-level rise and deltas: causes of change and human dimension implications. *Global and Planetary Change*, **50**, 63–82.

Felton, E.A. and Crook, K.A.W., 2003. Evaluating the impacts of huge waves on rocky shorelines: an essay review of the book 'Tsunami – The Underrated Hazard'. *Marine Geology*, **197**, 1–12.

Flemming, B.W., 2007. The influence of grain-size analysis methods and sediment mixing on curve shapes and textural parameters: implications for sediment trend analysis. *Sedimentary Geology*, **202**, 425–435.

Gehrels, W.R., Milne, G.A., Kirby, J.R., Patterson, R.T. and Belknap, D.F., 2004. Late Holocene sea-level changes and isostatic crustal movements in Atlantic Canada. *Quaternary International*, **120**, 79–89.

Hansom, J.D. and Hall, A.M., 2009. Magnitude and frequency of extra-tropical North Atlantic cyclones: a chronology from cliff-top storm deposits. *Quaternary International*, **195**, 42–52.

Hoffmann, T., Thorndycraft, V.R., Brown, A.G. et al., 2010. Human impact on fluvial regimes and sediment flux during the Holocene: review and future research agenda. *Global and Planetary Change*, **72**, 87–98.

Hopley, D., 2005. Coral reefs. In: M.L. Schwartz (Ed.), *Encyclopedia of Coastal Science*. Springer, Dordrecht, the Netherlands, pp. 343–349.

Inman, D.L. and Nordstrom, C.E., 1971. On the tectonic and morphological classification of coasts. *Journal of Geology*, **79**, 1–21.

Julian, M. and Anthony, E.J., 1996. Aspects of landslide activity in the Mercantour Massif and the French Riviera, southeastern France. *Geomorphology*, **15**, 275–289.

Marriner, N. and Morhange, C., 2007. Geoscience of ancient Mediterranean harbours. *Earth-Science Reviews*, **80**, 137–194.

Masselink, G. and Hughes, M.H., 2003. *Introduction to Coastal Processes and Geomorphology*. Arnold, London, UK.

Naylor, L.A., Stephenson, W.J. and Trenhaile, A.S., 2010. Rock coast geomorphology: recent advances and future research directions. *Geomorphology*, **114**, 3–11.

Nott, J., 2004. The tsunami hypothesis – comparisons of the field evidence against the effects, on the Western Australian coast, of some of the most powerful storms on Earth. *Marine Geology*, **208**, 1–12.

Orford, J.D., Forbes, D.L. and Jennings, S.C., 2002. Organisational controls, typologies and time scales of paraglacial gravel-dominated coastal systems. *Geomorphology*, **48**, 51–85.

Phillips, J.D. and Slattery, M.C., 2006. Sediment storage, sea level, and sediment delivery to the ocean by coastal rivers. *Progress in Physical Geography*, **30**, 513–530.

Ranasinghe, R. and Stive, M.J.F., 2009. Rising seas and retreating coastlines. *Climate Change*, **97**, 465–468.

Regard V., Saillard, M., Martinod, J. et al., 2010. Renewed uplift of the Central Andes Forearc revealed by coastal evolution during the Quaternary. *Earth and Planetary Science Letters*, **297**, 199–210.

Roy, P.S., Cowell, P.J., Ferland, M.A. and Thom, B.G., 1994. Wave dominated coasts. In: R.W.G. Carter and C.D. Woodroffe (Eds.), *Coastal Evolution: Late Quaternary Shoreline Morphodynamics*. Cambridge University Press, Cambridge, UK, pp. 121–186.

Schellart, W.P. and Rawlinson, N., 2010. Convergent plate margin dynamics: new perspectives from structural geology, geophysics and geodynamic modelling. *Tectonophysics*, **483**, 4–19.

Scott, T.W., Swift, D.J.P., Whittecar, G.R. and Brook, G.A., 2010. Glacioisostatic influences on Virginia's late Pleistocene coastal plain deposits. *Geomorphology*, **116**, 175–188.

Short, A.D., 1999. Global variation in beach systems. In: A.D. Short (Ed.), *Handbook of Beach and Shoreface Morphodynamics*. John Wiley & Sons, Chichester, UK, pp. 21–35.

Short, A.D. and Masselink, G., 1999. Embayed and structurally controlled beaches. In: A.D. Short (Ed.), *Handbook of Beach and Shoreface Morphodynamics*. John Wiley & Sons, Chichester, UK, pp. 230–250.

Short, A.D. and Woodroffe, C.D., 2009. *The Coast of Australia*. Cambridge University Press, Cambridge, UK.

Spencer, T. and Viles, H., 2002. Bioconstruction, bioerosion and disturbance on tropical coasts: coral reefs and rocky limestone shores. *Geomorphology*, **48**, 23–50.

Suanez, S., Fichaut, B. and Magne, R., 2009. Cliff-top storm deposits on Banneg Island, Brittany, France: effects of giant waves in the Eastern Atlantic Ocean. *Sedimentary Geology*, **220**, 12–28.

Syvitski, J.P.M., Kettner, A.J., Overeem, I. et al., 2009. Sinking deltas due to human activities. *Nature Geoscience*, **2**, 681–689, doi: 10.1038/NGEO629.

Tamura, T., Nanayama, F.K., Saito, Y., Murakami, F., Nakashima, R. and Watanabe, K., 2007. Intra-shoreface erosion in response to rapid sea-level fall: depositional record of a tectonically uplifted strand plain, Pacific coast of Japan. *Sedimentology*, **54**, 1149–1162.

Trenhaile, A.S., 2002. Rock coasts, with particular emphasis on shore platforms. *Geomorphology*, **48**, 7–22.

Walker, M., 2005. *Quaternary Dating Methods*. John Wiley & Sons, Chichester, UK.

Wolanski, E., 2007. *Estuarine Ecohydrology*. Elsevier, Amsterdam, the Netherlands.

Woodroffe, C.D., 2003. *Coasts: Form, Process and Evolution*. Cambridge University Press, Cambridge, UK.

4 Drivers: Waves and Tides

DANIEL C. CONLEY

School of Marine Science and Engineering, University of Plymouth, Plymouth, UK

4.1 Physical drivers of the coastal environment

Waves and tides are the physical drivers that provide the energy for essentially all changes in coastal geomorphology. These drivers are derivatives of, respectively, solar energy and the gravitational pull of the Sun and the Moon. Significantly, it is gradients in these energy sources that lead to the creation of the forcing agents. Gradients in heating between different parts of the Earth cause the winds to blow across the surface of the ocean, which in turn generates the waves that efficiently transport this energy to distant locations. Gradients in the gravitational potential of the Moon and the Sun lead to the dramatic daily tidal variations in the level of the sea surface, which not only expose different sections of coastal geomorphology to wave attack but also generate strong tidal currents that also work to shape the morphology of the coast. Driven by gradients themselves, the waves and tides, which force coastal processes all over the globe, exhibit strong spatial and temporal variability that gives rise to the great variety present along the world's coastlines. While ultimately driven by celestial forces that would appear to be beyond human influence, the details of where warming and cooling take place, as well the depths of the coastal basin where tides propagate, are affected by climate change, so that in the end the distribution of waves and tides around the face of the Earth can indeed be deeply influenced by climate change. In the following sections we will become familiar with the characteristics of these drivers, how they are generated, how they transmit energy to distant shores, and how they are affected by global climate change.

4.2 Waves

4.2.1 Importance and definitions

Waves represent physical processes that are tremendously efficient at transporting energy over long distances, typically without permanently disturbing the environment

Coastal Environments and Global Change, First Edition. Edited by Gerd Masselink and Roland Gehrels.
© 2014 John Wiley & Sons, Ltd. Published 2014 by John Wiley & Sons, Ltd. Companion Website: www.wiley.com/go/masselink/coastal

through which they travel. For example, electromagnetic waves transport energy from the Sun to the Earth, and seismic waves transmit the energy released by the movement of the tectonic plates throughout the globe. The waves we are interested in, however, are called surface gravity waves and these are one of the key drivers for shaping the morphology of the world's coastlines. Gravity waves represent a type of mechanical wave in which the Earth's gravity provides the restoring force that drives the dynamics of the waves. As wind blows across the surface of the ocean, it generates disturbances on the sea surface that render local patches of the surface out of hydrostatic equilibrium with the rest of the surface. This disequilibrium represents potential energy, and as gravity forces the water particles back into equilibrium, the potential energy is converted into kinetic energy, which is transferred to neighbouring particles, which in turn move out of equilibrium, radiating the disturbance outward from the initial disturbance. This repetitive propagation of energy, originally supplied by the wind and powered by the cyclic interchange between potential and kinetic energy, defines the dynamics of surface gravity waves. Waves travelling across the ocean in this fashion are generally considered non-dissipative. This means that the energy contained in the waves can be transmitted for long distances without loss, due to this near-perfect, gravity-driven, transformation of kinetic and potential energy. Studies of wave propagation (Snodgrass et al., 1966) have documented the transmission of energy from storms off New Zealand in the southwestern Pacific across 11,000 km of ocean to the shores of Alaska. It is only upon reaching shallow water that non-linear processes lead to a release of the energy, which is then available for driving sediment transport and the shaping of the coastline.

Waves are typically characterized by their wave height, wavelength and period (Fig. 4.1). The wave height H is defined as the elevation difference between the highest part of the wave, known as the crest, and the lowest part, or trough. For reasons that will soon become apparent, this elevation is also described by the wave amplitude a, which is equivalent to half of the wave height ($H/2$). The wavelength λ is the distance between two successive points of equivalent phase (specific part of the wave cycle), such as the distance between two crests or two troughs. The wave period T is the time required for the wave to repeat itself, for example the time between the appearance of two subsequent crests of a wave at a specific location. Another method of representing this information is the frequency f, which tells how many wave cycles occur in a period of time and is equal to the inverse of the period ($f = 1/T$). As has already been alluded to, the depth of the water through which surface gravity waves propagate is an important parameter and is typically represented by a lower case h, which is not to be confused with the upper case H for wave height.

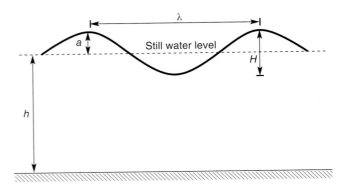

Fig. 4.1 Diagram detailing the definition of the various parameters used to describe a wave.

4.2.2 Wave theories

Many of the key features of surface gravity wave dynamics can be described mathematically by linearized wave theory based on potential flow. This treatment, which assumes a homogenous, incompressible fluid, and ignores friction and turbulence, is known as Airy theory in honour of the 19th century mathematician who first published the correct derivation of the theory. From this theory, we can derive an expression that describes the evolution of the sea surface (η) as a function of location in space (x, y) and time (t):

$$\eta(x,t) = a \sin\left(\frac{2\pi}{\lambda}x - \frac{2\pi}{T}t\right) \quad (4.1)$$

Here, the sea-surface elevation is defined as the distance, positive upwards, from the level of the sea in the absence of the wave and, for clarity, we assume that the wave propagates in the x-direction (the minus in eqn. 4.1 is necessary to define a wave that propagates in the positive x-direction). This equation describes how in a progressive wave, for a snapshot in time (hold t constant), the sea surface varies sinusoidally in space (Fig. 4.2a), and also for a fixed location in space (hold x constant) the sea surface varies sinusoidally in time (Fig. 4.2b). In order to make our expressions simpler, it is traditional to introduce two new quantities, the wave number k and the radian frequency σ. These quantities are related to our earlier parameters according to:

$$k = \frac{2\pi}{\lambda}; \; \sigma = \frac{2\pi}{T} = 2\pi f \quad (4.2)$$

Using these quantities, we can rewrite eqn. 4.1 as:

$$\eta(x,t) = a \sin(kx - \sigma t) \quad (4.3)$$

One of the results of Airy wave theory is that the frequency and wavelength are not independent of each other and only specific combinations of wavelength and period

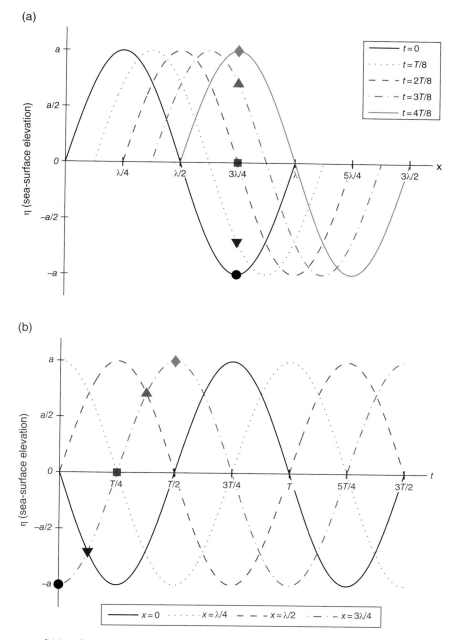

Fig. 4.2 Figure showing spatial (a) and temporal (b) evolution of waves (sinusoids) as expressed in eqn. 4.1. The different lines in (a) represent snapshots of the same wave form at different points in time, with each point separated from the previous by 1/8 period. Notice that the wave form moves in the positive *x*-direction. The different lines in (b) represent the temporal evolution of the sea surface at fixed locations in space. The different symbols plotted at the fixed location of 3/4λ in (a) are now plotted on the appropriate line in (b) for comparison.

can exist. The relationship between wavelength and period is known as the dispersion relationship and is expressed as:

$$\sigma = \sqrt{gk\tanh(kh)} \qquad (4.4)$$

Notice that the relationship between wavelength and period is governed by the strength of gravity *g*, reflecting its role as the restoring force in surface gravity waves. The dispersion relation uses the hyperbolic tangent tanh, which, in analogy to the trigonometric functions, is the

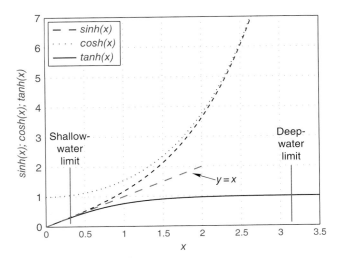

Fig. 4.3 Plot of the hyperbolic functions sinh, cosh and tanh. Notice how for small values of x, $\sinh(x) \approx \tanh(x) \approx x$, and that for large values of x, $\tanh(x) \approx 1$. The shallow-water wave limit $h < \lambda/20$ is equivalent to $x < 0.31$ and the deep-water limit is equivalent to $x > 3.1$.

ratio of the hyperbolic sine (sinh) to the hyperbolic cosine (cosh). The curves of these functions are displayed in Fig. 4.3. It is worth noting that for small values the hyperbolic tangent of an argument is very close to the argument, and that for large values the hyperbolic tangent of an argument is approximately 1.

The wave phase speed or celerity C is defined as the speed with which a point of constant phase, such as the crest, moves through the water. From the definitions of the wave parameters it can quickly be seen that $C = \lambda/T = \sigma/k$. If we consider the deep-water limit where the water depth is much larger than the wavelength so that $kh >> 1$ (and $\tanh(kh) = 1$), the deep-water phase speed is given by:

$$C_{deep} = \frac{\sigma}{k} = \frac{\sqrt{gk}}{k} = \sqrt{\frac{g}{k}} = \sqrt{\frac{g\lambda}{2\pi}} = \frac{gT}{2\pi} \qquad (4.5)$$

All three of the equations on the right-hand side are alternatives for expressing the deep-water phase speed, with the first being the classical expression and the last one expressing it in terms of wave period, which is commonly the quantity known about surface gravity waves. The key result of eqn. 4.5 is that in deep water the phase speed of waves is dependent on the wavelength (or period), with the longer waves travelling faster. This behaviour explains the origin of the name 'dispersion relationship', because in a collection of waves of different frequencies the waves will sort themselves out, with the longer waves propagating away faster than the shorter ones. These waves are said to be dispersive.

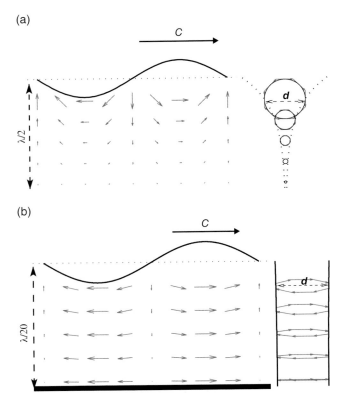

Fig. 4.4 Figure exhibiting the particle motions at different phases and depths under a surface gravity wave propagating to the right in deep (a) and shallow (b) water. The particle path at a single point integrated over an entire cycle is drawn to the right of each particle motion map and the physical significance of d_0 is indicated. Notice that d_0 is a function of depth for deep-water waves but is constant for shallow-water waves.

The motions of the water particles below a progressive surface gravity wave are called orbital motions, with the individual velocities called orbital velocities. Equations for these velocities are also part of the solution obtained from Airy theory. The equation for the horizontal orbital velocity in the direction of wave advance under a deep-water wave can be approximated as:

$$u_{deep}(x, z, t) = a\sigma e^{kz} \sin(kx - \sigma t) \qquad (4.6)$$

Notice that this equation is now also a function of z, the vertical coordinate. As the z-coordinate is defined with positive direction pointing upwards and the origin at the sea surface, the term e^{kz} will lead to horizontal orbital motions decreasing exponentially with depth. It is also worth noting that the horizontal orbital motion is in phase with the sea-surface elevation (eqns 4.3 and 4.6). This means that when the sea surface is highest (lowest), the strongest forward (backward) velocities occur (Fig. 4.4a). At the zero-crossings (when the sea-surface elevations are at mean sea level), there is no horizontal motion. The equivalent equation for vertical velocity is:

$$w_{deep}(x, z, t) = a\sigma e^{kz} \cos(kx - \sigma t) \qquad (4.7)$$

The vertical motions are a quarter wave cycle out of phase with the sea surface and the horizontal velocities, but the magnitude (strength of the strongest velocities) is the same as the horizontal velocities (Fig. 4.4a).

Particle paths are defined as the path a particle of fluid at a fixed point in space (x and z constant) would take over the course of a wave cycle. These can be derived by summing together all the velocities during the cycle, and for a deep-water wave this results in a circular particle path (Fig. 4.4a). In deep water, the diameters of these paths d_O decrease exponentially with depth, rapidly approaching zero, which indicates no orbital motion. The orbital motions half a wavelength below the surface ($h = \lambda/2$) are a mere 4% of those at the surface, and at one wavelength below the surface ($h = \lambda$) they are tenths of a percent. The quantity d_O is also known as the horizontal displacement distance, and represents the maximum horizontal distance that a particle of water will be transported during a wave cycle. Because of the circular particle paths in deep water this is also the vertical distance of displacement, and the equation for this quantity depends on the depth below the surface and is expressed as:

$$d_O = He^{kz} \qquad (4.8)$$

The rapidly decreasing orbital motions below a deep-water wave mean that such waves will not affect the bottom, so are of little interest to coastal evolution other than as a mechanism to transport energy from distant storms. The same is not true for shallow-water waves, which are defined as progressive surface gravity waves in which the water depth is much less than the wave length, so that $kh << 1$ (and $\tanh(kh) \approx kh$). In this case, the shallow-water phase speed is given by:

$$C_{shallow} = \frac{\sigma}{k} = \frac{\sqrt{gk^2 h}}{k} = \sqrt{gh} \qquad (4.9)$$

and the horizontal orbital velocity is given by:

$$u_{shallow}(x, z, t) = \frac{H}{2}\sqrt{\frac{g}{h}} \sin(kx - \sigma t) \qquad (4.10)$$

and the horizontal displacement distance in shallow water is expressed as:

$$d_O = \frac{H}{\sigma}\sqrt{\frac{g}{h}} = \frac{HT}{2\pi}\sqrt{\frac{g}{h}} \qquad (4.11)$$

One of the major differences between shallow- and deep-water waves is that shallow-water waves are no longer dispersive, because the wave speed depends only on the water depth (eqn. 4.9). A note of caution is warranted here as this is true only if the waves under consideration all meet the shallow-water wave limit (provided below). Other major differences are that the diameter of the orbital motions are no longer a function of depth below the water surface (see Fig. 4.4; eqn. 4.10), and that the vertical orbital motions are greatly reduced or absent, resulting in elliptical particle paths (see Fig. 4.4). This means that in shallow water, the wave orbital motions reach the seabed where they are capable of mobilizing and transporting sediment.

The limit for the validity of the deep-water approximations is typically reached when the water depth is greater than half a wavelength ($h > \lambda/2$), and the limit for shallow-water waves is for water depths that are less than a twentieth of the wavelength ($h < \lambda/20$). These limits represent the range for which errors in using the approximations for the hyperbolic functions are less than 4%. Unfortunately, these two limits exclude many conditions that are of interest for application to coastal geomorphology, meaning that the exact solutions, which represent intermediate wave conditions, must be applied. The forms of these equations are all provided in Table 4.1. The resulting orbital motions and particle paths for shallow waves can be observed in Fig. 4.4b, where reduced, but finite, vertical motions at the top of the water column result in oval-shaped particle paths but reduce to purely horizontal orbital motions at the seabed. In shallow water, the horizontal displacement distance appears constant, with depth clearly having the potential to affect the seabed. Intermediate waves exhibit particle path behaviour in between these two extremes.

Two further assumptions required to derive Airy theory are: (1) the low steepness assumption, which states that the wave height is small in comparison to the wavelength ($H << \lambda$); and (2) the low relative wave height assumption, in which the wave height is assumed to be small relative to the water depth ($H << h$). The latter assumption is particularly restrictive as waves approach shallow water (shoal), but even the former assumption can be violated for the largest and most energetic waves. An indication of the inaccuracy of the linear theory can be obtained from observing shoaling waves in nature, which exhibit wave profiles that are quite clearly non-sinusoidal. In fact, real waves often exhibit asymmetries of two distinct types. The first is a vertical asymmetry, where the crest is more peaked and narrow than the trough, which is broad and shallow. This type of asymmetry is commonly referred to as wave skewness (Fig. 4.5a), and results in orbital motions under the crest of the wave that are of higher velocity but shorter duration than the slower but longer-lived return flows under the trough. This type of profile is observed under higher waves in the deep ocean, and waves in the early stages of shoaling. The

Table 4.1 Table providing both the deep- and shallow-water approximations, as well as the full expression, for various wave properties based on Airy wave theory. Due to the dispersion relation, these expressions are not necessarily unique and where alternate forms are possible, the expression that favoured the use of frequency has been selected.

Quantity	Deep water ($h \geq \lambda/2$)	Intermediate depth (full expression)	Shallow water ($h \leq \lambda/20$)
Surface elevation (η)	$\eta(x,t) = \dfrac{H}{2}\sin(kx - \sigma t)$	$\eta(x,t) = \dfrac{H}{2}\sin(kx - \sigma t)$	$\eta(x,t) = \dfrac{H}{2}\sin(kx - \sigma t)$
Horizontal orbital velocity (u)	$u(x,z,t) = \dfrac{H}{2}\sigma e^{kz}\sin(kx - \sigma t)$	$u(x,z,t) = \sigma\dfrac{H}{2}\dfrac{\cosh[k(z+h)]}{\sinh(kh)}\sin(kx - \sigma t)$	$u(x,z,t) = \dfrac{H}{2}\sqrt{\dfrac{g}{h}}\sin(kx - \sigma t)$
Vertical orbital velocity (w)	$w(x,z,t) = -\dfrac{H}{2}\sigma e^{kz}\cos(kx - \sigma t)$	$w(x,z,t) = -\sigma\dfrac{H}{2}\dfrac{\sinh[k(z+h)]}{\sinh(kh)}\cos(kx - \sigma t)$	$w = 0$
Hydrostatic pressure (p)	$p(x,z,t) = \rho g\left(\dfrac{H}{2}e^{kz}\sin(kx - \sigma t) - z\right)$	$p(x,z,t) = \rho g\left(\dfrac{H}{2}\dfrac{\cosh[k(z+h)]}{\sinh(kh)}\sin(kx - \sigma t) - z\right)$	$p(x,z,t) = \rho g\left(\dfrac{H}{2}\sin(kx - \sigma t) - z\right)$
Dispersion relation	$\sigma = \sqrt{gk}$	$\sigma = \sqrt{gk\tanh(kh)}$	$\sigma = k\sqrt{gh}$
Phase velocity (C)	$C = \dfrac{g}{\sigma} = \sqrt{\dfrac{g}{k^2 h}}$	$C = \dfrac{g}{\sigma}\tanh(kh)$	$C = \sqrt{gh}$
Group velocity (C_g)	$C_g = \dfrac{1}{2}\dfrac{g}{\sigma} = \dfrac{1}{2}\sqrt{\dfrac{g}{k^2 h}}$	$C_g = \dfrac{1}{2}\left[1 + \dfrac{2kh}{\sinh(2kh)}\right]\dfrac{g}{\sigma}\tanh(kh)$	$C_g = \sqrt{gh}$
Horizontal orbital diameter (d_0)	$d_0 = He^{kz}$	$d_0 = H\dfrac{\cosh[k(z+h)]}{\sinh(kh)}$	$d_0 = \dfrac{H}{kh} = \dfrac{H}{\sigma}\sqrt{\dfrac{g}{h}}$

(a)   (b)

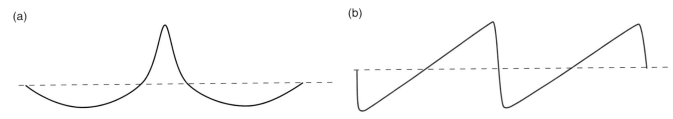

Fig. 4.5 Schematic diagram of wave surface profiles that demonstrate skewness (a) and asymmetry (b). Waves typically exhibit skewness first and will develop asymmetry as their non-linearity increases.

second type of asymmetry is a front-back asymmetry (Fig. 4.5b), which is characterized by a steep wave face as the crest approaches and a more gentle profile on the back of the crest. The resulting wave orbital motions exhibit much higher accelerations during the offshore to onshore flow transition than during the onshore to offshore transition. This type of front-back asymmetry, which is called, surprisingly enough, 'asymmetry', is observed under waves as they approach breaking, and always occurs in addition to wave skewness.

In order to be able to represent waves that violate the linear theory assumptions and to be able to describe the various asymmetries observed in nature, a collection of non-linear wave theories have been developed since the publication of Airy theory. The non-linear theory most similar to Airy theory is Stokes theory, which relaxes, but does not eliminate, the low steepness and relative wave height approximations. This theory, first published by the

famous mathematician and fluid dynamicist George Gabriel Stokes, involves a solution that takes the form of a series of expansions in which the analytical expression is the sum of a series of sinusoids that are harmonics of each other. The equation for sea-surface elevation correct up to third-order under the Stokes theory is:

$$\eta(x,t) = a_1\sin(kx - \sigma t) - a_2 k\cos(2kx - 2\sigma t) \quad (4.12)$$

A plot of this profile can be seen in Fig. 4.6, where it is clear from the higher, but narrower, crest and reduced, but broader, trough that Stokes theory is capable of producing wave skewness. In theory, the Stokes wave can be extended to an infinite series of terms, and in practice as many as five terms are commonly applied (Fenton, 1985).

As suggested earlier, Stokes wave theory begins to break down in shallow water. There are two properties that provide insight as to whether the simpler theories are

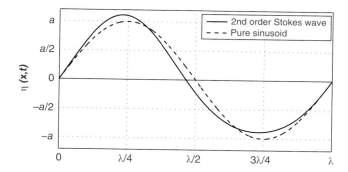

Fig. 4.6 Plot of wave-surface elevation as obtained from Stokes wave solution (solid line) and from pure sinusoid (dashed line). Notice how Stokes wave has a higher crest and a shallower trough than the sinusoid, and that the spatial/temporal extent of the crest is significantly shorter than that of the trough. The Stokes wave exhibits skewness, but not asymmetry.

appropriate. The first is relative wave height H/h and the second is relative wavelength λ/h, where relative is defined in reference to the local water depth h. When either of these parameters grows large, the assumptions inherent in Stokes theory are being violated. However, the relative wavelength is considered the more stringent of these conditions and a combined parameter, the Ursell number Ur, is defined as the product of the relative height and the relative wavelength squared, and is given by:

$$Ur = \frac{H\lambda^2}{h^3} \qquad (4.13)$$

When the Ursell number grows larger (order 100), Stokes theory is no longer considered valid and higher order theories such as solitary wave or cnoidal wave theory are called for. A discussion of these theories is beyond the scope of this book but thorough treatments can be found in Dingemans (1997).

4.2.3 Wave generation

Anyone who has experienced a light breeze blowing across a placid body of water and observed the generation of small ripples on the surface of the water is aware that it is the wind that generates the surface gravity waves that this chapter discusses. Exactly how this occurs is still not completely resolved, but most of the details can be explained by the combined theories of Miles (1957) and Phillips (1957). The Phillips theory, which is the dominant mechanism in early wave growth, suggests that the wind field subjects the water surface to random travelling pressure fluctuations, but the water can only respond to those fluctuations that have spatial and temporal characteristics consistent with the physics inherent in wave

dispersion relations (e.g. eqn. 4.4). The presence of these undulations in the water surface will feed back into the wind field to enhance these components in the pressure fluctuations. This results in a linear growth in wave energy in a process that involves positive feedback. Once an initial waveform exists on the sea surface, the Miles theory predicts boundary layer separation in the lee of the wave crest, leading to a recirculation eddy in the trough of the wave, which reinforces the wave orbital velocity kinematics and leads to an exponential growth in wave energy. Given continued wind, this growth is limited only by wave breaking.

What is important to recall is that the processes discussed earlier do not necessarily lead to a single frequency of waves, but potentially to a wide distribution of frequencies. The exact distribution of frequencies that are present in the sea surface are continually changing. In a process described by Longuet-Higgins (1969), it is believed that shorter wavelength waves preferentially steepen and break on the crest of longer waves, leading to a continuous transfer of energy from higher frequency to lower frequency waves. The frequency content of the wave field is described using a wave energy spectrum (Fig. 4.7) that gives the distribution of wave energy either as a function of frequency alone, or as a function of both frequency and direction of propagation (i.e. directional wave spectrum). While the nature of the wave spectrum is critically important for many applications, for convenience sake the wave field is commonly described through a single set of parameters related to wave height and period. The most commonly used height parameter is the significant wave height H_{sig}. For historical reasons related to the original visual observations of wave height, this parameter was defined as the average height of the 1/3rd highest waves. However, it has long been known (e.g. Kinsman, 1985) that if a wave field has a Rayleigh distribution of heights, then this statistical description of wave heights is equivalent to four times the standard deviation of the sea-surface elevation. This is the definition most commonly applied when using instruments to continuously measure the motion of the sea surface. Other commonly utilized wave heights are the mean wave height H_{mn} and the root-mean-squared wave height H_{rms}. While similar definitions exist for wave period, one of the most common parameters is the peak period T_p, which is the period associated with the most energetic or 'peak' frequency in the wave spectra.

The earliest models for the prediction of wave conditions focused on predicting these individual wave parameters. One of the most commonly utilized techniques is that developed during and immediately following World War II by Sverdrup and Munk (1947), whose work was later updated by Bretschneider (1952, 1958). It is commonly referred to as 'significant wave technique' or 'SMB method' in recognition of the contributions of Sverdrup,

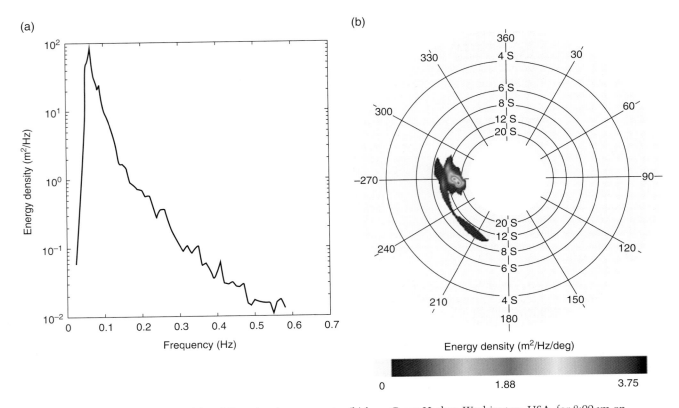

Fig. 4.7 Plot of wave energy spectra (a) and directional wave spectra (b) from Grays Harbor, Washington, USA, for 8:09 am on 30 March 2012. The plots show how the energy in the wave field is distributed across frequencies alone (a) or frequency and direction (b) for this period of time. The figure is assembled from products delivered by the Coastal Data Information Program (CDIP) (http://cdip.ucsd.edu/).

Munk and Bretschneider. This method recognizes that the nature of the wave conditions generated by a storm are largely dependent on three characteristics of the storm, namely: (1) the velocity of the wind U expressed as a wind stress factor; (2) the fetch length F, or the distance over which this wind blows; and (3) the duration D, which represents the period of time for which the wind blows. In the absence of limits on fetch and duration, there is an upper limit on the wave height and period that may be generated by a particular wind speed; these wave conditions are known as the 'fully developed seas' or 'fully arisen seas'. However, more commonly there is a limit to the fetch length and/or the duration for which the wind acts, and this produces seas of an inferior height and period. Wave conditions arising from such a situation are known as 'fetch limited' or 'duration limited', depending on which characteristic provided the most stringent limitation. While the SMB method can be expressed by a series of empirical, highly non-linear analytical relations, it is most commonly applied using graphical representations of a family of curves called nomographs (Fig. 4.8).

Wave predictions derived from the SMB method provide only a basic, parametric description of the wave field and lack practically any description of the distribution of energy with frequency and direction. A desire for more complete predictions of wave fields led to a series of studies that sought to improve on the SMB methodology in order to be able to describe the complete frequency spectrum. One of the first such efforts was the Pierson-Moskowitz spectra (Pierson and Moskowitz, 1964), or PM spectra, which used weather ship data from the North Atlantic to derive a relationship for the frequency spectra of fully developed seas. This algorithm was therefore a function only of the wind speed, leading to a one-to-one relation between wave height and peak period. This is a commonly applied relation that is poorly supported by observations (Fig. 4.9). The JONSWAP spectrum (Joint North Sea Wave Project; Hasselmann et al., 1973, 1976) was developed to account for fetch limitation and the observation that waves never are completely developed fully and continue to evolve through non-linear interactions between the waves. The JONSWAP spectra, which applies a correction factor to the PM spectra, is a

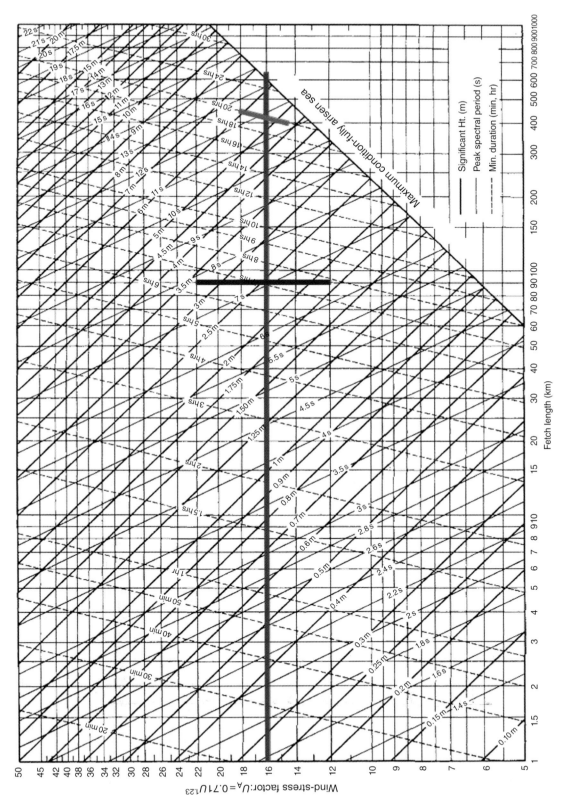

Fig. 4.8 Example wave nomograph used to make parametric wave predictions based on wind speed, fetch and duration. Examination of the wave parameters around the intersection of the horizontal and vertical lines indicates that under wind conditions equivalent to a wind stress factor of 16 where wave growth is limited by a fetch of 90 km, waves with a period of 7 s and 2.5 m wave height would be generated. The intersection of the horizontal and oblique line provides the wave parameters for the waves generated under a 20 h duration limitation. The fully developed conditions are those present at the intersection of the wind stress factor line and the fully developed limit line. (Source: Masselink et al. 2011.)

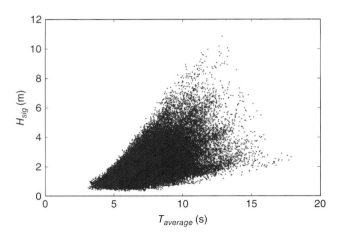

Fig. 4.9 Scatter plot of average wave period $T_{average}$ versus significant wave height H_{sig} for five years of data (2005–2009) at Grays Harbor, Washington, USA. Each point represents a half-hourly estimate. While a general trend may be suggested by the data, it is clear that a one-to-one relation does not exist. (Source: Data from CDIP, http://cdip.ucsd.edu/).

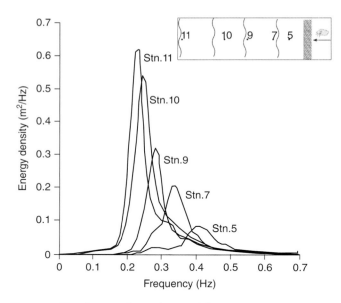

Fig. 4.10 Plot showing the evolution of the wave-energy spectra observed at the JONSWAP (Joint North Sea Wave Project) experiment. The results demonstrate that the further downwind the measurements, the higher the energy content, the lower the frequency, and the narrower the bandwidth of the waves generated.

two-parameter model dependent on both wind speed and fetch, and was created following a clever set of observations in the North Sea (Fig. 4.10). Finally, the TMA spectral model (Bouws et al., 1985) represents a further refinement of the wave spectral model predictions to

account for the effects on limited water depth in the region of wave generation.

4.2.4 Wave propagation and shoaling

As the locally generated waves, containing a broad spectrum of waves, propagate out of the region of generation, they are no longer affected by the wind and, instead, the wave characteristics begin to evolve on the basis of wave physics and the environment through which they travel. Once waves have travelled out of the region of generation, the major changes that occur in deep water are due to wave 'dispersion'. While individual waves propagate at their phase speed, the energy in a packet of waves travels at the group velocity. The group velocity can be represented by the independent notation C_g, but it is common practice to present the group velocity as a function of the phase speed, as in:

$$C_g = Cn \tag{4.14}$$

where the scaling parameter n is a function of frequency and water depth, which can be expressed as:

$$n = \frac{1}{2}\left[1 + \frac{2kh}{\sinh(2kh)}\right] \tag{4.15}$$

The deep-water value for this relation is $n = 0.5$ and the shallow-water value is $n = 1.0$.

The significance of the differing values for n in deep and shallow water is that as wave energy travels out of the region of generation, the energy is distributed across a relatively broad distribution of frequencies. As this energy progresses, the energy contained in the lower frequency and longer wavelength waves will travel faster, moving out in front of the energy contained in the slower, higher frequency and shorter wavelength waves. This process of dispersion results in the development of wave fields with relatively narrow-banded spectra, with the longer wavelength waves (or larger peak period) preceding the shorter wavelength waves (or shorter peak period). The term 'swell' is used to refer to this kind of wave condition in which the waves have travelled out of the region of generation and the spectrum has evolved through a process of dispersion to result in well-sorted wave fields that exhibit narrow-banded spectra. The equivalent term for the heterogeneous, broad-spectrum conditions observed in the region of wave generation and growth is 'sea'. While these terms are used to describe two distinct conditions, it is quite common to observe both conditions simultaneously where pre-existing swell waves propagate into a region of active wave generation. This leads to bimodal wave spectra, i.e. spectra with two peaks.

While dispersion results in the biggest changes to wave fields during propagation across deep water there are other less dramatic processes that also contribute to their evolution. One such process has to do with the directional spreading of the wave field. Unsteadiness in the generating wind field means that waves leaving the zone of generation will typically have a small, but finite, distribution of directions about the mean direction of propagation. As these waves propagate, this will lead to a broadening of the wave front, which, due to the conservation of energy, requires a reduction in the energy contained in each unit of wave crest. This quantity, the mean wave energy density E_{wv}, can be expressed for a wave of height H as:

$$E_{wv} = \frac{1}{8}\rho g H^2 \qquad (4.16)$$

and has the units of energy per square area. It is clear from this relationship that the spreading of the wave front must result in a concomitant reduction in wave height. Additional factors that can alter the wave field during wave propagation include internal friction (viscous damping) and wave-wave interactions (e.g. Ardhuin et al., 2009). The level of damping from internal friction is a rapidly decreasing function of wave period, and so it is significant only for higher frequency waves and will occur relatively close to the region of wave generation.

The above processes lead to a transformation of wave fields as they travel across the ocean, but the most dramatic and rapid changes occur as the waves approach a coastline and encounter intermediate and shallow water. The different changes that occur as a wave field travels through shallow water are collectively known as 'shoaling' processes, and these changes occur because the wave kinematics are now constrained to occur in an increasingly limited vertical space due to the proximity of the seabed. We have already discussed the changes in wave shape that occur as waves shoal, but there are other fundamental changes to the wave properties that occur during shoaling. To determine what these changes are, it is necessary to understand that the one invariant property of a wave field is the wave period or frequency. Using this axiom, the next step to deriving these changes is to recognize the fact that, under non-dissipative conditions, the wave power P is conserved. Wave power is often referred to as 'wave energy flux', and recalling the discussion about the velocity with which energy propagates (section 4.2.4), it should be clear that the relation for P is:

$$P = E_{wv}C_g = \frac{1}{8}\rho g H^2 Cn \qquad (4.17)$$

which has the units of average power per unit length of wave crest. Locations characterized by large values of wave power are favoured sites for the installation of renewable wave-energy devices (Box 4.1).

Referring to eqns 4.9 and 4.15, it can be confirmed that C and n are depth-dependent, so that when waves travel

---

CONCEPTS BOX 4.1   Wave renewable energy

In order to reduce the climate changes induced by anthropogenic release of greenhouse gases, the search for renewable energy sources that promise a sustainable future is currently taking place worldwide. The waves that approach the shores of the world's coastlines represent an enormous potential source of energy that is currently dissipated largely through wave breaking. Estimates of the global wave-energy potential range from 1 to 10 terawatts (TW), which is a significant amount when considering that the current mean global consumption of electricity is about 2 TW. While this resource is not distributed equally around the globe, there are favourable locations for wave-energy collection on all the inhabited continents of the world (Fig. 4.11).

Despite the clear benefits of harvesting wave energy, there are still numerous technical issues to be resolved, a fact that is borne out by the wide range of devices that are proposed to collect this energy and convert it into electricity. However, practically all devices that are at an advanced stage of development can be grouped into three categories. The first category is the oscillating water column type device, where the displacement of the surface of the ocean compresses or expands a controlled volume of air, which is then used to drive an energy-generating turbine. While such devices can be floating, structures fixed to the shoreline are more common and make up a large portion of the full-scale wave-energy devices currently deployed. The second group of devices might be called oscillating bodies, where wave-induced differential motion between two or more bodies provide the energy-generating potential. These devices are typically placed in open water, although they may float on the surface or be attached to the bottom. The final category is the over-topping device, in which wave overwash is trapped in a closed basin and then drained through a low-head generating turbine. Most designs for this type of device envision large platforms moored at sea, although an alternative design incorporates the design into structures fixed at the shoreline. Whatever the basic design concept, successful devices must have costs that allow them to

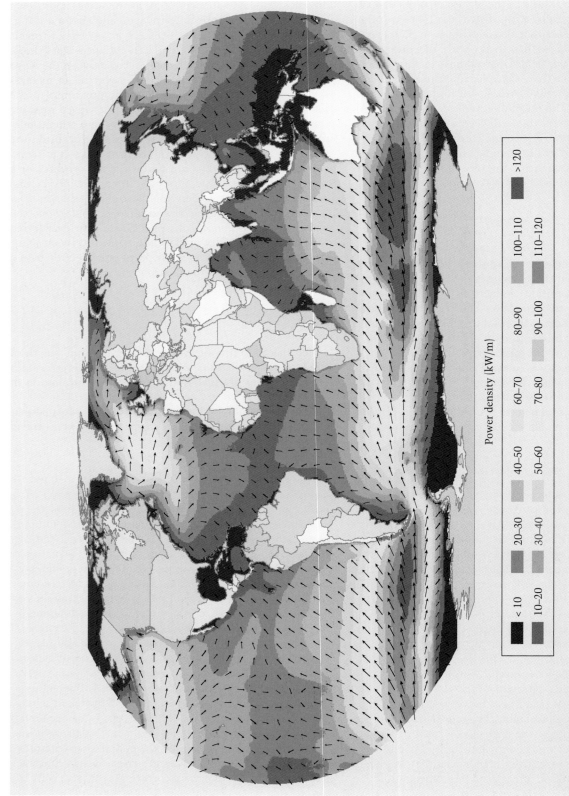

Power density (kW/m)

< 10	20–30	40–50	60–70	80–90	100–110	>120
10–20	30–40	50–60	70–80	90–100	110–120	

Fig. 4.11 Map of annual mean wave-power density (in kW/m) represented by colour, and annual mean best direction as calculated from six years of wave model results. Figure adapted from Gunn and Stock-Williams (2012; fig. 1), who estimate a global wave-power resource of 2.11 TW. (Source: Gunn and Stock-Williams 2012. Reproduced with permission of Elsevier.) For colour details, please see Plate 13.

generate electricity for a competitive price and at the same time be able to survive and generate electricity under the wave-condition extremes that are present at any location. This is a challenge that is currently proving difficult to meet, although steady progress is being made.

When wave energy does become a viable alternative source of energy, there are environmental issues that will have to be considered. The removable of energy from waves propagating in the ocean must affect whatever processes those undisturbed waves contributed to.

Such processes can range from ocean mixing to current generation to the transport of sediment and the shaping of morphology. Whether those effects in fact lead to significant impacts on the environment will be very much site- and project-specific. Whether those impacts are negative or positive is also an open question, although the current expenditure on engineering works designed to reduce the amount of wave energy that reaches the shoreline would suggest that there is plenty of scope for the latter.

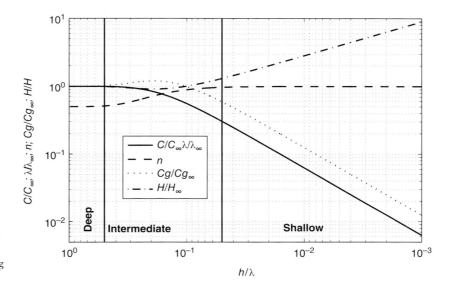

Fig. 4.12   Plots demonstrating the evolution of relative wave phase speed (C), wavelength (λ), group velocity (Cg), wave height (H) and scaling parameter n, as a function of water depth h.

from one water depth into another, these quantities will change. If P is to be conserved, it is then clear that H will also have to change. In fact, a shoaling coefficient κ_s, which gives the relationship between wave height at one depth relative to another depth, can be defined by setting the wave power at each depth in eqn. 4.17 equal to each other and then rearranging the terms as:

$$\kappa_s = \frac{H_b}{H_a} = \sqrt{\frac{C_a n_a}{C_b n_b}} \qquad (4.18)$$

where the subscripts a and b refer to locations with different depths. It is common to define this shoaling parameter in terms of local conditions and deep-water conditions, which can be done by using the subscript 0 to refer to deep-water conditions, dropping the subscript for local conditions and substituting eqns 4.5, 4.9 and 4.15 into eqn. 4.18. The final result is then:

$$\kappa_s = \frac{H}{H_0} = \sqrt{\frac{gT}{4\pi Cn}} \qquad (4.19)$$

Figure 4.12 presents a graph displaying the behaviour of various wave parameters for normally incident waves as they enter shallow water, as predicted by linear wave theory. Notice that while shallow-water wave heights are larger than deep-water wave heights, there is a region in intermediate depth water where heights decrease. This is due to the increase in the wave group velocity in this same region.

Another significant process of wave transformation that occurs in shallow and intermediate water is wave refraction. Refraction, which is a phenomenon observed in waves of all kinds, can be defined as the change in direction of waves obliquely encountering a gradient in wave phase speed. For water waves, this occurs most commonly when waves travel across depth contours that are not parallel to the wave crest. This process is relatively straightforward to understand if one considers the wave front depicted in Fig. 4.13. As this wave approaches the shoreline, the section of wave crest on the left of the figure encounters the shallow contours earlier than sections of the crest on the right. The lower phase speed of the shallow water means that this part of the crest will slow down,

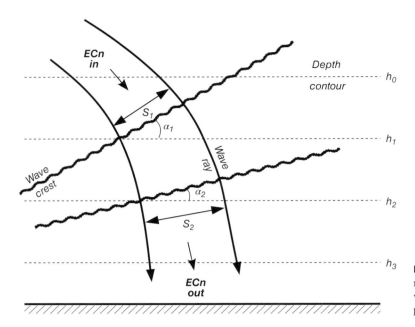

Fig. 4.13 Schematic diagram of depth-based wave refraction. Wave rays are vectors perpendicular to the wave crest, which indicate the direction of wave propagation. See text for further explanation.

Fig. 4.14 Photo of wave crests approaching a curved shoreline in a crescentic bay, demonstrating how initially straight waves will refract to approach the shoreline approximately shore normal. (Source: Photograph by D. Conley.)

resulting in a clockwise rotation of the wave front. Clearly, the rotation would be counter-clockwise had the wave approached the depth contours from the other direction. In fact, the process of refraction works to steer waves so that the direction of wave advance is increasingly aligned perpendicular (normal) to bathymetric contours (Fig. 4.14). Quantitatively, the degree of bending under refraction can be determined for all types of waves using Snell's Law, which states that:

$$\frac{\sin(\alpha_a)}{C_a} = \frac{\sin(\alpha_b)}{C_b} \qquad (4.20)$$

where, as in eqn. 4.18, the subscripts a and b refer to locations with different depths.

For water waves, α_a is the angle between the wave crest and the depth contour in the deeper water and C_a is the phase speed at this depth, while α_b and C_b are the corresponding quantities in shallower water. As the relation between depth and phase speed is known, the above relation can be used to solve for the new direction of travel of a wave that approaches a depth contour with a known direction of travel. Sequentially performing this calculation to determine the path of wave propagation is known as 'ray tracing' and represents one of the oldest methods of wave modelling.

Wave refraction also has implications for the wave height, because as the crest of the wave bends towards the depth contours, the wave crest will effectively spread out (see Fig. 4.13). In order to conserve the wave power transmitted by the wave, this spreading of the wave crest must also result in a compensatory diminishing of the wave height. Using some relatively simple geometrical arguments and equating the wave power from deep water to shallow water, a wave refraction coefficient κ_r, similar to the shoaling coefficient in eqn. 4.19, can be defined as:

$$\kappa_r = \frac{H}{H_0} = \sqrt{\frac{\cos(\alpha_0)}{\cos(\alpha)}} \qquad (4.21)$$

The shallow-water wave height H can be determined by multiplying the deep-water wave height H_0 by both κ_s and κ_r. These two terms will have opposing effects, but, generally, the shoaling coefficient is much stronger, resulting in a significant increase in wave height as waves shoal.

4.2.5 Wave measurement

Visual wave measurements have a long history and continue to this day for various volunteer-based programs or ship-board observations, but quantitative studies of such observations show that, while longer term patterns are similar to sensor-based measurements, the individual observations have a clear bias. It appears that human observers consistently ignore the smaller amplitude waves and the shorter period disturbances, so that visual observations overestimate both height and period. Thus, a short description of sensor-based techniques shall be presented. The most basic wave measurements are typically point measurements such as wave staffs, pressure sensors and wave buoys. Wave staffs are surface piercing sensors that provide the only direct measurements of the displacement of the sea surface. Unfortunately, they are surface piercing devices, which require rigid mounting at the air-water interface. Thus, they are rarely used in the field, but they are often the sensor of choice for water surface elevation measurements in wave basins on account of their high accuracy and sampling rate capability. Pressure sensors are commonly used for field measurements where sub-sea water pressure is used to infer the movement of the sea surface. Modern pressure sensors also have good accuracy and sampling capabilities, and the potential to utilize bottom-mounting systems offers multiple logistic benefits. Unfortunately, due to the exponential decay with depth of the wave-generated pressure signal (eqns 4.6 and 4.7), pressure sensors have significant depth limitations and pressure-based measurements of waves is limited to shallower water. To overcome the depth limitation of pressure-based measurements, and thereby provide deep-water sensing capability, surface-following wave buoys have been developed. The sensing elements in such buoys are either accelerometers or tilt sensors. The measurements from these sensors can be processed through integration to provide changes in the sea-surface elevation, which then permits a determination of the wave-energy spectrum.

In general, the point measurements of sea-surface disturbance described above are incapable of distinguishing the direction of travel of the wave causing the disturbance. Even for a single frequency purely sinusoidal waveform, the sea-surface disturbance at a point is the same no matter whether the wave approaches from east, west, north or south. To resolve this directional information, a minimum of three independent measurements is required, with four or more typically employed for redundancy. This is typically achieved through the use of arrays of the point sensors previously described, but can also be collections of heterogeneous measurements such as combined pressure and current measurements. An increasingly common technique for collecting directional wave information involves the use of acoustic velocity profilers. These instruments can use combinations of velocity measurements and reflections of the acoustic beams off the sea surface to arrive at estimates of the directional wave spectra in a variety of methods. These sensors also have depth limitations, which are due to the increasing separation of the acoustic beams with distance from the sensor, but it is not as limiting as pressure sensors.

The deployment and maintenance of sensors on the sea surface is a matter of not insignificant logistical difficulty, which involves considerations not only related to the harsh environmental conditions but also to the conflicting requirements of other users of the ocean. No doubt this has helped to motivate the development in recent decades of multiple forms of remote-sensing techniques for wave measurements. In addition, such remote-sensing techniques generally provide spatial distribution of wave measurements in contrast to the single location measurements of all the previously discussed *in situ* measurements. While land-based measurements are now being developed (Box 4.2), most remote-sensing development has come in the area of satellite-based measurements. Techniques have been developed to estimate the significant wave height in the ocean from the shape of the returns from satellite altimeters, and validation of such measurements suggest accuracies for estimates of significant wave height in the order of 10%. Limitations to such measurements include the fact that the effective footprint of the measurements is in the order of 10 km × 10 km and satellite return periods of approximately 10 days. Despite these limitations, the development of global wave-height climatologies for the first time represents a major advance in wave measurement. The increasing deployment of synthetic aperture radars (SAR) result in effective sensor footprints on the sea surface in the 1 m range, which means that directional wave spectra may be estimated from sequential images of patches of the sea surface. The fine resolution of these sensors means wave information may be collected closer to the coast, although assumptions of stationarity in the image space prohibit estimates in locations of rapidly changing bathymetry.

METHODS BOX 4.2   Wave measurements by radar

The logistical difficulties associated with deploying and maintaining *in situ* wave sensors along with the point measurement nature of such instruments have provided a strong motivation to develop remote-sensing technologies that can overcome these shortcomings. In recent years, this has led to the development of terrestrial-based techniques employing radar to measure waves. There are two basic approaches for measuring waves using radar. One technique adapts the high-frequency (HF) radar systems [frequency $O(10–50\,MHz)$], which were originally developed to measure surface currents over a large spatial area. Such systems derive the current strength from changes in the Doppler shift of the radar signal reflected off surface gravity waves through Bragg scattering. In order to use these devices to determine the wave directional spectra, it is necessary to accurately record the spectral shape of the scattered radar energy, which is an

inherently more demanding requirement. With the use of multiple sites that contain both directional transmit antennae and multi-element receive antennae it is possible to estimate independent directional estimates over a large area (Fig. 4.15).

The so called X-band radar technique, which has reduced spatial coverage [$O(km)$] but enhanced spatial [$O(m)$] and temporal [$O(s)$] resolution, is based on devices that operate at standard navigation radar frequencies [$O(10\,GHz)$]. Returns from these images give a representation of the instantaneous roughness of the sea surface, and can be processed to provide wave spectral and directional information. Some success has been achieved in retrieving amplitude information through the application of empirical relations. This technique has the attractive feature that it can potentially be a dual-use technology on devices that have been obtained for other functions.

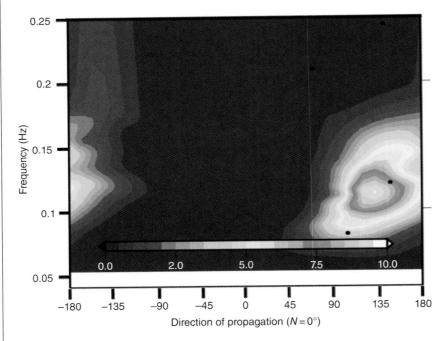

Fig. 4.15   Wave directional spectra as estimated from high-frequency (HF) radar measurements, and photos of a 16-element receive antenna (lower left panel) and a four-element transmit antenna (lower right panel) from a Wellen Radar (WERA) high-frequency radar installation on the north coast of Cornwall, UK. (Source: Photographs by D. Conley.) For colour details, please see Plate 14.

4.2.6 Long waves

While the wind waves (or sea) and swell discussed in the preceding sections represent the most easily perceived ocean wave phenomenon in the world's oceans, a myriad collection of waves are present, many of which have considerably larger wavelengths than typical wind waves. Hidden below the surface of the ocean are gravity waves, which propagate due to vertical gradients in density. Known as internal waves, they have amplitudes on the scale of $O(10\,m)$ and periods on the scale of hours. A type of surface gravity wave that is also relatively unnoticeable in the deep ocean are tsunami (see Chapter 5), which have very long wavelengths $O(100\,km)$ and therefore behave as shallow-water waves. When tsunami arrive at the continental shelf, the resulting shoaling processes lead to extreme wave heights, which give these waves the terrifying potential for destruction for which they are known. Oceanic (and atmospheric) waves of a completely different type include the vast planetary scale waves known as Rossby waves, which rely on the latitudinal gradients of Coriolis force as a restoring force and only travel from east to west. Rossby waves have wavelengths and periods in the order of hundreds of kilometres and months, respectively. Rossby waves travel slowly, $O(km/day)$, but their potential to transport great quantities of heat can have fundamental importance on global weather patterns (Jacobs et al., 1994). Another type of large-scale wave that requires gravity, Coriolis force and a lateral boundary to exist are Kelvin waves. While Kelvin wave phenomenon can be observed in internal waves, in the atmosphere and along the equator, it is their role in facilitating the regular daily fluctuations of the sea-surface elevation known as the tide that is of most interest in the current context. We shall examine these waves more closely in section 4.3.

4.2.7 Wave climate and response to global climate change

Wave climate is the name given to the statistical description of wave conditions experienced by a location over a fixed period of time, typically a year. This description is usually broken down into distributions of wave height, period and direction. While the full wave climate is of critical importance for many applications (see Box 4.1), it is often the extreme waves that are of significance for morphologic evolution, as these are the events that are responsible for the most dramatic changes to coastal morphology. Section 4.2.3 has already provided the background necessary to understand where the major global climate change driven alterations to wave climate will come from. It should be clear that any systematic alterations to wind speed during storms will have a knock-on effect to the generated waves as well. To a lesser extent, systematic changes to the spatial extent of storms (fetch) or temporal length (duration) may also affect wave climate where that climate represents waves that are near the threshold of being fully developed.

As will be covered more fully in Chapter 5, climate scientists now believe that future climate scenarios in a globe with elevated carbon dioxide levels will result in more of the most intense and destructive tropical storms, even if the total number of tropical storms decreases (e.g. Bender et al., 2010). The result of such a scenario, with the higher wind speeds characteristic of such storms, would be an increase in the frequency of occurrence of the extreme waves generated by such storms. This increase in the frequency of extreme waves with larger heights and longer periods will be of great significance to locations where tropical-storm generated waves represent the most energetic forcing conditions, but what of other locations? While predictions for changes in the frequency and intensity of extra-tropical storms are not currently available, satellite observations of wind speed over the ocean have indicated a global increase in wind speeds over the past 20 years on an average and extreme (>90%) basis (Young et al., 2011). While the rate of increase varies across the globe, significant increases of greater than 2.5% per decade are seen in larger areas of the world's oceans. Indicative of the complex relation between waves and wind, a concomitant increase in mean wave heights has not been detected. However, significant increases in the extreme wave heights have been observed at high latitudes, with this increase tapering out in equatorial regions. Experts are still discussing whether these changes are driven by a warming Earth, but these observations have been supported by other studies using wave-buoy observations, all of which are consistent with an increase in the intensity of extra-tropical storms.

It is worth noting that any such changes in wave climate are not likely to be universal and will most definitely exhibit spatial variations. In a study of wave climate over the past 25 years on the Pacific coast of the USA, Allan and Komar (2006) observed a strong latitudinal variation in wave climate changes, with the largest significant increase occurring in the north off Washington state; then smaller, but significant, increases off Oregon and northern California; and no significant increase observed off southern California. This type of variation is due to both proximity to location of storm paths as well as to differences in origin for local wave climate.

Another driver for changes in wave climate derived from global climate change has to do with rising sea level. As discussed in Chapter 2, all scenarios for future global climate change include projections for increasing sea level. As is apparent from the discussion on shoaling, changing sea level has the potential to affect the path waves follow as they approach the coast and shift regions of concentration of wave energy. A particularly sensitive example of this is around coral reef systems when reef growth cannot keep pace with sea-level rise and the increased water depth above the reef crest results in propagation of elevated levels of wave energy across the reef flat into susceptible coastal landforms in the lee of the reef. This sequence of events is believed to be responsible for observed early Holocene sedimentary deposits, which require significantly higher energy conditions than currently experienced at the location of deposition.

In a future of increased sea level and higher energy waves, coastal landforms will be under attack from both waves and currents, which will reach the coast at a higher elevation, and also from increased wave run-up, which will carry the erosive energy of the waves further up the beach. Which of these processes will dominate? While this is a complex question, which depends on multiple variables, Ruggiero (2012) attempted to answer it for the local observed wave and sea-level increases in the US Pacific Northwest. Even in regions where the observed local rates of sea-level change are as great as those from climate change projections, this study suggested that the dominant process was the increase in run-up heights from the more energetic wave climate. The importance of this wave climate driven increase in total water level (TWL) is underscored by the result that regions of the study that were emergent relative to mean sea level were still effectively experiencing increasing TWL because of the wave run-up considerations. Whether this relationship continues into the future and/or is applicable to other locations depends on the rate of wave energy increase, the rate of sea-level change, and the beach slope, with flatter beaches being more sensitive to absolute sea-level changes.

4.3 Tides

4.3.1 Tidal characteristics

While the recognition of beauty in the world's coasts is almost a universally recognized aesthetic, there is a special magic reserved for those who live near or visit coastal areas that experience strong tidal variations. The regular cyclical filling and emptying of large basins, or the acceleration, deceleration and subsequent flow reversal of roiled tidal currents, has drawn the attention of human observers since perhaps prior to the advent of conscious thought. Considering the fascination that the motions of celestial bodies held for early humans, it is not hard to imagine that connections between these motions and tidal fluctuations were detected from earliest times. This supposition is supported by written records as old as 4000 years from the ancient Indian Vedas, which contain discussions of this relation. This ancient knowledge notwithstanding, it was only with the development of Newton's theory of gravitational attraction that a theory

that accounted for the majority of features of tidal variations could be developed.

The examination of tidal records from different locations around the globe (Fig. 4.16) reveals a number of common features of tides that any tidal theory must be able to explain. The tidal records from South Pass, Los Angeles, USA, represent the tidal signal that is instinctively the easiest to understand. This pattern exhibits one high-water and one low-level tide every day, completing an entire tidal cycle once every 24 hours. This type of tide, the 'diurnal tide', is by no means the most common tide. That role is reserved for the 'semi-diurnal tide', which has roughly two tidal cycles per day and a tidal period of just over 12 hours. The tides at Otago Harbour, New Zealand, are a relatively good example of a pure semi-diurnal tide, but most locations exhibit a 'diurnal inequality', in which one of the two high tides in a day is significantly larger than the other, as in Baracoa, Cuba. As can be observed in Hinkley Point, UK, and South Pass, all locations exhibit high-and low-water levels that vary in a similar manner over a 28-day lunar cycle. The particularly strong tides that occur every 14 days are known as spring tides, and the weakest tides, which occur at the mid-point between the spring tides, are known as neap tides. In locations such as the Kuril Islands, Russia, where the diurnal inequality is particularly strong, the neap-spring beating can lead to a condition where one of the semi-diurnal tides is indistinguishable for a period of time and such locations are said to experience mixed diurnal and semi-diurnal tides. This rich collection of tidal patterns observed around the world presents a significant challenge that any theory that seeks to explain the origin of the tides must be able to account for. The first theory that was able to do so is commonly referred to as the equilibrium theory of the tides.

4.3.2 Equilibrium tides

The equilibrium theory is based on the approximations that the Earth is a liquid-covered sphere and that the surface of that liquid planet is in equilibrium with the forces acting on it. For the equilibrium tide, the relevant force is gravitational attraction. Examining the diagram in Fig. 4.17a, it is easy to understand how the gravitational attraction of the Moon on the Earth will tend to lead to a bulge in the liquid surface of the Earth towards the Moon, and that as the Earth spins about its axis during the day, this bulge will stay fixed relative to the Moon, so that an observer on a fixed point on the Earth will experience a daily rising and lowering of the sea surface exactly as in a diurnal tide. What is not so immediately clear from this explanation is why semi-diurnal tides are so common on Earth. To understand this, it is necessary to remember that the Moon does not truly orbit about the Earth, but instead both the Moon and the Earth orbit about their

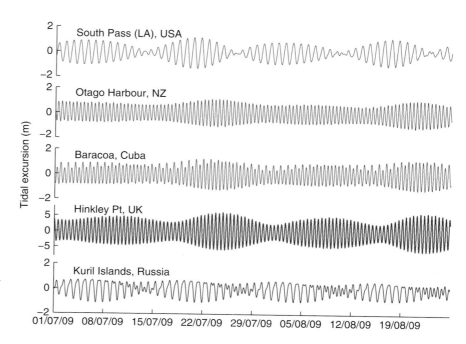

Fig. 4.16 Tidal predictions for multiple locations around the world. Predictions are for a 56-day window commencing on 1 July 2009. Predictions were made using the program xtide (http://www.flaterco.com/xtide/).

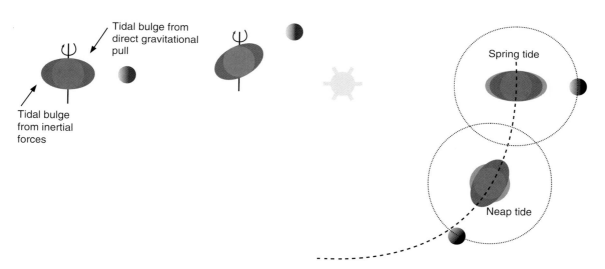

Fig. 4.17 Schematics to illustrate the tide-generating processes.

common centre of mass, which actually lies towards the surface of the Earth about 4671 km above the centre of the Earth. Thus, the Earth can be understood to revolve around this off-centre point so that, in the frame of reference of the Earth, the liquid on the side of the Earth far from this point will experience a 'centrifugal force', much like the water in a bucket that is spun in circles around an observer's shoulder. The result of this will be a bulge of water on the side of the Earth opposite to the Moon, which turns out to be of the same magnitude as the bulge on the side

facing the Moon. Many purists disdain the reference to centrifugal force when explaining tides, as this is really only an apparent force, used to describe the apparent effects of true forces in a universal frame of reference when observed in a rotating frame of reference. Nonetheless, this apparent force has very real results that are well within the reader's experience and shall suffice for current purposes. Any reader desiring a more thorough derivation of the tide-generating forces is referred to a complete text on tides, such as that by Pugh (1987). With two bulges following the Moon's orbit around the Earth, the cause of semi-diurnal tides is now apparent. The diurnal inequality arises from the fact that the plane of the Moon's orbit about the Earth is rarely perpendicular to the axis of the Earth's rotation, so that the stationary observer on the Earth's surface passes through thicker and thinner sections of the two bulges (Fig. 4.17b), and the inequality between these two sections changes as the inclination of the Moon's orbit with respect to the Earth's axis of rotation changes. It can be observed that there exist locations at high latitudes where the nature of the inequality will appear to be a diurnal tide.

In order to understand the spring-neap beating of the tides, it is necessary to recall that the Sun also exerts gravitational attraction on the Earth. Although the Sun is close to 30 million times as massive as the Moon, it is approximately 400 times further away from the Earth, relative to the Moon. Because gravitational attraction is directly proportional to the mass of each body, but inversely proportional to the cube of the distance between two bodies, the resulting solar-derived tidal forces are roughly half the lunar-derived tidal forces. Thus, the solar tides are smaller than lunar tides, but not undistinguishable and they exhibit all the features previously described for the lunar tides. When the Earth, the Moon and the Sun are all aligned (Fig. 4.17c), the tidal contributions from both systems will sum, leading to exceptionally strong tides (springs); when the Sun and the Moon are at right-angles with respect to the Earth, the tides will be in opposition, leading to particularly weak tidal signals (neaps). The former condition is that associated with full and new Moon, while the latter describe the conditions of quarter Moon.

4.3.3 Dynamical considerations

The equilibrium theory has now provided a comprehensive explanation for all of the tidal characteristics introduced in section 4.3.1, but consideration of all the ramifications of this theory reveals serious shortcomings in the theory. In particular:

• If the oceans were in constant equilibrium with the attraction of the Moon and Sun, then the time of occurrence of the different phases of the tide would be constant for all locations at a fixed longitude.
• Similarly, equilibrium would mean that tidal amplitudes were very close to constant for all locations of specified latitude, and the occurrence of diurnal/mixed/semi-diurnal tides would all be dependent only on latitude.
• Finally, tidal equilibrium would require that spring tides occur precisely at full or new Moon.

That all these conditions are not observed, thereby indicating that the tides are not exactly in equilibrium with the Moon and Sun, is not after all particularly surprising for multiple reasons. Firstly, when the idealized picture of the tides is considered, it is clear that the tidal bulges represent a wave with a wavelength equal to half the circumference of the globe. Such a large wavelength would imply that the tide would propagate as a shallow-water wave with a phase speed equal to eqn. 4.10. Considering the average depth of the ocean is 4000 m, this gives a phase speed of just under 200 m/s, which is clearly too slow to keep up with the greater than 450 m/s speed of the rotation of the Earth at sea level at the equator. In addition, the fact that the Earth is only 70% covered in ocean clearly challenges the basic assumptions of the equilibrium theory.

The modern understanding of the tides is that they can be represented by wave systems that are forced by the gravitational attractions described in the equilibrium theory and which circulate like standing Kelvin waves (Fig. 4.18) in virtual basins. Careful examination of Fig. 4.18 indicates that there is a point in the centre of the basin where the sea surface is immobile and about which the wave propagates (anti-clockwise in the northern hemisphere and clockwise in the southern hemisphere), with no fluid motion at that point. Tides in the global ocean can be represented by a series of these cells, which are called 'amphidromes' (Fig. 4.19). Each amphidrome has an immobile (in terms of tidal motion) point in the centre, which is called the amphidromic point, and radiating from these points are a series of 'co-tidal' lines, which represent the locus of all points in the amphidrome with the same phase. This 'spoke' structure helps to understand the dynamics of the amphidromic system by providing an impression of how the tide propagates around the amphidrome, which only partly resembles the idealized square used in Fig. 4.18. The colour contours, which are generally normal to the co-tidal lines, represent amplitude bands of the amphidromic system. In general, the tidal amplitude increases with distance from the amphidromic point, and the tidal phase increases in an anti-clockwise (clockwise) sense in the northern (southern) hemisphere.

4.3.4 Tidal analysis and prediction

While recognition of the tide-generating forces and identification of the amphidromic patterns of tidal propagation provides for a high level of understanding of the tides, the precise tidal response of specific locations, particularly in the coastal ocean, is too strongly dependent on local factors, such as water depth and topographic flow restrictions, to render predictions based solely on dynamical considerations practical. Nonetheless, as every beachcomber and fisherman knows, reliable predictions of tidal water-level changes and tidal currents are quite common.

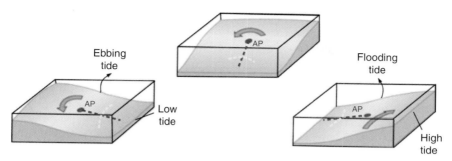

Fig. 4.18 Diagram illustrating a Kelvin wave propagating about an amphidromic point (AP) in a basin. The diagram is for the northern hemisphere where the Coriolis force makes the wave propagate in a cyclonic (anti-clockwise) fashion. The red lines represent lines of constant phase that radiate from the amphidromic point and the yellow lines indicate contours of constant amplitude that are normal to the lines of constant phase. (Source: Gross and Gross 1993. Reproduced by permission of Pearson Education, Inc.)

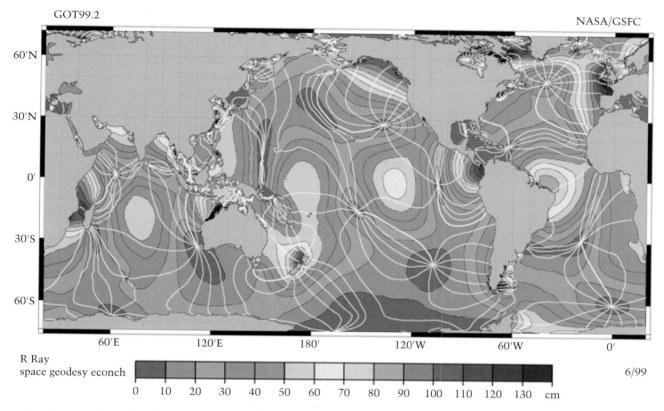

Fig. 4.19 A map of the M2 tidal constituent in which the amplitude of this component is indicated by colour and the white lines represent co-tidal lines with 30° phase difference. (Source: R. Ray; NASA – Goddard Space Flight Center; NASA – Jet Propulsion Laboratory; Scientific Visualization Studio; Television Production NASA-TV/GSFC.) For colour details, please see Plate 15.

While there are now multiple techniques used to provide this kind of prediction, the most common methodology involves harmonic analysis and prediction. Harmonic analysis recognizes that tides are driven by contributions from the different celestial bodies and their interactions. Each one of these contributions has a specific individual frequency and is known as a tidal constituent (Table 4.2).

Harmonic analysis uses recorded tidal elevation changes to determine the local amplitude and phase of each tidal constituent. As the orbits of celestial bodies and the shape of the ocean basins change imperceptibly on human time-scales, once the nature of the local tidal constituents are known, they can then be used to predict tidal elevations far into the future. While most official tidal predictions are based on tidal constituents that were calculated from 19 years of observations (known as a tidal epoch), high-accuracy predictions can be generated from observation periods as short as a year or less. Nonetheless, a comparison of recorded sea-surface elevations and tidal predictions will always exhibit differences. This difference between observed elevations and accurate tidal predictions is known as the 'non-tidal residual', which represents contributions from stochastic processes such as barometric pressure changes and wind- and wave-driven set-up and set-down.

4.3.5 Tidal currents

The orbital velocities of a tidal wave propagating in an ocean 4000 m deep is in the order of centimetres per second, but that number changes as the water depth decreases and as the tidal amplitude increases. Add to this natural flow restrictions, such as narrow straights or prominent headlands, and tidal currents can achieve peak values approaching 10 m/s at very specific locations around the globe. Increasingly, attempts are made to harness this energy through installation of renewable tidal-energy devices (Box 4.3). While it is

Table 4.2 Name, period in hours, relative amplitude (M2 = 1), type and origin of the most important tidal constituents for the equilibrium tide. While the International Hydrographic Organization Tidal Committee recognizes more than 140 long-term, diurnal and semi-diurnal constituents alone, high-quality tidal predictions can typically be achieved including relatively few constituents. In shallow water, and other situations where non-linear interactions between primary constituents are strong, compound constituents such as the M4 component can become important. (Source: Adapted from Pugh 1987. Reproduced with permission from John Wiley & Sons.)

Tidal constituent	Period (hours)	Amplitude (relative to M2)	Type	Origin
M2	12.42	1.0000	Semi-diurnal	Principal lunar
S2	12.00	0.4652	Semi-diurnal	Principal solar
K1	23.93	0.5842	Diurnal	Principal lunar-solar
01	25.82	0.4151	Diurnal	Principal lunar
P1	24.07	0.1931	Diurnal	Principal solar
N2	12.66	0.1915	Semi-diurnal	Larger elliptical lunar
MF	327.90	0.1723	Long term	Lunar monthly
K2	11.97	0.1267	Semi-diurnal	Declinational lunar-solar
MM	661.30	0.0909	Long term	Lunar semi-monthly
SSA	4383.00	0.0802	Long term	Solar semi-annual
Q1	26.87	0.0794	Diurnal	Larger elliptical lunar
M4	6.21	NA	Compound	

CONCEPTS BOX 4.3   Tidal renewable energy

As the search for green forms of energy accelerates, many historic forms of harnessing nature are being revisited. Windmills have been used for centuries to grind grain and to pump water and are currently generating a significant fraction of the world's electricity budget. Similarly, tidal impoundments have also been used for milling around the North Atlantic coastline since the early Middle Ages, and today harvesting of tidal energy is poised to become another important form of clean electricity generation. The great potential for tidal energy is amplified by the high density of water, which increases the energy potential over wind by more than a thousand-fold. In addition, the almost perfect predictability of tidal currents is a highly attractive feature, which is lacking in most alternative forms of green energy (solar, wind and waves). Today, practically all tidal-energy developments are focused on either potential energy capture or tidal-stream technology.

Potential energy capture is achieved either by constructing barrages that permit the complete isolation of the head of a tidal estuary or the construction of an impoundment area in the side or centre of such an estuary. Low-head turbines built into the walls of these developments generate electricity, as water from the high-water side drains through the turbine to the low-water side. While the potential exists for two-way generation on flood and ebb tide, engineering considerations have resulted in most current project schemes to opt for one-way generation. Clever actions such as active pumping into the reservoir at high tide can increase the energy output of these schemes. Potential energy capture is not merely a theoretical construct and existing developments at La Rance in France and at Sihwa Lake in Korea, along with experimental installations at Annapolis Royal in Nova Scotia and Murmansk in Russia have been providing electricity to consumers for as long as 45 years. Clearly, potential energy capture is most attractive in macro-tidal locations, where the difference between high and low tide is greatest, and projects have been proposed in the Bay of Fundy in Canada, the Bristol

Channel in the UK, and in Russia and Alaska. The proposed Bristol Channel tidal barrage alone is estimated to be able to provide 5–10% of the UK's electricity needs.

Tidal-stream energy capture refers to the use of devices to harvest the kinetic energy present in rapidly flowing tidal currents. Most such devices are essentially the underwater equivalent of wind turbines, although there are alternative designs that have produced encouraging results. All such devices are designed to extract kinetic energy from strong tidal currents, where the available power varies with the cube of the velocity. Current technologies typically require velocities greater than 2 m/s to achieve full potential and, due to the cubic relation, power generation decreases rapidly for slower velocities. Tidal-stream technology is a late entrant into the marine renewable energy field, originally lagging behind wave energy and tidal barrages, but rapid development has occurred in recent years. The double turbine 1.2 MW SeaGen S (Fig. 4.20) has been deployed in Strangford Lough in Northern Ireland since 2008 and is claimed to

have generated over 3 GWh of electricity in that time (www.marineturbines.com). During a similar period, the Roosevelt Island Tidal Energy (RITE) Project demonstrated the world's first operation of a grid-connected tidal turbine array in the East River running through the centre of New York City (www.theriteproject.com). In addition to generating power, these projects have both made significant contributions to the development of tidal energy by performing detailed studies on the environmental impacts of tidal-stream energy projects. In these studies, no significant impact has been observed so far on local bird and benthic communities, although minor behavioural changes on seal and fish populations have been noted, as they exhibit a pattern of avoidance of the relatively slow-moving turbines. A remaining and unresolved question relates to how the installation of large arrays of turbines will affect local as well as far-field tidal patterns. This concern is perhaps even more relevant for tidal impoundment schemes, and in particular barrages.

Fig. 4.20   Photo of the Seagen tidal turbine device deployed in the Strangford Lough in Northern Ireland since 2008. The device can generate 1.2 MW of energy and since its installation has delivered more than 6 GWh of carbon-neutral energy to the electrical grid. (Source: By Ardfern (Own work) [CC-BY-SA-3.0 (http://creativecommons.org/licenses/by-sa/3.0) or GFDL (http://www.gnu.org/copyleft/fdl.html)], via Wikimedia Commons.)

very much the local geography that determines the magnitude and direction of these currents, they arise due to tidal propagation, which is responding to celestial forcing. This suggests that tidal currents can also be analysed by

harmonic analysis, which is indeed true. Harmonic analysis of currents, however, is complicated by the fact that tidal currents can vary vertically and are free to flow in directions that may vary during the tidal cycle.

4.3.6 *Global change effects on tides*

Greenhouse gases and a warmer Earth will not have an effect on the gravitational attraction of celestial bodies. This means that the tidal potential on the surface of the Earth will be unaltered in the face of global climate change. However, we have already seen (section 4.3.3) that the actual tidal dynamics represent a balance of the tide-generating potential and the physical constraints that determine how the Earth can respond to this potential. Global change effects on tides will come about through changes that alter the ocean's ability to respond to the forcing. For the most part, such changes will arise through sea-level rise, which will have numerous knock-on effects including increased tidal propagation speed 4.10, reduction in tidal dissipation through bottom friction, and alterations of basin shapes. When considering projections for the next century, even if an increase in mean sea level of 1 m is assumed, the associated change in phase speed in the deep ocean will be in the order of 0.01%, which is close to negligible. Thus, it is clear that changes in tidal behaviour are likely to be most noticeable on the shallower sections of the coastal ocean, where such a rise represents a much greater relative change. The form that such a change will take is a complex question that will be highly dependent on the specifics of the location and is certainly spatially variable. The most significant changes will occur in locations where changes in the tidal characteristics will either push a region closer to resonance or pull it further away. For example, in a study of the palaeo-tides in the Gulf of Maine over the past 7000 years, Gehrels et al. (1995) demonstrated that the M2 tidal range increased as mean sea level increased all across the Gulf, more than doubling in range over the period of the study. Using sea-level rise scenarios suggested by the Intergovernmental Panel on Climate Change (IPCC) for the next century, they estimated that the M2 amplitude would increase a further 0.3–1.9%. It is worth remembering that an increase in tidal amplitude translates into an increase in extreme water levels, which is in addition to any sea-level change.

The tidal changes that occurred and are predicted in the Gulf of Maine are the result of a coastal basin being pushed closer to resonance by an increasing sea level. This is not a behaviour that can be assumed to be general for all locations. A study of the response of tides on the northwest European Continental Shelf (Pickering et al., 2012) to a 2 m sea-level change predicts increased tidal amplitude through much of the North Sea, but significantly reduced amplitude in the Gulf of St Malo and the Bristol Channel. This study also highlighted that changes in tidal properties derived from sea-level rise can extend to alterations in the shape of the tides and alterations in the strength of the neap-spring cycle.

4.4 Summary

Waves and tides represent the forcing functions for coastal morphologic change. Both are manifestations of celestially derived energy, with waves being derived from wind energy that has been transmitted from distant locations of generation, and tides being the response to gradients in gravitational attraction on the surface of the ocean. Both phenomena travel relatively unimpeded in the open ocean, but as they approach the coastal ocean and shallow, constrained waterways, they begin to lose energy. This dissipation of energy in the coastal ocean is what ends up driving processes of erosion and deposition, which shape the morphology of the coast.

- For a given water depth, there is a fixed relation between wave periods and wavelengths; the wave dispersion relation specifies what this relation is.
- Linear wave theory provides a good estimate for wave phase and group velocities, the orbital motions below a progressive wave, and the pressure field in the water column below a wave. Non-linear theories are required for accurate descriptions in waves of high steepness or shallow water.
- The environmental factors that determine the characteristics of generated wave fields are the wind speed, duration and fetch. While wave dynamics will not change in a warmer Earth, the above factors may well change, and current studies suggest that the most extreme waves will become more common.
- The equilibrium theory of tides provides a convenient explanation for various features of tides, but dynamic theory is required to explain the actual tides in the ocean's tidal basins. Changes to the geometry of those basins caused by a rising sea level will therefore change the tides.
- Waves and tides represent vast reservoirs of carbon-free solar and celestial energy, which are so far largely untapped but are the focus of an accelerating level of research and investment.

Key publications

ABP Marine Environmental Research Ltd., 2011. *Atlas of UK Marine Renewable Energy Resources*, http://www.renewables-atlas.info.
  [*One of the more comprehensive examples of strategic assessment of wave and tidal reserves*]

Dean, R. and Dalrymple, R.A., 1991. *Water Wave Mechanics for Engineers and Scientists*. World Scientific Publishing, Singapore.
  [*A recent text on water wave mechanics that has quickly become a standard reference among coastal engineers and scientists*]

Kinsman, B., 1984. *Wind Waves: Their Generation and Propagation on the Ocean Surface*. Prentice Hall, Englewood Cliffs, NJ, USA.

[*The classic reference on the generation and propagation of waves on the surface of the ocean*]

Pugh, D.T., 1987. *Tides, Surges and Mean Sea-Level: A Handbook for Engineers and Scientists*. Wiley, Chichester, UK.

[*A highly accessible but comprehensive text on tidal theory, no longer in print but an electronic version is readily available on the internet*]

References

Allan, J.C. and Komar, P.D., 2006. Climate controls on US West Coast erosion processes. *Journal of Coastal Research*, **22**, 511–529.

Ardhuin, F., Chapron, B. and Collard, F., 2009. Observation of swell dissipation across oceans. *Geophysical Research Letters*, **36**, L06607, doi: 10.1029/2008GL037030.

Bender, M.A., Knutson, T.R., Tuleya, R.E. et al., 2010. Modeled impact of anthropogenic warming on the frequency of intense Atlantic hurricanes. *Science*, **327**(5964), 454–458. doi: 10.1126/science.1180568.

Bouws, E., Gunther, H., Rosenthal, W. and Vincent, C., 1985. Similarity of the wind wave spectrum in finite depth water. *Journal of Geophysical Research*, **90**(C1), 975–986.

Bretschneider, C.L., 1952. The generation and decay of wind waves in deep water. *Transactions of the American Geophysical Union*, **33**, 381–389.

Bretschneider, C.L., 1958. Revision in wave forecasting deep and shallow water. *Proceedings of the 6th International Conference on Coastal Engineering*, ASCE, pp. 30–67.

Dingemans, M.W., 1997. *Water Wave Propagation Over Uneven Bottoms. Part 1: Linear Wave Propagation*. World Scientific Singapore.

Fenton, J.D., 1985. A fifth-order Stokes theory for steady waves. *Journal of Waterway, Port, Coastal and Ocean Engineering*, **111**, 216–234.

Gehrels, W.R., Belknap, D.F., Pearce, B.R. and Gong, B., 1995. Modeling the contribution of M(2) tidal amplification to the Holocene rise of mean high water in the Gulf of Maine and the Bay of Fundy. *Marine Geology*, **124**, 71–85.

Gunn, K. and Stock-Williams, C., 2012. Quantifying the global wave power resource. *Renewable Energy*, **44**, 296–304.

Hasselmann, K., Barnett, T.P., Bouws, E. et al., 1973. Measurements of wind-wave growth and swell decay during the Joint North Sea Wave Project (JONSWAP). *Deutschen Hydrographischen Zeitschrift*, **8**, 1–95.

Hasselmann, K., Ross, D.B., Muller, P. and Sell, W., 1976. A parametric wave prediction model. *Journal of Physical Oceanography*, **6**, 200–228.

Jacobs, G.A., Hurlburt, H.E., Kindle, J.C. et al., 1994. Decade-scale trans-Pacific propagation and warming effects of an El Niño anomaly. *Nature*, **370**, 360–363.

Kinsman, B., 1965. *Wind Waves: Their Generation and Propagations on the Ocean Surfaces*. Prentice-Hall, Englewood Cliffs, NJ, USA.

Longuet-Higgins, M.S., 1969. A nonlinear mechanism for the generation of sea waves. *Proceedings of the Royal Society A*, **311**, 371–389.

Miles, J.W., 1957. On the generation of surface waves by shear flow. *Journal of Fluid Mechanics*, **3**, 185–204.

Phillips, O.M., 1957. On the generation of waves by turbulent wind. *Journal of Fluid Mechanics*, **2**, 417–445.

Pickering, M.D., Wells, N.C., Horsburgh, K.J. and Green, J.A.M., 2012. The impact of future sea-level rise on the European Shelf tides. *Continental Shelf Research*, **35**, 1–15, doi: 10.1016/j.csr.2011.11.011.

Pierson, W.J. and Moskowitz, L., 1964. A proposed spectral form for fully developed wind seas based on the similarity theory of S.A. Kitaigorodskii. *Journal of Geophysical Research*, **69**, 5181–5190.

Pugh, D.T., 1987. *Tides, Surges and Mean Sea-Level: A Handbook for Engineers and Scientists*. Wiley, Chichester, UK.

Ruggiero, P., 2012. Is the intensifying wave climate of the U.S. Pacific Northwest increasing flooding and erosion risk faster than sea-level rise? *Journal of Waterway, Port, Coastal, Ocean Engineering*, **139**(2), 88–97, doi: 10.1061/(ASCE)WW.1943-5460.0000172.

Snodgrass, F.E., Groves, G.W., Hasselmann, K.F., Miller, G.R., Munk, W.H. and Powers, W.H., 1966. Propagation of ocean swell across the Pacific. *Philosophical Transactions of the Royal Society A*, **259**(1103), 431–497, doi: 10.1098/rsta.1966.0022.

Sverdrup, H.U. and Munk, W.H., 1947. *Wind, Sea, and Swell: Theory of Relations for Forecasting*. H.0. Publication Number 601, US Navy Department, Hydrographic Office.

Young, I.R., Zieger, S. and Babanin A.V., 2011. Global trends in wind speed and wave height. *Science*, **332**, 451–455, doi: 10.1126/science.1197219.

5 Coastal Hazards: Storms and Tsunamis

ADAM D. SWITZER

*Earth Observatory of Singapore, Nanyang Technological University,
Nanyang Avenue, Singapore*

5.1 Coastal hazards

Although the coastal zone constitutes a relatively minor portion of the Earth's land area, it accommodates more than 60% of the world's population, with many more billions of people relying on the resources these regions provide. These figures are impressive even at first glance. However, when one considers that many coasts around the world are inaccessible or not comfortably habitable, only then does the true importance of the coast in terms of human vulnerability become apparent. This is particularly the case along subtropical and temperate coasts as they have above-average concentrations of people and economic activity.

People, industries, infrastructure and ecological systems along coasts are vulnerable to a number of natural hazards, some of which have, or will, become more serious with changing climate. Climate change is likely to affect rainfall and climate patterns, potentially intensifying and changing seasonal patterns and frequency of storms. Sea-level rise associated with climate change will also impact on, and reduce the stability and relative height of, natural barriers to marine inundation, such as by eroding dune systems.

Despite a long history of disasters, and the projections of increasing hazards under a changing climate, coastal development continues unabated, leaving many communities at risk of enormous catastrophic losses as cities grow at unprecedented rates even where historical events clearly indicate a considerable risk (Fig. 5.1). All coastal communities must live with a certain risk of coastal hazards, but measures ranging from coastal engineering to studies of human behaviour can minimize the potential for fatalities and economic loss, and therefore must be undertaken.

Coastal Environments and Global Change, First Edition. Edited by Gerd Masselink and Roland Gehrels.
© 2014 John Wiley & Sons, Ltd. Published 2014 by John Wiley & Sons, Ltd. Companion Website: www.wiley.com/go/masselink/coastal

Fig. 5.1 The coastal megacity of Hong Kong in southern China lies in the path of seasonal typhoons with the potential to cause immense damage to the coastal infrastructure that rests predominately on low-lying reclaimed land. (Source: By Lichunngai (Own work) [CCO], via Wikimedia Commons.) Inset: Historical photos like this one of the *Tymeric*, a large ship sunk in Hong Kong in September 1937, provide examples of past events. The unnamed typhoon (tropical cyclone) sunk numerous ships, fishing boats and killed an estimated 11,000 people as it struck southern China. A similar event in southern China today would likely cause losses of more than $US50 billion. (Source: Reproduced with permission of D. Hayden.)

The previous three chapters have outlined the main long-term drivers of coastal evolution, namely sea-level change, geology and sediment dynamics, and the long-term actions of waves and tides. This chapter investigates storms (including extratropical storms and tropical cyclones) and tsunamis, which are the primary coastal hazards that can generate marine inundation events, and describes their effects on the coast and coastal populations. This chapter will not deal with other coastal hazards such as coastal land subsidence. Sections on co-seismic subsidence and subsidence due to sediment loading or to groundwater extraction are found in Chapters 2, 3 and 12.

5.1.1 Coastal vulnerability to storm and tsunami hazards

Over the last decade, numerous notable coastal disasters have occurred as a result of storms (tropical cyclones) and tsunamis. The Indian Ocean (2004), Samoan (2009), Chilean (2010) and Japanese (2011) tsunamis, as well as the tremendous impacts of tropical cyclones including Hurricane Katrina in New Orleans, USA (2005), Cyclone Nargis in Myanmar (2008), Cyclone Yasi in Australia and Hurricane Irene in northeastern USA (2011), have highlighted the vulnerability of the coastal zone and its infrastructure, inhabitants, economies and ecological systems. The damages arising as a result of these and other less extreme events are significant (Table 5.1).

Increasing socio-economic pressures, extensive human alteration of the landscape, and the continued over-exploitation of coastal resources have reduced the resilience of the coastal system to both short- and long-term environmental change. In addition to anthropogenic pressure, the coastal system now also faces warmer global temperatures and accelerating sea-level rise (see Chapter 17). When combined with the potential for an increase in extreme events, one can only assume that these factors will further exacerbate the potential for coastal flooding and destruction from marine inundation events.

Despite an increase in the understanding and awareness of coastal hazards and a demonstrated global increase in efforts to estimate and manage coastal risks, disasters associated with coastal hazards are still occurring. In fact, storms and tsunamis are still causing significant socio-economic damage, in many cases more serious than ever before (e.g. Tōhoku Tsunami of March 2011). In order to prepare for future coastal change and manage the coasts of today, scientists, consultants, engineers and decision-makers have been forced to investigate the relationship between changing natural and socio-economic conditions, and to rethink hazard response and adaptation strategies for sustainable coastal living.

Tropical cyclones (also called hurricanes in North America and the Caribbean, and typhoons in Asia) (Fig. 5.2) are some of the most economically disruptive natural disasters that occur on Earth. According to the US National

Table 5.1 The 10 costliest cyclones (hurricanes) in the USA by insured losses, 1980–2011 (in $US millions). (Source: Adapted from Munich Reinsurance Company, Geo Risks Research, NatCatSERVICE 2012.)

Rank	Date	Hurricane	Locations	Overall losses ($US millions)	Insured losses ($US millions)	Fatalities
1	25–30 Aug. 2005	Katrina	USA: Louisiana, New Orleans, Slidell; Mississippi, Biloxi,Pascagoula, Waveland, Gulfport	125,000	62,200	1,322
2	6–14 Sep. 2008	Ike	USA; Cuba; Haiti; Dominican Republic; Turks and Caicos Islands; Bahamas	38,000	18,500	170
3	23–27 Aug. 1992	Andrew	USA: Florida, Homestead; Louisiana; Bahamas	26,500	17,000	60
4	7–21 Sep. 2004	Ivan	USA; Cayman Islands; Grenada; Jamaica; Trinidad and Tobago; Venezuela	23,000	13,800	125
5	19–24 Oct. 2005	Wilma	USA; Bahamas; Cuba; Haiti; Jamaica; Mexico	22,000	12,500	42
6	20–24 Sep. 2005	Rita	USA: Florida, Keys; Louisiana, Lake Charles, Holly Beach, Cameron, New Orleans; Mississippi; Texas	16,000	12,100	10
7	11–14 Aug. 2004	Charley	USA: Florida; Cuba; Jamaica; Cayman Islands	18,000	8,000	36
8	22 Aug.–2 Sep. 2011	Irene	USA; Canada; Carribean	7,400	5,600	55
9	1–9 Sep. 2004	Frances	USA; Bahamas; Canada; Turks and Caicos Islands; Cayman Islands	12,000	5,500	50
10	14–22 Sep. 1989	Hugo	USA: Virgin Islands, Puerto Rico; Antigua and Barbuda; Montserrat; Guadeloupe	9,600	5,100	116

Oceanic and Atmospheric Administration (NOAA), the average insured losses from hurricanes in the USA exceed $US5 billion per year. In the USA, emergency managers at all levels rely on NOAA's weather and climate data. For hurricanes in the USA, this information comes from satellite imagery and Doppler radar, which improve forecasts, accurately identify the stretch of coastline under storm warning, and reduce the total cost of evacuation. A study of the potential for improved satellite-based tropical cyclone forecasts by Centrec Consulting Group (2007) suggested that efforts to provide more effective actions to protect property and to enable evacuation of individuals residing in the path of the storm could be valued at $US450 million.

The impacts of tsunamis and storms have reshaped the history of many cities around the world over the last century, as tsunami waves, cyclonic high winds, storm surges and flooding have caused widespread devastation (Fig. 5.3). In addition to people and infrastructure, ecosystems and economies are also impacted by the landfall of storms and tsunamis. For example, research by Chambers et al. (2007) estimated that Hurricane Katrina in 2005 killed or severely damaged 320 million large trees across 20,000 km² of forest across the southern USA. Additionally, coastal food plantations, fisheries (e.g. fish, oyster and prawn (shrimp) farms) and harvesting infrastructure (e.g. boats, processing and storage facilities) can also be severely damaged by storms and tsunamis.

5.1.2 Assessing risk

Coastal vulnerability assessment has emerged as a key concept for understanding the impacts of climate change and natural hazards. Such assessments are essential for developing adequate risk-management strategies. Coastal vulnerability associated with coastal hazards relates to the susceptibility of both the natural system and coastal societies to these hazards. The vulnerability of a coastal community can be regarded as the sum of risks to its social, economic and physiographic properties, and is inversely related to the natural and social coping and adaptive capacity to adverse impacts, i.e. a society's resilience. Assessing coastal vulnerability is an important prerequisite to determine the 'where', 'why' and 'how' questions related to inundation risk. Only when armed with such knowledge can government agencies take steps to prepare for and reduce the risk of coastal disasters (e.g. where to place nuclear power plants; Fig. 5.4).

There are a variety of coastal vulnerability assessment methods available, often encompassing a broad range of sectorial or multidisciplinary applications at a variety of spatial scales. Despite huge advances in the quality of data (e.g. tidal data, wave models, satellite imagery, digital terrain models) and assessment (e.g. sophisticated flooding models), there remains a limited usefulness

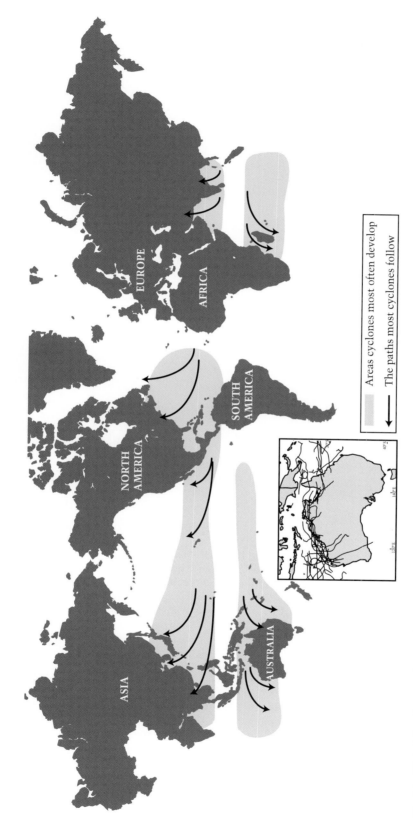

Fig. 5.2 Global distribution of tropical cyclone areas. Cyclones are generated over warm tropical seas and move in arcuate tracks toward the poles. The global map provides an overview of where most cyclones develop and their likely paths. (Source: Nilfanion, via Wikimedia Commons and NASA 2006.) The inset shows actual cyclone tracks from 2005 to 2010 over northern Australia. (Source: Pacific Tropical Cyclone Data Portal (beta), 30 July 2012; www.bom.gov.au.)

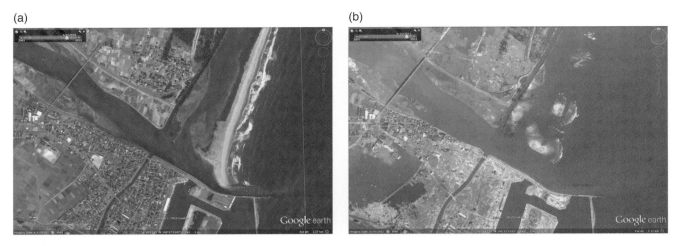

Fig. 5.3   Google Earth images of (a) before and (b) after the 11 March 2011 tsunami at Sendai Fujitsuka township in northeast Japan. Much of the infrastructure has been destroyed. Note the destruction of the sandbar. (Source: (a) Google Earth, © 2010 GeoEye. (b) Google Earth, © 2011 GeoEye.)

Fig. 5.4   (a) The Quinshan Nuclear Power Plant, in Zhejiang, China, and (b) the Madras Atomic Power Station, Kalpakkam, near Chennai, India, are two of many coastal nuclear power plants in Asia. As many Asian countries have frantically built coastal nuclear facilities in recent decades, it is apparent that in many cases they have made little use of science to determine whether these areas are safe. More than 30 plants in operation or under construction in Asia are at risk of one day being hit by a very large (Category 5) cyclone and/or tsunami. (Source: (a) Atomic Energy of Canada Limited [Attribution], via Wikimedia Commons. (b) Google Earth, © 2012 GeoEye.)

for integrated assessment methods. In particular, most vulnerability assessment methodologies are designed to investigate potential geophysical changes (e.g. erosion) or economic loss (e.g. insurance). In contrast, very few assessment schemes look at coastal resilience or consider the impacts of social disturbance in planning for long-term recovery.

In terms of coastal hazards, there are three main aspects of vulnerability (risk) to consider: (1) the potential for economic damage (economic risk); (2) the potential for mortality or social disruption (social risk); and (3) the potential for landscape disturbance or natural habitat destruction (geo-ecological risk).

5.2   Extratropical storms and tropical cyclones

5.2.1   *The anatomy of storms and cyclones*

Before we discuss how storms affect coasts, we need to look briefly at how they are formed. In the most simple of terms, storms are caused by the meeting of cold and warm air, and the compulsion for warm air to rise above cold (convection). This creates areas of low pressure at the Earth's surface where the warm air has diverged upwards. When warm air rises, its water vapour condenses into clouds, releasing heat, which further warms the air, forcing it to rise even higher, and continuing the cycle. Both

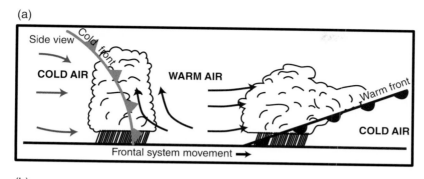

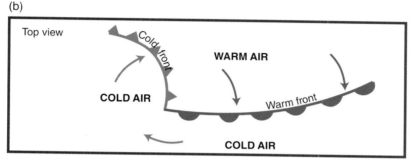

Fig. 5.5 Extratropical (shown here) and sub-polar cyclones differ from the tropical cyclones in appearance and cloud structure. Extratropical cyclones encompass a class of storms with many names. Although they are sometimes referred to as 'cyclones', this is imprecise; the term cyclone applies to numerous types of low-pressure areas. In this case, 'extratropical' is a descriptor that signifies that this type of cyclone usually occurs outside the tropics in the middle latitudes. Such systems are often described in lay terms as 'depressions' or 'lows'. When they occur over the ocean, they can generate large ocean swells and storm surges as they cross the land. (Source: Earth Observatory of Singapore.)

warm and cold air rush towards the low-pressure area (convergence) trying to 'fill in' the low, generating winds at the Earth's surface. The Earth's rotation causes an apparent force (the Coriolis effect) that pulls the winds to the right (creating counter-clockwise rotation) in the northern hemisphere and to the left (creating clockwise rotation) in the southern hemisphere.

In temperate regions (30–60° north and south of the equator), extratropical storms arise from the collision of cold and warm 'fronts' (the boundaries between air masses of different temperatures), over an area of several hundreds to thousands of kilometres (Fig. 5.5). When a slow-moving warm front moves towards an area of relatively colder air, the warm air slides relatively easily up over the colder air. However, when a faster-moving cold front hits an area of relatively warmer air, the warm air present is forced quickly upwards over the cold front, creating an area of low pressure at the Earth's surface.

A *tropical cyclone* is a generic term for a storm with an organized system of thunderstorms that are not based on a frontal system. They form in the region between 30° north and south of the equator. In contrast to the temperate regions, the tropics do not have fronts, as the ambient temperature gradient is usually relatively weak between air masses. Instead, localized patches of warm air, only

several tens of kilometres wide, are caused by the heating of surface air due to evaporation from warm ocean waters and subsequent condensation of this water vapour in the air column. The heating of the surface air results in convection, creating areas of low pressure at the surface. The winds caused by convergence into this low-pressure zone cause greater evaporation, further intensifying the process. In some instances, smaller tropical storms may merge, and low pressure is intensified due to increasing rotation as result of greater horizontal convergence. As these systems intensify, larger-scale tropical depressions, or in some cases tropical cyclones, can result (Fig. 5.6). The intensity of tropical cyclones is commonly measured using the Saffir-Simpson scale (Category 1 to 5), a measure of wind speed and the likely damage such wind speeds will cause (Fig. 5.7). Characteristics of tropical cyclones include a central low-pressure zone (the 'eye' – an area characterized by light winds and fair weather) around which high-speed winds (at least 34 knots) circle.

5.2.2 Vulnerable coasts and the storm cycle

The passage of storms over the ocean can generate strong ocean swells that can strike the coast as high surf waves (see Chapter 4). The degree to which a coast is

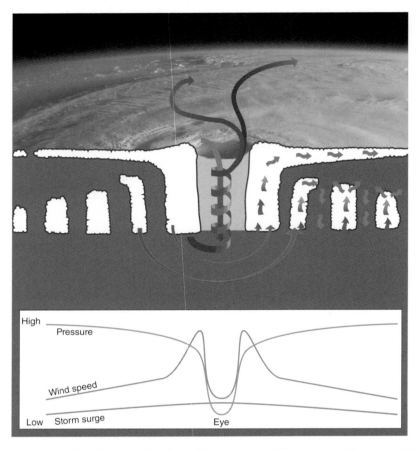

Fig. 5.6 In contrast to other storms (cyclones), tropical cyclones have a recognizable structure of an eye with radiating rain (cloud) bands. This figure provides a schematic summary of the main components of a tropical cyclone, including rain bands and eyewall, as well as showing the pressure, wind speed and storm surge gradients across the storm. (Source: Adapted from Earth Observatory, NASA 2003.) For colour details, please see Plate 16.

open to wind-generated waves is a primary factor in determining the equilibrium 'energy state' of the system. A coastline where wave energy is predominantly high is considered as a high-energy coast. In contrast, embayed coasts in sheltered seas are usually low-energy coasts. This background energy condition plays a vital role in determining the degree to which a coastline is open to wave influence, and thus susceptible to storm hazards.

Beaches are dynamic sedimentary systems that experience phases of erosion and accretion over a range of timescales (see Chapter 7) in response to smaller-scale (short-term) events, such as storms, waves, tides and winds, or in response to large-scale (long-term) events (e.g. El Niño Southern Oscillation; Box 5.1) and geomorphic processes. All coastlines are naturally dynamic, and cycles of erosion are often an important feature of their

physiographic evolution. On most beaches, wind, waves and currents move and rearrange unconsolidated sediments in the beach system, sometimes resulting in rapid changes in the position of the shoreline.

Short-term changes are often seasonal, with erosion mostly occurring when storms that generate erosional wave regimes are more frequent. For example, many temperate beaches erode in winter, when stronger and more frequent storms occur, and renew (accrete) in summer. In addition to seasonal erosion, rapid erosional episodes may also be produced by high-magnitude storms, such as tropical cyclones or intense low-pressure systems. The degree of erosion that occurs within a particular erosional phase is highly variable, but is usually linked to the magnitude and frequency of storms that impact the system, as well as to the equilibrium energy state. For example, during a high-magnitude (1 in 100 year) storm, high-energy waves

Saffir-Simpson Hurricane Scale

Category **1**

Winds 119 – 153 kph (74 – 95 mph)
Damage to mobile homes and vegetation.
Some downed trees. Some power outages.
Loose items become projectiles.

Category **2**

Winds 154 – 177 kph (96 – 110 mph)
Severe damage to mobile homes. Some damage to roofs, doors, windows of houses. Many downed trees. Widespread power outages.

Category **3**

Winds 178 – 209 kph (111 – 130 mph)
Structural damage to houses. Large trees downed. Airborne debris will kill or injure. Flooding near coast. Damage could extend well inland.

Category **4**

Winds 210 – 249 kph (131 – 155 mph)
Severe damage to buildings. Nearly all trees downed. Flooding extends far inland. Mass evacuations of areas 10 km from coast.

Category **5**

Winds > 249 kph (155 mph)
Widespread catastrophic damage to buildings, some complete building failures. Mass evacuations of areas up to 16 km from coast.

EARTH OBSERVATORY of SINGAPORE

Fig. 5.7 Schematic summary of the Saffir-Simpson hurricane (tropical cyclone) wind scale. This scale ranges from 1 to 5, based on the intensity at the indicated time. The scale was originally developed by wind engineer Herb Saffir and meteorologist Bob Simpson for US hurricanes. The scale provides examples of the types of damage and impacts expected with winds of the indicated intensity. In general, damage rises by about a factor of four for every category increase. The determining factor in the scale is the maximum sustained surface wind speed (peak 1-minute wind at the standard meteorological observation height of 10 m over unobstructed exposure) associated with the cyclone. (Source: Adapted from Palm Beach Post 2012.)

may erode a considerable way into the dune system (see Chapter 8). Such events can impact areas that lie outside of the normally active zone of accretion and erosion, and form structures such as erosion 'scarps' (Fig. 5.8). In many cases, the sand that was stored in the dune and beach is carried offshore, where it accumulates into sandbars. If enough sand accumulates offshore, these sandbars intercept large waves before they reach shore, reducing their impact on the coastline. This is an example of morphodynamic feedback (see Chapter 1).

Storm sequencing is also a key driver of beach evolution and change, as a series of lower-magnitude storms that occur in quick succession may produce a similar degree of erosion to a single higher-magnitude storm. This is because the intervening periods are too short for the constructive background wave regime to return a significant amount of sediment back to the shoreline (shoreline accretion) following erosion.

Generally, beach recovery begins soon after the storm or swell system passes: as the high-energy waves become

CONCEPTS BOX 5.1   El Niño Southern Oscillation (ENSO)

The El Niño/La Niña Southern Oscillation (ENSO) is a climatic cycle that occurs across the Pacific Ocean roughly every three to seven years, and which lasts for nine months to two years at a time. The mechanism for this oscillation is not completely understood. During the El Niño phase, the western Pacific experiences cooler sea-surface temperatures, resulting in high sea-surface pressure (and a subsequent decrease in frequency and intensity of rain and storms), while in the eastern Pacific warmer sea-surface temperatures lead to low sea-surface pressure (resulting in an increased number of storms). During La Niña, this cycle is reversed, with the western Pacific experiencing warmer surface temperatures and the eastern Pacific's surface temperatures cooling.

The resultant change in average weather affects coastlines through longer erosional phases, which occur in addition to beach erosion from normal seasonal storms. For example, La Niña has been correlated with erosional phases along the east coast of Australia due to a higher frequency of storms (Short and Trembanis, 2000), but lower wave heights and storm activity in the western USA (Storlazzi and Griggs, 2000; Allan and Komar, 2006). In contrast, El Niño events result in phases of positive sediment budget (or beach growth) in eastern Australia linked to the lower frequency of coastal storms, while the western USA experiences significant increases in storminess and wave energy.

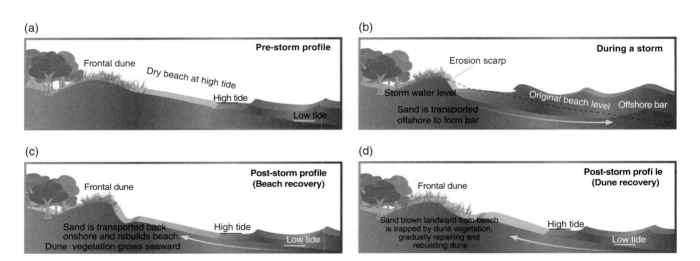

Fig. 5.8 Schematic diagram of the storm cycle on a high-energy beach. (a) The original pre-storm profile. (b) Under storm attack, the beach profile changes as sediment is removed offshore and an erosional scarp is formed in the beach. (c) Post-storm (in fair weather conditions), lower-energy waves return sediment to the beach. (d) Following the establishment of a new beach profile, onshore winds can carry sediments further onshore to form a new frontal dune. (Source: Author, with Earth Observatory of Singapore.)

smaller and the background, constructive wave pattern returns. During these conditions (known as fair-weather), smaller waves slowly dismantle the offshore sandbars and return sediments to the beach (Fig. 5.8). Although some sand may have been permanently washed away from the beach system into deep water by the storm, eventually the beach and the dunes will regenerate to closely match their pre-storm profile. Most of the sand transported offshore during storms and stormy seasons is eventually reincorporated into the dune. Beach accretion is generally a much slower process than beach erosion. It may take several months to years for a beach to return to its pre-storm condition after one major storm or several smaller storms in quick succession.

5.2.3   Disequilibrium in the storm cycle

While the normal storm cycle allows a beach system to equilibrate with changing conditions, very large storm and tsunami events can cause irreversible changes to the coastal profile. This concept is also known as 'disequilibrium', as the changes to the coastal environment are too large for normal coastal accretion processes to reverse. For example, in the low-energy coastlines of the Atlantic and Gulf coasts of the USA, the systems are dominated by a micro-tidal regime, dominant offshore winds, characteristically low wave energy and a shallow, gently sloping coastal shelf. Such low-energy conditions lead to the

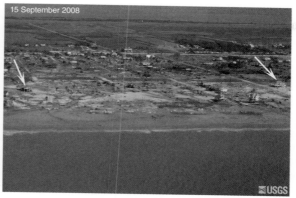

Fig. 5.9 Oblique aerial photographs of the Bolivar Peninsula, Texas, USA, before and after Hurricane Ike (a tropical cyclone), which struck the coast on 13 September 2008. The arrows mark features recognizable in each image. (Source: USGS 2008.)

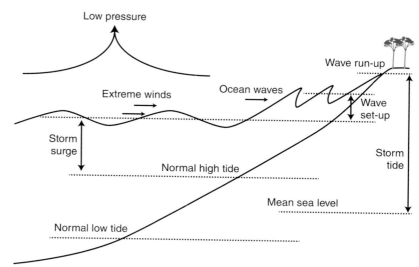

Fig. 5.10 Schematic diagram of the storm surge set-up. The storm surge is the combined effect of water that is pushed toward the shore by the force of the onshore winds caused by the storm, along with the superimposed temporary rise in sea level caused by low atmospheric pressure. The rise in water level can cause severe flooding in coastal areas, particularly when the storm tide coincides with the normal spring high tides. (Source: Author, with Earth Observatory of Singapore.)

formation of extensive low-lying barrier islands (see Chapter 9). However, these low-energy coasts experience seasonal extreme events associated with tropical cyclones (locally known as hurricanes), and, in the case of the east coast, extratropical storms (locally known as nor'easters). These high-energy events are not in equilibrium with the ambient setting and often result in significant coastal modification, including frequent storm-dominated overwash events (see section 5.4) that can result in the permanent modification/disruption of the coastline and coastline processes (Fig. 5.9). In these settings, temporarily elevated sea levels and storm waves move significant sediment loads out of the dune system and into the coastal plain (see section 5.4).

One particular coastal disequilibrium event capable of significant damage is known as a storm surge. A storm surge is the sum of two components. The first component is generated by the inverse relationship between water level and barometric pressure – in essence, the less the atmosphere pushes down on the water surface (e.g. because of low-pressure areas during storms), the higher the water surface rises (see Fig. 5.6). The second component is the shoreward mass-transport of water in steep, high waves, resulting in water piling up against the coast. The storm surge is the difference between the observed level of the sea surface and the astronomical tide that would have occurred in the absence of the storm (Fig. 5.10). As a general rule, the low barometric pressure allows a hydrostatic rise in the water level of approximately 0.1 m for each 10 hPa (mb) that the central pressure of the system is lower than the surrounding pressure. Importantly, as a low-pressure system moves into shallow coastal waters, the shape of the nearshore seabed (bathymetry) and coastline shape (morphology) modify the surge and can cause substantial amplification and attenuation of its height.

5.3 Tsunamis

Tsunamis are often mistakenly called 'tidal waves', although they are unrelated to tides. They are also often referred to as 'seismic sea waves,' although they are not purely associated with seismicity (earthquakes). A tsunami event is not usually the result of one wave, but is more often the term applied to a series of long-wavelength water waves.

In the most general of terms, a tsunami can be considered as a wave in a water body that is caused by the sudden and rapid displacement of a large volume of that body of water. Tsunamis usually occur in the ocean, but they can also occur in large lakes and bays. In Japan, where the term originated (tsunami is the Japanese term for 'harbour wave'), tsunamis are common due to the occurrence of frequent, large, offshore earthquakes. Tsunamis have the potential to flood the coastal zone with enormous volumes of water, causing immense damage to coastal communities (Box 5.2).

Although tsunamis are generally associated with submarine earthquakes, there are many potential causes (Fig. 5.13). Rarer secondary causes include underwater explosions (including violent volcanic eruptions and detonations of underwater nuclear devices), terrestrial and submarine landslides and other mass geological movements, meteorite impacts, and other disturbances above or below the water. Although earthquake-generated tsunamis are the most common, the highest run-up (a measurement of the height of the water onshore observed above normal water level) for a historical event is attributed to a landslide-generated event. The Lituya Bay event in Alaska in 1958 generated a tsunami that had a run-up of more than 500 m, i.e. the waves continued to flow onshore up to a height of over 500 m above normal sea level. In comparison, the explosion and collapse of Krakatau in 1883 created waves in excess of 45 m at Merak, a port in northern Java, Indonesia.

5.3.1 Tsunamis in the ocean

Tsunamis are commonly characterized by a very long wavelength L of several hundred kilometres and very low amplitudes in the order of 1 m when travelling in deep water (Murty, 1977). Because of their low amplitudes, tsunamis often remain undetected in the open ocean, although their long wavelengths mean that they can affect the whole water column. Tsunamis are shallow water waves and, according to linear wave theory (see Chapter 4), the velocity c of tsunami waves (celerity) only depends on the water depth h:

$$c = \sqrt{gh} \tag{5.1}$$

where g is the acceleration of gravity.

The deeper the water, the larger the velocity of the tsunami waves. Tsunamis move through the deep ocean

---

CASE STUDY BOX 5.2 The Tōhoku tsunami

The Tōhoku tsunami was generated by the Great East Japan earthquake. The magnitude 9.0 megathrust earthquake occurred off the east coast of Japan on 11 March 2011. The earthquake triggered a very large tsunami, in places generating wave run-ups more than 35 m high that struck the Japanese coast minutes after the quake, travelling up to 10 km inland.

The Japanese government has confirmed that there were more than 15,000 deaths and estimates that the cost of the earthquake and tsunami could exceed $US300 billion, making it the most expensive natural disaster on record. Additionally, the clean-up after the Fukushima Nuclear Power Plant incident, which occurred as a result of the earthquake and tsunami inundation (Fig. 5.11), has been estimated to cost between $US13 billion and $US250 billion on top of this sum.

The tsunami was generated by vertical slip on a rupture located approximately 65 km offshore from the east coast of Tōhoku. Following the main earthquake, the Japan Meteorological Agency issued a warning for a 'major' tsunami (>3 m). The largest tsunami run-ups recorded were at Ryōri Bay, ōfunato (~30 m), Tarō, Iwate (~37.9 m) and Omoe peninsula (38.9 m).

The tsunami took 10–30 minutes to reach the coast. The tsunami inundation was graphically shown on television. This first image showed Sendai Airport with waves sweeping away cars and planes as the tsunami moved inland. The second graphic imagery was filmed in and around Sendai Airport by a news helicopter, showing residents in vehicles trying to escape the approaching waves, which contained considerable debris.

In Japan, a similar scenario to the 2004 Indian Ocean earthquake and tsunami was played out, where damage by the tsunami was much more deadly and destructive than was the actual earthquake. There are several reasons why Japan – arguably the best-prepared country in the world in terms of tsunami preparedness – still experienced a high death toll from the tsunami. The primary reasons are the geographic proximity of the source, and the unexpectedly large size of the tsunami waves.

Fig. 5.11 Tsunami waves from the 2011 Tōhoku tsunami engulf the coastal defences in front of Fukushima Nuclear Power Plant in Japan. At the time of the earthquake and tsunami, three of the six reactors in the plant were undergoing maintenance and were not operating. The remaining reactors were shut down automatically after the earthquake, and the residual heat of the fuel was being cooled using emergency generators. The tsunami disabled the emergency generators cooling the reactors, causing partial nuclear meltdowns and a series of explosions. These events caused large-scale evacuations and considerable concern for worker safety and food and water supplies. On the International Nuclear Event Scale, the events rated level 7 (major release of radioactive material with widespread health and environmental effects requiring implementation of planned and extended countermeasures). Such events highlight the vulnerability of coastal infrastructure, including nuclear power plants, to tsunami and storm surges. (Source: AP Photo/Tokyo Electric Power Co. 2011. Reproduced with permission.)

Although many of the coastal cities were defended by tsunami walls, most were constructed to deal with much lower tsunami heights, and did little to defend the cities against the surge. Some media outlets suggested that residents may have been complacent and relied on the tsunami defences. This remains uncertain, but it is clear that many people caught in the tsunami are likely to have thought they were located in areas that were high enough above sea level to be safe.

The tsunami waves generated also propagated to the southeast (Fig. 5.12), and shortly after the earthquake, the Pacific Tsunami Warning Centre (PTWC) in Hawaii issued tsunami watches for locations in the Pacific before then raising the warning to cover the entire Pacific Ocean. Later, the West Coast and Alaska Tsunami Warning Centre issued a tsunami warning for the coastal areas of most of California, all of Oregon, and the western part of Alaska, plus a tsunami advisory covering the Pacific coastlines of most of Alaska, and all of Washington and British Columbia, Canada. In California and Oregon,

waves up to 2.4 m hit many harbours, causing more than $US10 million worth of damage to boats and harbours. Tsunami waves up to 1 m high also hit Vancouver Island in Canada. On the other side of the Pacific, waves up to 0.5 m were recorded in the Philippines where several houses were destroyed, and on the coast of Jayapura, Indonesia. $US4 million in damages were also recorded in Wewak, Papua New Guinea. In the central Pacific, Hawaii reported significant damage to private properties, including considerable damage to the Four Seasons Resort Hualalai. On Midway Atoll's reef inlets and Spit Island, 1.5 m was submerged, killing more than 100,000 nesting seabirds. Along the eastern Pacific coast, small surges were reported in Mexico, although little or no damage was reported. Further south, Peru and Chile reported waves of up 1.5 m and 3 m respectively, and damage to several homes in both countries. In the Galapagos Islands, a 3 m tsunami wave arrived 20 hours after the earthquake, causing considerable damage to buildings and coastal infrastructure.

(a)

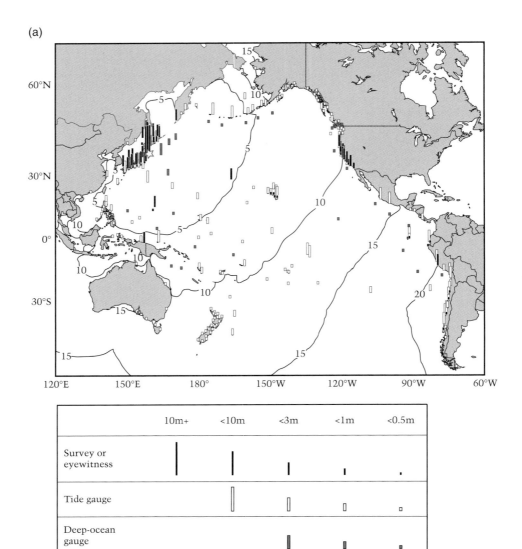

	10m+	<10m	<3m	<1m	<0.5m
Survey or eyewitness	❙	❙	❙	∎	.
Tide gauge		▯	▯	▯	▫
Deep-ocean gauge			▮	▮	∎

Tsunami travel times at 5-hour contour intervals.

(b)

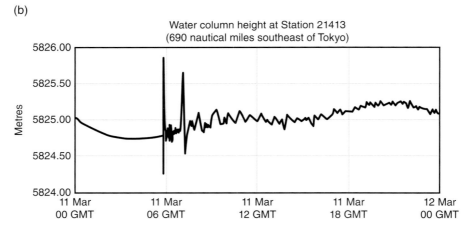

Fig. 5.12 (a) The map shows the tsunami source (indicated by a star) and water heights from tide gauges, deep-ocean gauges and eyewitness reports throughout the entire Pacific region overlaid onto a tsunami travel time base map (data current as of 18 March 2011). (b) Water column height on 11 March 2011 at DART (Deep-ocean Assessment and Reporting of Tsunamis) station 21413, which is situated 690 nautical miles southeast of Tokyo. (Source: (a) NOAA Research 2011. (b) Adapted from NOAA Research 2011b.)

at up to 800 km/hr, about as fast as a large passenger aircraft. As tsunami travel at great speeds, they can cross the expanse of large oceans rapidly. For example, the Tōhoku tsunami of 11 March 2011 generated waves that struck the Japanese coast less than 20 minutes after the initial earthquake. Waves moving in the other direction travelled across the Pacific Ocean, reaching the western USA coast a little over 12 hours after the event (see Fig. 5.12).

5.3.2 Tsunamis at the coast

In the deep ocean, tsunamis may appear only a metre or so high. If that is the case, why are they so big and destructive at the coast? The reason is the volume of water involved and the very long wavelengths. As a tsunami approaches the shoreline and enters shallower water, the front of the wave starts to slow down due to friction with the ocean bottom, and its amplitude grows. Tsunami wavelengths are several orders of magnitude longer than normal waves, and therefore the wave amplitude has the potential to become much taller (see Chapter 4). As the front of the wave slows, the back of the wave is still moving much faster. In order to accommodate this change in dynamics, the wave rapidly grows in height. This process is called shoaling and is also common to ocean waves. In many cases (usually in shallow shoreline settings), the upper part of the wave will be moving faster than the lower part, which is experiencing greater friction with the sea floor. The wave will then rise precipitously and break much like a normal ocean wave (see Chapter 4). Tsunamis that broke like regular ocean waves were observed in India and Thailand after the 2004 Indian Ocean tsunami and in northeast Japan after the Tōhoku earthquake. On coastlines where the shoreface is steep (e.g. coral atolls), the wave may not have time to be slowed by friction with the bottom, and hence will arrive as an unbroken wave or surge similar to a quickly rising tide. Such surges were observed in Sri Lanka and the Maldives as a result of the 2004 Indian Ocean tsunami. Although there are hydrodynamic differences between the velocity of breaking and surging tsunamis at the shore, wave arrival in both cases is usually followed by several minutes of inundation as the remainder of the tsunami arrives.

A tsunami is generally composed of a series of individual waves, called a wave train. As the tsunami wave trains are not of equal size, the destructive force and damage may be compounded as successive waves reach shore. Each wave in a tsunami wave train is affected by local bathymetry as it nears the coast. As a result, the danger from a tsunami is not likely to have passed with the first wave as, depending on the situation, danger from further wave arrivals can persist for hours.

5.3.3 Tsunami drawdown

Like all waves, tsunami waves have a crest and a trough (see Chapter 4). The first arrival of a tsunami is commonly, but not always, associated with a temporary lowering of sea level as coastal water is drawn seaward and the sea floor is exposed. This unusual retreat of seawater is an important warning sign of a tsunami, and the first peak will typically arrive with an enormous volume of water several minutes later. Recognizing this phenomenon and reacting to it correctly (by evacuation) can save lives. In the 2004 Indian Ocean tsunami, Tilly Smith, a young English girl (who was then only 10 years old) was credited with saving almost a hundred foreign tourists at Maikhao Beach, Phuket (Thailand), by raising the alarm minutes before the arrival of the tsunami. The young girl had learnt about tsunamis in a geography lesson two weeks prior. She recognized the receding waters and what she described as 'frothing bubbles' on the surface of the sea. Luckily, Tilly shared her knowledge with her parents, who warned the staff at the hotel, along with others on the beach. Quick action by her parents and hotel staff meant that the beach was evacuated before the tsunami arrived. This was one of the few beaches on Phuket Island that reported no casualties.

5.3.4 Tsunami warning systems

Tsunami warning systems are used to detect tsunamis and issue warnings with the aim of preventing the loss of life and property. Unfortunately, tsunami warning systems are primarily based on the initial detection of an earthquake. Hence, other tsunami-generating mechanisms (as shown in Fig. 5.13) are likely to be underestimated in terms of timing and magnitude, or not detected at all. In this section we will primarily concern ourselves with earthquake-generated tsunami, as this is how most warning systems are focused. In most tsunami warning systems, the initial advisory is based on the earthquake epicentre and magnitude and the associated potential for a tsunami.

Although there is little hope of a cost-effective way of defending a coastline against a very large tsunami originating close to the coast (e.g. Tōhoku, 2011), for those further away from the tsunami source there are tsunami warnings that allow people to seek higher ground and routes of escape. For example, the Pacific Tsunami Warning Center (PTWC), headquartered in Hawaii and Alaska, maintains a web of seismic equipment and water-level gauges to identify tsunamis at sea, and provides timely warnings of potential transoceanic tsunamis. Similar regional systems are in existence or are proposed to protect coastal areas worldwide. Most tsunami warning systems consist of two components: (1) a network of sensors that detect tsunamis; and (2) a

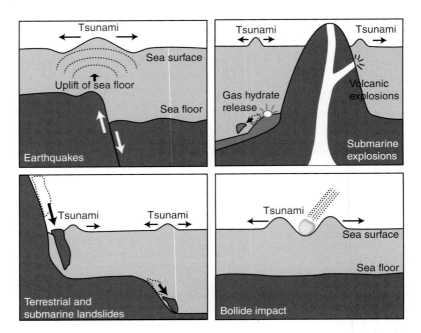

Fig. 5.13 Schematic diagram of the potential mechanisms for generating tsunamis. Although earthquakes are the dominant and best known of the tsunami generation mechanisms, it is other mechanisms such as terrestrial and submarine landslides, along with submarine volcanic explosions or flank collapses, that have produced the largest events in the historic and pre-historic (geological) records. (Source: Earth Observatory of Singapore.)

communications infrastructure to issue alerts that may permit the evacuation of areas judged to be at risk. Such systems need a complementary program of educational efforts targeted at both residents and visitors so that people are prepared and know what to do in the case of a warning.

There are two main types of tsunami warning systems: (1) international; and (2) regional. Both systems rely on the properties of seismic waves generated by earthquakes. Seismic waves move through the Earth at much greater speeds than tsunamis move across the ocean. It is therefore possible to detect an earthquake and assess the potential for a tsunami usually in less than 30 minutes. This gives time for a possible tsunami forecast to be made, and warnings (if warranted) to be issued to threatened areas. Unfortunately, until a reliable model is able to predict which earthquakes will produce significant tsunamis, this approach will produce many more false alarms than true warnings requiring significant action. In the current operational system of most tsunami warning systems, the seismic alerts are used to send out initial watches and warnings. Then, data from observed sea-level height (either shore-based tide gauges or Deep-ocean Assessment and Reporting of Tsunamis (DART) buoys) are used to verify the existence of a tsunami.

The DART system is an array of stations currently deployed in the Atlantic, Pacific and Indian Oceans (other DART buoys are planned for the South China Sea) (Fig. 5.14). These stations give information about tsunamis while they are still far offshore. Each station consists of a seabed bottom pressure recorder that detects the passage of a tsunami by measuring the pressure of the water column and relating this to the height of the sea surface. The data are then transmitted to a surface buoy via sonar. The surface buoy radios the information to a tsunami warning centre, such as PTWC, via satellite. The system has considerably improved the forecasting and warning of tsunamis throughout the world, as realized by the warnings issued to Pacific Ocean nations following the 2011 Tōhoku tsunami.

5.4 Overwash

One major aspect of storm and tsunami study is the concept of overwash. Overwash is the flow of water and sediment over a beach crest that does not directly return to the initial water body (ocean, sea, bay, or lake) from which it originated. Overwash begins when the run-up level of waves, usually coinciding with an associated storm surge, exceeds the local beach or dune crest height (Fig. 5.15). As the water level rises in response to a combination of waves and surge, the beach or dune crest is breached and the area behind is inundated. In many cases this initial inundation

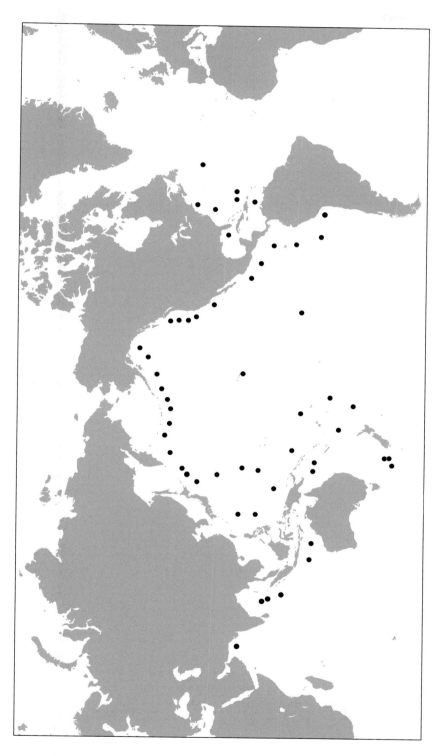

Fig. 5.14 Map showing the location of the Deep-ocean Assessment and Reporting of Tsunamis (DART) buoy network. The initial buoys were put out by the US agency National Oceanic and Atmospheric Administration (NOAA) in 2001. Every station consists of a floating buoy moored to a monitoring instrument on the ocean floor. This sensor measures temperature and pressure every 15 s, detecting tsunamis by the pressure changes they cause. The device converts pressure readings into estimates of sea-surface height, giving researchers an idea of how big the coming waves will be. These monitoring devices relay their readings to the surface buoys, which beam the information via satellite in real time. Researchers combine this information with data from tide gauges along various coasts to gain an integrated assessment of any incoming tsunami. (Source: Data from NOAA Research 2011c.)

(a) Washover plain

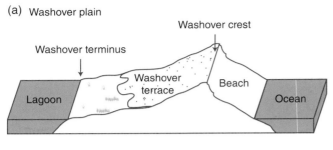

(b) Washover lobe

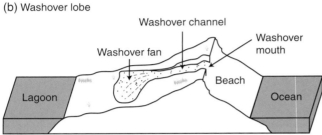

Fig. 5.15 In barrier systems, overwash by wave and surge run-up can be simply categorized in terms of the relative elevations of water level and the barrier, the frequency of overtopping waves, and the excess wave run-up. Overwash is the transport of seawater and sediment over a dune system by elevated wave and water levels primarily caused by storm surges and tsunamis. Washover refers to the morphology and sediments generated by overwash flows. The main washover types found are: (a) washover plains; and (b) washover lobes. (Source: Adapted from Matias et al. 2010. Reproduced with permission of Elsevier.)

occurs as a sheet of water and sediment (called sheetwash) runs over (overwashes) the barrier.

Although severe overwash primarily occurs in association with large storms or tropical cyclones, it is also caused by tsunami events. It is particularly common on barrier-island coasts, but can also occur on the margins of lakes and on low-profile coasts where sandy spits and gravel or shingle beaches are dominant.

Overwash is also often referred to as marine inundation, although many other terms, for example 'catastrophic saltwater inundation events' (Goff et al., 2001), are found in the literature. Although the terms are often used interchangeably, we will consider overwash as distinct from washover. Washover refers to the water and sediments deposited as a result of overwash, i.e. the sediment deposited inland of a beach crest by overwash flow.

5.4.1 Overwash impacts

In terms of coastal morphology, washover deposits contribute significantly to the evolution and sediment budget of barrier islands (see Chapter 9), and overwash is believed to be a major process in the retreat mechanism (landward

sedimentation processes, also known as transgression) of some coastal barriers in response to sea-level rise. Sediments that are transported by overwash can be deposited onto the upper beach or as far as the back-barrier bay, estuary or lagoon. Overwash can also enter small estuaries and back-barrier channels that typically run normal (perpendicular) and parallel to the coast behind the protection of barrier islands or narrow low-lying barriers. The inundation, landward sediment transport, and erosion as a result of overwash, can affect coastal management, primarily through the loss or damage of property and infrastructure such as roads, car parks, amenities, recreational areas and navigation channels. Secondary problems include loss of property function (e.g. hotels) and the loss of the natural protection afforded by barriers or dunes.

The catastrophic nature of overwash, and its frequent occurrence in some areas of the world, indicate that where these processes occur they must be accounted for in the long-term management of coastal communities. Although overwash events can be devastating, not all are disastrous. Smaller events occur regularly on low-lying sand spits and barrier islands where the dunes are low or absent. For many coasts, the likelihood of overwash events is also tide-dependent; if the storm surge coincides with high tide, the water level and, hence, potential for overwash is greater. A deflated (eroded) post-storm coastal system that has not had time to return to equilibrium may also have increased vulnerability should an overwash event occur.

In barrier systems, overwash by wave and surge run-up can be simply categorized in terms of the relative elevations of water level and the barrier, the frequency of overwashing waves, and the excess wave run-up, ΔR. The quantity ΔR is defined as:

$$\Delta R = R + S - dc \qquad (5.2)$$

where R is the wave run-up height, S is the storm surge height, and dc is the elevation to the dune crest from the mean water level (or from some other common datum).

One key factor in determining the response of a dune or barrier to overwash is the degree of freeboard available. Freeboard refers to the difference between the elevation of the barrier crest and the run-up limit of waves (see Orford et al., 2003; and Fig. 1.16). This difference can be directly related to the occurrence of overwash events. In the case of positive freeboard (i.e. where the dune crest is higher than the run-up limit), only the seaward face of the barrier is affected by wave action and the barrier or dune remains proud. In the case of zero freeboard, the run-up marginally overtops the crest of the barrier or dune, commonly resulting in an increase in crest elevation. Overwash occurs in the case of negative freeboard, which is the most destructive of the freeboard states. Overwash can cause a lowering of the barrier crest, which if prolonged may lead to the crest becoming lower than mean sea level, or in more destructive phases, overwash can even result in a breaching of the barrier. Where the barrier crest becomes lower than the mean sea level, this will result in permanent flooding of the coastal zone.

5.4.2 *Washover deposits*

Overwash by storms and tsunamis of various magnitudes results in morphologically different deposits. Figure 5.15 shows a schematic plan-view over a typical dune line being subjected to overwash with the common washover deposit types. In cases where overwash waves are small and infrequent, small-scale rearrangement of the surface sediments occurs. If overwash events are more frequent and/or larger bores result, the resulting deposits usually vary according to the local topography. Where dunes are relatively high, but uneven, overwash usually exploits existing gaps or lower areas in the foredune line, funnelling through the breach (gaps) and spreading out behind the dunes. Washover plains are wide, low-lying, denuded areas including a crest, a terrace and a terminus. Washover lobes are conspicuous features that cut the dune field in specific places and consist of a mouth, a channel and a fan. Unfortunately, there is considerable inconsistency in the naming of washover morphologies (Matias et al., 2010). For example, washover plains have been named washover ramps, sheets, flats and terraces. Washover lobes have also received different names, for example, hurricane channels, overwash features, dune terrace morphologies, washover fans and perched fans. One problem with these naming inconsistencies is that coastal planners and managers are left confused by the terminology applied by coastal geomorphologists.

In the final stages of overwash, the velocity of the bore usually decreases and the suspended sediment is deposited as a washover plain or lobe. Where the beach crest is low and uniform, or in locations where numerous overwash sites occur in one beach system, the individual washover fans may coalesce to form a washover terrace or apron. If overwash extends into a back-barrier lagoon, the deposit can appear as a subaqueous washover delta or layer of sand in the lagoonal strata (Leatherman, 1976).

In tsunami events or very large storm surges, the entire longshore segment of a beach can be subject to large-scale inundation, at which time a continuous flow of water streams over the dune crest. This can be referred to as sluicing overwash (Orford et al., 2003) and it is common where coastal dunes are low (e.g. barrier spits). In the event of sluicing overwash, sediment is commonly carried into the back of the dunes and deposited on the coastal plain or in back-barrier lagoons. In extreme cases (e.g. large tsunamis or huge storm surges), sluicing overwash can cause severe erosion of the back-barrier environments and can

precipitate considerable modification of the landscape, including the rapid and complete removal of the dune.

5.5 Palaeostudies of coastal hazards

While tsunami and cyclone warning systems allow us to monitor and predict the immediate impact of modern events, it is useful to study the history of a coastal site, or the prehistoric record preserved in the geological record. Analysis of recurrence intervals of past events allows us to provide better long-term predictions. Whilst the occurrence of tropical storms and cyclones is seasonal, and hence reasonably predictable, information on the biggest storm (cyclone) to ever have affected a coast is still valuable. On many coasts, similar research can and should be applied to past tsunamis. At almost all locations on the planet, the only way to adequately answer these questions is to turn to the geological record, although historical documents can also provide supplementary information.

5.5.1 *Reviewing historical documents*

There is an immediate and obvious need to assess the risk of catastrophic inundation events along many of the world's coasts. The first step in this exercise is usually to look at historical records, some of which have already been successfully used to reconstruct storm and tsunami characteristics. For example, Atwater et al. (2005) modelled tsunami heights that were inferred from written records in Japan and suggested that a ~ M9 megathrust earthquake off northwest North America occurred on 26 January 1700.

On some coasts, the use of historical records to provide information about past storm and tsunami occurrence is a relatively simple exercise as the historical record is not very long. On other coasts, such as in China, Japan or coasts bordering the Mediterranean Sea, the historical records are much longer and more detailed. The material is assessed and cross-checked to remove errors and misinterpretations, to refine early work and ensure a more robust dataset. Unfortunately, as researchers step back further in time, those studying the historical record are often frustrated by inconsistencies in descriptions, inaccuracies in translation between different languages, calendars and location names (see discussion in Lau et al., 2010).

5.5.2 *Using the geological record*

On many coasts, the lack of detailed historical records along with the low frequency of large overwash events make it difficult to adequately assess the recurrence intervals of storms and tsunamis. The Indian Ocean (26 December 2004) and Tōhoku (11 March 2011) tsunamis poignantly showed that catastrophic overwash events

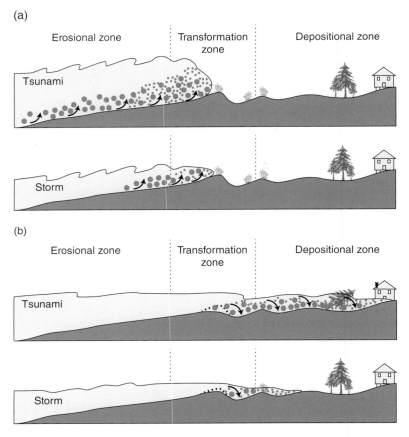

Fig. 5.16   Comparison of tsunami and storm processes. Tsunamis generally erode sediment from further offshore than do storms (a), and careful analysis of faunal and mineral content in overwash deposits (b) may allow the differentiation between storm and tsunami deposition (see Switzer et al., 2005; Switzer and Jones, 2008a, 2008b; Phantuwongraj and Choowong, 2012). (Source: Adapted from Phantuwongraj and Choowong 2012.)

are too infrequent for their hazard to be characterized by historical records alone. The geological and geomorphological record provides opportunities to assess storm and tsunami hazards more fully. The sand and boulder deposits left by storm and tsunami overwash can help to assess a variety of environmental parameters about a particular storm or tsunami event. Such deposits can yield information that allow the investigation of water depths and velocities of past inundations, the estimation of source locations (of tsunamigenic events), and the mapping of the likely inundation distance of future events, to name but a few.

Dating deposits using techniques such as radiocarbon dating and optically stimulated luminescence (OSL) can help develop detailed chronological histories that allow the analysis of the recurrence intervals of past events. Such information is vital to coastal planning and can be used to guide mitigation efforts that may reduce losses from future overwash events. Research into the geological record of palaeo-washover deposits has led to the

discovery of records spanning many thousands of years in numerous locations around the world. One notable example is the study of Nanayama et al. (2003) on the coast of Hokkaido, Japan. Here, tsunami-deposited sand sheets (some extending several kilometres inland) show that large tsunamis inundated the coast on average every 500 years, between 2000 and 7000 years ago. Such records provide clear evidence of large-scale palaeo-tsunamis in the region, and information from these palaeo-events could have been used to help guide management and planning procedures for the unfortunate events that were to occur on 11 March 2011.

To correctly interpret the local history of storms and tsunamis, earth scientists must be able to distinguish between the type of washover deposit (tsunami or storm) found in the geological record. Comparative studies of both types of deposit reveal some differences in sedimentology, stratigraphy, faunal composition, and inland height and extent. Although several researchers have proposed criteria to distinguish between these two types of deposits, it is apparent that each deposit must be carefully considered in the context of its regional setting.

There are many physical, sedimentological and geochemical techniques that can be used to identify washover deposits, but there is no single analytical technique that can unambiguously identify palaeo-tsunami from palaeo-storm deposits. In all palaeo-washover studies, local geomorphic and stratigraphic field criteria must be applied first. What can be compiled is a list of commonalities, where combining field observations and the analysis of geomorphic, sedimentological, geochemical, and palaeontological signatures may enable positive identification.

Generally, overwash events that deposit sand cause a decrease in the total organic matter within coastal plain sediments. Additionally, salinity changes caused by short-lived marine inundation events can cause notable changes in the assemblages of ostracods, diatoms, foraminifera, pollen and aquatic plants. The internal sedimentology (e.g. bedding), the regional physiography (e.g. height above sea level) and the three-dimensional distribution (e.g. landward extent, changes in thickness and regional continuity), can also assist in determining the origin of washover deposits. The existence of marine fossils or exotic (deep-water) heavy mineral assemblages within a terrestrial environment may indicate transport via a tsunami as opposed to via a storm (Fig. 5.16).

5.5.3 Overwash studies: successes and limitations

To date, the majority of work on palaeo-washover deposits has been conducted in sub-tropical and temperate latitudes where there are extensive salt marshes and coastal lagoons. The Indian Ocean tsunami of 2004 and Cyclone Nargis in 2008 indicated a real need for research on palaeo-washover in tropical environments. Unfortunately, such environments come with a myriad of challenges, including poor deposit preservation, intensive bioturbation and seasonal flooding. Many events also occur in countries that have socio-economic or political issues, and/or limited or difficult access, or few resources for such studies (e.g. Bangladesh, Myanmar, Indonesia). Despite these challenges, long-term washover records, when found, can provide an opportunity to evaluate the recurrence of storms and tsunamis that are large enough to leave lasting sedimentary signatures.

As dating is fundamental to the determination of recurrence intervals, it is important that coastal stratigraphers understand the limitations of the geological and analytical uncertainties in estimating event ages. In some cases, these uncertainties may be as long as the recurrence intervals, but new analytical approaches, such as the stratigraphic ordering of calibrated radiocarbon age distributions and the summing of probability density

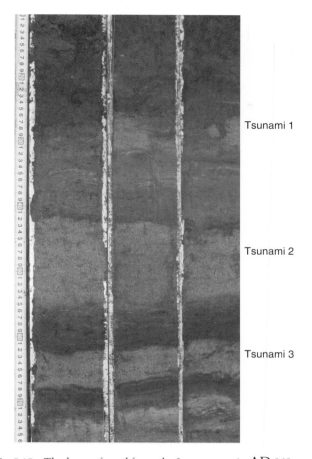

Fig. 5.17 The layer of sand from the Jogan event in AD 869 (Tsunami 1) is overlain by volcanic tephra dated to an event in AD 915. The two sand layers below probably represent previous tsunami events over several thousand years that are vertically confined by much finer coastal-plain sediments. Work such as that of Minoura et al. (2001) pioneered the use of deposits as indicators of storms and tsunamis. (Source: Photo © J. Goff.)

functions of dates, have helped to narrow uncertainties of event timing.

5.6 Integrating hazard studies with coastal planning

While palaeo-washover deposits provide a mechanism for extending the historical record of storms and tsunamis, and for studying overwash events for which there are no written or oral records, transferring this information to well-designed coastal planning is often problematic. For example, it is tragically apparent that although geological evidence for previous large tsunamis near Tōhoku had been identified in the past (Fig. 5.17), this science did not have a sufficient impact on coastal management of what

(a)   (b)

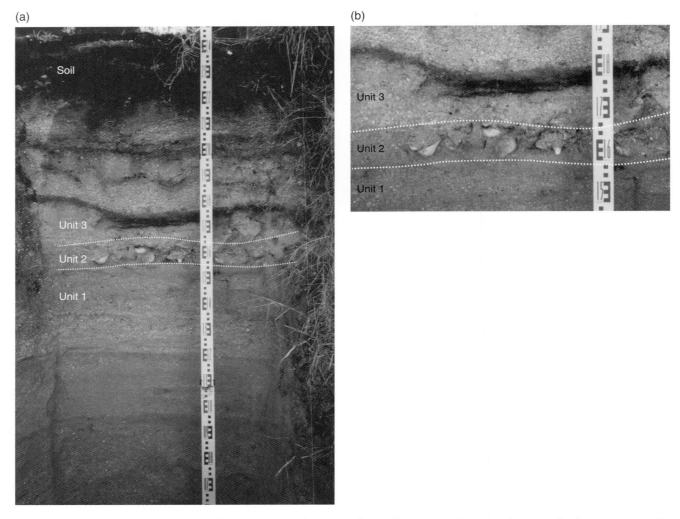

Fig. 5.18 Excavated face (a) showing a high-energy marine deposit on the southeast Australian coast (Unit 2). The deposit is vertically confined by an underlying sandy beach deposit (Unit 1) and an overlying soil (Unit 3) (b). The presence of marine fauna, articulated bivalves and marine sand indicate a marine origin for the deposit. Although the deposit is found 2 km inland and up to 2.4 m above sea level, a lack of definitive evidence for a tsunami or storm has limited the usefulness of the implications. (Source: A. Switzer.)

was once a beautiful city (see Fig. 5.3). Similarly, if the coastal washover stratigraphy of various locations around the northeast Indian Ocean (Thailand, Sri Lanka, India) had been carefully studied before 26 December 2004, signs of giant prehistoric tsunamis similar to the 2004 event could have been discovered (e.g. Jankaew et al., 2008). In many instances, the geological and geomorphological records of large overwash events are deemed interesting scientifically, but fail to transfer to meaningful action by local and regional governments. It is only when this information is used as a basis for signage, evacuation maps and emergency planning, that lives may be saved in future events of a similar magnitude.

In addition to providing information on the occurrence and frequency of coastal hazards in the stratigraphic record, detailed analysis of the facies, thickness, grain-size changes, faunal content and chronological framework within the deposits has the potential to provide information on the dynamics and recurrence interval of inundation – a valuable tool for hazard mapping and evacuation planning. The investigation of washover deposits is not necessarily straightforward, and there is no 'recipe' for distinguishing between storms and tsunamis in the geological record. The site-by-site approach to these studies, and the fact that such work requires more interpretation than does the

Driving research aim:
Safer coasts through increased understanding

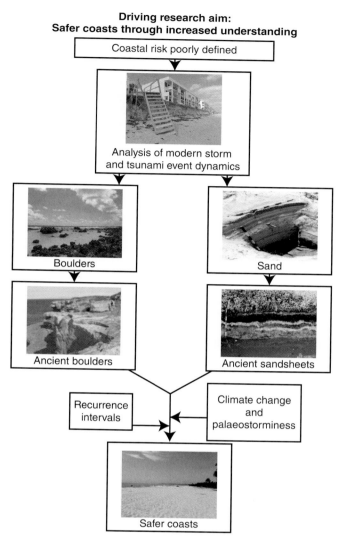

Fig. 5.19 Graphic representation of the systematic approach using modern and palaeo-washovers to investigate recurrence intervals of catastrophic events. (Source: Photos: Analysis, Ty Harrington [Public domain], via Wikimedia Commons. Sand, C. Gouramanis. Boulders, Y.S. Lee. Ancient sandsheets, Kruawun Jankaew. Ancient boulders, A. Switzer. Safer coasts, R. Rush. Diagram: Earth Observatory of Singapore.)

importance to coastlines that have short historical records and is useful for future coastal and hazard mapping and planning (Fig. 5.19).

5.7 Cyclones in a warmer world

The last decade has seen considerable debate and conjecture on whether the characteristics of tropical cyclones have changed, or will change, in a warming climate — and if so, how (see also Chapter 4). Many studies offer conflicting opinions. Knutson et al. (2010) suggested that this inconsistency is most likely due to amplitude fluctuations in the frequency and intensity of tropical cyclones that complicate the detection of statistically significant trends in multi-decadal records. This variability adds considerable complexity to the detection of long-term trends and their potential attribution of intensity changes to rising levels of atmospheric greenhouse gases. This complexity, combined with a lack of reliable historical records of tropical cyclones, means it remains uncertain whether past changes in tropical cyclone activity have exceeded the variability expected from natural causes. Despite the inherent uncertainties, current future projections based on theory and high-resolution models of climate dynamics consistently indicate that greenhouse warming will cause the globally averaged intensity of tropical cyclones to shift towards stronger storms. Interestingly, Knutson et al. (2010) also noted that existing modelling studies consistently project decreases in the globally averaged frequency of tropical cyclones.

Recently, important progress has been made in higher-resolution modelling that provides improved simulations of global storm frequencies and further support for theoretical expectations for a globally averaged increase in tropical cyclone intensity and rainfall. There have also been considerable increases in the quality and consistency of tools used to study the dynamics and statistics of tropical cyclone activity. Such advances have increased our confidence that tropical cyclone frequency is likely to either remain essentially the same or decrease. They also noted that although no increase in total storm count is projected, a future increase in the globally averaged frequency of the strongest tropical cyclones is more likely than not. Knutson et al. (2010) also stated that despite some suggestive observational studies, they could not at the time conclusively identify anthropogenic signals in past tropical cyclone data. They then qualified this with the statement that substantial human influence on future tropical cyclone activity cannot be ruled out, and pointed out that such change could arise from several mechanisms (including oceanic warming, sea-level rise and circulation changes). Unfortunately, in the absence of a detectable change, society is dependent on a combination of observational, theoretical and modelling studies to assess the

historical record, often means the true 'hazard' message may be lost in the scientific arguments of whether a deposit is a tsunami deposit or a storm deposit (Fig. 5.18). If one was to step away from the scientific arguments of storm versus tsunami and look at the basic premise of coastal hazard and planning assessment, then a deposit's genesis becomes less relevant (Switzer et al., 2011). The pure fact that a 'marine' deposit is found in an overwash setting yields basic information on the recurrence of 'large marine overwash events'. That in itself is of key likely changes in tropical cyclone activity given the likelihood of future changes in climate.

5.8   Summary

Coastal hazards such as extratropical storms, tropical cyclones and tsunamis can be major forces in shaping beach dynamics and geomorphology. Billions of people live in and rely on the coastal zone, and coastal development continues at unprecedented rates, despite numerous major coastal disasters in recent years and their likely increase in frequency and intensity under a changing climate. Historical and geological studies of past marine inundation events can provide information on recurrence intervals, intensity, vulnerability and risk, which can and should be integrated into coastal management and risk plans to reduce damage and mortality.

In particular, this chapter focusses on:
• The risks to society from coastal hazards.
• The causes, properties and structure of storms and tsunamis.
• What happens when storms and tsunamis hit the coast.
• How risks can be reduced, managed or included in plans.
• Using records of past occurrences to provide predictions for future impacts.
• How this information can be used to reduce risks to coastal communities.

Key publications

Daniels, R.J. Kettl, D.F. and Kunreuther. H. (Eds.), 2006. On Risk and Disaster: Lessons from Hurricane Katrina. University of Pennsylvania Press, Philadelphia, PA, USA.
[*This book has 20 papers that deal with Hurricane Katrina in 2005. The diverse spread of papers examined provide lessons and commentaries on assessing, perceiving and managing risks from future disasters*]
Donnelly, C., Kraus, N. and Larson, M., 2006. State of knowledge on measurement and modeling of coastal overwash. *Journal of Coastal Research*, **22**, 965–991.
[*An excellent summary of the state of play in overwash studies. Essentially a PhD student's literature review with a focus on the engineering aspects of overwash*]
Knutson, T.R., McBride, J.L., Chan, J. et al., 2010. Tropical cyclones and climate change. *Nature Geoscience*, **33**, 157–163.
[*Summary paper from an expert team of cyclone-climate researchers convened by the World Meteorological Organization (WMO) to assess the causes of past changes in cyclone activity. The team concluded that it remains uncertain if past changes in any tropical cyclone activity (frequency, intensity, rainfall, etc.) exceed the natural variability*]
Kundu, A. (Ed.), 2007. Tsunami and Nonlinear Waves. Springer, Berlin, Germany.

[*This book provides a detailed theoretical and mathematical analysis of the fundamentals of tsunamis, with a focus on the generation and dynamics of tsunamis. Specific results from the 2004 Indian Ocean tsunami are used to highlight the nature of tsunami waves and their links to non-linear phenomena*]
Terry, J.P., 2007. Tropical Cyclones: Climatology and Impacts in the South Pacific. Springer, New York, USA.
[*This book describes the behaviour, dynamics and characteristics of cyclones and their landscape impacts in the South Pacific, and investigates the broad range of disturbance effects cyclones have on the physical environments of the Pacific Islands*]
Tsunemasa, S., Yoshinobu, T., Minoura, K. and Yamazaki, T. (Eds.), 2010. Tsunamiites – Features and Implications. Elsevier, Amsterdam, the Netherlands.
[*This book is an overview of recent developments in the sedimentology of tsunami deposits (also called tsunamiites). It also covers some problems that need additional investigation (e.g. boulders), as well as indicating some direction of future tsunami deposit research*]

References

Allan J.C. and Komar P.D., 2006. Climate controls on US west coast erosion processes. *Journal of Coastal Research*, **22**, 511–529.
Atwater, B.F., Musumi-Rokkaku, S., Satake, K., Tsuji, Y., Ueda, K. and Yamaguchi, D., 2005. The Orphan Tsunami of 1700: Japanese clues to a parent earthquake in North America. *USGS Professional Paper*, **1707**, 3–123.
Centrec Consulting Group, 2007. *An Investigation of the Economic and Social Value of Selected NOAA Data and Products for Geostationary Operational Environmental Satellites (GOES).* A report to NOAA's National Climatic Data Center. Centrec Consulting Group, Savoy, IL, USA. Available at: http://www.centrec.com/resources/reports/GOES%20Economic%20Value%20Report.pdf
Chambers, J.Q., Fisher, J.I., Zeng, H., Chapman, E.L., Baker, D.B. and Hurtt, G.C., 2007. Hurricane Katrina's carbon footprint on US Gulf Coast forests. *Science*, **318**, 1107.
Goff, J., Chagué-Goff, C. and Nichol, S., 2001. Palaeotsunami deposits: a New Zealand perspective. *Sedimentary Geology*, **143**, 1–6.
Jankaew, K., Atwater, B.F., Sawai, Y. et al., 2008. Medieval forewarning of the 2004 Indian Ocean tsunami in Thailand. *Nature*, **455**, 1228–1231.
Knutson, T.R., McBride, J.L., Chan, J. et al., 2010. Tropical cyclones and climate change. *Nature Geoscience*, **3**, 157–163.
Lau, A.Y.A., Switzer, A.D., Dominey-Howes, D., Aitchison, J.C. and Zong, Y., 2010. Written records of historical tsunamis in the northeastern South China Sea: challenges associated with developing a new integrated database. *Natural Hazards and Earth System Sciences*, **10**(9), 1793–1806.
Leatherman, S.P., 1976. Quantification of overwash processes. PhD thesis, University of Virginia, Charlottesville, VA, USA.
Matias, A., Ferreira, O., Vila-Concejo, A., Morris, B. and Alveirinho Dias, J., 2010. Short-term morphodynamics of non-storm overwash. *Marine Geology*, **274**, 69–84.

Minoura K., Imamura, F., Sugawara, D., Kono, Y. and Iwashita, T., 2001. The 869 Jogan tsunami deposit and recurrence interval of large-scale tsunami on the Pacific coast of northeast Japan. *Journal of Natural Disaster Science*, **23**, 83–88.

Murty, T.S., 1977. *Seismic Sea Waves: Tsunamis*. Bulletin of the Fisheries Resources Board of Canada, No. 198. Department of Fisheries and the Environment, Fisheries and Marine Service, Ottawa, Canada

Nanayama, F., Satake, K., Furukawa, R. et al., 2003. Unusually large earthquakes inferred from tsunami deposits along the Kuril trench. *Nature*, **424**, 660–663.

Orford, J.D., Jennings, S.C. and Pethick, J., 2003. Extreme storm effect on gravel-dominated barriers. Coastal Sediments '03. In: R.A. Davis (Ed.), *Proceedings of the International Conference on Coastal Sediments 2003*. CD-ROM published by World Scientific Publishing Corporation and East Meets West Productions, Corpus Christi, TX, USA.

Phantuwongraj, S. and Choowong, M., 2012. Tsunamis versus storm deposits from Thailand. *Natural Hazards*, **63**, 31–50.

Short, A.D. and Trembanis, A., 2000. Beach oscillation, rotation and the Southern Oscillation, Narrabeen Beach, Australia. *Proceedings of the 27th International Conference on Coastal Engineering*, **1** , 2439–2452.

Storlazzi C.D. and Griggs G.B., 2000. Influence of El Niño–Southern Oscillation (ENSO) events on the evolution of central California's shoreline. *Geological Society of America Bulletin*, **112**, 236–249.

Switzer, A.D. and Jones, B.G., 2008a. Large-scale washover sedimentation in a freshwater lagoon from the southeast Australian coast: sea-level change, tsunami or exceptionally large storm? *The Holocene*, **18**, 787–803.

Switzer, A.D. and Jones, B.G., 2008b. Setup, deposition, and sedimentary characteristics of two storm overwash deposits, Abrahams Bosom Beach, Southeastern Australia. *Journal of Coastal Research*, **24**, 189–200.

Switzer, A.D., Pucillo, K., Haredy, R.A., Jones, B.G. and Bryant, E.A., 2005. Sea level, storm, or tsunami: enigmatic sand sheet deposits in a sheltered coastal embayment from southeastern New South Wales, Australia. *Journal of Coastal Research*, **21**, 655–663.

Switzer, A.D., Mamo, B.L., Dominey-Howes, D. et al., 2011. On the possible origins of an unusual (mid-late Holocene) coastal deposit, Old Punt Bay, southeast Australia. *Geographical Research*, **49**, 408–430.

6 Coastal Groundwater

WILLIAM P. ANDERSON, JR.

Department of Geology, Appalachian State University, Boone, NC, USA

6.1 Introduction

Coastal groundwater is an important component of the global hydrological cycle. In general, it is water that begins as recharge to land-based aquifers and ends as discharge to marine waters. This water can take a variety of paths that result in residence times of days to weeks for wave- and tide-induced circulation to thousands of years for deeper circulations, such as those on the continental shelf. Although river discharge to the oceans has long been considered to be the dominant contributor of water and nutrients to the nearshore and continental-shelf environments, studies have demonstrated that the coastal groundwater zone may also serve as a significant contributor of water and nutrients (Moore, 1996). In some locations, groundwater flux to the oceans, known as 'submarine groundwater discharge' (SGD), may exceed riverine discharge (Moore, 2010).

The terrestrially derived groundwater of the coastal zone flows seaward under natural conditions because of its high elevation relative to the oceans. In locations affected by human activities, such as pumping for water supply, this natural circulation may be modified and flow may be landward. The shallow coastal groundwater originates as precipitation-derived recharge to shallow aquifers, and it circulates at relatively short temporal scales. Because of its origin on land, this water is a pathway for natural and human-derived contaminants to the nearshore environment. As this shallow groundwater discharges in the nearshore environment, it interacts with the ocean. This region is not a static zone. Rather, it is a dynamic zone that is affected by natural temporal variations common to all coastlines. On short timescales, waves, tides (Li et al., 1999) and extreme events such as hurricanes (Anderson and Lauer, 2008) modify groundwater pathways, greatly complicating conditions in the subterranean estuary, and creating a mixing zone between groundwater and recirculating seawater. On longer timescales, this zone may be affected by seasonal, interannual and interdecadal variations in flow rates (Michael et al., 2005; Anderson and Emanuel, 2010), or at millennial timescales by rising sea levels.

6.2   The subterranean estuary

Research into the interaction of fresh and saline groundwater in coastal zones began in the 19th century, with three separate studies that promoted the idea of a sharp interface between saline and fresh groundwater. The first conceptual model (Dupuit, 1863) acknowledged that fresh groundwater discharge exists in dynamic equilibrium with saline groundwater because it receives fresh recharge at the surface that raises water levels and promotes seaward flow. This model required several simplifications, including: (1) horizontal flow toward the sharp interface; (2) static marine groundwater; and (3) fresh groundwater discharge at the shoreline. Later studies based on observations of coastal aquifers in Europe (Badon-Ghyben, 1889; Herzberg, 1901) also assumed that marine-based groundwater is static and occupies a lower position due to its greater density (ocean water is 2.5% denser than freshwater). Based on the density difference, the Ghyben-Herzberg principle suggests that a sharp interface between fresh and saline groundwater exists at depth and can be approximated using:

$$z = \left(\frac{\rho_f}{\rho_s - \rho_f} \right) h = Gh \qquad (6.1)$$

where ρ_f is the density of freshwater (~1000 kg/m³), ρ_s is the density of seawater (~1025 kg/m³), G is the Ghyben-Herzberg ratio, h is the height of the water table above

(a)

(b)

Fig. 6.2   (a) Hatteras Island, USA. This beach comprises relatively uniform, medium-grained sand. There is relatively little topography on the island and conditions are microtidal. (b) Hallsands, Devon, UK. This beach comprises coarse, rounded gravel with slate shingles. The beach has relatively steep topography and conditions are macrotidal. (Source: (a) Cynthia Liutkus-Pierce. (b) Bill Anderson.)

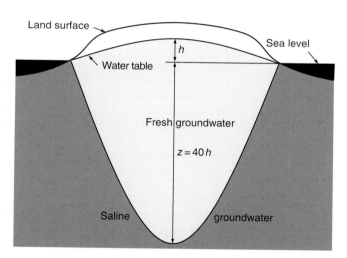

Fig. 6.1   Early conceptualization of coastal groundwater assumed static marine groundwater and a sharp interface between fresh and saline groundwater. Note that this conceptualization requires all discharge to occur at the shoreline. The Ghyben-Herzberg principle demonstrates that the depth to the sharp interface must be 40 times the height of the water table above sea level.

sea level, and z is the depth below sea level to the sharp interface. Using these densities and eqn. 6.1, the theoretical depth to the interface should be 40 times the height of the water table above sea level (Fig. 6.1). As an example, consider Hatteras Island on the Outer Banks barrier-island chain of North Carolina, USA (Fig. 6.2a). The mean maximum water-table elevation in this aquifer h is approximately 3.0 m. Using the Ghyben-Herzberg ratio, the theoretical depth to the sharp interface z is

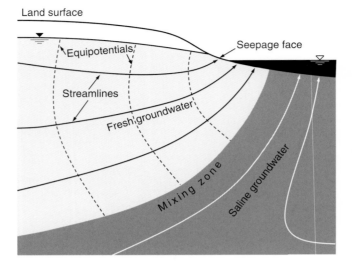

Fig. 6.3 Cooper (1959) produced a more complex conceptualization of the saline water/freshwater interface than the sharp interface model. Because coastal groundwater is dynamic, he noted that a mixing zone must exist between fresh and saline groundwater. (Source: Adapted from Cooper 1959. Reproduced with permission of John Wiley & Sons.)

Fig. 6.4 Example of a seepage face at low tide in Maceley Cove, Devon, UK. The seepage face at this site ranged from 10 to 20 m in width and decreased in size with the rising tide. In addition, the braided channels formed at the beginning of low tide and incised over the course of the ebbing tide. (Source: Photograph by Bill Anderson.)

approximately 120 m. Given that the aquifer thickness is only 24 m, however, there are obviously some problems with this simple conceptual model.

Hubbert (1940) questioned the validity of the Ghyben-Herzberg principle and suggested that fresh groundwater must discharge into a seepage zone within the saline water body. Two studies in the late 1950s used Hubbert's idea to bring more realism to the view of the mixing zone between fresh and saline water. Cooper (1959) noted that marine groundwater is not static but instead is dynamic. Taking a cue from Hubbert (1940), he described a 'zone of diffusion' between the freshwater discharging to a seepage face and recirculating seawater below the sea floor (Fig. 6.3). Seepage faces are often visible along shorelines, especially at low tide (Fig. 6.4). The offshore extent of the seepage face can be calculated with knowledge of aquifer permeability and freshwater discharge rates (Glover, 1959) using:

$$x_0 = -\frac{GQ}{2K} \tag{6.2}$$

where G is the Ghyben-Herzberg ratio, Q is the rate of discharge per unit length of shoreline, and K is the hydraulic conductivity. The variable Q can be considered as the rate at which topographically driven groundwater flow is exiting the aquifer. Also, the negative sign is included to denote that the seepage zone extends offshore a distance

of x_0. It should be noted that this equation assumes that there is a lack of tidal oscillations, which increase the complexity of the seepage face. This equation is most applicable to microtidal conditions. With macrotidal conditions, as shown in the seepage face picture in Fig. 6.4, the simple static seepage face is complicated by a migrating shoreline; thus, groundwater seepage will not only include land-derived freshwater, but also recirculated seawater.

Consider the Hatteras Island aquifer once again. Pumping tests indicate that the hydraulic conductivity of the aquifer is 21.5 m/day and it has a thickness of 24 m. The maximum water-table elevation of 3 m drops to sea level at a distance of 1500 m, giving a gradient of 0.002. By applying Darcy's law (Box 6.1), we can estimate the discharge per unit length to be about 1.0 m³/day/m of shoreline. According to eqn. 6.2, then, the seepage face will extend approximately 0.93 m offshore. Because Hatteras Island has microtidal conditions, this value may be a good approximation. Consider also the Hallsands barrier in Devon, UK (Fig. 6.2b). The gradient in this aquifer is much larger: 4.5 m of elevation is lost by the groundwater within 100 m of lateral distance, giving a gradient of 0.045. The gravel barrier comprising the barrier is also much more permeable, with an estimated hydraulic conductivity of 100 m/day over a thickness of 2 m. Once again applying Darcy's law, the discharge per unit length of shoreline is 9 m³/day/m and the calculated offshore extent of the seepage face is 1.8 m. Given that conditions at Hallsands are

METHODS BOX 6.1 Darcy's law

Henry Darcy (1803–1858) was a French civil engineer who is famous for designing and constructing a water supply for the city of Dijon, France, in the mid-1800s. Late in his life, he wrote a report, *Les Fontaines Publiques de la Ville de Dijon* (1856), detailing the construction of that water supply (Deming, 2002). He also devoted much of his time to hydraulic experimentation, which he described in the appendix of his report. These efforts led to the discovery of what we now call Darcy's law in his honour. This law describes the flow of water through porous media. It is a constitutive law analogous to Ohm's law for the conduction of electrical current and Fourier's law for the conduction of heat.

Figure 6.5 shows a sketch of Darcy's experimental setup. The variables h_1 and h_2 represent hydraulic head, which is the total energy at the upper and lower ends of the column. Taking the difference between these values and dividing by the distance L between the points gives the hydraulic gradient. The variable Q represents the volume of discharge leaving the column.

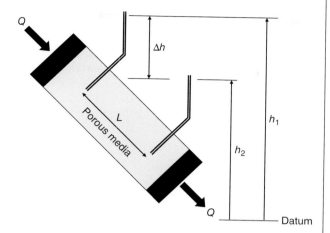

Fig. 6.5 Henry Darcy performed a series of column experiments with a variety of porous media. This diagram shows a conceptualization of his experiment. The column was filled with a material, and then volumetric flow through the column Q was measured at different hydraulic gradients $\Delta h/L$.

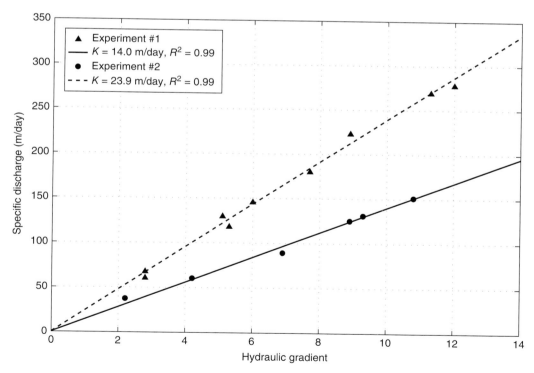

Fig. 6.6 Experimental data are shown for two of the porous media tested by Darcy. The linear relationship between hydraulic gradient and specific discharge led to the discovery of the constitutive property of the flow of water through porous media that has been named Darcy's law in his honour.

Darcy conducted a series of column experiments in which he used various sediment types, or porous media, and applied a variety of hydraulic gradients to each sediment sample. The data, shown in Fig. 6.6 for two types of porous media, suggest that the discharge is directly proportional to the hydraulic gradient and the cross-sectional area of the column, or:

$$Q \propto A \frac{\Delta h}{L} \qquad (6.3)$$

If we replace the proportionality with a constant K we get the basic form of Darcy's law:

$$Q = -KA \frac{\Delta h}{L} \qquad (6.4)$$

In this equation, K is the hydraulic conductivity, a parameter that depends on both the porous medium and fluid properties of the system. Other forms of the equation include the specific discharge q:

$$q = \frac{Q}{A} = -K \frac{\Delta h}{L} \qquad (6.5)$$

and the average linear velocity v_x:

$$v_x = \frac{q}{n} = -\frac{K}{n} \frac{\Delta h}{L} \qquad (6.6)$$

where n is the porosity of the porous medium. Darcy's law is an essential tool for coastal hydrogeologists because this law, when combined with conservation of mass, allows for the development of the groundwater flow equation and the study of flow distribution within coastal aquifers.

---

macrotidal, the calculated width is likely not to be a good estimate.

It was not until the 1990s that the complexity of the interface between fresh and saline groundwater in coastal zones became more apparent. In a classic paper, Moore (1996) describes enriched ^{226}Ra in the coastal waters of the South Atlantic Bight of North America. The important component of this paper is that: (1) this enrichment must come from onshore groundwater; and (2) it extends tens of kilometres offshore. Thus, the simple nearshore focus of early coastal zone groundwater studies was missing the larger picture. Moore (1999) goes on to describe further this mixing zone and coins the term 'subterranean estuary', which he defines as a coastal aquifer where groundwater derived from land drainage measurably dilutes seawater that has invaded the aquifer through a free connection to the sea. In other words, the function of the coastal groundwater zone is akin to the processes that occur in a surface-water estuary. In the subterranean estuary, chemical reactions take place between groundwater and the aquifer materials, and these reactions are modified and enhanced by geothermal heating of recirculating groundwater, topographically driven flow from the land (which may vary at seasonal and longer timescales), tidal oscillations, wave run-up, storm events and millennial-scale variations in sea level.

The subterranean estuary is much more complex than suggested by the mixing zone of Cooper (1959). In fact, there are four distinct components to this estuary that vary in magnitude, depending on coastal properties such as wave climate, tidal oscillations, aquifer heterogeneity and permeability, and land-derived groundwater flow rates (Fig. 6.7). Topographically driven groundwater flow from

land balances the density-driven circulation of saline groundwater from the ocean, as in the conceptual model of the traditional mixing zone; however, two other components also exist. Wave run-up and tidal oscillations induce the formation of an upper saline plume in the intertidal zone (Robinson et al., 2007a). This plume of saline water lies typically at the level of the mean high tide. A freshwater discharge tube separates this upper plume from the saline groundwater beneath the ocean. It is through this freshwater discharge tube that the topographically driven groundwater flow exits the aquifer at the seepage face. The majority of seepage occurs at low tide, especially under spring-tide conditions, when the gradient is at a maximum.

The subterranean estuary may only be a small portion of the overall interaction between groundwater and surface water in a coastal region. On many coastlines globally, groundwater-surface water interaction may extend well out onto the continental shelf (Fig. 6.8) or well inland. A study by Wilson (2005) used groundwater flow, heat, and solute transport modelling results to demonstrate that the zone of interaction between land and sea-based groundwater may extend for tens of kilometres on either side of the shoreline. In addition, geothermal convection, driven by thermal gradients between cold, deep seawater and warmer shelf sediments may drive groundwater flow from the continental slope landward tens of kilometres onto the continental shelf. Coastal heterogeneity complicates things further. In areas with coastal wetlands, the fresh/saline boundary may actually move to the landward edge of the wetlands, well inland of what is traditionally considered to be the zone of land-sea interaction.

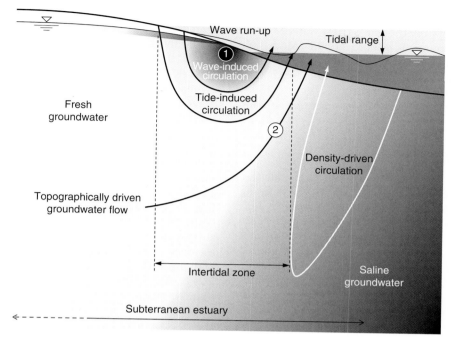

Fig. 6.7 Conceptual model of the subterranean estuary, showing the next level of complexity beyond the standard mixing zone model. In tidally affected shorelines, two features emerge in the fresh/saline boundary: (1) an upper saline plume; and (2) a freshwater discharge tube. Note that the landward boundary of the subterranean estuary depends on the limit of tidal influence. (Source: Adapted from Robinson et al. 2007b with permission from John Wiley & Sons.)

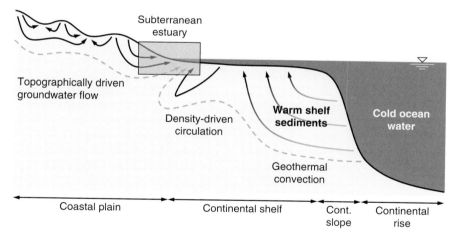

Fig. 6.8 The larger picture of land-coast groundwater interaction extends tens to hundreds of kilometres onto the continental shelf and beyond. In addition to the topographically driven flow and density-driven circulation of the subterranean estuary, geothermal convection through the continental shelf sediments also provides groundwater discharge to the continental shelf. (Source: Adapted from Wilson 2005 with permission from John Wiley & Sons.)

6.3 Submarine groundwater discharge (SGD)

6.3.1 Measurement of SGD

Probably the most common techniques of SGD measurement involve the use of tracers and numerical groundwater flow and solute transport models. Much of the SGD literature involves the use of tracer techniques for SGD rate estimation at regional scales. Thermal tracers have been used in several studies, but chemical tracers are much more widely used. By far the most common chemical tracer in the SGD literature is ^{226}Ra, which is naturally enriched in groundwater relative to ocean water. ^{226}Ra decays to ^{222}Rn with a half-life of 1620 years, making it a useful tracer for estimating the rate of freshwater SGD in the nearshore environment (Moore, 1996). Modelling techniques are becoming more widely used in the study of SGD. These studies range from small-scale, local studies of individual coastal aquifers (e.g. Li et al., 1999; Anderson and Emanuel, 2010) to regional-scale studies (e.g. Wilson, 2005).

6.3.2 SGD in the global hydrological cycle

Up until the 1990s, little was known about the relative rates of the various components of SGD. Several defining works in the mid-1990s set off a wave of research into rates of SGD. Improved modelling and measurement techniques since that time have allowed more detailed studies to be conducted. An important paper that likely started SGD research used radium isotopes as tracers to estimate that rates of SGD in the South Atlantic Bight of the southeastern USA must equal 40% of the total riverine discharge in the same region (Moore, 1996). Given that: (1) the assumption had always been that rivers are the primary pathways for the transport of weathering products and nutrients to the coastal zone; and (2) previous estimates had shown that SGD is at most 0.01 to 10% of riverine discharge (Church, 1996), this was a substantial discovery. Several researchers called the 40% estimate into question. Noting that the land-derived component of SGD is limited by the rate of aquifer recharge to the land, Younger (1996) concluded that only 4% of SGD can come from freshwater discharge across the land/ocean boundary. This result suggests that 96% of SGD is actually marine water that has recirculated through the nearshore environment with wave run-up and tidal oscillations.

Li et al. (1999) performed the first modelling study of the effect of waves and tides at the nearshore scale and found that the recirculation of marine water with waves and tides is larger than the land-derived freshwater component. They derived an equation to reflect the fact that there are several components to SGD. Their equation for discharge to the oceans as SGD, or D_{SGD}, is:

$$D_{SGD} = D_n + D_w + D_t \qquad (6.7)$$

where D_n is the net groundwater discharge from topographically driven groundwater flow from the continents, D_w is the wave-induced component of circulation, and D_t is the tide-induced component of circulation. It should be noted that D_w and D_t are both oscillating properties and therefore contain both recharge and discharge components. Each of the arrows discharging to the nearshore in Fig. 6.7 represents one of the components of eqn. 6.7. An additional term could also be included to represent the density-driven circulation further from shore as represented by the white arrow in the same figure.

The discrepancy here is likely to be due to the scales of observation and the definitions of what actually constitutes SGD. Moore (1996) operates at a regional scale and estimates SGD at this large scale using the mass balance of radium isotopes. Younger (1996) bases his estimate on the mass balance of nearshore fluxes and assumptions about the size of coastal aquifers relative to coastal streams, which he suggests are several orders of magnitude larger. His estimates, however, rely on the assumption that most of the coastal groundwater discharges not at the coast but instead to coastal streams as baseflow. Eqn. 6.7 represents primarily SGD at the nearshore scale, while Moore (1996) estimates SGD at the continental-shelf scale. Moore (2010) recently suggested that his 40% estimate of SGD relative to riverine discharge is too low. Looking at the kilometre, regional (South Atlantic Bight), and ocean (Atlantic Ocean) scales, the new estimate is that the flux of SGD is three times that of riverine fluxes. In addition to scale independence, the data also suggest that SGD must not only exist at the shoreline, but also throughout the entire continental shelf.

SGD is a tangible property of groundwater that has an obvious effect on sea level. There are components of terrestrial groundwater other than SGD, however, that indirectly contribute to sea level in locations that may be far from the coast. These are detailed nicely by Lettenmaier and Milly (2009). Some of these components enhance aquifer storage and have the net effect of reducing groundwater contribution to sea-level change. These include: (1) the direct injection of treated wastewater into aquifers for storage purposes; (2) infiltration from surface reservoirs; and (3) infiltration of irrigation water derived from surface water supplies. Other components act to reduce groundwater storage, thereby increasing groundwater contribution to sea-level change. These include: (1) declines in recharge due to urbanization; and (2) groundwater mining. As Lettenmaier and Milly (2009) point out, quantifying groundwater's contribution to sea-level change is a challenge for the hydrological community, both due to its perceived importance to the balance of the global hydrological cycle and the difficulty in accurately measuring changes in groundwater storage.

6.4 Controls on SGD variability

The contribution of each of the components of coastal groundwater flow (i.e. topographically driven flow, tidal and wave circulation, density-driven circulation) relative to stream contributions has been debated for years and probably reflects the method of estimating SGD, the hydrogeology of the coastal aquifers that were studied, and the boundary conditions at each study site (e.g. wave climate, tidal amplitudes). This section explores the various factors that may influence rates of SGD.

6.4.1 Spatial variations

Clearly, spatial heterogeneity in coastal aquifers will play a large role in the contribution of SGD relative to streams at various scales. It should be noted that Moore (1996, 2010) did his research in the South Atlantic Bight of North America, an area comprising permeable barrier island and karst aquifers, and relatively few low-permeability zones. Layered coastal plain aquifers in this region also likely contribute groundwater inputs well offshore where they outcrop on the continental shelf. Had this work been carried out in areas with less productive coastal aquifers, such as the fractured bedrock coastal aquifers of the northeastern USA, groundwater flux as a function of riverine flux may have been quite different.

Figure 6.9 shows several coastal hydrogeological conditions, each comprising a different material and, therefore, a different ability to conduct terrestrial groundwater to the coast or recirculate tidal oscillations and wave energy (Box 6.2). As Moore (2010) notes, most of the estimates of the relative contributions of various parameters to SGD, such as those proposed by Li et al. (1999), employ simple models in their calculations. Coastlines, however, are notoriously heterogeneous and anisotropic. They may comprise a mixture of aquifer materials, aquifer conditions (e.g. confined, unconfined), tidal ranges, and wave climates, among other parameters, and each of these variations will affect the proportional contributions to SGD.

The composition of coastal aquifers will vary over a wide range of materials, from karst limestones to fractured granites. Coastal conditions will also vary in continuity, from long barrier-island systems to pocket beaches comprising gravel that are separated by headlands of relatively impermeable crystalline rocks. All of these variations will affect the sources of SGD. For example, karst aquifers such as those that dominate the Yucatan Peninsula in Mexico and The Burren, County Clare, Ireland (Fig. 6.9a), are highly transmissive aquifers. In karst regions, it is likely that sizable rivers are rare or non-existent, and that groundwater dominates as the primary mode of flux between the land and the ocean. In karst regions with low wave climate and low tidal range, the land-derived component of SGD may be quite large.

Beaches comprising gravel, although extremely permeable, may not transmit as much topographically derived groundwater flow to the nearshore environment. Depending upon the underlying material on which the beach is built, gravel beaches may be quite efficient at transmitting tidal and wave energy deep into the aquifer, but lack a sizable topographically derived component.

These beaches often comprise a thin veneer of gravel over less-permeable material, such as the peat-gravel barrier systems of southwest England, or gravel beaches overlying bedrock along the northeastern coast of the USA or the Black Sea coast of Turkey (Fig. 6.9b).

Sandy coastal aquifers, which exist throughout the world and often comprise thick sequences of permeable sediment, may exhibit a variety of conditions, such as the 100 km long barrier-island chain along the east coast of the USA or the kilometre-scale beaches of the northern Dominican Republic coast (Fig. 6.9c). These aquifers are of moderate permeability and therefore do not transmit wave and tidal energy as efficiently into the aquifer as do karst materials or gravel (see Box 6.2). Depending on their connection to land-based aquifers, sandy coasts may have high or low terrestrial contributions to SGD.

At the opposite extreme to the previous three examples are coastlines dominated by relatively impermeable bedrock, such as the rocky cliffs of Maine, USA, or the Côte de Granit Rose of Brittany, France (Fig. 6.9d). These coastal aquifers may have small rates of SGD from any source of contribution (land, waves, tidal oscillations) due to the lack of aquifer-ocean interactions. Fractured bedrock aquifers tend to have a low permeability that relies on transmissive fractures to conduct groundwater flow. As these fractures are relatively discontinuous along a cliff face, such as at coastal headlands, the interactions between the ocean and nearshore groundwater at these locations tends to be limited.

Larger-scale continuity will also help to control the contribution of SGD to the nearshore and continental-shelf environments relative to stream discharge. Long sequences of relatively homogeneous barriers, such as the gravel beaches of South Devon, UK (Fig. 6.11a), may contribute a consistent rate of SGD from all sources to the nearshore per unit length of coastline. Discontinuous coastlines, such as the coasts of Cornwall, UK, and the Pacific Northwest of the USA, where pocket beaches are scattered along a bedrock coast, will have large variations in SGD per unit length of coastline, with a variety of sources (Fig. 6.11b). For example, the beaches in these systems will likely interact with wave setup and tidal oscillations, which will contribute much of the local SGD. The relatively low permeability of the bedrock aquifers supporting the pocket beaches, however, means that little contribution to SGD will derive from terrestrial sources.

All of the previously described conditions affect SGD primarily in the nearshore environment. Keeping in mind the regional extent of ocean-groundwater interactions, however, spatial variations may not only extend alongshore, but also well offshore. In the case of fractured bedrock coastal aquifers, the offshore extent of ocean-groundwater interaction may be limited. In contrast, coastal-plain type settings, such as exist in the Atlantic

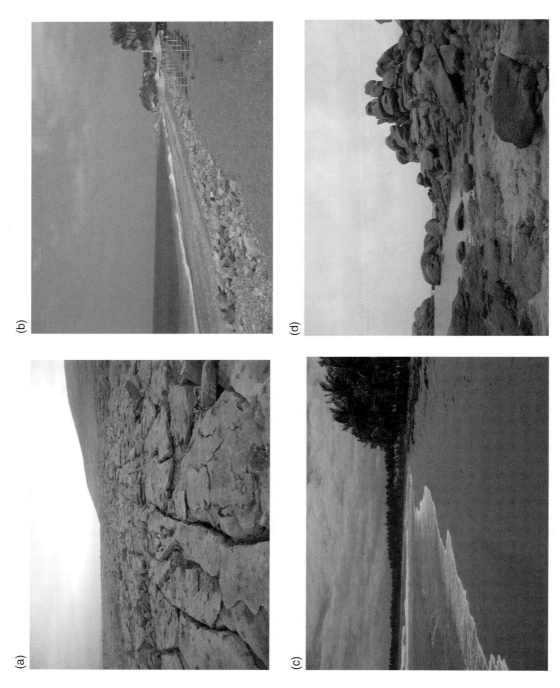

Fig. 6.9 Coastal hydrogeology will play a large role in the interaction between terrestrial and marine groundwater. (a) Karst aquifers are highly permeable and allow for easy interaction between land and sea (The Burren, County Clare, Ireland). (b) Gravel beaches, while not as permeable as karst, have high rates of submarine groundwater discharge (SGD) (Black Sea, Inebolu, Turkey). (c) Sandy beaches are less permeable and, at the lower end of their permeability range, may supply orders of magnitude less SGD to coastal waters (North Coast, Dominican Republic). (d) Regions of bedrock are relatively impermeable and have limited rates of SGD (Côte de Granit Rose, Ploumanac'h, Brittany, France). (Source: Photographs by Bill Anderson.)

CONCEPTS BOX 6.2 Tidal oscillations in groundwater

Coastal aquifers are unique in that they are bordered by a marine system that fluctuates in a predictable and periodic fashion. As we have already seen in this chapter, the magnitude of these oscillations can lead to a variety of morphologies in the subterranean estuary as well as leading to oscillating rates of submarine groundwater discharge (SGD).

Jacob (1950) derived a simple analytical solution to the oscillating boundary condition in which the amplitude of the fluctuation induced in the aquifer at any distance from the shoreline can be calculated. Although his derivation is for an idealized confined aquifer, it can be used in any coastal aquifer as an estimate as long as the tidal oscillations are small relative to the aquifer's thickness (Erskine, 1991). The equation for this predicted oscillation in the aquifer when divided by the tidal amplitude is:

$$TE = \frac{H_x}{H_0} = \exp\left(-x\sqrt{\frac{\pi S}{t_0 T}}\right) \qquad (6.8)$$

where TE is the tidal efficiency, H_0 is the tidal amplitude in the sea, H_x is the tidal response in the aquifer at a distance x from the shoreline, S is the storativity of the aquifer, T is the transmissivity of the aquifer, and t_0 is the period of the tidal oscillation. Transmissivity is the product of the

hydraulic conductivity of the porous medium and the thickness of the aquifer. An equation of similar form can be used to calculate the lag time t_τ it would take for a particular phase of the tidal oscillation, such as high tide, to penetrate a distance x into the aquifer. The equation is:

$$t_\tau = x\sqrt{\frac{t_0 S}{4\pi T}} \qquad (6.9)$$

where all parameters are the same as in eqn. 6.8.

Figure 6.10 shows calculations for three aquifer materials: gravel, sand and silt. All three aquifers have the same thickness of 10 m and the same storativity of 0.001. The upper panel (Fig. 6.10a) shows tidal efficiency. The highly permeable gravel easily transmits the tidal signal. Even 500 m from the shoreline, tidal efficiency is 30%. The low-permeability silt, however, essentially loses all tidal oscillation only 70 m from the shoreline. Time lags show a similar trend (Fig. 6.10b). Not only is the gravel's tidal efficiency high, but the time lag 500 m from shore is only 0.1 days. In other words, it only takes 2.4 h for the tidal signal to penetrate 500 m into the aquifer. The tidal signal in the silt, at the other extreme, requires 0.4 days, or 9.6 h, to penetrate only 70 m into the aquifer.

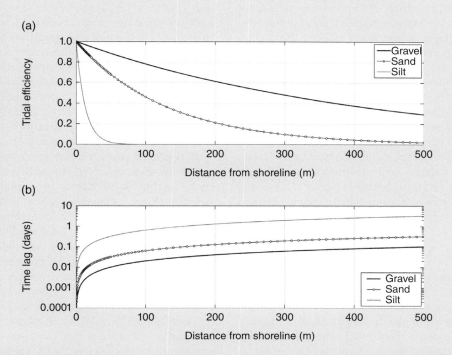

Fig. 6.10 (a) Tidal efficiency depends on the permeability of the porous medium in the aquifer. The highly permeable gravel has a high efficiency relative to the less-permeable silt. (b) The time lag for the tidal signal to penetrate into the aquifer also varies with permeability. Time lags are low for gravels relative to sands and silts.

(a)

(b)

Fig. 6.11 The continuity of beaches and the materials comprising them are both important factors for submarine groundwater discharge (SGD). (a) A long, permeable barrier beach (Slapton Sands, Devon, UK). (b) A short, discontinuous beach separated by impermeable bedrock (Ecola State Park, Oregon, USA). (Source: Photographs by Bill Anderson.)

and Gulf Coasts of the southeastern USA, may have extensive offshore interactions. For example, consider the hypothetical cross-section of a coastal plain–continental shelf system shown in Fig. 6.12. Nearshore groundwater-ocean exchange takes place at the land-ocean interface. This is the location at which the subterranean estuary develops in response to topographically driven groundwater flow, tidal oscillations, and wave run-up. Broad, layered regional aquifer systems, like coastal-plain settings, however, not only have the nearshore to consider, but also groundwater-ocean interactions at places where the deeper aquifers intersect with the bathymetry. This may occur on the continental shelf, but it is more likely to occur on the continental slope, where groundwater-ocean interactions take place. It is also in these regions that mixing between fresh and saline waters develops in response to hydraulic conditions. Other factors, such as pumping, will also affect the exchange of groundwater between the various layers. Vertical exchange may also take place between these aquifers with vertical flow through confining layers.

6.4.2 Temporal variations

Not all coastal aquifers contain all of the components of the conceptual model of the subterranean estuary (see Fig. 6.7). In the previous section, it was demonstrated that spatial variations in aquifer permeability modify the subterranean estuary from its theoretical shape. Temporally varying boundary conditions in the nearshore, such as wave setup and tidal oscillations, can also have a strong impact on rates of SGD and subterranean estuary morphology. They may also, when interacting with nearshore

groundwater, have an effect on sediment transport and beach morphology. For example, areas of low wave climate or low tidal amplitudes will likely not form an upper saline plume and all topographically driven groundwater flow will exit through a wider seepage face. Most of these temporally varying boundary conditions have short timescales of variation, from seconds (e.g. waves), to hours (e.g. tidal cycle), to weeks and months (e.g. spring-neap tidal cycle). Rates of SGD can also respond to longer-timescale variations, from yearly (seasonal variations), to interannual (e.g. El Niño Southern Oscillation, with a period of 2–6 years), to interdecadal (e.g. Pacific Decadal Oscillation, with a period of 10–25 years). The subterranean estuary may also be affected by long-term sea-level variations at the millennial scale (see section 6.6).

Waves and wave run-up are important at localized scales. In microtidal regions with a high wave climate, such as the Outer Banks barrier-island chain on the Atlantic Coast of North Carolina, USA, they may be the dominant factor in the development of the subterranean estuary and rates of SGD. Field studies of wave influence on the subterranean estuary and SGD are difficult to employ because of the complexity of the field situation such as wave energy and wave complexity (e.g. deep-water waves, shallow-water waves and wave run-up), and the added complexity of high-frequency waves superimposed on low-frequency tidal oscillations. As a result, most research on the influence of waves involves analytical and numerical modelling methods applied to simple two-dimensional cross-sectional model domains.

Xin et al. (2010) provide a good example of this type of study. They use numerical simulations to quantify the

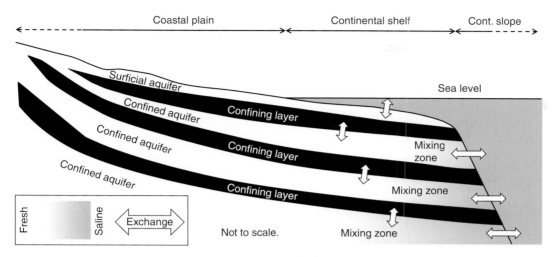

Fig. 6.12 At the regional scale, groundwater–ocean exchange may extend far offshore, even reaching the continental slope where layered aquifers outcrop. This type of layering is typical of coastal plain–continental shelf aquifer systems. Mixing between fresh and saline waters also occurs within these aquifers and varies in response to many factors (e.g. pumping, permeability).

effect of waves on rates of SGD, building on the earlier groundwater flow and solute transport model of Robinson et al. (2007a). Waves are high-frequency boundary effects, and because of this high frequency, there tends to be a rapid attenuation of the wave energy in the groundwater domain. In addition, waves produce instantaneous pore-water flow in response to the shape of the wave and wave run-up. Wave run-up (Fig. 6.13) causes a groundwater over-height in which water levels are higher under the swash zone (Turner and Masselink, 2012). This bulge in the nearshore water table moves in response to variations in tidal phase and also depends on the height of water in the backbarrier lagoon, if present. The phase-averaged influence of the wave process on the subterranean estuary is infiltration at the upper limit of wave run-up and discharge at the breaking point of waves. In other words, waves produce an upper saline plume (see Fig. 6.7) similar to that developed in response to tidal oscillations (Xin et al., 2010). The range of this influence can cover a large cross-shore distance in locations with a flat beach profile.

Tidal oscillations are another temporally varying forcing factor that is a critical component in the development of the subterranean estuary. These oscillations are especially important in areas with macrotidal conditions and moderate wave climate, such as the peat-gravel barrier systems of Devon, UK. At this location, the moderate wave climate produces maximum inshore wave heights of 2.6–3.0 m at a one in 50 year recurrence interval (SCOPAC, 2003). Unlike waves, tides are relatively low-frequency boundary conditions that result in a complex mixing zone. Numerical studies have demonstrated that in tidally dominated coastal aquifers, an upper saline plume will form at the upper tidal limit and will be separated from the

Fig. 6.13 Waves and wave run-up play an important role in the development of the subterranean estuary. Wave run-up, pictured here in Duck, North Carolina, USA, raises the elevation of mean sea level landward of the ocean and induces an upper saline plume similar to that developed in response to tidal oscillations. (Source: Photograph by Bill Anderson.)

primary mixing zone by a freshwater discharge tube (see Fig. 6.7).

The characteristics of the subterranean estuary under these conditions will depend to a large extent on the tidal properties at a particular site. For example, Fig. 6.14 shows simulated subterranean estuaries that develop under tidal and non-tidal conditions. The upper panel (Fig. 6.14a) shows the simulated mixing zone for a macrotidal gravel aquifer. Under tidal forcing, an upper saline plume develops at the

(a)

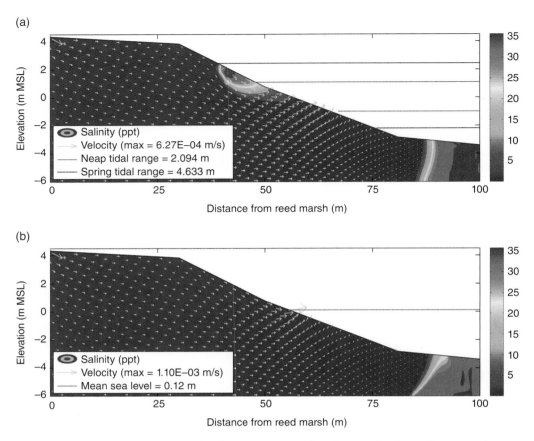

Fig. 6.14 Tidal oscillations are the key ingredient to developing an upper saline plume. This figure shows simulation results for a homogeneous gravel aquifer. (a) Simulations of a tidally oscillating ocean and a homogeneous gravel aquifer produce the theoretical components of the subterranean estuary. The high permeability of the gravel aquifer, coupled with a strong topographically driven groundwater component, produces a wide freshwater discharge tube. A second mixing zone develops at the saline wedge. (b) Tidal oscillations are not included in this simulation. Note that groundwater discharges around mean sea level (MSL) and only one mixing zone develops at the saline wedge. For colour details, please see Plate 17.

limit of the high-tide line in response to infiltration. A freshwater discharge tube separates the upper saline plume from the primary mixing zone. Note that the lower mixing zone is below the low-tide limit. This occurs because of the development of a seepage face, which extends offshore due to the energy from topographically driven groundwater flow. The lower panel of the figure (Fig. 6.14b) shows a non-tidal simulation for the same aquifer. Because tides are not simulated, no upper saline plume develops. Instead, all that develops is the primary mixing zone between the topographically driven land-derived groundwater flow and the density-driven groundwater flow in the nearshore.

Robinson et al. (2007a) undertook an extensive numerical study of the effect of tidal oscillations on the development of the subterranean estuary. Their findings suggest that the behaviour of the subterranean estuary is similar to that of surface estuaries: they are affected by both land and ocean-based forcings. For the subterranean estuary, this

means that there exists a balance between the discharge of topographically driven groundwater flow from land and the recirculating seawater that moves through the nearshore due to tidal oscillations. Robinson et al. (2007a) also suggest that subterranean estuaries should be classified in a manner similar to the classification of surface estuaries: stratified, partially stratified and well-mixed. Stratified systems develop in coastal aquifers that are microtidal or lack tidal variations. These systems do not develop an upper saline plume and mixing occurs along the saline wedge (Fig. 6.14b). Partially stratified systems have two mixing zones, one in the upper saline plume and one along the saline wedge. The freshwater discharge tube under these conditions lacks mixing except along its edges, which are potentially zones for geochemical reactions. The authors also suggest that the geochemical conditions in the two mixing zones could be completely different. Well-mixed systems develop in response to either limited land-derived groundwater flow or as a result of macrotidal conditions. In these systems, the freshwater discharge tube is lost and there is only a small zone separating the upper saline plume from the saline wedge. Although they acknowledge that waves play a role in the formation of the upper saline plume at shorter timescales than those for tidal oscillations, they do not assess the influence of waves on the stratification potential within subterranean estuaries.

Waves and tides play an important role in the development of a region's coastal groundwater conditions, and these conditions may be enhanced by a combination of strong wave and tidal forcings. The model output of the base case simulation of Robinson et al. (2007a), which utilizes a tidal amplitude of 1 m and does not include waves, indicates that tidally driven recirculation accounts for approximately 45% of SGD, with approximately 19% produced by density-driven flow. Similar modelling of waves (wave amplitude of 0.2 m, wave period of 10 s) without tides suggests that waves account for 49% of SGD (Xin et al., 2010), which is similar in scale to tidal contributions. The natural condition of both waves and tides enhances their role, but is not additive, due to the modulation of the signals; the combined contribution of simulated waves and tides is 61% of SGD. It should be noted that these results represent a complex interaction between wave climate, tidal conditions, beach morphology and the hydraulic properties of the aquifer. Thus, it does not apply to all coastal aquifers.

Waves and tides operate on relatively short timescales of seconds to hours, although the total spring/neap cycle operates at a bi-weekly timescale. There are also longer-scale factors that affect the subterranean estuary and rates of SGD at scales of days to months to years. Consider, for example, storm events, which can be conceptualized as a large, long-period wave. As detailed in Chapter 5, storm surge, which is induced by low barometric pressures and wind, coupled with waves and tides, may occur during winter extra-tropical storms, tropical storms and hurricanes. Storm surge may greatly expand the swash zone, sending saline waters well inland of their normal wave- and tide-induced variations, resulting in overwash events. The presence of this saline water produces a temporary recharge of saline water into the fresh land-based portion of the subterranean estuary, resulting in a lateral expansion of the mixing zone. This can be thought of as a form of saltwater intrusion from above. Thus, in areas prone to overwash events, the theoretical subterranean estuary morphology may never be realized. Instead, the subterranean estuary may be working its way toward dynamic equilibrium with waves and tidal oscillations between each storm event. Areas suffering from frequent overwash events may perennially have higher saline levels in nearshore aquifers than would be predicted theoretically.

As an example, consider the sandy barrier-island aquifer of Hatteras Island. Coastal North Carolina frequently experiences hurricanes. In fact, it is not uncommon for this stretch of the Atlantic coast to receive two or more hurricanes in a single year. Thus, there can easily be an additive effect of multiple hurricanes within the coastal aquifer because there may not be enough time for a storm's influence to work its way through the aquifer prior to the arrival of the next storm event.

Hurricane Emily struck Hatteras Island in August and September 1993, sending a storm surge of 3 m over the northern third of the island, which averages approximately 1.5 m in elevation. This overwash event produced a saline recharge event, which raised total dissolved solids above 500 mg/l at the local water-treatment plant. The source of the raw water for the treatment plant was a well-field consisting of 42 wells, all of which were inundated by the storm surge. Anderson and Lauer (2008) used the raw water data to calibrate a groundwater flow and solute transport model of the Hurricane Emily overwash event. They then utilized the calibrated model to predict the residence time of the storm event and to see its effect on the morphology of the subterranean estuary. Their simulation results indicate that the pulse source of saline water from Pamlico Sound, which was the source of the overwash, produces a wave of saline infiltration into the aquifer (Fig. 6.15). The infiltration of the saline plume gradually works its way deeper into the aquifer as it is diluted by freshwater recharge from above. One year after the overwash event, the aquifer has become nearly filled with elevated salinity levels throughout its thickness. The simulations also showed that even after 10 years of freshwater recharge and dilution of the pulse saline source, the subterranean estuary had not returned to its pre-storm morphology. Given that multiple overwash events would have struck the island during this decade, it is unlikely that the subterranean estuary of Hatteras Island ever approaches its theoretical morphology.

Longer temporal-scale variations at yearly to multiannual scales may also affect the morphology of the subterranean estuary and rates of SGD, but these effects are primarily due to the topographically driven flow component of the subterranean estuary. Seasonal variations occur at the annual scale, but interannual variations in climate, such as the El Niño Southern Oscillation (ENSO), which varies at 2–6 year scales, and interdecadal variations in climate, such as the Pacific Decadal Oscillation (PDO), which varies at 10–25 year scales, have also been demonstrated to affect coastal aquifers.

Seasonal variations primarily affect recharge rates and, therefore, the land-derived component of the subterranean estuary. For example, during late winter and early spring, aquifer recharge rates tend to be higher because of lower temperatures and low rates of evaporation and transpiration, or evapotranspiration. Thus, water levels, and consequently groundwater flow rates, tend to be higher during

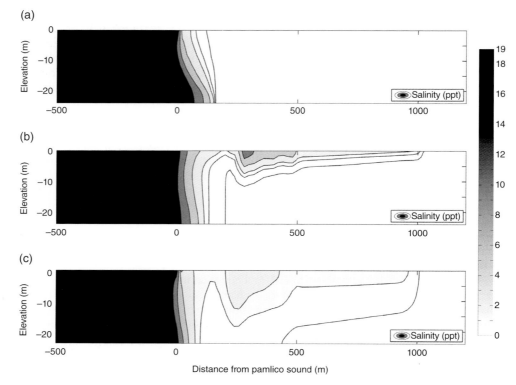

Fig. 6.15 Storm events that produce saline overwash may have long residence times in coastal aquifers. This figure shows simulation results produced by Anderson and Lauer (2008). Hatteras Island lies between 0 and 1200 m in the figure. Pamlico Sound is at negative distances. (a) Pre-storm simulations indicate the morphology of a theoretical subterranean estuary without tides. (b) Overwash during Hurricane Emily in 1993 extended approximately 1 km inland. One month after the storm, approximately one-third of the aquifer has elevated salinity levels. The upward deflection of the isochlors around 200 m occurs due to a dune that runs perpendicular to the plane of the simulation. (c) After one year of dilution, the aquifer still has not returned to pre-storm levels, and nearly the entire thickness of the aquifer over the region of the overwash event has elevated salinity levels. (Source: Data from Anderson and Lauer 2008.)

this time of year. During the late summer and early autumn, water levels typically drop to their lowest levels of the year due to reduced summer recharge rates that occur in response to high temperatures and correspondingly high levels of evapotranspiration. Michael et al. (2005) used a field study of Waquoit Bay, Massachusetts, USA, to suggest that seasonal variations in recharge on land affect rates of SGD in the subterranean estuary. They document the seasonality with a time series of hydraulic gradients and seepage meter flux measurements in the field. They also use generic groundwater flow and solute transport simulations to explain the lag times that occur between high fluxes in the land-based aquifer and high rates of SGD in the nearshore.

While seasonal variations in recharge are common to many coastal aquifers, some regions are subjected to strong interannual and interdecadal variations in climate that may mask the seasonal variations. For example, many of the coastal aquifers of the southeastern USA are heavily influenced by ENSO. Anderson and Emanuel (2010)

demonstrate that recharge rates to Hatteras Island do not show a seasonal periodicity, but rather significantly correlate with the interannual oscillations of ENSO. They then use a modelling study to simulate the effect of these interannual recharge variations on rates of SGD. They find that not only does the topographically driven flow component of SGD show significant correlation with ENSO, but also that anomalies, or variations, in these rates between the wet (El Niño) and dry (La Niña) phases of ENSO vary by up to 35%. The results of this study suggest that the biogeochemistry of the nearshore zone may vary considerably at these temporal scales.

6.5 Human influences

Martinez et al. (2007) suggest that 40% of the world's population lives within 100 km of the coast. As detailed in Chapter 17, 20 million of these people live below the normal high-tide line, and 200 million are exposed to coastal flooding during storm events (Nicholls, 2010). This human forcing on coastal groundwater processes, especially in the context of rising populations and global climate change, will likely continue to increase over the next century. As much as coastal armouring has caused shifts in coastal sediment budgets and morphology, increasing usage of coastal groundwater by a larger populace will dramatically affect coastal hydrogeological budgets. This will, in turn, affect the morphology of the subterranean estuary and rates of SGD. The changing dynamics of the coastal groundwater flow system, then, could ultimately have widespread implications to nearshore biogeochemistry over the next century.

Increasing pumping of groundwater resources in the coastal zone is a natural consequence of increasing populations. People need an easy and reliable source of potable water, and often coastal aquifers are highly productive and shallow, making their exploitation relatively simple and economical. Pumping of coastal aquifers, however, has consequences, the most obvious of which is pumping-induced saltwater intrusion. This occurs in response to the changing hydrodynamics of the subterranean estuary, including the creation of a cone of depression around a pumping well that reverses the natural hydraulic gradient inland (Box 6.3). Figure 6.16 shows simulations of the same gravel aquifer that has been shown throughout this chapter. The initial conditions in the upper panel (Fig. 6.16a) are the same as those shown in Fig. 6.14. The simulated pumping rate is 270 m³/day. The figure displays several snapshots of the modified subterranean estuary at one day, five days and one month after the start of pumping. The simulation shows that there are clear effects on the subterranean estuary as a result of groundwater extraction, even in a highly permeable gravel aquifer. After one day of pumping it is obvious that the upper saline plume is growing, relative to that developed under dynamic

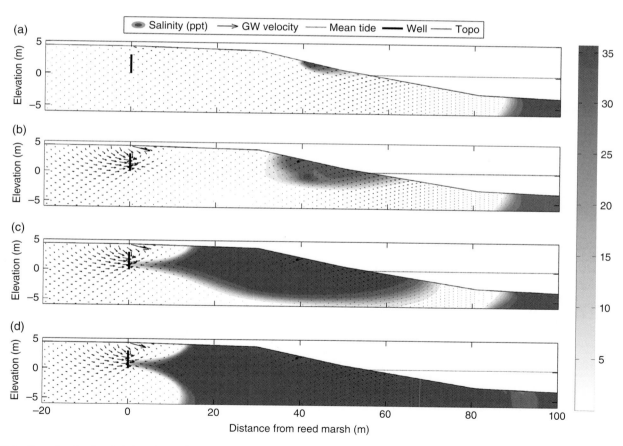

Fig. 6.16 Pumping in the vicinity of coastlines can have a dramatic effect on the subterranean estuary. This figure shows a simulation of the influence of a pumping well on the gravel aquifer of previous simulations. The pumping rate of the well, which is located at 0 m, is 270 m³/day. Conditions are shown for (a) pre-pumping conditions, as well as (b) one day, (c) five days, and (d) one month after the beginning of pumping. GW, groundwater.

CONCEPTS BOX 6.3   Saltwater intrusion

The abstraction of groundwater from wells creates a cone of depression that under ideal conditions is centred symmetrically around the pumping well. The difference in the elevation of the cone of depression at a particular radial distance from the pumping well and the pre-pumping elevation of the water table at that point is called 'drawdown'.

Figure 6.17 shows pumping effects in a sandy barrier-island setting with a hydraulic conductivity of 25 m/day. Drawdown at radial distances up to 250 m from the pumping well is shown in the upper panel (Fig. 6.17a) for two pumping rates. Although the process of forming a cone of depression is dynamic, this figure shows conditions at quasi-steady state. Note the

difference in drawdown caused by the two pumping rates, which differ by an order of magnitude. Note also the increase in the slope of the cone of depression with proximity to the pumping well, a process necessitated by the fact that the cross-sectional area of the aquifer has decreased and more gradient is required to conserve mass.

The middle and lower panels of the figure (Fig. 6.17b, c) show the application of these two pumping rates to the sandy barrier-island aquifer with a fairly strong shoreward gradient. In the middle panel of the figure (Fig. 6.17b), the pumping well lies only 100 m from the shoreline. The proximity of the well to the shore means that the cone of depression intersects the saline water

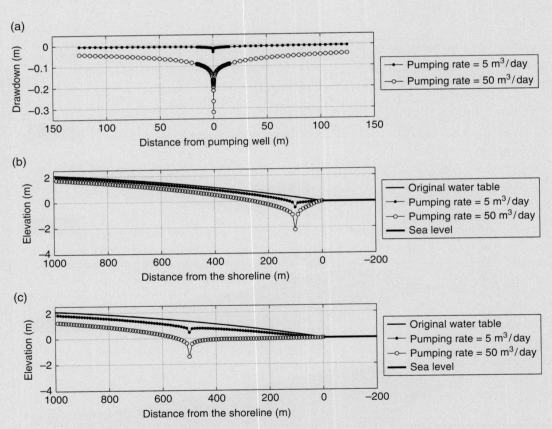

Fig. 6.17 (a) The upper panel shows simulated cones of depression around a pumping well at pumping rates of 5 and 50 m³/day. (b) The pumping rates of the upper panel have been applied to a sandy coastal aquifer. In this panel, the pumping well lies 100 m from the shoreline. The strong reversed gradient induced by pumping proximal to the shoreline would likely cause rapid contamination of the pumping well. (c) In the lower panel, the well lies at a distance of 500 m from the shoreline. Although the gradient in the nearshore is not as strong, there is more effect on the aquifer at distances further into the aquifer because the water table is below sea level more than 500 m inland. This situation will also cause changes in the subterranean estuary.

in the sea. The reversed gradient that this produces means that there will be a landward migration of the mixing zone inland, even at low pumping rates, and likely contamination of the well. The lower panel (Fig. 6.17c) applies the same pumping rates to a well located 500 m from the shoreline. At this location,

there is less effect on nearshore gradient, especially at low pumping rates; however, note that pumping at this location has more influence on the regional water table, where at 1000 m from the shoreline the water table has been reduced by 1 m compared with non-pumping conditions.

equilibrium. Although there is little change in the saline wedge, the upper saline plume is more easily altered by the changed flow field due to its proximity and the influence that the cone of depression is having. Five days after the onset of pumping, saltwater 'upconing' has occurred, and the wellfield is now drawing in saline water. Note that the freshwater discharge tube continues to exist, although the saline wedge is beginning to work its way upgradient. Finally, after one month of pumping, the effect is more severe: SGD in this section of the coast has stopped and saltwater intrusion is replacing the formerly fresh portion of the aquifer.

The conditions simulated in Fig. 6.16 assume that boundary conditions affecting the aquifer, such as rates of recharge, are constant. However, the influence of pumping may be exacerbated during times of low recharge rates (e.g. summer months, drought). This also does not take into account the potential effects of storm events, which may cause saltwater intrusion from above with the occurrence of overwash. Anderson (2002) simulates the effect of overwash processes on aquifer water quality at a municipal wellfield on Hatteras Island with a simple analytical solution. Multiple storm events are simulated by adding together various storms using linear superposition. The simulations demonstrate that not only is pumping during benign conditions a threat to the water quality of the coastal aquifer, but overwash events may provide a much quicker and long-lasting threat to coastal groundwater supplies.

It is not only saltwater intrusion due to freshwater extraction that is a threat to the coastal zone. Many regions extract moderately saline groundwater from deeper zones and employ desalinization techniques to produce the freshwater resource. For example, consider once again Hatteras Island. After the salinization of the surficial aquifer following Hurricane Emily in 1993, much of the island shifted its water supply to a mixture of relatively fresh, shallow groundwater with relatively saline, deep groundwater. This saline water source is then treated with reverse-osmosis desalinization processes to make the water potable. The consequence, of course, is a broader influence of pumping, because now both deep and shallow aquifers are being exploited. An unintended consequence is the disposal of the brines that are the by-product of the water-treatment process. Where can this wastewater be

properly disposed without contaminating the marine environment?

Groundwater abstraction is not confined to the nearshore and shallow aquifers. Well inland, especially in the layered aquifers of coastal-plain settings, fresh groundwater is withdrawn from deep aquifers for use in irrigation and industrial processes. In many cases, these productive aquifers are pumped at high rates and the resulting cone of depression is large, thereby promoting reversed gradients and the flow of saline groundwater inland. Aquifer consolidation may also occur, exacerbating rates of sea-level rise due to the drop in elevation promoted by consolidation. Figure 6.12 shows a layered coastal-plain aquifer. With pumping of a deep aquifer, saline water that has already penetrated the aquifer due to density-driven flow will be induced inland in response to reversed hydraulic gradients. Also, depending on the relative permeability of the confining layers, saline groundwater may also be induced to flow across these low-flow confining layers. The result may be the migration of the mixing zone well inland of the marine boundary and the loss of a fresh groundwater supply. Over-pumping may also cause subsidence, which in coastal settings will result in a change in relative sea level. Consider the tide-gauge record for Manila, Philippines, in Fig. 2.12. A gradual rate of sea-level rise is apparent at the beginning of the record from 1900 to 1960; however, beginning in 1960, a combination of sediment deposition, reclamation, and subsidence due to over-pumping results in an accelerated rate of sea-level rise. Increasing human access to coastal groundwater resources globally will result in similar effects in many locations worldwide in the next century.

Overuse of the coastal groundwater resource is not the only human threat to the subterranean estuary, because rising populations produce more of the by-products of modern life: contamination. One of the principal mechanisms of contamination of coastal aquifers is septic contamination. While a large portion of the world's population lives near the coast, many coastal areas do not have proper treatment facilities for human waste. This may be due to a lack of means for treatment, but often it is due to low gradients that prevent easy transport to wastewater treatment facilities through gravity-driven sanitary sewers. As

Fig. 6.18 This picture shows the aftermath of erosion from winter extratropical storms along the Outer Banks barrier-island chain of North Carolina, USA. Note that the house now lies in the swash zone. Potential septic contamination, in addition to saline intrusion, is a reality all along this coastline. (Source: Photograph by Roland Gehrels.)

a result, many coastal regions rely on septic systems for waste storage and disposal. Given the shallow water table, highly permeable aquifers and dynamic boundaries common to coastal regions, these areas are prone to septic contamination, especially in the wake of storm events (Fig. 6.18). The shallow, permeable aquifers of coastal regions are also prone to more traditional contamination such as oil spills and run-off from the urban infrastructure. Again, the high permeability and shallow water table common to coastal aquifers, and the increasing likelihood of accidental spills with increasing populations, makes these locations prone to contamination. The ultimate destination of these contaminants once they get into coastal aquifers is the subterranean estuary and, ultimately, discharge to the nearshore marine environment through SGD.

6.6 Influence of global climate change

Increasing coastal populations are not the only threat to coastal aquifers. Global climate change will also have an effect on conditions within the subterranean estuary, especially as relative sea level rises and precipitation rates change.

The most obvious consequence of a warming Earth is a change in relative sea level (Chapter 2). This can potentially affect the location of the subterranean estuary because as sea level rises, the saline wedge will migrate as well. In areas dominated by macrotidal conditions, the

location of the upper saline plume will also migrate up the beach to higher elevations. The balance between the onshore and offshore components, however, is the most interesting response that will be expected in the aquifer. A rising sea level means that, conditions onshore being equal, the gradient driving topographically driven groundwater flow will be reduced. Thus, not only will the saline wedge migrate inland, but there will also be less energy in the freshwater portion of the aquifer to offset the saline wedge, and SGD rates will decline.

Perhaps an unexpected and potentially more important consequence of global climate change will be the terrestrial-based variations that will affect the topographically driven component of flow. Many climate simulations predict that precipitation rates will drop in many locations. Falling rates of precipitation mean that recharge rates on land will decline, thereby reducing fresh groundwater flow to the coast. The energy contained in the land-based portion of coastal aquifers balances the energy offshore, keeping the position of the saline wedge close to the shoreline and generating the freshwater discharge tube in macrotidal environments. Reduced water-table elevations in the mainland aquifers, however, will reduce this energy, leading to a migration of the saline wedge inland and a likely reduction in the magnitude of the freshwater discharge tube. With further lowering of the water table and reduced gradients, the energy in the aquifer may become low enough to completely eliminate the freshwater discharge tube altogether. As with rising sea levels, there will also be greater contamination potential with storm events.

Global climate change has, of course, been a reality throughout Earth's history, and it operates at much longer timescales than have yet been discussed. Most notably, global climate change has had a prominent effect on coastal aquifers during transitions from glacial to interglacial periods. Rising sea levels during these transitions to interglacials have meant that coastal aquifers are continually responding to changing boundary conditions, thereby putting the subterranean estuary in a constant state of disequilibrium. Recent research suggests that freshwater marshes lying behind barrier beaches record and preserve the record of sea-level rise because the coastal water table rises in tandem with sea level. Anderson and Gehrels (2011) collected organic samples from basal peats in Hallsands, Devon, UK, and compared their resulting ages and depths to regional sea-level curves. They found that relative sea levels predicted by the freshwater peat samples mirror relative sea levels predicted by glacial isostatic adjustment (GIA) models (Bradley et al., 2011) and traditional salt-marsh techniques (Fig. 6.19). Stratigraphical investigations of coastal freshwater marshes will further enhance our understanding of rates of sea-level rise during the present interglacial.

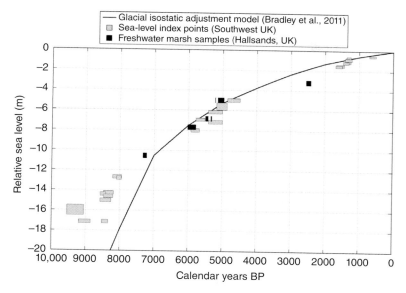

Fig. 6.19 Coastal aquifers may preserve longer-scale sea-level variations. Anderson and Gehrels (2011) collected basal peat samples from a coastal freshwater reed marsh in Devon, UK. The surface of this marsh has tracked the rise in sea level over the past 7000 years. Radiocarbon ages on peat samples (black boxes), when plotted in an age-altitude diagram, line up with both the sea-level curve predicted by a glacial isostatic adjustment model (Bradley et al. 2011; black line) and with regional sea-level index points (grey boxes).

6.7 Summary

Coastal groundwater flow systems provide an important connection between terrestrial and marine waters in the form of the subterranean estuary. Operating at a variety of timescales, groundwater discharging to the nearshore marine environment, known as submarine groundwater discharge, provides a source of both water and nutrients, and in some locations may exceed riverine inputs. Groundwater discharging to the marine environment affects nearshore biogeochemical processes, such as oxidation-reduction reactions and ion desorption. Coastal groundwater may also be more regional in scope, encompassing much of the continental shelf. Short- and long-term processes, such as tidal oscillations and sea-level rise, respectively, also affect the circulation of groundwater in this region, and may contribute to the development of complex subterranean estuary morphologies. Storm processes that produce overwash can affect the salinity distribution, thereby preventing the subterranean estuary from reaching its theoretical morphology. Human activities that impact on coastal groundwater include coastal contamination and over-pumping. These activities may become more important in the coming decades due to population growth and the shift of populations toward the world's coastlines.

- Early study of coastal aquifers noted that a mixing zone must exist between fresh terrestrial groundwater and saline submarine groundwater. Early conceptual models suggested that this zone was static. Over time, however, enhanced research techniques have shown that this mixing region, now known as the subterranean estuary, is quite complex.

- The subterranean estuary responds to a range of influences, including land-based groundwater flow, tides and waves, and density-driven marine recirculation. All of these factors combine to determine a particular subterranean estuary's morphology. The primary components of this morphology may include an upper saline plume, a freshwater discharge tube and a saline wedge. Between these primary components, zones of mixing are present.

- Groundwater discharging at the shoreline is referred to as submarine groundwater discharge (SGD). SGD rates, which are important to nearshore environments, will vary spatially (e.g. local and regional aquifer heterogeneity) as well as temporally (e.g. waves, tidal oscillations, spring/neap tidal oscillations, interannual oscillations).

- Approximately 40% of the world's population lives within 100 km of coastlines, 200 million of which are exposed to coastal flooding during storms. Pumping of coastal groundwater to supply this population affects coastal groundwater dynamics and ultimately the subterranean estuary. Over-pumping may lead to saltwater intrusion and a loss of the water supply for future use.

- Global climate change has affected, and will continue to affect, coastal aquifers. Sea-level change over the past 10,000 years has been recorded in freshwater peats lying behind barrier beaches. Continued sea-level rise means that the subterranean estuary is always in dynamic equilibrium. Changing climate may also lower terrestrial groundwater flow, thereby changing relative amounts of SGD and, ultimately, nearshore biogeochemistry.

Key publications

Cooper, H.H., Jr., 1959. A hypothesis concerning the dynamic balance of fresh water and salt water in a coastal aquifer. *Journal of Geophysical Research*, **64**(4), 461–467.
[*The first comprehensive description of the fresh/saline boundary as a zone of diffusion*]

Darcy, H., 1856. *Les Fontaines Publiques de la Ville de Digon*. Victor Dalmont, Paris, France.
[*This is a seminal paper on groundwater hydrology that provides the first demonstration of Darcy's law*]

Moore, W.S., 1996. Large groundwater inputs to coastal waters revealed by ^{226}Ra inputs. *Nature*, **380**, 612–614.
[*This is the first paper to demonstrate that groundwater contributions to oceans are on par with river contributions*]

Moore, W.S., 1999. The subterranean estuary: a reaction zone of ground water and sea water. *Marine Chemistry*, **65**, 111–125.
[*This paper defines the morphology of the fresh/saline mixing zone with a term that is now a standard part of the coastal groundwater lexicon*]

Robinson, C., Li, L. and Barry, D.A., 2007. Effect of tidal forcing on a subterranean estuary. *Advances in Water Resources*, **30**, 851–865.
[*This paper demonstrates, with simulations, that the subterranean estuary morphology is influenced by the rate of land-based groundwater flow and tidal forcing*]

References

Anderson, W.P., Jr., 2002. Aquifer salinization from storm overwash. *Journal of Coastal Research*, **18**(3), 413–420.

Anderson, W.P., Jr. and Emanuel, R.E., 2010. Effect of interannual climate oscillations on rates of submarine groundwater discharge. *Water Resources Research*, **46**, W05503, doi: 10.1029/2009WR008212.

Anderson, W.P., Jr. and Gehrels, W.R., 2011. Short and long temporal controls on peat/gravel barrier systems, H34E-01. 2011 Fall Meeting, American Geophysical Union, San Francisco, CA, USA, 5–9 December 2011.

Anderson, W.P., Jr. and Lauer, R.M., 2008. The role of overwash in the evolution of mixing zone morphology within barrier islands. *Hydrogeology Journal*, **6**, 1483–1495, doi: 10.1007/s10040-008-0340-z.

Badon-Ghyben, W., 1889. *Nota in verband met de voorgenomen putboring nabij Amsterdam*. Koninklijk Instituut Ingenieurs Tijdschrift, The Hague, the Netherlands.

Bradley, S.L., Milne, G.A., Shennan, I. and Edwards, R., 2011. An improved glacial isostatic adjustment model for the British Isles. *Journal of Quaternary Science*, **26**(5), 541–552, doi: 10.1002/jqs.1481.

Church, T.M., 1996. An underground route for the water cycle. *Nature*, **380**, 579–580.

Cooper, H.H., Jr., 1959. A hypothesis concerning the dynamic balance of fresh water and salt water in a coastal aquifer. *Journal of Geophysical Research*, **64**(4), 461–467.

Darcy, H., 1856. *Les Fontaines Publiques de la Ville de Digon*. Victor Dalmont, Paris, France.

Deming, D., 2002. *Introduction to Hydrogeology*. McGraw Hill, New York, USA.

Dupuit, J., 1863. *Études théoriques et pratiques sur le mouvement des eaux dans les canaux découverts et à travers les terrains perméables* (2eme edn). Dunot, Paris, France.

Erskine, A.D., 1991. The effect of tidal fluctuation on a coastal aquifer in the UK. *Ground Water*, **29**(4), 556–562.

Glover, R.E., 1959. The pattern of fresh-water flow in a coastal aquifer. *Journal of Geophysical Research*, **64**(4), 457–459.

Herzberg, A., 1901. Die Wasserversorgung einiger Nordseebäder. *Journal für Gasbeleuchtung und Wasserversorgung*, **44**, 815–819.

Hubbert, M.K., 1940. The theory of ground-water motion. *Journal of Geology*, **48**, 785–944.

Jacob, C.E., 1950. Flow of ground-water. In: H. Rouse (Ed.), *Engineering Hydraulics*. Wiley, New York, USA, pp. 321–386.

Lettenmaier, D.P. and Milly, P.C.D., 2009. Land waters and sea level. *Nature Geoscience*, **2**, 452–454.

Li, L., Barry, D.A., Stagnitti, F. and Parlange, J.-Y., 1999. Submarine groundwater discharge and associated chemical input to a coastal sea. *Water Resources Research*, **35**(11), 3253–3259.

Martínez, M.L., Intralawan, A., Vázquez, G., Pérez-Maqueo, O., Sutton, P. and Landgrave, R., 2007. The coasts of our world: ecological, economic and social importance. *Ecological Economics*, **63**(2–3), 254–272, doi: 10.1016/j.ecolecon.2006.10.022.

Michael, H., Mulligan, A.E. and Harvey, C.F., 2005. Seasonal oscillations in water exchange between aquifers and the coastal ocean. *Nature*, **436**, 1145–1148, doi: 10.1038/nature03935.

Moore, W.S., 1996. Large groundwater inputs to coastal waters revealed by ^{226}Ra inputs. *Nature*, **380**, 612–614.

Moore, W.S., 1999. The subterranean estuary: a reaction zone of ground water and sea water. *Marine Chemistry*, **65**, 111–125.

Moore, W.S., 2010. A reevaluation of submarine groundwater discharge along the southeastern coast of North America. *Global Biogeochemical Cycles*, **24**, GB4005, doi: 10.1029/2009GB003747.

Nicholls, R.J., 2010. Impacts of and responses to sea-level rise. In: J.A. Church, P.L. Woodworth, T. Aarup and S. Wilson (Eds), *Understanding Sea-Level Rise and Variability*. Wiley-Blackwell, Chichester, UK, pp. 17–51.

Robinson, C., Li, L. and Barry, D.A., 2007a. Effect of tidal forcing on a subterranean estuary. *Advances in Water Resources*, **30**, 851–865.

Robinson, C., Gibbes, B., Carey, H. and Li, L., 2007b. Salt-freshwater dynamics in a subterranean estuary over a spring-neap tidal cycle. *Journal of Geophysical Research*, **112**, C09007, doi: 10.1029/2006JC003888.

SCOPAC (Standing Conference on Problems Associated with the Coastline), 2003. *Lyme Bay and SE Devon Sediment Transport Study, 2003. Start Point to Berry Head*. Available at www.scopac.org.uk.

Turner, I.L. and Masselink, G., 2012. Coastal gravel barrier hydrology – observations from a prototype-scale laboratory experiment (BARDEX). *Coastal Engineering*, **63**, 13–22, doi: 10.1016/j.coastaleng.2011.12.008.

Wilson, A.M., 2005. Fresh and saline groundwater discharge to the ocean: a regional perspective. *Water Resources Research*, **41**, W02016, doi: 10.1029/2004WR003399.

Xin, P., Robinson, C., Li, L., Barry, D.A. and Bakhtyar, R., 2010. Effects of wave forcing on a subterranean estuary. *Water Resources Research*, **46**, W12505, doi: 10.1029/2010WR009632.

Younger, P.L., 1996. Submarine groundwater discharge. *Nature*, **382**, 121–122.

7 Beaches

GERBEN RUESSINK[1] AND ROSHANKA RANASINGHE[2]

[1]*Department of Physical Geography, Faculty of Geosciences, Institute for Marine and Atmospheric Research Utrecht, Utrecht University, Utrecht, The Netherlands*
[2]*Department of Water Science Engineering, UNESCO-IHE, Delft, The Netherlands*

7.1 Introduction

7.1.1 Setting

Beaches can be defined as accumulations of sand or gravel found along marine, lacustrine and estuarine shorelines, and deposited by waves and wave-induced flows. Sandy beaches account for about 20% of the world's coasts. Gravel beaches occupy approximately 10% of the global coastline and are found particularly in high-latitude areas. The beach constitutes the upper part of the shoreface (Fig. 7.1) and can be subdivided into three sub-systems:

(1) Subtidal zone: This zone extends from the low-water mark to typical water depths of 5–10 m. It is here that waves incident from deep water shoal and break. The breaking-wave zone is also called the surf zone, and in its outer part the breaking waves change rapidly in appearance (shape). Further onshore, in the inner region, all breaking waves have transformed in sawtooth-shaped bores that propagate to the shore without any major change in shape. During wave breaking, the motion of the incident waves is transformed into motions of different types and scales. These include small-scale turbulence, about 20–200 s long infragravity waves that are related to the grouped structure of the wind-generated waves, and steady flows, including alongshore, rip and bed return (undertow) flow currents. The topography of subtidal sandy beaches may be simple, but may also contain one or more alongshore ridges known as sandbars.

(2) Intertidal zone: This zone is located between the low- and high-tide marks. Here, the inner surf-zone bores finally expand their energy and run up and down the beachface as swash and backwash. When the tide range is sufficiently large and/or the incident waves are rather low, even shoaling waves can be active in the intertidal zone. With a decrease in bed gradient and an increase in offshore wave height, infragravity waves become increasingly important in swash and backwash action. These low-frequency waves do not suffer from wave breaking as much as the incoming sea and swell waves, and can, therefore, dominate the swash during storms. The intertidal zone may also display sandbar topography.

Coastal Environments and Global Change, First Edition. Edited by Gerd Masselink and Roland Gehrels.
© 2014 John Wiley & Sons, Ltd. Published 2014 by John Wiley & Sons, Ltd. Companion Website: www.wiley.com/go/masselink/coastal

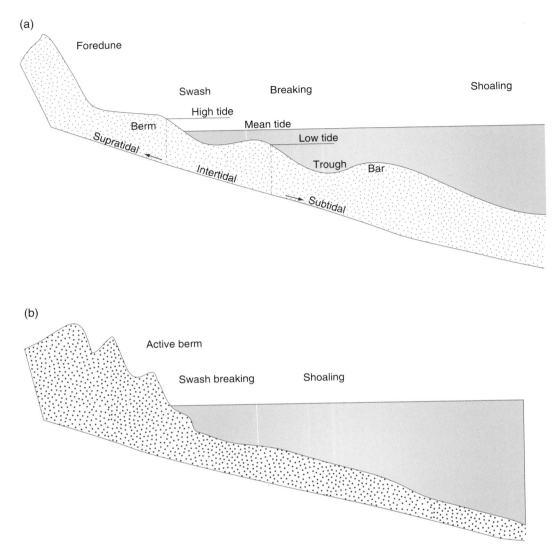

Fig. 7.1　Typical cross-shore profile for: (a) a gently sloping sandy beach; and (b) a steep gravel beach, showing characteristic morphological features and hydrodynamical zones. (Source: Adapted from Davidson-Arnott 2010. Reproduced with permission.)

(3) Supratidal zone: This zone is the part of the cross-shore profile between the high-tide level and some physiographic change such as a dune or a cliff, or permanent vegetation. It is normally above the level of the highest swash, except during storms, and is dominated by wind-driven processes. During prolonged periods of low-wave activity, sand may accumulate at the inter-supratidal boundary to form a berm (see Fig. 1.8). It protects the coastal dunes and cliffs from erosion during the early phase of higher-wave conditions. When berm erosion is pronounced, the beach profile may become scarped. In particular, gravel beaches can have multiple berms at different locations, even quite high on the profile (storm berms).

In the terminology adopted here, the beach is synonymous to the nearshore zone. Beaches can be long and uninterrupted for several tens of kilometres ('open beaches'), can be part of the seaward side of barrier islands, or can be located between headlands ('pocket beaches').

7.1.2　Scales of nearshore morphology

Morphological patterns abound on natural beaches. Figure 7.2 provides a graphical impression of the spatial and temporal scales of morphological classes commonly observed in the nearshore zone, from small-scale ripples to

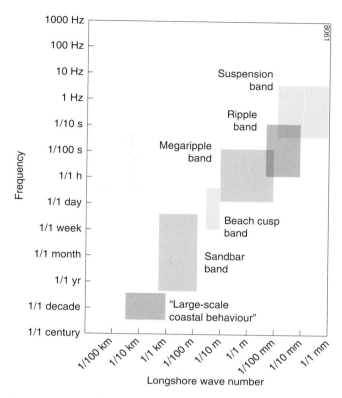

Fig. 7.2 Temporal and spatial classification of typical beach features. Longshore wave number is the reciprocal of wavelength. (Source: Adapted from Holman 2001. Reproduced with permission from Springer Science + Business Media.)

large-scale coastal behaviour. It is intriguing to see that the classes span six orders of magnitude in space, from centimetres to tens of kilometres. The temporal scales are approximately linearly coupled to the spatial scale (see also Fig. 1.7). Wave-induced ripples found in the shoaling zone can modify their appearance in a matter of seconds as they adjust their height in response to the passage of lower and higher waves in a wave group. Larger-scale sandbars change in position and appearance in response to the storms. The response of an entire coastal section to a rise in sea level or increased storminess may take decades or longer.

Wave ripples (Fig. 7.3a) are the smallest-scale patterns commonly observed on beaches, with typical wavelengths of 10–100 cm and height-over-length ratios (ripple steepness) of 0.2 or less. Wave ripples are roughness elements that reduce the energy of waves and currents, and they also locally modify sediment transport rate and direction. There are numerous terminologies or classification schemes for the various types of ripples observed. One of the most widely adopted schemes is Clifton's (1976) and comprises three types of ripples: (1) orbital ripples with a wavelength proportional to the wave orbital diameter; (2) anorbital ripples with a wavelength proportional to the grain-size diameter; and (3) suborbital ripples with intermediate wavelengths. The change in wave characteristics across the nearshore results in a cross-shore progression of ripple types over a beach. Based on scuba-diver observations, Clifton (1976) found asymmetric ripples in the shoaling zone, succeeded by a zone of cross-ripples and lunate megaripples just seaward of the breaker zone. Beneath breaking waves, the seabed was planar, as the bed shear stress was sufficiently high to prevent bed forms from forming (Fig. 7.4). A similar succession of bed states, albeit with somewhat different ripple types, has also been observed temporally at a single location in response to temporal variability in wave forcing (Hay and Mudge, 2005).

Beach cusps (Fig. 7.3b) represent alongshore rhythmic patterns in the swash zone of steep, coarse-sand or gravel beaches. An individual beach cusp comprises two topographic highs, or horns, and a single topographic low, or bay. The alongshore spacing between consecutive horns is in the order of several tens of metres and is proportional to the horizontal extent of the swash motion, with a constant of proportionality of about 1.5. Beach-cusp formation is a prime example of positive feedback (Werner and Fink, 1993). The wave uprush decelerates over the horns and is deflected from the horns into the adjacent bay, where it piles up and accelerates to return seaward as a narrow jet, or 'mini-rip'. The patterns of deceleration and acceleration result in sediment deposition on the horns and erosion from the bays. Thus, the beach cusps modify the swash in such a way that their height increases. At some point, negative feedback takes over to stabilize the cusp patterns. The sequence of positive and then negative feedback between morphology and hydrodynamics is characteristic of self-organization (see also Box 1.1).

Sandbars are the largest of the patterns found in the nearshore and have been observed both on the subtidal and intertidal beach (Wijnberg and Kroon, 2002). Intertidal bars can be found in tidal ranges as large as 6 m, while subtidal sandbars are mostly limited to tideless to mesotidal settings. Subtidal sandbar morphology is located in water depths of less than 10 m and can assume a large variety of configurations, including shore-obliquely oriented transverse bars, crescentic bars, and longshore uniform bars (Figs. 7.3c–d). All types can extend for up to several tens of kilometres along the shore. Crescentic and alongshore bars can occur singularly or in multiples of up to four or five bars. When the bar morphology is characterized by a dominant alongshore wavelength (i.e. spacing of rips), this is referred to as rhythmic morphology. In multiple-bar settings, the dominant wavelength in outer bars is usually larger than in inner bars. Also, rips tend to be less regularly spaced in inner than in outer bars. The dynamics of intertidal and subtidal sandbars is predominantly governed by breaking waves and associated flows.

(a)

(b)

(c)

(d)

Fig. 7.3 Examples of morphological patterns typical of sandy beaches: (a) ripples; (b) beach cusps; (c) alongshore sandbars; and (d) crescentic sandbar. The sandbars in (c) and (d) are located beneath the white high-intensity bands that are induced by the persistent wave-breaking above the sandbar crests (see Box 7.2). (Source: (a) Gerben Ruessink. (b) Giovanni Coco, courtesy of NIWA. (c) and (d) Ian Turner.) For colour details, please see Plate 18.

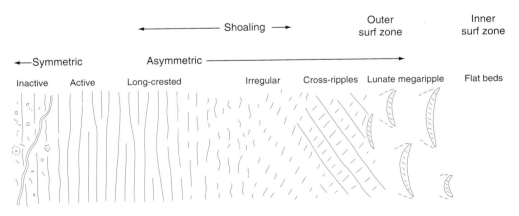

Fig. 7.4 Cross-shore evolution of ripple types based on scuba-diver observations on planar beaches. See text for further explanation. (Source: Adapted from Clifton 1976.)

Sandbars are highly important morphological features from a societal point of view. Firstly, subtidal sandbars are shallow parts in a cross-shore profile and thus force wave breaking to occur away from the shoreline, providing a means of coastal protection. Secondly, subtidal sandbars are rather voluminous and are thus significant in coastal budget studies. Thirdly, sandbars continuously modify their position in response to changes in the wave forcing. The change in sandbar position, and also the presence and location of rip channels, constitute the major control on nearshore bathymetric variability on the timescales of weeks to years. This is precisely the timescale most relevant to coastal zone managers. Accordingly, many measures to mitigate coastal erosion nowadays involve modifying sandbar characteristics, such as increasing their height by artificial sand dumping.

7.1.3 Scope of this chapter

This chapter provides a description of the nearshore zone, including its governing wave and wave-driven processes and their feedback with surf-zone morphology through sediment transport. This chapter also examines the potential effects of global-change induced accelerated sea-level rise and changes in the number and magnitude of severe storms, and how coastal management may assist in dealing with the associated coastal problems. Our focus is on sandy, wave-dominated, open beaches. The dominant processes in gravel beach dynamics are reviewed in Buscombe and Masselink (2006).

7.2 Nearshore hydrodynamics

7.2.1 Wave breaking

Shoaling waves increase in wave height and decrease in celerity until the water particle velocity at the crest exceeds the wave celerity. This limit is associated with a limiting wave steepness, which is given by:

$$\left[\frac{H}{L}\right]_{max} = 0.142 \tan h(kh) \qquad (7.1)$$

where H is the wave height, L is the wavelength, k is the wave number $(2\pi/L)$, and h is the water depth. Using shallow-water approximations, eqn. 7.1 reduces to $\gamma = H_b/h_b = 0.88$, where H_b is the breaking wave height, h_b is the breaker depth, and γ is the breaker criterion. For solitary waves, the value of γ has been found to range between 0.73 and 1.03, although the 0.78 value determined by McCowan (1894) is most commonly adopted. However, laboratory studies have indicated that the value of γ can depart significantly from these theoretical values, depending on the beach slope and deep-water wave steepness, and also on

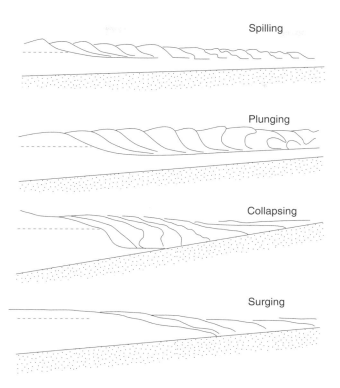

Fig. 7.5 The four different breaker types. (Source: Adapted from Davidson-Arnott 2010. Reproduced with permission.)

the wind speed and direction. All of these γ values are for monochromatic waves. In a natural environment where waves are irregular, the value of γ can vary between approximately 0.3 and 0.8.

Breaking waves can take different forms, depending on the beach slope and wave characteristics. In general, three different breaker types have been defined: spilling, plunging, and surging breakers. An intermediate breaker type between plunging and surging breakers has also been identified as collapsing breakers. The different breaker types are shown in Fig. 7.5. Battjes (1974) classified the three main breaker types in relation to the Iribarren number $\xi = \tan \beta / \sqrt{H/L}$ ($\tan \beta$ = beach slope) as follows:

Spilling	$\xi_\infty < 0.5$	$\xi_b < 0.4$
Plunging	$0.5 < \xi_\infty < 3.3$	$0.4 < \xi_b < 2.0$
Surging	$\xi_\infty > 3.3$	$\xi_b > 2.0$

The subscripts ∞ and b denote deep water and breaking conditions, respectively.

This classification is purely based on uniform waves in laboratory tests. The relatively few attempts to verify these ξ ranges in the field have not been very successful (Komar, 1998). Although the trends were found to be correct (i.e. breaker type changes from spilling to plunging to surging as ξ increases), the limit values given above should be treated with some caution.

Waves transport energy and momentum as they propagate. The wave energy E is given by:

$$E = \frac{1}{8}\rho g H^2 \qquad (7.2)$$

where ρ is the density of water and g is the gravitational acceleration. In Newtonian physics, momentum equals mass multiplied by velocity, that is, a mass being transported or a mass flux. Thus, momentum q can be expressed as $q = \rho \bar{u}$, where $\bar{u}$ = (cross-shore, alongshore, vertical) velocity vector. Momentum is therefore a vector quantity of which the direction is the same as that of the velocity vector.

As waves propagate, the momentum of a fluid element can change, resulting in a wave-induced momentum flux, which is given by ρu^2. According to Newton's second law, any change of momentum results in a force acting upon the fluid element. Hence, wave-induced momentum flux results in a wave force acting in the opposite direction. The wave-induced momentum flux is referred to as 'radiation stress', which is conventionally defined as 'the excess momentum flux due to the presence of waves' (Longuet-Higgins and Stewart, 1964).

Radiation stress is a vector quantity with both normal and shear components. For an oblique wave approaching an alongshore-uniform coast at an angle of θ to the shore normal, these components can be expressed (using linear wave theory) as:

Normal component in the cross-shore (x) direction

$$S_{xx} = \left(n - \frac{1}{2} + n\cos^2\theta\right)E \qquad (7.3)$$

Normal component in the along-shore (y) direction

$$S_{yy} = \left(n - \frac{1}{2} + n\sin^2\theta\right)E \qquad (7.4)$$

Shear components $S_{xy} = S_{yx} = \left(n\cos\theta\,\sin\theta\right)E \qquad (7.5)$

In eqns 7.3 to 7.5, n is the ratio between the wave speed and the wave group speed, and $n = 0.5$ in deep water and $n = 1$ in shallow water. The shear components act in the direction of the axis given by the first subscript, but exert themselves on a plane normal to the direction of the axis given by the second subscript.

For shore-normal waves, $\theta = 0$ and thus:

$$S_{xx} = \left(2n - \frac{1}{2}\right)E \qquad (7.6)$$

$$S_{yy} = \left(n - \frac{1}{2}\right)E \qquad (7.7)$$

$$S_{xy} = S_{yx} = 0 \qquad (7.8)$$

which in shallow water $(n = 1)$ reduces to $S_{xx} = 1.5E$ and $S_{yy} = 0.5E$.

Wave forces in the cross-shore direction are governed by eqn. 7.3 and in intermediate water depths prior to wave breaking the value of n increases in the direction of wave travel. Therefore, S_{xx} increases in the shoreward direction (prior to breaking), resulting in an equal and opposite wave force acting in the seaward direction. This offshore-directed force is balanced by a small hydrostatic pressure gradient, resulting in a slightly lower (than the still water level, SWL) surface elevation at the shoreward end of the shoaling zone (prior to wave breaking). This small lowering of the surface elevation just prior to the breaker location is referred to as wave set-down (Fig. 7.6). The opposite phenomenon occurs in the surf zone (after

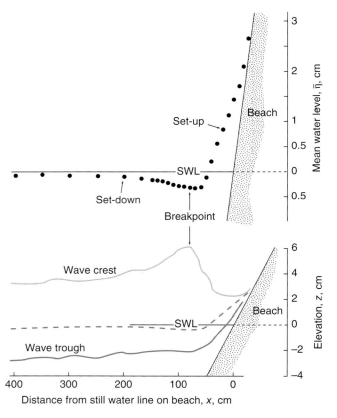

Fig. 7.6 Laboratory experiments of wave set-down and set-up. The location of the maximum set-down is located just landward of the breakpoint. SWL, still water level. (Source: Adapted from Bowen 1968. Reproduced with permission of John Wiley & Sons.)

waves break). When waves break, energy is dissipated rapidly, resulting in a shoreward decrease in S_{xx}. This results in an onshore-directed wave force, which must then be balanced by a hydrostatic pressure gradient such that the water level is higher (than the SWL) at the shoreline. This raising of the water level in the surf zone is referred to as wave set-up (Fig. 7.6).

Analytical expressions for wave set-down and set-up can be derived by considering the cross-shore force balance:

$$F_x = -\frac{dS_{xx}}{dx} = pg\left(h_0 + \bar{\eta}\right)\frac{d\bar{\eta}}{dx} \qquad (7.9)$$

where F_x is the wave force in cross-shore direction, h_0 is the still water depth, and $\bar{\eta}$ is the wave set-down/up. It should be noted that outside the surf zone $dS_{xx}/dx > 0$, while inside the surf zone $dS_{xx}/dx < 0$. Substituting $S_{xx} = 1.5E$ and $\gamma = H_b/h_b$ it can easily be shown that the set-down just prior to wave breaking is:

$$\bar{\eta}_{setdown} = -\frac{1}{16}\gamma H_b \qquad (7.10)$$

The water level inside the surf zone varies such that:

$$\frac{d\bar{\eta}}{dx} = -\frac{\tfrac{3}{8}\gamma^2}{\left(1 + \tfrac{3}{8}\gamma^2\right)}\frac{dh_0}{dx} \qquad (7.11)$$

with a maximum wave set-up $\bar{\eta}_{max}$ (relative to SWL) of $\tfrac{5}{16}\gamma H_b$ at the shoreline. With $\gamma = 0.5$ (for irregular waves) this returns a maximum shoreline set-up of about $0.16H_b$. This compares well with the maximum shoreline set-up measured along southern California beaches.

When waves break, energy is dissipated. The characteristics of wave energy dissipation are governed by the underlying beach profile. If the beach profile is steep, waves break practically at the shoreline and immediately run up the beach. In such situations, there is virtually no surf zone, and all wave energy is dissipated in the breaking process and in a highly turbulent swash zone. On the other hand, if the beach slope is mild (such as on intermediate or dissipative beaches), the wave energy will be gradually dissipated across a wide surf zone. If the beach slope is mild and uniform, waves (of different height) will break at all depths, resulting in fairly uniform wave energy dissipation across the width of the surf zone. However, if the profile consists of pronounced bars and troughs, then there will be more waves breaking over the bars and little or no breaking over the deeper troughs. In such cases, waves that break over the shallow bar might re-form over the deeper trough area and then shoal and break again, either at an inner bar

or on the slope close to the shoreline (shore-break). Note that this results in multiple set-up/down zones.

Due to the irregularity of wave forcing, at any given position in a natural surf zone there are some waves that are broken or breaking (characterized by white foam), and some that are still unbroken. The proportion of broken or breaking waves increases progressively from the seaward end of the surf zone to the shoreline. Field experiments conducted at Torrey Pines Beach, California, USA, have shown that wave heights in the inner surf zone are independent of deep-water wave heights, but are limited by the local water depth such that $H_{rms} \approx 0.42h$ (Thornton and Guza, 1982). The inner surf zone where this stable value of H_{rms} (and thus energy) occurs is referred to as the 'energy saturated' zone.

7.2.2 Wave-generated nearshore currents

Wave-generated currents can be classified into three main types (Fig. 7.7): longshore currents, cross-shore currents and rip currents.

7.2.2.1 Longshore currents

The wave-induced force F_y due to horizontal gradients in radiation stresses in the longshore direction is given by:

$$F_y = -\left(\frac{\partial S_{yy}}{\partial y} + \frac{\partial S_{yx}}{\partial x}\right) \qquad (7.12)$$

For a longshore uniform coast the $\frac{\partial}{\partial y}$ term is zero and hence eqn. 7.12 reduces to:

$$F_y = -\left(\frac{\partial S_{yx}}{\partial x}\right) \qquad (7.13)$$

This alongshore force is balanced by the bed shear stresses resulting from the generation of an alongshore current. Note that for shore-normal waves $S_{yx} = 0$, and hence there is no wave-driven longshore current under shore-normal waves.

By substituting eqn. 7.5 for S_{yx} in eqn. 7.13 and introducing wave celerity C, eqn. 7.13 can be re-written as:

$$F_y = -\frac{\partial}{\partial x}\left[ECn\cos\theta\left(\frac{\sin\theta}{C}\right)\right] \qquad (7.14)$$

From Snell's law (wave refraction) $\sin\theta/C$ is a constant. Also, in the nearshore zone, it is reasonable to assume that θ is small (usually < 10°), and thus $\cos\theta \approx 1$. Using these two approximations, and eqns 7.2 and 7.13, it can be shown that:

$$F_y = -\frac{5}{16}\frac{\sin\theta_\infty}{C_\infty}\rho\gamma^2\left(gh\right)^{3/2}\frac{dh}{dx} \qquad (7.15)$$

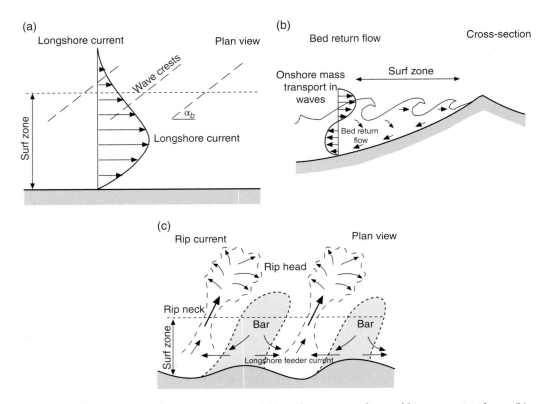

Fig. 7.7 Three main types of wave-generated nearshore currents: (a) longshore current due to oblique wave incidence; (b) cross-shore mass flux and undertow (bed return flow); and (c) rip currents due to onshore flow over bars and alongshore setup gradients resulting from alongshore non-uniform morphology. (Source: Masselink and Hughes 2003. Reproduced with permission from Hodder Education.)

By equating the above wave force to a quadratic bed shear stress, an expression for the cross-shore varying longshore current can be obtained as:

$$v(x) = -\frac{5}{16} g\pi h \frac{\gamma}{c_f} \frac{\sin\theta_\infty}{C_\infty} \frac{dh}{dx} \quad (7.16)$$

where c_f is a friction factor and $0 < x < x_b$ (x_b is the surf-zone width). This cross-shore distribution is only valid from the breaker line to the shoreline, and predicts that the longshore current is zero just outside the breaker line, suddenly increases to its maximum value at the breaker line, and then gradually decreases to zero across the surf zone (the dashed line in Fig. 7.8). However, in nature, the cross-shore profile of the longshore current is smoother. This is due to the lateral dispersion of momentum via horizontal eddies. This phenomenon is often referred to as 'horizontal (or lateral) mixing'. Each line shown in Fig. 7.8 represents a different level of lateral mixing, represented by the parameter P. The level of lateral mixing increases with increasing P, with $P = 0$ representing no lateral mixing resulting in the saw-tooth shape of the longshore current profile given by eqn. 7.16 (the dashed line in Fig. 7.8).

7.2.2.2 Cross-shore currents

The momentum of an unbroken propagating wave $q_{unbroken}$, which is in the direction of wave propagation, is limited to the area between the wave trough and wave crest and can be expressed as:

$$q_{unbroken} = \int_0^\eta \rho u \, dz \quad (7.17)$$

As explained earlier, momentum can be considered as a flux of mass, or a mass flux. This mass flux can be visualized in an Eulerian sense by considering waves travelling past a pole fixed on the seabed. As each wave goes past the pole, the pole will bend in the direction of wave propagation (onshore) during a small part of the wave period. Thus, if velocities were measured above SWL, recordings will only be made during a small part of the wave period, but all velocities will be in the onshore direction. If velocity measurements were made just below SWL (between SWL and wave trough level), recordings will be made during a longer part of the wave period and some of the measured velocities will be in the offshore direction. However, the time-averaged velocity

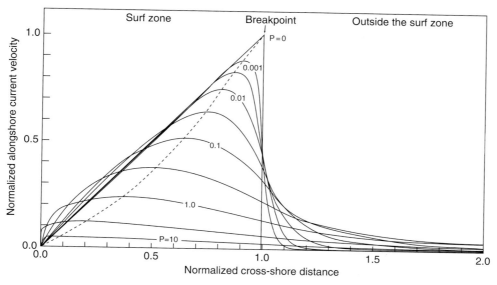

Fig. 7.8 The cross-shore distribution of the alongshore current on a planar beach. Each line represents a different level of lateral mixing (indicated by the P-value), with $P = 0$ representing no lateral mixing. Note that the surf zone has been normalized such that the breakpoint is located at $x = 1$. The dashed line connects the locations of maximum alongshore-current velocity. (Source: Adapted from Longuet-Higgins 1970. Reproduced with permission of John Wiley & Sons.)

over the wave period at this location will still be in the onshore direction. Thus, there is a net onshore mass flux between the wave crest and the wave trough. Mass flux can also be explained in Lagrangian terms. Considering linear wave theory it can be seen that a water particle will travel onshore faster while it is under the wave crest than when it travels offshore while it is under the wave trough. Thus, particle paths will not be entirely closed, resulting in a residual onshore movement of water particles. This residual transport is known as Stokes drift, which when integrated over the water depth is exactly the same as the mass flux described above in Eulerian terms.

By substituting for orbital velocity from linear wave theory, it can be shown from eqn. 7.17 that the mass flux per unit surface area for unbroken waves is:

$$q_{unbroken} = \frac{\rho g a^2}{2C} = \frac{E}{C} \qquad (7.18)$$

where a is the wave amplitude (half the wave height). For breaking waves, the surface roller energy also contributes to the mass flux, thus resulting in a much larger onshore mass flux. The surface roller is the layer of air-water mixture that moves onshore in the upper water column. The surface roller results in a delay of wave energy dissipation due to the temporary storage of energy and momentum in the roller. The mass flux under breaking waves $q_{breaking}$ is:

$$q_{breaking} = \frac{E}{C} + \frac{\alpha E_r}{C} \qquad (7.19)$$

where E_r is the wave energy associated with the roller, and α is a constant that can vary between 0.22 and 2 (a value of 1 is commonly adopted).

As the coast is a closed boundary across which no water can be transported, for a strictly alongshore-uniform coast, all of the mass flux has to be returned offshore. This is facilitated by a net offshore-directed flow below the trough level (Fig. 7.7b). This offshore-directed return flow is known as the undertow, or the bed return flow, and when integrated over the water column must exactly balance the mass flux. Therefore, the depth-averaged undertow velocity U_w can be expressed as:

$$U_w = \frac{q}{\rho h_t} = \frac{Ek}{\rho \omega h_t} \qquad (7.20)$$

where q is the mass flux (unbroken or breaking waves), $\omega = 2\pi/T$ (T is the wave period), and h_t is the water depth below trough level. As the mass flux for unbroken waves is relatively small, the undertow is very small outside the surf zone. However, for breaking waves the mass flux increases by about two orders of magnitude due to the influence of the surface roller; typical undertow velocities are about 0.2–0.3 m/s (Masselink and Black, 1995).

Rip currents are usually associated with relatively deep, cross-shore oriented channels from the shoreline to beyond the breakpoint. These channels are separated alongshore by shallow alongshore bars or transverse bars. Offshore-directed rip currents are a component of the horizontal cell-circulation patterns that are generated due to a combination of onshore flow over shallow bars and alongshore set-up gradients.

On an alongshore-uniform plane beach, for shore-normal waves, the cross-shore force balance is given by eqn. 7.9: the cross-shore gradient in momentum flux (radiation stress) is exactly matched by the set-up gradient. In this case, there can be no net cross-shore flows. However, on an alongshore non-uniform barred beach (e.g. Fig. 7.3d) in addition to wave set-up at the bar there will also be an onshore flow over the bar. In this case the radiation stress gradient is matched by a combination of wave set-up and shear stress. The partitioning of the radiation stress gradient into wave set-up and onshore flow is a function of many parameters, including wave height, water depth over bar, bar crest width and seaward slope of the bar.

The onshore flow over the bar flows landward until it encounters the closed boundary of the coast, and then turns alongshore, forming a rip feeder current. When this feeder current meets another feeder current flowing in the opposite direction (formed due to the same mechanism, but over an adjacent bar), they merge and turn offshore, forming a strong rip current that flows offshore in the channel between the two bars. This rip cell circulation pattern is further enhanced by alongshore set-up gradients due to the alongshore alternating bar-channel-bar morphology. Waves break on the shallow bars, but do not break in the deeper channel in between bars. Thus, at the cross-shore location of the bar, set-up will only occur at the bar locations. This will result in higher surface elevations at the bars and a lower surface elevation at the channel, resulting in a tendency for water to flow from the bar area to the channel. The combination of these two mechanisms results in rip cell circulations. Under moderate wave energy conditions, rip cells are mostly confined to the surf zone, while under high-energy wave conditions the circulation cells may not always be closed, leading to rip currents that extend far beyond the surf zone. Under oblique wave incidence, a longshore current is super-imposed on the mechanisms described above and thus the described rip cell circulation patterns (which are for shore-normal wave incidence) will be skewed in the downdrift direction (Fig. 7.7c).

7.3 Surf-zone morphology

7.3.1 Intertidal sandbars

Depending primarily on the wave conditions and tidal range, and to a lesser extent on the nearshore gradient, intertidal bars can attain a number of shapes (Masselink et al., 2006). Slip-face bars (Fig. 7.9a), characterized by a well-defined landward-facing slip-face and a more gentle

Fig. 7.9 Intertidal sandbars: (a) a slip-face bar (Castricum, the Netherlands); (b) low-amplitude ridges (Cleveleys, Fylde Coast, northwest England); and (c) sand waves (Truro, Massachusetts, USA). (Source: (a) Jacobs 2005. Reproduced with permission of Beeldbank.rws.nl/ Rijkswaterstaat. (b) Courtesy of S. Ilic. (c) Moore et al. 2003. Reproduced with permission of Elsevier.)

seaward slope, generally occur on beaches with an overall mild slope (typically 2°) subject to variable wave conditions and a micro- to mesotidal tide range. Their height can become quite pronounced, with the elevation difference between the bar crest and the shoreward-located trough sometimes exceeding 1 m. They generally form during post-storm conditions around the low-water line and migrate onshore to merge to the supra-tidal beach after several weeks or months of prolonged low-energy wave conditions. High-energy wave conditions associated with a surge that submerges the entire intertidal zone for several tides interrupt the onshore migration and may destroy the bar. Onshore bar migration is most pronounced when surf-zone bores propagate over the crest in water depths of a few decimetres at most. The excess water brought onshore by the bores is channeled alongshore to nearby rip channels that cut through the bar approximately every 100 m. The net onshore flow over the bar takes the sand stirred by the surf-zone bores from seaward to landward of the crest, causing onshore bar migration.

Coastal environments subjected to a macrotidal range tide and low-to-intermediate wave conditions generally contain a series of shore-parallel low-amplitude ridges (two to six; Fig. 7.9b), collectively referred to as ridge-and-runnel morphology. Their cross-shore spacing is approximately 100 m and, just as with slip-face bars, they are intersected by small rip channels. Low-amplitude ridges are generally less than 1 m in height and are more or less symmetric in cross-shore shape. Because of the large tidal range, the swash and surf zone translate rapidly across the profile and are not stationary for a sufficient period of time to modify the ridges significantly. Accordingly, the ridges are rather immobile features, in marked contrast to slip-face bars. Early research indicated that ridge location roughly corresponds to the positions on the intertidal profile where the water level is stationary for the longest time, but more recent research has falsified this claim (Masselink and Anthony, 2001).

A third type of intertidal sandbars, referred to as intertidal sand waves, is typically found in low wave energy settings and very gentle intertidal slope (<0.5°). Often, intertidal sand waves grade offshore into subtidal bars. The number of bars may range from four to 20 (Fig. 7.7c). Their heights are generally less than 0.5 m, making them even more subdued than low-amplitude ridges. Their cross-shore spacing varies between 50 and 150 m. Crest lines are usually shore-parallel, but sinuous forms and bifurcations have also been observed.

7.3.2 *Subtidal sandbars: cross-shore migration*

Subtidal sandbars are highly dynamic and can migrate in the onshore and offshore direction in response to changing wave conditions (Wijnberg and Kroon, 2002). Offshore sandbar migration is generally observed during high wave events and can attain values of 20–30 m/day. The large

waves break on the sandbar, generating a strong offshore-directed current (undertow) that is largest near or just onshore of the bar crest. Here, also, most sediment is brought in suspension. The resulting gradients in sediment transport (Fig. 7.10a) cause erosion landward and deposition seaward of the bar crest. Consequently, the sandbar moves offshore and often the height difference between the bar crest and trough increases. The location of wave breaking and of maximum undertow and sediment suspension move seaward with the sandbar. This feedback between hydrodynamics, sediment transport and sandbar morphology causes the sandbar to move offshore until the storm subsides and the waves no longer break on the sandbar.

Onshore sandbar migration is generally observed during moderate-energy wave conditions and is substantially slower than offshore migration; observed migration rates rarely exceed a few metres per day. The physical processes underlying onshore sandbar migration are not well understood; however, there is increasing consensus in the coastal scientific community that it is related to wave non-linearity (Box 7.1). As depicted in Fig. 7.10b, wave non-linearity causes onshore transport over the entire sandbar, peaking at the bar crest. The resulting gradients in sediment transport result in erosion seaward and deposition landward of the bar crest, resulting in onshore sandbar migration. The location of maximum wave non-linearity and sediment suspension move landward with the sandbar. Again, the feedback between the waves, sediment transport, and sandbar morphology causes a continuation of the migration. At the same time, the sandbar will diminish in height. With the arrival of the next episode of large waves, the onshore bar migration ceases and the sandbar migrates offshore once more.

When low-to-moderate-energy wave conditions persist for a sufficiently long period of time, an onshore-migrating sandbar will ultimately weld to the beach to form a berm. On beaches with a pronounced seasonality in the wave climate, such as are found on parts of the west coast of the USA, this results in a seasonal cycle of a 'winter' and 'summer' profile. During the low-energy summer, existing sandbars move onshore to weld with the beach, building a wide, non-barred beach with a pronounced berm. The subsequent winter waves erode the beach to form one or more subtidal sandbars.

While variability in sandbar location on timescales of storms to seasons is forced by temporal variability in wave conditions, long-term monitoring programmes have revealed the presence of longer-term trends in location without any corresponding variability in the wave forcing (Ruessink et al., 2003). In particular, multiple subtidal sandbars may show a cyclic behaviour, with a generation phase just seaward of the low-tide line, a net seaward migration through the nearshore, and a decay phase at the seaward end of the nearshore (Fig. 7.12). The time period between successive decay phases is approximately constant on a single site, but has been reported to show remarkable inter-site variability, from about one year on a beach in Japan facing the

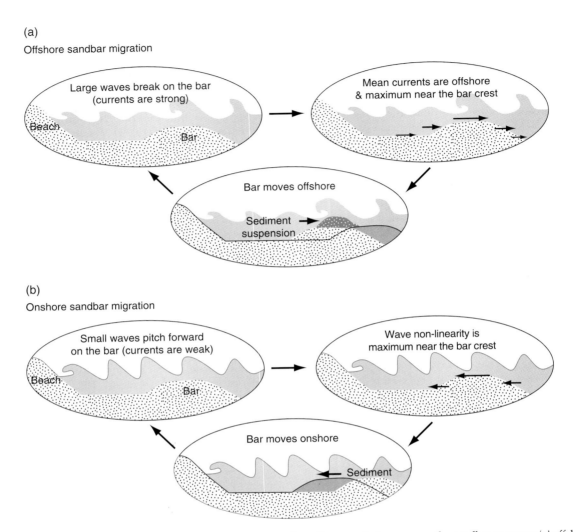

Fig. 7.10 Feedback between waves, currents, sediment transport and sandbar morphology causes the sandbar to move: (a) offshore when high waves break on the bar crest; and (b) onshore when intermediate, non-linear waves propagate across the sandbar. (Source: Adapted from Hoefel and Elgar 2003. Reproduced with permission of American Association for the Advancement of Science and S. Elgar.)

Pacific Ocean, to more than 15 years along some parts of the Dutch coast (Ruessink et al., 2003). The net offshore movement of the bars does not cause a net loss of sand offshore. Instead, it has been hypothesized that the sand of the decaying bar is moved onshore due to wave non-linearity, or is lost due to longshore sediment transport. The outer bar plays a key role in the continuation of the cycle. Once it starts to decay, the next bar disattaches from the shore and starts its journey offshore, and a new sandbar is generated just seaward of the low-tide level. The gentle increase in water depth over the crest of a decaying bar allows larger waves to reach the next sandbar, increasingly favouring conditions that result in offshore, rather than onshore, bar migration. The onset of net offshore bar migration can be alongshore coherent for several kilometres, but this

obviously requires the water-depth variability along the crest of the decaying bar to be minimal. In cases where this variability is not negligible, the next inner bar will loosen from the shore at some locations, but not at others. Ultimately, this will cause the alongshore disconnection of the initially uniform inner bar and the realignment of the disconnected bars with other sandbars (Fig. 7.13), a phenomenon referred to as bar switching (Shand et al., 2001).

Not all subtidal sandbar systems exhibit cycles of long-term net offshore migration. On some beaches, bars migrate progressively onshore. In settings where low-wave conditions persist for several months or longer, such as in parts of the Mediterranean, the outer bar is only active during episodic, short-duration storms. This very occasional activation of the outer bar is likely to be insufficient to cause

CONCEPTS BOX 7.1   Wave non-linearity and sand transport

During wave shoaling and breaking, waves become increasingly non-linear as they enter shallow water. This change in wave shape is of profound importance to sand transport in the nearshore. As a rule of thumb, shoaling waves are skewed, with high, short crests and shallow, long troughs. This is also reflected in the orbital motion near the bed: the onshore stroke of the wave is stronger but of shorter duration than the offshore stroke (Fig. 7.11). As a consequence, substantially more sand is brought into suspension during the onshore stroke of the wave than during the offshore stroke. Therefore, the onshore transport under the wave crest will exceed the offshore transport under the wave trough. Further onshore, in the surf zone, the wave skewness decreases; instead, the waves become asymmetric, and resemble forward-leaning sawtooth waves (Fig. 7.11).

The temporal derivative of the near-bed orbital motion, the acceleration, under sawtooth waves is skewed, with larger accelerations during the transition from offshore to onshore velocity. Although the precise mechanism is still under debate, acceleration skewness causes more sand suspension during the onshore stroke of the wave, and hence onshore transport. Also, the time difference between the moment of maximum offshore velocity and the next flow reversal is small compared to the time needed for the stirred sand to settle back to the bed. Thus, it is possible that some sand stirred during the offshore wave stroke persists into the onshore wave stroke, and thus contributes to onshore sand transport. In contrast, maximum onshore velocity happens early in the onshore wave stroke of the wave (Fig. 7.11). Accordingly, most of the stirred sand has settled by the time of the reversal to offshore velocity (Ruessink et al., 2011) and does not contribute to offshore sand transport. Hoefel and Elgar (2003) have argued that acceleration skewness is the key mechanism to onshore sandbar migration (see Fig. 7.10b).

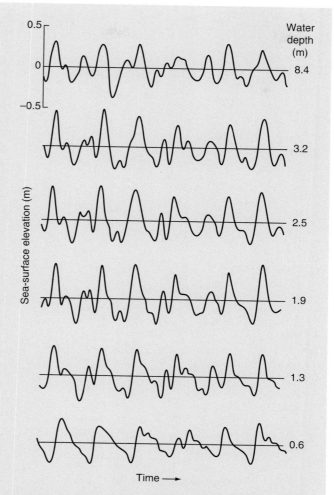

Fig. 7.11 Sea-surface elevation versus time for about 90 s. The wave shape changes from sinusoidal through skewed to forward-leaning asymmetric at the shallowest location. (Source: Adapted from Elgar and Guza 1985. Reproduced with permission of Cambridge University Press.)

---

any net long-term trend. The reasons for inter-site differences in long-term sandbar behaviour are not understood.

So far, we have explored the evolution of existing sandbars. But which processes lead to the formation of sandbars in the first place? This question remains unresolved despite considerable research and modelling efforts. The most likely hypothesis is known as the breakpoint hypothesis. Seaward of the breakpoint, sand is transported onshore because of wave non-linearity (see Box 7.1), while within the surf zone, sand is transported offshore due to the bed return flow. Thus, the breakpoint represents a point of sediment convergence. Once the bar has formed, the waves immediately shoreward of the bar crest will no longer be breaking. Instead, they will reform, shoal, and break again

further onshore, giving rise to yet another zone of sediment convergence and a sandbar. The number of sandbars that can form on a cross-shore profile likely depends inversely on the overall beach slope (Short and Aagaard, 1993).

7.3.3 *Subtidal sandbars: alongshore non-uniform dynamics*

Subtidal sandbars almost always exhibit significant alongshore morphological variability. The most obvious indicators of alongshore morphological variability are bar/shoal and rip channel patterns (e.g. see Fig. 7.3d). The alongshore variability of bar/shoal and rip channel patterns is conveniently encompassed by the 'beach state'. The beach

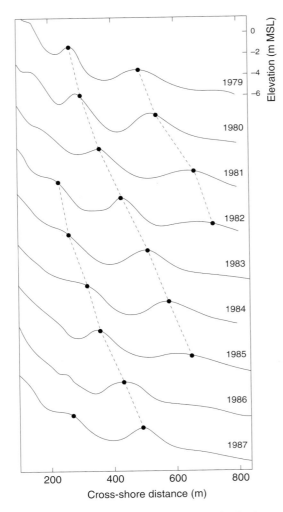

Fig. 7.12 Annual cross-shore profiles at Noordwijk, the Netherlands, showing multi-year cycles of bar generation, net offshore migration, and bar decay. The dotted lines connect crests of the same sandbar. At Noordwijk, the time period between consecutive bar decays is about four years. MSL, mean sea level.

state can thus be considered as a bulk parameter that qualitatively describes the predominant features of alongshore morphological variability at a given time.

Wright and Short (1984) have presented the most widely accepted sequential beach classification scheme based on beach states (Fig. 7.14). The three main beach states identified in this scheme are dissipative, intermediate and reflective. The intermediate beach state is further subdivided into four states: low tide terrace (LTT), transverse bar and rip (TBR), rhythmic bar and beach (RBB) and long-shore bar trough (LBT). Wright and Short (1984) employed a morphodynamic database spanning six years, which consisted of visual observations of surf zones of a number of beaches. These observations, in conjunction with measurements of wave parameters, were used to derive an empirical relationship between beach state and a dimensionless sediment fall velocity Ω, given by:

$$\Omega = \frac{H_b}{wT} \qquad (7.21)$$

Here, H_b is the breaker height, T is the wave period, and w is the sediment fall velocity. Reflective beach states are expected to occur when $\Omega < 1$, intermediate states when $1 < \Omega < 6$, and dissipative states when $\Omega > 6$. Of these three main beach states, the intermediate states are the most commonly observed and most temporally and spatially variable states.

The above classification scheme, which was based on observations at micro-tidal single-barred beaches, was later extended by Short and Aagaard (1993) to micro-tidal multi-barred beaches. The most notable phenomenon in the double-bar scheme is that the outer bar is never down-state of the inner bar (i.e. if the inner bar is RBB, then the outer bar will never be TBR). Masselink and Short (1993) examined the effect of tide range of beach states and introduced another dimensionless parameter, the relative tide range, RTR, defined as the ratio of the mean spring tide range to the breaker height. The beach scheme of Wright

Fig. 7.13 An example of bar switching at Wanganui, New Zealand. The high-intensity bands signal bar crests (see Box 7.2). In the foreground, the bars have been marked 1–3 and in the distance 1'–3'. In (b), bar 2 has realigned with bar 1' and bar 3 is realigning with bar 2'. (Source: Shand et al. 2001. Reproduced with permission of Elsevier.)

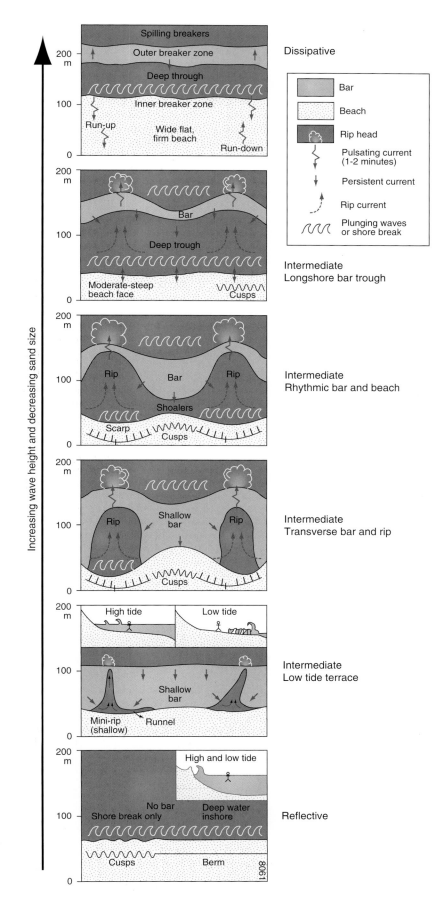

Fig. 7.14 Wright and Short's (1984) beach classification scheme, also known as the Australian beach model. (Source: Adapted from Wright and Short 1984. Reproduced with permission of Elsevier.) For colour details, please see Plate 19.

and Short (1984) is valid for RTR < 3. For Ω < 1, the reflective state grades with increasing RTR into low-tide terrace beaches with and finally without rips; for larger Ω, the intermediate beach types shift toward a bar-rip morphology at the low-tide level to finally a flat, ultra-dissipative beach; and, for the highest Ω, the barred dissipative extreme grades into a non-barred and finally ultra-dissipative beach. Masselink and Short's (1993) work thus highlights that all wave-dominated beaches in all tidal ranges

can be classified with knowledge of Ω and RTR (e.g. Scott et al., 2011).

Based on ARGUS video-imaging techniques (Box 7.2), Lippmann and Holman (1990) and Ranasinghe et al. (2004a) gained detailed insight into micro-tidal, intermediate beach-state dynamics at two different single-barred beaches: Duck, North Carolina, USA, and Palm Beach, Sydney, Australia, respectively. These studies showed that beaches continuously cycle through the four intermediate

---

METHODS BOX 7.2 Video remote-sensing

Over the past decades, optical remote-sensing has become an indispensible tool for monitoring the nearshore environment with high temporal resolution (days) over extended periods (years). The shore-based, automated video systems in the Argus programme (Holman and Stanley, 2007) have played a key role in popularizing nearshore imaging. An Argus system typically comprises four or five digital cameras (Fig. 7.15a) mounted on a high viewpoint, such as apartment building, hotel, cliff or lighthouse, and pointed obliquely along the beach to yield an 180° view of the nearshore. Each daylight (half-)hour an Argus system routinely collects a single snapshot and a 10-minute time-exposure image (Fig. 7.15b, c) for every camera. The individual images can be rectified to a top-view and merged into a single plan-view of the nearshore (Fig. 7.15d).

While a snapshot image shows individual breaking waves, a time-exposure image – the mathematical time-mean of all images collected with 1 or 2 Hz over a 10-minute period – shows a breaker region. This region

is the preferred location of wave breaking in the surf zone, in clear contrast to the darker alongshore lines where waves are not breaking. As waves break primarily because of limiting depth, the breaker regions are optical signatures of subtidal sandbars and associated morphological features, such as rip channels. The 10-minute averaging period is a compromise between the minimum period required to average over sufficient images to remove the signal of individual breaking waves and the maximum period required to capture true changes in breaker position owing to tidal variability. Intertidal sandbars are clearly visible in low-tide images.

Alternative video products are so-called pixel time series, that is, time series of image intensity collected at an individual pixel and sampled at 2 Hz. Several hydrodynamic variables, such as wave celerity, wave direction and alongshore current speed, can be estimated when time series are simultaneously collected at spatially extensive arrays of pixels. In essence, such pixel arrays act as optical equivalents of arrays of *in situ* instruments.

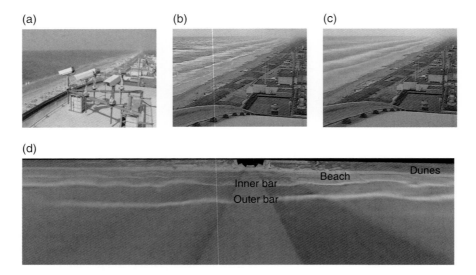

Fig. 7.15 (a) Argus station at Noordwijk, the Netherlands, with an example of (b) a snapshot, (c) a 10-minute time-exposure image from one of its cameras, and (d) a five-camera merged plan-view of (alongshore x cross-shore) 6000×1500 m.

beach states in response to changing wave conditions. A typical cycle observed at Palm Beach is shown in Fig. 7.16. The nearshore morphology is reset by storm events resulting in an LBT state, regardless of the beach state prior to the storm. When the wave energy decreases, the nearshore morphology sequentially evolves into RBB, TBR and LTT states. However, the occurrence of a storm at any stage of this evolutionary cycle would result in the re-set of the nearshore morphology to an LBT state. Higher beach states, which are associated with high-energy wave conditions, persist for shorter durations than do lower beach states, which are associated with low-energy wave

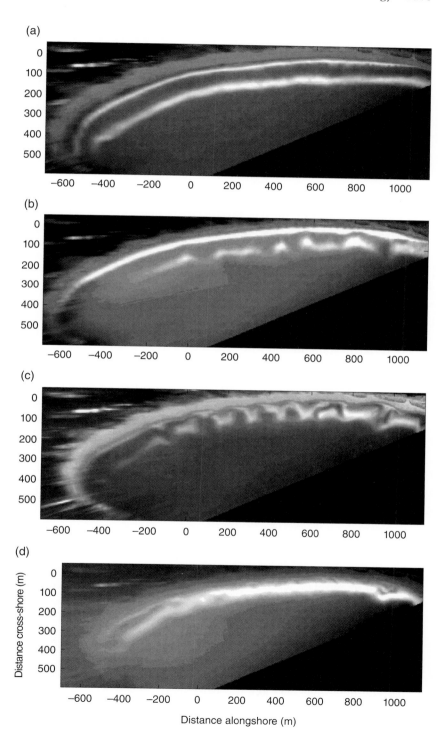

Fig. 7.16 Rectified time-exposure images of a full transitionary cycle of intermediate beach states at Palm Beach, Sydney, Australia: (a) longshore bar trough (LBT) on 8 May 1996; (b) rhythmic bar beach (RBB) on 15 May 1996; (c) transverse bar and rip (TBR) on 25 May 1996; and (d) low tide terrace (LTT) on 11 June 1996. (Source: Ranasinghe et al. 2004a. Reproduced with permission of Elsevier.)

conditions. Up-state transitions (e.g. RBB to LBT) are mainly associated with storm events and occur less frequently than do sequential down-state transitions (e.g. RBB to TBR) associated with accretive wave conditions. Additional video observations from the Gold Coast, Australia, further demonstrate that nearshore variability in the bar decreases during obliquely incident waves with intermediate height (Price and Ruessink, 2011).

During energetic wave conditions, particularly during storms, enhanced offshore sediment transport and/or alongshore transport can result in the complete re-working of the nearshore morphology and the emergence of LBT morphology during so-called reset events. Shortly after the reset event, wave energy decreases sharply and alongshore variability starts to re-emerge in the alongshore bar morphology. This re-emergence has puzzled scientists for decades. Why would an alongshore-uniform sandbar develop crescentic patterns or become dissected by rip channels after a storm? The present-day view on the causative mechanism for the LBT to RBB transition is that of self-organization, examined in great detail by Falques et al. (2000) (see also Box 1.1, Fig. 1.10 and Fig. 1.11). A small perturbation, characterized by a specific alongshore wavelength, is added to an otherwise alongshore-uniform bathymetry and its subsequent evolution is then examined. The perturbation results in minute variations in wave breaking, and hence causes cell-circulation patterns with onshore flow where the water depth is shallower than average and offshore flow where it is slightly deeper (section 7.2). In combination with offshore-increasing sediment transport rates, the circulation patterns cause sediment deposition on the shallow parts that thus grow into horns, and sediment erosion in the deeper parts to produce bays. The positive feedback of the irregularities into the flow causes the shallower parts to become shallower and the deeper parts to become deeper. The rate at which the initial perturbation becomes more pronounced turns out to depend on its wavelength, and it is expected that the wavelength that grows fastest corresponds to the one observed in nature. In general, self-organization models predict an increase in the fastest growing wavelength with wave height (for unbarred beaches) or the distance between the shoreline and the sandbar crest. While in nature the wavelength in outer bars is indeed often larger than in inner bars, most of the observations have unsuccessfully attempted to correlate wave height and wavelength. The fastest growing wavelength does not depend solely on wave height, but also on the water depth above the bar, the bed slope, and the cross-sectional area of the shoreward located bar trough, with other more complicated parameters potentially playing a role as well. Later (non-linear) models demonstrate that negative feedback starts to become important once the bed patterns reach a specific height. In particular, the downslope sediment transport by gravity acts to slow down and ultimately stop the growth of the emerging bed patterns.

Once initial alongshore variability is established in LBT morphology, it rapidly develops into rhythmic crescentic bar shapes resulting in RBB morphology through the above-mentioned feedback between wave breaking, mean flows, sediment transport and the evolving bathymetric variability. The data presented for Duck and Palm Beach by Lippmann and Holman (1990) and Ranasinghe et al (2004a), respectively, both show that the LBT to RBB transition is in fact the fastest inter-state transition, indicating that LBT is the most unstable beach state. The RBB to TBR transition also occurs relatively rapidly, albeit slower than the LBT to RBB transition. Based on their analysis of conditions at Palm Beach, Ranasinghe et al. (2004a) showed that a very small decrease in wave height (~2.5% decrease) is sufficient to trigger an RBB to TBR transition. Ranasinghe et al (2004a) also investigated the causative mechanisms for this transition via process-based numerical modelling and concluded that the hydrodynamic processes governing this transition consist of onshore flow over the bar crest, alongshore feeder currents and rip currents. These hydrodynamic processes result in: (a) formation of shoals in the lee of the longshore bars due to onshore sediment transport caused by strong onshore flow over the bars; (b) formation of shoreward extensions of the longshore bars at their extremities due to accretion at the confluence of feeder and rip currents; and (c) the combination of these two types of shoals and their eventual welding onto the beach. The rip currents in a fully established TBR morphology operate in the deep and well-defined transverse channels. The TBR state is the most stable beach state, making it the modal state at most high wave energy, microtidal beaches. The TBR to LTT transition appears to require a more significant decrease in wave height compared with the RBB to TBR transition. Ranasinghe et al (2004a) indicated that the TBR to LTT transition can only occur under obliquely incident waves and that it is a negative feedback mechanism that consists of: (a) erosion of the transverse shoals and the deposition of the eroded sediment in the rip channels by longshore sediment transport, thus diminishing the longshore variability in morphology; and (b) weakening of the rip circulation due to the decreasing longshore morphological variability.

The above process descriptions are based on observations from single-barred beaches. However, to some extent, these processes are also valid for both the outer and inner bars of a double-barred system. There are, however, some phenomena that are particular to double-barred systems. One of these, explored by Castelle et al. (2010), is the occasional striking coupling between inner- and outer-bar patterns (Fig. 7.17). This coupling suggests that self-organization processes on an inner bar either do not exist or are largely overwhelmed by some other mechanism. Video observations of the temporal evolution of rhythmic patterns in

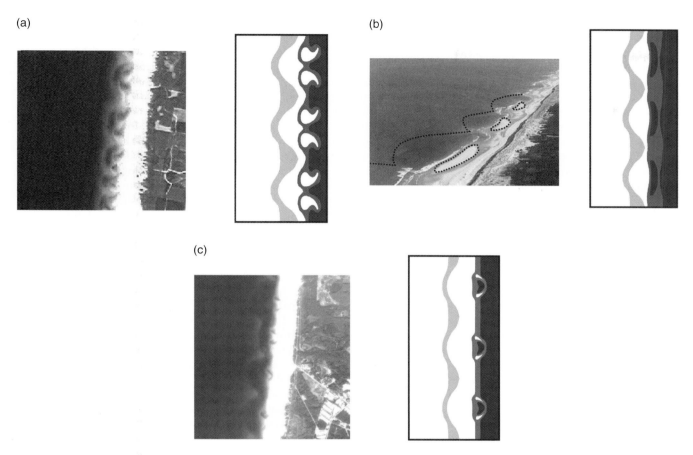

Fig. 7.17 Various examples in which the alongshore variability in the inner bar is forced by the variability in the outer bar. (Source: Castelle et al. 2007. Reproduced with permission of Elsevier.)

the double-sandbar system at the Gold Coast, Australia, highlight that immediately after a storm, inner- and outer-bar patterns are formed that grow independently of each other, highlighting the role of self-organization. However, at some point, the water-depth variability along the outer bar becomes so pronounced that it leads to wave-height variations at the inner bar due to differential wave breaking and refraction. The resulting cell-circulation patterns then enforce the growth of inner-bar patterns having the same length scale as the outer-bar patterns, instead of the smaller-scale features expected from self-organization processes alone. This suggests a temporal change in the dominance of self-organization into morphological coupling.

7.4 Anthropogenic activities

Most beaches provide safety from coastal flooding and support multiple human activities, including tourism and recreation, nature conservation and fishing. Interactions between two or more beach activities are often conflicting and can disrupt natural beach dynamics.

For example, the construction of jetties to allow for safe navigation into a harbour can lead to erosion of nearby beaches, endangering human infrastructure and coastal safety. Management is required to plan and co-ordinate the different activities, with the overall aim being to safeguard long-term sustainable development. Proper management is becoming increasingly important due to increased human occupation of the coastal zone and the anticipated adverse effects of global climate change on coastal evolution (section 7.5). Even the smallest coastal problem brings in many different stakeholders, including administrative organizations (councils, government agencies), scientists and various interest groups (residents, tourists, environmental organizations). The resulting complexity of coastal management issues and the various spatial and temporal scales often involved require an integrated, multidisciplinary approach for coastal zone management to be effective. Such an approach was born in 1992 during the Earth Summit of Rio de Janeiro and is referred to as Integrated Coastal Zone Management (ICZM) or Integrated Coastal Management (ICM).

Communication is of utmost importance to the success of any ICZM project. Davidson et al. (2007) introduced the Frame of Reference approach and paid particular interest on how coastal scientists (should) interact with coastal managers (Fig. 7.18). At the highest level, the coastal managers need to define clearly the management issues they face. With the issues clearly laid out, the frame of reference is invoked. As a starting point, the managers need to define the issues in terms of a strategic objective, in essence the long-term management vision for the sustainable development of the coastal zone. To make the strategic objective practical, one or more operational objectives have to be defined. The operational objective is achieved in a cyclic four-stage process that involves information collection (stage 1), decision-making (stage 2), intervention implementation (stage 3), and evaluation (stage 4). An essential ingredient of the first stage is to transform the relevant data and, sometimes, model results into a simple set of issue-related parameters that accurately and adequately describe the current state and evolution of a coastal system in relatively simple terms. Such parameters, referred to as coastal state indicators (CSIs), are meant to facilitate a management decision directly and to ensure effective communication between coastal managers, policy-makers, scientists and stakeholders. Poor or ineffective communication has indeed been a major reason why some coastal management projects were not as successful as intended. Scientists often felt that their knowledge was not implemented to solve practical problems, while coastal managers often claimed that the scientific knowledge does not come in the form

required for practical use. CSIs are thus meant to provide a common language among the various parties involved. The fourth, evaluation, stage ensures that ICZM is an iterative process. The success or failure of an implemented intervention to meet the operational objective can lead to refinements or complete modifications of the intervention, or to a revision of the benchmarks.

For many beaches, the major management issue is coastal protection, with the strategic objective 'to guarantee a sustainable safety level' and the operational objective 'to maintain the coastline seaward of a particular location'. Over most of the last century, the construction of large structures made of rocks, concrete or timber was the most commonly implemented intervention to combat coastal erosion. These hard coastal-protection structures, discussed in detail in Reeve et al. (2004), include shore-parallel breakwaters and shore-attached groynes. All types of hard structures are meant to separate the cause from the effect. Breakwaters, for example, are designed to dissipate part of the incident energy, reducing the direct impact of storm waves on beaches. Groynes are meant to trap sediment transported alongshore to locally maintain a sufficiently wide beach. Often, groynes trap so much sand that downdrift beaches become sediment starved and are eroded severely. This leads to the extension of hard structures to beaches that initially did not have any problem. The adverse effects of coastal structures on adjacent beaches, coupled with the increasing significance of the coastal zone for ecology, economy and recreation, has recently resulted in a shift towards coastal-protection methods based on the principle of 'Building with Nature'.

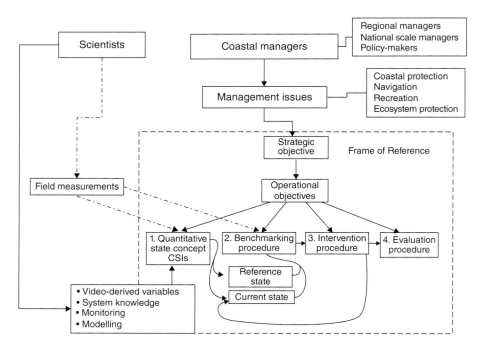

Fig. 7.18 Recommended model describing interaction between coastal scientists and managers according to the Frame of Reference approach discussed in the text. This model was designed in the framework of the EU-funded CoastView project, which aimed to support coastal zone management by means of video systems (see Box 7.2). CSIs, coastal state indicators. (Source: Davidson et al. 2007. Reproduced with permission of Elsevier.)

These new methods involve the artificial deposition, or nourishment, of naturally occurring material on dunes, the intertidal beach, or the shoreface (Box 7.3). Natural processes (waves, currents and wind) are expected to redistribute the nourished material across and along the coast, resulting in coastal protection or even growth. In contrast to hard structures, nourishments do not remove the actual cause of coastal erosion; instead, the nourished sediment takes the place of the otherwise eroded local sediment. Thus, sediment will continue to be lost and, accordingly, nourishments will have to be repeated at regular intervals.

The placement of sediment on the dry and wet beach is the most common type of nourishment. Extensive guidelines for its design can be found in Dean (2002). Generally, the size of the nourished material is slightly coarser than that of the local beach sediment. Finer sediments tend to be rapidly lost offshore, while much coarser material results in a steep beach that is unattractive for recreational purposes. Irrespective of the size of the material used, the first few post-nourishment storms will erode large quantities of sediment from the nourishment to form a more natural coastal profile. This rapid initial loss of sediment is often perceived by the public as a failure of the nourishment scheme; however, the eroded sediment often resides just seaward of the waterline and still contributes to coastal protection. Nonetheless, the rapid initial loss, coupled with the undesirable closing of the beach during the low wave energy recreational season for the implementation of the nourishment itself, has resulted in an increased popularity of sediment nourishment on the shoreface. Along the Dutch coast, for example, shoreface nourishments are nowadays the dominant type of coastal protection, with a placement of 10–15 million cubic metres of sand per year. The design of shoreface nourishment relies predominantly on practical experience and, in order to play safe, most shoreface nourishments contain up to twice the amount of sediment that is expected to be eroded from a stretch of coast during the anticipated nourishment lifetime.

Shoreface nourishments are either designed as breaker or feeder berms in relation to their intended lee and feeder effects, respectively (Fig. 7.19). A breaker berm will result in increased wave dissipation. The corresponding shoreward reduction in wave height and alongshore flow velocities will result in increased deposition of sediment supplied from alongshore. Thus, the beach shoreward of a breaker berm will grow because of the capture of sediment moving alongshore. On the unprotected downdrift side of the nourishment, the alongshore flow and associated alongshore sediment transport will increase in magnitude, leading to an inevitable erosion of sediment. A wider beach is also expected for a feeder berm. In this case, however, the sediment of the berm itself is the main source of beach widening. Processes contributing to this onshore transport can be wave non-linearity (see Box 7.1) or the onshore flow over the nourishment related to cell-circulation patterns. As the water depth over the nourishment will be slightly less than at similar distances from the shore elsewhere, the enhanced wave breaking will result in a larger set-up and hence drive a cell-circulation pattern with rip currents at the alongshore heads of the nourishment. These head effects are one of the main potential problems with shoreface nourishments. However, extensive monitoring of various nourishment projects along the Dutch coast (e.g. Box 7.3) has not provided any indication that the heads are locations of

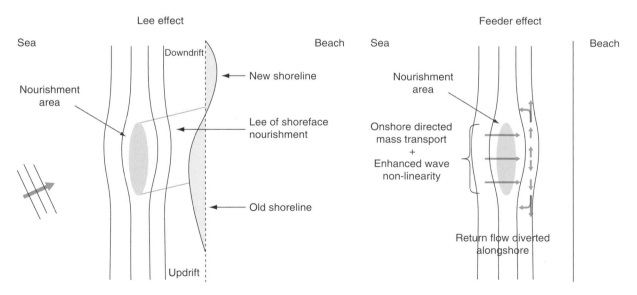

Fig. 7.19 Lee and feeder effects of a shoreface nourishment on its adjacent surroundings. (Source: Adapted from Van Duin et al. 2004. Reproduced with permission of Elsevier.)

CASE STUDY BOX 7.3   Shoreface nourishment at Noordwijk Beach, The Netherlands

In 1990, the Dutch government adopted the concept of dynamic preservation of the coastline. This concept aims at combining safety against flooding with sustainable preservation of other coastal functions. The safety component implies that the coastline will at least be maintained at its 1990 position. Sand nourishment within or seaward of the subtidal sandbars has been chosen as the primary means to mitigate persistent coastal erosion. The annual volume of nourished sand amounted to some 7 million cubic metres between 1990 and 2000, and had risen to 10–15 million cubic metres by 2010. From a safety perspective, the dynamic preservation policy is viewed as highly successful. While in 1992, about 35% of the Dutch sandy beaches suffered from persistent erosion, this number had dropped to 15% by 2000. However, there is now growing concern about the potentially negative effects of nourishments on ecosystems. In particular, it is feared that the reduced degree of dune erosion has suppressed the natural dynamics in the entire dune area and, accordingly, has resulted in habitat degradation and loss of biodiversity.

One of the largest and best-monitored nourishments was implemented in 1998 in front of the beach town Noordwijk aan Zee. The nourishment had a total volume of 1.7 million cubic metres, an alongshore length of about 3 km, and was positioned some 500 m seaward of the outer bar (Fig. 7.20) in 6–7 m water depth. Using a data set of Argus time-exposure images collected between 1998 and 2004, Ojeda et al. (2008) observed how the nourishment migrated more than 300 m onshore before losing its integrity in 2003. This onshore migration likely reflects the nourishment's feeder function. Intriguingly, the nourishment halted the net offshore migration of the sandbars (section 7.3.2) observed prior to the nourishment. Even though the nourishment was no longer discernible after 2003, the sandbars remained stagnant at their original pre-nourishment location. Because the sandbars elsewhere did continue migrating offshore, the originally alongshore-uniform bars broke up and realigned with other sandbars. Such a series of events, known as bar switching, can also been observed under natural conditions (section 7.3.2), but it is clear that the switching at Noordwijk was nourishment-induced. The Argus images further revealed that the nourishment did not result in more, or more pronounced, rip channels, and thus did not influence swimming safety. Despite expectations and the observed onshore migration of the nourishment, the subaerial beach did not widen. Apparently, the onshore-moving sand remained within the sandbar zone and did not end up on the beach.

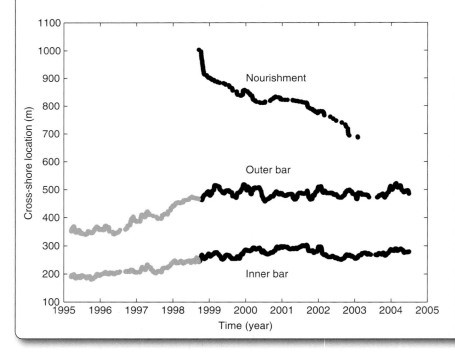

Fig. 7.20 Time series of alongshore-averaged location of the shoreface nourishment, and the outer and inner subtidal sandbar at Noordwijk, the Netherlands. The intertidal beach is located near 100 m. All estimates were based on the Argus video system at Noordwijk (see Box 7.2 and Fig. 7.15). Prior to the nourishment (grey dots), both the inner and outer bar migrated net offshore as part of an approximately four-year long net-offshore migration cycle (see Fig. 7.12). The nourishment can be seen to migrate onshore since its implementation in 1998 to finally disappear in 2003. The onshore migration reflects the feeder function of the nourishment. (Source: Adapted from Ojeda et al. 2008. Reproduced with permission of Elsevier.)

reduced swimming safety. Sometimes the functioning of shoreface nourishment changes with time. For example, extensive studies of the Terschelling nourishment in the Netherlands have shown that this nourishment initially acted as a feeder berm. This led to unprecedented growth of a more landward-located sandbar that started to act as a breaker berm. It is obvious that placement depth is crucial for shoreface nourishments to contribute to coastal protection. As a rule of thumb, nourishments built inside or just offshore of the nearshore sandbar zone are reported as successes, while nourishment projects involving placing material much farther offshore are generally perceived to be disappointments.

7.5 Climate change

7.5.1 Potential future impacts

Global climate change is expected to affect the hydrodynamic boundary conditions that determine both short-term (days) and long-term (decadal) beach evolution. In particular, climate change is expected to result in accelerated sea-level rise (SLR), and modifications in the number, magnitude and geographical location of severe storms, with associated changes in the wave and storm-surge climate. The latest projections given by the Intergovernmental Panel on Climate Change (IPCC, 2007) indicate SLR projections from 0.18 to 0.79 m, including an allowance of 0.2 m for uncertainty associated with ice-sheet flow (Chapter 2). However, IPCC (2007) also states that larger values of SLR cannot be excluded. Future tropical cyclones are expected to become more intense, with larger peak wind speeds and heavier precipitation. Extra-tropical storm tracks are expected to move poleward, with associated changes in wind, wave, precipitation and temperature patterns (Chapter 5). Also, changes in mean sea-ice conditions may modify fetch conditions, in particular at high latitudes

(Chapter 14). The projected climate-change driven variations in natural phenomena that govern coastal zone processes are very likely to have a significant impact on coastal processes. Any negative impacts of climate change in the coastal zone are certain to have massive socio-economic impacts worldwide. The potential first-order climate change impacts that may be felt along the world's beaches can be summarized as shown in Table 7.1.

Accelerated SLR will cause widespread coastline retreat (recession), although local response will be governed by total sediment budgets. The commonly used Bruun rule (Box 7.4) predicts an upward and landward movement of the coastal profile in response to SLR (Fig. 7.21), suggesting a recession of 50–200 times the SLR, depending on the local beach slope. However, the use of the Bruun rule to obtain site-specific estimates of coastal recession due to SLR is a topic of much debate. In fact, following a comprehensive review of the Bruun rule, Ranasinghe and Stive (2009) concluded that the Bruun rule is generally unsuitable for obtaining exact and site-specific predictions of coastal recession due to SLR. They further recommended that Bruun rule estimates should only be considered in a broadly indicative sense and should not be directly used in coastal planning/management. Apart from causing coastline recession, SLR could also have other negative impacts, such as permanent inundation of unprotected low-lying coastal land and decreasing the efficacy of existing coastal-protection structures (e.g. overtopping of breakwaters, groynes, seawalls, dykes). In extreme cases, an existing effective coastal-protection structure might turn into a coastal-erosion hazard due to SLR.

Any climate-change driven variation in the mean (or dominant) wave direction could lead to major erosion on the updrift side, and comparable accretion on the downdrift side, of pocket beaches, resulting in permanent re-orientation (or rotation) of such beaches. Another phenomenon particularly relevant to pocket beaches is the El Niño Southern Oscillation (ENSO), which has been

Table 7.1 Potential first-order climate change impacts on beaches.

Potential impact	Main driver
Permanent inundation of low-lying land and increased flood height	Sea-level rise
Extreme and sudden erosion and inundation due to failure of coastal defences	Sea-level rise, increasing intensity and frequency of storms, increase in storm surge
Coastline recession	Sea-level rise, changes in alongshore sediment transport gradients
Increased storm erosion	Increasing intensity and frequency of storms, increase in storm surge, changes in storm direction and wave period
Erosion/accretion due to permanent re-alignment of embayed beaches	Changes in mean offshore wave direction
Changes in cyclic erosion/accretion patterns at pocket beaches due to changes in El Niño Southern Oscillation (ENSO) driven beach rotation	Changes in mean and storm wave characteristics
Increased periodic inundation due to increased wave run-up	Sea-level rise, increasing intensity and frequency of storms, increase in storm surge

CONCEPTS BOX 7.4   The Bruun rule

The method most commonly used to estimate coastal recession due to sea-level rise (SLR) is the simple two-dimensional mass conservation principle known as the Bruun rule (Bruun, 1962). Essentially, the Bruun rule predicts a landward and upward displacement of the cross-shore seabed profile in response to a rise in the mean sea level (Fig. 7.21), and is expressed as:

$$R = \frac{LS}{B+h} \qquad (7.22)$$

where h is the maximum depth of exchange of material between nearshore and offshore (i.e. depth of closure), L is the horizontal distance from the shoreline to depth h, B is the berm or dune elevation estimate for the eroded area, S is sea-level rise, and R is the horizontal extent of coastal recession.

Although its usefulness as a predictive tool has been a controversial issue for decades, coastal scientists and engineers have been routinely using the Bruun rule for

almost five decades, mostly due to its simplicity and the lack of any other easy-to-use alternative method. One of the main criticisms regarding the Bruun rule is the uncertainty associated with recession estimates obtained using this method. While the Bruun concept implies that SLR will result in the movement of sand from the berm or dune to the submerged nearshore profile, the actual physical processes by which this dune erosion will occur are not explicitly accounted for in this concept. Therefore, by necessity, the Bruun rule requires the specification of several input parameters that are associated with significant uncertainty. The Bruun rule expresses coastal recession due to SLR as simply the product of the SLR and the active profile slope. An application of the Bruun rule to several beaches in Sydney, Australia, which took into account the uncertainties associated with both input parameters (i.e. SLR and active profile slope), has shown that the potential variability of recession estimates obtained for this area can be as high as 4000% (Ranasinghe and Stive, 2009). The assumption that all sand transport occurs perpendicularly to the coastline is another severe limitation of the Bruun rule. Thus, the Bruun rule does not accommodate any three-dimensional variability, which is common along natural coastlines, precluding its application in areas adjacent to headlands or engineering structures, lagoon and estuary inlets, deltas, and areas with significant gradients in alongshore sediment transport. An alternative process-based, probabilistic approach to obtain recession estimates that are suitable for risk informed coastal zone management is presented in Ranasinghe et al. (2012).

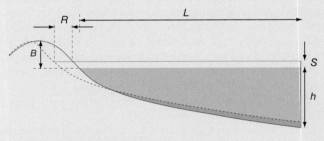

Fig. 7.21   Schematic diagram showing the Bruun rule for coastal recession. (Source: Ranasinghe et al. 2009.)

linked to the cyclic rotation of pocket beaches (Ranasinghe et al., 2004b). Measurements at Narrabeen Beach and Palm Beach, two pocket beaches located near Sydney, Australia, have indicated shoreline changes exceeding 50 m (over a period of 2–5 years) due to ENSO-driven beach rotation. Thus, any climate-change driven variations in the ENSO phenomenon is likely to result in changes in the magnitude and frequency of this cyclic rotation phenomenon, which may result in more intense and more frequent erosion/accretion cycles on the world's many pocket beaches. ENSO also plays a role in controlling the hydrology of beaches on barrier islands (Anderson and Emanuel, 2010).

An increase in the frequency of storm occurrence and/or storm intensity will undoubtedly result in more severe coastal erosion. The situation will be further exacerbated by a concurrent increase in storm surge. Indeed, increased storm erosion may have a more damaging impact than

the slow, steady effect of SLR. Coastal setback lines that currently only take into account, for example, the one in 100 year storm event estimated using historical data, need to be re-evaluated using future projected storm and surge characteristics. The combination of SLR, increased storm wave height, and increased storm surge will also result in more instances of dune erosion, overwash (either by run-up or inundation), and, in extreme cases, breaching and complete destruction of dunes. This will present major threats to coastal communities located in low-lying coastal zones that depend on the stability of coastal dunes as a primary defence mechanism (e.g. the Dutch coast and barrier Islands along the east coast of the USA). Furthermore, any increases in storm wave heights, occurrence frequency and/or storm surge might render existing coastal-protection structures such as offshore breakwaters and seawalls ineffective, and in some extreme cases detrimental.

7.5.2 Quantification of climate change impacts: future directions

Beaches are highly dynamic and are continually adjusting to subtle changes in hydrodynamic forcing. The feedback between hydrodynamics and morphology is highly non-linear and scale-dependent, both temporally and spatially. The exact response of a given stretch of the coast to a given set of environmental forcing functions will also depend to a large extent on site-specific geomorphic features. Climate change processes can result in variations in some or all of these environmental phenomena. However, due to the site-specificity of the composite impact of climate change at local scales, it is impossible to predict climate change impacts on the coastal zone at local scales (<10 km) without a comprehensive local-scale study that takes into account the highly complex and non-linear forcing-response mechanisms and site-specific geomorphology.

Technically, a carefully selected and validated suite of mathematical models could be used to quantify the above-mentioned climate change impacts. However, there are large uncertainties associated with not only the various models, but also with the forcing (i.e. greenhouse gas (GHG) emissions scenario). Any conscientious effort to quantify climate change impacts on coasts should therefore include the quantification of the range of uncertainty associated with model predictions. This can be achieved via ensemble modelling.

A thorough coastal climate change impacts study would ideally follow the broad structure shown in Fig. 7.22. In the suggested structure, GHG emissions scenario uncertainty is accounted for by considering several IPCC Special Report on Emissions Scenarios (SRES) scenarios (e.g. A1, A1F1, B2 – IPCC SRES 2000, see section 1.3.2). In selecting the SRES scenarios, it should be ensured that both low- and high-emissions scenarios are included in the ensemble. Several atmospheric ocean general circulation models (AOGCM) (e.g. Mark3.5, GFDLCM2.1, ECHAM5.0) are owned and operated by large research organizations around the world. These AOGCMs are forced in line with the various IPCC SRES scenarios and provide output consisting of time series of various climate variables (e.g. surface temperature, ocean temperature, atmospheric pressure, precipitation and wind; note that mean sea level is usually obtained by post-processing AOGCM output) on a global grid at a fairly coarse resolution of about 200 km. As different AOGCMs give somewhat different projections, there is a significant uncertainty associated with AOGCM output. In the study structure suggested in Fig. 7.22, the uncertainty associated with AOGCMs is accounted for by considering output from several AOGCMs.

As AOGCM output is generally available at a resolution that is inappropriate for direct use in local/

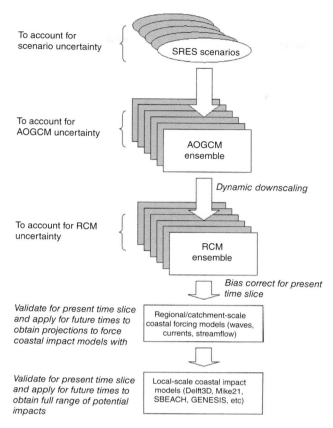

Fig. 7.22 Suggested structure for a coastal climate change impact quantification study. Abbreviations are defined in the text.

regional-scale coastal impact models, this output needs to be downscaled to a finer resolution. This is best achieved via dynamic downscaling, where the coarse-grid AOGCM output is used to drive regional-scale models (regional climate models, RCMs). Typically, the RCMs are nudged towards the parent AOGCM at regular intervals to ensure that RCM output is consistent with AOGCM output over long timescales. RCMs generally provide output at 2–50 km resolutions. As there is model uncertainty associated with RCMs also, the use of several RCMs to account for this uncertainty is suggested in the study structure shown in Fig. 7.22. A necessary step prior to using the RCM output in regional/catchment-scale forcing is bias correction of RCM output. This can be achieved by comparing RCM output for the present time slice (e.g. 1980–2000) with concurrent field measurements of relevant climate variables (e.g. wind, air pressure, temperature, precipitation), and applying correction techniques to the RCM output as required. The bias-corrected RCM output can then be used to drive validated regional-scale models to obtain future

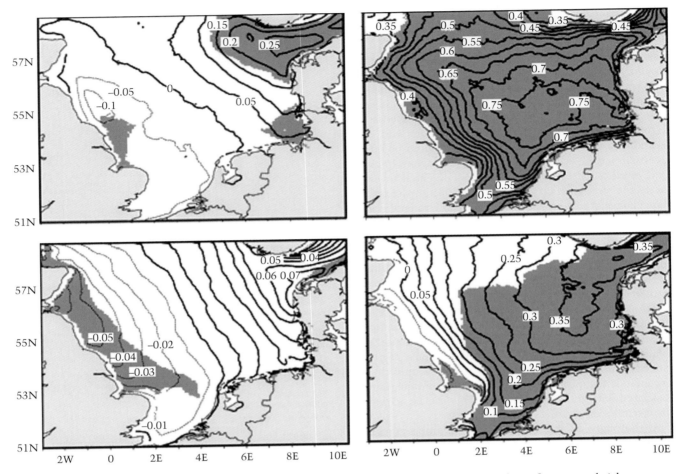

Fig. 7.23 Change in 50th (left) and 99th (right) percentile wind speed, in metres per second (top), and significant wave height, in metres (bottom). Change represents the difference between the 2071–2100 and the 1961–1990 simulation. (Source: Grabemann and Weisse 2008. Reproduced with permission from Springer Science + Business Media and the author.)

projections of forcing parameters (waves, surge levels, etc.) that are relevant for coastal processes. Finally, the projected forcing conditions thus obtained can be used to drive appropriate validated coastal impact models (e.g. Delft3D, Mike21, GENESIS) to obtain projections of climate change driven physical impacts on coasts.

At the time of writing, there are many studies on possible future changes in storm characteristics, but the number of studies on regional wave-climate changes is minimal. Figure 7.23 demonstrates the results of a study for the North Sea, based on two models and three emission scenarios. The figure indicates that future changes in wave heights for this area are expected to be small, but statistically significant, especially along the Dutch-German-Danish coast. All existing studies on wave-height changes under anthropogenic climate change are based on not more than a few transient climate simulations of 10–30 year time slices of the 21st century due to the high

computational demand of a single simulation. The resulting time series are too short to accurately assess return levels of storms that are sufficiently intense to result in major beach and dune erosion. Figure 7.23, for example, demonstrates the 99th percentile of significant wave height, which implies wave conditions that happen for three to four days each year. Significant beach and dune erosion along the North Sea coast is restricted to events with 1/10-year or smaller return values; the Dutch policy on coastal safety is even based on 1/10,000-year return values. The accurate assessment of such extreme events requires a large ensemble of simulations with a single model, each started with slightly different initial conditions (e.g. Sterl et al., 2008).

The ensemble modelling approach suggested in Fig. 7.22 will provide a number of different projections of the coastal impact under investigation. The range of projections will account for GHG scenario uncertainty,

AOGCM uncertainty, and RCM uncertainty. If necessary, the uncertainty induced by the regional/catchment-scale forcing model and the coastal impact model can also be included in this approach, at a significant computing cost. The range of coastal impact projections thus obtained can then be statistically analyzed to obtain not only a best estimate of coastal impacts but also the range of uncertainty associated with the projections, which will enable coastal managers/planners to make risk-informed decisions.

7.6 Summary

This chapter provides an overview of key hydrodynamic processes (waves, currents) and landforms of predominantly sandy beaches. Importantly, the beach is here defined as the nearshore zone, in which waves transform to ultimately end up as swash/backwash motions on the beach face. The feedback between the transforming waves, the resulting currents and sand transport, and the evolving landforms is crucial in order to understand nearshore evolution at all scales, both temporally and spatially. This chapter further highlights the impacts of anthropogenic activities and climate change on nearshore evolution, and illustrates how this evolution can be monitored and quantified. The focus in coastal climate change studies has traditionally been on accelerated sea-level rise, but possible changes in storm-wave characteristics may impact beaches more profoundly. Process-based morphodynamic models are useful tools to estimate this impact; the assessment of the uncertainty associated with these estimates must be an integral part of any coastal climate change study.

- Beaches are defined as accumulations of sand or gravel found along marine, lacustrine and estuarine shorelines. Morphological patterns, from centimetre-scale wave-induced ripples to sandbars of $O(100\,\text{m to 1 km})$, are inherent to almost all beaches and are highly variable, both spatially and temporally.
- Cross-shore beach profiles can be subdivided conveniently based on tidal exposure, into a subtidal, intertidal and supratidal zone. Each zone is associated with a distinct set of hydrodynamic processes. Wave breaking and the resulting mean flows (undertow, horizontal cell circulations, and longshore currents) form the dominant hydrodynamics in much of the subtidal zone. The intertidal zone contains the swash and backwash motions, while the supratidal zone is flooded solely during high-energy wave events.
- Subtidal sandbars are among the largest morphological patterns found on beaches. Often, their spatial and temporal variability constitutes the largest source of depth variation on timescales of weeks to years. The feedback between waves, wave-induced currents, sedi-

ment transport and the sandbar(s) causes the sandbars to move offshore rapidly (20–30 m/day) during storms and onshore slowly (1–3 m/day) during intermediate wave-energy conditions. Morphodynamic feedback also causes variability in sandbars (e.g. crescentic shapes and rip channels) to form, to grow and, ultimately, to disappear. The overall morphological appearance of a beach (sandbars, rip channels, etc.), in combination with the morphodynamic process regime (reflective versus dissipative), allows beaches to be classified into distinct beach types.

- Most beaches provide safety from coastal flooding and support multiple human activities, including tourism and recreation, nature conservation and fishing. Human activities need to be carefully managed, especially when there are conflicting interests. The 'frame of reference' approach is meant to guide coastal managers, policy-makers, scientists and stakeholders through the complicated process of integrated coastal zone management.
- Climate change is likely to affect the external hydrodynamic forcing of beaches in a significant way. Attention has traditionally focused on accelerated sea-level rise but it is likely that changes in the location of storm tracks and in the intensity, return frequency and duration of intense storms will affect beach evolution more profoundly. The quantification of climate-change induced beach change demands a model cascade from global to local models. An important point is to quantify the uncertainty due to different emission scenarios and to the use of different climate, wave and coastal-impact models.

Key publications

Battjes, J.A., 1988. Surf-zone dynamics. *Annual Review of Fluid Mechanics*, **20**, 257–291.
[*Somewhat dated, but otherwise excellent overview of surf zone dynamics*]

MacMahan, J.H., Thornton, E.B. and Reniers, A.J.H.M., 2006. Rip current review. *Coastal Engineering*, **53**, 191–208.
[*Up-to-date review on rip current dynamics*]

Masselink, G. and Puleo, J., 2006. Swash zone morphodynamics. *Continental Shelf Research*, **26**, 661–680.
[*Overview of swash zone hydrodynamics, sediment transport processes and beachface response*]

Weisse, R. and von Storch, H., 2010. Past and future changes in wind, wave and storm surge climates. In: *Marine Climate and Climate Change*. Springer Verlag, Berlin, Germany, pp. 165–203.
[*Regional study into effect of climate change on wave conditions*]

Wright, L.D. and Short, A.D., 1984. Morphodynamic variability of surf zones and beaches: a synthesis. *Marine Geology*, **56**, 93–118.
[*Benchmark paper on beach morphodynamics – probably the most cited paper in the sandy beach literature*]

References

Anderson, W.P. and Emanual, R.E., 2010. Effect of interannual climate oscillations on rates of submarine groundwater discharge. *Water Resources Research*, **46**, W05503, doi: 10.1029/2009WR008212.

Battjes, J.A., 1974. Surf similarity. *14th International Conference on Coastal Engineering*, American Society of Civil Engineers, pp. 466–480.

Bowen, A.J., Inman, D.L. and Simmons, V.P., 1968. Wave 'set-down' and set-up. *Journal of Geophysical Research*, **73**, 2569–2577.

Bruun, P., 1962. Sea-level rise as a cause of shore erosion. *Journal of the Waterways and Harbours Division*, **88**, 117–130.

Buscombe, D. and Masselink, G., 2006. Concepts in gravel beach dynamics. *Earth-Science Reviews*, **79**, 33–52.

Castelle, B., Bonneton, P., Dupuis, H. and Sénéchal, N., 2007. Double bar beach dynamics on the high-energy meso-macrotidal French Aquitanian coast: a review. *Marine Geology*, **245**, 141–159.

Castelle, B., Ruessink, B.G., Bonneton, P., Marieu, V., Bruneau, N. and Price, T.D., 2010. Coupling mechanisms in double sandbar systems. Part 1: Patterns and physical explanation. *Earth Surface Processes and Landforms*, **35**, 476–486.

Clifton, H.E., 1976. Wave-formed sedimentary structures – a conceptual model. In: R.A. Davis Jr. and R.L. Ethington (Eds), *Beach and Nearshore Sedimentation*. Society of Economic Paleontologists and Mineralogists, Special Publication No. 24, pp. 126–148.

Davidson, M., van Koningsveld, M., de Kruif, A. et al., 2007. The CoastView project: developing Coastal State Indicators in support of coastal zone management. *Coastal Engineering*, **54**, 463–475.

Davidson-Arnott, R., 2010. *Introduction to Coastal Processes and Geomorphology*. Cambridge University Press, Cambridge, UK.

Dean, R.G., 2002. *Beach Nourishment: Theory and Practice*. Advanced Series on Ocean Engineering, Vol. 18. World Scientific, Singapore.

Elgar, S. and Guza, R.T., 1985. Observations of bispectra of shoaling surface gravity waves. *Journal of Fluid Mechanics*, **161**, 425–448.

Falqués, A., Coco, G. and Huntley, D., 2000. A mechanism for the generation of wave-driven rhythmic patterns in the surf zone. *Journal of Geophysical Research*, **105**, 24071–24088.

Grabemann, I. and Weisse, R., 2008. Climate change impact on extreme wave conditions in the North Sea: an ensemble study. *Ocean Dynamics*, **58**, 199–212.

Hay, A.E. and Mudge, T., 2005. Principal bed states during SandyDuck97: occurrence, spectral anisotrophy, and the bed state storm cycle. *Journal of Geophysical Research*, **110**, C03013, doi: 1029/2004JC002451.

Hoefel, F. and Elgar, S., 2003. Wave-induced sediment transport and sandbar migration. *Science*, **299**, 1885–1887.

Holman, R.A., 2001. Pattern formation in the nearshore. In: G. Seminara and P. Blondeaux (Eds), *River, Coastal and Estuarine Morphodynamics*. Springer Verlag, Berlin, Germany, pp. 141–162.

Holman, R.A. and Stanley, J., 2007. The history and technical capabilities of Argus. *Coastal Engineering*, **54**, 477–491.

IPCC (Intergovernmental Panel on Climate Change), 2007. *Climate Change 2007: The Physical Science Basis*. Contribution of Working Group I to the Fourth Assessment Report of the Intergovernmental Panel on Climate Change. Cambridge University Press, Cambridge, UK.

Komar, P.D., 1998. *Beach Processes and Sedimentation* (2nd edn). Prentice Hall, Upper Saddle River, NJ, USA.

Lippmann, T.C. and Holman, R.A., 1990. The spatial and temporal variability of sandbar morphology. *Journal of Geophysical Research*, **95**, 11575–11590.

Longuet-Higgins, M.S., 1970. Longshore currents generated by obliquely incident sea waves. *Journal of Geophysical Research*, **75**, 6778–6801.

Longuet-Higgins, M.S. and Stewart, R.W., 1964. Radiation stresses in water waves: a physical discussion, with applications. *Deep-Sea Research*, **11**, 529–562.

Masselink, G. and Anthony, E.J., 2001. Location and height of intertidal bars on macrotidal ridge and runnel beaches. *Earth Surface Processes and Landforms*, **26**, 759–774.

Masselink, G. and Black, K.P., 1995. Magnitude and cross-shore distribution of bed return flow measured on natural beaches. *Coastal Engineering*, **25**, 165–190.

Masselink, G. and Hughes, M.G., 2003. *Introduction to Coastal Processes and Geomorphology*. Hodder Arnold, London, UK.

Masselink, G. and Short, A.D., 1993. The effect of tide range on beach morphodynamics and morphology: a conceptual model. *Journal of Coastal Research*, **9**, 785–800.

Masselink, G., Kroon, A. and Davidson-Arnott, R.G.D., 2006. Morphodynamics of intertidal bars in wave-dominated coastal settings: a review. *Geomorphology*, **73**, 33–49.

McCowan, J., 1894. On the highest wave of permanent type. *Philosophical Magazine*, **38**, 351–358.

Moore, L.J., Sullivan, C. and Aubrey, D.G., 2003. Interannual evolution of multiple longshore sandbars in a mesotidal environment, Truro, Massachusetts, USA. *Marine Geology*, **196**, 127–144.

Ojeda, E., Ruessink, B.G. and Guillén, J., 2008. Morphodynamic response of a two-barred beach to a shoreface nourishment. *Coastal Engineering*, **55**, 761–770.

Price, T.D. and B.G. Ruessink, 2011. State dynamics of a double sandbar system. *Continental Shelf Research*, **31**, 659–674.

Ranasinghe, R. and Stive, M.J.F., 2009. Rising sea and retreating coastlines. *Climatic Change*, **97**, 465–468.

Ranasinghe, R., Symonds, G., Black, K. and Holman, R., 2004a. Morphodynamics of intermediate beaches: a video imaging and numerical modeling study. *Coastal Engineering*, **51**, 629–655.

Ransinghe, R., McLoughlin, R., Short, A. and Symonds, G., 2004b. The Southern Oscillation Index, wave climate, and beach rotation. *Marine Geology*, **204**, 273–287.

Ranasinghe, R., Callaghan, D. and Stive, M.J.F., 2012. Estimating coastal recession due to sea level rise: beyond the Bruun rule. *Climatic Change*, **110**, 561–574.

Reeve, D., Chadwick, A. and Fleming, C., 2004. *Coastal Engineering: Processes*, Theory and Design Practice. SPON Press, Abingdon, UK.

Ruessink, B.G., Wijnberg, K.M., Holman, R.A., Kuriyama, Y. and van Enckevort, I.M.J., 2003. Intersite comparison of interannual nearshore bar behavior. *Journal of Geophysical Research*, **108**, doi: 10.1029/2002JC001505.

Ruessink, B.G., Michallet, H., Abreu, T. et al., 2011. Observations of velocities, sand concentrations, and fluxes under velocity-asymmetric oscillatory flows. *Journal of Geophysical Research*, **116**, C03004, doi: 10.1029/2010JC006443.

Scott, T., Masselink, G. and Russell, P.E., 2011. Morphodynamic characteristics and classification of beaches in England and Wales. *Marine Geology*, **286**, 1–20, doi: /10.1016/j.margeo. 2011.04.004.

Shand, R.D., Bailey, D.G. and Shepard, M.J., 2001. Longshore realignment of shore-parallel sand-bars at Wanganui, New Zealand. *Marine Geology*, **179**, 147–161.

Short, A.D. and Aagaard, T., 1993. Single and multi-bar beach change models. *Journal of Coastal Research*, Special Issue, **15**, 141–157.

Sterl, A., Severijns, C., Dijkstra, H. et al., 2008. When can we expect extremely high surface temperatures? *Geophysical Research Letters*, **35**, L14703, doi: 10.1029/2008GL034071.

Thornton, E.B. and Guza, R.T., 1982. Energy saturation and phase speeds measured on a natural beach. *Journal of Geophysical Research*, **87**, 9499–9508.

Van Duin, M.J.P, Wiersma, N.R., Walstra, D.J.R., van Rijn, L.C. and Stive, M.J.F., 2004. Nourishing the shoreface: observations and hindcasting of the Egmond case, the Netherlands. *Coastal Engineering*, **51**, 813–837.

Werner, B.T. and Fink, T.M. 1993. Beach cusps as self-organized patterns. *Science*, **260**, 968–971.

Wijnberg, K.M. and Kroon, A., 2002. Barred beaches. *Geomorphology*, **48**, 103–120.

Wright, L.D. and Short, A.D., 1984. Morphodynamic variability of surf zones and beaches: a synthesis. *Marine Geology*, **56**, 93–118.

8 Coastal Dunes

KARL F. NORDSTROM

Institute of Marine and Coastal Sciences, Rutgers – the State University of New Jersey, New Brunswick, NJ, USA

8.1 Conditions for dune formation

8.1.1 Aeolian transport

The movement of sand by wind results from momentum transfer from air to sediment (Sherman and Hotta, 1990). Simple deterministic models have been developed to quantify this concept in order to predict aeolian transport rates across ideal surfaces. One of the most frequently used is the equation of Bagnold (1936):

$$q = C(\rho_a/g)(D/D_r)^{0.5} u_*^3 \qquad (8.1)$$

where q is the rate of sediment transport, C is an empirical constant ranging from 1.5 (nearly uniform sand) to 2.8 (sand of a very wide range of grain sizes), ρ_a is air density, g is acceleration due to gravity, D is grain diameter, D_r is a reference grain diameter of 0.25 mm, and u_* is shear velocity. Of note is that the sediment transport rate according to eqn. 8.1 is proportional to the velocity cubed and that the equation disregards the existence of a threshold velocity. Models of aeolian transport are primarily based on assumptions that: (1) the wind field is unidirectional, fully turbulent, uniform and steady, implying that the vertical velocity profile can be readily defined; (2) sediments available for entrainment and transport are uniform in size and composition; and (3) the surface is planar, horizontal, dry and unobstructed (Sherman et al., 1998). Not surprisingly, deterministic models are often poor predictors of aeolian transport monitored in the field. A glance at sand blowing on a beach (Fig. 8.1) reveals considerable spatial and temporal differences over the short term over distances of only a few metres. These differences are due to the complexity of natural conditions related to local differences in moisture content of sediments, salt crusts, grain-size characteristics, beach litter and micro-topography

Fig. 8.1 Sand streamers and aeolian bedforms on the beach near the River Dyfi, Wales, UK. Due to their location within the intertidal zone or on the backshore seaward of the upper limit of storm-wave uprush, aeolian deposits such as shown here are temporary. (Source: Photograph by Karl Nordstrom.)

Fig. 8.2 Dune landward of an estuarine beach at Sandy Hook, New Jersey, USA. The distinction between wind-created dunes and wave-created beach ridges is blurred where wind deposits form a veneer on wave deposits, especially on narrow reflective beaches. (Source: Photograph by Karl Nordstrom.)

(Sherman et al., 1998; Wiggs et al., 2004; Baas and Sherman, 2005; Davidson-Arnott et al., 2005). Human actions, such as raking or driving on beaches and deploying sand fences, add even more local variability.

8.1.2 Potential for dune building

Quantification of aeolian transport rates may be complicated at small spatial scales, but the net result of this transport is clearly revealed in the dunes landward of the backshore. A dune may be defined as a mound or ridge formed by the deposition of wind-blown sand, but the definitions used for management purposes may differ dramatically (see section 8.3). A smaller, temporary feature composed of wind-blown sand is usually called an aeolian bedform. Dunes may vary in size from the less than 1 m high forms landward of estuarine beaches (Fig. 8.2) to forms over 100 m high. Many of the high dunes are landward of large sand sources (Fig. 8.3).

Coastal dunes will form where there is a source of sand, a wind strong enough to entrain and move the sand, and a means of causing the sand to be deposited by blocking it or reducing the speed of winds. The shape, size and type of dunes, and the natural environments that form on them, depend on many local factors, including storm frequency and magnitude, which determine the effect of wave and wind characteristics, and the type of vegetation and its health and vigour. The conditions for dune development are readily met on most beaches composed of fine- to medium-sized sand, but dunes are absent from many locations, even those with wide sandy beaches. The lack of dunes in developed areas is due primarily to human actions such as beach raking and grading. Lack of dunes in

Fig. 8.3 Dune at Oregon Dunes National Recreation Area, USA, looking toward the beach. This ridge is migrating inland and extending the dune field into forested land. Such landward-migrating dunes are often referred to as transgressive dunes. (Source: Photograph by Karl Nordstrom.)

undeveloped areas is usually related to the narrowness of beaches, which limits fetch length for wind-blown sand and exposes incipient dunes to wave attack.

8.1.3 Processes of dune formation

Sand transported landward from the beach by aeolian processes is the principal sediment input to coastal dunes, and the width of the beach is critical in providing sediment to the dune (Nickling and Davidson-Arnott, 1990; Houser, 2009). Dissipative beaches are wide and flat, and provide good sources of wind-blown sand. Narrow reflective beaches are less effective. The geomorphological and bio-logical processes forming natural foredunes are reviewed in Hesp (1989, 1991, 2002). Sand blown across unvegetated portions of sandy beaches may accumulate at the seaward-most vegetated surface and at the base of existing dunes. Alternatively, sand may collect at the wrack lines that form on the backshore, especially the upper-storm wrack line (Fig. 8.4) that provides a relatively high barrier to transport. Vegetation on the beach can grow seaward from an estab-lished dune or grow as discrete clumps of pioneer vegeta-tion right on the beach. These patches of vegetation often take on a linear, shore-parallel form as a result of reworking by wave swash. Wrack delivered by wave swash is also deposited as a shore-parallel line. Wrack lines consist of natural and cultural litter, and contain seeds and rhizomes of coastal vegetation and nutrients that aid in growth of new vegetation (Godfrey, 1977; Ranwell and Boar, 1986).

The new dunes that form on the backshore seaward of established dunes are called incipient (or embryo) dunes. Space and time are important to allow incipient dunes to survive erosion by wave uprush during small storms and to increase in size to create a foredune ridge large enough to survive storms of annual or greater frequency and mag-nitude. Erosion during subsequent storms can eliminate the seaward portion of an established foredune and create a vertical erosional scarp. On coasts with a nearly bal-anced or positive sediment budget, post-storm deposition will replace sediment on the beach, and the seaward por-tion of the foredune will begin to reform as a dune ramp

Fig. 8.4 Beach wrack deposited at Ocean City, New Jersey, USA, after a winter storm. Post-storm accretion by swell waves will create a wider and dryer beach berm, increasing the potential for aeolian transport to the wrack. Seeds, rhizomes and nutrients within the wrack will cause vegetation to grow during warmer months and contribute to sand trapping and growth of the foredune at this seaward location. (Source: Photograph by Karl Nordstrom.)

Fig. 8.5 Aeolian transport forming a ramp covering a wave-eroded scarp at Fire Island, New York, USA. The ramp replaces sediment seaward of the dune and facilitates delivery of sediment inland, contributing to growth of the crest and landward migration, and ultimately a loss of sediment from the beach system. (Source: Photograph by Karl Nordstrom.)

(Fig. 8.5). The dune ramp is both a source of sand and a conduit for delivery of sediment inland (Aagaard et al., 2004). Dune building thus proceeds as a series of intermit-tent phases of accretion and erosion. Under natural condi-tions, establishment of the morphology and vegetation assemblages of foredunes can take up to 10 years (Woodhouse et al., 1977; Maun, 2004).

Wave erosion diminishes the width of the beach and removes sand from the dune, so the long-term beach sedi-ment budget is important in determining the availability of sediment and the frequency at which the dune is attacked (Psuty, 1988; Hesp, 2002). If the beach sediment budget is strongly negative, dune erosion is relatively fre-quent and the dune crest may be overwashed, with sedi-ment displaced inland (Fig. 8.6a). On shores where the volume of sediment supplied to the beach is balanced or slightly negative, beach width is moderate and dune scarp-ing occurs during large storms. Subsequent aeolian action will replenish the sand in the dune and the dune will be characterized by a high, wide foredune ridge (Fig. 8.6b). On shores where the volume of sand delivered to the beach exceeds outputs, shoreline progradation occurs, resulting in formation of multiple foredune ridges separated by lower swales (Fig. 8.6c). The higher ridges usually repre-sent locations where the foredune was stable for a rela-tively long time and it built up over many sand-blowing events. High rates of progradation tend to produce more dune ridges that are lower in elevation.

Blowouts (Fig. 8.7) form in dunes following destruc-tion of stabilizing vegetation by disease, drought, grazing or trampling. These features can grow laterally by dis-rupting the vegetative cover and undermining substrate.

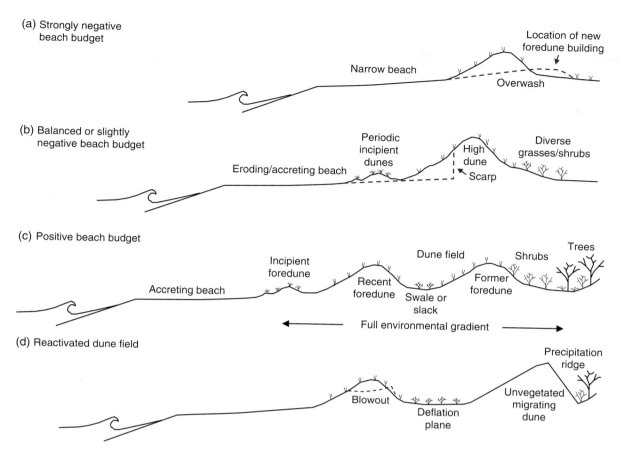

Fig. 8.6 Scenarios for development of coastal dunes, based on beach and dune sediment budgets. (a) A narrow eroding beach leads to frequent dune erosion and overwash. New incipient dunes may form on the wrack lines delivered landward of the former crest, leading to an irregular, low and hummocky foredune that contributes to further overwash. (b) A balanced or slightly negative sediment budget leads to frequent wave erosion and scarp formation on the front of the dune followed by rebuilding by aeolian transport, resulting in a high foredune ridge. (c) Beach progradation allows new foredunes to grow seaward, leaving multiple foredune crests and intervening moister swales (or slacks) with different habitat value. Wide dune fields can be characterized by a complete environmental gradient, from the few species that can survive the stresses on the beach to the species-rich environment farther landward where trees can also survive. (d) A dune field can also be reactivated due to loss of stabilizing vegetation. The former dune surface can be deflated down to the water table, and an unvegetated ridge can migrate into adjacent areas, including developed lands. (Source: Adapted from Nickling and Davidson-Arnott 1990.)

The gaps they create in higher dunes can accelerate wind speeds locally, thereby enhancing transport of sand and salt aerosols farther inland, thus reversing vegetation succession (Hesp, 1991). Blowouts may also deflate down closer to the location of the water table where they can evolve into species-rich moist slacks (Doody, 2001). Blowouts that form low points in the foredune can increase the likelihood of wave overwash. Blowouts can occur under natural conditions, but the likelihood of their formation is often increased as a result of human activities. Widespread destabilization of surface vegetation can lead to formation of a migrating ridge of bare sand (see Fig. 8.3) that can bury established vegetation and cultural features landward of it (Fig. 8.6d). Migrating

dune ridges are now less common than in the past due to widespread stabilization programmes.

8.1.4 Dune fields

Dune fields several kilometres wide (Fig. 8.8) are most common where sand from wide, shallow nearshore shelves or near river-mouths was made available in the past due to storm flooding or periods of alternating sea levels. Activation and reactivation of dune fields can be enhanced by either rising or falling sea level. Evaluations of dune fields in Europe reveal alternate phases of dune building, erosion and remobilization associated with sea-level changes following the past Pleistocene glaciation (Bird, 1990; Christiansen et al., 1990;

Fig. 8.7   Blowout in southeast Queensland, Australia. Blowouts are wind-eroded troughs or saucer-shaped depressions that provide conduits for wind-blown sand and storm-wave uprush. Managers frequently made attempts to stabilize blowouts in the past, but their value in rejuvenating landscapes and increasing diversity of habitats is becoming increasingly recognized. (Source: Photograph by Karl Nordstrom.)

Fig. 8.8   Dune field west of Haarlem, the Netherlands. Many formerly active dune fields are now stabilized, but reveal the hummocky topography created during periods of dune mobility. The hummocky terrain can result in great variations in growing conditions for vegetation and result in high landscape diversity and aesthetic appeal. (Source: Photograph by Karl Nordstrom.)

Klijn, 1990). Low sea-level phases during the Pleistocene exposed much of the continental shelves to wind erosion. Glacial deposits and outwash added considerable volumes of sand-size sediment to be blown inland and deposited in locations that are now landward of the coastline. Marine transgressions carried sediment from the shelf shoreward, building sandy beaches and eroding coastal formations. Wave erosion of dunes, followed by aeolian transport off the unvegetated slopes of eroded dunes and off the beaches, delivered additional sediment inland. Periods of stability or intervening

regressions allowed time for sediment to accumulate seaward. Periods of widespread dune building are often revealed in clearly delineated shore-parallel ridges formed by vigorous vegetation growth. Remobilization of stabilized dunes can take the form of shore-perpendicular blowouts and parabolic dunes or shore-parallel transgressive dune ridges. All of these features provide conduits for landward transport.

Landward migration of coastal dunes in the Netherlands in the past is considered to be primarily triggered by coastal erosion and flooding, assumed to have been induced by a rising sea level and increased storm-surge frequency (Klijn, 1990). In contrast, the three most recent dune-building periods in Denmark are associated with lower sea levels, the most recent being between 1450 and 1750, falling within the period known as the Little Ice Age (Christiansen et al., 1990; Box 8.1). Chronology of dune formation differs in different countries, even within Western Europe (Christiansen et al., 1990; Klijn, 1990). It is likely that many of the discrepancies in dates of activation are due to regional constraints, including human-induced changes, rather than broad-scale climatic effects. A combination of pedological,

CONCEPTS BOX 8.1   Dune formation during the Little Ice Age

Episodes of significant aeolian sand accumulation during the Little Ice Age are documented for many locations in northwestern Europe (Wilson et al., 2001). The Little Ice Age occurred in the northern hemisphere between AD 1300 and 1900, with specific beginning and ending dates varying considerably for different regions. This period was not a single phase of sustained cold, but a series of warm and cold climate anomalies that varied in degree and importance regionally. Differences occurred in relative sea level, degrees of storminess and even dominant wind direction. For example, movement of polar waters to the south during cold phases favoured easterly air flow across the British Isles, whereas warm phases re-established westerly flows; these reversals contributed to dune development on both east-facing and west-facing shores (Wilson et al., 2001). Dune mobility was increased by removal of stabilizing vegetation by human action for use as feed, fuel or roof cover (Christiansen et al., 1990). These activities were likely enhanced during colder phases, as growing seasons for crops became restricted and the demand for firewood increased. It is not always clear whether the increased dune activity is associated with initiation of new dunes landward of the beach (resulting from increased intensity of coastal storms or flooding) or due to reactivation of existing dunes (resulting from removal of stabilizing vegetation by human action).

biological, anthropogenic and climatic factors appears to be the cause of changes rather than a single factor.

8.2 Dunes as habitat

Vegetation influences dune morphology, but morphology also determines subsequent growing conditions for vegetation. Disturbance is critical to the composition and richness of vegetation (Moreno-Casasola, 1986), and the kinds and levels of disturbance differ with distance from the water. Richness is diminished near the beach, where few species can tolerate the stresses of sand mobility and salt spray (Moreno-Casasola, 1986). Pioneer plants (e.g. *Cakile edentula*) that are tolerant of salt spray and sand blasting form incipient dunes on the backshore, and grasses (e.g. *Ammophila* spp. and *Spinifex* spp.) form foredune ridges (Hesp, 1989). Well landward of the foredune, protection from salt spray and sand inundation favours growth of woody shrubs, with trees and upland species even farther landward (Fig. 8.9). On natural dunes of coasts with a relatively balanced sediment budget, the transition from pioneer beach plants to fully mature forests can extend over cross-shore environmental gradients of hundreds to thousands of metres, depending on the frequency and magnitude of winds that drive the physical stresses (McLachlan, 1990). This gradient can be truncated by elimination of the seaward dune sub-environments by wave erosion, causing woody shrubs or trees to be closer to the water than they would originally grow.

Natural foredunes are inherently dynamic and fragmented (García-Novo et al., 2004; Martínez et al., 2004), and they undergo exchanges of sediment, nutrients and biota, follow cycles of accretion and erosion, and retain diversity and complexity even as they change shape and position (Doody, 2001). Wind-blown sand can have a stimulating positive effect on the growth of dune-building plants and prevent their degeneration (Maun, 2004). Bare areas and low areas within dunes provide a variety of habitats, and poorly vegetated dry areas high in the dune are not as barren as often perceived. Invertebrate fauna thrive on the open and dry dunes that are often warmer than surrounding areas (Doody, 2001). Low areas sheltered by dune crests can allow species less tolerant of wind spray and salt stress to grow close to the water. Slacks are centres of diversity within both mobile and stabilized dunes (Grootjans et al., 2004).

The optimum condition for maintaining species diversity in dune environments may be one where different sections are evolving at different rates or stages, providing many different degrees of sediment movement and ground cover. This conception of dunes as dynamic systems is in sharp contrast to the idea that bare sand and migrating dunes are bad and must be stabilized, which guided past plans for managing them. Dune mobility is not a problem for species adapted to the dune. A dune managed as a

Fig. 8.9 Dune field preserved as an undeveloped enclave on the developed shore of Avalon, New Jersey, USA. Vegetation species differ with distance landward of the beach due to increased sheltering from salt spray and inundation by seawater and wind-blown sand. Grasses occur near the beach, yielding to woody shrubs then trees farther inland. (Source: Photograph by Karl Nordstrom.)

dynamic system can be more resistant to erosion, cheaper to maintain, have greater natural values, and be more sustainable over the long term than a fixed dune (García-Novo et al., 2004; Martínez et al., 2004).

8.3 Dunes in developed areas

Dunes, like many other natural features, have been eliminated in many locations by human actions, but many of these alterations occurred so long ago that they have been forgotten. Few visitors to the coast realize that many farm fields, pine plantations or developed portions of many coastal towns and cities are on former dune fields. Even in many parks and nature reserves, effects of past human interventions persist, and few dune systems in these areas can be considered entirely natural, although most visitors would not realize it. The basic ways dunes can be modified by human actions are identified in Table 8.1. Not all of these activities totally destroy dunes, but the resulting dunes can depart dramatically from the kinds of landforms that would evolve by natural processes alone. Even the most benign activities may have to be controlled to ensure that dunes continue to provide natural and human benefits.

Foredunes that are allowed to exist in developed areas are usually lowest where residents demand easy access and direct views of the water from shorefront homes and highest where safety is the principal value, and sand fences and vegetation plantings are used to augment natural trapping rates. Dunes in human-occupied areas often consist of a single linear ridge. The dunes are usually narrow, because the landward extent is restricted by human infrastructure. Wider dunes exist in developed areas where accretion has occurred updrift of long shore-parallel structures, such as jetties, providing that regulations have

Table 8.1 Principal ways that dunes are altered by human actions. (Source: Adapted from Nordstrom 2000.)

Eliminating for alternative uses
Constructing buildings, transportation routes, promenades, golf courses
Mining

Altering through use
Pedestrian trampling and off-road vehicle use
Harvesting and grazing
Extracting oil, gas, water
Military activities

Reshaping (grading)
Piling up sand to increase flood protection levels
Removing sand that inundates facilities
Eliminating dunes to create wider, flatter recreation beaches
Breaching, lowering or eliminating dunes to facilitate access or views of the sea

Altering dune mobility
Constructing shore protection structures that change locations of accretion and scour
Introducing resistant sediment into dunes
Clearing the beach of litter
Stabilizing dunes using sand fences or vegetation plantings
Remobilizing dunes by removing vegetation

Altering external conditions
Damming or mining streams that supply sediment to beaches
Introducing pollutants that affect dune vegetation

Creating or changing habitat
Nourishing beaches and dunes
Adding species to increase diversity
Introducing or removing exotic vegetation

prevented new development from encroaching. Dunes in developed areas generally have less topographic variability than do their natural counterparts when they are built using sand fences or earth-moving equipment, because they usually are designed to a common standard. Dunes created by bulldozing or dumping from trucks have poorly defined internal stratification, and they may contain sediments that are too coarse or too fine to be aeolian deposits. Once dunes are created and shaped according to human needs, attempts usually are made to protect them in place, and they become less mobile than natural dunes.

8.3.1 Degradational activities

The reason dunes are not found in many areas of human habitation is not because of natural constraints, but because they have been removed or prevented from forming. Where dunes do exist, they are usually managed to fulfill only a few human-use functions, which may be incompatible with natural functions. Whether dunes become mere artifacts or are restored to more naturally functioning systems depends on human perceptions of values for coastal resources and the perceived role of natural components in providing these values. Developers attempt to build as close to the water as possible so as to provide homeowners and visitors ready access to beaches, eliminating portions of the foredune, or even all of it (Fig. 8.10). Beach nourishment can provide the opportunity for restoration of dunes and the habitats they provide (Fig. 8.10d), but dune restoration is not always carried out on nourished beaches.

Use of beaches solely for recreation can result in the total elimination of dunes by grading them and raking the beach to provide views of the sea in order to maintain the beach as a recreation platform. Raking to remove wrack eliminates this habitat and the large quantities of invertebrates contained within it, with a resulting decrease in biodiversity, and it prevents formation of incipient dunes that could grow into naturally evolving foredunes. Vehicles used on beaches can kill fauna and disperse organic matter in drift lines, thereby destroying young dune vegetation and losing nutrients (Godfrey and Godfrey, 1981), and the unnatural landscape image can undermine attempts to instill an appreciation of the shore as a natural environment. Exotic vegetation was often planted in dunes in the past because it was seen to be effective for stabilizing dunes, valuable economically, or attractive. Exotic species can form mono-specific stands of vegetation that outcompete native species and decrease biodiversity and authenticity of vegetation, interfere with successional processes, create an undesirable landscape appearance, and reduce the value of land for conservation. Exotics have different sand stabilization effects and can change the morphology and mobility of landforms as well (Cooper, 1958; Avis, 1995).

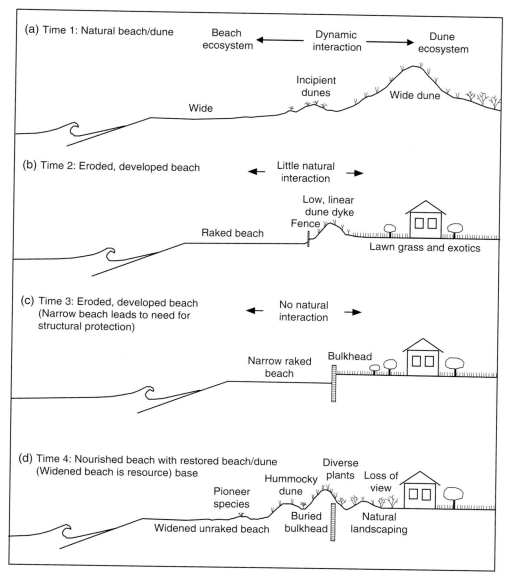

Fig. 8.10 Frequently encountered scenarios of coastal change in developed areas. (a) Time 1 represents the conditions prior to development. (b) Time 2 depicts elimination of the former dune to facilitate construction close to the sea and raking of the beach to facilitate recreation. Private lots are often landscaped according to suburban ideals, and exotic vegetation and lawns replace natural vegetation. Beach erosion continues, leading to the need to protect infrastructure using a dune that is now smaller and farther seaward than the former foredune. (c) As the beach becomes narrower (Time 3), the dune cannot survive to provide protection, leading to the construction of a bulkhead or seawall. Elimination of the beach can lead to calls for beach nourishment. (d) Time 4 reveals how beach-dune habitat can be restored, although this scenario rarely occurs. Nourished beaches in developed areas are more often managed as a wider expanse of raked beach, and private lots retain their suburban appearance.

8.3.2 Value of dunes

Coastal sand dunes provide important ecosystem values and services (Table 8.2), but many existing uses of dunes are not compatible with the natural values they provide. Reshaping a dune to facilitate home construction, or bulldozing it down to facilitate access to the sea or retain views from shorefront buildings, effectively eliminates all of its natural values. Excessive driving or walking in the dunes can destroy vegetation and interfere with use by fauna. Accordingly, direct use of dunes should be kept to low levels.

Dunes that are allowed to survive or are intentionally built up in developed areas are managed principally for their value in protecting infrastructure from erosion and flooding. Dunes protect human structures by providing sediment to replace erosion losses on the beach during storms. The sediment removed from dunes during storm wave attack is incorporated into the beach and helps dissipate storm waves. The portions of the dunes that survive wave attack can provide a physical barrier to wave uprush (e.g. compare Fig. 8.6a and b). The vegetation on the surface helps prevent erosion by winds, and the roots and rhizomes of vegetation and filaments of fungi below the surface help resist wave erosion of the sediment contained in the dune. Dunes managed for shore protection are usually built and maintained as stable, linear features, often with a consistent design height meant to provide protection against flooding (Fig. 8.11). These dunes may be

Table 8.2   Values and services provided by coastal dunes.

Protecting human structures (by providing sediment, physical barriers or resistant vegetation)
Providing sites for passive recreation (nature appreciation, aesthetic or therapeutic opportunities)
Filtering pollutants
Retention area for groundwater
Ecological niches for plants adapted to dynamic conditions
Refuge areas for invertebrates, birds, small mammals
Nest or incubation sites
Food for primary consumers and higher trophic levels (scavengers and predators)
Synergistic benefits of additional habitat types, ecotones and corridors for faunal migrations
Intrinsic value

Fig. 8.11   Flat-topped, linear dune dyke built for shore protection at Avalon, New Jersey, USA. The coarser sediments on the surface reveal the non-aeolian origin of the landform, and the regular pattern of the vegetation reveals that it was planted. (Source: Photograph by Karl Nordstrom.)

better termed 'dune dykes' than true dunes. They can be considered dunes if they occupy the position that a dune would occupy on a shoreline subject to only natural processes, providing they have similar form and functions (natural and human) to natural dunes and are modified by aeolian action, even if this modification is restricted to a surface veneer (Nordstrom and Arens, 1998). Dunes that are built to these characteristics could retain all of the values and services of a natural dune, despite human modifications, if proper restoration and conservation measures are taken.

8.4   Dune restoration and management

Effective dune management requires developing programmes for: (1) identifying remaining natural environments and establishing them as new nature reserves; (2) protecting existing reserves from human-induced damage; and (3) restoring degraded areas and reinstating natural dynamic processes. Managing landforms to achieve their full potential involves evaluating them on multiple criteria, related to ecological, geomorphological and social factors (Table 8.3), and allowing these landforms freedom to evolve with minimal human input. Steps in restoring

Table 8.3   Guidelines for coastal management applicable to dune environments. (Source: Nordstrom 2008.)

Establish a natural strip between human development and the water line
Prevent construction on the beach and foredune
Require environmentally compatible recreational uses

Protect human lives and settlements
Restrict development close to water
Favour dune formation to provide a barrier against overwash and flooding

Preserve natural coastal dynamics
Establish large conservation areas for nature protection
Restrict structures that would interfere with sediment transfers to beach and dune
Prevent habitat fragmentation and create and maintain ecological corridors

Provide sustainable and environmentally friendly tourism and development
Ensure the carrying capacity of the environment is not exceeded
Orient tourism toward conservation goals
Increase environmental awareness of tourists

Protect endangered or threatened biota and landscapes
Give preference to endangered or threatened species and habitat types
Prohibit damaging activities, or require mitigation or compensation
Conduct restoration projects
Prevent the introduction of exotic species

Fig. 8.12 Dune constructed using bulldozers and sand fences at Lavallette, New Jersey, USA. Many of the sand-trapping fences seen on the landward side of the dune are not needed for dune building. (Source: Photograph by Karl Nordstrom.)

Fig. 8.13 The dune field created on the nourished beach at Ocean City, New Jersey, USA. This site is in an intensively developed segment of the shore that was heavily used by tourists. Its conversion to a naturally functioning habitat (described in Box 8.2) demonstrates the effectiveness of allowing nature to evolve by ceasing beach raking. The dune ridge at left was built by emplacing sand fences and planting beach grass. The bulk of the dune to the right occurred through natural deposition, unaided by fences or plantings. (Source: Photograph by Karl Nordstrom.)

dunes in locations where they have been eliminated or severely degraded involve: (1) obtaining stakeholder acceptance for new dunes or larger dunes; (2) restoring the dunes; (3) allowing these new landforms to function like natural dunes by allowing some form of dynamism; (4) controlling subsequent negative human actions; and (5) favouring dune evolution through time using adaptive management. If time and space are not available, natural constraints can be partially overcome by aiding natural processes using sand fences or vegetation plantings or by using earth-moving equipment (Fig. 8.12). There is no single way to restore a coastal dune, and many strategies have been employed (Martínez et al., 2013).

8.4.1 Favouring dune building by natural aeolian processes

Beach nourishment provides a wider sediment source for aeolian transport and greater protection for newly formed dunes from wave erosion. A dune that is allowed to evolve by natural processes on a nourished beach (Fig. 8.13; Box 8.2) will have the internal stratification, topographic variability, surface cover and root mass of a natural dune. The wide cross-shore gradient of physical processes can allow for a suite of distinctive habitats, from pioneer species on the seaward side to woody shrubs and trees on the landward side. The initial nourishment project can provide the necessary sand volume and space for dunes to form, but maintenance nourishment is required to retain dune integrity, given subsequent wave-induced erosion. Given the long time frame for restoration of the morphology and vegetation of foredunes under natural conditions, it may be desirable to use earth-moving equipment, sand fences or vegetation plantings to create a dune ridge for initial

protection (Fig. 8.13, foreground) and allow a more natural dune to evolve gradually seaward of it.

8.4.2 Depositing fill directly

Dunes constructed by dumping sediment directly and reshaping it using earth-moving equipment are usually built to optimize a flood-protection function and have a designed shape with little topographic diversity (see Fig. 8.11). The grain-size characteristics, rates of change, and characteristic vegetation of dunes built by mechanical placement differ from dunes created by aeolian deposition. Bulldozed dunes can provide habitat similar to natural dunes if actions are taken to enhance this value during and after construction. Well-sorted sands that resemble dune sediments can be used if suitable borrow areas are found. Patchiness of habitats can be increased by creating an undulating crest, resulting in local differences in drainage and wind speed, and converting the landward boundary from a line to a zone (Nordstrom, 2008). Subsequent deposition of wind-blown sand on bulldozed dunes can create surface characteristics similar to a natural dune if no barriers to aeolian transport are created seaward of it.

8.4.3 Using sand fences

Fencing materials include canes and tree branches that are inserted directly into sand, and wooden slats, plastic and jute fabric that are attached to fence posts. Using

CASE STUDY BOX 8.2   Allowing dunes to evolve on nourished beaches

Events at Ocean City, New Jersey, USA, reveal the advantages of allowing natural aeolian deposition to occur on a nourished beach, following artificial construction of an initial foredune (see Nordstrom, 2008; Fig. 8.13). The site illustrates how naturally functioning landforms and vegetation, with their inherent dynamism, can evolve after beaches and dunes are initially built to protect human structures. A federal project involved using fill dredged from a nearby inlet to create a wide beach berm. The municipality then placed two rows of sand-trapping fences 5 m apart and planted the space between only with American beachgrass (*Ammophila breviligulata*). The nourished beach provided an excellent source of wind-blown sediment and protected the foredune from wave damage during small storms. Additional fences were placed on the seaward side of the dune to encourage horizontal rather than upward growth, so shorefront residents could retain views of the sea. Designation of nesting sites for piping plovers (*Charadrius melodus*) by the state endangered-species programme resulted in prohibition of beach raking, leading to local colonization of the backshore by plants and growth of incipient dunes that survived several winter storm seasons and grew into new foredune ridges seaward of the dunes maintained by sand fences. The seaward portion of the dune is dynamic, while the landward crest built by the sand fences and *Ammophila* plantings remain as a protection structure. The presence of natural seed sources nearby has enabled succession to occur within the evolving dune field, even as some portions of the dune remained dynamic. Time and space are critical to the evolution of vegetation gradients after planting of the initial stabilizer. The artificially widened beach and use of sand fences and vegetation plantings to help create and stabilize the initial dune crest at Ocean City allowed the dune to quickly provide its human-use function as protection for buildings and infrastructure against flooding. Once the dune achieved its protective function, it could be allowed to evolve as a naturally functioning environment.

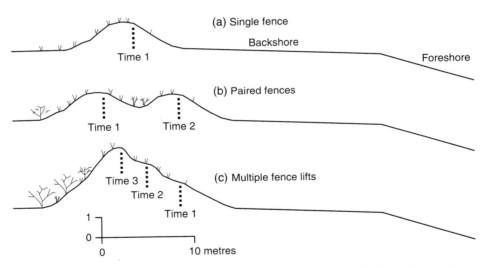

Fig. 8.14 Differences in dune topography and vegetation growing conditions associated with alternative sand-fence configurations. Each time period represents at least a year of accumulation at a fence. A single straight fence (a) produces a single linear ridge that provides limited protection from salt spray and overwash. Placement of an additional fence seaward (b) can create an intervening swale that is partially protected from winds and salt spray and closer to the water table, enabling growth of species less tolerant of the stresses of a beach environment. Deployment of multiple fence lifts (c) can create a higher dune that provides better protection against flooding, sand inundation and salt spray and allows more species, including woody shrubs, to thrive landward of the crest.

sand fences with different characteristics and configurations can result in considerable variety in topography and vegetation. Straight fences placed parallel to the shore appear to provide the most economical method of building protective dunes (CERC, 1984; Miller et al., 2001).

Zig-zag fences can create wider dunes with more undulating crest lines and more gently sloping dune faces, which are closer to the shapes of natural dunes. Single straight fences produce a linear ridge (Fig. 8.14). Paired fences can create a broader-based foredune with a rounded crest (if placed close together) or multiple crests with intervening swales (if spaced apart). Multiple lifts of fences (Fig. 8.14) can create a higher dune with much greater volume (CERC, 1984; Mendelssohn et al., 1991; Miller et al., 2001). Sand accumulation efficiency and morphological changes depend on fence porosity, height, inclination, scale and shape of openings, wind speed and direction, number of fence rows, separation distance between fence rows, and placement relative to existing topography. Fencing with a porosity of about 50% can fill to capacity in about a year where appreciable sand is moving, but vegetation plantings may still be required to stabilize the surface and establish a natural trajectory.

Sand fences are not needed where a dune of adequate size already exists, where they would trap sand in unnatural configurations, or where they cannot be buried, such as in vegetated portions of the dune or on narrow beaches where sources of wind-blown sand are restricted. Sand-trapping fences are often used to prevent users from trampling the dune. Fences used for purposes other than building dunes or preventing sand inundation can remain as conspicuous intrusions in the landscape, acting as physical boundaries to movement of fauna and reminders of the artificial nature of the dunes (see Fig. 8.12).

8.4.4 Using vegetation

Vegetation reduces wind speed, traps sand, stabilizes surface sediment, provides habitat, and improves aesthetic appeal. The characteristics of the vegetation are important in providing many of the goods and services provided by coastal dunes (see Table 8.2). Basic dune-stabilizing vegetation for use by managers in shorefront municipalities and by residents should be easy to propagate, harvest, store and transplant with a high survival rate, be commercially available at local nurseries at relatively low cost, and be able to grow in a variety of microhabitats on a spatially restricted foredune (Feagin, 2005). The species that are most useful in building a foredune rapidly react positively to sand burial. When sand deposition diminishes, these species can degenerate, but successional species may then take over. Degeneration of initial dune-building species is not a problem where the loss is coincident with colonization by later successional species. Species more typical of sheltered environments may be planted, but this may not be necessary because they will eventually opportunistically colonize the dune by themselves (Feagin, 2005).

8.4.5 Restricting negative human actions

Management actions in natural reserves may only be required to control human impacts, not natural processes. Restoration actions may be confined to removal of problematic exotic species and removal of vestiges of past human activities that bear no relationship to present use. Day-use recreation and nature appreciation are appropriate human uses. Beach fill may be a reasonable option to temporarily protect isolated pockets of valued natural resources down-drift of developed areas, where sediment inputs are reduced by past human actions, such as construction of jetties or groyne fields. Use of fences and vegetation plantings to stabilize mobile portions of dunes should be confined to locations where valuable infrastructure is threatened by sand inundation but cannot be relocated.

In intensively developed areas, the natural dynamism of the dune will be perceived as negative. The limit to how much dynamism can be tolerated by shorefront stakeholders is often related to the distance between the dune crest and the nearest human infrastructure. Foredunes can be allowed to erode or be breached by storm waves if there is space and the foredune is not the only protection (Arens et al., 2001), but dunes in most developed areas are narrow and close to human facilities and may not be allowed to evolve solely by natural processes. Ongoing human efforts may be required to allow portions of the dune to be mobile and evolve through time, while the overall feature remains intact as a barrier against storm wave runup. Many of the methods taken to create dunes where residents and tourists do not want them or increase the dynamism of stabilized dunes represent departures from past practices, so it is likely that most dune restoration projects will be experimental and small scale and will require a strong public information component until the feasibility of large-scale future projects is demonstrated.

Many of the degradational activities identified in section 8.3.1 can be prevented or reduced in their impact by good management programmes. Alternatives to widespread raking to manage wrack include: (1) selectively removing the cultural litter and leaving the natural litter; (2) leaving the highest storm wrack line on the backshore while raking the beach below it; (3) restricting cleaning operations to the tourist season; and (4) leaving longshore segments unraked to develop as natural enclaves (Nordstrom, 2008). Sand-trapping fences could be restricted after creation of the first dune ridge that functions as the core around which the natural dune can evolve. Symbolic fences may be used instead of sand-trapping fences to control pedestrians while allowing the sand blown landward to build up dune height and volume without interference to movement of fauna.

Driving could be restricted to only a few segments along a beach or to portions of the beach. Shore-perpendicular access could be restricted to a few prescribed dune crossings. There is little reason to allow private vehicles on beaches in developed municipalities when road networks exist landward and where undeveloped enclaves are so small that beaches can be reached on foot. Municipal vehicles used for patrolling the beach and for trash removal can be restricted as well.

Many public agencies now recognize the adverse effects of exotic vegetation on biodiversity, authenticity and succession and have made attempts to eliminate problematic species in conservation areas. The problem with invasion of exotics into dunes in developed areas from nearby residential and commercial properties is less well known. Achieving restoration goals on dunes on private lots is important but virtually ignored in coastal management programmes.

8.4.6   Maintaining dune environments

The first step in building new dunes where they have been eliminated may be to get any dune-like feature on the ground, even if it is low, narrow, linear, and fixed in position. Eventually, these dunes may evolve into functional habitats with resource value and aesthetic appeal (Fig. 8.15). Gaining acceptance for larger, more naturally functioning dunes requires demonstrating their value by providing examples of good management practice and implementing public education programmes. Education programmes should specify the role of human actions in both the degradation and restoration of natural environments. Once dunes have been re-established, monitoring and adaptive management are important in assessing: (1) whether restoration goals have been achieved; (2) what kind of follow-up actions are required; and (3) how plans can be modified to achieve better future projects. Recognition of the many values of coastal dunes in both natural and human contexts will help ensure that the full resource potential of dunes is reached. Examples of good dune management programmes exist, even at the local level, where most dune degradation occurs (Box 8.3).

Fig. 8.15   Avalon, New Jersey, USA. The older ridge to the left is a dyke created by fill- and earth-moving equipment but has developed a vegetation cover that obscures its artificial origin and provides habitat and aesthetic value. The species diversity on the landward ridge was increased by the protection provided by the high crest on the right, which was created using multiple lifts of sand fences. (Source: Photograph by Karl Nordstrom.)

8.5   Effects of future climate change

Changes in climate affect storm paths, storm severity, sea-level rise and conditions for growth of coastal vegetation, as well as human actions taken in response to these changes. Many of the future rates of environmental change will fall within limits of past periods of natural changes, but they are likely to exceed the recent rates to which humans have become accustomed. Climate changes may not create different kinds of problems, but are likely to intensify existing problems. Of special concern is the projected increase in rate of sea-level rise, which will increase rates of erosion and reduce beach widths. Beach sediment budgets will become more negative and erosion rates of foredunes will increase. Formerly high, relatively stable foredunes (see Fig. 8.6b) may be converted to smaller, more ephemeral forms (see Fig. 8.6a). Dune fields that extend across coastal barriers will be vulnerable to wave erosion on both the sea and bay sides.

Changes in the interior portions of dune fields are likely to be related to growth conditions for vegetation that will, in turn, affect dune mobility. Some dune fields will become more mobile and some less mobile, and the species inventories will change. Some studies predict dramatic future changes in specific dune fields. A simulation of the Coto Doñana National Park in Spain indicates an extension of the dry season, a decrease in summer soil moisture, and an increase in potential evaporation in the winter, leading to an increase in sand drifts; climate in Slowinski National Park in Poland may change from a sub-continental to a more Atlantic-type climate, with an increase in winter temperatures, a decrease in occurrence of frost, and an increase in rainfall, extending the vegetation cover over barren parts of the dunes and decreasing wind erosion (van Huis, 1989).

Managers of dune conservation areas may initially attempt to use human actions to maintain prior conditions, but a broader spatial view of coastal resources may be more effective. There is little reason to attempt to maintain vulnerable species within their former habitat if they can migrate with the climate zones to a new viable conservation site. A more global view of changes in resources is required to recognize that displacement, per se, may not be a problem.

Many of the more dramatic effects of climate change and sea-level rise may be obscured in developed areas by human efforts to protect property and maintain a stable or predictable resource base (Titus, 1990). The potential scale of human influences on dunes easily exceeds the scale of likely change resulting from climatic alteration, and resultant changes may be largely determined by human activity. The types of responses and scales will differ greatly due to differences in national, regional and local economies. Where little money is available for

CASE STUDY BOX 8.3   A municipal programme for dune management

Actions in the Borough of Avalon, New Jersey, USA, reveal how dunes can be protected and restored at the municipal level, which is where much of the degradation of dunes occurs (see Nordstrom, 2008). A disastrous storm in 1962 resulted in an aggressive programme for building dunes using sand fences and vegetation plantings and acquisition of undeveloped lots. Dunes are now found along the entire ocean-front of the municipality and provide many of the values of natural dunes, even where they are artificially created (Fig. 8.15). The borough purchased a 1.5-km long undeveloped segment (see Fig. 8.9) soon after the storm occurred, which the owners were willing to sell. This conservation area provided space for landforms to evolve naturally and a location for testing environmentally compatible management strategies, such as suspending use of sand fences and raking, which continued in other portions of the borough. Use of fences and raking ceased in the undeveloped segment in 1991. By 2009, foredune crest heights in the naturally evolving area were 1 m lower than in the raked and fenced area, but dune and beach volumes were greater, and the dune field was wider (72 m versus 40 m) and had an additional ridge. Suspension of raking increased the number and type of habitats for flora and fauna by increasing the number of dune crests and swales that create topographic variability. The new dune that evolved seaward now compares favourably in volume

with dunes built with fences, but it has a gentler seaward slope and fewer restrictions to cross-shore movement of sediment and biota and greater diversity of vegetation. The site has greater natural value and human-use value for nature appreciation and education, although the lower height of the dune makes it more vulnerable to still-water flooding (Nordstrom et al., 2012).

Education of residents and landowners in Avalon has been crucial in the acceptance of dunes as an important resource. Frequent meetings between representatives of the borough and landowners are held because of the rapid turnover in resident population. Stakeholders are encouraged to be active in the community, to educate themselves about coastal protection, and to attend meetings of the Chamber of Commerce, Realtors Association, and Land and Homeowners Association every year. A borough newsletter and flood hazard information are regularly mailed to property owners. Weekly guided tours of dunes in the dedicated undeveloped area are offered by a nearby university research and education facility. A municipal historical museum traces the history of the dunes from the earliest photos from the 1890s to the present. Dune stewardship is incorporated into the 4th grade curriculum, and students participate in planting dunes in the spring. The mayor and director of public works conduct periodic television interviews discussing the significance of dunes and the value of actions to manage them.

maintaining infrastructure in the face of increased erosion or dune mobility, the likely scenario would be to allow the coast to evolve, and accommodate the changes by elevating buildings on pilings and converting usable surfaces to those that can tolerate more frequent inundation. Landward migration of the beach and seaward portion of large dune fields could be tolerated, but this migration would eventually result in exposure and loss of habitats that evolved in more stable landward areas. The role of dunes in developed areas is unclear under this accommodation scenario. The severe restriction to beach width will limit the value of the beach as a source of sand for dunes and as a space for them to form, but the value of dunes in providing protection to remaining buildings would increase and human actions to preserve or restore them could be increased. In any case, erosion/accretion cycles in the dunes are likely to occur more rapidly than in the past.

Raising the substrate subject to inundation by placing sand on the beach and under buildings and roads is an alternative to more passive measures, but the cost of dredge and fill projects and environmental problems in

borrow areas may prohibit this option from being implemented on a large scale (Titus, 1990). Where affordable, this scenario would allow dunes to form and survive near the water, but they would still be modified by human actions where they migrated into landward infrastructure.

Where the level of development justifies the expense, seawalls and dykes can provide protection, just as in past centuries. Coasts protected by artificial structures would undergo increase in water depths and continued or accelerated beach loss. The beach could be maintained by artificial nourishment, although at increasingly greater cost. Dune preserves, if included landward of protection structures, would be cut off from the beach and exchanges of sediment and biota would be prevented. It is more likely that dune preserves that extend for any distance alongshore would be incorporated into protection strategies as naturally functioning barriers to flooding, but attempts may be made to modify the foredunes to reduce the likelihood that they would breach.

It is easy to envision many of the elements of these scenarios occurring in a single location through time. Individual houses will be destroyed and removed; others will be raised as they are threatened; individual buildings and roads may be constructed or reconstructed on substrate built to higher elevations than the surrounding unimproved terrain; seawalls, bulkheads and dykes will be built and rebuilt in segments until they eventually may form a continuous barrier. Many of these human alterations are likely to be incremental because of the difficulty of funding large-scale projects. The changes in storm frequency and severity, sea-level rise, and growing conditions are likely to result in dunes with different morphology and habitat value from those that exist today, but the conversion will occur slowly.

8.6 Summary

Coastal dunes form almost anywhere there is a sandy beach, but their sizes and shapes differ in response to regional differences in climate conditions, sediment budgets, vegetation growth and human actions. Dunes are dynamic features that undergo periods of accretion and erosion, and stability and instability. The landforms and habitats we see in dune conservation areas today are transient features that should be allowed to evolve by natural processes. Climate change may displace them from their present locations, but this may not be a problem if the habitats can be accommodated elsewhere. The acquisition of additional dune fields for conservation purposes is an important goal, but these dunes will not be pristine and will require additional restoration actions. Dunes in developed areas are spatially restricted and often perceived solely as protection structures and managed as static landforms. These dunes can provide more of their potential functions and services if at least some portions of them are allowed to evolve naturally. Natural processes can be aided through beach nourishment, bulldozing, fencing and planting programmes, and degradational practices can be controlled by restricting raking or driving on beaches and over-use of sand fences.

• Dunes are not fragile features; they can form in nearly any sandy beach environment.
• Dunes are mobile and change size, shape and location in response to changes in wave erosion, wind conditions and vegetation growth, all of which are affected by climate change.
• Human actions have played a key role in the past by eliminating dunes for construction, destabilizing dunes by destroying vegetation cover, and stabilizing dunes through use of sand fences and vegetation plantings.
• Human actions will be increasingly important in creating and maintaining dunes in the future as population pressure and rising sea levels restrict the space available for coastal landforms to evolve.

Key publications

Doody, J.P., 2001. *Coastal Conservation and Management: An Ecological Perspective.* Kluwer Academic Publishers, Dordrecht, the Netherlands.
[*Very accessible overview of ecological aspects of dune management, suitable for non-ecologists*]

Hesp, P.A., 2002. Foredunes and blowouts: initiation, geomorphology and dynamics. *Geomorphology*, **48**, 245–268.
[*Very comprehensive review of foredunes and blowouts*]

Nordstrom, K.F., 2008. *Beach and Dune Restoration.* Cambridge University Press, Cambridge, UK.
[*Comprehensive and up-to-date text on all aspects of restoring beach and dune systems*]

Sherman, D.J. and Hotta, S., 1990. Aeolian sediment transport: theory and measurement. In: K.F. Nordstrom, N. Psuty and B. Carter (Eds), *Coastal Dunes: Form and Process.* John Wiley & Sons, Chichester, UK, pp. 17–37.
[*Technical review of aeolian sediment transport processes*]

Wilson, P., Orford, J.D., Knight, J., Braley, S.M. and Wintle, A.G., 2001. Late-Holocene (post-4000 years BP) coastal dune development in Northumberland, northeast England. *The Holocene*, **11**, 215–229.
[*Very nice case study of coastal dune development during the second part of the Holocene*]

References

Aagaard, T., Davidson-Arnott, R., Greenwood, B. and Nielsen, J., 2004. Sediment supply from shoreface to dunes: linking sediment transport measurements and long-term morphological evolution. *Geomorphology*, **60**, 205–224.

Arens, S.M., Jungerius, P.D. and van der Meulen, F., 2001. Coastal dunes. In: A. Warren and J.R. French (Eds), *Habitat Conservation: Managing the Physical Environment.* John Wiley & Sons, London, UK, pp. 229–272.

Avis, A.M., 1995. An evaluation of the vegetation developed after artificially stabilizing South African coastal dunes with indigenous species. *Journal of Coastal Conservation*, **1**, 41–50.

Baas, A.C.W. and Sherman, D.J., 2005. Formation and behavior of aeolian sand streamers. *Journal of Geophysical Research*, **110**, F03011, 1–15.

Bagnold, R.A., 1936. The movement of desert sand. *Proceedings of the Royal Society A*, **157**, 594–620.

Bird, E.C.F., 1990. Classification of European dune coasts. In: W. Bakker, P.D. Jungerius and A. Klijn (Eds), *Dunes of the European Coasts: Geomorphology-Hydrology-Soils.* Catena Supplement 18, pp. 15–24.

CERC (Coastal Engineering Research Center), 1984. *Shore Protection Manual.* US Army Corps of Engineers, Ft. Belvoir, Virginia, USA.

Christiansen, C., Dalsgaard, K., Møller, J.T. and Bowman, D., 1990. Coastal dunes in Denmark. Chronology in relation to sea level. In: W. Bakker, P.D. Jungerius and A. Klijn (Eds), *Dunes of the European Coasts: Geomorphology-Hydrology-Soils.* Catena Supplement 18, pp. 61–70.

Cooper, W.S., 1958. *The Coastal Sand Dunes of Oregon and Washington. Geological Society of America Memoir* **72**.

Davidson-Arnott, R.G.D., MacQuarrie, K. and Aagaard, T., 2005. The effect of wind gusts, moisture content and fetch length on sand transport on a beach. *Geomorphology*, **68**, 115–129.

Doody, J.P., 2001. *Coastal Conservation and Management: An Ecological Perspective.* Kluwer Academic Publishers, Dordrecht, the Netherlands.

Feagin, R.A., 2005. Artificial dunes created to protect property on Galveston Island, Texas: the lessons learned. *Ecological Restoration*, **23**, 89–94.

García Novo, F., Díaz Barradas, M.C., Zunzunegui, M., García Mora, R. and Gallego Fernández, J.B., 2004. Plant functional types in coastal dune habitats. In: M.L. Martínez and N.P. Psuty (Eds), *Coastal Dunes, Ecology and Conservation.* Springer-Verlag, Berlin, Germany, pp. 155–169.

Godfrey, P.J., 1977. Climate, plant response, and development of dunes on barrier beaches along the U.S. east coast. *International Journal of Biometeorology*, **21**, 203–215.

Godfrey, P.J. and Godfrey, M.M., 1981. Ecological effects of off-road vehicles on Cape Cod. *Oceanus*, **23**, 56–67.

Grootjans, A.P., Adema, E.B., Bekker, R.M. and Lammerts, E.J., 2004. Why young coastal dune slacks sustain a high biodiversity. In: M.L. Martínez and N.P. Psuty (Eds), *Coastal Dunes, Ecology and Conservation.* Springer-Verlag, Berlin, Germany, pp. 85–101.

Hesp, P.A., 1989. A review of biological and geomorphological processes involved in the initiation and development of incipient foredunes. *Proceedings of the Royal Society of Edinburgh*, **96B**, 181–201.

Hesp, P.A., 1991. Ecological processes and plant adaptations on coastal dunes. *Journal of Arid Environments*, **21**, 165–191.

Hesp, P.A., 2002. Foredunes and blowouts: initiation, geomorphology and dynamics. *Geomorphology*, **48**, 245–268.

Houser, C., 2009. Synchronization of transport and supply in beach-dune interaction. *Progress in Physical Geography*, **33**, 733–746.

Klijn, J.A., 1990. The younger dunes in the Netherlands: chronology and causation. In: W. Bakker, P.D. Jungerius and A. Klijn (Eds), *Dunes of the European Coasts: Geomorphology-Hydrology-Soils.* Catena Supplement 18, pp. 89–100.

Martínez, M.L., Maun, M.A. and Psuty, N.P., 2004. The fragility and conservation of the world's coastal dunes: geomorphological, ecological, and socioeconomic perspectives. In: M.L. Martínez and N.P. Psuty (Eds), *Coastal Dunes, Ecology and Conservation.* Springer-Verlag, Berlin, Germany, pp. 355–369.

Martínez, M.L., Hesp, P.A. and Gallego-Fernandez, J. (Eds), (2013) *Coastal Dune Restoration.* Springer-Verlag, Heidelberg, Germany.

Maun, M.A., 2004. Burial of plants as a selective force in sand dunes. In: M.L. Martínez and N.P. Psuty (Eds), *Coastal Dunes, Ecology and Conservation.* Springer-Verlag, Berlin, Germany, pp. 119–135.

McLachlan, A., 1990. The exchange of materials between dune and beach systems. In: K.F. Nordstrom, N.P. Psuty and R.W.G. Carter (Eds), *Coastal Dunes: Form and Process.* John Wiley & Sons, Chichester, UK, pp. 201–215.

Mendelssohn, I.A., Hester, M.W., Monteferrante, F.J. and Talbot, F., 1991. Experimental dune building and vegetative stabilization in a sand-deficient barrier island setting on the Louisiana coast, USA. *Journal of Coastal Research*, **7**, 137–149.

Miller, D.L., Thetford, M. and Yager, L., 2001. Evaluating sand fence and vegetation for dune building following overwash by Hurricane Opal on Santa Rosa Island, Florida. *Journal of Coastal Research*, **17**, 936–948.

Moreno-Casasola, P., 1986. Sand movement as a factor in the distribution of plant communities in a coastal dune system. *Vegetatio*, **65**, 67–76.

Nickling, W.G. and Davidson-Arnott, R.G.D., 1990. Aeolian sediment transport on beaches and coastal sand dunes. In: R.G.D. Davidson-Arnott (Ed.), *Proceedings of the Symposium on Coastal Sand Dunes.* National Research Council Canada, Ottawa, Canada, pp. 1–35.

Nordstrom, K.F., 2000. *Beaches and Dunes of Developed Coasts.* Cambridge University Press, Cambridge, UK.

Nordstrom, K.F., 2008. *Beach and Dune Restoration.* Cambridge University Press, Cambridge, UK.

Nordstrom, K.F. and Arens, S.M., 1998. The role of human actions in evolution and management of foredunes in The Netherlands and New Jersey, USA. *Journal of Coastal Conservation*, **4**, 169–180.

Nordstrom, K.F., Jackson, N.L., Freestone, A.L., Korotky, K.H. and Puleo, J.A., 2012. Effects of beach raking and sand fences on dune dimensions and morphology. *Geomorphology*, **179**, 106–115.

Psuty, N.P., 1988. Sediment budget and dune/beach interaction. *Journal of Coastal Research*, **SI3**, 1–4.

Ranwell, D.S. and Boar, R., 1986. *Coast Dune Management Guide.* Institute of Terrestrial Ecology, NERC, Monkswood, Huntingdon, UK.

Sherman, D.J. and Hotta, S., 1990. Aeolian sediment transport: theory and measurement. In: K.F. Nordstrom, N. Psuty and B. Carter (Eds), *Coastal Dunes: Form and Process.* John Wiley & Sons, Chichester, UK, pp. 17–37.

Sherman, D.J., Jackson, D.W.T., Namikas, S.L. and Wang, J., 1998. Wind-blown sand on beaches: an evaluation of models. *Geomorphology*, **22**, 113–133.

Titus, J.G., 1990. Greenhouse effect, sea level rise, and barrier islands: case study of Long Beach Island, New Jersey. *Coastal Management*, **18**, 65–90.

van Huis, J., 1989. European dunes, climate and climatic change, with case studies of the Coto Doñana (Spain) and the Slowinski (Poland) National Parks. In: F. van der Meulen, P.D. Jungerius and J.H. Visser (Eds), *Perspectives in Coastal Dune Management.* SPB Academic Publishing, The Hague, the Netherlands, pp. 313–326.

Wiggs, G.F.S., Baird, A.J. and Atherton, R.J., 2004. The dynamics of moisture on the entrainment and transport of sand by wind. *Geomorphology*, **59**, 13–30.

Wilson, P., Orford, J.D., Knight, J., Braley, S.M. and Wintle, A.G., 2001. Late-Holocene (post-4000 years BP) coastal dune development in Northumberland, northeast England. *The Holocene*, **11**, 215–229.

Woodhouse, W.W., Seneca, E.D. and Broome, S.W., 1977. Effect of species on dune grass growth. *International Journal of Biometeorology*, **21**, 256–266.

9   Barrier Systems

SYTZE VAN HETEREN

Geological Survey of the Netherlands, Utrecht, The Netherlands

9.1   Definition and description of barriers and barrier systems

In discussing barriers and barrier systems, it is important to consistently distinguish the individual landform (barrier) from multiple inter-related barriers within the context of an overall coastal setting (barrier system).

Barriers are elongated, wave-, tide- and wind-built ridges that are composed predominantly of unconsolidated sand and gravel, and protect the adjacent mainland from open-water processes. Most of them are oriented parallel to the general shoreline trend. Their crests are above high-tide level, distinguishing them from bars. Barriers are potentially mobile, and usually separate marine or lacustrine open-water bodies from marshes, lagoons and estuaries that may be connected to sea or lake by tidal inlets.

Barrier systems are series of barriers and their associated environments along a coastal compartment defined on the basis of a common setting or sediment source. They may contain series of similar barriers, but more commonly include a range of barrier types. On the one hand, barrier systems are a composite of various smaller-scale coastal environments, including beaches (Chapter 7), coastal dunes (Chapter 8), salt marshes and tidal flats (Chapter 10), mangrove shorelines (Chapter 11), tidal inlets and tidal deltas (Chapter 12), and the shoreface. On the other hand,

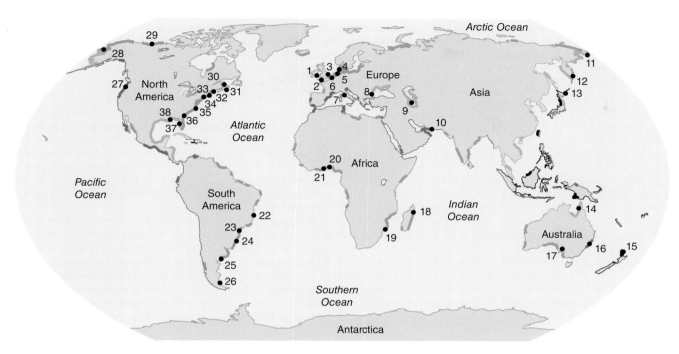

Fig. 9.1 Barrier-system distribution, and position of barriers or barrier systems mentioned in Chapter 9. Thin lines denote trailing-margin (blue), lake and marginal-sea (green), collision (red) and island-arc (black) coasts, respectively. Thick lines with the same colour code denote barrier systems along these coasts. 1, Carnsore barriers, Ireland; 2, Chesil Beach, England; 3, northern Norfolk barriers, England; 4, Holmsland and Skallingen barriers, Denmark; 5, German Bight; 6, North Holland, Netherlands; 7, Feniglia tombolo, Italy; 8, Danube Delta barriers, Romania and Ukraine; 9, Caspian Sea; 10, Khor Kalamat spit, Pakistan; 11, Chukchi Peninsula, Russia; 12, eastern Kamchatka, Russia; 13, Notsuke Peninsula, Japan; 14, western Cape York, Australia; 15, Pakiri, New Zealand; 16, Tuncurry, Australia; 17, Lefevre Peninsula, Australia; 18, Ambodiampana barrier, Madagascar; 19, Ujembje barrier, Mozambique; 20, Lekki barrier, Nigeria; 21, eastern Ghana and Togo barriers; 22, Doce strand plain, Brazil; 23, Paranaguá, Brazil; 24, Rio Grande do Sul, Brazil; 25, Caleta Valdés barrier, Argentina; 26, El Páramo spit, Argentina; 27, Willapa barrier, USA; 28, Kasegaluk barrier, USA; 29, Beaufort Sea barriers, USA and Canada; 30, Malpeque barriers and Buctouche spit, Canada; 31, Nova Scotia barriers, Canada; 32, Saco Bay, USA; 33, Horseneck Beach, USA; 34, Fire Island, USA; 35, Bogue Banks, Shackleford Banks and Outer Banks, USA; 36, Georgia Bight, USA; 37, Tampa Bay, USA; 38, northeastern Gulf of Mexico, USA. For colour details, please see Plate 20.

barrier systems are part of or associated with larger-scale coastal environments such as deltas (Chapter 13) and estuaries (Chapter 12). Where immediately relevant to barriers, aspects of these environments will be discussed in this chapter.

Barrier systems fringe about 15% of the world's seas and oceans (Fig. 9.1), and they are equally common along lake shores. They occur on all continents except Antarctica (Table 9.1), and at all latitudes. Their distribution is very uneven, however. This uneven distribution is a function of tectonic setting (Glaeser, 1978). About three-quarters of barrier coastline occupies low-relief coastal plains on trailing-edge continental margins (including marginal seas). Relative sea-level (RSL) rise during the Holocene, which has characterized many barrier systems in the northern hemisphere, has resulted in the submergence of extensive coastal-plain areas and in redistribution of vast amounts of sediment stored on drowning continental shelves. The presence of barrier systems along other types of coastline, such as collision coasts, is a function of local or regional sediment availability. Deltas are important sources for sandy barrier systems. High-gradient streams and eroding paraglacial bluffs are important sources for coarse-clastic systems. Chenier plains are common downdrift of low-gradient tropical and subtropical river mouths.

9.2 Classification

9.2.1 Barriers

Barrier planform can be classified using several parameters:
- attachment versus detachment,
- tide- versus wave-dominance,
- island length versus inlet width,
- open-water versus fetch-limited exposure,
- drift- versus swash-alignment, and
- single- versus multiple-barrier planform.

Table 9.1 Characteristics of barriers and barrier systems, as reported in the literature. For some locations, missing tidal and wave characteristics were added, using ranges derived from satellite data. The transport volumes should be interpreted as first-order assessments rather than exact quantities.

Location	Tidal range (m)	Wave height (m)	Landward transport (m³/yr)	Longshore transport (m³/yr)	Material	Type
Carnsore, SE Ireland	2.2–3.2	0.9–1.7			Sand and gravel	Baymouth barrier, baymouth spit
N Norfolk, E England	2.2–3.2	0.2–0.4		<350,000	Sand and gravel	Mixed-energy barrier islands, recurved spit
Chesil Beach, S England	1.1–1.5	3.3–4.2		3500–4700	Gravel and sand	Welded barrier
Skallingen, Denmark	1.5	1.0	90,000	70,000 (natural); 640,000 (dredging-related)	Sand	Barrier spit
W Netherlands	1.3–2.0	1.3–1.8		500,000–600,000 (gross)	Sand	Beach-ridge plain
Danube Delta, Romania	0.1	0.8		700,000–1,900,000	Sand	Baymouth barriers, baymouth spits, hooked spits, beach-ridge plains
Beaufort Sea, N Alaska (USA)	0.3	<1		100,000	Sand and gravel	Baymouth spits and barriers, wave-dominated barrier islands, hooked and recurved spits
Willapa, Washington (USA)	2.5–3.0	2.1	400,000	400,000–2,300,000	Sand	Recurved spit
Malpeque, Prince Edward Island (Canada)	1.2	1.0–1.5	91,100	40,000–200,000	Sand	Mixed-energy barrier islands, recurved spits
Buctouche spit, New Brunswick (Canada)	0.5–1.0	0.3		8000–56,000 (net)	Sand	Recurved spit
Fire Island, New York (USA)	1.0–1.3	1.5–2.0		230,000–460,000	Sand and gravel	Wave-dominated barrier island
Outer Banks, North Carolina (USA)	0.6–1.0	0.69	337,000	590,000–720,000	Sand	Wave-dominated barrier islands
NW Florida and SE Alabama (USA)	0.43	1.0	−47,000–58,000	17,000–223,000	Sand	Wave-dominated barrier islands, recurved spit, welded barrier, beach-ridge plains
Paranaguá, Brazil	2.2	0.7		300,000–1,100,000	Sand	Islands and spits (beach-ridge plain)
Caleta Valdés, Argentina	3.1	2.0	Little	>100,000	Sand and gravel	Looped barrier, recurved spit
El Páramo spit, S Argentina	5.7–6.6	1.0			Sand and gravel	Recurved spit
E Ghana and Togo	1.0–1.2	0.5–1.5		1,200,000–1,500,000	Sand	Baymouth barrier and spits, beach-ridge plains
Tuncurry, New South Wales (Australia)	1.5	2.0–2.5			Sand (and gravel)	Baymouth spits and barriers, welded and pocket barriers, beach-ridge plain
Lefevre Peninsula, S Australia	2.4	0.5		30,000–80,000	Sand	Developed spit
Pakiri, New Zealand	2.0–2.5	1.4		5000	Sand	Baymouth spit, recurved spit, welded barrier

As a result of the interplay among these parameters, there is significant variability in barrier shape (Fig. 9.2; Fig. 9.3) and size (see also Zenkovich, 1967). Attached barriers include welded barriers, pocket barriers, baymouth barriers, looped barriers, cuspate barriers, tombolos and double tombolos, and barrier spits. Detached barriers form islands.

Welded barriers, pocket barriers and baymouth barriers are attached to the mainland at both ends and concave seaward in planform. Welded barriers smooth stretches of coast with relatively minor indentations, protecting narrow back-barrier areas. Pocket barriers occupy somewhat larger indentations, sheltering small back-barrier areas. They may have a single, small tidal inlet. Baymouth

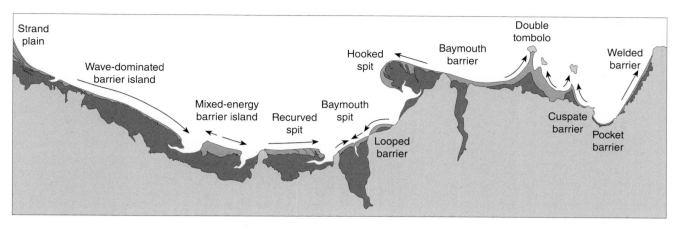

Fig. 9.2 Barrier-planform classes. Attached forms include welded barriers, pocket barriers, cuspate barriers, double tombolos, baymouth barriers, and various types of spit. Detached forms are mixed-energy and wave-dominated islands. Strand plains are characterized by multiple-barrier planform in progradational systems. Arrows denote littoral drift.

barriers commonly develop as accreting barrier spits close off relatively large bodies of open water that occupy medium-sized indentations. Such indented coastlines are typical of collision coasts, as well as of paraglacial trailing-edge and marginal-sea coasts. Additional conditions conducive to the formation of attached barriers are high wave energy and limited sediment supply.

Whereas welded barriers, pocket barriers and baymouth barriers tend to straighten the coastline, looped and cuspate barriers do the opposite. Looped barriers protect semi-circular or elliptically shaped back-barrier areas, and may form where the distal end of an accreting barrier spit intersects the mainland coast at an acute angle. Cuspate barriers have a concave plan view protruding from the general coastal alignment and close off triangular-shaped back-barrier areas. They form on the lee side of islands or shoals that are close to the mainland. Unlike tombolos, they are not attached to these highs. Double tombolos have a plan view similar to that of cuspate barriers, and may develop as spits anchored to an island accrete to the mainland. Tombolos are single sediment ridges.

Barrier spits are attached to the mainland at one end, commonly where the shoreline shows an abrupt directional change. They can be simple, hooked or recurved, and are present in the same indented settings as most other attached barrier types. Wave-driven longshore currents and strong tidal currents are instrumental in their development. Spits have slightly concave to highly convex planforms, and may show distinct recurved ridges and intervening swales near their distal ends (compound planform). Swales may be filled with aeolian sand, but they are usually turned into wetlands as RSL rises.

The planform of detached barriers (barrier islands) in coastal-plain settings is mainly a function of wave and tidal regime (Davis and Hayes, 1984). Wave-dominated and mixed-energy types can be distinguished (Fig. 9.4). Wave-dominated barrier islands, the most common type, may be hundreds of kilometres long, such as along Rio Grande do Sul in Brazil. Tidal inlets are widely spaced and tidal deltas are relatively small. Washovers are abundant. Mixed-energy barrier islands are short, and usually have a drumstick shape. The wide ends of islands, directly downdrift of tidal inlets, commonly protrude farther seaward than their updrift counterparts. They are marked by one or more ridges representing former coastlines. Tidal inlets are relatively stable and closely spaced, and tidal deltas are large. Tide-dominated barrier islands do not exist. Under these conditions, attached barriers characterize barrier systems.

Because barrier-island planform is governed by the ratio between wave and tidal energy, rather than by tidal regime alone, systems with large tidal ranges may be wave-dominated, whereas systems with very small tidal ranges may be tide-dominated (Davis and Hayes, 1984). Barriers fringing the entrance to the Minas Basin in the Canadian Bay of Fundy (Nova Scotia) experience a tidal range of about 10 m, but are shaped by storm waves that have grown in a long fetch. The opposite is true along the west-peninsular coast of Florida. Here, tidal range is less than 1 m, but mixed-energy barriers have developed under the influence of low wave energy and large tidal basins. Along coasts dominated by low wave and tidal energy, small temporal changes in either parameter may trigger significant changes in barrier planform. The morphology of deltaic barrier islands is not necessarily indicative of present-day hydrodynamic processes and does not conform to the relationship with wave and tidal energy as found for coastal-plain barrier islands (Stutz and Pilkey, 2011). They are classified on the basis of island length and inlet width.

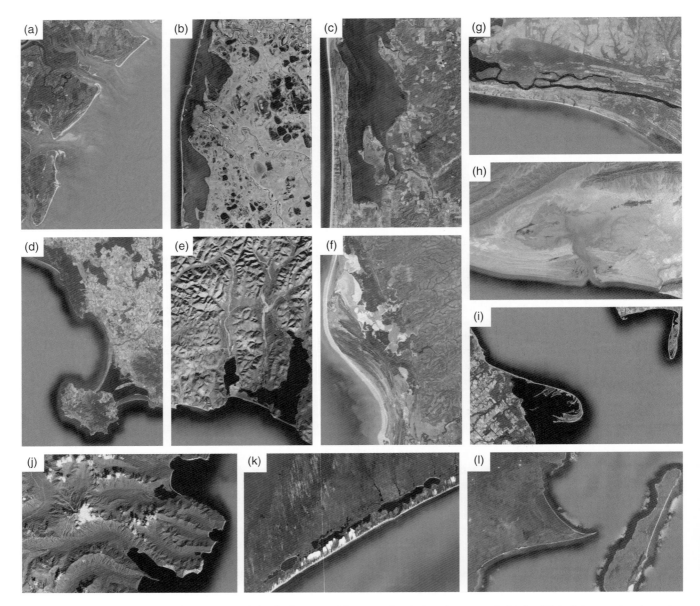

Fig. 9.3 Examples of different barrier types: (a) tide-dominated barrier islands fringing Georgia Bight, USA; (b) wave-dominated barrier island fronting Kasegaluk Bay, Alaska; (c) recurved spit protecting Willapa Bay, Washington, USA; (d) double tombolo at Feniglia, Italy; (e) baymouth barriers at Chukchi Peninsula, Russia; (f) looped barrier at Cape York, Australia; (g) strand plain fronting Lekki Lagoon, Nigeria; (h) baymouth spits at Khor Kalamat, Pakistan; (i) hooked spit at Notsuke Peninsula, Japan; (j) pocket barriers occupying indentations in the eastern Kamchatka coast, Russia; (k) welded barrier at Ujembje, Mozambique; (l) cuspate barrier at Ambodiampana, Madagascar. Images show areas of about 30 × 45 km. See Fig. 9.1 for locations. (Source: Google Maps. (a–e, g–j) © Google, TerraMetrics. (f) © Google, NASA, TerraMetrics. (k) and (l) © Google.) For colour details, please see Plate 21.

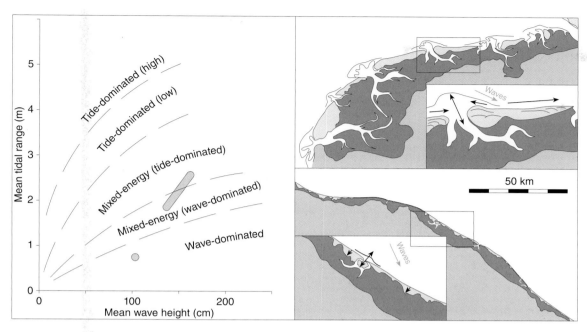

Fig. 9.4 Barrier-island classification as a function of wave and tidal regime. (Source: Hayes 1979. Reproduced with permission of Elsevier.) Mixed-energy barriers are commonly short, separated by prominent tidal inlets, and marked by protruding updrift termini. The stronger the tide dominance, the shorter and more stunted the barriers. Wave-dominated barriers are commonly narrow and long, with few small inlets. Wave and tide characteristics of examples on the right are shown in the diagram on the left (grey bar and dot). Analysis of an increasing number of barrier islands during the past decades has shown that the boundary lines drawn to distinguish the different barrier types do not have universal validity (Stutz and Pilkey 2011).

Not all barrier islands occur in open-water settings and have a seaward or lakeward coastline affected by high-energy processes. Some form and change in sheltered coastal re-entrants, behind open-ocean barrier islands, on flood-tidal deltas, or on river deltas sheltered by offshore islands. Others are surrounded by land, in areas characterized by RSL fall, and become islands only during spring tides and storm surges. The behaviour of such fetch-limited barrier islands is governed by high-energy events. They are much smaller than their open-ocean counterparts.

The development of drift- or swash-alignment (Fig. 9.5) is primarily a function of local topography and bathymetry, and of the size and distribution of sediment sources (Forbes et al., 1995). Drift-aligned barriers have an orientation oblique to the crests of the prevailing incident waves; they are controlled mainly by longshore sediment transport, forming where sediment supply is plentiful. They are part of open systems, marked by sediment exchange with adjacent coastal cells (compartments that contain the sources, transport paths and sinks of sediment). Swash-aligned barriers have an orientation parallel to the crests of the prevailing incident waves. They are part of closed systems, marked by little or no sediment exchange with adjacent coastal cells. They form where sediment supply is limited. In general, barrier islands are drift-aligned, barrier spits may be either drift- or swash-aligned, and all other attached barriers tend toward swash-alignment.

Although many barrier spits and barrier islands are characterized locally by multiple ridges, they are still considered to have a single-barrier planform. Multi-barrier planform (Fig. 9.6) involves some mechanism of long-term coastal progradation, commonly into the shallow waters of gently sloping shoreface zones. Series of ridges form strand plains, in which only the most seaward ridges function as barriers. Farther landward, swales are no longer occupied by typical back-barrier environments. Three types of strand plain are distinguished: the beach-ridge plain, the foredune-ridge plain, and the chenier plain. They can all extend tens to hundreds of kilometres parallel to shore.

The planform of barriers reflects their internal structure (Fig. 9.7). In retrogradational barriers, sediments deposited in a seaward setting overlie sediments that accumulated farther landward. From bottom to top, typical sequences include back-barrier, barrier and shoreface elements. Tidal-delta and overwash deposits form the majority of the preserved barrier sediment in transgressive settings. In South Carolina, USA, most of the barrier sand is stored in ebb-tidal deltas. In Denmark, the Holmsland barrier is

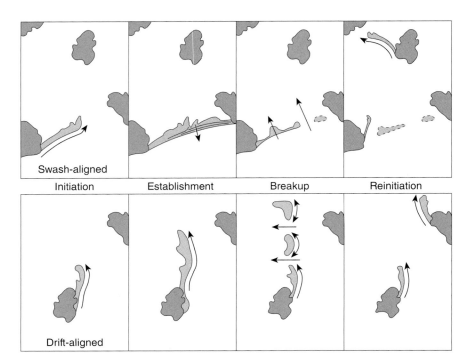

Fig. 9.5 Typical development of drift- and swash-aligned coarse-clastic barriers nourished by finite sediment sources. They go through phases of initiation, establishment, breakup and reinitiation in a regime of relative sea-level (RSL) rise. Temporal transitions between drift- and swash-alignment are common. Arrows denote main sediment-transport directions. The example shown is from southeastern Canada. (Source: Adapted from Forbes et al. 1995. Reproduced with permission of Elsevier.)

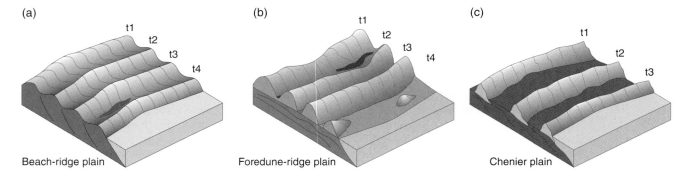

Fig. 9.6 Multiple-barrier planforms, with temporal development denoted by t1 to t4. Beach-ridge plains (a) consist of sandy or gravelly wave-formed ridges that may contain abundant shells and shell fragments. Individual beach ridges are underlain by sand and alternate with swales of intertidal shelly sand. If present, the cover of aeolian sand is thin and does not overprint the beach-ridge morphology. Foredune-ridge plains (b) consist of series of ridges formed by wind and stabilized by vegetation. Both beach-ridge plains and foredune-ridge plains may be overlain by large dunes that tend to obliterate the well-defined record of barrier progradation. In areas of limited exposure to waves, beach-ridge plains may grade or change abruptly into chenier plains (c), which consist of sandy and shelly ridges overlying muddy deposits. The scale of these features varies. Ridge spacing may be as little as 10 m to more than 1000 m.

composed almost entirely of overwash deposits. Inlet fill is a significant component of transgressive barrier sequences where tidal prisms (water volumes transported through inlets during ebb and flood) are large, distances between inlets are short, and inlet migration is significant. The frequent exposure of back-barrier sediments on present-day beaches, particularly following storm surges, indicates that near-complete removal of supratidal barrier elements is common in retrogradational settings. In progradational barriers, sediments deposited in a landward setting overlie sediments that accumulated farther seaward. From bottom to top, typical sequences include shoreface, barrier and

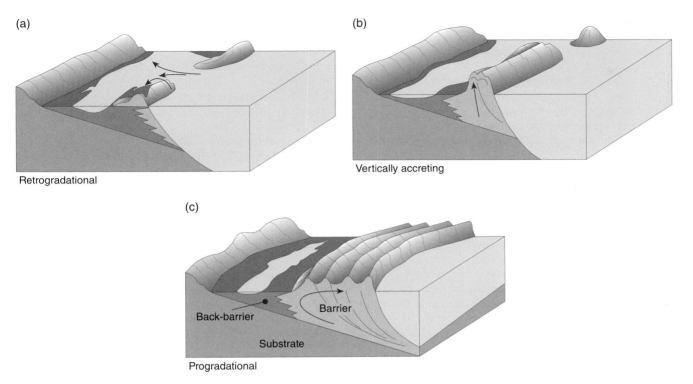

Fig. 9.7 Internal structure of barriers and their associated planform. (a) In retrogradational systems, barriers migrate landward over back-barrier deposits. (b) In vertically accreting systems, barriers migrate very little, showing planform stability during thousands of years regardless of sea-level variability. (c) In progradational systems, barriers migrate seaward, capping thick accumulations of shoreface sediments. Many barrier sequences contain both transgressive and regressive elements, reflecting temporally changing and spatially varying ratios between changing accommodation space (the space in which sediments may accumulate) and sediment supply. The scale of these barrier types varies, particularly in progradational settings. Perpendicular to the coast, some coastal wedges are as little as 100 m wide and only a few metres thick, whereas others are tens of kilometres wide and tens of metres thick.

back-barrier elements. Usually, supratidal aeolian sediments mark the present barrier surface.

A final element in the classification of barriers is grain size. Sand is by far the most common barrier sediment. Gravel dominates many paraglacial barriers in mid- and high latitudes and some carbonate barriers in low latitudes. Coarse-clastic barriers differ from their sandy counterparts by steeper, reflective equilibrium profiles, allowing easy percolation of wave swash, and by well-developed swash ramps that facilitate barrier-crest accumulation in response to low-magnitude run-up overwash. Dunes are commonly less developed. Mud, although typical of chenier coasts, does not usually comprise the barrier lithosomes themselves.

9.2.2 Barrier systems

Regional barrier systems can be classified on the basis of degree of compartmentalization, reflecting the relative importance of antecedent topography and geology, sediment abundance and size, and exposure to wave and tidal

energy (FitzGerald and Van Heteren, 1999). Sediment-starved 'isolated' barrier coastlines are characterized by short, widely spaced barriers. Small, localized updrift and offshore sources are associated with short barriers along 'clustered headland-separated' barrier coastlines. Various amounts of sediment from larger updrift and offshore sources or directly from rivers result in longer barriers along 'wave-dominated mainland-segmented', 'mixed-energy mainland-segmented', 'wave-dominated inlet-segmented' and 'mixed-energy inlet-segmented' barrier coastlines.

On an even larger scale, systematic patterns of different barrier types have been recognized as well, but no terminology has been proposed to classify these systems. Along the East Coast of the USA, several coastal compartments consist of a cuspate spit, an eroding headland, barrier spits and long barrier islands, and short barrier islands. Large-scale patterns along the Georgia Bight and the German Bight reflect shore-parallel variability in tidal and wave regime. With tidal range decreasing and wave energy increasing away from the apex of these bights, barrier morphology changes from tide-dominated to wave-dominated. The apex itself is commonly marked by an absence of barriers (Hayes, 1979).

9.3   Barrier sub-environments

Is a barrier system just a series of subaerially exposed sandy or gravelly landforms that protect back-barrier areas and the mainland from open-marine or -lacustrine processes? Not for coastal morphologists. Along the shore, it includes tidal inlets, tidal deltas and barrier platforms. In the cross-shore domain, it also includes several of the following sub-environments: the shoreface and nearshore zones, beaches, beach ridges, cheniers, coastal dunes, washovers, tidal flats and channels, bays, lagoons, estuarine basins, marshes and mangrove swamps (Fig. 9.8). These sub-environments occupy set positions relative to each other, and show consistent morphodynamic inter-relationships. In the analysis of large-scale coastal evolution, the cross-shore sequence of morpho-sedimentary sub-units was defined as a 'coastal tract' by Cowell et al. (2003a).

The shoreface is the most seaward environment of a barrier system. Its boundaries with the inner shelf (seaward) and surf zone (landward) vary spatially and temporally in response to changing hydrodynamic conditions. The shoreface is a critical environment whose behaviour has a major impact on other sub-environments. It acts as a source or sink of barrier sediment, and its changing morphology is an important factor in the hydrodynamic regime impacting barriers.

The dissipative or reflective nature of the nearshore zone and beach affect the amount of wave energy that reaches barrier coastlines under fair-weather and storm conditions. In addition, both of these sub-environments store sediment during and after storm surges, slowing down dune scarping and speeding up post-storm recovery of barriers. When truncated because of shifting sediment cells, beach ridges and cheniers are important sources for barriers developing downdrift.

Coastal dunes include stationary foredunes adjacent to the open water and mobile forms located farther inland. Foredunes protect barrier and back-barrier areas farther landward from open-marine wave impact and flooding, and their morphology determines the likelihood and location of overwash. Sediment eroded from foredunes during storm surges may form a buffer on the beach, reducing the impact of further storm waves. Aeolian landforms landward of the foredunes may limit overwash to back-barrier areas, forming additional barriers for water penetrating the foredunes during storm surges. Because of their large sediment volume, they reduce barrier mobility. Washovers are an important element in landward barrier migration (see Chapter 5), and a source for dune development.

Tidal inlets (see Chapter 12) function as major pathways for water and sediment transport between open-water and back-barrier elements of barrier systems, particularly in retrogradational settings. Tidal deltas provide protection for barrier coastlines and temporarily store sediment that may benefit barriers at a later point in time. Finally,

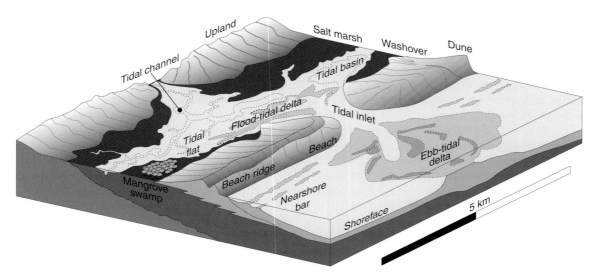

Fig. 9.8   Barrier-system sub-environments. Open-marine elements include the shoreface, ebb-tidal deltas and nearshore bars. Back-barrier elements include flood-tidal deltas, tidal flats and channels, tidal basins, marshes and mangrove swamps. They are separated by the barrier proper, which consists of beaches, beach ridges, dunes and washovers. Tidal inlets may connect the open-marine and back-barrier sub-environments. All elements occupy set positions relative to each other, and show consistent morphodynamic inter-relationships. The cross-shore succession of morphosedimentary sub-units is termed a 'coastal tract' (Source: Cowell et al. 2003a). Scale is approximate.

back-barrier sub-environments have an effect on fronting barriers by influencing accommodation space, tidal prisms, and landward migration rate of the backside of those barriers.

9.4 Theories on barrier formation

Most present-day barriers have moved to their present positions from farther seaward. They may have formed when sea-level fall slowed down or ended at the culmination of the last glacial period, allowing waves and currents to create supratidal sediment accumulations, or when the rising sea drowned outer shelves and initiated a period of major onshore sediment transport. Subsequent size changes, merger and disintegration have erased every trace of their origin. So is the question how barriers and barrier systems formed initially trivial? Heated debates in the coastal literature, fuelled by a lack of observations on barrier formation during lowstand conditions, prove otherwise.

Three observed mechanisms for barrier formation under various present-day conditions are: drowning of coastal ridges, spit development, and emergence of nearshore bars (Fig. 9.9). Barrier formation through drowning of coastal ridges has been observed best along the Caspian Sea shoreline. Here, new barriers are formed during periods of RSL rise following lowstands, by detachment of storm-related beach ridges from more irregular inner shorelines fringing the mainland. Spit elongation has been observed along many high-relief coastal regions, and along low-gradient coastal plains where these are

marked by a significant change in shoreline orientation. When breached, spits change into barrier islands. Observations of bar emergence in response to hydrodynamic processes and wind action primarily concern small barriers in protected settings, such as the large ebb-tidal-delta complexes at the mouth of Tampa Bay in Florida, USA. During fair-weather periods, wave-generated currents transport sand over shoals in environments characterized by abundant sediment, gentle gradients and low wave energy. In the absence of storm surges, wind processes redistribute sand on widening embryonic islands. Once the resulting ridges are stabilized by vegetation, they may survive moderate storm surges.

9.5 Modes of barrier behaviour

McBride et al. (1995) identified eight shore-normal and shore-parallel types of barrier behaviour (Fig. 9.10). In a shore-parallel direction, spits and islands may accrete and migrate very rapidly, both updrift and downdrift. The recurved spit of Shackleford Banks in North Carolina, USA, for example, accreted 1.5 km during a 40-year time span. Shore-parallel barrier migration is intricately linked with inlet behaviour. In a shore-normal direction, progradational, stable (usually vertically accreting) and retrogradational behaviour are the main modes of movement. Commonly, the life span of progradational barriers is limited to a few thousand years, as back-barrier areas tend to be filled rapidly. In the western Netherlands, mean progradation rates were hundreds of metres per century for

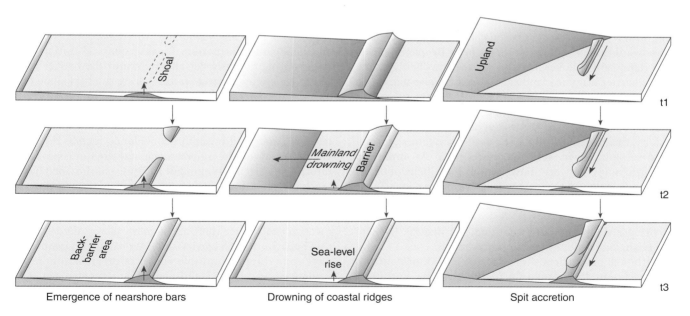

Fig. 9.9 Mechanisms of barrier formation: emergence of bars (left), drowning of coastal ridges (centre), and spit development (right) are shown from top (t1) to bottom (t3).

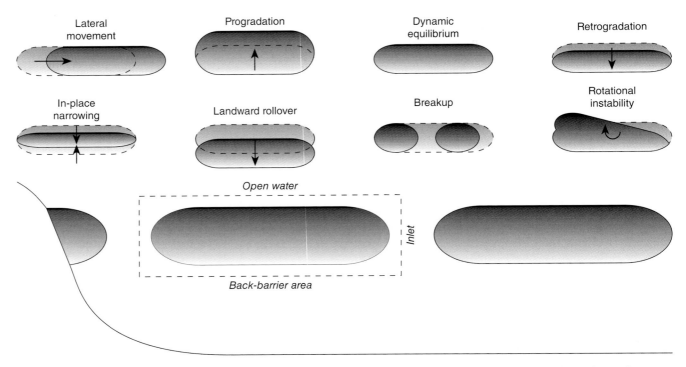

Fig. 9.10 Modes of barrier behaviour. If a barrier shows minimal net change along its back-barrier shoreline and significant change along the open-water shoreline, behaviour is classified as lateral movement, progradation, dynamic equilibrium or retrogradation. If both shorelines are moving, behaviour is classified as in-place narrowing, landward rollover, breakup or rotational instability. (Source: Adapted from McBride et al. 1995. Reproduced with permission of Elsevier.)

thousands of years. Stable barriers may be in dynamic equilibrium (little shoreline change) or experience in-place narrowing (shorelines move toward each other). In-place narrowing is common in coarse-clastic barriers that accrete vertically under increasingly rare overtopping events in a regime of RSL rise. Retrogradational barriers are marked by a landward-migrating seaside shoreline or by overall recession (landward rollover). In the rollover mechanism, the whole barrier migrates landward. Barriers must narrow to some critical width and height before they are subject to rollover. Rates may be tens of metres per year, determined both by the magnitude and frequency of extreme events and by the nature of the back-barrier system. Subtidal lagoons slow down rollover more than do intertidal marshes. Some barriers show non-uniform behaviour. In rotational instability, neap-berm development, swash-bar welding and berm-ridge development cause local prograda-tion on one end of the barrier while other parts of the same barrier are stable or retrogradational. Drumstick barriers and changing barrier orientation may result.

Although all barriers show some degree of discontinu-ous behaviour linked mainly to the periodic impact of extreme events, their temporal changes are generally grad-ual and continuous when considered on a decadal time-scale. When barriers are nourished by small sediment sources that become exhausted, gradual shoreline change

is punctuated by events of rapid reorganization (overstep-ping), particularly in a regime of rapid RSL rise (see Fig. 9.5). In Nova Scotia (Canada), initially drift-aligned coarse-clastic barriers are transformed into swash-aligned forms when sediment supply decreases. These swash-aligned barriers are vulnerable to catastrophic breakup and landward reformation. Breakup may be initiated by the scouring of washover channels into the barrier core. Other causes are seepage failure along the backslope of a barrier, and backwash drawdown and percolation on the beachface. Overstepping is common in paraglacial coarse-clastic barriers. Although overstepping of sandy barriers has been inferred from indirect evidence such as the pres-ence of shoals on the inner continental shelf, it is probably rare. Shoals originally thought to represent drowned barri-ers have been found to migrate landward in a process called transgressive submergence (Penland et al., 1988).

Barrier systems do not experience uniform behaviour through time. Since there are so many factors determining this behaviour, each barrier and even barrier segment may have gone through multiple phases of progradation, retro-gradation and relative stability during their existence, only partly preserved in their present-day stratigraphy and mor-phology. A commonly recognized large-scale pattern in low and middle latitudes has three main phases (Box 9.1): (1) barrier formation and retrogradation following the Last

CONCEPTS BOX 9.1 Phases in large-scale barrier development governed by a shifting balance between accommodation space and sediment availability

Barriers do not show consistent behaviour during their lifespans. Although they may experience extensive periods of uniform progradation, stability or retrogradation, changes in various drivers of barrier development can tip the balance between accommodation space and sediment supply at any time. Such balance shifts occur on all spatial scales. Although present-day morphology and stratigraphy form only incomplete geological records, they are essential in identifying and explaining past changes in barrier behaviour.

Barrier morphology can be used to reconstruct shifting coastline positions through time. Many beach-ridge series, for example, show evidence of alternating periods of erosion and accretion. Seaward bulges may be related to river outlets or tidal inlets, and ridge truncations may be related to changing coastline orientation. Stratigraphy helps to distinguish between progradational and retrogradational barrier elements that do not have a distinct surface expression.

Using morphological and stratigraphical data, the overall, long-term behaviour of barrier systems can be visualized in diagrams showing accommodation space and sediment availability. Systems showing similar behaviour through time can be grouped into categories.

The southeastern Australia type (A1–A5) is marked by an initial phase of retrogradation during rapid early Holocene relative sea-level (RSL) rise and abundant sediment supply, followed by progradation during middle- to late-Holocene RSL highstand and fall and abundant but decreasing sediment supply. Renewed retrogradation took place during a final phase of RSL stability and dwindling sediment supply. The western Netherlands type (B1–B6) is characterized by major retrogradation during rapid early Holocene RSL rise, progradation during decelerating middle- to late-Holocene RSL rise, and renewed retrogradation during continued RSL rise and dwindling sediment supply. A final phase of renewed progradation is the result of present-day beach and shoreface nourishment. The Mississippi type (C1–C6) behaves less uniformly. Retrogradation during rapid early Holocene RSL rise is followed by localized progradation during decelerating middle- to late-Holocene RSL rise and high fluvial sediment supply, and localized retrogradation during dwindling sediment supply following delta abandonment. Areas with multiple phases of delta activation and abandonment exhibit cycles of progradation and retrogradation (Fig. 9.11).

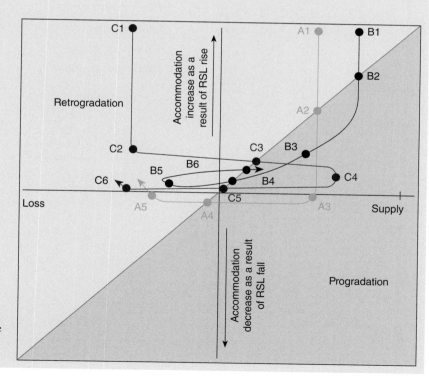

Fig. 9.11 Multi-phase barrier-system behaviour for three types of barrier coast: southeastern Australia type (A1–A5), western Netherlands type (B1–B6), and Mississippi type (C1–C6). RSL, relative sea level. (Source: Adapted from Cowell et al. 2003b.)

Glacial Maximum lowstand, in conditions of rapid RSL rise; (2) subsequent progradation and back-barrier infilling as RSL decelerated or stabilized and newly submerged sediment was transported from the shoreface to the shoreline; and (3) renewed retrogradation under continued RSL stability (or slow rise/fall) as sediment sources were less and less accessible, ending up beyond the reach of wave action on a steepening shoreface.

9.6 Drivers in barrier development and behaviour

9.6.1 Introduction

The development and behaviour of barriers and barrier systems are governed by a range of processes and parameters, whose interplay results in a wide variety of dynamic barrier forms.

Most factors affecting barrier behaviour are spatially and temporally variable on different scales (Fig. 9.12). As a result, no two barriers are alike. Sedimentary processes and associated morphological change take place on varying temporal (instantaneous, event, engineering, geological) and spatial scales (see Fig. 1.7).

The natural variability and autonomous trends that result from the interplay of drivers in barrier behaviour are captured in the analysis of large-scale coastal behaviour. This integrated analysis provides a framework encompassing all factors responsible for barrier evolution over the complete spectrum of spatial and temporal scales (Roy et al., 1994), including the influence by humans.

9.6.2 Antecedent topography and substrate lithology

Antecedent highs function as anchoring or stabilization points for perched barrier islands, which have a core of non-barrier sediment, and for attached barriers such as spits. The highs may be visible as promontories or as hills fronted on the seaside by barriers. Alternatively, they may be buried completely by modern barrier sediment.

Modelling experiments and field observations (from the Caspian Sea) show a relationship between barrier type and dimension on the one hand and the gradient of the coastal plain and continental shelf on the other (Fig. 9.13). In the case of gentle offshore slopes (in the order of 1 : 10,000), passive flooding dominates the coastal area and little sediment is redistributed. Under somewhat steeper offshore gradients (in the order of 1 : 1000), bars formed in the breaker zone are gradually transformed into barriers. Bottom friction reduces wave energy, limiting barrier size and transferred sand volumes. Where gradients are in the order of 1 : 100, storm-built beach ridges turn into large barriers. Back-barrier width is a function of gradient. With gradients steeper than 1 : 100, beaches weld directly to the mainland, and welded or mainland barriers without back-barrier areas are formed.

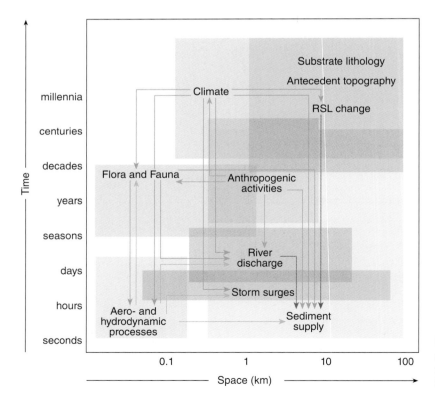

Fig. 9.12 Interlinked drivers of barrier behaviour, and associated temporal and spatial scales. RSL, relative sea level. For colour details, please see Plate 22.

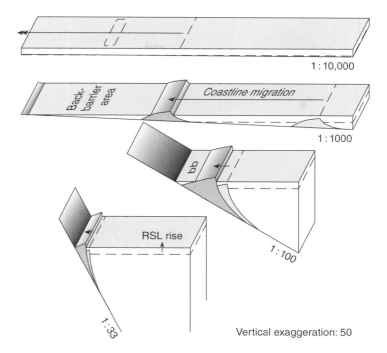

Fig. 9.13 Transgression during RSL rise and no net sediment input as a function of substrate gradient. Barriers and back-barrier areas are present when gradients are in the order of 1 : 100 to 1 : 1000. RSL, relative sea level; bb, back-barrier area. (Source: Adapted from Roy et al. 1994.)

Generally, longshore and cross-shore barrier-coastline migration rates increase as the gradient of the coastal plain and continental shelf decreases. In large embayments, such a decrease is present from the periphery to the apex. Since barriers can extend very rapidly in shallow water when sufficient sediment is available, a progradational planform is common along low-gradient, relatively sediment-rich shelves. However, progradation rates decelerate as expanding barriers reach increasingly deeper water. Substrate gradient also influences barrier evolution by governing how much sediment is within reach of the upper erosional zone of the shoreface (Fig. 9.14).

On the basis of regional bathymetry and topography, a partition can be made into low-lying coastal-plain coasts, more irregular embayed coasts, and relatively steep, cliffed or protruding coasts, each with a distinctive barrier-coastline morphology. Coastal plains with limited relief are dominated by strand plains or by inlet-segmented barriers. Such morphology was much more common at times of lower sea level than it is today because many outer shelves are less rugged than their inner counterparts (Roy et al., 1994). Embayed coasts are characterized by mainland-segmented barrier coastlines and by clusters of headland-separated barriers, depending on the distribution and volume of sediment sources and the size of embayments. Rugged cliffed or protruding coasts may be home to isolated barriers in sediment-starved settings.

Smaller-scale topographic and bathymetric features modify waves and currents and thus affect erosion, transport and deposition on individual barriers. The locations of barrier islands and associated inlets may be constrained by the presence of palaeovalleys. Upon submergence, large drainage basins form coastal bays with sizable tidal prisms, even when the tidal range is small. Retrogradational barriers typically form along shorelines that are either straight or convex seaward because these coasts focus wave energy and promote sediment transport away from the barrier. Bluffs fringing the backshore may favour coastal set-up by restricting submergence, resulting in downwelling and offshore sediment transport.

In determining coastline curvature, and therefore orientation relative to prevailing and dominant wind and waves, antecedent topography has an additional, indirect control on barrier morphology. In long basins, for example, conditions are favourable for the dominance of wind-forced waves from the directions with the longest fetch. Asymmetric hooks will build and migrate in two dominant directions and along two sides of these basins. When the basins are narrow enough, hooks on opposing sides may attract each other. Upon their merger, small oval sub-basins are formed (Ashton et al., 2009; see Fig. 1.10).

Substrate lithology affects barrier compositions, and influences barrier-system behaviour because it determines erodibility and ease of sediment transport. It also controls the resistance of anchor points (including beachrock and aeolianites) and channel margins, and thus the mobility of barriers and associated inlets, and influences the rate of shoreface and shoreline erosion (high for erodible sandy substrates, lower for lithified muds). Finally, it creates

208   SYTZE VAN HETEREN

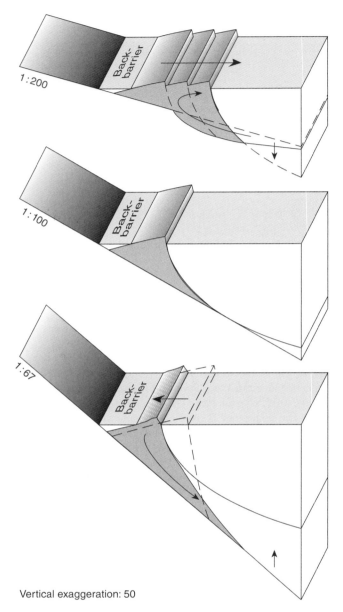

Vertical exaggeration: 50

Fig. 9.14 Influence of substrate gradient on sediment fluxes. In general, the shoreface supplies material to barriers when its slope is gentler than would be produced by the interaction of waves and currents. As the shoreface erodes, water depth increases and bottom sediments remain out of reach of wave and current action under increasingly extreme conditions. With steeper gradients, barrier sediment tends to be transported offshore, to the shoreface and shelf. (Source: Adapted from Roy et al. 1994.)

differential seabed roughness and, thus, hydrodynamic regime. Recent studies in North Carolina, USA, have suggested a link between barrier erosion and gravel outcrops seen on the updrift flanks of shore-oblique bars (Fig. 9.15).

9.6.3 Relative sea-level change

The decelerating rise in eustatic sea level following the Last Glacial Maximum and its stability during the middle to late Holocene has been one of the dominant factors in the evolution of present-day barrier systems. Where RSL change has mimicked the decelerating rate of eustatic sea-level rise, barrier formation at or near the shelf edge has usually been followed by an initial period of slowing retrogradation and a subsequent phase of progradation. Vertical land movements may have overprinted the eustatic component to a varying degree, however. Subsidence and compaction at rates of many metres per century commonly have had more extreme effects on barrier behaviour than eustatic sea-level change or crustal movements. In areas of uplift, RSL fall reduces barrier longevity. Although new beach ridges may form periodically as coastlines migrate seaward, long-term RSL fall limits the time during which each beach ridge serves as primary barrier. The strand plains formed by series of beach ridges and intervening swales form relatively stable barrier systems.

The influence of the rate of RSL change on barrier behaviour is not yet understood in detail. Few measurements exist and there is a strong overprint by other drivers. Evidence from eastern Canadian and northwest European gravel barriers suggests that barrier-retrogradation rate correlates positively with the rate of RSL rise on a subdecadal timescale. Qualitative proof for the sensitivity of barriers to the rate of RSL change comes from the Chinese coast, where transitional areas between subsidence and uplift belts, with a dominant eustatic sea-level component, are preferential areas for the formation of barrier systems. Some numerical models suggest that rapid RSL rise favours the stepwise retrogradation of barriers. Others produce fairly continuous retrogradation when RSL is rising rapidly, and intermittent landward migration when RSL is rising slowly. Evidence from the Caspian Sea in Asia, an interesting 'laboratory' because of metre-scale RSL fluctuations during the past few centuries, shows that continuous and discontinuous behaviour both occur, reflecting the roles of barrier dimensions, sediment supply, and sediment type.

RSL change affects sediment supply to, and flux within, barrier systems in various ways. Observed and modelled effects include: maximized accessibility of potential sediment sources when RSL varies most extensively; decreasing sediment supply as the rate of RSL rise decelerates (Forbes and Syvitski, 1994); a role switch for the shelf from sink to source during RSL fall; and reworking of sand eroded from barrier fronts into transgressive dunes during RSL rise. Finally, RSL change affects sediment supply indirectly, by changing the gradients and courses of rivers.

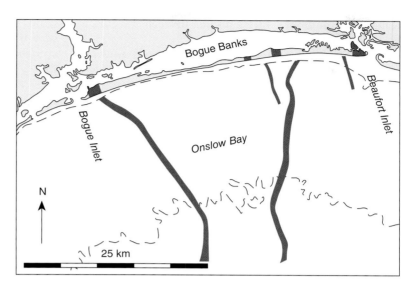

Fig. 9.15 Link between the distribution of relict channel fills and barrier-coastline behaviour in North Carolina, USA. Erosion hotspots on Bogue Banks, denoted in dark grey, are thought to be caused by the interaction of relatively coarse channel-fill sediments with the hydrodynamic regime. This gravelly material, a surface expression of the underlying geology, produces sorted bedforms that alter incident wave energy. (Source: Adapted from Browder and McNinch 2006. Reproduced with permission of Elsevier.)

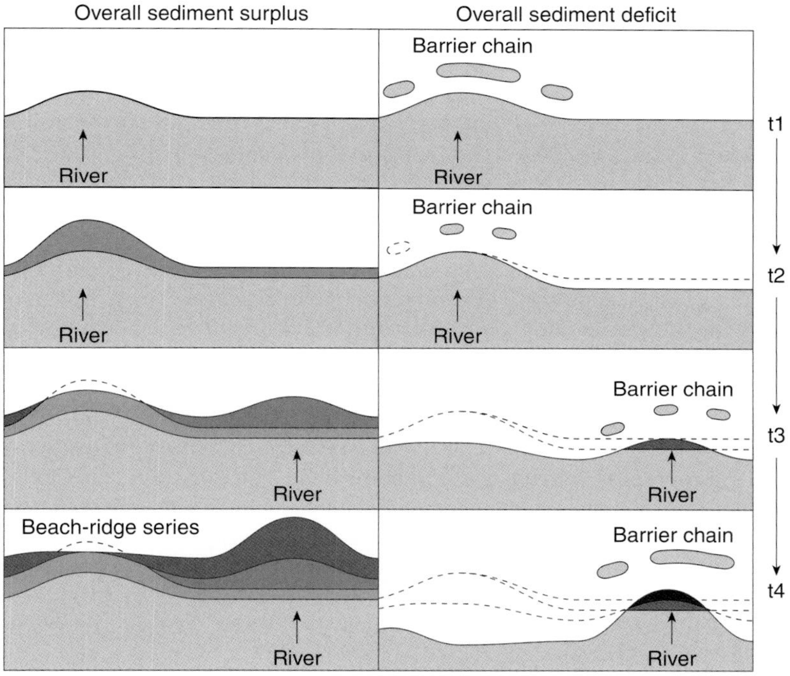

Fig. 9.16 Influence of periodic shifts of sediment sources, such as rivers, on barrier behaviour. Development in time is denoted by t1 to t4. In a scenario of overall sediment surplus, these shifts may be reflected in beach-ridge orientations. In a scenario of overall sediment deficit, they are more commonly reflected in the growth and decay of barrier chains. Note that rivers also affect barrier behaviour indirectly, by their influence on coastal currents.

9.6.4 Sediment sources, sinks and fluxes

In combination with other drivers, sediment supply controls barrier behaviour by filling accommodation space, and by determining the grain size of barrier sediment.

As a result of differences in sediment abundance, retrogradational and progradational barriers coexist side by side under identical RSL change. A non-uniform balance between sediment demand and supply does not just result in spatial barrier variability, but also in temporal change. Barrier behaviour is known to reflect periodic shifts of sediment sources such as rivers (Fig. 9.16).

Sediment type has several effects on barrier behaviour. Strong seepage in coarse-grained barriers, for example, may translate into a lack of distinct tidal passes (Carter and Orford, 1984). Different transport rates (and separate arrival) of sand and gravel eroded from updrift sources during extreme events may cause cyclic transformation from gentle sand-rich to steep gravel-rich barrier surfaces.

Barrier sediments are derived from various sources, including rivers, palimpsest and modern deposits on shoreface and inner shelf, tidal deltas, updrift barrier sections, updrift erosional headlands, and marine organisms. Many of these sources are part of the coastal tract, and are on the providing end of sediment redistribution that benefits barrier growth.

The volume of sediment supplied by modern rivers is determined in part by drainage-basin size and climate. Much of the fluvial sediment is contributed episodically during high-discharge events. On a larger scale, temporal changes associated with avulsion-related distributary switching are a main driver of barrier behaviour. Because modern rivers supply relatively little sediment to most barrier systems, with contributions decreasing rapidly away from river-mouths, there is no clear link between present-day sediment discharge and barrier abundance and size.

Many shoreface and inner-shelf areas in the world are marked by a surficial unit of palimpsest (showing attributes of both an earlier and the present depositional environment) sediment that has been reworked from relict deposits of earlier, commonly continental, depositional environments by waves and currents, including inlet scour. Voluminous lowstand fluvial deposits preserved and reworked offshore serve as major sources for barrier growth in a process termed shoreface bypassing, even near modern deltas. Shoreface bypassing was instrumental for widespread barrier progradation in the middle Holocene (Fig. 9.17).

Tidal deltas of degrading and closed inlets, modified by waves and tidal currents, are an additional offshore source. Alternating beach ridges and plains in the western Netherlands barrier system have been thought to reflect periodic increases in sediment supply from ebb-tidal deltas as inlets were closed one by one (Beets et al., 1992). The wide parts of some southern Portuguese barriers owe their shape to the incorporation of flood-tidal deltas after inlet migration or closure (Pilkey et al., 1989).

Voluminous barriers provide considerable scope for internal re-organization. Such cannibalization occurs when sediment supply diminishes at updrift barrier sections. It is common in spits, where sediment is eroded from the ocean-facing coastline to maintain longshore growth. Cannibalization may also occur among adjacent barriers, but net losses of sediment from updrift barriers may not be directly linked with net gains of sediment on adjacent downdrift barriers because sediment by-passing at inlets is episodic and inefficient.

Erosion of updrift bluffs and continental dune fields supplies sand and gravel to attached barriers on a local scale. When headland bluffs are eroded, adjacent lows may fill up and become fronted by baymouth barriers, straightening the coast. Sources vary in size, but are finite. In mid- to high latitudes, material eroded from tills, glaciolacustrine deposits, fluvial sediments, and outwash material is a particularly important source for paraglacial barriers. In arid environments, material eroded from ergs builds barriers downdrift. Bluff erosion is commonly a gradual process, but landslides may introduce larger volumes into coastal sediment cells at once. The lifespan of bluff-fed barrier systems depends in large part on the volume and erodibility of the sediment source.

The reworking of carbonate skeletal sediments formed by marine organisms contributes significantly to some low-latitude barriers such as those fringing the Persian Gulf. At present, barriers in South Florida, USA, are supplied with sediment from siliciclastic deposits preserved offshore and with carbonate sediments.

Aside from sources, sediment sinks also play an important role in barrier behaviour. On the one hand, back-barrier areas, tidal inlets and estuaries trap sediment that therefore is unavailable to nourish the barrier coastline. On the other hand, sediment eroded from the primary barrier coastline is redistributed and ends up in transgressive dunes, tidal basins, neighbouring coastal compartments, the shoreface and inner shelf, and in submarine canyons that extend sufficiently shoreward to intercept nearshore sediment-transport cells.

Rates of sediment transport and supply range widely (see Table 9.1). In the short term, sediment supply may be governed by extreme events such as landslides and jökulhlaups (glacial outburst floods). During the sudden drainage of glacier-dammed lakes, discharges may reach $100,000\,m^3/s$. Jökulhlaups provide the majority of sediment available to some Icelandic barriers during a single event. In the long term, net cross-shore sediment-transport rates are up to $100\,m^3/m/yr$, related in part to weather conditions. Cross-shore fluxes have been found to account for approximately three-quarters of the total annual barrier-coastline change for El Niño years along the Washington (USA) coast, but for only one-quarter (with longshore sediment transport dominant) on a decadal timescale (Ruggiero et al., 2010). Net longshore sediment-transport rates are up to millions of cubic metres per year. An average multi-year movement of approximately 1400 metric tons of gravel a day has been measured at Caleta Valdés Barrier Spit in Argentina. Sediment transport also varies spatially, as reflected in differential barrier behaviour. The drivers responsible for this transport, provided that sediment is available, are wind, waves, longshore and cross-shore currents, and tides.

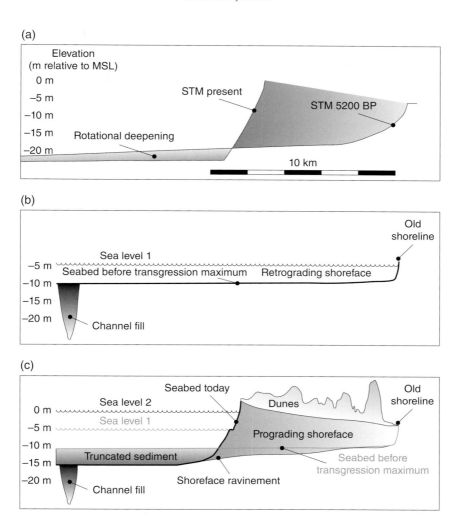

Fig. 9.17 Shoreface translation model (STM) simulation (a) of seaward shoreface translation and rotational shoreface deepening for the western Netherlands coast, as compared to a morphostratigraphic reconstruction based on core and seismic data. During the middle-Holocene transgression maximum, when relative sea level (RSL) was about 5 m lower than today, the coastline was located some 10 km landward of its present position (b). This palaeocoastline can be recognized as a very small beach ridge formed in a position sheltered from large-scale wave action. Remnants of even older tidal-channel fills are present at or near the seabed at 14–16 m below present-day mean sea level (MSL), in a zone that extends up to 12 km seaward of the present coastline. The upper 3–6 m of the tidal-channel fills, adjacent tidal-basin deposits and underlying Pleistocene river deposits have been truncated by later shoreface erosion. The degree of truncation could be constrained on the basis of the limited preservation of isolated channel fills characterized by a lack of shallow tributaries. It is likely that part of the eroded sediment was used for up to two-thirds of coastal progradation following the middle-Holocene transgression maximum, around 6000 years ago (c). (Source: Cowell et al. 2003b. Adapted with permission from The Coastal Education & Research Foundation, Inc.)

9.6.5 *Wind, waves, longshore and cross-shore currents, and tides*

Air and water flow generated by wind, waves and tides result in sediment fluxes that influence barriers by adding or removing sand and gravel (Fig. 9.18). The direct influence of wind is limited to intertidal and supratidal areas, but wind also drives waves and currents (including the non-astronomical component of tides), and therefore affects the subtidal zone as well. The influence of waves and tides is limited to intertidal and subtidal areas. Jointly, they may generate sediment-circulation cells. Subtidal sediment movement on the shoreface and in the nearshore zone is important because it adds or removes sediment from barrier systems.

Wind strength and direction are primary elements in sediment mobilization and transport on barrier coasts, especially when sparsely or non-vegetated intertidal and supratidal areas are laterally extensive. Wind is most effective during times of drought and has little impact when

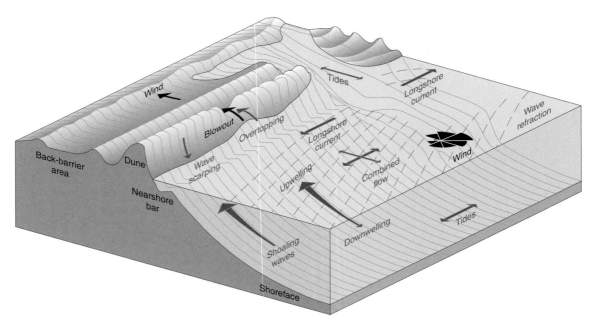

Fig. 9.18 Hydrodynamic (in grey) and aerodynamic (in black) processes affecting barrier behaviour. (Source: Adapted from Stive and De Vriend 1995. Reproduced with permission of Elsevier.)

source areas such as beaches and shoals are wet because of rain or flooding. Persistent onshore winds produce embryo dunes, foredunes, blowouts and transgressive dune fields. Under the influence of wind action, blowouts in foredunes may erode down to below the threshold level for wave overtopping, and function as corridors of sand transport, not only to transgressive dune fields but also to back-barrier areas. Aeolian transport increases during storms, as reflected in slipface-migration rates of transgressive dunes.

Energy imparted to coastal waters and sediments by wind is a very important geological agent controlling sediment transport and evolution of barrier systems. Wind-generated waves and the associated sediment transport are a function of wind speed and direction, fetch (open-water distance, also on the back-barrier side), width of area affected by fetch, duration of the wind event over a given area, and bathymetry (including shelf width). Barrier-coastline exposure to waves is a function of shoreline orientation, presence or absence of offshore islands and ledges, and nearshore gradients. Differential wave refraction leads to more aggregate wave energy at headlands than in bays.

Direct wave action dominates hydrodynamics from the middle and lower beachface downward to the middle shoreface. Its influence peaks in the nearshore zone and decreases toward the lower shoreface, where the wave base associated with major storms is located. Impact on barrier systems includes entrainment of bottom sediments and onshore sediment flux driven by shoaling waves, erosion of frontal dunes by incident waves, and overtopping of low

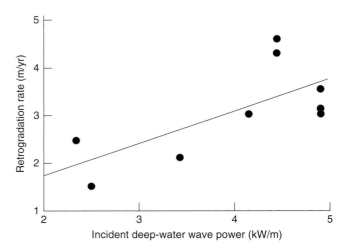

Fig. 9.19 Landward migration rate of barrier islands fringing the Canadian Beaufort Sea as a function of offshore wave energy. (Source: Héquette and Ruz 1991. Adapted with permission from The Coastal Education & Research Foundation, Inc.)

barrier sections or even entire barriers. Retrogradation rates of barrier islands fringing the Canadian Beaufort Sea show a correlation with incident deep-water wave energy (Fig. 9.19). Direct wave action also creates the bar-and-trough morphology of the nearshore zone, as well as swash bars and berms that weld onto barriers. Finally, it erodes back-barrier surfaces during extremely high tides. Continuous reworking by wave action results in a general separation of size, shape and mineralogical fractions in barrier sediment. Fractionation has been documented under storm conditions along Horseneck barrier in Massachusetts, USA, where sand is moved offshore, whereas gravel is moved onshore.

Aside from direct effects, wave action governs sediment transport by resultant cross-shore and longshore currents. The transport directions and rates of sediment stirred up from the seabed by incident waves are commonly controlled by these currents. Documenting current patterns and the associated sediment-transport paths on the shoreface is of critical importance to analyzing, explaining and predicting barrier behaviour.

When near-coastal water levels experience wind-driven set-up or set-down, downwelling or upwelling currents may generate cross-shore sediment transport. Such transport takes place between the beach and the lower shoreface, but to shallower depths in non-storm conditions than during extreme events. Strong onshore (and some shore-parallel) winds push water toward the shore, raising the level of the sea surface. Return flow from this wind set-up takes place at the base of the water column and potentially transports sediment seaward. Other processes, such as wave refraction, modify or overprint the downwelling and upwelling currents, depending on the meteorological and hydrodynamic conditions. Net long-term cross-shore transport is the residual of large and complicated offshore and onshore fluxes of sediment, which vary considerably under storm, moderate-energy and fair-weather conditions.

Indirect evidence for net sediment transport from shoreface to barrier is common. It comes from sediment-budget calculations where onshore transport is needed to explain barrier-coastline behaviour, current measurements, long-term bathymetric time series, behaviour- and process-based models, and provenance studies. Heavy-mineral characteristics of beach sand are testament to the importance of cross-shore transport in the progradation of the voluminous western Netherlands barrier system. Little direct evidence exists for onshore sediment transport, because cross-shore fluxes are difficult to measure across large areas and during time intervals long enough to capture fair-weather periods as well as storms of different magnitude. At Skallingen in Denmark, onshore bar movement in the mid-outer surf zone has been linked to a dominance of wave-induced sediment transport during storm conditions (Aagaard et al., 2004). When large undertow velocities dominate, offshore sediment transport results.

Permanent sediment loss from barrier systems may occur when sediments are swept offshore by mega-rips during extreme events and settle in water too deep for onshore-directed currents to return the sediments to the shoreface. The importance of this process may be limited. Many open-ocean shelves contain little modern sand, and sediment from the storm-dominated inner shelf of Texas, USA, collected immediately before and after a major hurricane,

suggests that most of the sand eroded from the adjacent barrier system was not transported beyond the shoreface.

Littoral drift or longshore sediment transport is powered by processes associated with breaking waves. The primary direction of longshore currents is controlled by the local shoreline orientation in combination with the direction of wave approach. Fining of sediment is common where regionally uniform longshore transport redistributes sediment from local updrift sources. Spatially variable longshore-transport directions, common along undulating or indented coastlines, complicate longshore grain-size patterns. Temporal variability, reflecting the dynamic nature of the forcing wind and waves, does the same.

The direct effects of littoral drift on barrier systems are commonly preserved in barrier morphology. They include shore-parallel elongation of barrier spits and barrier-island tips, and shore-perpendicular drumstick-shape widening of spit and island termini near large inlets. Here, wave refraction around ebb-tidal deltas leads to local reversals of the regional longshore-current direction. The location where swash bars attach to the beach is controlled by the amount of overlap of the ebb-tidal delta with the downdrift inlet shoreline. In many locations, past directions of longshore transport are reflected in the orientation of beach-ridge and foredune-ridge series.

When angles between wave crests and coastlines are small, longshore sediment transport tends to smooth indentations, closing off small inlets and nourishing spits that accrete from headlands across embayments. Drift-aligned barriers dominate under these conditions. When these angles are large, minor perturbations along originally straight coastlines may develop into cuspate barriers and hooked (or flying) spits. Constructive feedback at initially small coastline irregularities culminates in single or multiple cusps or in an alternation of swash- and drift-aligned barrier segments.

Tidal currents are a function of astronomical forcing and back-barrier dimensions. Ebb- and flood-tidal deltas are the main morphological elements reflecting tidal action in barrier systems, but both inlet-related shore-perpendicular and marine shore-parallel tidal flow have a direct effect on barrier shape as well. Where inlets cut barrier coastlines, tidal currents may extend barriers up to kilometres in a shore-normal direction. Longshore barrier growth may be a result of beach erosion by ebb-tidal currents on the back-barrier side.

Along with morphology, tidal range determines the size of the area where waves and currents can rework the substrate of intertidal areas. When all other factors are similar, a large tidal range will expose a more extensive area to wave and current action than will a small tidal range. In the subtidal zone, the effectiveness of wind and wave action diminishes and tidal-current activity increases as the tidal range increases. When tidal range is large and shoreface and nearshore slopes are gentle, waves cannot break in a concentrated area for a long period of time.

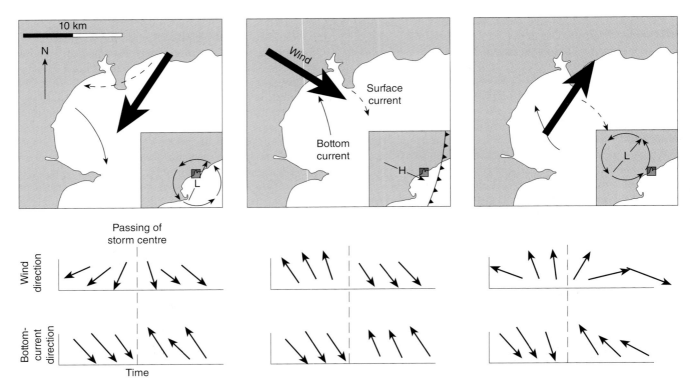

Fig. 9.20 Influence of different storm types on sediment transport toward and away from the Saco Bay barrier system in Maine, USA. Depending on storm characteristics, there may be either net onshore or net offshore transport. Onshore transport occurs when downwelling abates during the waning stages of storms (Stive et al. 1999). (Source: Hill et al. 2004. Reproduced with permission of Elsevier.)

9.6.6 Storm surges

Storm surges induced by wind and low pressure interact with astronomical tide, wave set-up and wave run-up to form storm tides that have a strong, instantaneous impact on barrier systems, especially in areas with wide, gently sloping shelves. They may leave barriers temporarily or permanently vulnerable to subsequent storm surges.

Storm surges affect the sediment budgets of entire barrier systems, from shoreface to back-barrier zone, even when the storms responsible for them do not make landfall. The main sand and gravel fluxes lead to internal reorganization. There is increasing field and modelling evidence that net sediment loss, traditionally thought to be widespread, is limited in many areas (Fig. 9.20). Rather, storms may cause inner-shelf and lower-shoreface erosion and onshore transport, as evidenced by the presence of fossil clam shells and rock fragments covered by soft corals on post-storm barriers in North Carolina, USA. They are transported landward from more than 1 km seaward of the modern surf zone. In the Caspian Sea, storm surges are the only mechanism capable of winnowing enough coarse-grained sediment from the shoreface to construct barriers (Kroonenberg et al., 2000). Storm surges also cause the

most dynamic erosion of updrift sediment sources, such as bluffs, thus providing additional material.

Barrier behaviour during storm surges is a result of erosional effects (including dune scarping, beach erosion, overwash-channel incision, and inlet formation) and depositional processes (including berm, beach-ridge and chenier formation, development of perched fans, washover-terrace construction, and sedimentation by sheetwash) (Morton and Sallenger, 2003). In the 'swash' regime, run-up is confined to the foreshore. In the 'collision' regime, wave run-up reaches the base of the foredune ridge. In the 'over-wash' regime, wave run-up overtops the berm or, if present, part of the foredune ridge. In the 'inundation' regime, the storm surge is sufficient to submerge much of the barrier.

When surges and accompanying set-up and run-up encounter high and continuous foredunes that prevent overtopping, they cause severe dune and beach erosion (Fig. 9.21). Barrier widths may be reduced by many metres in just a few hours. Eroded sediment is transported along-shore or temporarily stored in the nearshore zone or on the upper shoreface where it may dissipate wave energy. Dune scarping has not been shown to accelerate long-term coastal erosion. To the contrary, some barriers with large foredunes and inland dunes have lower rates of historical

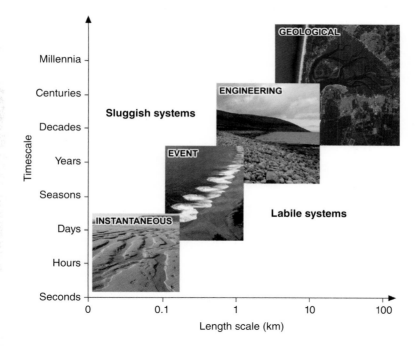

Plate 1 Relationship between spatial and temporal scales of coastal systems. Sluggish and labile systems are those that respond relatively slow and fast, respectively. (Source: Adapted from Cowell and Thom 1994. Imagery © 2013 Terrametrics. Map data © 2013 Google.)

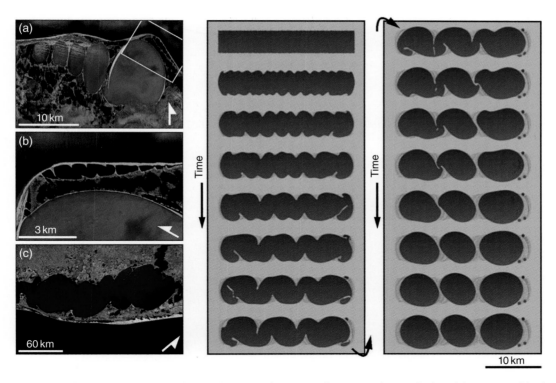

Plate 2 Natural examples of enclosed water bodies with cuspate features and segmented water bodies. (a) Laguna Val'karkynmangkak, Russia; (b) inset of (a); and (c) Lagoa Dos Patos, Brazil. The results of a numerical simulation of the formation of cuspate features and segmented water bodies are shown in the right panels. (Source: Ashton et al. 2009. Reproduced with permission of the Geological Society of America.)

Coastal Environments and Global Change, First Edition. Edited by Gerd Masselink and Roland Gehrels.
© 2014 John Wiley & Sons, Ltd. Published 2014 by John Wiley & Sons, Ltd. Companion Website: www.wiley.com/go/masselink/coastal

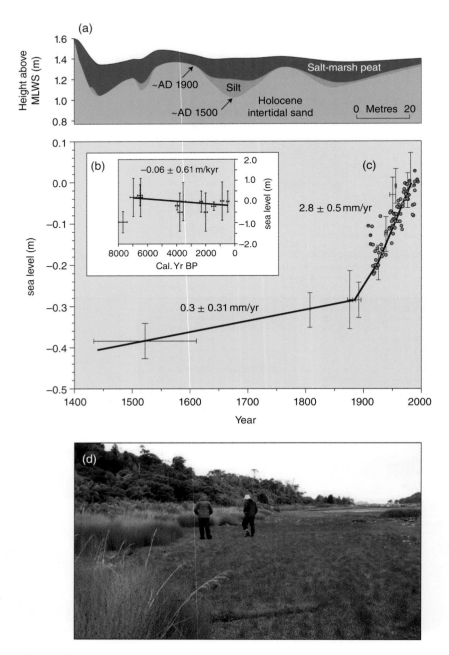

Plate 3 Deriving sea-level history from salt-marsh stratigraphy. (a) Stratigraphy of a salt marsh in southern New Zealand. The marsh developed in the past half millennium on a substrate of late Holocene intertidal sands. The sands were deposited in the middle and late Holocene when sea level was slightly higher than present. MLWS, mean low water spring. (b) Since about AD 1900, accumulation has been very rapid as a consequence of the accommodation space provided by the sharp sea-level acceleration (c). The crosses in (b) and (c) represent dated samples of shells and plant material, respectively, which can be related to former sea levels (with vertical and age uncertainties). Different coloured dots in (c) represent annual measurements of sea level from two nearby tide gauges. Cal. Yr BP, calibrated years before present. (d) The photo shows an overview of the marsh, which can be found on the Catlins coast in southeastern New Zealand, near the village of Pounawea. (Source: Adapted from Gehrels et al. 2008. Reproduced with permission of John Wiley & Sons.)

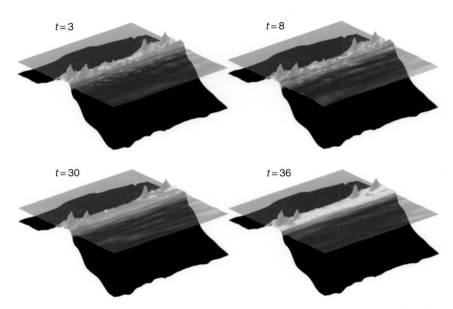

Plate 4 Numerical simulation using the XBeach model of the impact of Hurricane Ivan (the largest of five hurricanes to strike to US coast in 2004) on a 2-km wide section of Santa Rosa Island, a narrow barrier island in Florida. The Gulf of Mexico is in the lower right and the back barrier bay in the upper left of all four panels. The simulation was run over 36 hours and the four panels represent the water-level variation and barrier morphology at different time steps. The maximum storm surge and significant wave height used for forcing the model were 1.75 m and 7 m, respectively, and were attained at t = 18 hours. Note the extensive overwashing at t = 30 hours and the complete destruction of the dunes in the central part of the modelled region. (Source: McCall et al. 2010. Reproduced with permission of Elsevier.)

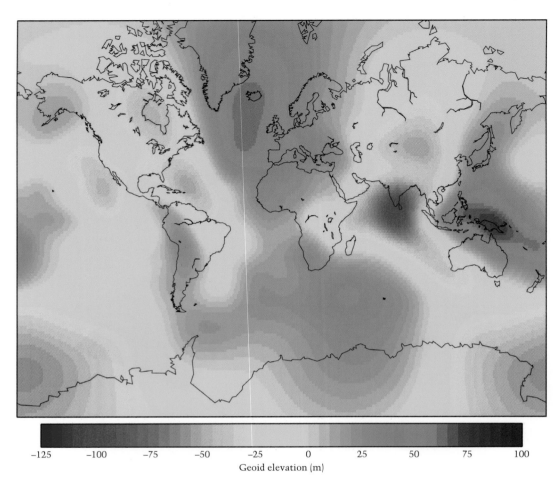

Plate 5 Map of geoid height relative to the reference ellipsoid. In general, these heights are a few 10s of metres, but they can reach ~100 m in some regions due to the existence of strong lateral variations in sub-surface rock density. The geoid model shown here is EGM2008 (Source: Pavlis et al. 2012), which is available via http://earth-info.nga.mil/GandG/wgs84/gravitymod/egm2008/index.html.

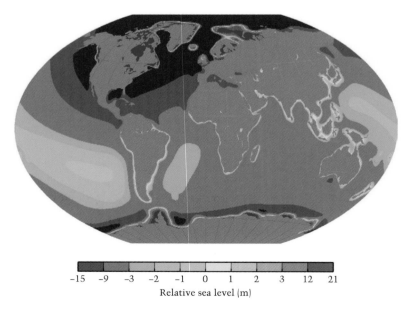

Plate 6 Relative sea level at 7 kyr BP from a glacial isostatic adjustment (GIA) model. Note that the model eustatic value is about −5 m at this time.

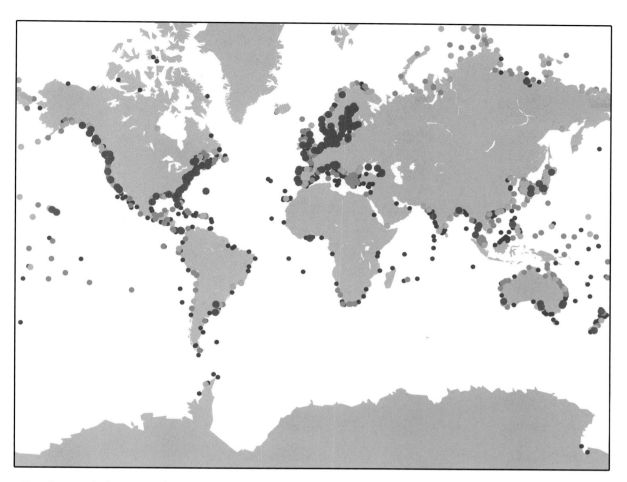

Plate 7 Map showing the location and time span of data for the Revised Local Reference network of tide gauges. The different colours indicate the period over which a given gauge has been operational: purple, less than 20 years; green, 20–40 years; orange, 40–60 years; red, greater than 60 years. These gauges are regularly levelled relative to a local, land-based benchmark to check for vertical motion of the structure the tide gauge is mounted on (e.g. a pier). (Source: Data from Permanent Service for Mean Sea Level, from Holgate et al. 2013.)

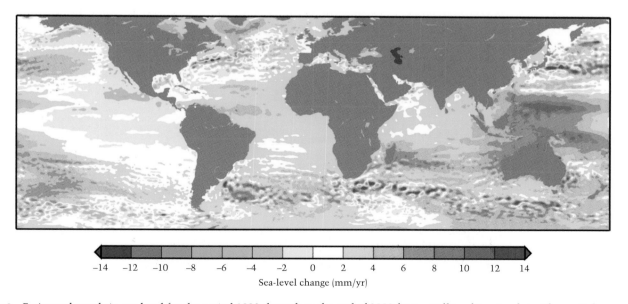

-14 -12 -10 -8 -6 -4 -2 0 2 4 6 8 10 12 14

Sea-level change (mm/yr)

Plate 8 Estimated trends in sea level for the period 1993 through to the end of 2009 from satellite altimeter data. The spatial variability in rate is a few centimetres and thus is an order of magnitude larger than the global mean rise (3.2 mm/yr) over the same period. Note that the altimeter satellite missions only measure ocean height between latitudes 66° north and south, hence the truncation of the map. (Source: Data from AVISO: www.aviso.oceanobs.com)

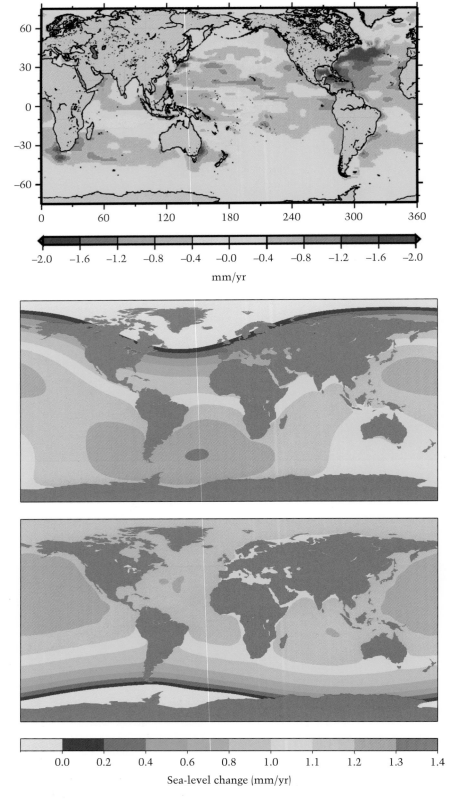

Plate 9 Top frame shows an estimate of the rates of sea-level change for the period 1950– 2003 due to changes in ocean temperature and salinity. (Source: Berge-Nguyen et al. 2008. Reproduced with permission from Elsevier.) The bottom two frames show the modelled relative sea-level response assuming a mass loss equivalent to 1 mm/yr of glacio-eustatic rise for the Greenland (middle frame) and West Antarctic (bottom frame) ice sheets. These model results include the influence of changes in gravity and Earth rotation. Note that the response of the solid Earth and gravity field result in a sea-level fall in the vicinity of the ablating ice sheet and an enhanced sea-level rise (relative to the glacio-eustatic value) in the far field. One consequence of this is that low-latitude regions will experience the greatest sea-level rise associated with melting of the polar ice sheets. (Source: Adapted from Milne et al. 2009. Reproduced with permission of Nature Publishing Group.)

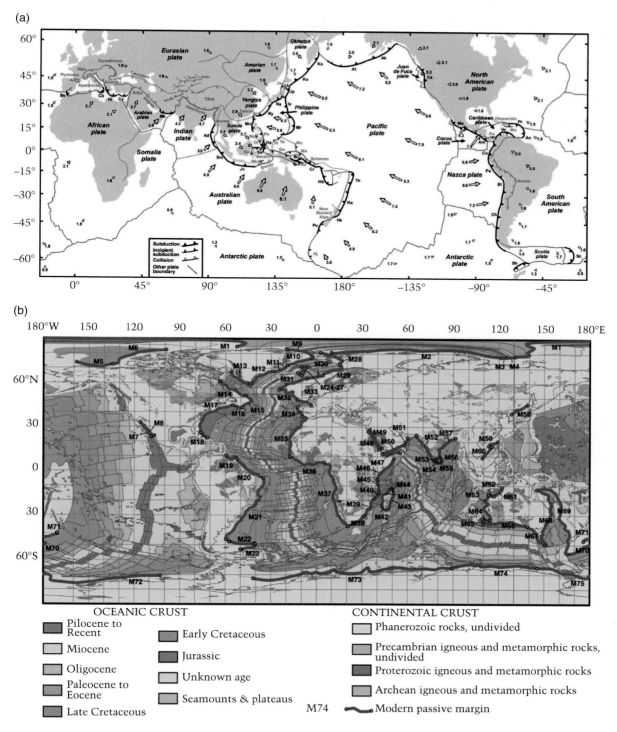

Plate 10 Global tectonic maps showing: (a) active; and (b) passive plate margins. Each type of plate margin has a fundamental influence on coastal geomorphology. In (a), plates and all subduction zones on Earth (including incipient subduction zones), collision zones and the velocities of the major plates are shown (Source: Schellart and Rawlinson 2010. Reproduced with permission of Elsevier.). The subduction zones are: Ad, Andaman; Ak, Alaska; Am, Central America; An, Lesser Antilles; At, Aleutian; Be, Betic-Rif, Bl, Bolivia; Br, New Britain; Cb, Calabria; Ch, Chile; Co, Colombia; Cr, San Cristobal; Cs, Cascadia; Cy, Cyprus; Ha, Halmahera; Hb, New Hebrides; Hk, Hikurangi; Hl, Hellenic; Iz, Izu-Bonin; Jp, Japan; Jv, Java; Ka, Kamchatka; Ke, Kermadec; Ku, Kuril; Me, Mexico; Mk, Makran; Mn, Manila; Mr, Mariana; Na, Nankai; Pe, Peru; Pr, Puerto Rico; Pu, Puysegur; Ry, Ryukyu; Sa, Sangihe; Sc, Scotia; Sh, South Shetland; Sl, North Sulawesi; Sm, Sumatra; To, Tonga; Tr, Trobriand; Ve, Venezuela. Incipient subduction zones: Ct, Cotobato; Gu, New Guinea; Hj, Hjort; Js, Japan Sea; Mo, Muertos; Ms, Manus; Mu, Mussau; Ne, Negros; Ph, Philippine; Pl, Palau; Pn, Panama; Sw, West Sulawesi; We, Wetar; and Ya, Yap. The green circles in (b) divide margins into age sectors. (Source: Bradley 2008. Reproduced with permission of Elsevier.)

(a)

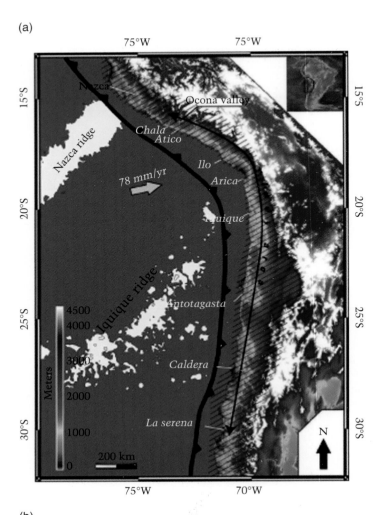

(b)

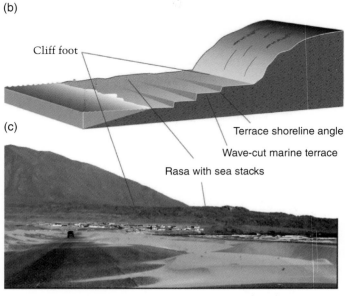

Cliff foot

(c)

Terrace shoreline angle

Wave-cut marine terrace

Rasa with sea stacks

Plate 11   Effects of neotectonic uplift on an active margin coast. The map (a) shows the topography of the Central Andes, the rate of migration of the Pacific plate and the subduction zone associated with tectonic uplift of this active margin. Most of the Pacific coast of the Central Andes, between 15°S and 30°S, displays a wide (a couple of kilometres) planar feature, gently dipping seawards and backed by a cliff. It is now widely accepted that the inner edges of these terrace features, called 'rasa', correspond to sea-level highstands. The presence of these rasa argues for a recent and spatially continuous uplift of the margin over this 1500 km long coast. A rasa is a wave-cut surface limited at its continental edge by a cliff foot. (b) General sketch of a rasa, which locally can be occupied by marine terraces. (c) Example of a rasa and cliff at Tanaka, southern Peru (15.75°S); the cliff foot is at ~300 m above mean sea level (MSL). (Source: Regard et al. 2010. Reproduced with permission of Elsevier.)

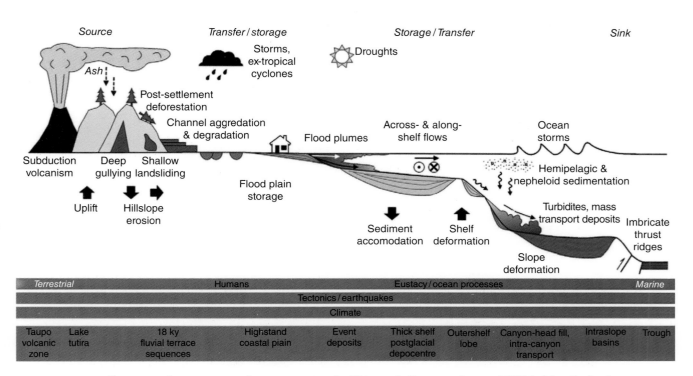

Plate 12 Diagram illustrating the source-to-sink concept, using the Waipaoa Sedimentary System (WSS) in New Zealand as an example. (Source: Carter et al. 2010. Reproduced with permission of Elsevier.)

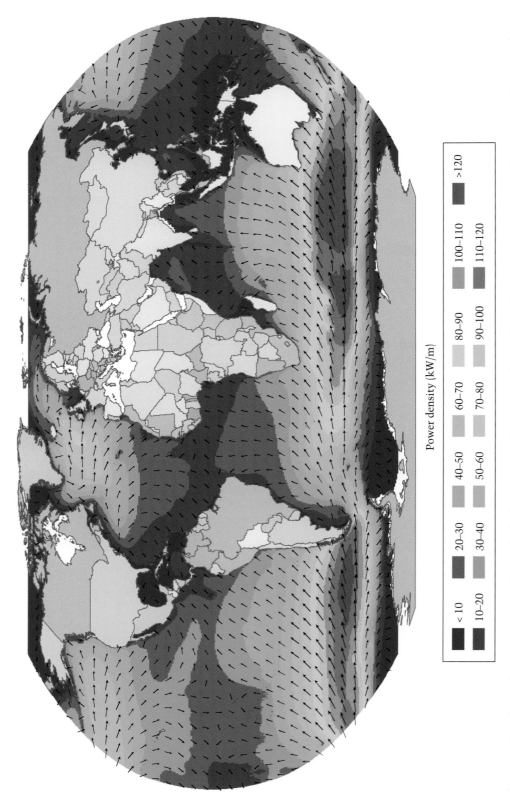

Power density (kW/m)

<table>
<tr><td>■</td><td>< 10</td><td>■</td><td>20–30</td><td>■</td><td>40–50</td><td>■</td><td>60–70</td><td>■</td><td>80–90</td><td>■</td><td>100–110</td><td>■</td><td>>120</td></tr>
<tr><td>■</td><td>10–20</td><td>■</td><td>30–40</td><td>■</td><td>50–60</td><td>■</td><td>70–80</td><td>■</td><td>90–100</td><td>■</td><td>110–120</td><td></td><td></td></tr>
</table>

Plate 13 Map of annual mean wave-power density (in kW/m) represented by colour, and annual mean best direction as calculated from six years of wave model results. Figure adapted from Gunn and Stock-Williams (2012; fig. 1), who estimate a global wave-power resource of 2.11 TW. (Source: Gunn and Stock-Williams 2012. Reproduced with permission of Elsevier.)

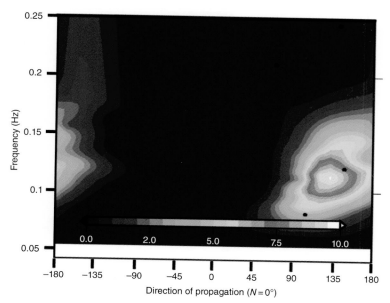

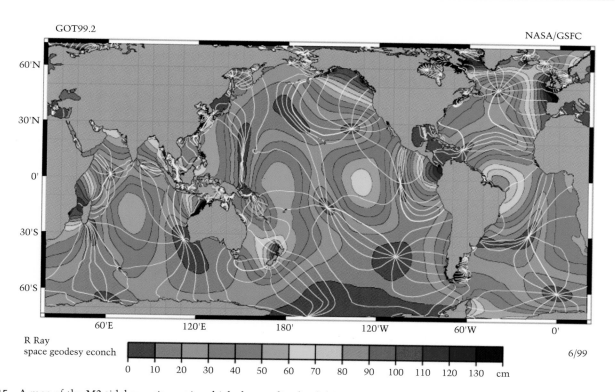

Plate 14 Wave directional spectra as estimated from high-frequency (HF) radar measurements, and photos of a 16-element receive antenna (lower left panel) and a four-element transmit antenna (lower right panel) from a Wellen Radar (WERA) high-frequency radar installation on the north coast of Cornwall, UK. (Source: Photographs by D. Conley.)

Plate 15 A map of the M2 tidal constituent in which the amplitude of this component is indicated by colour and the white lines represent co-tidal lines with 30° phase difference. (Source: R. Ray; NASA – Goddard Space Flight Center; NASA – Jet Propulsion Laboratory; Scientific Visualization Studio; Television Production NASA-TV/GSFC.)

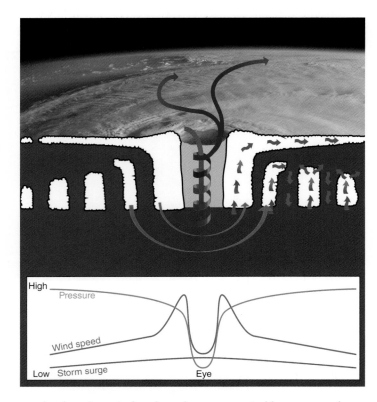

Plate 16 In contrast to other storms (cyclones), tropical cyclones have a recognizable structure of an eye with radiating rain (cloud) bands. This figure provides a schematic summary of the main components of a tropical cyclone, including rain bands and eyewall, as well as showing the pressure, wind speed and storm surge gradients across the storm. (Source: Adapted from Earth Observatory, NASA 2003.)

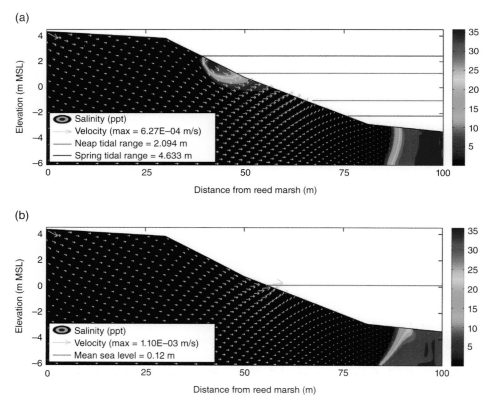

Plate 17 Tidal oscillations are the key ingredient to developing an upper saline plume. This figure shows simulation results for a homogeneous gravel aquifer. (a) Simulations of a tidally oscillating ocean and a homogeneous gravel aquifer produce the theoretical components of the subterranean estuary. The high permeability of the gravel aquifer, coupled with a strong topographically driven groundwater component, produces a wide freshwater discharge tube. A second mixing zone develops at the saline wedge. (b) Tidal oscillations are not included in this simulation. Note that groundwater discharges around mean sea level (MSL) and only one mixing zone develops at the saline wedge.

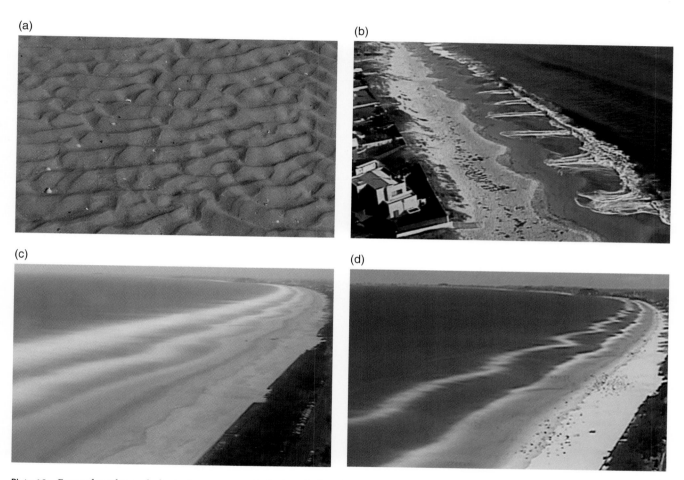

(a)

(b)

(c)

(d)

Plate 18 Examples of morphological patterns typical of sandy beaches: (a) ripples; (b) beach cusps; (c) alongshore sandbars; and (d) crescentic sandbar. The sandbars in (c) and (d) are located beneath the white high-intensity bands that are induced by the persistent wave-breaking above the sandbar crests (see Box 7.2). (Source: (a) Gerben ruessink. (b) Giovanni Coco., courtesy of NIWA. (c) and (d) Ian Turner.)

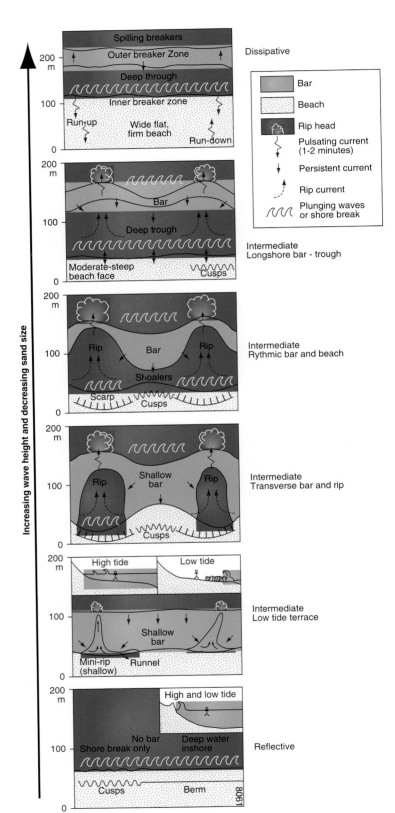

Plate 19 Wright and Short's (1984) beach classification scheme, also known as the Australian beach model. (Source: Adapted from Wright and Short 1984. Reproduced with permission of Elsevier.)

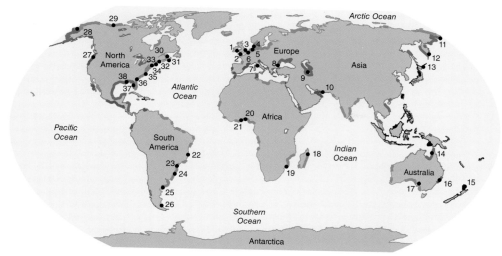

Plate 20 Barrier-system distribution, and position of barriers or barrier systems mentioned in Chapter 9. Thin lines denote trailing-margin (blue), lake and marginal-sea (green), collision (red) and island-arc (black) coasts, respectively. Thick lines with the same colour code denote barrier systems along these coasts. 1, Carnsore barriers, Ireland; 2, Chesil Beach, England; 3, northern Norfolk barriers, England; 4, Holmsland and Skallingen barriers, Denmark; 5, German Bight; 6, North Holland, Netherlands; 7, Feniglia tombolo, Italy; 8, Danube Delta barriers, Romania and Ukraine; 9, Caspian Sea; 10, Khor Kalamat spit, Pakistan; 11, Chukchi Peninsula, Russia; 12, eastern Kamchatka, Russia; 13, Notsuke Peninsula, Japan; 14, western Cape York, Australia; 15, Pakiri, New Zealand; 16, Tuncurry, Australia; 17, Lefevre Peninsula, Australia; 18, Ambodiampana barrier, Madagascar; 19, Ujembje barrier, Mozambique; 20, Lekki barrier, Nigeria; 21, eastern Ghana and Togo barriers; 22, Doce strand plain, Brazil; 23, Paranaguá, Brazil; 24, Rio Grande do Sul, Brazil; 25, Caleta Valdés barrier, Argentina; 26, El Páramo spit, Argentina; 27, Willapa barrier, USA; 28, Kasegaluk barrier, USA; 29, Beaufort Sea barriers, USA and Canada; 30, Malpeque barriers and Buctouche spit, Canada; 31, Nova Scotia barriers, Canada; 32, Saco Bay, USA; 33, Horseneck Beach, USA; 34, Fire Island, USA; 35, Bogue Banks, Shackleford Banks and Outer Banks, USA; 36, Georgia Bight, USA; 37, Tampa Bay, USA; 38, northeastern Gulf of Mexico, USA.

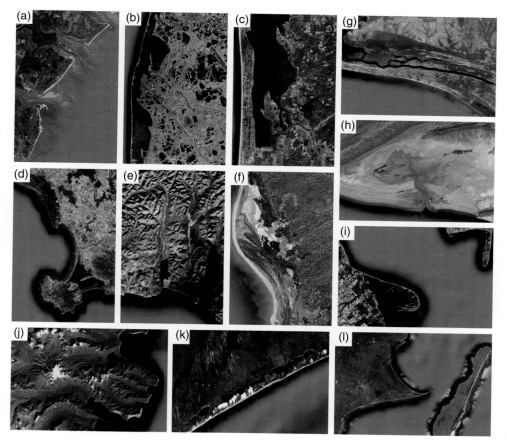

Plate 21 Examples of different barrier types: (a) tide-dominated barrier islands fringing Georgia Bight, USA; (b) wave-dominated barrier island fronting Kasegaluk Bay, Alaska; (c) recurved spit protecting Willapa Bay, Washington, USA; (d) double tombolo at Feniglia, Italy; (e) baymouth barriers at Chukchi Peninsula, Russia; (f) looped barrier at Cape York, Australia; (g) strand plain fronting Lekki Lagoon, Nigeria; (h) baymouth spits at Khor Kalamat, Pakistan; (i) hooked spit at Notsuke Peninsula, Japan; (j) pocket barriers occupying indentations in the eastern Kamchatka coast, Russia; (k) welded barrier at Ujembje, Mozambique; (l) cuspate barrier at Ambodiampana, Madagascar. Images show areas of about 30×45 km. See Fig. 9.1 for locations. (Source: Google Maps. (a–e, g–j) © Google, TerraMetrics. (f) © Google, NASA, TerraMetrics. (k) and (l) © Google).

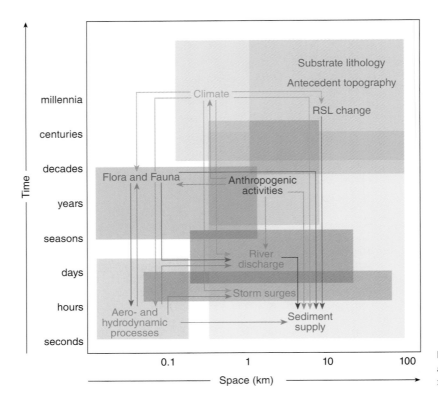

Plate 22 Interlinked drivers of barrier behaviour, and associated temporal and spatial scales. RSL, relative sea level.

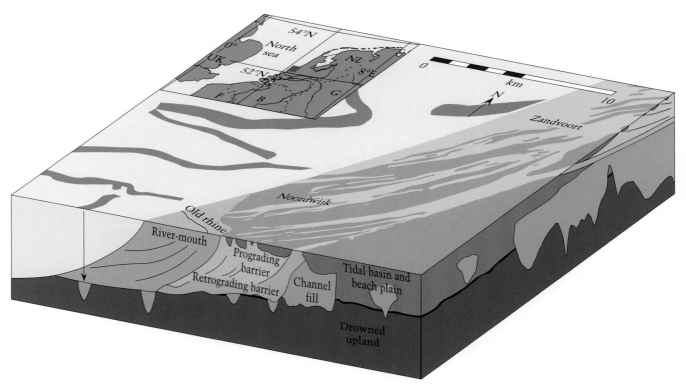

Plate 23 Preservation of barrier morphostratigraphy in the western Netherlands (excluding the frontal dunes and the belt of parabolic, transgressive dunes formed during the last millennium). All that remains of an early Holocene retrogradational barrier-island system that existed offshore the present coastline is a set of partially truncated tidal-channel fills that occupied the associated back-barrier areas and tidal inlets (in blue). The mid-Holocene retrogradational barrier system culminated in a low beach ridge that formed under low wave energy on the landward side of a wide and gently sloping shoreface. The mid- to late-Holocene progradational barrier system has been preserved as a series of beach ridges and beach plains, truncated on its seaward side by rivers and by renewed coastline retrogradation during the late Holocene. Much of the surface expression of the mid- and late-Holocene ridges has been erased by human activities. The location is marked by the red box in the inset.

Plate 24 The 33-km long seawall constructed on Saemangeum estuary and completed in 2010. The engineering structure will reclaim 400 km² estuarine land through drainage and infilling. (Source: NASA Earth Observatory, http://earthobservatory.nasa.gov/IOTD/view.php?id=7688).

(a)

(b)

(c)

Plate 25 (a) Mangroves show a range of adaptations to the inhospitable intertidal environment in which they flourish, and that contribute to the ecosystem services they provide. These include distinctive root systems, such as the prop roots of *Rhizophora*, seen in this stand of *Rhizophora* on Low Isles on the Great Barrier Reef, Australia. (b) Vivipary is a significant adaptation as a result of which mangroves are able to establish and expand in muddy intertidal environments. The fruit of several of the Rhizophoraceae, such as *Bruguiera*, germinate while still on the tree. When they then detach, they are able to float and establish in favourable habitats, rapidly developing rootlets that anchor them to the mud. (c) The southernmost mangroves in the world occur at Corner Inlet in Victoria, Australia. Here, *Avicennia* is stunted, reaching little more than a metre in height, and merges into salt-marsh vegetation, which is seen in the foreground. (Source: Photographs by (a) Catherine Lovelock. (b) Colin Woodroffe. (c) Roland Gehrels.)

(a)

(b)

Plate 26 (a) Scrub mangrove setting – a stand of *Ceriops* in Hinchinbrook Channel, northern Queensland, Australia, in which individual tree growth is stunted. (b) Carbonate settings – mangroves are not limited to muddy continental shorelines, but also occur in carbonate environments. These mangroves have developed on a reef platform in Torres Strait, Australia. (Source: Photographs by (a) Catherine Lovelock and (b) Javier Leon.)

(a)

(b)

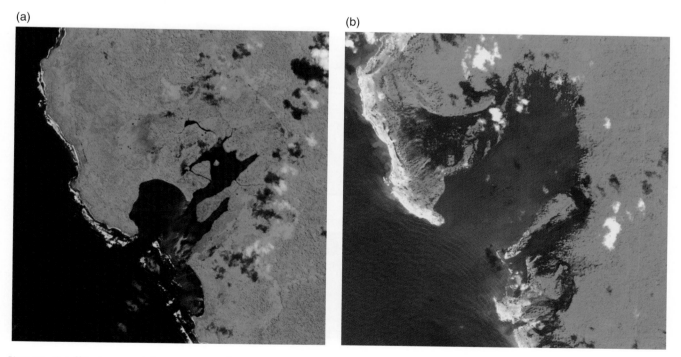

Plate 27 Satellite images of the west coast of Kirchall Island in the Nicobar Islands, Indian Ocean, before (a) and after (b) the Boxing Day tsunami that devastated this coast on 26 December 2004. Extensive areas of mangrove forest were swept away on this coast by the tsunami. (Source: CNES 2004/Distribution Spot Image/Processing CRISP.)

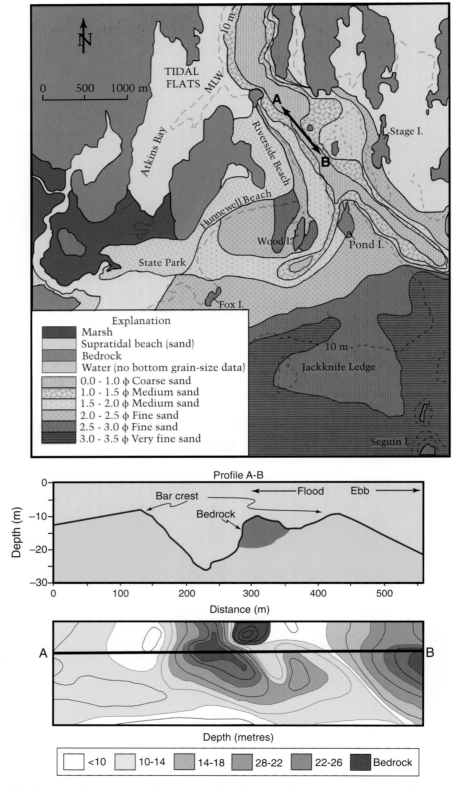

Plate 28 Grain size and bathymetric data indicate that the mouth of the Kennebec River estuary along the central Maine coast, USA, is dominated by a seaward sediment transport. The channel is floored by coarse to medium sand that is sourced from the river. Two large seaward-oriented transverse bars demonstrate that sand is exported from this estuary. MLW, mean low water.

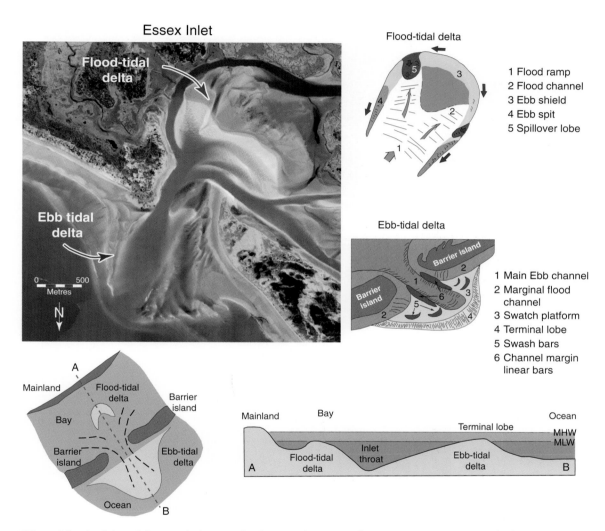

Essex Inlet

Flood-tidal delta

Flood-tidal delta

1 Flood ramp
2 Flood channel
3 Ebb shield
4 Ebb spit
5 Spillover lobe

Ebb tidal delta

Ebb-tidal delta

Barrier island

Barrier island

1 Main Ebb channel
2 Marginal flood channel
3 Swatch platform
4 Terminal lobe
5 Swash bars
6 Channel margin linear bars

0   500
Metres

N

A

Mainland

Flood-tidal delta

Barrier island

Bay

Barrier island

Ebb-tidal delta

Ocean

B

Mainland   Bay   Ocean

Terminal lobe
MHW
MLW

A   Flood-tidal delta   Inlet throat   Ebb-tidal delta   B

Plate 29 Ebb- and flood-tidal models. Aerial photograph of Essex Inlet, Massachusetts, USA. MHW, mean high water; MLW, mean low water. (Source: Hayes 1979. Photograph: FitzGerald 1996.)

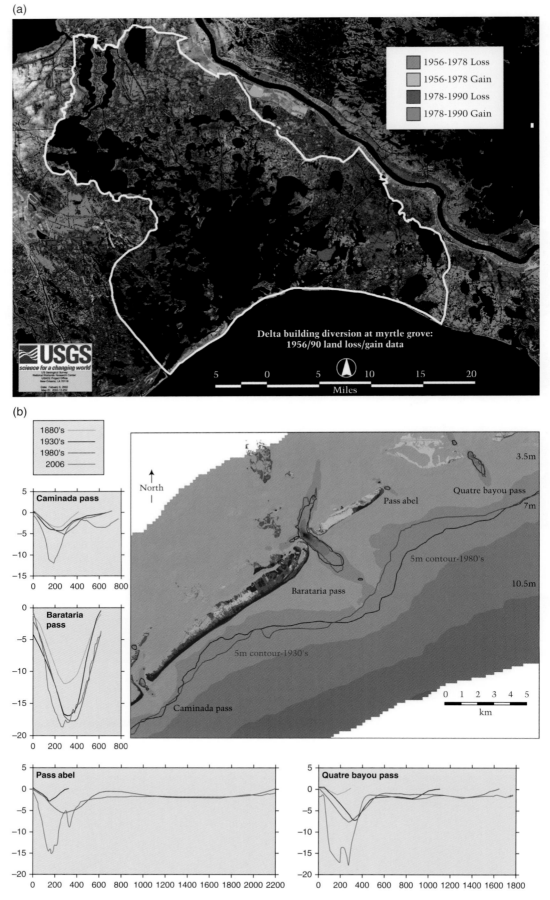

Plate 30 Wetland loss. (a) Barataria Bay, located along the central Louisiana coast, has experienced extensive wetland loss during the past 60 years (Source: Barras 2006), (b) resulting in greater tidal exchange through the tidal inlets. This increasing tidal prism has enlarged the size of the tidal inlets, resulting in the movement of sediment offshore to the ebb deltas and the loss of sand along the barrier chain.

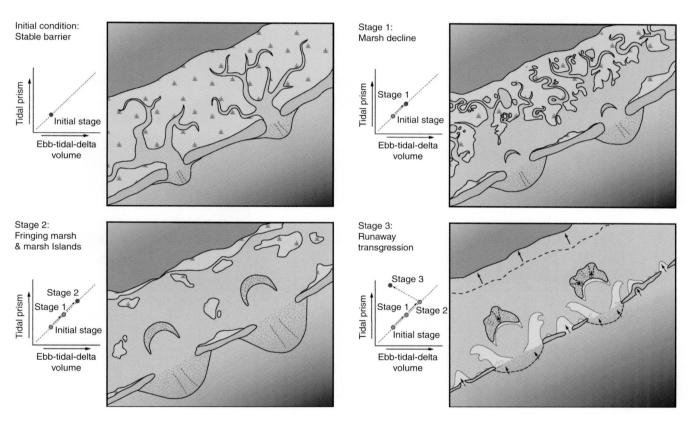

Plate 31 Conceptual evolutionary model of a mixed-energy barrier coast to a regime of accelerated sea-level rise. (Source: FitzGerald et al. 2006.)

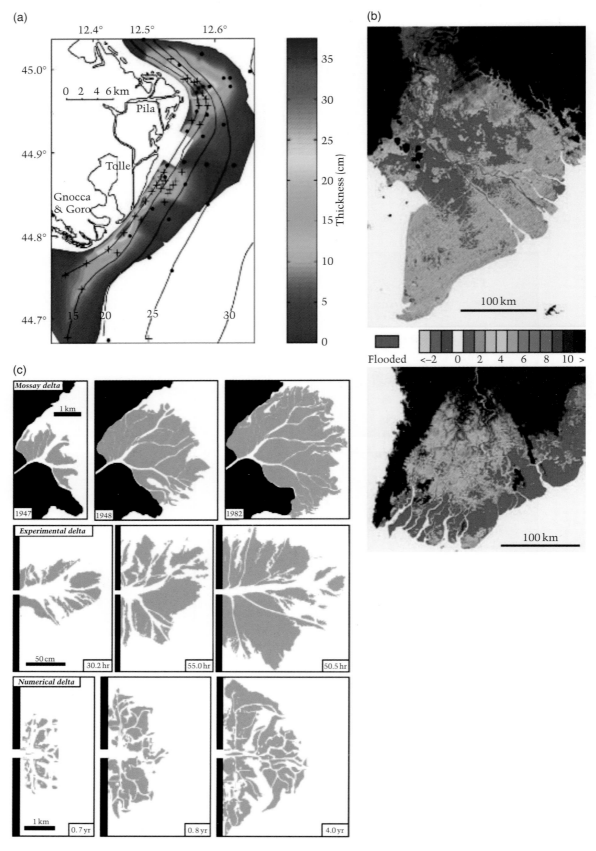

Plate 32 A selection of methods used in delta studies. (a) Digital X-radiography of Po River flood deposition on the shoreface; (b) SRTM (Shuttle Radar Topography Mission) and MODIS (Moderate Resolution Imaging Spectroradiometer) imaging of the Mekong (top) and Irrawaddy (bottom) deltas; (c) time series imagery of deltas. (Source: (a) Wheatcroft et al. 2006. Reproduced with permission from Elsevier. (b) Adapted from Syvitski et al. 2009. Reproduced with permission of Nature Publishing Group. (c) Wolinsky et al. 2010. Reproduced with permission from John Wiley & Sons.)

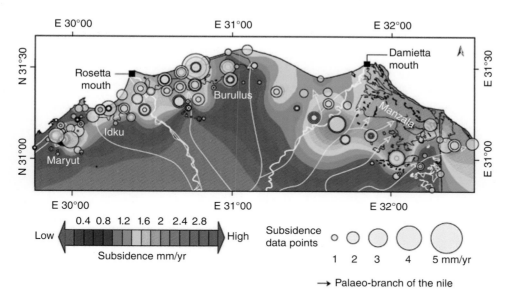

Plate 33 Subsidence in the Nile delta in the course of the Holocene. (Source: Adapted from Marriner et al. 2012. Reproduced with permission from the Geological Society of America.)

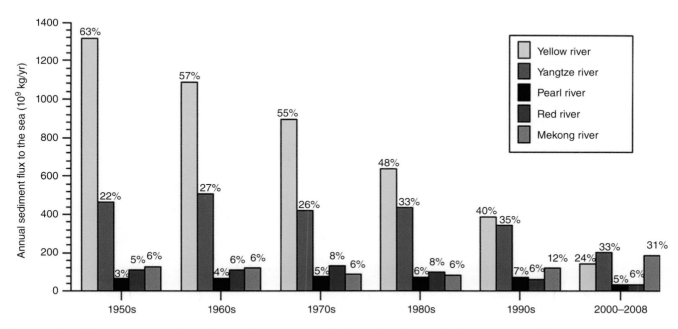

Plate 34 Decadal variations in sediment flux to the western Pacific Ocean from five major rivers in East and Southeast Asia. (Source: Wang et al. 2011. Reproduced with permission of Elsevier.)

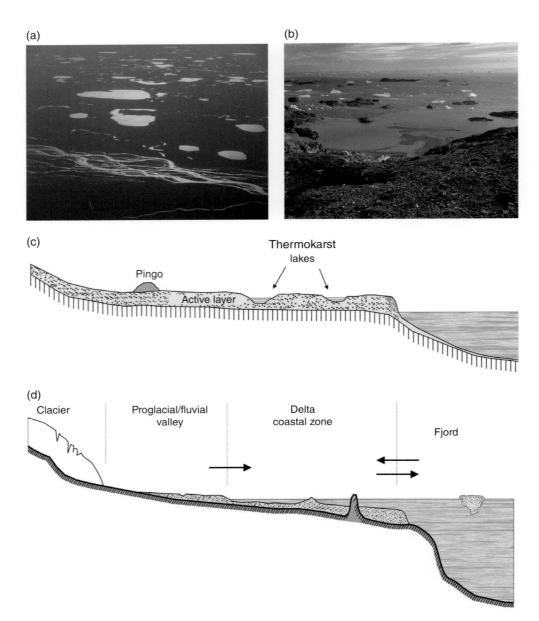

Plate 35 Coastal landscapes on high-latitude coasts. (a) Thermokarst lakes and a braided river on the Alaskan coastal plain. (b) Delta at high tide in a high-relief area, Sermilik, southeast Greenland. The distance from the delta mouth towards the glacier is c. 2 km. Note the icebergs and plume of sediment in the fjord. Schematic coastal landscapes for: (c) a low-relief area; and (d) a high-relief area (Source: Data from Nielsen 1994). Transects in a low-relief area are typically in the order of 10–1000 km, like those over tundras with major rivers on coastal plains in Siberia and Canada. The transects in a high-relief area are typically small and in the order of kilometres, like in eastern Greenland where glaciers are close to open water.

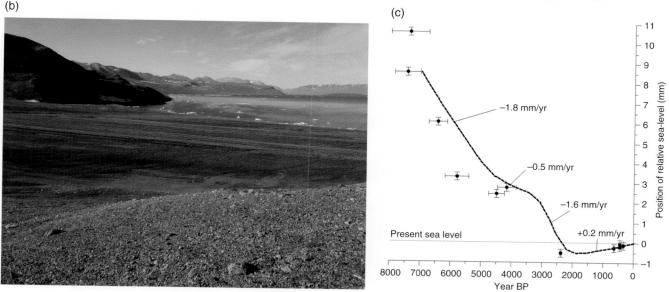

Plate 36 (a, b) Beach-ridge plain at Grønnedal, northeast Grønland, with Holocene beach ridges from the present water level up to 36 m above mean sea level. (c) Typical sea-level curve at the same location in northeast Greenland. The crosses are optically stimulated luminescence (OSL) dating on the lower beach ridges. (Source: Adapted from Pedersen et al. 2011. Reproduced with permission of John Wiley & Sons.)

Plate 37 Variety of rock coast morphologies in differing geological settings and rock types: (a) granitic coast, Putuo Island, East China; (b) cliff in greywacke sandstone, Catlins Coast, southeast coast South Island, New Zealand; (c) sandstone plunging cliff, Sydney Harbour, Australia; (d) sloping shore platform developed in Blue Lias limestone, Glamorgan Heritage Coast, South Wales, UK; (e) horizontal shore platform developed in Jurassic sandstone, Curio Bay, Southland, New Zealand; (f) horizontal platform developed in greywacke sandstone, Victoria, southeastern Australia; (g) sloping platform developed in mudstone with inactive sea cliff behind, Kaikoura Peninsula, South Island, New Zealand; and (h) cliff developed in consolidated glacial till, Lake Pukaki, South Island, New Zealand. (Source: Photographs by Wayne Stephenson.)

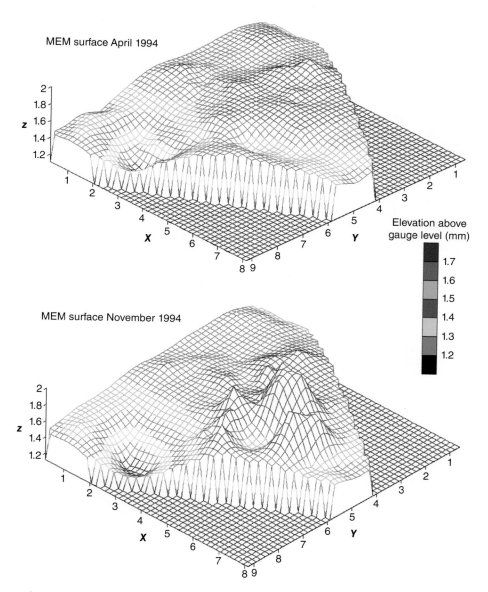

Plate 38 Wire frame surface plots from a traversing micro-erosion meter (MEM) site. The upper plot is the surface in April 1994 and the lower plot is the same surface in November 1994. Note surface swelling evident at the lower right corner of the surface indicated by the dark blue peaks that have risen up. Elevations are in millimetres and relative to the MEM gauge zero level.

Plate 39 Examples of anthropogenic transformation of reef islands and approaches to shoreline stabilization in developing countries: (a) densely populated and modified capital island of Malé, Maldives; (b) modified tourist island shoreline and reef, Maldives; (c) sandbag revetment, Tarawa atoll, Kiribati; (d) coral cobble seawall, South Tarawa atoll, Kiribati; (e) failed sandbag revetment, Bairiki island shoreline, Tarawa atoll, Kiribati; and (f) failed sandbag revetment, South Tarawa atoll, Kiribati. (Source: Photographs by Paul Kench.)

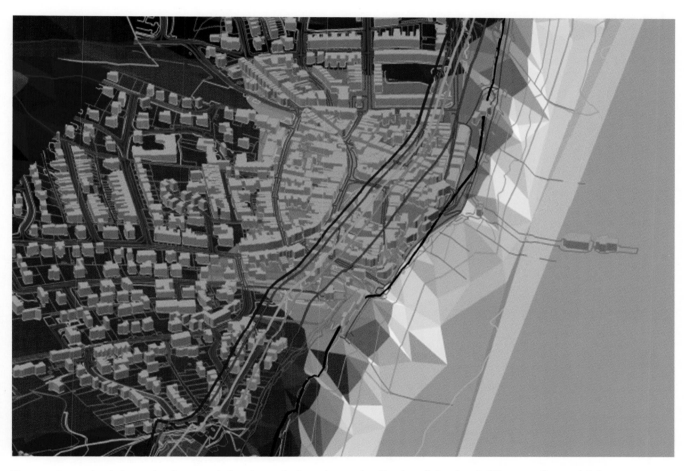

Plate 40 Improving information for coastal planning and adaptation: a visualization of simulated cliff retreat if coastal defences were removed at Cromer, Norfolk, UK. The lines show change every 10 years. Note that the recession decreases. (Source: Dawson et al. 2009. Reproduced with permission of Springer-Verlag Dordrecht.)

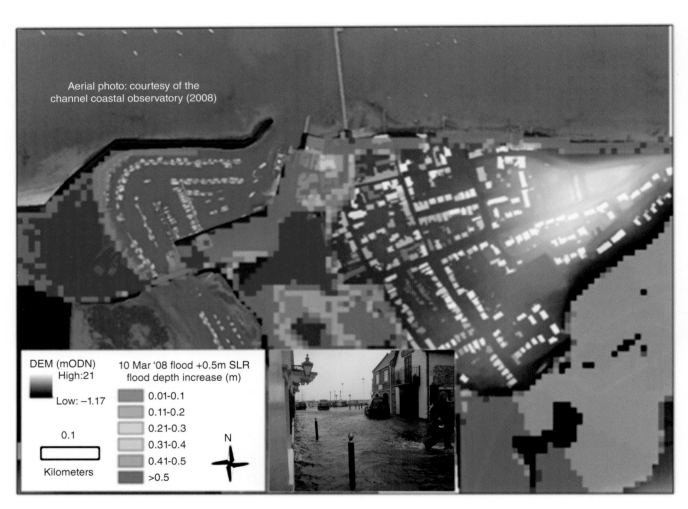

Plate 41 Flood simulations of Yarmouth, Isle of Wight, UK, using the LISFLOOD-FP inundation model. This shows the increase in flood depth due to a 0.5-m rise in sea-level, assuming a repeat of the flood of 10 March 2008 (see inset photograph) (Source: Yarmouth Harbour Commissioners 2008), see Wadley et al. 2012.

Fig. 9.21 Dune scarping in North Holland (western Netherlands) during a 2007 storm surge. Foredune, which slopes to the upper left, is more than 10m high. (Source: Photograph by Marcel Bakker.)

erosion than do areas with small dunes (Houser et al., 2008). Along prograding barriers, buried storm-surge scarps are marked by concentrations of heavy minerals or coarse particles.

Overwash, discussed in more detail in Chapter 5, is the supratidal flow of sediment-laden water across barriers when wave run-up or storm-surge levels exceed the elevations of at least some lows in foredune ridges. This water does not directly return to the adjacent open-water area from which it originated. Run-up overwash is linked to local frontal-dune breaches, and inundation overwash (or sheetwash) is linked to breaches across large stretches of coastline. Overwash is directed landward, except when originating in back-barrier areas.

Overwash produces a variety of morphological changes in barriers, which fall into three categories. (1) Crest accumulation, deposition of perched fans, and barrier narrowing are dominant when dunes prevent overwash from crossing an entire barrier. (2) Barrier rollover takes place when foredunes are discontinuous, undulatory or absent. (3) Barrier breaching and development of large washover channels into (ephemeral) inter- and subtidal storm passes or inlets are most common in narrow barriers.

Overwash is instrumental in the landward migration of barriers. Through this mechanism, barriers subject to erosion maintain their integrity. New supratidal areas and habitat are created on the back-barrier side while existing habitat, property and infrastructure are destroyed on the seaside.

Washover is the sediment deposited by overwash (Fig. 9.22) when water overtopping gaps in barriers spreads laterally, and friction and percolation further reduce its energy, particularly in dry sand. The process is explained in more detail in Chapter 5.

Some beach ridges and cheniers are constructed by multiple high-energy events. As these ridges are formed at high water levels, they are generally persistent. Storm surges are also instrumental in the distal extension of gravelly beach ridges at the cost of the updrift coast, as sediment-transport cells extend farther downdrift during extreme events than during fair-weather periods.

Barriers recover from the effects of storm surges in different stages that may take decades to complete. Typically, recovery starts with the transport of overwashed sediment back to the coastline, where it may seal storm channels. Onshore sand migration in swash bars, foreshore deposition, and rapid advancement of the berm crest happen next. Subsequent aggradation and progradation of berm crests is followed by foredune construction over and slightly seaward of erosional escarpments, and dune stabilization by vegetation (Morton et al., 1995). Non-vegetated sandy washover fans are temporary reservoirs for eventual aeolian redistribution of sand. In the case of prevailing offshore winds, much sand is returned to beaches. Monitoring data from retrograding USA barriers show recovery to positions consistent with the long-term (100+ years) trend, suggesting that storm surges have little effect on long-term barrier erosion (Fig. 9.23). Recovery is not possible, however, when storm surges force some radical change in barrier morphology, such as during overstepping.

The geomorphological impacts of tsunamis, on barriers may be similar to those of storm surges, with beach erosion or levelling, destruction of barriers, and the formation of dune scarps. Generally, distinguishing storm-surge and tsunami deposits in barrier records is difficult (see Chapter 5). Clearly recognizable tsunami marker horizons are typically located away from barriers, either farther inland or in barrier-starved coastal regions.

9.6.7 River discharge

Rivers affect barrier systems not only by supplying sediment, but also by runoff-related processes. Their discharge into the open-marine environment stimulates ebb domination, may create inlets, drives inlet migration, and acts as jetties that trap sediment supplied by longshore currents. When fluvial runoff varies significantly, inlets may go through cycles of opening and closure. Inlets may also be relocated when river-bank erosion results in barrier breaching in one place and reduced tidal flow is unable to keep open the original inlet farther updrift or downdrift. Entrapment of sediment reworked from the inner shelf explains the formation of beach-ridge plains associated with several Brazilian river-mouths that have been classified as wave-dominated deltas (Dominguez et al., 1992).

Fig. 9.22 Overwash (a and b) at an artificial gap in the western Netherlands frontal dune during a 2007 storm surge. The resulting washover (c and d) contained a lot of debris and litter. (Source: Photographs by (a) Marcel Bakker. (b) Johan Bos. Reproduced with permission. (c) and (d) Sytze van Heteren.)

9.6.8 *Climate*

Climate affects barriers through direct effects of rainfall, winds and waves, currents, and water and air temperature. Humic tropical lowlands, for example, are not conducive to dune development, and their barrier systems are dominated by mud made available by intense chemical weathering. In such settings, cheniers form downcurrent from river mouths during low mud influx accompanying long-term dry periods. Uniquely arctic processes affecting barrier behaviour include ice push, ice rafting, thermal erosion of permafrost areas, thaw subsidence, and fetch limitation due to ice shelves (see Chapter 14).

Long-term patterns of climate change (such as global warming and shifting high-pressure cells) and shorter-term periodicity (such as El Niño and La Niña) govern sea and groundwater level and storminess. They also influence type and resilience of vegetation, weathering, discharge and sediment supply by rivers, aeolian transport, lithification of sediment (beachrock and aeolianite formation), and availability of carbonate skeletal material.

Barrier systems register climate fluctuations on all timescales. Individual El Niño events are linked to increased overwash frequency. Series of El Niño events are thought to be responsible for beach-ridge formation along the coast of Peru. Twentieth-century coastline accretion and erosion in southeastern Australia have been correlated to decadal variability in rainfall, river discharge and storminess. Regional dune mobility

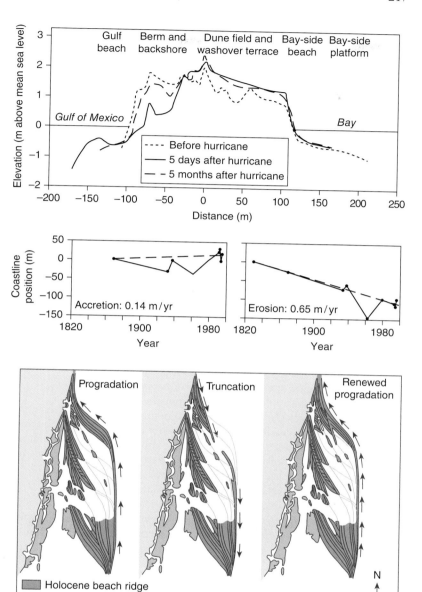

Fig. 9.23 Hurricane erosion and post-storm recovery, as shown by a series of profiles from a western Florida barrier (top) and by average coastline change of a New York barrier (bottom). The bottom panels show that extreme events do not necessarily have long-lasting effects. For the New York barrier, storm-related erosive events have changed the long-term coastline trend in neither accretionary nor eroding systems. (Source: (Top) Stone et al. 2004. Reproduced with permission of Elsevier. (Bottom) Zhang et al. 2002. Reproduced with permission from University of Chicago Press.)

Fig. 9.24 Beach-ridge series at the Doce strand plain in east-northeastern Brazil. Truncations are linked to climate-related inversions in longshore drift (arrows). Dark and light grey lines denote progradation- and erosion-related palaeocoastlines. (Source: Dominguez et al. 1992. Reproduced with permission of Elsevier.)

occurs during century-scale phases of increased aridity or windiness, such as in northwestern Europe during the Little Ice Age. Millennial changes in atmospheric circulation induce inversions in the dominant direction of littoral drift, as reflected in the orientation of truncated Brazilian beach-ridge series (Fig. 9.24).

Numerous modern barriers behave differently than in the recent past as a result of climate change, or are expected to be affected in the coming decades. Accelerating RSL rise forces many barrier coastlines landward and increases accommodation space at increasing rates, but also has an indirect effect. The rising base level on which storm surges are superposed, shortens the

return period of peak water levels associated with extreme events. Low-lying barrier islands on the Mid-Atlantic coast of the USA, for example, are particularly prone to increased flooding as a result of rising peak water levels.

Changing wind patterns also influence barrier development. Increases in the frequency and strength of hurricanes and other storms are expected to result in intensifying coastal erosion. Barrier systems fringing the northeastern Gulf of Mexico were relatively stable from the 1970s to the mid-1990s, when tropical cyclone activity was low, but have experienced significant erosion since 1995, the start of a period of pronounced storminess (Stone et al., 2004).

In affecting wave climate in high-latitude areas, changing sea-ice cover has a strong, indirect influence on arctic barrier systems. The farther and longer sea ice retreats from the shore during the summer, the longer the fetch over open water and the stronger the wave action. This effect has resulted in increased vulnerability of many existing Alaskan and Siberian barriers, and simultaneously may create the conditions needed for the accumulation of sand and gravel to form new barriers along Canada's arctic coastline.

9.6.9 Flora and fauna

Plants are intricately linked to the behaviour of barrier systems. This relationship is most obvious in coastal dune fields (see Chapter 8). Vegetation is particularly sensitive to changing island topography, physical impact of waves and overwash currents, groundwater level and salinity, precipitation, availability of nutrients in soil and air, and exposure to salt spray. Vegetation cover is weakened by fire, overgrazing, or excessive trampling by animals or man. Barriers typically show a succession of vegetation. Dune grasses, which are capable of dealing with rapid burial by aeolian or overwash sediment, dominate foredunes in most areas. They are replaced by shrubs, trees and bunch grasses once sand supply to a dune is halted. In humid tropical areas, vines occupy the upper beach, shrubs and rain forest cover the main ridges, and mangroves characterize swamps in swales and in back-barrier areas.

Various plant species are eco-engineers, modifying the abiotic barrier environment by biological activity. In the subtidal zone, sea grass stabilizes the seabed, thus influencing the availability of sand and gravel along barrier coasts. In intertidal areas, algal mats stabilize tidal flats and algal fronds increase the buoyancy of gravel particles. Depending on the climate setting, mangroves or marsh plants trap subaqueous sediment and cause dissipation of storm waves. Supratidally, dune grasses and other vegetation trap wind-blown sediment and stabilize dune fields as a result of soil formation. Finally, coastal forests reduce the impact of hurricane-force winds.

The most important effects of vegetation on supratidal barrier behaviour are related to foredune formation and prevention of blowout development. Vegetation is the key to creating increasingly high coastal dunes that are progressively more resistant to removal by wave action. Blowouts are generally initiated where dune vegetation is damaged or destroyed. In back-barrier areas, marsh disintegration has the most prominent effect on barrier behaviour. Expanding open-water areas increase the vulnerability of barriers on their back-barrier sides.

Animals can stabilize and destabilize sediment, and may affect hydrodynamic processes. Shellfish such as oysters trap sediment in open-marine water and in back-barrier areas. When present as colonies, they may help

Table 9.2 Anthropogenic influence on barriers and barrier systems.

Increased sediment supply or reduced sediment demand
Dredge-spoil dumping
Shoreface and beach nourishment
Downdrift groyne and jetty construction
Nearshore breakwater construction
Updrift diversion of river outlets
Stabilization of dunes with vegetation and fertilizers
Deforestation
Agriculture
Reduced sediment supply or increased sediment demand
Levee construction
Hydrocarbon extraction
Mining of sand and gravel from barriers and adjacent open water
River-bed excavation
Dredging
Seawall construction
Updrift groyne and jetty construction
Downdrift diversion of river outlets
Construction of dams in the lower reaches of rivers
Measures reducing soil erosion
Overpumping of groundwater
Overloading by constructions in lowlands

dissipate wave energy impacting adjacent coastlines. Grazing and digging rabbits are well-known destabilizers in dune areas, and cattle may destroy both barrier and back-barrier vegetation by grazing and trampling.

9.6.10 Processes and impacts of anthropogenic activities

Many barrier systems are strongly influenced by anthropogenic activities (Table 9.2).

One of the leading causes for the shift from sediment surplus to sediment deficit in many barrier systems is the multiple damming of major source rivers. Sediment-load reductions of 95% in the Ebro (Spain), 98% in the Nile (Egypt), and 25–30% in the Danube (Romania/Ukraine) have been reported. Near river mouths, this effect may be exacerbated by overpumping of groundwater, which led to subsidence rates of 3–4 m per century in the Italian Po delta.

Traffic and overgrazing destroy vegetation, making dunes vulnerable to erosion and to overwash, which frequently occurs at access trails and roads to the beach. The current practice of foredune stabilization, although reducing sediment loss and vulnerability to overwash in many areas, limits barrier migration and thus may cause barriers to narrow in the long term.

A final consequence of anthropogenic activities is segmentation of barrier coastlines. Long impermeable jetties and deep navigation channels disrupt littoral-drift patterns and lead to extreme longshore variability in barrier behaviour.

9.7 Barrier sequences as archives of barrier behaviour

Information regarding annual to multi-decadal responses of present-day barriers to changes in interacting coastal variables is abundant. It forms the basis for our understanding of barrier behaviour. Morphological and stratigraphical data supplement observations of modern-day process-response relationships because they cover combinations of conditions (rapid RSL change/extreme storm surges) that are different from, or more extreme than, those captured in field measurements or represented in monitoring series. Knowledge of barrier stratigraphy is essential in evaluating numerical hindcasts in which different evolutionary histories produce identical final morphologies. As a proxy of past patterns and processes driving coastal-environmental change on centennial to millennial timescales, barrier stratigraphy is also invaluable as direct model input. It is a key element in predicting future barrier behaviour (Stolper et al., 2005).

The distillation of quantitative process-response links from barrier stratigraphy requires identification of unambiguous marker horizons and boundaries that represent drivers of barrier behaviour. Visible indicators include lag deposits of shells or gravel, garnet concentrations, and changes in sedimentary structures. They are supplemented by micropalaeontological information. Their indicative meaning is determined on the basis of observations in present-day barrier-system environments. A second step involves calculation of sediment volumes that can be tied to barrier behaviour resulting from identified past events and changing variables. Well-constrained volumes can be used in the assessment of sedimentation rates and source contributions, but not until events and changes involving known sediment volumes are tied to time in a final step.

The main problem associated with the use of barrier stratigraphy in the long-term analysis of drivers is related to preservation. Barrier behaviour leaves only fragmentary records, with sediments deposited in a subaqueous environment having a better preservation potential than sediments deposited in a supratidal environment (Fig. 9.25). Erosive events are at best represented by hiatus or diastems.

9.8 Lessons from numerical and conceptual models

9.8.1 Data and concepts

To further our understanding of barrier behaviour, especially for changes that do not leave a morphological or stratigraphical record, we need numerical models. Incorporation of well-constrained process-response relationships from measurements and historical reconstructions has steadily improved the performance of these models. Although information regarding annual to decadal responses of present-day barriers to changes in interacting coastal variables is abundant, it is biased toward supratidal, intertidal and shallow subtidal areas that are relatively easy to monitor. Suitable input data from deep subtidal areas and from barrier stratigraphy, both needed to model long-term barrier behaviour, are still in short supply. On the geological timescale, numerical models are increasingly capable of reproducing large-scale barrier behaviour as distilled from stratigraphical sequences (Cowell et al., 2003a, 2003b). On the engineering timescale, we are faced with a shortage of accurate monitoring data from the deep subtidal environment. During the past decades, improving qualitative concepts have proven to be very useful, but they are increasingly stretched beyond their limitations.

Constancy of hydrodynamic regime, longshore uniformity of coastal cells, and a balance with modern-day processes, which are essential requirements of the equilibrium-profile concept, are not always met. More often than not, the deeper parts of coastal tracts are out of equilibrium with actual hydrodynamic processes, not only because coastal tracts experience external forcing such as sea-level change, but also because not enough time has passed for hydrodynamic forces to cause complete adjustment of out-of-equilibrium drowning landscapes (Cowell et al., 2003a, 2003b). Observations suggest adjustment times of hours around the shoreline to millennia near the inner shelf (Stive and De Vriend, 1995). Systematic export of sand from the shoreface to the shelf, a core element of the Bruun rule (which states that a typical concave-upward beach profile erodes sand from the beachface and deposits it offshore to maintain constant water depth when RSL rises), is difficult to explain by known processes and cannot be applied uniformly. Where the shoreface has gentle slopes, net onshore transport of sand is known to have occurred across the globe (Inman and Dolan, 1989). The concept of a quantifiable depth of closure, where there is no significant transport of sediment between shallower and deeper parts of the shoreface, is used in the vast majority of coastal sediment-budget analyses. This concept, based primarily on bathymetric monitoring in the absence of process measurements, is widely considered to be erroneous. It would require a zone of minor sediment flux between two areas marked by significant sediment transport. Even immeasurably small changes in bed level translate into large volumes, because the lower shoreface and adjacent part of the inner shelf occupy large areas.

9.8.2 Model output relevant to barrier-system behaviour

Concerted modelling and observational efforts for some well-studied locations, including the Columbia River coastal cell (Washington), North Carolina, USA, the western Netherlands, and southeastern Australia, have

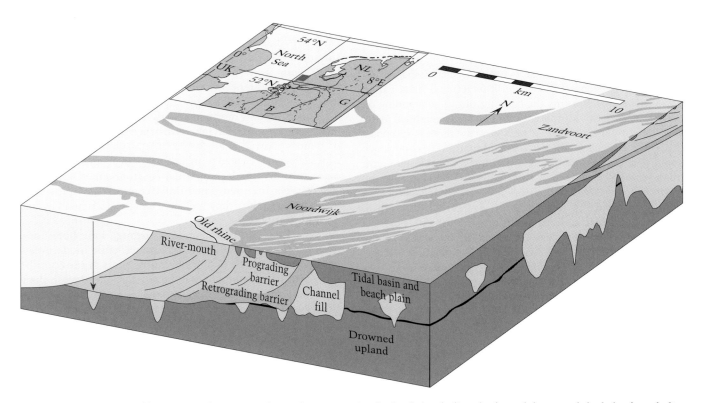

Fig. 9.25 Preservation of barrier morphostratigraphy in the western Netherlands (excluding the frontal dunes and the belt of parabolic, transgressive dunes formed during the last millennium). All that remains of an early Holocene retrogradational barrier-island system that existed offshore the present coastline is a set of partially truncated tidal-channel fills that occupied the associated back-barrier areas and tidal inlets (in blue). The mid-Holocene retrogradational barrier system culminated in a low beach ridge that formed under low wave energy on the landward side of a wide and gently sloping shoreface. The mid- to late-Holocene progradational barrier system has been preserved as a series of beach ridges and beach plains, truncated on its seaward side by rivers and by renewed coastline retrogradation during the late Holocene. Much of the surface expression of the mid- and late-Holocene ridges has been erased by human activities. The location is marked by the red box in the inset. For colour details, please see Plate 23.

produced new insights explaining barrier-system changes as reflected in morphology and stratigraphy:
- small gradient differences may produce major variability in barrier-coastline migration during RSL rise (**STM**, Shoreface Translation-barrier Model);
- grain size of supplied sediment has a considerable effect on shoreface morphology (**BARSIM**, BARrier SIMulation);
- lithified substrates lead to lower rates of barrier transgression than do erodible, sand-rich substrates (**GEOMBEST**, GEOmorphic Model of Barrier, Estuarine, and Shoreface Translations);
- shoreline changes are sensitive to directional changes in the incident waves, including those caused by interannual climate fluctuations such as major El Niño events (**UNIBEST**-TC, UNIform BEach Sediment Transport-Time-averaged Cross-shore model);
- inland topography is the main parameter influencing millennial and longer-term coastal retrogradation under

conditions of RSL rise (model based on the **Exner** equation); and
- variations in incident wave height and wave period affect entire barriers during storm surges, whereas variations in the surge-level gradient between ocean and back-barrier areas affect only the amount of deposition on the back side of barriers and in back-barrier areas (**XBeach**, eXtreme Beach behaviour).

Current numerical models address limited sets of process-response relationships for limited timescales. In modelling barrier behaviour on engineering timescales (the most relevant in coastal-zone management), it is common to either upscale results of models for short-term processes or downscale results of models for long-term changes. Behaviour-oriented, probabilistic techniques appear to be better suited to modelling barrier behaviour on these engineering timescales than deterministic approaches. Uncertainty assessments become increasingly indispensable as the time periods for which predictions of non-linear barrier systems are made are extended from a single decade to the entire 21st century. A main source of uncertainty is the lack of data input from the deep subtidal environment.

9.9 Coastal-zone management and global change

Many barriers, especially those fringing mid-latitude coastlines, are densely populated and protect areas of great economic value. In light of their vulnerability to flooding and hurricane-force winds, and given the increasing impact of residential, industrial and recreational development, sound coastal-zone management is essential. As part of such management, barriers must be considered in a wider environmental context.

Accelerating rates of RSL rise and increased storminess, related to global warming, have a direct impact on many of the world's barrier systems. Some of the major problems faced by society are coastline retrogradation and recurrent flooding. These problems are amplified by diminishing sediment supply as sand and gravel sources are either becoming exhausted or more and more protected from erosion by man-made structures. Key questions faced by coastal-zone managers are:
• How will barrier systems and individual barriers respond to RSL rise, the rising frequency of storm surges, and various anthropogenic activities?
• What measures can be taken to reduce the risk of future barrier loss and flooding?

Answering these questions requires knowledge of barrier systems on all timescales. Analyses of long-term and large-scale evolution provide useful boundary conditions for process-response relationships that shape regional barrier-system morphodynamics. They help circumvent the problem that predictions for one barrier may have little bearing on others, and show how positive developments of one barrier may have negative impacts on the next. Thus, an understanding of large-scale coastal behaviour is invaluable in regional and national decision-making and in developing realistic management strategies prior to major interventions. Strategies can be optimized when probabilistic shoreline-change models fed by high-quality field data (including accurate sediment budgets) are used to compare the predicted effects of different management alternatives (Ruggiero et al., 2010).

Short- and medium-term process-response understanding is important for near real-time forecasting and risk assessment of expected storm impacts, which rely primarily on measurements from previous extreme events. Optimized storm-surge and run-up models with detailed pre-storm bathymetry and morphology are fed by real-time data and forecasts of storm parameters. Observations of consistent vulnerability to storm surges help coastal managers involved in the repair and rebuilding of coastal infrastructure. Some high-risk areas may not be redeveloped, and closely spaced constructions with massive foundations may be replaced by widely spaced structures with small footprints (Morton, 2002).

Coastal-zone managers focus increasingly on keeping human activity compatible with natural barrier morphodynamics. This approach may allow for barrier systems that are expected to migrate landward. Coastal managers might formulate setback zones to ensure that the main source areas remain free to erode. The concept of a stationary coastline may be feasible in areas with abundant sand and the financial resources to transport it to barriers, especially when these have major economic value. It may not be feasible, however, in rapidly subsiding areas with low sediment supply, especially in light of climate change.

Increasingly, soft solutions are being used rather than hard structures, even for large-scale problems. Stationary hard structures require stringent design criteria and are commonly over-dimensioned, in part because of uncertain future outlooks. High cost-benefit ratios result. Soft solutions, such as large-scale shoreface nourishments, include the use of natural processes in coastal protection. So-called sand engines allow long-term redistribution of sand to barrier beaches and dunes by cross-shore and longshore processes (Box 9.2). Important advantages are cost effectiveness and the possibility to adjust sediment volumes along the way.

9.10 Future perspectives

Changes over decadal time are the most relevant to coastal managers and engineers, but are the least known in terms of driver influence on barrier behaviour. Reliable, quantitative predictions of barrier behaviour at timescales of decades and longer require a focus on engineering-scale studies, downscaling of patterns distilled from morphostratigraphical studies to provide boundary conditions and constrained freedom ranges for shorter-term behavioural models, and upscaling of patterns observed in short-term process-response studies.

Major challenges for coastal morphologists and modellers are:
• tackling non-linearity and therefore limited predictability of dynamics that dominate decadal-scale barrier-system behaviour;
• establishment of accurate and high-resolution temporal and spatial control in morphostratigraphical studies; and
• development of innovative instrumentation for the monitoring of coastal processes in challenging environments and under extreme weather conditions.

CONCEPTS BOX 9.2   Sand engines to mimic mid-Holocene barrier progradation under rapid relative sea-level rise

Throughout the world, examples are known of extensive barrier progradation at times of significant relative sea-level (RSL) rise. Such progradation is possible when sand volumes available for natural barrier development are even larger than the accommodation space created by the rising sea level. In the western Netherlands, for example, the barrier coastline prograded more than 10 km between about 5500 and 1500 years ago, when RSL was still rising at an average rate of 0.1 m per century. At the beginning of this progradation, the rate of RSL rise was about 0.25 m per century, comparable to that of today.

At present, newly created accommodation space is no longer filled with naturally supplied sediment, as the sand sources are either exhausted or beyond the reach of everyday wave action. To keep the coastline in place, the Dutch government has established a nourishment policy in which sand is extracted from the inner shelf and dumped on the beach or on the upper shoreface. This policy, along with a range of hard structures, has stopped the coastal erosion that formed an increasing threat to the western Netherlands. Traditionally, these nourishments have been small, in the order of several million cubic metres. To minimize environmental impact, infrequent mega-nourishments will likely replace small-scale nourishments in the near future. These so-called sand engines are created by placing large volumes of sediment in a place from where it can be redistributed by natural hydro- and aerodynamic processes to build engineering structures for coastal safety, while creating new opportunities for nature.

Each sand engine will be about 20 million cubic metres. The nourished material is expected to be gradually redistributed by waves, currents and wind over a period of 10–20 years. The feasibility of the sand-engine concept is tested along the southern part of the western Netherlands coast. In 2011, 21.5 million cubic metres of sand extracted 10 km offshore was used to build a hook-shaped peninsula that extends 1 km seaward from the pre-existing coastline and 2 km along the shore, covering a surface area of 1 km^2 (Fig. 9.26). Models suggest that about 0.35 km^2 of new beach and dunes will have formed in one or two decades (right panel of Fig. 9.26).

Extending the coast seaward by putting large volumes of sand within reach of waves and currents that may transport the sediment toward the shore at a time of accelerating RSL rise mimics the natural mid-Holocene barrier progradation. By adding material to the entire open-marine part of the coastal tract, a new long-term sediment source is put into place that will benefit part of the western Netherlands coastline during the next decades.

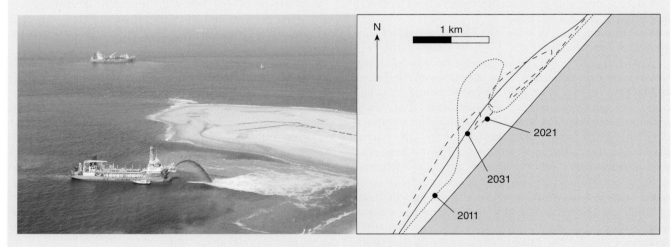

Fig. 9.26   Left: Construction of a sand engine along the western Netherlands coastline. A trailing suction hopper dredger deposits dredged sand by rainbowing it in the nearshore zone on 11 April 2011. The existing beach is visible in the lower right of the image. Right: Modelled development of the Delfland sand engine created along the western Netherlands coast in 2011. An initial hook-shaped peninsula will be modified into a slight bulge in the overall coastline by 2031. (Source: (Left) Rijkswaterstaat/Joop van Houdt. www.dezandmotor.nl. (Right) Provincie Zuid-Holland, The Netherlands Houdt. www.dezandmotor.nl.)

The different response time for different components of the coastal tract, from millennia on the lower shoreface to hours on the beach, complicates the identification of stability thresholds that may induce non-linear barrier behaviour. Linear extrapolation of instantaneous-scale activity does not necessarily equal barrier response on engineering or geological timescales. Minor changes in a single factor may lead to major change. A relatively small increase

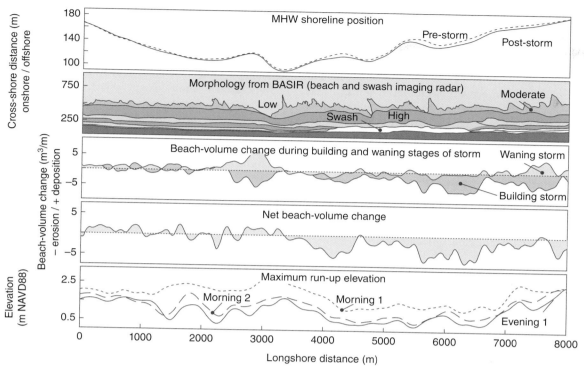

Fig. 9.27 Near-simultaneous observations from the Coastal Lidar And Radar Imaging System (CLARIS) during a storm event impacting the North Carolina barrier coast, USA. MHW, mean high water. NAVD88, North American Vertical Datum of 1988. Increased application of these and comparable systems will result in a database that will allow coastal morphologists to establish quantitative links between coastal drivers and barrier response. (Source: Adapted from Brodie and McNinch 2009. Reproduced with permission from the American Geophysical Union.)

in surge magnitude, for example, is sufficient to trigger breaching and subsequent destruction of coarse-clastic barriers made vulnerable by long periods of stability and vertical accretion. In another feedback mechanism, a steepening shoreface reduces wave dissipation, thus prompting further increases in gradient.

Quantification of long-term process-response relationships to predict 21st-century barrier behaviour is only possible when accurate volume and flux estimates are complemented by reliable age constraints. Dating of sedimentary sequences and event markers has become increasingly feasible with the introduction and rapid improvement of new techniques such as optical dating. Sediment budgets for entire coastal tracts are commonly far from perfect. Slow fluxes between shoreface and barrier, and rapid fluxes during extreme events, are still particularly difficult to monitor in the field.

Further innovations in instrumentation are needed to measure and monitor process-response relationships under these challenging conditions. Recently, the monitoring of overwash has become easier with commercially available acoustic and optical instrumentation

that can be employed in a wide range of flow speeds and in very shallow water. Airborne lidar significantly improves the capability to collect time series of coastal topography and shallow-water bathymetry. The newly developed Coastal Lidar And Radar Imaging System (CLARIS) is now providing near-simultaneous observations of beach topography, nearshore bathymetry, and wave run-up during fair-weather and high-energy events for large stretches of coastline (Fig. 9.27), reducing our reliance on *in situ* measurements from vulnerable instrument platforms. Finally, scale models of increasing size allow state-of-the-art monitoring under controlled conditions, which is instrumental in quantifying process-response relationships that cannot yet be measured in the field.

9.11 Summary

Barrier-system dynamics are a function of antecedent topography and substrate lithology, RSL changes, sediment availability and type, climate, vegetation type and cover, and various aero- and hydrodynamic processes during fair-weather conditions and extreme events. Global change has an influence on many of these drivers. Accelerating eustatic sea-level rise modifies the speed and direction of RSL change and reduces the return period of extreme events, changing rainfall and vegetation patterns alter river-discharge characteristics, shifting and strengthening ocean-current and wind patterns are reflected in increased storminess, and anthropogenic activities have reduced the natural supply of sediment to barriers. As a result, many modern barriers behave differently today than in the past. With a shifting balance between sediment supply and accommodation space created or eliminated by RSL change, more and more barrier systems are becoming subject to persistent erosion, retrogradation and recurrent flooding. When faced with these issues, coastal managers and engineers need to weigh economic and safety aspects against environmental and morphodynamic concerns. In selecting and designing measures to reduce the risk of barrier loss and coastal flooding and in keeping human activity compatible with changing natural barrier morphodynamics, they need quantitative process-response input from coastal geomorphologists and modellers.

• Predictions of barrier-system behaviour over decadal time are the most relevant to coastal managers and engineers, but for this timescale, driver influence on barrier behaviour is still insufficiently known. New engineering-scale studies need to be integrated with downscaled patterns from morphostratigraphical studies and upscaled patterns from short-term process-response studies.

• A second gap in barrier-system knowledge concerns coastal processes in challenging environments and under extreme weather conditions. Monitoring their dynamics requires constant innovation in the development of field instruments.

• Accurate temporal and spatial control is indispensable when morphostratigraphical studies are to be used as input for risk assessments and other decision-making criteria relevant to policy-makers, coastal engineers, and other end users. It helps constraining the magnitudes and recurrence intervals of past extreme events and decadal-scale changes.

• Numerical models are needed, not only to predict future coastal change, but also to further our understanding of barrier behaviour. Sensitivity analyses are particularly useful in explaining erosive developments, most of which do not leave a morphological or stratigraphical record. They may also shed light on the non-linearity of decadal-scale barrier-system behaviour.

Key publications

Beets, D.J. and Van der Spek, A.J.F., 2000. The Holocene evolution of the barrier and the back barrier basins of Belgium and the Netherlands as a function of late Weichselian morphology, relative sea-level rise and sediment supply. *Geologie en Mijnbouw/Netherlands Journal of Geosciences*, **79**, 3–16.
[*Excellent overview of long-term barrier behaviour as a function of various drivers for one of the most extensively studied barrier systems in the world*]

Belknap, D.F. and Kraft, J.C., 1985. Influence of antecedent geology on stratigraphic preservation potential and evolution of Delaware's barrier systems. *Marine Geology*, **63**, 235–262.
[*Well-illustrated early account of barrier behaviour in light of antecedent topography and subsurface lithology*]

Hine, A.C., 1979. Mechanisms of berm development and resulting beach growth along a barrier spit complex. *Sedimentology*, **26**, 333–351.
[*Early study linking barrier morphology and behaviour to sedimentary processes*]

Lindhorst, S., Betzler, C. and Hass, H.C., 2008. The sedimentary architecture of a Holocene barrier spit (Sylt, German Bight): swash-bar accretion and storm erosion. *Sedimentary Geology*, **206**, 1–16.
[*High-quality illustration of barrier development using state-of-the-art ground-penetrating-radar images*]

List, J.H., Sallenger, A.H., Jr., Hansen, M.E. and Jaffe, B.E., 1997. Accelerated relative sea-level rise and rapid coastal erosion: testing a causal relationship for the Louisiana barrier islands. *Marine Geology*, **140**, 347–365.
[*Critical assessment of standard concepts and assumptions used by coastal engineers to predict barrier-coastline change*]

Orford, J.D. and Carter, R.W.G., 1995. Examination of mesoscale forcing of a swash-aligned, gravel barrier from Nova Scotia. *Marine Geology*, **126**, 201–211.
[*Thorough account of non-linear aspects of modern barrier behaviour along paraglacial coarse-clastic coasts*]

Roy, P.S., Thom, B.G. and Wright, L.D., 1980. Holocene sequences on an embayed high-energy coast: an evolutionary model. *Sedimentary Geology*, **26**, 1–19.
[*Well-illustrated analysis of indented barrier coasts and prelude to the influential 1994 chapter on wave-dominated coasts in Bill Carter and Colin Woodroffe's book on coastal evolution (see References)*]

Swift, D.J.P., 1975. Barrier-island genesis: evidence from the central Atlantic shelf, eastern U.S.A. *Sedimentary Geology*, **14**, 1–43.
[*Comprehensive study of the important link between shelf processes and barrier behaviour*]

References

Aagaard, T., Davidson-Arnott, R., Greenwood, B. and Nielsen, J., 2004. Sediment supply from shoreface to dunes: linking sediment transport measurements and long-term morphological evolution. *Geomorphology*, **60**, 205–224.

Ashton, A.D., Murray, A.B., Littlewood, R., Lewis, D.A. and Hong, P., 2009. Fetch-limited self-organization of elongate water bodies. *Geology*, **37**, 187–190.

Beets, D.J., van der Valk, L. and Stive, M.J.F., 1992. Holocene evolution of the coast of Holland. *Marine Geology*, **103**, 423–443.

Brodie, K.L. and McNinch, J.E., 2009. Measuring bathymetry, runup, and beach volume change during storms: new methodology quantifies substantial changes in cross-shore sediment flux. Poster: AGU Fall Meeting, San Francisco, CA, USA, 14–18 December (abstract: adsabs.harvard.edu// abs/2009AGUFMNH11A1111B).

Browder, A.G. and McNinch, J.E., 2006. Linking framework geology and nearshore morphology: Correlation of paleo-channels with shore-oblique sandbars and gravel outcrops. *Marine Geology*, **231**, 141–162.

Carter, R.W.G. and Orford, J.D. (1984) Coarse clastic barrier beaches: a discussion of the distinctive dynamic and morphosedimentary characteristics. *Marine Geology*, **60**, 377–389.

Cowell, P.J., Stive, M.J.F., Niedoroda, A.W. et al., 2003a. The coastal-tract (part 1): a conceptual approach to aggregated modelling of low-order coastal change. *Journal of Coastal Research*, **19**, 812–827.

Cowell, P.J., Stive, M.J.F., Niedoroda, A.W. et al., 2003b. The coastal-tract (part 2): applications of aggregated modelling of lower-order coastal change. *Journal of Coastal Research*, **19**, 828–848.

Davis, R.A., Jr. and Hayes, M.O., 1984. What is a wave-dominated coast? *Marine Geology*, **60**, 313–329.

Dominguez, J.M.L., Bittencourt, A.C.S.P. and Martin, L., 1992. Controls on Quaternary coastal evolution of the east-northeastern coast of Brazil: roles of sea-level history, trade winds and climate. *Sedimentary Geology*, **80**, 213–232.

FitzGerald, D.M. and Van Heteren, S., 1999. Classification of paraglacial barrier systems: coastal New England, USA. *Sedimentology*, **46**, 1083–1108.

Forbes, D.L. and Syvitski, J.P.M, 1994. Paraglacial coasts. In: R.W.G. Carter and C.D. Woodroffe (Eds), *Coastal Evolution*. Cambridge University Press, Cambridge, UK, pp. 373–424.

Forbes, D.L., Taylor, R.B., Orford, J.D., Carter, R.W.G. and Shaw, J., 1991. Gravel-barrier migration and overstepping. *Marine Geology*, **97**, 305–313.

Forbes, D.L., Orford, J.D., Carter, R.W.G., Shaw, J. and Jennings, S.C., 1995. Morphodynamic evolution, self-organisation, and instability of coarse-clastic barriers on paraglacial coasts. *Marine Geology*, **126**, 63–85.

Glaeser, J.D., 1978. Global distribution of barrier islands in terms of tectonic setting. *Journal of Geology*, **86**, 283–297.

Hayes, M.O., 1979. Barrier island morphology as a function of tidal and wave regime. In: S.P. Leatherman (Ed.), *Barrier Islands:*

From the Gulf of St. Lawrence to the Gulf of Mexico. Academic Press, New York, USA, pp. 1–27.

Héquette, A. and Ruz, M.-H., 1991. Spit and barrier island migration in the southeastern Canadian Beaufort Sea. *Journal of Coastal Research*, **7**, 677–698.

Hill, H.W., Kelley, J.T., Belknap, D.F. and Dickson, S.M., 2004. The effects of storms and storm-generated currents on sand beaches in Southern Maine, USA. *Marine Geology*, **210**, 149–168.

Houser, C., Hapke, C. and Hamilton, S., 2008. Controls on coastal dune morphology, shoreline erosion and barrier island response to extreme storms. *Geomorphology*, **100**, 223–240.

Inman, D.L. and Dolan, R., 1989. The Outer Banks of North Carolina: budget of sediment and inlet dynamics along a migrating barrier system. *Journal of Coastal Research*, **5**, 193–237.

Kroonenberg, S.B., Badyukova, E.N., Storms, J.E.A., Ignatov, E.I. and Kasimov, N.S., 2000. A full sea-level cycle in 65 years: barrier dynamics along Caspian shores. *Sedimentary Geology*, **134**, 257–274.

McBride, R.A., Byrnes, M.R. and Hiland, M.W., 1995. Geomorphic response-type model for barrier coastlines: a regional perspective. *Marine Geology*, **126**, 143–159.

Morton, R.A., 2002. Factors controlling storm impacts on coastal barriers and beaches – a preliminary basis for near real-time forecasting. *Journal of Coastal Research*, **18**, 486–501.

Morton, R.A. and Sallenger A.H., Jr., 2003. Morphological impacts of extreme storms on sandy beaches and barriers. *Journal of Coastal Research*, **19**, 560–573.

Morton, R.A., Gibeaut, J.C. and Paine, J.G., 1995. Meso-scale transfer of sand during and after storms: implications for prediction of shoreline movement. *Marine Geology*, **126**, 161–179.

Penland, S., Boyd, R. and Suter, J.R., 1988. Transgressive depositional systems of the Mississippi delta plain: a model for barrier shoreline and shelf sand development. *Journal of Sedimentary Petrology*, **58**, 932–949.

Pilkey, O.H., Neal, W.J., Monteiro, J.H. and Dias, J.M.A., 1989. Algarve barrier islands: a noncoastal-plain system in Portugal. *Journal of Coastal Research*, **5**, 239–261.

Roy, P.S., Cowell, P.J., Ferland, M.J. and Thom, B.G., 1994. Wave-dominated coasts. In: R.W.G. Carter and C.D. Woodroffe (Eds), *Coastal Evolution*. Cambridge University Press, Cambridge, UK, pp. 121–186.

Ruggiero, P., Buijsman, M., Kaminsky, G.M. and Gelfenbaum, G., 2010. Modelling the effects of wave climate and sediment supply variability on large-scale shoreline change. *Marine Geology*, **273**, 127–140.

Stive, M.J.F. and De Vriend, H.J., 1995. Modelling shoreface profile evolution. *Marine Geology*, **126**, 235–248.

Stive, M.J.F., Cloin, B., Jimenez, J. and Bosboom, J., 1999. Long-term cross-shoreface sediment fluxes. In: N.C. Kraus and W.G. McDougal (Eds), *Coastal Sediments '99*. American Society of Civil Engineers, pp. 505–518.

Stolper, D., List, J.H. and Thieler, E.R., 2005. Simulating the evolution of coastal morphology and stratigraphy with a new morphological-behaviour model (GEOMBEST). *Marine Geology*, **218**, 17–36.

Stone, G.W., Liu, B., Pepper, D.A. and Wang, P., 2004. The importance of extratropical and tropical cyclones on the short-term
evolution of barrier islands along the northern Gulf of Mexico,
USA. *Marine Geology*, **210**, 63–78.

Stutz, M.L. and Pilkey, O.H., 2011. Open-ocean barrier islands:
global influence of climatic, oceanographic, and depositional
settings. *Journal of Coastal Research*, **27**, 207–222.

Zenkovich, V.P., 1967. *Processes of Coastal Development*. Oliver
and Boyd, Edinburgh, UK.

Zhang, K., Douglas, B. and Leatherman, S., 2002. Do storms cause
long-term beach erosion along the U.S. East Barrier Coast?
Journal of Geology, **110**, 493–502.

10 Tidal Flats and Salt Marshes

KERRYLEE ROGERS AND COLIN D. WOODROFFE

School of Earth and Environmental Sciences, University of Wollongong, Wollongong, NSW, Australia

10.1 Introduction

Along most coasts, fine-grained sediments are winnowed away by wave and current action, and landforms are rocky, or composed of sand or gravel. However, there are substantial sections of coast that are dominated by muddy sediments, either in sheltered locations where low-energy marine processes dominate, or where the supply of silt and clay-sized sediment is so large that there is a positive sediment budget. These muddy coasts are distinctive for several reasons. Firstly, fine sediment behaves differently from sand and gravel; it takes a long time to settle, but interactions between grains, such as flocculation, accelerate deposition and promote the cohesion of mud once deposited, meaning it requires significantly higher energy to resuspend. Secondly, these muddy sediments support significant biological activity, including organisms within

the sediment (infauna) that bioturbate the sediments, organisms on the surface (benthic epibiota) that form mats and help bind the sediment, and important macrophyte communities in the upper intertidal zone. Thirdly, the accumulation of mud provides a sedimentary record of the gradual accretion that has occurred on these coastlines, providing the opportunity for palaeoenvironmental reconstruction of past habitats and the way in which they have responded to altered boundary conditions, such as sea-level change (see Chapter 1).

10.2 Tidal flats

Tidal flats are low-gradient landforms that occur within the intertidal zone, being alternately exposed at low tide and inundated at high tide. There have been various attempts to classify tidal flats (Dyer, 1998; Semeniuk, 2005), but the term has been used to refer to many quite different environments. Tidal flats can be regarded as components of other coastal settings (e.g. deltas or macrotidal estuaries; discussed Chapters 12 and 13) or they can be the dominant landform on some open coasts (e.g. the west coast of Korea, and the Kerala coast of India).

Tidal flats contrast with the predominantly sandy shoreface of higher-energy coastlines in a number of respects. Firstly, they are generally recognized because of the muddy nature of the upper intertidal zone, though tidal flats may also be composed of sand or gravel in areas with extreme tidal range, such as the Bay of Fundy on the Atlantic coast of North America. Secondly, they are usually of much lower gradient, appearing almost horizontal, often for several kilometres. Perhaps most distinctive are the wetlands that are associated with these coasts, comprising mangrove forests in the tropics, and salt marshes across many temperate and high-latitude coasts (as well as occurring in association with mangroves on tropical and some temperate coastlines). Salt marshes are important coastal ecosystems, and they are examined in detail later in this chapter, but first it is necessary to describe the broader tidal-flat setting in which these landforms occur.

10.2.1   Tidal-flat setting

Tidal flats are most extensive on, but are not limited to, low-energy, sheltered, muddy coastlines that experience a large tidal range. They rarely develop on open coasts with vigorous wave action and may transition to wave-dominated beaches under specific conditions related to increasing wave energy and decreasing tidal range (Masselink and Short, 1993). However, there are exceptions. Where enormous quantities of fine sediment disgorge to the sea, tidal flats develop despite relatively exposed conditions, and higher energy. The Amazon River carries a large suspended sediment load, and much of this fine sediment is carried along the open coast, resulting in broad mudflats, backed by mangroves, prograding along the northeastern coast of South America between the Amazon and Orinoco Rivers, and beyond. Similar open bight mudflats occur to the west of the Mississippi River in the Gulf of Mexico, accumulating on the shores of western Louisiana and eastern Texas. A broad plain has built up during the mid to late Holocene, termed a chenier plain because of the chenier ridges that mark past shorelines stranded across it (these are shelly ridges dominated by the live oak, *Quercus virginiana*, called 'chene', hence the name chenier). The Louisiana chenier plain appears to have prograded during periods when the Mississippi delta supplied mud along its western margin. The mudflats underwent episodic erosion, resulting in reworking of coarse sediments and concentration of shell hash into shore-parallel ridges. The broad

plain with its intermittent shell ridges provides an important palaeoenvironmental archive.

Elsewhere, fine-grained sediments that accumulate in tidal flats have been derived from other sources. They may be eroded from coastal deposits, particularly where these are poorly consolidated glacial moraines or other Pleistocene sediments on the margins of estuaries, as in the Bay of Fundy or the Severn River Estuary in southwest England. The source of sediments comprising tidal flats may be marine or terrestrial; for example, around the North Sea, strong tidal currents have concentrated mud into the Wadden Sea, and muds deposited in the Wash are derived partly from the rapid erosion of the adjacent cliffs on the east coast of England.

These tidal flats are siliciclastic in composition. However, there are also extensive tidal flats composed of carbonate sediment. Most of this is biogenic, derived from the breakdown of the skeletal remains of calcareous organisms. In the tropics these are typically corals and coralline algae, whereas in temperate regions they are likely to be foraminifera, bryozoans and molluscs. In the Bahamas on the Great Bahama Bank, the largest example of a carbonate bank in the world, broad tidal flats are composed of ooids (spherical grains of concentric carbonate layers), as well as pelletal and grapestone aggregates and chemically precipitated carbonate muds.

10.2.2   Geomorphological and ecosystem functioning

In sedimentological terms, tidal flats are sediment sinks. However, even this generalization has its exceptions; for example, the extensive tidal flats off Jiangsu in eastern China are composed of the vast quantities of sediments delivered by the Huanghe (Yellow) River, which discharged directly to the open Yellow Sea coast in the past. Since the diversion of the Yellow River to its new course in Bohai Bay in 1855, where it continues to build new mudflats, the Jiangsu tidal flats have been experiencing erosion. Here, the tidal flats have become a sediment source, a further example of the cyclicity that can be associated with the dynamics of large rivers and their deltas.

Coasts are classified as macrotidal where the spring tidal range exceeds 4 m; mesotidal where it is 2–4 m; and microtidal where it is less than 2 m. Tidal flats are most extensive in areas of large tidal range, primarily because the greater amplitude of the tide exposes much broader areas between the highest and lowest spring tide maxima. However, landforms that might be considered tidal flats develop in any tidal range; intertidal surfaces can even occur in embayments with small tidal range. In their statistical classification of attributes of mudflats, Dyer et al. (2000) found tidal range, wave exposure and slope to be key factors. Modelling also indicates that tidal flats are likely to be more extensive where there is abundant sediment and a large tidal range (Liu et al., 2011). Classic tidal-flat sequences occur in the following megatidal settings (megatidal refers to a tidal range >6 m; each of these examples experience tides in excess of 10 m): the Bay of Fundy; the Bay of Mont St Michel in northwest France; and King Sound in northwestern Western Australia. Even in these extensive tidal-flat regions, where tidal currents are at their most effective in concentrating muds, there are abundant signs of mudflat erosion.

Although by definition tidal flats are intertidal, there are important supratidal flats that occur landward of them, and which share a common origin. Supratidal is generally used to refer to landforms that occur above the elevation reached by the highest tide, although many of the broad flats that are no longer actively inundated by the highest tide need not necessarily be at a higher elevation. In some situations, there is a broad continuum between the active tidal flats and those that were formerly influenced by tides, although they may still be flooded by extreme storm events or freshwater from runoff. Along the arid Trucial Coast of the Persian Gulf, intertidal carbonate environments grade into the broad hypersaline calcareous environments known as 'sabkha'. These sabkhas develop in response to accumulation of sediments within existing waterbodies such as lakes or lagoons, or the deflation of sediment surfaces, thereby enabling periodic inundation and concentration of carbonates and sulphides within sediments. In the semi-arid regions of northern Australia, there are extensive salt flats where intensive evaporative processes ensure vegetation is limited or absent.

However, elsewhere there are significant ecosystems on the upper margins of tidal flats, occupying the associated wetland plains with mangroves or salt marsh. Seagrass can occur seaward, whereas algal or microbial mats can also cover much of the mid and lower intertidal zone. All of these are productive ecosystems and support herbivores that contribute to terrestrial food webs.

In contrast to sandy shorefaces and the sand and gravel beaches that characterize open coasts, the fine-grained sediment that settles in the relatively sheltered tidal flats supports a rich organic community. The sediments are less reworked than the biotically poorer sandy environments. Muds flocculate, which accelerates their settling, and once deposited, muds become cohesive. The biota also contribute to stabilizing the surface of tidal flats; biodeposition of faeces can occur, altering bed roughness, but organic products, such as extracellular polymeric substances (EPS), produce mucus that can modify surface cohesion (Le Hir et al., 2007). On the other hand, biological activity modifies the mud surface through the activities of organisms in the sediment (infauna), which is termed bioturbation, reducing the strength of the upper few centimetres.

The lower sections of the tidal flat may contain molluscs. Many of the bivalve species that occur there live within the sediment, but contiguous beds of sub-fossil articulate bivalves may be seen where erosion has exposed them; loose valves can be concentrated into a shell pavement, or winnowed into an incipient shell ridge or bank. Such intertidal shell ridges are dynamic and undergo gradual landward reworking. When these ridges become stranded in the upper intertidal or supratidal zone, they are preserved as cheniers.

Oyster beds, in some cases forming oyster reefs, may be important in some tidal flats. The rich invertebrate fauna, including polychaetes and crustaceans, are food resources for wading birds. Roebuck Bay in northern Western Australia, for example, comprises macrotidal carbonate tidal flats, with a rich invertebrate infauna (more than 1000 macro-invertebrates per square metre), which in turn are the food resources supporting dense shorebird numbers (often 100,000 individual waterbirds, and occasionally up to 300,000). Roebuck Bay has extensive areas that are exposed at low tide, and the benthic food stocks available to migratory waders before they leave Australia are considerable, exceeding those of other known shorebird departure sites in the Wadden Sea or Mauritania (Tulp and de Goeij, 1994). At least two million shorebirds use the extensive intertidal flats of the Yellow Sea in China and Korea.

10.2.3 Sediments and sedimentology of tidal flats

Early detailed studies of the zonation of muddy environments and the stratigraphy of tidal-flat sediments were undertaken in the Netherlands (Postma, 1961) and in the Wash in eastern England (Evans, 1965). In contrast to sandy coasts, which tend towards a concave 'equilibrium profile' with the shoreface getting progressively deeper with distance from the water line and sediment becoming finer, tidal-flat sediments tend to become finer landwards. In the stratigraphy of regressive coastal settings this is seen as a fining-upward sedimentary sequence. The low tidal flat, exposed only at lowest tides, is generally sandy, comprising various channel environments, although there can be significant gravel deposits in macrotidal settings like the Severn River estuary, which forms the boundary between England and Wales, and the Bay of Fundy. A generalized morphostratigraphic framework is illustrated in Fig. 10.1. Tidal flats can exhibit more variability at local level than might appear likely when comparing superficially similar tidally dominated embayments.

The tidal-flat surface may comprise various surface morphologies and sedimentary structures. Those surfaces that are exposed for long periods of time develop desiccation cracks. These are particularly prominent across supratidal hypersaline flats, but also typify the upper intertidal zone where there is insufficient biological activity to

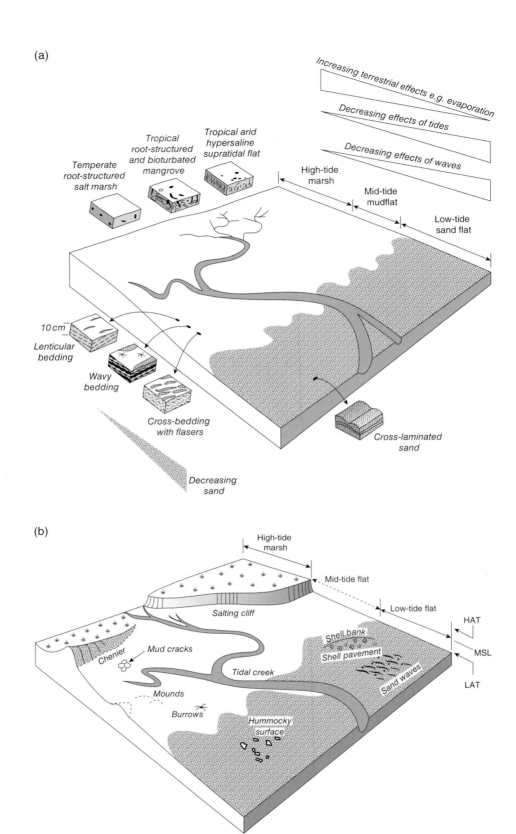

Fig. 10.1 Generalised morphostratigraphic description of a tidal flat. (a) Sedimentology of a typical tidal flat, relating sediment types to marine and terrestrial processes; and (b) geomorphology with typical sedimentary features and their relationship to highest astronomical tide (HAT), mean sea level (MSL) and lowest astronomical tide (LAT). (Source: Adapted from Semeniuk 2005. Reproduced with permission of Springer Science + Business Media.)

disrupt the drying process. Lower in the intertidal zone, biological activity results in scours and feeding excavations (such as by sting rays), but also greater topographic roughness as a result of burrowing and mounds (e.g. through the activity of shrimps and other crustaceans, or mud skippers). In the more sandy regions, complex patterns of ripples develop, with major fields of bedforms of greater than 60 cm wavelength, termed megaripples.

Geomorphic elements across the lower parts of tidal flats have been little studied. These are low-gradient areas, but tidal-flat slope is likely to vary across the zones, and locally in response to tidal channels. Depositional features include sand shoals and shell ridges. An erosional cliff is frequently found at the landward margin of the tidal flat (or the seaward margin of the salt marsh), and it is generally considered to result from concentration of wave activity or current rates of sea-level rise exceeding contemporary sedimentation rates (Kirwan et al., 2011). Associated with the slumping of the small cliff, undercutting the salt-marsh vegetation, it is not uncommon to see mud balls of consolidated mud, generally disc-shaped (Fig. 10.2).

In addition to surface features, such as desiccation cracks, or surface mounds or scars, there are sometimes other features that relate to groundwater. In some places there may be bubble structures. They are formed by seawater, rainwater or groundwater seepage from adjacent land. These structures have been little studied, but hydrology and salinity gradients are likely to exert controls on biota, and seepage may affect the geochemistry and diagenesis of sediments (Semeniuk, 2005). Salinity increases higher on the upper tidal flat, and higher in the tidal range, through evaporation and macrophyte transpiration.

Sedimentary structures that develop where they are not bioturbated include cross-laminated and ripple-laminated sand. Sand is transported across tidal flats by traction as bedload, in contrast to mud (silt and clay-sized particles), which is transported by tidal currents as suspended load (Fig. 10.3). The different behaviour results in complex laminated sediments, including lenticular and flaser bedding. In the case of sand, it is necessary for bed velocities to exceed the shear stress needed to entrain the particular grain size. Sand can be transported landwards until the velocity decreases and grains are deposited. Flood-tide velocities tend to exceed ebb-tide velocities, with this asymmetry becoming greater beneath tidal currents with distance from

Fig. 10.2 Salting cliff eroded into the seaward margin of salt marsh in East Mersea, southeastern England. (Source: Photograph by Bob Jones.)

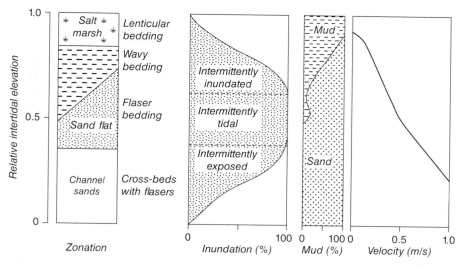

Fig. 10.3 Typical variation in sedimentary characteristics, inundation frequency, mud content and tidal-current velocity across the intertidal zone. (Source: Adapted from Amos 1995 and Woodroffe 2003.)

the sea, favouring the gradual landward movement of sand until the threshold beyond which sand can no longer be transported and at which deposition takes place.

10.2.4 Mud dynamics

The behaviour of mud is more complex and more difficult to understand than that of coarser grains. Tidal currents transport mud in suspension and waters are always highly turbid in these muddy environments. There is a high suspended sediment concentration near the bed and often a transition to lower concentrations in the upper water column; the transition is called a lutocline. The accumulation of fine sediments attests to the effectiveness of the tides in importing mud and its rapid accretion in these settings. Postma proposed what has been called the 'settling and scour lag' hypothesis to explain this influx of mud into the Dutch tidal flats (Postma, 1961). Mud deposition occurs on the upper tidal-flat environments at slack water during high tide. However, at peak tidal stage (see Box 10.1), after the velocity has dropped below the entrainment threshold, there is still water influx, and sediment will remain in

suspension due to low settling velocity, continuing to be carried landward. On the ebb tide, the water at these most landward extremes does not reach the velocities required to entrain the deposited muds. There is therefore a lag after velocity drops at high tide (settling lag), with the mud in suspension carried further landward, and there is a lag before velocities recur on the ebb (scour lag), meaning it is less likely these sediments will be entrained again on the outgoing tide. The net effect is an import of sediment.

Mud dynamics in these environments are much more complex than captured by the settling and scour lag idea. Several factors favour the gradual accumulation of mud in the upper intertidal zone. Tidal currents effectively bring in large concentrations of fine-grained sediment, particularly where they develop asymmetries in which flood velocities exceed ebb velocities. In addition, the currents in mid tidal flats are generally sufficient to ensure continual reworking of sediments, such that mud is re-suspended and winnowed from these sandier environments, ensuring that it does not accumulate as efficiently there. Re-suspension of fine sediment requires a greater velocity of flow than its settling velocity. Consolidation of sediment once it has been

---

CONCEPTS BOX 10.1   Tidal creeks

Tidal creeks are a distinctive feature of the intertidal zone, and extend across tidal flats, salt marshes and mangrove forests. They are the conduits through which much of the marsh is flooded during the incoming tide and drained during the outgoing ebb tide. They are the primary routes by which sediment and nutrients are exchanged, living organisms enter and leave the marsh, and detrital matter is exported. Creek systems are dendritic; a major channel crossing the adjacent tidal flat bifurcates into successively smaller, sinuous tidal creeks, with distance into the marsh interior. Superficially resembling river channels, the flow in tidal creeks is bidirectional and the volume of water discharged is a function of the tidal prism, being greater on spring tides than on neap tides (Fig. 10.4).

On neap tides, the creeks fill as the tide rises, but the water does not overflow the banks and inundate all of the marsh surface. On spring tides, the water does overtop the creek banks and flows across the marsh surface. The velocity of flow in the creek accelerates when the bank is overtopped and tidal waters flood across the salt marsh. At high tide, the flow decelerates, and at slack water (or shortly after, as water can continue to flow into the more distant parts of the high marsh even though it has reached the peak of the tide over the adjacent tidal flat), inflow ceases. As the tide falls, the marsh surface drains because of the negative gradient on the water surface. In those studies that have measured flow

velocity, there is often a peak rate of flow when these overmarsh tides drain off the marsh surface (French and Stoddart, 1992). If ebb velocities exceed flood velocities, the creeks are considered ebb-dominated.

There may be subtle lateral erosion, particularly on meander bends, but erosion, transport and deposition are balanced, and cross-sectional areas of creeks

Fig. 10.4 A salt-marsh tidal creek in the marshes of Ho Bugt, western Denmark, seen at low tide. Spring tides overtop the banks and inundate the marsh. (Source: Katie Szkornik.)

decrease uniformly with distance into the marsh (Fig. 10.5). Tidal creeks appear to have reached a short-term dynamic equilibrium such that over time they neither silt up nor enlarge (Allen, 2000). Over longer timescales, there seem to be cycles of creek expansion and abandonment associated with the autocyclic nature of estuarine sedimentation (Allen and Haslett, 2002).

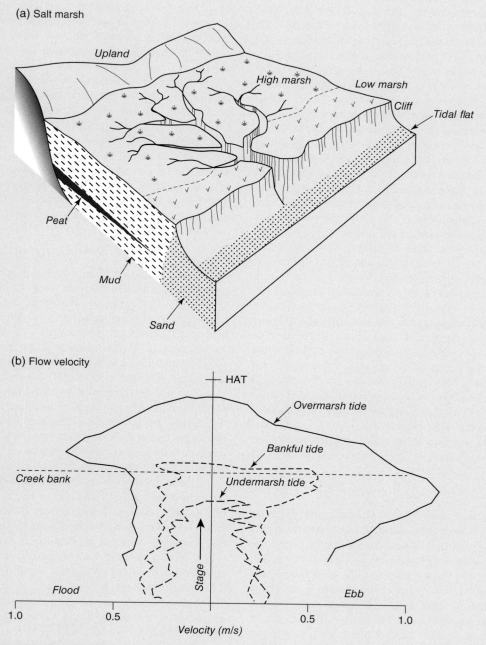

Fig. 10.5　Schematic illustration of tidal-creek system, showing: (a) a dendritic creek network traversing a marsh; and (b) the flow velocities during tides that do not overtop the banks (undermarsh), reach the banks (bankfull), and overtop them (overmarsh). Flow velocities illustrated with respect to highest astronomical tide (HAT). (Source: Adapted from Woodroffe 2003. Reproduced with permission of Cambridge University Press.)

deposited is also very important, and is increased through compaction, pore-water reduction, and biological factors.

The movement of sand and the contrasting behaviour of mud result in tidal bedding across much of the mid tidal flat. The sand becomes rippled (however, ripples visibly exposed at low tide may represent patterns of sediment movement that occurred only during the ebb tide, which may have reshaped ripples formed under the flood tide). These may be draped by mud (mud drapes) deposited at slack water, giving rise to sand-mud laminae. The limit of sand movement will differ under different tides. Only the spring tides may have strong enough bed currents to move sand grains, so sand is less abundant in the most landward settings, giving rise to isolated sand lenses (lenticular bedding). Lower in the intertidal zone, sand and mud may be more equal (wavy bedding), and sand-dominated bedding with inter-ripple mud lenses (flaser bedding) is more likely seaward of this zone. Complex lamination patterns can result, including tidal bundles of muds (representing the neap-spring cycle), differentiated in terms of cross-bedding and thickness (Boersma and Terwindt, 1981). Slope is often greatest at the sand-mud boundary at the limit of bedload transport, in the mid tidal flat.

Tidal flats contain conspicuous tidal channels and tidal-creek systems. These dendritic networks are important conduits for the transport of water, sediment, nutrients and organic detritus, and their role is examined in Box 10.1. Creek systems are typically tapering from large channel cross-sections on the lower flats to narrower and shallower creeks within salt marshes or mangroves. In these, it is frequently the case that ebb flows exceed flood flow velocities; nevertheless mud is still accumulating in such systems as shown by detailed studies in the Dollard Estuary in the Netherlands, despite ebb-dominated channel flow (Christie et al., 1999). Whereas creeks are confined by vegetation in the salt marshes, they are less impeded and may migrate across the mid tidal and low tidal flats, although few studies have examined their dynamics.

10.2.5   Post-depositional change

Broad coastal plains have formed through past tidal-flat accumulation. Many of these are now important locations for industrial, agricultural or settlement-related economic activity. The history of deposition is partially preserved in the sediments below these plains. Significant stratigraphic preservation means that we probably know more about how tidal flats form and evolve through study of the sedimentary sequences beneath them than through direct process study of these landforms. Post-depositional changes and diagenetic changes can be important. For example, the gradual subsidence of some plains, accentuated by groundwater extraction, can exacerbate the risks that threaten communities in these areas as a result of sea-level rise associated with global warming.

There is also another factor, often overlooked, that presents management challenges in these settings. Although the muddy sands deposited beneath tidal flats provide important near-horizontal landscapes, favoured for settlements, and for coastal infrastructure such as airports, subtle geochemical changes in the sediment are likely to have occurred. These include the development of carbonate nodules, as well as precipitates such as gypsum. Iron sulphides, initially derived from the sulphate in seawater, undergo redox reactions that result in jarosite, which can in turn lead to complex oxidation products when these sediments are exposed. Staining of sediments dark grey to black is typical with iron sulphide, these iron compounds become orange and brown when oxidized. Oxidation results in highly acidic conditions, known as acid-sulphate soils. These problems are not experienced until the substrates, termed potential acid-sulphate soils, are exposed to the air, instigating oxidation and the release of acidic waters. Such acidification has been associated with fish kills, and other environmental impacts, in many areas where reclamation has been undertaken.

10.2.6   Erosion and re-suspension

Incoming flood tides initiate a bore over some tidal flats that re-suspends sediment. Tidal currents can be rapid where the tidal range is large, but, although tidal dynamics are clearly important in shaping these environments, it has become increasingly apparent that wave activity also plays a significant role. Waves generally do not break on the broad mudflats that occur in open bights, and these muddy accumulations are very effective at dissipating wave energy. However, locally generated wind waves can result in re-suspension of sediment, as seen during empirical studies in the Humber estuary, northern England (Paterson et al., 2009). Waves are particularly important during storm events. Even rain can cause greater turbidity over broad tidal flats, increasing suspended sediment concentrations on the ensuing tide.

In these enormous sediment sinks, erosion can be an important factor. For example, in the Severn estuary the bed is largely stripped of sediment. Salt marshes here are limited to narrow and contracting fringes, and sea-level rise averaging 4 mm/yr appears likely to have further erosional effects on tidal flats that have undergone degradation at rates of approximately 16 mm/yr (Kirby, 2010). There has been recent research that indicates that the shape of tidal flats may provide clues about dominance of aggradation or erosion. Kirby (2000) suggested that accretionary mudflats are typically convex-up and that erosional flats are concave-up. The convexity of tidal flats enhances ebb domination. Bearman et al. (2010) propose that the convex shape of tidal flats increases with increasing tidal range and sediment load, while the concave shape is enhanced by increasing tidal-flat width, wave energy and grain size. Detailed monitoring of the dynamics of tidal-flat surface changes was undertaken by O'Brien et al. (2000), who showed that the surface was higher in summer than in winter, due partly to algal binding of sediments in summer, and partly to the winter storminess, which winnows the flats, resulting in erosion. Such storms tend to mask the sediment cycling due to tidal rhythms. Greater water movement on rough days inhibits settling, and suspension of mud is greater during gales, with fluid mud re-eroded on successive spring tides (Kirby, 2010).

Recent research is beginning to demonstrate morphodynamic feedbacks across tidal flats and interrelationships with associated wetlands. The hypothesis that there might be an equilibrium morphology towards which a tidal flat trends, or that there might be a preferred shape related to the dominant operative processes, reflects thinking that has been central to much research on better-studied sandy coasts for many decades (see Chapter 1). Convex profiles mean that larger proportions of the intertidal zone are at elevations higher in the tidal range and therefore amenable to colonization by vegetation and able to become wetland.

The processes of vegetation colonization and the geomorphology of salt marshes are considered next. However, the numerous exposures of peat layers in the margins of the Severn estuary, examined in detail through many meticulous studies by J.R.L. Allen, indicate that there have been cycles of erosion, mudflat progradation, salt-marsh colonization, and subsequent demise and erosion. Such studies, as well as examination of cliffing along the seaward margins of other marshes, reinforce the autocyclic nature of change that such systems experience. In addition, modelling has linked the temporal extent of marsh erosion with the tidal-flat sediment budget, showing their interdependence on sedimentation and sea-level tendency. When erosion occurs in muddy sediments, the mobilized mud may be lost from the system, as opposed to morphodynamics on sandy coasts, where it can be returned over ensuing months. Nevertheless, it appears that marshes, including both mangrove and salt marsh, are a component of the tidal-flat system; evolution of these components may be modulated by other factors and is closely linked to subtidal channels. For example, numerical modelling of long-term evolution of salt marshes and tidal flats indicates that vegetation plays a critical role in the redistribution of eroded sediments across the intertidal area, while tidal-flat shape and vegetation may indicate the rate of conversion of marshes to tidal flats (Mariotti and Fagherazzi, 2010).

10.3 Salt marshes

Coastal salt marsh is a significant ecosystem occupying the upper margins of the intertidal zone. They may be defined as vegetated tidal flats, but are considered here as a separate geomorphological system, as the presence of vegetation leads to morphodynamical processes that are quite distinct from those acting lower in the intertidal zone. Salt marshes consist largely of intertidal communities that are dominated by flowering plants, including herbs, grasses, rushes and shrubs. They are typically located on soft, muddy substrates within estuaries and embayments, or along low-energy coastlines. However, some plant species typical of salt marshes may also occur on exposed headlands, where they are subjected to salt spray. Salt marsh can also develop where a gently shelving coast is combined with high concentrations of suspended sediment, such as the chenier plain coast of southwestern Louisiana, USA. Salt marsh is estimated to cover 38 million hectares worldwide (Woodwell et al., 1973), but details about the distribution and composition of salt marsh remain relatively poor on the coasts of Asia, South America and Australia, and there is an urgent need to review the global extent of salt marsh.

Salt marshes have been regarded as highly productive ecosystems, and they are also increasingly valued as carbon sinks due to their ability to sequester carbon from the atmosphere within their soils. Salt marshes reportedly sequester carbon at an average rate of $210\,g/m^2/yr$ and globally store at least 430 Tg of carbon in the upper 50 cm of marsh soils (Chmura et al., 2003). The value of carbon stored in tidal wetland soils, including both salt marsh and mangrove, may be greater than any other natural ecosystem for three main reasons: (1) they continue to sequester carbon through vertical accretion; (2) the decay of organic carbon is slow due to anoxic conditions; and (3) the release of methane to the atmosphere is limited due to saline conditions (McLeod et al., 2011). Salt marshes provide numerous ecosystem services, including raw materials and food, coastal protection and erosion control, water purification, maintenance of fisheries, carbon sequestration, tourism, recreation, education and research (Barbier et al., 2011).

10.3.1 Salt-marsh ecology

Salt-marsh plants have developed various adaptations to the numerous stressors they encounter in occupying the soft substrates of low-energy coastlines, estuaries and embayments. The most obvious challenges to plant growth in these habitats are the highly saline soils and anoxic conditions associated with flooding from saline and brackish waters.

The majority of plants that occupy salt marshes are referred to as halophytes as they characteristically grow in saline soils of varying salinities; however, glycophytes (plants that grow in non-saline soils and may be damaged by high salinity levels) may establish at higher marsh elevations. As the elevation of a salt-marsh surface increases, the influence of tidal flooding decreases and the flux of salt into the salt marsh also decreases. However,

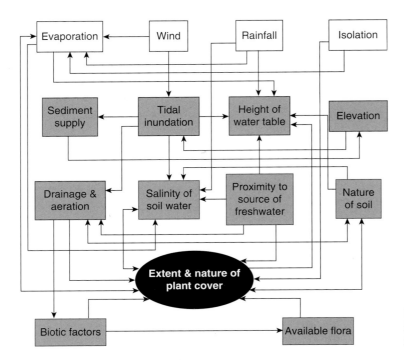

Fig. 10.6 The interactions between environmental factors and vegetation in salt marshes. (Source: Adapted from Adam 2009.)

the salinity of the soil does not follow a similar relationship. Soil salinity within low-elevation salt marshes that receive frequent immersion largely reflects the salinity of the tidal waters. At higher marsh elevations that receive less frequent immersion from tidal waters, soil salinity is influenced by flooding, groundwater influences and the prevailing climatic conditions (Fig. 10.6). During drier periods, higher evapotranspiration can increase soil salinity and salt crusts may form. Where rainfall and run-off limit the development of hypersaline conditions at upper marsh elevations, intermediate marsh elevations often exhibit hypersaline conditions. This is particularly evident at lower latitudes that are exposed to high solar radiation, and may result in characteristic salt-pan or saline-flat development. The development of highly saline soils can occur even when flooding water is brackish, as long as evapotranspiration is high.

The presence of salt in soils has significant implications for the uptake of water and nutrients by salt-marsh plants. High soil salinity may result in reduced water uptake and damage cell membranes – causing increased cell permeability and loss of nutrients to the soil (Alongi, 1998). The growth response of salt-marsh plants to salinity varies widely between species and even genotypes. Adam (1990) indicates that regulation of salt content within plant biomass can be facilitated by the following:
• Ion exclusion: halophytes may exclude salt from being absorbed by roots.

• Growth and/or succulence: sodium and other ions may be diluted within halophytes by directing sodium uptake to new biomass or increasing succulence so that salt concentration within the foliage remains relatively constant.
• Secretion: halophytes may secrete a salt solution through salt glands on the outside of leaves or may accumulate salt within bladder cells on salt hairs.
• Redistribution of ions and/or loss of plant biomass: halophytes may accumulate salt within plant biomass that is shed and replaced when salt has accumulated at high concentrations.
• Reduction in transpiration: many salt-marsh plants exhibit particular physiology, such as low leaf surface : volume ratios or use of C_4 photosynthetic pathways, that limit transpiration rates and the uptake of toxic ions such as salt.

Due to waterlogging from tidal flooding, salt-marsh soils are frequently anaerobic and salt-marsh plants must adapt to the lack of oxygen and the presence of toxins, such as hydrogen sulphide, associated with anaerobic conditions. To cope with anaerobic soil conditions, aerenchyma – channels and cavities filled with air – may form within the roots of salt-marsh plants and facilitate the exchange of oxygen between shoots and roots. Salt-marsh plants that are unable to form aerenchyma, such as *Atriplex portulacoides*, which reportedly has root porosity not exceeding 5%, are intolerant of waterlogged soils and may be restricted to parts of the tidal marsh with well-aerated soils, such as creek banks (Adam, 1990).

Despite the ability to develop root aerenchyma in response to anaerobic soil conditions, the movement of interstitial water greatly affects salt-marsh development. Frequent flooding and drainage flushes toxins, such as salts and sulphides, and replenishes nutrients and oxygen within the soil profile (Alongi, 1998).

10.3.2 Salt-marsh vegetation and biogeography

Salt marshes are commonly divided into low marsh and high marsh on the basis of elevation and plant zonation. The lower limit of low marsh is defined as the seaward margin of vegetation within the intertidal zone. The upper limit of high marsh is commonly marked by a sharp elevation change between the marsh vegetation and the hinterland, or artificial structures, such as roads or buildings. Alternatively, the upper limit can be defined on the basis of the highest tide, and an ecotone may separate the salt marsh and hinterland. Halophytes commonly colonize low and mid elevation salt marshes, while the upper marsh may be colonized by both halophytes and glycophytes, with the proportion of each varying in response to tidal immersion (Adam, 1990). While salt marshes are typically dominated by halophytes, glycophyte diversity may be high; Adam (1990) reported from a survey of British salt marshes that halophytes constituted only 45 of 325 species of vascular plants recorded. The halophytic component of salt-marsh vegetation is dominated by few families: Poaceae, Chenopodiaceae, Juncaceae, Cyperaceae, Plumbaginaceae and Frankinaceae.

Salt-marsh distribution is largely dependent upon the availability of favourable local conditions (i.e. low-energy depositional coastlines with a gently sloping bedrock on which a tidal flat can develop), and the tolerance of salt-marsh species to salinity and hydroperiod. As a consequence, long-term temporal changes in the distribution of tidal flats and salt marsh that occurred during the Quaternary were driven primarily by sea-level change associated with changing climate. Modern salt marshes are regarded as being relatively young; having developed during the past few millennia when sea level has been close to its present level. Over shorter timescales (decades to centuries), salt-marsh biogeography is largely explained by the interplay between geomorphic processes, biological processes and human interference. Meso- and micro-scale variations in salt-marsh distribution relate principally to tolerances of salinity and wetting.

At the macro-scale, salt marshes have been classified into regional types on the basis of dominant vegetation or vegetation similarities (see for example Chapman, 1974; Frey and Basan, 1985; Adam, 1990), and these types generally correlate with latitude. In the tropics, competition with mangroves and persistent high soil salinities may act to reduce salt-marsh diversity. By contrast, cool summers and cold winters at high latitudes limit the diversity of salt marshes. A summary of the salt-marsh regionalization of Adam (1990) is provided below:

- **Tropical marshes.** Limited in extent by the dominance of mangroves. Salt flats are extensive due to the hypersaline conditions in the upper intertidal zone. Marshes may develop as a pioneer community seaward of mangrove, as a grassland associated with mangrove forests, or as a secondary community within disturbed mangrove areas.
- **Dry coast marshes.** Generally dominated by halophytic plants; glycophytic plants may be rare due to the dry conditions and high soil salinities for extended periods.
- **Arctic marshes.** Relatively species-poor and floristically uniform.
- **Boreal marshes.** Often regarded as an ecotone between arctic and temperate marshes, and slightly more diverse than arctic marshes.
- **Temperate marshes.** Neither too hot to allow extensive development of hypersaline soil conditions, nor physiologically too cold to limit germination and metabolic processes. Temperate marshes exhibit the greatest species diversity, and at a family and generic level are regarded to exhibit strong similarity. Adam (1990) identifies European, western North American, Japanese, Australasian and South African sub-groups, but argues that there is an overall similarity between northern hemisphere sub-groups and southern hemisphere sub-groups that may be explained by the origin of angiosperms following the separation of Laurasia and Gondwanaland.
- **West Atlantic marshes.** Occur over a wide range of climatic and tidal-range conditions, and are dominated by *Spartina* species, particularly *S. alterniflora* (Fig. 10.7).

Adam (1990) argues that the dominance of *Spartina* species in the West Atlantic may represent a relatively recent evolutionary event. Interestingly, *Spartina* species

Fig. 10.7 *Spartina alterniflora* salt marsh located on the Barataria-Terrebonne estuary, south Louisiana, USA. (Source: USDA-NRCS PLANTS database, http://plants.usda.gov/java/largeImage?imageID=spal_002_ahp.tif).

are globally regarded as invasive (Adam, 2009), and four exotic species have been recorded in San Francisco Bay, USA (Ayres et al., 2004). The fertile *S. anglica* and sterile *S. townsendii* are highly invasive and originated on the south coast of England as a result of spontaneous hybridization of the native *S. maritima* and the West Atlantic *S. alterniflora*. *Spartina anglica* is extremely tolerant of tidal submergence and has colonized mudflats. This colonization has displaced seagrass communities, *Salicornia* monocultures, and *S. maritima*, which is now regarded as rare in northern Europe (Adam, 1990). The tolerance of this hybrid is attested by its colonization of mudflats seaward of mangrove within Western Port Bay, Victoria, Australia.

Interestingly, salt marshes in temperate zones may co-exist and competitively interact with mangroves (Saintilan et al., 2009). While mangroves more commonly occupy tropical latitudes and their global distribution is largely controlled by physiological intolerance to frost (Woodroffe and Grindrod, 1991), salt-marsh vegetation does not exhibit a strong physiological preference for low temperatures (although individual species may exhibit intolerances to temperature extremes). Rather, salt marsh establishes on shorelines where mangrove establishment is precluded or growth is limited (Kangas and Lugo, 1990). Hence, salt marshes are more prevalent in temperate, sub-arctic and arctic zones. The species richness of both mangrove and salt-marsh species reflects their distribution: mangrove species richness is greatest at lower latitudes, whereas salt-marsh species richness is greatest in temperate zones.

Salt marsh and mangrove co-exist in southeastern Australia (Fig. 10.8), and analysis of archival aerial photographs has identified a trend of mangrove encroachment into salt marsh (Saintilan and Williams, 1999). Hypothesized causes include increased rainfall, cessation of agriculture and industry, thus enabling mangroves to re-colonize areas they previously inhabited, altered tidal regimes resulting from engineering works such as dredging or related to sea-level rise, increased sedimentation and associated nutrients, and subsidence or compaction. A strong relationship was identified between relative sea-level rise and the upslope migration of mangrove into salt-marsh communities at selected sites in southeastern Australia (Rogers et al., 2006); however, further research is required to establish the mechanisms behind this interaction at a local scale. In the West Atlantic-Gulf of Mexico region, periodic freezes have killed or damaged mangroves, allowing salt marsh to replace damaged mangroves within about four to five years (Stevens et al., 2006). Relatively mild winters and lack of frost since 2000 have allowed mangroves to expand into salt marsh in Florida and Louisiana, USA (Saintilan et al., 2009). Expansion of mangrove in coastal Louisiana has also been attributed to drought-induced dieback of *S. alterniflora* (McKee et al., 2004).

10.3.3 Salt-marsh zonation and succession

Salt marshes commonly exhibit patterns of variation in the composition of flora and fauna. The vertical arrangement of species in overlapping bands, often parallel to shorelines, is usually referred to as zonation. A further longitudinal pattern of zonation may also be evident along the length of an estuary. Although broad patterns of zonation have been described for decades (Chapman, 1974), the underlying causes of zonation are relatively poorly understood. Generally, species that are dominant on low marshes tend to exhibit broad tolerance to environmental gradients, particularly to salinity and water-logging. Consequently, these species may be observed at many elevations within a salt marsh. In contrast, upper-level salt marshes may act as a niche for species with narrow tolerance to environmental gradients (Adam, 1990). As a corollary, salt marshes tend to exhibit increasing species richness with elevation, and high marshes often support a mosaic of communities. However, there are many exceptions to this simple rule; for example, many *Spartina* species tend to colonize salt marshes only at low elevations. An alternative view of salt-marsh zonation is that the lower boundary of species distribution is limited by tolerance to environmental gradients, while the upper boundary is controlled by species competition (Adam, 1990).

There is considerable variation in zonation patterns at various spatial scales, including within an estuary, between estuaries at different latitudes, and between coastlines (Adam, 1990). In addition, due to the role of competition in ordering salt-marsh species distribution, it cannot be assumed that zonation along a gradient is indicative of

Fig. 10.8 Co-existing salt marsh (foreground) and mangrove (background) on the Tweed River, a sub-tropical tide-dominated estuary in northern New South Wales, Australia. (Source: Photograph by Kerrylee Rogers.)

community succession or that salt-marsh communities in upper-level marshes have developed through succession from vegetation communities at lower elevations (Gray, 1992). While models of marsh development may enforce the idea of a series of successions from early colonizers to mature communities, this model only holds true for the pioneering stages of marsh development. Current studies of salt-marsh morphodynamics promote equilibrium between tidal inundation, salt-marsh accretion and elevation (discussed later), which gives rise to zonation of species and communities, rather than succession from one community type to another (e.g. Friedrichs and Perry, 2001).

10.3.4 Geomorphological and ecosystem functioning

At the landform scale, salt-marsh geomorphology may be regarded as a subset of the geomorphology of tidal flats; both mangrove and salt marsh generally occur within the upper half of the tidal spectrum. Salt marshes, and indeed mangrove, are located where the action of storm and wind waves is reduced, primarily on low-energy coasts, in estuaries and shallow bays, or behind spits and barrier islands. Salt-marsh development is generally greater when wind and wave conditions promote sediment accumulation and the underlying geology allows the development of low-gradient plains within the intertidal zone. Salt marshes have been classified on the basis of their physical setting (Fig. 10.9) and include the following:

- **Open-coast marshes** are often poorly developed due to wave action. They are typically sandy systems with relatively exposed sandy tidal flats to seaward.
- **Deltaic marshes** occur on actively prograding deltas, such as the Atchafalaya delta that is building seaward as the Mississippi River seeks a new distributary to the sea. Small islands of mud accumulation provide favourable habitat for marsh colonization
- **Estuarine marshes** include marshes that fringe estuaries where muddy sediments accumulate, and lagoons that are partially or intermittently closed off at the mouth by spits, promontories or barriers and support extensive salt marshes.
- **Back-barrier marshes** are located on the sheltered landward side of open-coast barriers, or behind barriers or spits at the mouth of estuaries. Substrates may be sandy to muddy, and the barriers are often retreating landwards across the marsh, as is occurring on barrier islands such as Timbalier Island in Louisiana.
- **Embayment marshes** may fringe the edges of large, open tidal embayments with unobstructed entrances. Due to the strong marine influence, substrates are commonly dominated by sand.
- **Drowned-valley marshes** may develop in drowned river valleys, including rias, which are narrow bedrock-fringed embayments formed by submergence of a river valley, and fjords, resulting from glacial erosion, which forms a steep-sided narrow inlet; their development is generally restricted to low-lying areas where sediment may accumulate.

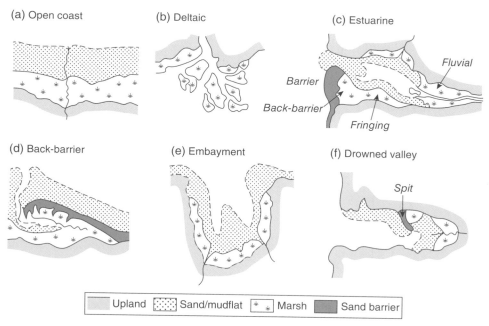

Fig. 10.9 Geomorphological settings in which salt marsh occurs. (Source: Adapted from Allen 2000 and Woodroffe 2003.)

10.3.5   The role of sedimentation and tides in salt-marsh maintenance

The rate and source of sediment supply, tidal regime, wind-wave climate and water-level changes are important determinants of the behaviour and development of salt marshes. Colonization of tidal flats by vegetation is initially dependent upon tidal-flat elevation being raised within the tidal prism through processes of sediment accumulation so that the surface is at a suitable elevation within the intertidal zone for marsh vegetation to survive. Once salt-marsh plants establish, further salt-marsh development is promoted by the *in situ* contribution of organic matter to the adjacent tidal flat, while the mechanical action of tides moving over plants acts to baffle tidal flows and promote the deposition of sediments. Deposited sediment and organic material consequently becomes trapped within the soil profile by the salt-marsh plants that bind them. Sediment supply and deposition are critical for the maintenance of most salt marshes.

Many salt marshes exhibit a strong relationship between sediment supply and distance to source of sediment (e.g. proximity to tidal creeks), elevation and deposition, whereby sediment accumulation is generally regarded to be inversely proportional to the hydroperiod, which is defined as the duration of inundation on a surface (Fig. 10.10). Hydroperiod is a function of the tidal range and surface elevation. Low marsh generally has a greater hydroperiod and greater opportunity for accumulating sediment than does high marsh. This relationship between sedimentation and hydroperiod is widely documented, and is largely caused by fewer flooding tides at higher elevation in the marsh and the reduced sediment availability within tidal waters that reach these heights, as much of the sediment has been deposited at lower elevations before the tide reaches the high marsh. A basic model of

sediment accumulation devised by Pethick (1981), and referred to as the negative feedback loop, follows this relationship; accretion increases the marsh elevation, causing a decrease in hydroperiod, and a subsequent decrease in the rate of sediment accretion.

Sediments on which salt marshes form are sourced from tidal waters that transport minerogenic material and organic material to the salt marsh, referred to as allochthonous sediment, or from the salt-marsh plants themselves that supply organic material, referred to as autochthonous sediment. Rarely is sediment sourced from the immediate upland slopes of salt marshes, although rivers may deliver silt and clay to estuaries and deltas, and shoreline erosion can also be an important source of sediment, as described earlier for tidal flats. Many salt marshes are primarily minerogenic, with sediment ranging from fine sands to clayey silts. In contrast, there are also organogenic salt marshes, such as the salt marshes of New England, USA, which form on a peat substrate from *in situ* processes of organic matter accumulation and development of underground biomass.

Terrigenous sources of minerogenic sediment include river catchments, erosion of coastal cliffs, and offshore sediment deposits. Salt marshes on rivers with emergent deltas, such as the Mississippi River, establish where major sediment inputs are from riverborne sediments derived from erosion within the catchments. Land-use in the surrounding catchment, catchment clearing and climatic regimes can strongly influence the quantity and quality of minerogenic sediment available for deposition. Salt marshes located on relatively open estuaries or open coasts may exhibit a predominance of marine sediments. In other settings, there may be no net input of sediment. For example, some margins of the Severn River in Britain are undergoing vigorous erosion, with sediment being reworked and transported into active environments (Allen, 1990). Salt marshes can also occur on carbonate sediments where the calcareous substrate is formed from the skeletal remains of organisms, such as molluscs and foraminifera.

Some salt marshes may be largely dependent on sediment inputs derived from storm activity. Storms have long been recognized as events that can cause significant geomorphological change to salt marshes. Storm surges may cause erosion, sediment redistribution and result in loss of organic material, and they may deliver significant volumes of sediment to salt marshes. For example, hurricanes along the Gulf of Mexico have supplied sediment to subsiding brackish marshes on the Mississippi River Delta, even to the degree that elevation gain from sediments transported to salt marshes during Hurricane Katrina exceeded the degree of subsidence of the substrate that had occurred in the two years following the storm (McKee and Cherry, 2009). Interestingly, elevation loss post-Hurricane Katrina was greater within a brackish marsh that received storm sediments with higher organic

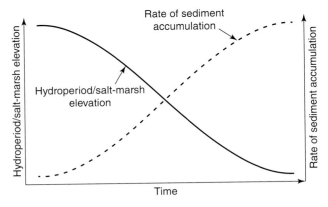

Fig. 10.10 Hypothetical model of the relationship between hydroperiod, which generally correlates with salt-marsh elevation, and the rate of sediment accumulation. (Source: Adapted from Adam 2009.)

matter and water content due to initial compaction of the storm layer and collapse of the root zone. The influence of storms on marshes is dependent upon the marsh location and type, and the storm intensity (Guntenspergen et al., 1995). Cahoon (2006) identified storm-related processes that may influence salt-marsh elevations and have long-lasting ecological consequences, including sediment deposition, sediment erosion, sediment compaction, soil shrinkage, root decomposition, root growth, soil swelling, and lateral folding of the marsh root mat.

10.3.6 Response to sea-level changes

The maintenance of salt-marsh elevation with respect to sea level, termed 'relative elevation', is integral to their long-term survival. An increase in relative sea level may submerge salt marsh unless the salt marsh increases its elevation proportionately. However, the processes controlling salt-marsh surface elevation are complex and may be divided into geomorphological processes, biological processes and hydrological processes. These processes operate in response to a range of drivers, including tides, sea-level rise, nutrient addition and climate. Figure 10.11 explores the interaction of geomorphological, biological and hydrological processes and their influence on the response of salt marshes to sea-level change.

The long-term evolution of salt marshes is largely attributed to the accumulation of minerogenic sediments that accumulate through tidal redistribution. In tide-dominated settings, salt marshes have reportedly kept pace with

sea-level changes, in some places for as long as 6000 years (Gehrels 1999), providing sediment archives from which sea-level changes can be reconstructed (see Chapter 2). Salt marshes may remain in a morphodynamic balance through geomorphological processes alone when rates of sediment accumulation are equivalent to, or exceed, rates of relative sea-level rise. Large-scale wetland loss in the interior regions of the Mississippi Delta has been attributed to marsh accretion being unable to keep pace with relative sea-level rise (Day et al., 2007). Since 1900, overbank flooding of the Mississippi River into the deltaic plain, and associated sedimentation in the interior of wetlands, has been limited by flood mitigation measures, such as additional construction of levees. This period of wetland loss contrasts markedly with the period in which the present Mississippi Delta plain formed 6000 to 1000 years ago. During this period of deltaic formation, sea levels were relatively stable, enabling riverine sediments to be deposited at the mouth of rivers and older distributaries, and by overbank flooding and crevasse formation (Day et al., 2007). McKee and Cherry (2009) found that salt marshes exhibited greater submergence from relative sea-level rise when accumulated sediments had a higher ratio of organic material to minerogenic material.

The organic material within salt-marsh soils is complex, and may include algae and microbial mats, living roots and rhizomes, structurally intact dead material, and organic material remaining after decomposition of dead vegetation. The contribution of organic material to salt marshes may be very high, for example the wet organic

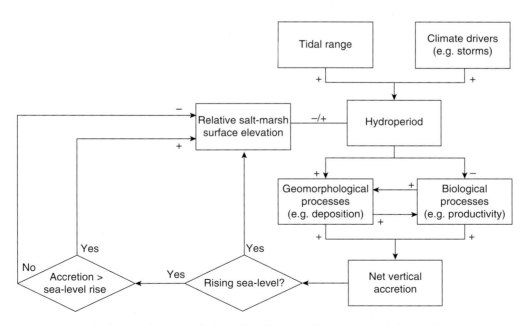

Fig. 10.11 Interactions between hydroperiod, geomorphological and biological processes, and their influence on vertical accretion and salt-marsh response to sea-level rise. (Source: Adapted from Reed 1990.)

matter to soil volume in Narragansett Bay in Rhode Island, Massachusetts, USA, is reportedly as high as 96% (Bricker-Urso et al., 1989). Recent research using surface elevation tables (SETs) has established the contribution of organic material to the maintenance of salt-marsh elevations (see Box 10.2) by enabling the disaggregation of

---

METHODS BOX 10.2   Exploring surface elevation dynamics of marshes using surface elevation tables

Surface elevation tables (SETs) have been used to explore sedimentation and the influence of below-ground processes on marsh elevation and stability. SETs are a precise non-destructive method for measuring relative changes in marsh-surface elevation. The SET consists of a stable vertical benchmark and a portable component. The portable component fits to the stable benchmark, and the horizontal arm on the portable component can be accurately levelled. Pins are lowered from the levelled horizontal arm and the length of pins above the horizontal arm is measured. Over time, SET sites are re-visited and re-measured, and changes in the measured pin length above the horizontal arm provide an indication of the relative change in the marsh-surface elevation.

Simultaneous measurements of vertical accretion and surface elevation enable the surface processes of sediment accumulation to be disaggregated from the below-ground processes influencing the soil volume. Such processes include autocompaction (compression of the sediment under its own weight and through loss of organic material), plant productivity, peat decomposition and groundwater flux. Accretion is commonly determined from measurements of sediment accumulation above an artificial marker horizon at sites adjacent to the SET, using distinctive layers, such as feldspar, in the substrate. Autocompaction of the soil volume between the marsh surface and the base of the stable benchmark (and in some cases uplift) can be determined by the difference between the elevation change measured using SETs and the amount of sediment accretion, determined from marker horizons (Fig. 10.12).

SETs were originally developed in the Netherlands (Schoot and de Jong, 1982) and have since been redeveloped by the United States Geological Survey. Currently, three versions of the SET are used for exploring marsh-elevation changes within the root zone (shallow-rod SET), within the zone of shallow subsidence (original SET, Fig. 10.12; Cahoon et al., 2002a), and within the zone of deep subsidence (deep-rod SET; Cahoon et al., 2002b).

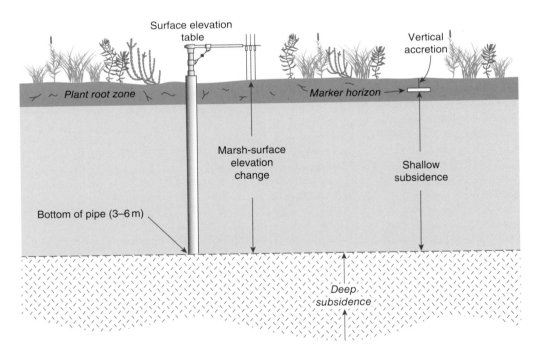

Fig. 10.12   Schematic representation of the surface elevation table (SET) and feldspar marker horizon. (Source: Adapted from Cahoon: www.pwrc.usgs.gov/set)

surface processes of sediment accumulation from below-ground processes that may influence soil volumes, including accumulation and/or decomposition of organic material.

Surface plant-litter accumulation and subsequent decomposition contribute to soil volume and relative elevation change. Highly organic sediments deposited during storms are considered to make a significant contribution to the stability and survival of salt marshes (Guntenspergen et al., 1995). Despite an implicit understanding of the significance of litter and organic material to salt-marsh stability, very little research has focussed on identifying the contribution of surface litter to salt-marsh substrate volume. Similarly, algal and microbial mats that form on salt marshes are highly productive (Zedler, 1980), and may make a significant contribution to the elevation of the wetland surface and sedimentation. However, the contribution of these algae and microbial mats to the overall wetland surface elevation remains largely unknown and requires further consideration.

The role of below-ground productivity in ensuring the stability of salt marshes against relative sea-level rise is becoming increasingly apparent. Variation in vertical accretion in a Terrebone Basin marsh, Louisiana, USA, was attributed to variable organic-matter accumulation and plant productivity (Nyman et al., 1993), indicating that the degree of submergence a marsh can tolerate may be dependent upon plant productivity. Similarly, the application of nutrients and associated increases in plant productivity are reported to result directly in increases in wetland elevation (Morris et al., 2002). Recent research has also indicated that elevated carbon dioxide enhanced photosynthesis and plant productivity of a salt-marsh species that uses a C_3 photosynthesis pathway, causing vertical translation of the wetland surface that was proportional to expansion of the root zone (Cherry et al., 2009). There is increasing awareness that enhanced productivity of C_3 salt-marsh species under conditions of greenhouse warming may contribute to off-setting the ecological impacts of sea-level rise on salt marshes.

Wetland hydrological processes influence wetland elevations and the response of salt marshes to sea-level rise through the delivery of sediments and organic matter to wetland surfaces, the regulation of plant growth and decomposition, and via water flux dynamics within the soil volume that can influence salt-marsh substrate elevations. Tidal range is considered to be an indicator of potential for sediment transport, whereby salt marshes experiencing high tidal ranges may exhibit higher rates of accretion than similar salt marshes within microtidal estuaries. Consequently, salt marshes with high tidal ranges are regarded to be less vulnerable to submergence from sea-level rise than those in low tidal-range settings (Kirwan and Guntenspergen, 2010).

Alterations to salt-marsh soil volume through water flux mechanisms may occur as a result of compression of the substrate, which causes lateral movement and uplift in adjacent areas (Cahoon, 2006), or shrinkage and swelling of the soil volume in response to tides, rainfall (or lack of) and groundwater depth fluctuations (e.g. Rogers and Saintilan, 2009). Shrinkage of soil volumes in response to longer-term perturbations, such as drought, may affect the capacity of wetlands to maintain their position within the tidal prism and adapt to sea-level rise. This may become particularly apparent should climate change projections of increased frequency and intensity of drought events hold true for some regions of the world.

10.3.7 Impact of future climate and sea-level change

Drivers of salt-marsh response to sea-level rise occur at a range of temporal and spatial scales. Climate change induced alterations to these drivers of elevation change are likely to have a significant influence on the future response of salt marshes to climate and sea-level change.

Alterations to the tidal regime associated with sea-level rise may alter relationships between tides, sedimentation and salt-marsh elevations. In addition, enhanced productivity and associated elevation increases from enhanced inorganic sediment trapping and *in situ* organic matter accumulation may also relate to inundation (Mudd et al., 2009). Engineering structures, such as levee banks, flood gates and culverts, further complicate the relationship between tides, accretion and wetland elevations, and may influence the capacity of salt marshes to respond to future sea-level rise through sedimentation and plant productivity (see for example the discussion on how these factors may relate to the Mississippi River Delta, USA, in Day et al., 2000). Models of wetland surface evolution suggest that salt-marsh surfaces may attain equilibrium with sea-level rise through sedimentation (e.g. Allen, 2000; Kirwan and Murray, 2007). Increased inundation frequency and duration associated with future sea-level rise may promote self-adaptation of salt marsh through elevation adjustment that is facilitated through the delivery of organic and mineralogic material or *in situ* processes that build elevation such as plant productivity, providing that salt-marsh elevation has sufficient time to equalize with water levels (Kirwan and Guntenspergen, 2010). Rates of sediment accumulation may also vary spatially in response to sediment source proximity, tidal range, the dominance of ebb- or flood-tides and marsh tendency towards maintaining an equilibrium marsh height (Friedrichs and Perry, 2001). Despite this research, the elevation response of salt marshes to high rates of sea-level rise is poorly understood and research is now being directed towards establishing the threshold of resilience for salt marshes, and to determine the rate of sea-level rise at which salt marshes become unstable due to insufficient sediment supply and organic matter accumulation.

Episodic events, such as hurricanes and storms, may influence the response of salt marsh to future sea-level rise. Mechanisms by which episodic storm events affect soil elevations are summarized by Cahoon (2006) and may include substrate disruption and sediment redistribution, loss of soil organic matter through tree mortality from high winds, lightning strikes and hail damage, delivery of sediment from terrigenous sources, compaction of soil volume from large storm tides, and changes in organic matter content of the soil volume due to incursions of seawater, which negatively influence growth, or estuarine flushing with freshwater, which may enhance plant productivity. As the intensity and frequency of these episodic events are projected to increase in association with climate change (see Chapter 5), the impact of these events on the ecology and geomorphology of salt marshes should be considered when determining the future resilience of salt marshes to climate and sea-level change. Similarly, there is some concern that enhanced intensity and frequency of long-term climatic perturbations (for example drought), as projected by the Intergovernmental Panel on Climate Change (IPCC), may affect the relationship between salt-marsh elevation and water levels, and ultimately influence the resilience of salt marshes to sea-level rise.

Salt-marsh resilience to sea-level rise is largely dependent upon the relationship between elevation, sedimentation (both allochthonous and autochthonous), and relative sea-level change. Analyses of the stratigraphy and chronology of Holocene coastal sedimentary sequences have been used to guide the development of salt-marsh morphodynamic models and provide insight into the resilience of coastal wetlands to projected sea-level rise in the 21st century. There appear to be many examples where sediment accumulation kept pace with Holocene sea-level rise and promoted salt-marsh development (Allen, 2000).

Considerable effort has been directed towards developing models that simulate salt-marsh morphology. The model of marsh morphodynamics developed by Allen (2000) attempts to integrate mineral and organic sediment contribution, sea-level rise and autocompaction of soils. This model equates the change in elevation of a marsh surface over a time step to the sum of mineral and organic sediment input, the change in mean sea level, and the height change attributed to autocompaction of sediments. Allen (2000) simulated marsh morphology under a range of sea-level conditions. These simulations were based upon the empirical work of Pethick (1981), who identified the negative feedback loop between accretion, elevation and hydroperiod (described earlier, see Fig. 10.10), and that the rate of mineral sediment input

into the modelled marsh decreased exponentially with elevation. Due to a lack of knowledge regarding plant productivity and decay across a marsh platform, early simulations incorporated a constant value of organic sediment input. This model satisfactorily established situations where a marsh surface may accrete up to the highest tide level (highest astronomical tide, HAT). Beyond this level, the mineral and organic sediment contribution from the marsh ceases, and the marsh surface undergoes an ecological transformation to freshwater vegetation, beneath which peat accumulates (Fig. 10.13). The stratigraphy of marshes throughout much of Europe and parts of North America indicates a sequence of interbedded peat layers, which record past episodes of alternation between salt marsh and 'fen' (as the freshwater wetlands dominated by birch and other wetland plants are called in eastern England), or other freshwater wetlands.

Figure 10.13 hypothesizes several different responses to various scenarios of sea-level change and human influence (e g. land reclamation and bank stabilization). The rate of accretion decreases as the marsh surface gets higher within the tidal range (a process of negative feedback because tidal inundation is less frequent and less sediment is imported to this elevation). If the level of the sea rises, the gradual accretion of the upper marsh can continue, and the marsh communities will not be replaced by freshwater wetlands underlain by peat. If the sea level fluctuates, then this may be recorded by episodes of fen peat accumulation, resembling those apparent in the stratigraphy. In many parts of the UK and northern Europe, embankments have been built and salt marshes 'reclaimed' from the sea (Allen, 2000). These marshes are now threatened by relative sea-level rise and erosion at lower elevations. In many cases, migration to higher elevations is limited by artificial structures, such as sea walls, which are designed to protect land from erosion or inundation. Coastal squeeze is the term used to describe this trend of loss at lower elevations and limited opportunity for migration to higher elevations. 'Managed realignment' has been adopted as a planning policy (as discussed further in Chapter 11), allowing tidal waters back into some former marshes so that they can revert to salt marshes, or providing opportunity for landward migration of marshes (Doody, 2013).

Analyses of resilience are now largely focussed on non linear-feedbacks between inundation, plant productivity, sediment accumulation (allochthonous and autochthonous) and salt-marsh elevations, and the incorporation of these into simulation models of response to sea-level changes (e.g. Kirwan et al., 2010). Results generally agree that the maximum rate of sea-level rise at which salt marshes maintain a morphodynamic balance are largely dependent on tidal range and suspended sediment

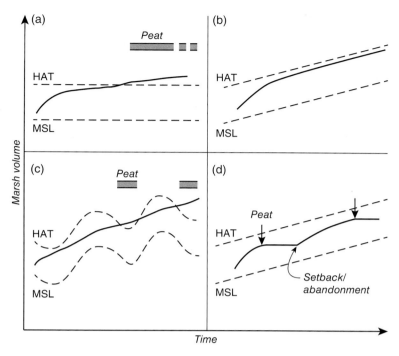

Fig. 10.13 Hypothesised salt-marsh morphodynamic responses to a range of sea-level changes. (a) Sea level (and other external boundary conditions) remain unchanged, and the salt-marsh surface accretes asymptotically until the substrate has built up to highest astronomical tide (HAT), and peat then accumulates beneath a 'fen' freshwater vegetation above the reach of salt water. (b) Constant sea-level rise; the salt-marsh substrate accretes, but continues to accrete at rates equivalent to sea-level rise and is not replaced by more landward 'fen' vegetation and a peaty substrate. (c) Irregular sea-level fluctuations through time; salt marsh accretes but the surface is above HAT at times, enabling peat to accumulate beneath freshwater vegetation that established above the muddy marsh substrate. (d) Human influence; an embankment is constructed as part of a land claim, after which the substrate ceases to accrete, until further sea-level rise overtops the first embankment, requiring setback and construction of a second embankment, or abandonment of the reclamation. Future sea-level rise threatens to overtop many such embanked and 'reclaimed' marshes in Europe and elsewhere, and a policy of 'managed realignment', by which tidal waters are allowed back into former salt marshes, is being trialled in several sites in the UK and Europe. MSL, mean sea level. (Source: Adapted from Allen 2000 and Woodroffe 2003.)

concentrations or sediment supply (French, 2006; Kirwan et al., 2010), with small influences driven by carbon dioxide concentrations, freshwater inputs, and pollutant inputs (e.g. Cherry et al., 2009). In accordance with this approach, Kirwan et al. (2010) indicate that salt-marsh resilience may be exhausted under rates of sea-level rise in the order of a few millimetres per year under conditions of low suspended sediment concentrations (~1–10 mg/l), while wetlands may adapt to sea-level rise in the order of several centimetres per year when suspended sediment concentrations are high (~30–100 mg/l). At rates of sea-level rise exceeding 20 mm/yr, only wetlands with tidal ranges exceeding 3 m and high suspended sediment concentrations (>30 mg/l) are likely to survive, while more typical wetlands with suspended sediment concentrations of 30 mg/l and tidal ranges of 1 m are projected to be drowned and replaced by habitats that represent lower intertidal or subtidal settings by the end of the 21st century.

10.4 Human influences

Despite our incomplete knowledge of tidal-flat and salt marsh behaviour and dynamics, these systems have been extensively modified by human activity. Many areas in northwestern Europe have undergone land claim, where embankments have been built to exclude the tide and 'reclaim' salt marsh and sometimes areas of tidal flat also. The extensive reclamation of the Dutch Zuiderzee has been the most extensive, although these have now been surpassed by Saemangeum, at the mouth of the Dongjin and Mangyeong Rivers in western Korea, where construction of a 33-km long seawall, begun in 1991, was completed in 2010, reclaiming 400 km² (Fig. 10.14). Other tidal flats around the world are increasingly threatened by declining river inputs of sediment, pollution, and other impacts.

Human manipulation of salt marshes is reported from as early as the Middle Ages, when they were used for fishing

Fig. 10.14 The 33-km long seawall constructed on Saemangeum estuary and completed in 2010. The engineering structure will reclaim 400 km² estuarine land through drainage and infilling. (Source: NASA Earth Observatory, http://earthobservatory.nasa.gov/IOTD/view.php?id=7688). For colour details, please see Plate 24.

and grazing. Since this time, humans have continued to use, and sometimes abuse, salt marshes on a large scale. A review of human-driven change of salt-marsh ecosystems by Gedan et al. (2009) identified seven categories of human impact on salt marshes:

(1) **Resource exploitation and extraction**: Marshes have been widely exploited for the fodder they provide for grazing livestock. Salt-marsh plants have also been used as animal bedding, thatch, rope, packing for pottery, insulation and mulch. Due to the apparent resilience of salt marshes to grazing, it has been proposed that they may have an evolutionary history that included grazing.

(2) **Land conversion**: Marshes have been converted to land for agricultural use; however, over the past century, conversion has increasingly occurred to accommodate urban expansion. Conversion for agricultural use largely involves the construction of ditches to promote soil drainage and construction of engineering structures (e.g. dykes) to mitigate flooding and inundation. Sulphide-rich soils that oxidize as a result of drainage can cause significant soil acidification, and acid drainage, as discussed earlier, can impact the health of biota. In addition, dykes can disrupt salinity gradients and cause significant shifts in ecology. Salt marshes have been, and continue to be, used for the production of salt, with conversion of flats to evaporative ponds. Conversion of salt marshes for urban and industrial expansion has largely resulted from filling with waste and soil, resulting in the loss of large salt-marsh areas.

(3) **Species introductions and invasive species**: As salt-marsh plants are tolerant of high salinities and waterlogging, many salt-marsh plants have been introduced to coastal areas, and in some cases this has caused significant displacement of native and endemic species and shifts in the ecological function of salt marshes. Non-native animals have also been introduced via ballast waters from ships.

(4) **Hydrological alterations**: Hydrological alterations attributed to ditching and restricting tidal exchange influence soil anaerobic conditions, salinity gradients and distribution of salt-marsh species. Ditching, or the establishment of networks of drainage ditches, has been undertaken to promote salt-marsh plant production for grazing and to control mosquito populations. The practice favours high-marsh biota due to the increased drainage, soil aeration and reduced salinity. Tidal restrictions limit tidal exchange upriver, creating brackish to freshwater conditions and causing community structure changes.

(5) **Pollution and eutrophication**: As salt marshes are depositional environments, they accumulate suspended sediments and associated nutrients and metals. These pollutants are readily absorbed and transformed by salt-marsh vegetation or precipitated into soils as metal sulphides. As most salt-marsh plants are nitrogen limited, nutrient addition may have a significant influence on plant productivity, alter the competitive interactions between species, and influence the zonation of salt marshes.

(6) **Changes in consumer control**: Relatively recent research is highlighting the influence of consumers on salt marshes. Human activity has been associated with the increased abundance of consumers, such as snow geese, marsh snail, and crabs. These enhanced populations may selectively consume salt-marsh resources, leading to salt-marsh denudation or ecosystem collapse.

(7) **Climate change**: Sea-level rise and increased storm intensity are the impacts of greatest concern in terms of how salt-marsh morphology will respond to climate change (as discussed earlier). However, climate change will also alter the global carbon cycle and carbon sequestration potential of salt marshes. Enhanced levels of atmospheric carbon dioxide will influence salt-marsh plant species differentially, with C_3 photosynthesizing plants likely to exhibit an enhanced growth response. Temperature changes may alter salt-marsh biogeography and enhance competition for resources between salt-marsh and mangrove species.

Human influences on salt marshes have altered the structure and function of salt marshes globally and limited the ecosystem services they provide. Management is now being directed towards minimizing human influences, ameliorating salt-marsh degradation, and restoring salt-marsh ecosystem services.

10.5 Summary

Tidal flats, salt marsh and other wetlands that are associated with them were regarded as wastelands in the early 20th century, and they were reclaimed, degraded and polluted. Towards the latter part of the century, their ecological values were increasingly realized, and there was greater emphasis on research and management of these productive ecosystems. The sedimentary processes by which mud is transported and deposited and the role that macrophytic vegetation and other biota play are the focus of ongoing research, and it is becoming increasingly clear that there are subtle feedback processes and patterns of system behaviour that have consequences beyond the local scale. Such systems appear especially vulnerable in the face of ongoing, and future accelerated, sea-level rise. Their sediments contain an incomplete record of past adjustments to former boundary conditions, and more detailed study of these, together with monitoring and modelling, offer many opportunities for further research to better understand these fascinating coastlines before they suffer further degradation, or complete loss from some coastal settings.

Key publications

Adam, P., 1990. *Saltmarsh Ecology*. Cambridge University Press, Cambridge, UK.
[*Comprehensive review of salt-marsh ecology, which is still largely relevant*]

Allen, J.R.L., 2000. Morphodynamics of Holocene salt marshes: a review sketch from the Atlantic and Southern North Sea coasts of Europe. *Quaternary Science Reviews*, **19**, 1155–1231.
[*Readily accessible and comprehensive review of morphodynamics of European salt marshes*]

Allen, J.R.L. and Pye, K., 1992. *Saltmarshes: Morphodynamics, Conservation, and Engineering Significance*. Cambridge University Press, Cambridge, UK.
[*Excellent introductory account of salt-marsh morphodynamics and their significance*]

Chapman, V., 1974. *Salt Marshes and Salt Deserts of the World*. Cramer, Lehre, Germany.
[*Early global account of salt marshes*]

Dyer, K.R., Christie, M.C. and Wright, E.W., 2000. The classification of intertidal mudflats. *Continental Shelf Research*, **20**, 1039–1060.
[*Readily accessible publication regarding intertidal classification*]

Perillo, G., Wolanski, E., Cahoon, D. and Brinson, M.M., 2009. *Coastal Wetlands: An Integrated Ecosystem Approach*. Elsevier, Amsterdam, the Netherlands.
[*A recent synthesis of coastal wetland knowledge, including ecosystems, physical processes, restoration and management, wetland dynamics and sustainability*]

Saintilan, N., 2009. *Australian Saltmarsh Ecology*. CSIRO Publishing, Collingwood, Australia.
[*A recent review of Australian salt-marsh research, including distribution, geomorphology and ecology*]

Semeniuk, V., 2005. Tidal flats. In: M.L. Schwartz (Ed.), *Encyclopedia of Coastal Science*. Springer, Dordrecht, the Netherlands, pp. 965–975.
[*Excellent overview of tidal-flat geomorphology and hydrology*]

References

Adam, P., 1990. *Saltmarsh Ecology*. Cambridge University Press, Cambridge, UK.

Adam, P., 2009. Australian saltmarshes in global context. In: N. Saintilan (Ed.), *Australian Saltmarsh Ecology*. CSIRO Publishing, Collingwood, Australia, pp. 1–21.

Allen, J.R.L., 2000. Morphodynamics of Holocene salt marshes: a review sketch from the Atlantic and southern North Sea coasts of Europe. *Quaternary Science Reviews*, **19**, 1155–1231.

Allen, J.R.L. and Haslett, S.K., 2002. Buried salt-marsh edges and tide-level cycles in the mid-Holocene of the Caldicot Level (Gwent), South Wales, UK. *The Holocene*, **12**, 303–324.

Alongi, D.M., 1998. *Coastal Ecosystem Processes*. CRC Press LLC, Boca Raton, FL, USA.

Amos, C.L., 1995. Siliclastic tidal flats. In: G.M.E. Perillo (Ed.), *Geomorphology and Sedimentology of Estuaries*. Developments in Sedimentology. Elsevier, Amsterdam, the Netherlands, pp. 273–306.

Ayres, D.R., Smith, D.L., Zaremba, K., Klohr, S. and Strong, D.R., 2004. Spread of exotic cordgrasses and hybrids (*Spartina* sp.) in the tidal marshes of San Francisco Bay, California, USA. *Biological Invasions*, **6**, 221–231.

Barbier, E.B., Hacker, S.D., Kennedy, C., Koch, E.W., Stier, A.C. and Silliman, B.R., 2011. The value of estuarine and coastal ecosystem services. *Ecological Monographs*, **81**, 169–193.

Bearman, J.A., Foxgrover, A.C., Friedrichs, C.T. and Jaffe, B.E., 2010. Spatial trends in tidal flat shape and associated environmental parameters in South San Francisco Bay. *Journal of Coastal Research*, **26**, 342–349.

Boersma, J.R. and Terwindt, J.H.J., 1981. Neap-spring tide sequences of intertidal shoal deposits in a mesotidal estuary. *Sedimentology*, **28**, 151–170.

Bricker-Urso, S., Nixon, S., Cochran, J., Hirschberg, D. and Hunt, C., 1989. Accretion rates and sediment accumulation in Rhode Island salt marshes. *Estuaries and Coasts*, **12**, 300–317.

Cahoon, D., 2006. A review of major storm impacts on coastal wetland elevations. *Estuaries and Coasts*, **29**, 889–898.

Cahoon, D.R., Lynch, J.C., Hensel, P. et al., 2002a. High-precision measurements of wetland sediment elevation: I. Recent improvements to the Sedimentation-Erosion Table. *Journal of Sedimentary Research*, **72**, 730–733.

Cahoon, D.R., Lynch, J.C., Perez, B.C. et al., 2002b. High-precision measurements of wetland sediment elevation: II. The rod Surface Elevation Table. *Journal of Sedimentary Research*, **72**, 734–739.

Chapman, V., 1974. *Salt Marshes and Salt Deserts of the World*. Cramer, Lehre, Germany.

Cherry, J.A., Mckee, K.L. and Grace, J.B., 2009. Elevated CO_2 enhances biological contributions to elevation change in coastal wetlands by offsetting stressors associated with sea-level rise. *Journal of Ecology*, **97**, 67–77.

Chmura, G.L., Anisfeld, S.C., Cahoon, D.R. and Lynch, J.C., 2003, Global carbon sequestration in tidal, saline wetland soils. *Global Biogeochemical Cycles*, **17**, 1111.

Christie, M.C., Dyer, K.R. and Turner, P., 1999. Sediment flux and bed level measurements from a macro tidal mudflat. *Estuarine Coastal and Shelf Science*, **49**, 667–688.

Day, J., Britsch, L., Hawes, S., Shaffer, G., Reed, D. and Cahoon, D., 2000. Pattern and process of land loss in the Mississippi Delta: a spatial and temporal analysis of wetland habitat change. *Estuaries and Coasts*, **23**, 425–438.

Day, J.W., Boesch, D.F., Clairain, E.J. et al., 2007. Restoration of the Mississippi Delta: lessons from Hurricanes Katrina and Rita. *Science*, **315**, 1679–1684.

Doody, J.P., 2013. Coastal squeeze and managed realignment in southeast England, does it tell us anything about the future? *Ocean and Coastal Management*, **79**, 34–41.

Dyer, K.R., 1998. The typology of intertidal mudflats. In: K.S. Black, D.M. Paterson and A. Cramp (Eds), *Sedimentary Processes in the Intertidal Zone*. Geological Society of London, Special Publication, pp. 11–24.

Dyer, K.R., Christie, M.C. and Wright, E.W., 2000. The classification of intertidal mudflats. *Continental Shelf Research*, **20**, 1039–1060.

Evans, G., 1965. Intertidal flat sediments and their environments of deposition in the Wash. *Geological Society of London Quarterly Journal*, **121**, 209–241.

French, J., 2006. Tidal marsh sedimentation and resilience to environmental change: exploratory modelling of tidal, sea-level and sediment supply forcing in predominantly allochthonous systems. *Marine Geology*, **235**, 119–136.

French, J.R. and Stoddart, D.R., 1992. Hydrodynamics of salt marsh creek systems: implications for marsh morphological development and material exchange. *Earth Surface Processes and Landforms*, **17**, 235–252.

Frey, R.W. and Basan, P.B., 1985. Coastal salt marshes. In: R.A. Davis (Ed.), *Coastal Sedimentary Environments*. Springer-Verlag, New York, USA, pp. 225–301.

Friedrichs, C.T. and Perry, J.E., 2001. Tidal salt marsh morphodynamics: a synthesis. *Journal of Coastal Research*, Special Issue, **27**, 7–37.

Gedan, K.B., Silliman, B.R. and Bertness, M.D., 2009. Centuries of human-driven change in salt marsh ecosystems. *Annual Review of Marine Science*, **1**, 117–141.

Gehrels, W.R., 1999. Middle and Late Holocene sea-level changes in eastern Maine reconstructed from foraminiferal saltmarsh stratigraphy and AMS 14C dates on basal peat. *Quaternary Research*, **52**, 350–359.

Gray, A.J., 1992. Saltmarsh plant ecology: zonation and succession revisited. In: J.R.L. Allen and K. Pye (Eds), *Saltmarshes: Morphodynamics, Conservation and Engineering Significance*. Cambridge University Press, Cambridge, UK, pp. 63–79.

Guntenspergen, G.R., Cahoon, D.R., Grace, J. et al., 1995. Disturbance and recovery of the Louisiana coastal marsh landscape from the impacts of Hurricane Andrew. *Journal of Coastal Research*, Special Issue, **21**, 324–339.

Kangas, P.C. and Lugo, A.E., 1990. The distribution of mangroves and saltmarsh in Florida. *Tropical Ecology*, **1**, 32–39.

Kirby, R., 2000. Practical implications of tidal flat shape. *Continental Shelf Research*, **20**, 1061–1077.

Kirby, R., 2010. Distribution, transport and exchanges of fine sediment, with tidal power implications: Severn Estuary, UK. *Marine Pollution Bulletin*, **61**, 21–36.

Kirwan, M.L. and Guntenspergen, G.R., 2010. Influence of tidal range on the stability of coastal marshland. *Journal of Geophysical Resesearch*, **115**, F02009.

Kirwan, M.L. and Murray, A.B., 2007. A coupled geomorphic and ecological model of tidal marsh evolution. *Proceedings of the National Academy of Sciences*, **104**, 6118–6122.

Kirwan, M.L., Guntenspergen, G.R., D'Alpaos, A., Morris, J.T., Mudd, S.M. and Temmerman, S., 2010. Limits on the adaptability of coastal marshes to rising sea level. *Geophysical Research Letters*, **37**, L23401.

Kirwan, M.L., Murray, A.B., Donnelly, J.P. and Corbett, D.R., 2011. Rapid wetland expansion during European settlement and its implication for marsh survival under modern sediment delivery rates. *Geology*, **39**, 507–510.

Le Hir, P., Monbet, Y. and Orvain, F., 2007. Sediment erodability in sediment transport modelling: can we account for biota effects? *Continental Shelf Research*, **27**, 1116–1142.

Liu, X.J., Gao, S. and Wang, Y.P., 2011. Modeling profile shape evolution for accreting tidal flats composed of mud and sand: a case study of the central Jiangsu coast, China. *Continental Shelf Research*, **31**, 1750–1760.

Mariotti, G. and Fagherazzi, S., 2010. A numerical model for the coupled long-term evolution of salt marshes and tidal flats. *Journal of Geophysical Research*, **115**, F01004.

Masselink, G. and Short, A.D., 1993. The effect of tide range on beach morphodynamics and morphology: a conceptual beach model. *Journal of Coastal Research*, **9**, 785–800.

McLeod, E., Chmura, G.L., Bouillon, S. et al., 2011. A blueprint for blue carbon: toward an improved understanding of the role of vegetated coastal habitats in sequestering CO_2. *Frontiers in Ecology and the Environment*, **9**, 552–560.

McKee, K.L. and Cherry, J.A., 2009. Hurricane Katrina sediment slowed elevation loss in subsiding brackish marshes of the Mississippi River Delta. *Wetlands*, **29**, 2–15.

McKee, K.L., Mendelssohn, I.A. and Materne, M.D., 2004. Acute salt marsh dieback in the Mississippi River deltaic plain: a drought-induced phenomenon? *Global Ecology and Biogeography*, **13**, 65–73.

Morris, J.T., Sundareshwar, P.V., Nietch, C.T., Kjerfve, B. and Cahoon, D.R., 2002. Responses of coastal wetlands to rising sea-levels. *Ecology*, **83**, 2869–2877.

Mudd, S.M., Howell, S.M. and Morris, J.T., 2009. Impact of dynamic feedbacks between sedimentation, sea-level rise, and biomass production on near-surface marsh stratigraphy and carbon accumulation. *Estuarine, Coastal and Shelf Science*, **82**, 377–389.

Nyman, J.A., DeLaune, R.D., Roberts, H.H. and Patrick, W.H.J., 1993. Relationship between vegetation and soil formation in a rapidly submerging coastal marsh. *Marine Ecology Progress Series*, **96**, 269–279.

O'Brien, D.J., Whitehouse, R.J.S. and Cramp, A., 2000. The cyclic development of a macrotidal mudflat on varying timescales. *Continental Shelf Research*, **20**, 1593–1619.

Paterson, D.M., Aspden, R.J. and Black, K.S., 2009. Intertidal flats: ecosystem functioning of soft sediment systems. In: G. Perillo, E. Wolanski, D. Cahoon and M.M. Brinson (Eds), *Coastal Wetlands: An Integrated Ecosystem Approach*. Elsevier, Amsterdam, the Netherlands, pp. 317–343.

Pethick, J.S., 1981. Long-term accretion rates on tidal salt marshes. *Journal of Sedimentary Research*, **51**, 571–577.

Postma, H., 1961. Transport and accumulation of suspended matter in the Dutch Wadden Sea. *Netherlands Journal of Sea Research*, **1**, 148–190.

Reed, D. J., 1990. The impact of sea-level rise on coastal salt marshes. *Progress in Physical Geography*, **14**, 465–481.

Rogers, K. and Saintilan, N., 2009. Relationships between surface elevation and groundwater in mangrove forests of southeast Australia. *Journal of Coastal Research*, **24**, 63–69.

Rogers, K., Wilton, K.M. and Saintilan, N., 2006. Vegetation change and surface elevation dynamics in estuarine wetlands of southeast Australia. *Estuarine, Coastal and Shelf Science*, **66**, 559–569.

Saintilan, N. and Williams, R.J., 1999. Mangrove transgression into saltmarsh environments in south-east Australia. *Global Ecology and Biogeography*, **8**, 117–124.

Saintilan, N., Rogers, K. and Mckee, K., 2009. Salt-marsh-mangrove interactions in Australia and the Americas. In:

G.M.E. Perillo, E. Wolanski, D.R. Cahoon and M.M. Brinson (Eds), *Coastal Wetlands: An Integrated Ecosystem Approach*. Elsevier, Amsterdam, the Netherlands.

Schoot, P.M. and de Jong, J.E.A., 1982. *Sedimentatie en erosiemetingen met behulp van de Sedi-Eros-Tafel (Set)*. Notitie DDMI-82.401. Rijkswaterstaat, The Hague, the Netherlands.

Semeniuk, V. 2005. Tidal flats. In: M.L. Schwartz (Ed.), *Encyclopedia of Coastal Science*. Springer, Dordrecht, the Netherlands, pp. 965–975.

Stevens, P.W., Fox, S.L. and Montague, C.L., 2006. The interplay between mangroves and saltmarshes at the transition between temperate and subtropical climate in Florida. *Wetlands Ecology and Management*, **14**, 435–444.

Tulp, I. and de Goeij, P., 1994. Evaluating wader habitats in Roebuck Bay (North-western Australia) as a springboard for northbound migration in waders, with a focus on great knots. *Emu*, **94**, 78–95.

Woodroffe, C.D., 2003. *Coasts: Form, Process and Evolution*. Cambridge University Press, Cambridge, UK.

Woodroffe, C.D. and Grindrod, J., 1991. Mangrove biogeography: the role of Quaternary environmental and sea-level fluctuations. *Journal of Biogeography*, **18**, 479–492.

Woodwell, G.M., Rich, P.H. and Hall, C.A.S., 1973. Carbon in estuaries. In: G.M. Woodwell and E.V. Pecan (Eds), *Carbon and the Biosphere*. US Atomic Energy Commission, Springfield, VA, USA, pp. 221–240.

Zedler, J., 1980. Algal mat productivity: comparisons in a salt marsh. *Estuaries and Coasts*, **3**, 122–131.

11 Mangrove Shorelines

COLIN D. WOODROFFE[1], CATHERINE E. LOVELOCK[2] AND KERRYLEE ROGERS[1]

[1]*School of Earth and Environmental Sciences, University of Wollongong, Wollongong, NSW, Australia*
[2]*The School of Biological Sciences, The University of Queensland, St Lucia, QLD, Australia*

11.1 Introduction

Mangroves are trees or shrubs that occur in the upper intertidal zone on many low-energy tropical shorelines. Globally they cover 137,760 km² (Giri et al., 2011). Salt marshes and other coastal wetlands may occur landwards of mangrove vegetation, and seagrass may be extensive seawards. Mangroves are not a single taxonomic group, but comprise a diverse range of plants with adaptations enabling survival in this otherwise inhospitable saline and anaerobic environment. Mangrove forests are highly productive ecosystems that support both terrestrial and marine biodiversity. They are important habitats for fish and crustaceans on which humans are dependent. They also provide many other ecosystem services; both direct, in terms of timber and fuel; and indirect, by supporting biodiversity, providing physical protection of coasts, retaining sediments, and regulating nutrient and carbon exchange between terrestrial and marine environments.

Mangrove forests are best developed where extensive near-horizontal topography occurs close to sea level. They cover substantial areas where there is a large tidal range; however, there are instances where isolated stands of mangroves persist inland where they are not influenced by tides. Wave energy has to be sufficiently low to allow establishment and growth of plants, but mature forests also act to attenuate wave energy.

11.2 Mangrove adaptation in relation to climate zones

Mangroves show distinctive adaptations to intertidal environments. Most mangroves have evolved mechanisms to tolerate salt, with some able to withstand salt concentrations in soils that are three times that of seawater. Salt tolerance is associated with particularly efficient use of water during photosynthesis and highly effective hydraulic systems (Ball, 1998; Lovelock et al., 2006). The majority of mangrove species have root systems that transport oxygen, enabling plants to respire and perform key metabolic processes (e.g. nutrient and water uptake

(a)

(b)

(c)

Fig. 11.1 (a) Mangroves show a range of adaptations to the inhospitable intertidal environment in which they flourish, and that contribute to the ecosystem services they provide. These include distinctive root systems, such as the prop roots of *Rhizophora*, seen in this stand of *Rhizophora* on Low Isles on the Great Barrier Reef, Australia. (b) Vivipary is a significant adaptation as a result of which mangroves are able to establish and expand in muddy intertidal environments. The fruit of several of the Rhizophoraceae, such as *Bruguiera*, germinate while still on the tree. When they then detach, they are able to float and establish in favourable habitats, rapidly developing rootlets that anchor them to the mud. (c) The southernmost mangroves in the world occur at Corner Inlet in Victoria, Australia. Here, *Avicennia* is stunted, reaching little more than a metre in height, and merges into salt-marsh vegetation, which is seen in the foreground. (Source: Photographs by (a) Catherine Lovelock. (b) Colin Woodroffe. (c) Roland Gehrels.) For colour details, please see Plate 25.

and salt exclusion) despite being anchored in saturated, non-porous soils depleted in oxygen. Above-ground root systems include pneumatophores, prop roots and buttresses (Fig. 11.1a), some of which provide structural support, and most of which are covered with lenticels that promote gas exchange. Many have viviparous propagules (Fig. 11.1b); for example, in many genera, seeds remain attached and germinate on the tree and then are buoyant during an aquatic dispersal phase (Tomlinson, 1986).

Mangroves are predominantly tropical, although they extend into sub-tropical and temperate regions along the eastern coasts of major continents where warm poleward-flowing ocean currents ameliorate low temperatures. Although salt marsh does occur in the tropics, it generally represents a minor component in comparison with mangroves, especially in the wet tropics, but becomes increasingly important towards the poleward limits of mangroves and in arid regions (Fig. 11.1c). Mangroves occur on arid coasts, but are limited by hypersaline groundwaters (Semeniuk, 1983). Temperature controls distribution in a broad sense; mean air temperatures of the coldest month must be at least 20 °C. Exposure to winter frosts constrains the latitudinal limit of mangrove distribution (Morrisey et al., 2010).

11.3 Mangrove biogeography

The global distribution of mangroves comprises two provinces, one centred in the West Indies and the other extending across the Indian Ocean and the western Pacific. The greatest diversity lies in Indonesia and northern Australia, where there are more than 30 species (Duke et al., 1998). Four tree species characterize West Indian mangroves. *Rhizophora mangle* (red mangrove) dominates the lower intertidal zone; *Avicennia germinans* (termed the black mangrove) occurs landward of *Rhizophora*; and *Laguncularia* and *Conocarpus* are found at the landward margin. New World mangrove associations include the succulent *Batis maritima* and the fern *Acrostichum aureum*. Grasses or sedges may cover freshwater wetlands landward of the mangroves, such as in the Everglades of Florida, USA. Mangrove wetlands in Central and South America are slightly more floristically complex. The genus *Pelliciera* occurs in wetter areas, and *Mora oleifera* is found upstream in estuaries. At the southern limit in Brazil, *Avicennia schaueriana* replaces *A. germinans*, and *Laguncularia* is often found as the seaward mangrove (Cunha-Lignon et al., 2009).

There are several species of *Rhizophora* in the Indo-Pacific, and also several other genera in the Rhizophoraceae family, such as *Ceriops* and *Bruguiera*. The patterning of mangrove species is complex and was first described in detail by Macnae (1968). There is often a seaward zone of *Sonneratia*, then a prominent zone of *Rhizophora*, frequently transitioning into *Ceriops* and *Bruguiera*, and a landward zone of other mangroves such as *Lumnitzera* and *Excoecaria*. There are also several species of *Avicennia*, the genus that demonstrates the broadest tolerance to environmental factors, especially salinity, as it is able to grow throughout the upper intertidal zone. Brackish communities of *Nypa fruticans* and *Heritiera littoralis* often dominate the transition to inland vegetation, including peat swamp forest, as in the Mahakam delta in Borneo (Anderson and Muller 1975).

In arid and semi-arid areas, such as Western Australia, there are extensive saline flats landward of mangroves, often colonized by cyanobacterial mats, with low shrubs and samphire (comprising six genera in the Chenopodiaceae) occurring on the interface with the supratidal zone (Thom et al., 1975; Lovelock et al., 2009). In the wet-dry tropics of northern Australia there are broad convex alluvial plains landward of mangroves, covered by seasonally inundated grass- and sedgelands. In the humid tropics of much of Malaysia and Indonesia, similar plains support domed accumulations of woody peat beneath peat swamp forest (Staub and Esterle, 1994).

11.4 Zonation and succession

Mangrove forests are some of the most conspicuously 'zoned' of plant communities. Zonation is also characteristic of salt marshes (Chapter 10), indicating the similarly strong environmental gradients across the intertidal zone. However, in contrast to the low herbs and grasses of salt marshes, some of which are annuals, mangrove trees are long-lived, and zonation persists for decades, even as the environmental factors that led to the establishment of the pattern may change (Lovelock et al., 2011). Zonation is due to differing physiological tolerance of species to inundation regimes (and associated variation in soil anoxia, salinity and nutrient availability), and to the success of propagule establishment, especially as a consequence of predation by herbivores (Feller et al., 2010).

Many intertidal organisms show zonation within the intertidal zone; for example, barnacles or limpets are frequently concentrated within narrow elevational zones on a rocky shoreline. Such zones are usually static, reflecting prevailing environmental conditions and the physiological response of the organisms. In the case of mangroves, however, there has been a widely held view, predicated on their vivipary and unique aerial root systems, that mangroves are 'trees that reclaim land from the sea' (Carlton, 1974). This view was widely promulgated by Davis (1940), who observed mangrove seedlings establishing over marine carbonate sediments in Florida. The Florida mangroves are underlain by mangrove peat, and it was envisaged that the zonation (*Rhizophora –
Avicennia –* freshwater sedges) represented a succession of species, with mangroves becoming progressively replaced by a more landward, climax plant community.

This view was challenged by Egler (1952), who realized that the several-metres thickness of mangrove peat underlying the wetlands exceeded the elevational range within which mangroves occurred, and must have been deposited as sea level had risen. On this basis, he considered that mangroves had expanded landwards into the sedgelands. This view was verified by other researchers who recognized that calcareous muds beneath the peats in Florida were of freshwater (not marine) origin (Scholl, 1964), and that the lower sections of the peat were also deposited in a freshwater environment (Spackman et al., 1966). This stratigraphic sequence (freshwater marl – freshwater peat – mangrove peat) is a transgressive sequence, deposited as sea level has risen relative to the coast of southern Florida during the late Holocene (Fig. 11.2). Later work has shown

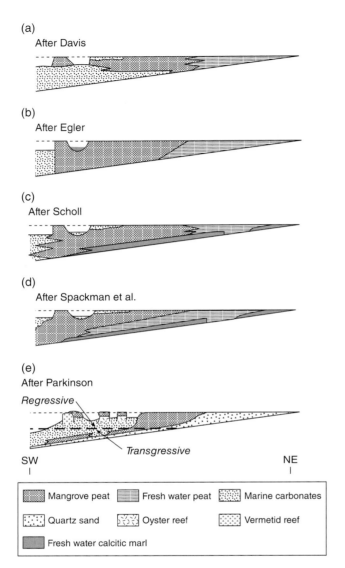

(a) After Davis

(b) After Egler

(c) After Scholl

(d) After Spackman et al.

(e) After Parkinson

Regressive

Transgressive

SW   NE

Mangrove peat   Fresh water peat   Marine carbonates

Quartz sand   Oyster reef   Vermetid reef

Fresh water calcitic marl

Fig. 11.2 Schematic representation of the stratigraphy and sequence of sedimentary environments in southwestern Florida, USA, as interpreted by successive researchers. (Source: Based on data from Davis 1940, Egler 1952, Scholl 1964, Spackman et al. 1966, and Parkinson 1989. Adapted from Woodroffe 1990. Reproduced with permission from Sage.)

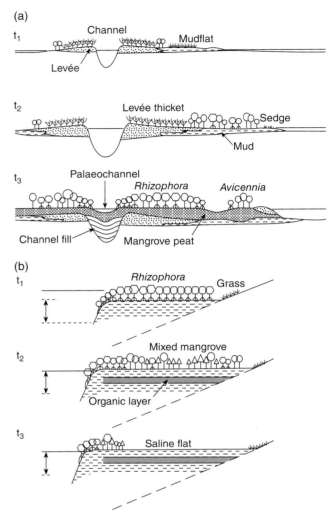

(a)

t_1 Channel   Mudflat

Levée

t_2 Levée thicket   Sedge

Mud

t_3 Palaeochannel   *Rhizophora*   *Avicennia*

Channel fill   Mangrove peat

(b)

t_1 *Rhizophora*   Grass

t_2 Mixed mangrove

Organic layer

t_3 Saline flat

Fig. 11.3 The association of mangrove habitats with landforms and the pattern of change that occurs as a consequence of the geomorphological development of deltaic-estuarine environments. (a) Three of the key stages in the evolution of distributaries in the Grijalva delta of Tabasco, Mexico. (b) Three of the key stages in the mid-late Holocene development of mangrove forests flanking the Ord River in Western Australia. Vertical arrows indicate spring tidal range. Stage 1 involves the development of side-channel or mid-channel flats that are rapidly occupied by monospecific mangrove stands, such as *Avicennia*. Stage 2 is dominated by processes of vertical accretion, and as the tidal flat approaches mean high water spring there is a reduction in tidal inundation, which is accompanied by the establishment of other mangrove species; some dieback of pioneer mangrove species may be evident and mangrove may form low shrub communities at infrequently inundated locations. By stage 3, most of the mangrove flat is above mean high water spring and mangroves have contracted to the margins of channels. (Source: (a) Adapted from Thom 1967. Reproduced with permission of John Wiley & Sons. (b) Adapted from Thom et al. 1975. Reproduced with permission of John Wiley & Sons.)

that as the rate of sea-level rise slowed, the transgressive trend gave way to a regressive trend, with mangroves colonizing carbonate banks in Florida Bay (Parkinson, 1989), the phenomenon that had triggered Davis' ideas.

In contrast to the view that mangrove zonation represents an obligate succession, Thom proposed, in a key paper on mangrove patterning in the Grijalva River delta in Tabasco, Mexico (Fig. 11.3a), that the distribution of mangrove species was an opportunistic response to landforms and microtopographic variability that influenced environmental conditions (Thom, 1967). He developed this concept

CONCEPTS BOX 11.1   Impacts of storms on mangrove shorelines

Intense tropical storms (called hurricanes in the West Indies, typhoons in Asia, and tropical cyclones in Australia) occur in many regions where there are mangroves, and have significant impacts on mangrove forests. Different species respond differently, depending on wetland characteristics (e.g. composition, health, stage of the tide, and tidal range) and particular storms (e.g. the direction of approach, wind speed, and storm-surge height). Impacts can be highly destructive, involving windthrow of trees, or defoliation. In some cases, trees remain in position; *Rhizophora*, for example, can act as a buffer because of their dense tangle of prop roots. This is an ecosystem service that is lost when mangrove forests are replaced by aquaculture ponds. Frequently, swathes of trees are toppled and the bare areas of substrate created may persist for decades without recolonization.

Severe storms, such as Cyclone Tracy that hit Darwin in 1974, can leave a legacy in the patterning of mangrove species (Woodroffe and Grime, 1999). In other cases, storms result in changes to bathymetry and hydrology, ultimately determining the plant community that can re-establish at the site (Cahoon et al., 2006). Intense storms can also be beneficial by providing inputs of freshwater, nutrient and sediments (Castañeda-Moya et al., 2010) that can enhance plant growth (Lovelock et al., 2011). Thus, storms may affect both surface (e.g. sediment delivery) and subsurface (e.g. root mortality, groundwater) processes (Cahoon et al., 2006).

An emerging view is that less-intense storms are important for sustaining the productivity of mangrove forests, which may otherwise become limited by high salinity, low nutrient availability and overwhelmed by sea-level rise. However, if the frequency of extremely intense storms increases with climate change, this may result in more drastic changes to mangrove communities in the future (Fig. 11.4).

Fig. 11.4 Devastation of mangroves on the coast of the Bay of Bengal in Bangladesh, as a consequence of Cyclone Sidr in November 2007. The stumps in the foreground are all that remains of mangrove trees, and these show that sediment has been stripped, unearthing the root system. Windthrow of mangroves in the background records the wind direction on this section of coast. (Source: Brian Jones.)

---

of the geomorphological development of mangrove habitats further through his work on the much more diverse mangrove forests along the margins of the Ord River (Fig. 11.3b) in Western Australia (Thom et al., 1975). Recognition of zones of mangrove species in the complex, species-rich mangrove forests of northern Australia, on shorelines that appear to be eroding, further supported the view that zonation of mangroves need not indicate a succession of vegetation associations advancing seaward (Semeniuk, 1980). High-resolution remote-sensing using multi-temporal photography and satellite imagery has detected changes in the extent of mangrove zones over time. In some cases, these involve mangrove expansion seaward over shallow marine mudbanks (Lucas et al., 2001), and in others, mangroves have shown substantial incursion into previously freshwater environments (Mulrennan and Woodroffe, 1998).

Replacement of one mangrove community by another over time is likely to occur in many mangrove forests, although this has rarely been observed. The life history of individual trees is of the order of several decades, making it difficult to monitor patterns of change. Only a few studies have revisited plots within mangrove forests to document the pattern of change (e.g. Devoe and Cole, 1998). Zonation that may be apparent to an observer who visits a forest has rarely been captured statistically, and may be less common than early descriptions implied (Feller et al., 2010). Evidence for long-term replacement of mangrove vegetation has been provided by analysis of pollen in sediment cores (Woodroffe et al., 1985, see later). However, even where sediment supply might appear ample, mangrove shorelines can demonstrate stability rather than vertical accretion and replacement of one community by a more landward one. The persistence of mangrove patterning over time indicates morphological resilience, as in Sherbro Bay in West Africa (Anthony, 2004), implying that such shorelines are in equilibrium with hydrodynamics and sediment transport through the system.

It is not always clear whether mangroves are successional, or whether on some coasts they represent a steady-state zoning of plant communities in response to an environmental gradient across the intertidal zone. Succession is often likely to be cyclic, where after disturbance or stand ageing, one mangrove forest is replaced by another (Lovelock et al., 2010). Succession during gap regeneration, perhaps from lightning strikes, may be an important process in the cycle of mangrove tree replacement (Duke, 2001), although gap formation may be more important for some species than others (Pinzón et al., 2003). There are also a series of disturbances that can interrupt or disrupt succession. Tropical storms are an episodic disruption (see Box 11.1). Geomorphological processes, such as the meandering of a tidal creek, can truncate a mangrove forest on an eroding outer bend, but provide new substrate suitable for colonization on accreting point bars on the inside of meanders.

11.5 Geomorphological setting and ecosystem functioning

Mangroves can occur in several geomorphological or environmental settings. Some of the most extensive mangrove regions are deltaic, associated with the extensive muddy environments that develop where large rivers bring sediment to the coast. Mangroves are often well-developed in estuarine environments where they are relatively sheltered, and are influenced primarily by tides. Mangroves may also develop in lagoonal settings or carbonate sedimentary environments (Fig. 11.5).

Mangrove forests have been classified for the purposes of describing differences in ecological function and for enhancing understanding of their responses to environmental change. Lugo and Snedaker (1974) and Lugo et al. (1976) defined five principal types of mangrove: overwash, fringe, basin, scrub and riverine (Fig. 11.6; a sixth type, hammocks (included in Fig. 11.6), was identified that is limited to isolated depressions in limestone underlying the Everglades). The classification has been widely adopted, as clear functional differences exist between the mangrove types.

• Fringing mangrove forests comprise the seaward forest zone (often dominated by *Rhizophora* or *Sonneratia*) that is flooded during each tidal cycle. These forests are often tall compared to the landward scrub stands. Fringing forests isolate the more landward habitats (scrub and basin forests) (Adame et al., 2009).

• Scrub mangrove comprises dwarfed stands, often of *Rhizophora* in the Caribbean and *Avicennia* and *Ceriops* in the Indo-Pacific, that appear to be constrained by salinity or nutrient availability due to limited tidal exchange and high evaporation (Lovelock et al., 2004). A broader range of mangrove species can adopt this growth form in the Indo-Pacific; Fig. 11.7 shows a stand of dwarfed *Ceriops*.

• Riverine mangrove forests occur along the banks of rivers, and have the largest trees and the highest productivity of all the types of mangrove forest, reflecting the low salinity of water that floods them and the high level of terrestrial inputs, which are often rich in nutrients (Adame et al., 2009).

• Overwash mangrove occurs on islands that are overtopped by the tide, as in the Bahamas and on the Belize barrier reef, or the low wooded islands of the Great Barrier Reef (Stoddart, 1980).

More recent nutrient and biogeochemical models, termed ecogeomorphic models, have recognized habitat types that are nested within a hierarchy of scales, using geomorphological settings at a regional scale, and ecological type at a more local scale (Twilley and Rivera-Monroy, 2009). Other researchers have examined how different habitats arise, and how differences in habitats are based on interactions among biological and physical components of the system (Feller et al., 2010). Such models have had application in restoring mangrove forests, assessing the goods and services that they provide, and interpreting their vulnerability to sea-level rise.

11.6 Sedimentation and morphodynamic feedback

The source and supply of sediment determine the nature of the substrate; an organic-rich mud is typical where there is both a large supply of sediment and prolific mangrove growth. By contrast, where terrestrial (clastic) sediments are limited or absent, as on isolated islands, mangroves grow on carbonate sediments, which are either biologically produced, as in the case of coral reefs (Fig. 11.7), or precipitated, as in the case of calcareous muds on the Great Bahama Bank. Successive stages of mangrove colonization and consolidation have been described in response to the evolutionary stage of reef-flat development on islands of the Great Barrier Reef (Stoddart, 1980), as well as in response to modern reef degradation (Alongi, 2008). Sediment accumulation beneath mangroves also incorporates organic matter produced by the mangroves themselves. In those systems where there is only limited inorganic sediment supplied from outside the wetland, the substrate is generally peaty, comprising the fibrous root material of mangroves. These highly organic peats characterize sediment-starved settings, such as limestone islands in the West Indies, the oceanward margin of the Everglades, and isolated islands in the Pacific Ocean.

The nature of the substrate beneath mangrove forests depends on sediment supply, comprising mud transported from outside, termed allochthonous, and *in situ* production, termed autochthonous (for example, mangrove-derived peat). Considerable volumes of mud can be advected into coastal wetlands from offshore by tidal processes. This is particularly true of macrotidal systems, such as those along the coast of northern and

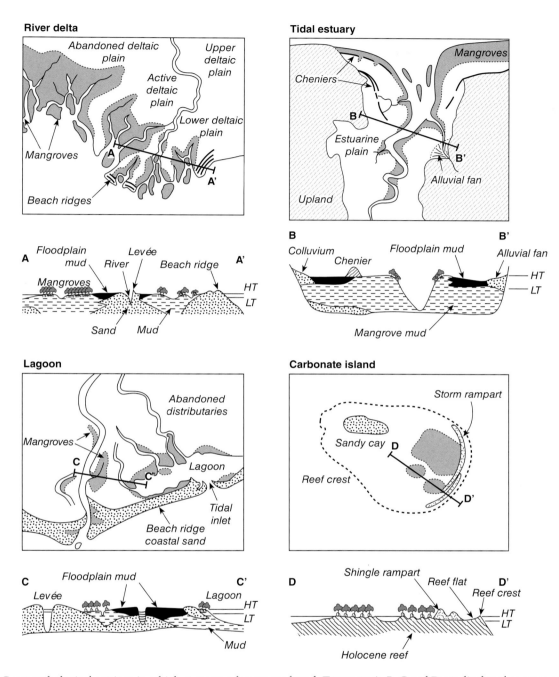

Fig. 11.5 Geomorphological settings in which mangrove forests are found. Transects A, B, C and D are displayed as cross-section diagrams. HT, high tide; LT, low tide. (Source: Adapted from Thom 1982; after Woodroffe 1992.)

northwestern Australia, where tidal flows are large enough to entrain considerable volumes of mud and carry them in suspension into mangrove forests where they are deposited when velocities are reduced (Wolanski, 2006). The effectiveness of tidal pumping is demonstrated on the Ord River, where reduction of river discharge as a result of dam construction to form Lake Argyle has been accompanied by rapid accumulation of tidal sediments in the upper part of the estuary (Wolanski et al., 2001). Accumulated sediments can be rapidly colonized by mangroves, increasing forest area (e.g. Lovelock et al., 2010).

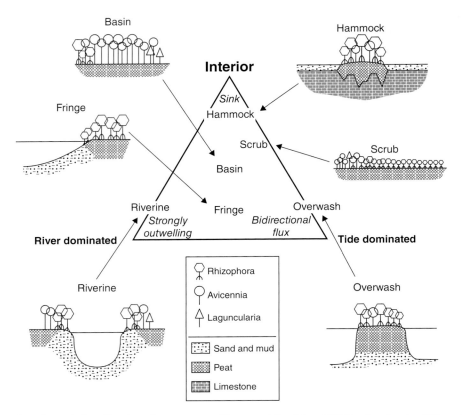

Fig. 11.6 The functional ecological classification of mangroves of Lugo and Snedaker (1974), whereby the different mangrove types are placed in a ternary context, based on river, tidal and interior influence. (Source: Adapted from Woodroffe 1992. Reproduced with permission from American Geophysical Union.)

Fig. 11.7 (a) Scrub mangrove setting – a stand of *Ceriops* in Hinchinbrook Channel, northern Queensland, Australia, in which individual tree growth is stunted. (b) Carbonate settings – mangroves are not limited to muddy continental shorelines, but also occur in carbonate environments. These mangroves have developed on a reef platform in Torres Strait, Australia. (Source: Photographs by (a) Catherine Lovelock and (b) Javier Leon.) For colour details, please see Plate 26.

Measurements of substrate accretion on the soil surface (including mineral sediments and biological accretion) in mangrove forests have not been undertaken as extensively as for salt marshes (see Chapter 10). For example, marker horizon and surface elevation tables (SET) have been used at only a few sites (Cahoon et al. 2006; see later in this chapter). However, a wide range of values is indicated (from 0 to 100mm/yr; Morrissey et al., 2010). Most sediment is trapped in fringing forests and riverine forests (Adame et al., 2009), and tolerance of high rates of sedimentation has been observed (Ellison, 1998). Tree root structure is an important factor in determining variation in sedimentation rates, with the most rapid potential rates accentuated by the presence of prop roots (Krauss et al., 2003). Accretion also showed a significant positive relationship with tidal range and duration of flooding, although the strength of this relationship varied over geomorphic settings (being strongest in embayments), suggesting significant differences in processes controlling surface elevation in different mangrove habitats (Cahoon and Lynch, 1997).

Detailed measurements of ground-surface elevation in mangroves demonstrate that surface elevation changes are only partially a record of sediment deposition, and that subsurface processes can lead to considerable variation in surface elevation both in sites with highly organic soils and those with mineral soils. In a review of the available data, Cahoon et al. (2006) found that surface accretion rates in mangroves exceeded sea-level rise, and increased with increasing rates of sea-level rise in selected coastal wetland settings in the Gulf of Mexico, the Caribbean, Central America, the Western Pacific, and Australia. Overall, mangroves with mineral soils had higher rates of surface elevation change than with peat soils, and root growth accounted for approximately 20% of increases in the height of the soil surface (Cahoon et al., 2006). A recent study in Micronesia (Krauss et al., 2010), found surface accretion rates were high (2.9 to 20.8mm/yr), but seven out of 13 sites showed mangrove soil surfaces declining in elevation (–0.6 to –5.8mm/yr), indicating significant shallow subsidence of soils likely due to compaction and biological processes.

Below-ground processes that influence surface elevation include contributions from root growth, groundwater changes within the substrate, and decomposition of organic matter. Root growth and decomposition are sensitive to environmental factors that affect plant growth (e.g. tidal inundation, nutrient availability, elevated carbon dioxide and salinity), thereby providing a strong biological signature on the adaptation of mangroves to sea-level rise. At this stage, there are insufficient studies to characterize the range of variation of sedimentation rates within a stand of mangroves, or between mangrove forests in different settings. It seems logical that sedimentation rates will increase as sea level rises, due to increased inundation frequency (Fig. 11.8). However, data are too sparse to determine whether this will give rise to stability of mangrove communities with sea-level rise, and models of the type proposed for salt marshes (see Chapter 10) have yet to be applied to mangrove shorelines.

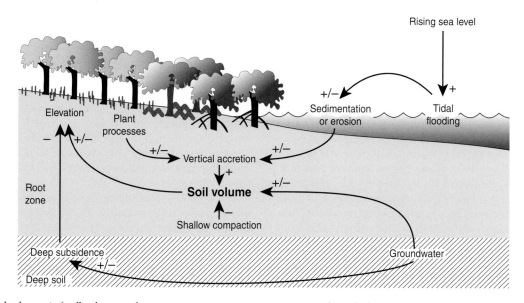

Fig. 11.8 Morphodynamic feedbacks on sedimentation in mangrove ecosystems. Relatively few studies have examined the contribution from organic and inorganic sediments, and the nature of the feedback, including the autocompaction of sediments. Many of these processes are likely to reflect the complexities that have been demonstrated for salt-marsh systems (see Chapter 10), but there have been fewer studies in mangrove forests. (Source: Adapted from Cahoon et al. 1999 with permission from Springer Science+Business Media B.V.)

11.7 Mangrove response to sea-level change

In many models, mangroves and other wetlands are considered to be in equilibrium with sea level, due to feedback processes linked to the relationship between water depth and accretion (Cahoon et al., 2006). As sea level rises, accretion increases, resulting in wetlands that keep pace with sea-level rise; conversely, if relative sea level falls (e.g. due to high rates of accretion), accretion declines, which also contributes to the equilibrium (Fig. 11.8). Some of the most recent models of the effects of sea-level rise on wetlands include both physical and biological components that represent the strong influence of the vegetation on sedimentation processes in wetlands (e.g. Kirwan and Murray, 2007). These models indicate that wetlands can 'keep-up' and even increase in area where sea-level rise is moderate and where sediment supply is sufficient. Large increases in the rate of sea-level rise and reductions in sediment supply lead to losses of wetland area, due to increased channel widths and the inability of accretion to match sea-level rise. Although these models do not take account of subsurface processes, and thus may overestimate the stability of wetlands, they indicate there may be thresholds in the ability of mangroves to keep pace with sea-level rise. Evidence of both equilibrium states and thresholds that have been exceeded can be found using palaeoenvironmental reconstructions. Additionally, palaeoenvironmental reconstruction can provide insights into the nature of the response of mangroves to sea-level rise, and may enable a clearer understanding of how such wetlands might respond to future environmental change.

Unfortunately, mangrove sediments do not preserve to the same extent as former coral reefs, so reconstruction of past mangrove shorelines is difficult. Late-glacial mangrove sediments have been identified on the broad Sahul and Sunda shelves, but these have been drowned by sea-level rise (Hanebuth et al., 2000). In shallower waters, and when rates of sea-level rise were slightly less, it seems that mangrove forests may have retreated landward. In this case, mangrove sediments evidently did not keep pace with sea-level rise, but mangrove re-established at a more landward location at higher elevation. Mangrove peat encountered on the lagoon floor of both the Belize barrier reef and the Great Barrier Reef indicates that mangrove sedimentation was unable to keep pace with sea-level rise in these situations (Stoddart, 1980; Woodroffe, 1990). Basal transgressive mangrove sediments, generally organic-rich peaty muds at the base of cores, are found beneath coastal plains in northern Australia (Woodroffe et al., 1993). The more landward of these mangrove sediments were able to keep pace with sea-level rise as it slowed prior to reaching present level, and the mangrove forests appear to have persisted into the 'big swamp' phase that characterized these estuarine plains (Fig. 11.9).

The Holocene evolution of a number of systems in northern Australia has been interpreted, primarily on the basis of stratigraphic and radiocarbon-dating analyses, with particular emphasis on depositional environments, but also including the evolution of channel form. Three phases of estuarine infill have been recognized.

(1) During the transgressive phase (8000–6800 years BP), sea-level rise drowned the valleys, resulting in the landward transgression of mangroves over terrestrial environments and the formation of broad, open embayments (Fig. 11.9). Initially, sediment supply could not keep pace with the rapid inundation of the landscape, resulting in the creation of a large area suitable for intertidal wetlands (termed accommodation space; see Chapter 1), and the catch-up transgressive sequence is recorded as a layer of organic-rich mangrove mud overlying pre-Holocene terrestrial sediments (Woodroffe et al., 1993).

(2) The big swamp phase (6800–5300 years BP) was marked by stabilizing sea level. Many estuarine systems were filled by 10–15 m of mangrove muds, indicating that the mangrove surface was able to keep up with the decelerating rate of sea-level rise over this period. This sedimentary sequence has been described from the South Alligator system in northern Australia in particular (e.g. Woodroffe et al., 1985), but has also been shown to occur in neighbouring systems and along the northern Australian coast (e.g. Thom et al., 1975; Woodroffe et al., 1993).

(3) The sinuous/cuspate phase (since 5300 years BP) is marked by a transition from mangrove to alluvial sediments as the floodplain matured and mangroves were replaced by grass and sedgeland vegetation. Pollen records indicate that as sediment accreted, lower-intertidal species (*Sonneratia*) were progressively replaced by mid-intertidal species (*Rhizophora/Ceriops*), higher-intertidal species (*Avicennia*) and freshwater floodplain species. In many cases, mangrove forests had been replaced by 4000 years BP, although in other places mangroves persisted until more recently (Clark and Guppy, 1988).

In other parts of the world, where sea level has fallen over the past few millennia, the mangrove substrate was left emergent, as shown, for example, by the highly oxidized muds found in coastal Malaysia (e.g. Geyh et al., 1979). In the case of the South Alligator and Daly rivers of northern Australia, the infilling of the mangrove swamps was also associated with the onset of channel migration throughout the estuarine and coastal plains, resulting in reworking of big-swamp sediments and lateral accretion of laminated channel margin and shoal sediments (e.g. Woodroffe et al., 1993).

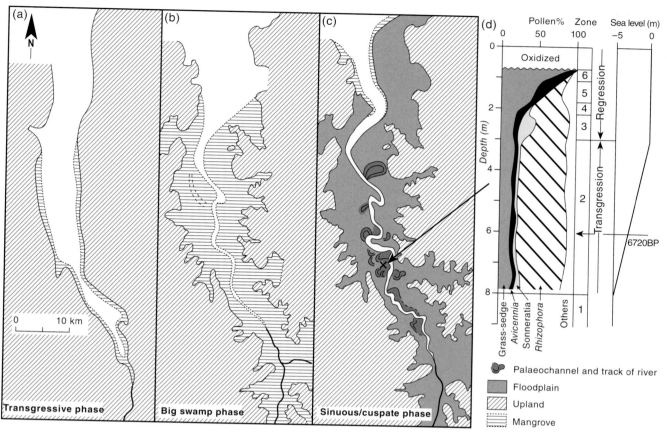

Fig. 11.9 Evolution of the South Alligator estuarine system, in the Northern Territory, Australia, during the mid- to late Holocene. (a) The prior valley was inundated around 8000 years BP during the transgressive phase; (b) extensive mangrove forests flourished across much of the area of the modern plains during the 'big swamp' phase, around 6000 years BP, as sea level stabilized; and (c) mangroves were replaced over most of the plains once the substrate had accreted beyond the intertidal zone, with deposition of alluvial clays and establishment of the grass-sedge freshwater wetland vegetation. (d) This broad long-term change is shown in a pollen diagram of the upper sediments of the plains, with the reduction in *Rhizophora* and *Avicennia* pollen and increase in Poaceae and Cyperaceae. (Source: Murray-Wallace and Woodroffe 2014. Reproduced with permission from Cambridge University Press.)

11.8 Human influences

Around much of the tropics, direct human activities in the coastal zone are already having a pronounced impact on mangrove coasts. Clearing of mangroves for aquaculture, agriculture and urban development has reduced the extent of mangroves globally by approximately 30%, exceeding rates of loss experienced by tropical forests (Spalding et al., 2010). In addition to clearing, reductions in the structural integrity of forests has also occurred as more valuable species are removed, which may also increase susceptibility to other stressors and reduce the ecosystem services provided (e.g. wave attenuation (Koch et al., 2009) and carbon sequestration (McLeod et al. 2011)).

Deforestation and land-use change in the terrestrial environment have in some instances led to increases in mangrove forest area as sediment delivered to the coast down rivers provides habitat for colonization. High sediment loads, while detrimental to subtidal tropical habitats (e.g. coral reefs and seagrass), can be beneficial to mangrove growth, and can increase the capacity for mangroves to keep pace with sea-level rise (Cahoon et al., 2006). However, high sediment loads are often associated with high nutrient loads, and while these may benefit mangrove growth, high levels of nutrients also enhance mortality in some settings (Lovelock et al., 2009).

The building of dykes, seawalls and other infrastructure in the coastal zone affects the current and future distribution of wetlands (Kirwan and Murray, 2008). Human-built barriers to upslope migration of mangroves with sea-level rise may result in large losses of mangrove area in the future due to coastal squeeze. Particularly insidious has been the destruction of mangroves in order to make way for construction of shrimp ponds (Alongi, 2002). Enormous expanses of forest have been lost to this process, in many cases for only a few years of aquaculture production before subsequent abandonment of the ponds, or their short-term use for salt production. After destruction of mangrove for these uses, it is rarely possible to re-establish mangrove forests again, due to a combination of poor management, eutrophication and acid sulphate soil development.

Mangrove forests are vulnerable to these direct human impacts, which are likely to be exacerbated by climate change and sea-level rise. However, there are also some positive aspects of human intervention in these systems. Recognition of the various goods and services that mangroves provide has led to many successful efforts to restore mangrove forests. Mangroves have been actively restored on many shorelines in

---

CASE STUDY BOX 11.2   Mangroves as protection in the event of a tsunami

On Boxing Day (26 December) in 2004, a massive earthquake off the coast of Sumatra caused a major tsunami that devastated Aceh and caused substantial damage around the coasts of Indonesia, Thailand, India and Sri Lanka, with tragic loss of life. A range of studies have implicated mangroves, as well as other coastal vegetation, as playing a significant role in dampening the effects of large waves, thereby protecting adjacent human settlements (e.g. Danielsen et al., 2005; Barbier, 2006). Comparison of the impact on villages where the natural mangrove fringe remained intact with those where it had been cleared implied that preserving mangroves had mitigated the loss (Kathiresan and Rajendran, 2005). In addition to attenuating wave energy by reducing friction, mangroves also played an important role in trapping debris and preventing people from being washed away (Paphavasit et al., 2009).

However, some of these conclusions are disputed, either on the grounds that there were insufficient data, the inadequacy of statistical analysis, or other factors, such as the relative position and elevation of settlements. Tsunami waves are quite different in period and wave energy from the wave types that have occurred in those studies that have examined attenuation, rendering the studies inconclusive (Alongi, 2008). Nevertheless, there are often other realistic expectations that lead to the restoration of mangroves, which are likely to be effective in reducing shoreline erosion during lesser wave conditions, increasing the resistance to storm-associated waves, and reducing damage to coasts (Duarte et al., 2013), such that mangrove vegetation may provide protection in the face of all but the largest tsunami (Fig. 11.10).

(a)   (b)

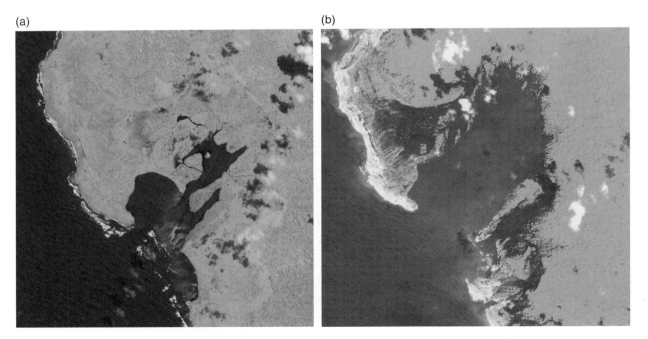

Fig. 11.10   Satellite images of the west coast of Kirchall Island in the Nicobar Islands, Indian Ocean, before (a) and after (b) the Boxing Day tsunami that devastated this coast on 26 December 2004. Extensive areas of mangrove forest were swept away on this coast by the tsunami. (Source: CNES 2004/Distribution Spot Image/Processing CRISP.) For colour details, please see Plate 27.

Fig.11.11 Mangrove forests fringing a tidal creek in northern Queensland, Australia, juxtaposed with residential properties. As sea level rises, the natural dynamics of intertidal wetlands, which would involve landwards encroachment by the upper-intertidal species, will come into increasing conflict with settlement and human constraints on land-use. This is termed coastal squeeze. (Source: Colin Woodroffe.)

the tropics for a range of purposes, from silviculture to shoreline protection (Ellison, 2000), although there remains debate about whether they provide protection against the most extreme events (see Box 11.2). As experience in establishing mangrove propagules in habitats where tidal, salinity and substrate conditions are favourable improves, there is an increasing potential to manage the distribution of these important ecosystems. Perhaps more than for most coastal systems addressed in this book, the expertise exists, or could be developed, to restore mangroves actively (e.g. Lewis, 2000), to establish future mangrove forests, and to conserve lands for future mangrove habitats on shorelines where they used to occur, or where they can establish as the sea level rises. A greater challenge revolves around whether there will be sufficient political, social, cultural or planning incentives to adopt such sustainable approaches to coastal management. Settlements and infrastructure built near the coast often means that the immediate hinterland behind mangroves forests has been embanked or reclaimed for alternative uses (Woodroffe, 1990), which prevents mangrove forests extending into these areas as the sea level rises, a conflict termed 'coastal squeeze' (Fig. 11.11). It remains doubtful whether society can show sufficient prescience to overcome the issue of coastal squeeze, and strategies such as managed re-alignment, practised already for salt marsh in Europe, can be extended to promote mangrove forests where there are presently other land-use challenges. Human impacts on mangrove ecosystems are being exacerbated by climate change, which further reduces the stability of mangrove forests and increases their vulnerability.

11.9 Impact of future climate and sea-level change

There is an emerging consensus that anthropogenic emission of greenhouse gases is causing global warming and as a consequence the sea is rising (see Chapter 1). Increasing carbon dioxide and temperature have direct effects on mangrove metabolism and ecosystem processes, which may have both positive and negative consequences for mangrove ecosystems (Lovelock and Ellison 2007; Gilman et al., 2008). Changes in storm intensity may affect mangrove distribution, as may a suite of other factors, but it seems clear that the major impact is likely to be from any acceleration of sea-level rise (Semeniuk, 1994).

The most fundamental response of mangrove shorelines to sea-level rise is incursion into more landward habitats. This has been seen in the stratigraphic record in Florida, where mangrove peat encroached over freshwater peat because of a gradual regional rise in sea level (see Fig. 11.2), and a more abrupt inundation of prior topography in the transgressive stage in the estuaries of northern Australia (see Fig. 11.9). This response, however, will not always be available in future, as it assumes that there is landward space in which mangrove forests can expand. Different mangrove hydrogeomorphic settings respond differently in terms of sediment accretion, which will be particularly important in relation to the regional rate of sea-level change (Krauss et al., 2010).

A major issue is coastal squeeze, described earlier (see Fig. 11.11). Perhaps less well investigated is the question of whether there is hinterland to invade. The presence of broad prograded plains, such as those developed on many coasts in the Indo-Pacific region, implies that rapid sea-level rise in these areas would lead to inundation of these plains. There has not been the detail of stratigraphic reconstruction to demonstrate whether such episodes have occurred in the past. It is clear that increased tidal processes will widen tidal channels (Wolanski and Chappell, 1996). The extent to which tidal flows will continue to be accommodated within the banks of tidal river systems depends on whether sedimentation builds up levées along the margins of these channels. Overtopping of such levées represents an important threshold in such estuarine systems, which, once exceeded, will result in widespread saltwater inundation of the plains that flank many of the tidal rivers of northern Australia, and which presently support extensive freshwater wetlands (Woodroffe, 2007).

Mangrove incursion into more landward salt-marsh habitats, and loss of salt marsh as a result, has been observed along a series of estuarine systems in southern New South Wales, Australia (Saintilan and Williams, 1999). This may relate to relative sea-level rise (Rogers et al. 2006), or be a more complex response to rainfall variability and anthropogenic pressures such as land-use change. Widespread inundation of plains might occur under future rapid sea-level rise, but more insidious might be saline incursion into freshwater wetlands through an expanding network of tidal creeks.

Fig.11.12 Expansion of tidal creeks across the plains of the Mary River in the Northern Territory, Australia. These plains were covered by freshwater wetlands, dominated by the paperbark, *Melaleuca*, until the tidal channels extended and deepened, enabling the incursion of saltwater and mangrove propagules into areas that previously did not receive tidal influence. The paperbark trees have been killed, the surface scarred by saltwater, and mangroves have established along the margins of the expanding creek system. (Source: Colin Woodroffe.)

Tidal-creek systems are a component of the natural dynamics of these complex coastal systems, but increased salinization through creek extension is occurring in several systems in northern Australia already (Mulrennan and Woodroffe, 1998). As the tidal channels extend and deepen, they create local increases in the tidal range, and transfer saltwater and mangrove propagules to regions previously protected from tidal influence (Fig. 11.12). While tidal-creek extension has not been demonstrated unambiguously to relate to sea-level change, observations of increased salinization from creek extension demonstrate the speed with which saline intrusion can proceed and is likely to become much more widespread as sea-level rise accelerates. Channel widening, as suggested in current models for salt-marsh response to sea-level rise (Kirwan and Murray, 2007), provide further support of the relationship between increased salinization through creek extension and sea-level rise. These complexities imply that the response of mangrove shorelines to sea-level change will differ between systems (Semeniuk, 1994). Under the most extreme conditions presently foreseeable, such as melting associated with the West Antarctic ice shelf, more rapid sea-level rise may challenge mangrove systems and trigger more rapid changes than currently seen. These changes will exacerbate threats already caused by human activities.

11.10 Summary

Mangrove forests fringe many tropical coastlines where there are low-lying plains at appropriate elevations,

extending into sub-tropical latitudes on the eastern margin of continents. Mangrove tree communities are relatively species-poor along Atlantic and East Pacific shorelines, however they also comprise more diverse forests associated with complex habitats around the margin of the Indian Ocean and throughout the west Pacific, and reach their greatest floristic and geomorphological diversity in southeast Asia and northern Australia. It is often possible to recognize a zonation of species; whether this indicates successional change depends on the environmental setting. Ecogeomorphic models have been developed that recognize various geomorphological settings within which there are ecologically distinct mangrove forests. Forest development and regeneration occur at both large and small (e.g. forest gaps) scales, giving rise to complexity in forest structure. Sedimentation can be rapid, and is a function of the type of substrate and the extent to which sediment is derived from the inland catchment (e.g. terrestrial mud) or the nearshore (e.g. marine carbonate), or is organic and derived from within the wetlands (e.g. fibrous peat). Mangrove shorelines have adapted to past patterns of climate and sea-level change, some of which have occurred more rapidly than at rates occurring at present, or anticipated in the immediate future. The fate of mangrove is dependent upon our ability to facilitate, as opposed to obstruct, adaptation of mangrove shorelines to sea-level rise.

- Mangrove systems can cope with substantial sediment loads, and sedimentation rates can be high and often spatially variable. Sediment accretion is a key process that contributes to mangroves keeping pace with sea-level rise.
- Mangrove forests, by comparison to other ecosystems that appear threatened by climate change, are relatively well adjusted to extreme environmental conditions, whether in terms of salinity tolerance or inundation by salt water.
- Sea-level rise, at the rates presently experienced or at accelerated rates being predicted, may be tolerated by mangrove forests. However, geomorphological adjustments are inevitable.
- Mangroves are generally well able to adapt to increasing sea level, but this will involve landward extension of mangroves into adjacent systems, which may no longer be possible due to modifications of the coastal plains (coastal squeeze), or desirable where loss of other habitats may occur (e.g. salt marsh, swamp forests).
- Anthropogenic pressures, combined with additional climate change, seem certain to lead to continued decline of mangrove ecosystems, unless the economic and ecological values of these systems are more broadly appreciated, and opportunities to preserve and extend their distribution by conservation and forest restoration are undertaken with greater urgency.

Key publications

Cahoon, D.R., Hensel, P.F., Spencer, T., Reed, D.J., McKee, K.L. and Saintilan, N., 2006. Wetland vulnerability to relative sea-level rise: wetland elevation trends and process controls. In: J.T.A. Verhoeven, B. Belman, R. Bobbink and D.F. Whigham (Eds), *Wetland Conservation and Management*. Springer-Verlag, Berlin, Germany, pp. 271–292.
[*Provides a summary of the current literature regarding the elevation adjustment of mangrove and sea-level rise*]

Duke, N.C., Ball, M.C. and Ellison, J.C., 1998. Factors influencing biodiversity and distributional gradients in mangroves. *Global Ecology and Biogeography Letters*, 7, 27–47.
[*Readily accessible publication regarding environmental gradients at different spatial scales*]

Gilman, E.L., Ellison, J. and Duke, N.C., 2008. Threats to mangroves from climate change and adaptation options. *Aquatic Botany*, 89, 237–250.
[*Reviews the impact of climate on drivers of mangrove change and provides a summary of adaptation options that build mangrove resilience*]

Krauss, K.W., Lovelock, C.E., McKee, K.L., Lopez-Hoffman, L., Ewe, S.M.L. and Sousa, W.P., 2008. Environmental drivers in mangrove establishment and early development: a review. *Aquatic Botany*, 89, 105–127.
[*Readily accessible review of drivers of mangrove establishment*]

Lugo, A.E. and Snedaker, S.C., 1974. The ecology of mangroves. *Annual Review of Ecology and Systematics*, 5, 39–64.
[*A comprehensive review of mangrove ecology that is still relevant*]

Perillo, G.M.E., Wolanski, E., Cahoon, D.R. and Brinson, M.M. (Eds), 2009. *Coastal Wetlands: An Integrated Ecosystem Approach*. Elsevier, Amsterdam, the Netherlands.
[*A comprehensive interdisciplinary book that reviews the current state of knowledge on coastal wetlands; Part V is focussed on mangroves*]

Robertson, A.I. and Alongi, D.M. (Eds), 1992. *Tropical Mangrove Ecosystems. Coastal and Estuarine Studies*. American Geophysical Union, Washington, DC, USA.
[*An interdisplinary compilation of the science of tropical mangrove ecosystems*]

References

Adame, M.F., Neil, D., Wright, S.F. and Lovelock, C.E., 2010. Sedimentation within and among mangrove forests along a gradient of geomorphological settings. *Estuarine, Coastal and Shelf Science*, 86, 21–30

Alongi, D.M., 2002. Present state and future of the world's mangrove forests. *Environmental Conservation*, 29, 331–349.

Alongi, D.M., 2008. Mangrove forests: resilience, protection from tsunamis, and responses to global climate change. *Estuarine Coastal and Shelf Science*, 76, 1–13.

Anderson, J.A.R. and Muller, J., 1975. Palynological study of a holocene peat and a miocene coal deposit from NW Borneo. *Review of Palaeobotany and Palynology*, 19, 291–351.

Anthony, E.J., 2004. Sediment dynamics and morphological stability of estuarine mangrove swamps in Shebro Bay, West Africa. *Marine Geology*, 208, 207–224.

Ball, M.C., 1988. Ecophysiology of mangroves. *Trees*, 2, 129–142.

Barbier, E.B., 2006. Natural barriers to natural disasters: replanting mangroves after the tsunami. *Frontiers in Ecology and Environment*, 4, 123–131.

Cahoon, D.R. and Lynch J.C., 1997. Vertical accretion and shallow subsidence in a mangrove forest of southwestern Florida, U.S.A. *Mangroves and Salt Marshes*, 1, 173–186.

Cahoon, D.R., Day, J.W.J. and Reed, D.J., 1999. The influence of surface and shallow subsurface soil processes on wetland elevation: a synthesis. *Current Topics in Wetland Biogeochemistry*, 3, 72–88.

Cahoon, D.R., Hensel, P.F., Spencer, T., Reed, D.J., McKee, K.L. and Saintilan, N., 2006. Wetland vulnerability to relative sea-level rise: wetland elevation trends and process controls. In: J.T.A. Verhoeven, B. Belman, R. Bobbink and D.F. Whigham (Eds), *Wetland Conservation and Management*. Springer-Verlag, Berlin, Germany, pp. 271–292.

Carlton, J.M., 1974. Land-building and stabilization by mangroves. *Environmental Conservation*, 1, 285–294.

Castañeda-Moya, E., Twilley, R.R., Rivera-Monroy, V.H., Zhang, K., Davis, S.E. and Ross, M., 2010. Sediment and nutrient deposition associated with Hurricane Wilma in mangroves of the Florida Coastal Everglades. *Estuaries and Coasts*, 33, 45–58.

Clark, R.L. and Guppy, J.C., 1988. A transition from mangrove forest to freshwater wetland in the monsoon tropics of Australia. *Journal of Biogeography*, 15, 665–684.

Cunha-Lignon, M., Coelho, C., Almeida, R. et al., 2009. Mangrove forests and sedimentary processes on the south coast of São Paulo State (Brazil). *Journal of Coastal Research*, Special Issue, 56, 1–5.

Danielsen, F., Sorensen, M.K., Olwig, M.F. et al., 2005. Asian tsunami: a protective role for coastal vegetation. *Science*, 310, 643.

Davis, J.H., 1940. The ecology and geologic role of mangroves in Florida. *Papers from Tortugas Laboratory*, 32, 307–412.

Devoe, N.N. and Cole, T.G., 1998. Growth and yield in mangrove forests of the Federated States of Micronesia. *Forest Ecology and Management*, 103, 33–48.

Duarte, C.M., Losada, I.J., Hendriks, I.E., Mazarrasa, I. and Marba, N., 2013. The role of coastal plant communities for climate change mitigation and adaptation. *Nature Climate Change*, 3, 961–968.

Duke, N.C., 2001. Gap creation and regeneration processes driving diversity and structure of mangrove ecosystems. *Wetlands Ecology and Management*, 9, 257–269.

Duke, N.C., Ball, M.C. and Ellison, J.C., 1998. Factors influencing biodiversity and distributional gradients in mangroves. *Global Ecology and Biogeography Letters*, 7, 27–47.

Egler, F.E., 1952. Southeast saline everglades vegetation, Florida: and its management. *Vegetatio*, 3, 213–265.

Ellison, A.M., 2000. Mangrove restoration: do we know enough? *Restoration Ecology*, **8**, 219–229.

Ellison, J.C., 1998. Impacts of sediment burial on mangroves. *Marine Pollution Bulletin*, **37**, 8–12.

Feller, I.C., Lovelock, C.E., Berger, U., McKee, K.L., Joyce, S.B. and Ball, M.C., 2010. Biocomplexity in mangrove ecosystems. *Annual Reviews of Marine Science*, **2**, 395–417.

Geyh, M.A., Kudrass, H.-R. and Streif, H., 1979. Sea-level changes during the late Pleistocene and Holocene in the Strait of Malacca. *Nature*, **278**, 441–443.

Giri, C., Ochieng, E., Tieszen, L.L. et al., 2011. Status and distribution of mangrove forests of the world using earth observation satellite data. *Global Ecology and Biogeography*, **20**, 154–159.

Hanebuth, T., Stattegger, K. and Grootes, P.M., 2000. Rapid flooding of the Sunda Shelf: a late-glacial sea-level record. *Science*, **288**, 1033–1035.

Kathiresan, K. and Rajendran, N., 2005. Coastal mangrove forests mitigated tsunami. *Estuarine Coastal and Shelf Science*, **65**, 601–606.

Kirwan, M.L. and Murray, A.B., 2007. A coupled geomorphic and ecological model of tidal marsh evolution. *Proceedings of the National Academy of Sciences*, **104**, 6118–6122.

Koch, E.W., Barbier, E.B., Sillman, B.R. et al., 2009. Non-linearity in ecosystem services: temporal and spatial variability in coastal protection. *Frontiers in Ecology and Environment*, **7**, 29–37.

Krauss, K.W., Allen, J.A. and Cahoon, D.R., 2003. Differential rates of vertical accretion and elevation change among aerial root types in Micronesian mangrove forests. *Estuarine Coastal and Shelf Science*, **56**, 251–259.

Krauss, K.W., Cahoon, D.R., Allen, J.A., Ewel, K.C., Lynch, J.C. and Cormier, N., 2010. Surface elevation change and susceptibility of different mangrove zones to sea-level rise on Pacific high islands of Micronesia. *Ecosystems*, **13**, 129–143.

Lewis, R.R., 2000. Ecologically based goal setting in mangrove forest and tidal marsh restoration. *Ecological Engineering*, **15**, 191–198.

Lovelock, C.E. and Ellison, J., 2007. Vulnerability of mangroves and tidal wetlands of the Great Barrier Reef to climate change. In: J. Johnson and P. Marshall (Eds), *Assessing the Vulnerability of the GBR to Climate Change*. Great Barrier Reef Marine Park Authority, Townsville, Australia, pp. 237–269.

Lovelock, C.E., Feller, I.C., McKee, K.L., Engelbrecht, B.M. and Ball, M.C., 2004. The effect of nutrient enrichment on growth, photosynthesis and hydraulic conductance of dwarf mangroves in Panama. *Functional Ecology*, **18**, 25–33.

Lovelock, C.E., Ball, M.C., Choat, B., Engelbrecht, B.M.J., Holbrook, N.M. and Feller, I.C., 2006. Linking physiological processes with mangrove forest structure: phosphorous deficiency limits canopy development, hydraulic conductance and photosynthetic carbon gain in dwarf *Rhizophora* mangle. *Plant and Cell Environment*, **29**, 793–802.

Lovelock C.E., Grinham, A., Adame, M.F. and Penrose, H.M., 2009. Elemental composition and productivity of cyanobacterial mats in an arid zone estuary in north Western Australia. *Wetlands Ecology and Management*, **18**, 37–47.

Lovelock, C.E., Sorrell, B., Hancock, N., Hua, Q. and Swales, A., 2010. Mangrove forest and soil development on a rapidly accreting shore in New Zealand. *Ecosystems*, **13**, 437–451.

Lovelock, C.E., Feller, I.C., Adame, M.F. et al., 2011. Intense storms and the delivery of materials that relieve nutrient limitations in mangroves of an arid zone estuary. *Functional Plant Biology*, **38**, 514–522.

Lucas, R., Ellison, J.C., Mitchell, A., Donelly, B., Finlayson, M. and Milne, A.K., 2001. Use of stereo aerial photography for quantifying changes in the extent and height of mangroves in tropical Australia. *Wetlands Ecology and Management*, **10**, 161–175.

Lugo, A.E. and Snedaker, S.C., 1974. The ecology of mangroves. *Annual Review of Ecology and Systematics*, **5**, 39–64.

Lugo, A.E., Sell, M. and Snedaker, S.C., 1976. Mangrove ecosystem analysis. In: B.C. Patten (Ed.), *Systems Analysis and Simulation in Ecology*. Academic Press, New York, USA, pp. 113–145.

Macnae, W., 1968. A general account of the fauna and flora of mangrove swamps and forests in the Indo-West-Pacific region. *Advances in Marine Biology*, **6**, 73–270.

McLeod, E., Chmura, G.L., Bouillon, S. et al., 2011. A blueprint for blue carbon: towards an improved understanding of the role of vegetated coastal habitats in sequestering CO_2. *Frontiers in Ecology and the Environment*, **9**, 552–560.

Morrisey, D.J., Swales, A., Dittmann, S., Morrison, M.A., Lovelock, C.E. and Beard, C.M., 2010. The ecology and management of temperate mangroves. *Oceanography and Marine Biology: An Annual Review*, **48**, 43–160.

Mulrennan, M.E. and Woodroffe, C.D., 1998. Saltwater intrusion into coastal plains of the Lower Mary River, Northern Territory, Australia. *Journal of Environmental Management*, **54**, 169–188.

Murray-Wallace, C.V. and Woodroffe, C.D., 2014. *Quaternary Sea-Level Changes*. Cambridge University Press, Cambridge, UK.

Paphavasit, N., Aksornkoae, S. and de Silva, J., 2009. *Tsunami Impact On Mangrove Ecosystems*. Thailand Environment Institute, Nonthaburi, Thailand.

Parkinson, R.W., 1989. Decelerating Holocene sea-level rise and its influence on southwest Florida coastal evolution: a transgressive/regressive stratigraphy. *Journal of Sedimentary Petrology*, **59**, 960–972.

Pinzón, Z.S., Ewel, K.C. and Putz, F.E., 2003. Gap formation and forest regeneration in a Micronesian mangrove forest. *Journal of Tropical Ecology*, **19**, 143–153.

Rogers, K., Wilton, K.M. and Saintilan, N., 2006. Vegetation change and surface elevation dynamics in estuarine wetlands of southeast Australia. *Estuarine, Coastal and Shelf Science*, **66**, 559–569.

Saintilan, N. and Williams, R.J., 1999. Mangrove transgression into saltmarsh environments in south-east Australia. *Global Ecology and Biogeography*, **8**, 117–124.

Scholl, D.W., 1964. Recent sedimentary record in mangrove swamps and rise in sea level over the southwestern coast of Florida: Part 1. *Marine Geology*, **1**, 344–366.

Semeniuk, V., 1980. Mangrove zonation along an eroding coastline in King Sound, northwestern Australia. *Journal of Ecology*, **68**, 789–812.

Semeniuk, V., 1983. Mangrove distribution in northwestern Australia in relationship to regional and local freshwater seepage. *Vegetatio*, **53**, 11–31.

Semeniuk, V., 1994. Predicting the effect of sea-level rise on mangroves in northwestern Australia. *Journal of Coastal Research*, **10**, 1050–1076.

Spackman, W., Dolsen, C.P. and Riegel, W., 1966. Phytogenic organic sediments and sedimentary environments in the everglades – mangrove complex. *Paleontographica Abteilung B Band*, **117**, 135–152.

Spalding, M., Kainuma, M. and Collins, L., 2010. *World Atlas of Mangroves*. Earthscan, London, UK.

Staub, J.R. and Esterle, J.S., 1994. Peat-accumulating depositional systems of Sarawak, east Malaysia. *Sedimentary Geology*, **89**, 91–106.

Stoddart, D.R., 1980. Mangroves as successional stages, inner reefs of the northern Great Barrier Reef. *Journal of Biogeography*, **7**, 269–284.

Thom, B.G., 1967. Mangrove ecology and deltaic geomorphology: Tabasco, Mexico. *Journal of Ecology*, **55**, 301–343.

Thom, B.G., 1982. Mangrove ecology: a geomorphological perspective. In: B.F. Clough (Ed.), *Mangrove Ecosystems in Australia, Structure, Function and Management*. ANU Press, Canberra, Australia, pp. 3–17.

Thom, B.G., Wright, L.D. and Coleman, J.M., 1975. Mangrove ecology and deltaic-estuarine geomorphology, Cambridge Gulf-Ord River, Western Australia. *Journal of Ecology*, **63**, 203–222.

Tomlinson, P.B., 1986. *The Botany of Mangroves*. Cambridge University Press, Cambridge, UK.

Twilley, R.R. and Rivera-Monroy, V.H., 2009. Ecogeomorphic models of nutrient biogeochemistry for mangrove wetlands. In G.M.E. Perillo, E. Wolanski, D.R. Cahoon and M.M. Brinson (Eds), *Coastal Wetlands: An Integrated Ecosystem Approach*. Elsevier, Amsterdam, the Netherlands, pp. 641–683.

Wolanski, E., 2006. The evolution time scale of macro-tidal estuaries: examples from the Pacific Rim. *Estuarine, Coastal and Shelf Science*, **66**, 544–549.

Wolanski, E. and Chappell, J., 1996. The response of tropical Australian estuaries to a sea level rise. *Journal of Marine Systems*, **7**, 267–279.

Wolanski, E., Moore, K., Spagnol, S., D'Adamo, N. and Pattiaratchi, C., 2001. Rapid, human-induced siltation of the macro-tidal Ord River Estuary, Western Australia. *Estuarine Coastal & Shelf Science*, **53**, 717–732.

Woodroffe, C.D., 1990. The impact of sea-level rise on mangrove shorelines. *Progress in Physical Geography*, **14**, 483–520.

Woodroffe, C.D., 1992. Mangrove sediments and geomorphology. In: D. Alongi and A. Robertson (Eds), *Tropical Mangrove Ecosystems*. American Geophysical Union, Coastal and Estuarine Studies, Washington, DC, USA, pp. 7–41.

Woodroffe, C.D., 2007. Critical thresholds and the vulnerability of Australian tropical coastal ecosystems to the impacts of climate change. *Journal of Coastal Research*, Special Issue, **50**, 469–473.

Woodroffe, C.D. and Grime, D., 1999. Storm impact and evolution of a mangrove-fringed chenier plain, Shoal Bay, Darwin, Australia. *Marine Geology*, **159**, 303–321.

Woodroffe, C.D., Thom, B.G. and Chappell, J., 1985. Development of widespread mangrove swamps in mid-Holocene times in northern Australia. *Nature*, **317**, 711–713.

Woodroffe, C.D., Mulrennan, M.E. and Chappell, J., 1993. Estuarine infill and coastal progradation, southern van Diemen Gulf, northern Australia. *Sedimentary Geology*, **83**, 257–275.

12 Estuaries and Tidal Inlets

DUNCAN FITZGERALD[1], IOANNIS GEORGIOU[2]
AND MICHAEL MINER[3]

[1]*Department of Earth and Environment, Boston University, Boston, MA, USA*
[2]*Department of Earth and Environmental Sciences, University of New Orleans, New Orleans, LA, USA*
[3]*Marine Minerals Program, Bureau of Ocean Energy Management, Gulf of Mexico Region,
New Orleans, LA, USA*

12.1 Introduction

This chapter is devoted to estuaries and tidal inlets – coastal systems that are closely related, but also have important differences. Both are waterways that provide navigation to ports, exchange of nutrients with the coastal ocean, and access to important feeding and breeding grounds for fin and shellfish. Estuaries most commonly occur at mouths of rivers, whereas tidal inlets are associated with barrier shorelines. The dimensions of tidal inlets and the distribution of their associated sand bodies are controlled by the volume and strength of tidal flow, sediment supply, and wave energy. Estuaries exist in a wide range of geological settings, including rias, glaciated valleys, structural basins, and bar-built systems. In the northern hemisphere, rising sea level following deglaciation flooded the incised valleys of river systems. These drowned valleys formed estuaries, except in cases where rivers with high sediment loads filled their channels (valleys) even though rising sea level continued to flood their valleys, resulting in the development of deltas (see Chapter 13). The distribution of estuarine-associated sand bodies is related to sediment availability and composition, wave energy, tidal range and tidal prism, and riverine discharge. Estuaries exist in the southern hemisphere, but their formation is complex due to a history of falling sea level during the past ~6 ka in the southern oceans. It should be noted that some estuaries are classified as tidal inlets. At the mouths of estuaries that are partially enclosed by sandy barriers, the dimension of the estuarine channel is commonly controlled by the saltwater tidal prism and thus these estuaries are also tidal inlets. These estuaries have small freshwater discharges compared with their saltwater tidal prism.

Understanding estuarine and tidal-inlet processes is not only important for the maintenance of navigable

Coastal Environments and Global Change, First Edition. Edited by Gerd Masselink and Roland Gehrels.
© 2014 John Wiley & Sons, Ltd. Published 2014 by John Wiley & Sons, Ltd. Companion Website: www.wiley.com/go/masselink/coastal

waterways, but also for the management of adjacent wetland and barrier shorelines. Estuaries are one of the chief exporters of nutrients to coastal oceans. Along with tropical rainforests, deltaic wetlands, and coral reefs, they are one of the world's most productive ecosystems, and are more productive than either the river itself or the coastal ocean that it empties into. Tidal inlets interrupt the long-shore transport of sediment, affecting both the supply of sand to the downdrift beaches and erosional-depositional processes along the inlet shoreline. The greatest magnitude of shoreline changes along barriers occurs in the vicinity of inlets, and these are a direct consequence of tidal-inlet processes.

This chapter examines the general morphodynamics, sedimentation trends, and hydrodynamics of estuaries and tidal inlets, and shows how these processes produce the facies architecture and large-scale bedding surfaces that define tidal-inlet and estuarine sedimentary deposits. We discuss how climate change and rising sea level may affect the character of estuaries, including their freshwater discharge and salinity, extent of associated wetlands, and sedimentation trends. In addition, we present a model of tidal-inlet and barrier-island evolution in a regime of accelerated sea-level rise (SLR).

12.2 Estuaries

12.2.1 Introduction

Cameron and Prichard (1963) proposed a definition for estuaries that has become widely accepted and is based on morphology and the mixing of ocean and river waters. According to their work, an estuary is a semi-enclosed coastal body of water with a free connection to the open sea within which saltwater is measurably diluted by freshwater coming from rivers or streams. Derived from the Latin word *aestuarium*, meaning tidal (*aestus*) inlet to the sea, estuaries are the transition from freshwater entering through fluvial input and the saline water of the open ocean. The interface between these waters results in a longitudinal density gradient, which controls circulation, mixing, and exchange dynamics of the estuary and the coastal ocean. In terms of sedimentation, Dalrymple et al. (1992) state that an estuary is the seaward portion of a drowned valley system, which formed during rising sea levels in the Pleistocene. The estuary receives sediment from fluvial and marine sources, and the resulting facies distribution is governed by the interaction of tides, waves and fluvial processes. In this section, we introduce different geological settings of estuaries and discuss their geomorphology, stratification and mixing, hydrodynamics, sedimentation processes, facies distribution, and impacts of climate change and rising sea level.

12.2.2 Geomorphic settings

In a geomorphic scheme, estuaries can be classified as bar-built, fjord, tectonic and ria-type, with coastal plain estuaries being an important type of a ria estuary (**Fig. 12.1**).

Bar-built estuaries are similar to drowned valleys and were once open embayments that gradually became enclosed due to high sedimentation rates and active littoral processes near the mouth. Typically, they develop along gently sloping plains on passive or tectonically stable margins and marginal seas. Longshore transport near the mouth promotes the formation of shore-parallel sandbars, spit platforms and sand spits, which commonly attach to adjacent headlands. One to multiple inlets sustain tidal exchange with the coastal ocean. For example, the Algarve in Portugal has as many as seven inlets along a 50 km stretch of the barrier coast. Bar-built estuaries tend to be shallow, rarely exceeding 10 m in depth. They have seasonal fluvial input and, potentially, a large sediment carrying capacity. Examples of bar-built estuaries can be found near mid-latitudes on the eastern USA coast (Pamlico Sound, North Carolina; Barnegat Bay, New Jersey; St John, Florida), the Gulf of Mexico (Laguna Madre, Texas; Barataria Bay, Louisiana), Hawkesbury, Australia, and in the Amazon and Nile River regions.

Fjords are found in high latitudes and are associated with deeply eroded glaciated valleys. They tend to be long (several kilometres) and deep (hundreds of metres), with small width-to-depth ratios (~1 : 10). Fjords have U-shaped cross-sections, with steep-sided walls and commonly a submerged rock bar or a sill near the mouth associated with the terminal moraine deposits. Water mass exchange near the mouth is governed by a buoyancy balance of upstream freshwater input arriving at the sill and saltwater input introduced by tidal oscillations. The source of freshwater can be from an active glacier or river input. Examples are found in Norway, Iceland, Puget Sound (Washington State, USA), British Colombia (Canada), Alaska, New Zealand, Greenland, Scandinavia, Antarctica and Chile.

Tectonic estuaries are formed due to block faulting near an ocean boundary. The depression developed through tectonism is filled by the ocean; rising sea level can aid in this process. Freshwater input in from one or several rivers. Examples of such estuaries include the San Francisco Bay in the USA, Manukau Harbour in Auckland, New Zealand, the Oro Province of Papua New Guinea, and several estuaries in northwest Spain.

Ria estuaries occur at the mouths of river valleys that have been drowned by rising sea level. Although the river valley may have been flooded by rising sea level during the

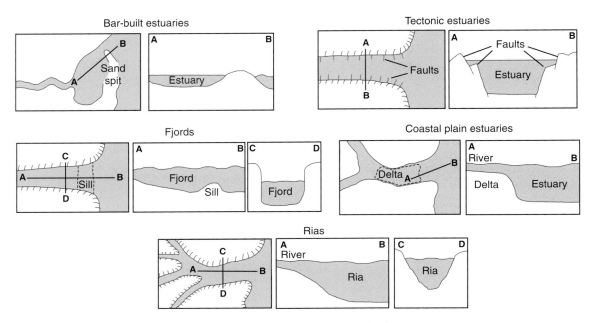

Fig. 12.1 Geomorphic classification of estuary types. (Source: Adapted from Pritchard 1952.)

last deglaciation, the valley itself may be much older, particularly those along bedrock coasts. For example, several rias along the central coast of Maine (USA), Ireland, Brazil and New Zealand owe their origin to fluvial erosion during the past hundreds of thousand to millions of years. Commonly, bedrock structures exert a strong control on estuary morphology.

Coastal plain estuaries are a special type of ria estuary that formed between 18,000 and 6000 years ago, following a large rise in eustatic sea level of more than 100 m, marking the end of the last stage of Pleistocene glaciation. Preceding this event during the lowstand of sea level, rivers extended their courses across the continental shelf, while at the same time incising and widening their valleys. Most estuaries resemble river courses, although they usually have very large width-to-depth ratios. Typically, these estuaries are shallow, with depths in the order of tens of metres, widths of several kilometres, and a channel that gradually widens and deepens towards the mouth. Some examples of coastal plain estuaries include the Thames and Gironde in western Europe, Delaware and Chesapeake Bay on the East Coast of the USA, St Lawrence and Miramichi in eastern Canada, and the Orange River in South Africa.

In a recent study of 96 estuaries along the coast of England and Wales, Prandle et al. (2006) found commonalities among channel characteristics for different geomorphic estuary types. Rias tend to be short, deep and steep-walled with low river discharges. Coastal plain estuaries are

relatively long and funnel-shaped, with broad V-shaped cross-sections and wide intertidal areas. Bar-built estuaries tend to be short, shallow and have low river inflow. Prandle et al. (2006) also discuss the correlation that exists between tidal amplitude and river flow, suggesting that estuaries evolve in response to changes in these parameters and other secondary controls, by adjusting their depth or length. This response will impact the dynamics and balance between the import and export of sediments. Their results suggest that the morphologic evolution of estuaries lies in the relative balance of tidal amplitude versus river flow.

12.2.3 Salinity stratification

Estuaries can be categorized by the degree of vertical stratification, varying from highly stratified estuaries to those that are well-mixed, with little or no vertical salinity gradients (Fig. 12.2). The degree of stratification is influenced by coastal morphology, channel geometry, freshwater input, tidal range near the mouth, and propagation characteristics of the tidal wave. For example, shallow bar-built estuaries are often well-mixed, whereas deep fjords can be highly stratified.

Cameron and Pritchard (1963) classified estuaries based on the ratio between tidal flow and river flow. Tidal flow is defined as the average influx of saltwater during the incoming tide, and river flow is the volume of freshwater influx during a half tidal cycle. This analysis yields estuary types (explained later) that vary from a low ratio (ratio ≤ 1), such as a salt-wedge estuary, to strongly and partially stratified (ratio ≤ 10–10^3), and, finally, to weakly stratified or

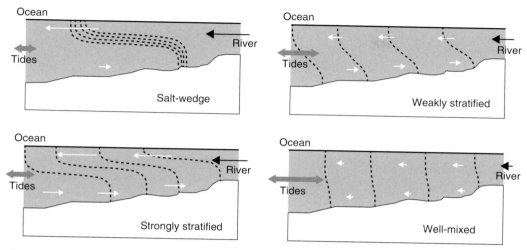

Fig. 12.2 Classification of estuaries based on the vertical structure of salinity. Different types of stratification (solid black lines) are controlled by the relative strength of tidal versus riverine flow as illustrated by the length of the arrows in the four models. (Source: Adapted from Valle-Levinson 2010.)

vertically mixed (ratio $\geq 10^3$). A classic salt-wedge estuary occurs at the mouth of the Mississippi River (USA), where Gulf of Mexico saltwater enters along the base of the channel in a wedge-shaped geometry while at the same time the large freshwater volume of the Mississippi flow exits the estuary at the surface, a process that maintains the vertical stratification. The actual balance between these two water masses varies with tidal stage, seasonal changes in river flow, wind-induced set-up, and tidal amplitude. Moreover, some estuaries loose their salt-wedge nature completely during periods of low river flow and become partially stratified. Examples of partially mixed estuaries include the James River in the Chesapeake Bay, USA, Miramachi, New Brunswick, Canada, and Itajai, Santa Catarina, Brazil.

As tidal forcing increases, thereby dominating river forcing, the resulting current-induced turbulence causes mixing of the entire water column. In these estuaries, salinity varies more horizontally than vertically, leading to a moderately stratified condition. Often the mechanism for salt flux is through steady (sub-tidal) shear dispersion, varying as much as an order of magnitude from the salt flux produced during the transition from neap to spring tides. The relatively wide and shallow nature of these estuaries makes them respond to wind stresses quickly, contributing further to the dynamics of salt flux transport mechanisms, mixing and exchange processes.

Well-mixed estuaries occur where tidal mixing is strong enough to overcome vertical stratification. These estuaries are more complex, and in the absence of vertical stratification, flow and exchange are largely governed by horizontal and transverse salinity gradients. When these estuaries are sufficiently broad, the Coriolis force may segregate the fresh- and saltwater masses during the flood

tide, causing them to flow in opposite directions (Masselink and Hughes, 2003). The incoming floodwaters are steered to the right (northern hemisphere) or to the left (southern hemisphere), creating a unique distribution of salinity in the estuary. This regime is modified further by the bathymetry of the estuary and the presence of deep natural or anthropogenic channels (navigation or dredged canals), which can cause axial convergence and secondary circulation that results in turbulent mixing, resulting in horizontal and transverse salinity gradients.

12.2.4 Circulation and sedimentation

Circulation in an estuary is controlled by the balance of riverine freshwater inflow, tidal amplitude, wind, and wave-induced currents. Secondary currents are also produced through mixing of salt- and freshwater; these water masses have different densities due to differences in salinity and temperature. The less-dense riverine freshwater typically flows over the more-dense saltwater, producing strong stratification and the characteristic saltwater wedge-type estuary. Seasonal changes in discharge and water temperatures of the river, particularly during early spring compared to summer temperatures, can strongly influence density gradients (e.g. in the Mississippi River). Molecular diffusion (movement of salt molecules from high to low concentrations) and turbulent mixing (movement of water parcels between fresh and saline regions), such as that produced by eddies, are two fundamental processes that lessen salinity gradients (Masselink and Hughes, 2003).

Large freshwater input by rivers to estuaries creates a net seaward outflow of water at the surface of the estuary, which can be responsible for discharging fine sediment. At the same time, the entrainment of underlying saltwater by freshwater outflow produces a mass deficiency of saltwater, inducing landward saltwater flow into the estuary. The transition from freshwater to saltwater, termed the halocline or pychnocline, takes place over a fraction of the water column, and is stable in salt-wedge estuaries. Often, the salt wedge is stationary or moves short distances landward and seaward with tidal movement, especially in microtidal settings. Although tides may reverse flows daily, the residual current in the region of the saltwater wedge is commonly responsible for net landward bedload sediment transport and an introduction of coarse sediment into the estuary, particularly in salt-wedge and partially mixed estuaries (Fig. 12.2). High discharges, storms and strong tide-generated currents can break down this type of circulation, producing a more vertically mixed estuary.

Fine particles, such as clays, have low settling velocities due to their small size, and are typically moving seaward near the surface of the estuary and settle slowly. Clays are composed of silica and alumina, and have appreciable quantities of other cations (Na^+, K^+, Ca^{2+}, Mg^{2+}) and can absorb certain anions (NO_3^-, SO_4^{2-}). This quality increases their cohesion, causing clay particles to flocculate at specific salinity thresholds whereby freshwater-laden suspended sediment interfaces with saltwater (Mehta and McAnally, 2008). At these locations, positively charged clay particles attract negatively charged anions, leading to the attraction of cations followed by other clay particles. This process continues until clay floccules are formed, with increased masses causing their settling to the bottom. The location where clay particles or flocs settle rapidly to the bottom, resulting in increased sedimentation rates, is termed the estuarine turbidity maximum (ETM), and usually occurs in the upper reaches of estuaries (Fig. 12.3; Dalrymple and Choi, 2007).

In ETM regions, concentrations of suspended sediment are much higher compared to waters landward or seaward and range from 0.1 to 20g/l, depending on the estuary type and tidal amplitude. High suspended sediment concentrations are often associated with macrotidal estuaries, such as the Tamar and the Thames estuaries (UK), or the Fly River (Papua New Guinea), whereas low concentrations are commonly associated with microtidal estuaries, such as the Hawkesbury, Australia. Within the ETM zone, the suspended sediment concentration can change appreciably, and is affected by the velocity and turbulence of river flow, tidal amplitude, depth of the estuary, and dynamics of the salinity structure. In the Scheldt Estuary, on the border of Belgium and the Netherlands, concentrations as high as 0.28g/l were reported, consisting of flocs with sizes up to 500μm. Settling velocities up to 11mm/s contribute as much as 95% of the measured sedimentation flux (Manning et al., 2007). Uncles (2002) demonstrated sediment concentrations ranging from 0.2 to 1.5g/l in the Tamar estuary (UK). This is consistent with other results that show that during the flooding tide a sudden peak in the suspended sediment concentration occurs as saltwater moves landward, occupying a greater portion of the water column and thereby intercepting the fine clay particles moving seaward.

12.2.5 Sedimentary regimes

Estuaries are highly efficient sediment traps because of their relatively low-energy conditions compared to adjacent fluvial and marine environments (Biggs and Howell, 1984). Although salinity levels have typically been used to define the limits of an estuary (e.g. Pritchard, 1967), geologically it is the physical processes (fluvial-wave-tidal) and relative dominance of each that produce distinctive estuarine depositional environments and characteristic lithofacies. Thus, geological estuarine models define the estuarine limits based on tidal influence instead of salinity (Dalrymple and Choi, 2007). Conceptual geological models have been introduced to explain differences in the morphology, sedimentary processes, and facies distribution in estuaries, which are tied closely to the relative contribution of wave, tidal and fluvial processes in controlling sedimentation (Dalrymple et al., 1992; Boyd et al. 1992). In their schemes, the estuary is divided into a landward zone dominated by riverine sedimentation, a seaward zone dominated by wave or tidal processes, and an intermediate zone of mixed fluvial-marine influence. The Dalrymple et al. (1992) and

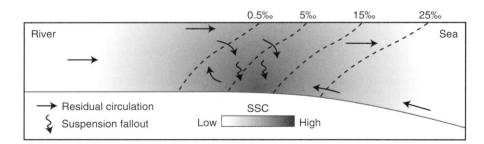

Fig. 12.3 Location of the estuarine turbidity maximum (ETM), showing the residual estuarine circulation and net sedimentation in the ETM. SSC, suspended sediment concentration. (Source: Adapted from Dalrymple and Choi 2007. Reproduced with permission of Elsevier.)

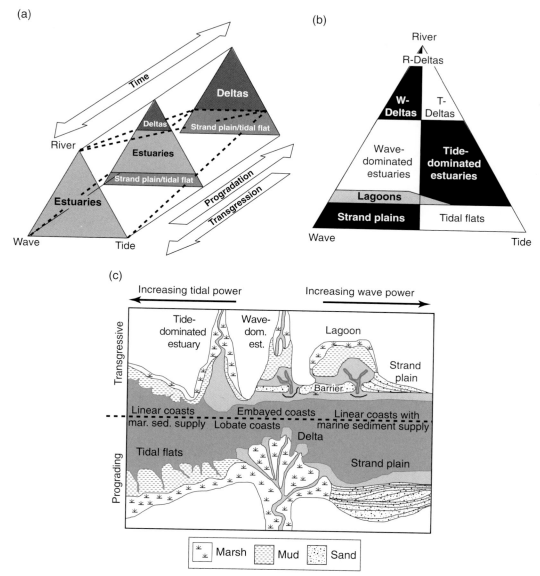

Fig. 12.4 Schematic morphologic and evolutionary diagrams for depositional coasts. (a) The evolution of estuaries versus deltas/ strandplains coasts is a function of sediment supply and sea level changes. (b) Constructive coasts can be classified according to the relative dominance of riverine, tidal, or wave processes. (c) The conceptual morphological model illustrates the types of depositional systems for transgressive and regressive regimes, given wave versus tidal dominance. Theestuary models are most appropriate for coastal plain, drowned river valley settings. Through time, an abundant sediment supply leads to estuarine infilling and, ultimately, delta formation. (Source: Harris et al. 2002; after Dalrymple et al. 1992 and Boyd et al. 1992.)

Boyd et al. (1992) model is presented as a ternary classification (Fig. 12.4), similar to those proposed by Coleman and Wright (1975) and Galloway (1975) for deltas, and Hayes (1979) and Davis and Hayes (1984) for barrier coasts.

The estuarine model has been expanded from the conventional two-dimensional ternary diagram to a three-dimensional prism to capture the estuarine evolution through time in response to changes in relative sea level and sediment supply (Fig. 12.4; Harris et al., 2002). Testing of these models in field locations worldwide has shown that the majority of clastic estuarine systems conform to this classification scheme (e.g. Harris et al., 2002). However, exceptions have been presented, especially in the case of bedrock or antecedent geology controlling morphology (Fenster and FitzGerald, 1996; FitzGerald et al., 2000) and fluvial dominance (Cooper, 1993).

Because estuarine morphology and facies development are dependent upon the interaction between river and marine processes, Dalrymple et al. (1992) proposed two idealized end-member models of estuarine sedimentation: tide-dominated and wave-dominated. Fluvial energy decreases down-estuary as hydraulic gradient decreases as the stream reaches base level, whereas marine energy decreases up-estuary. Three zones within most tide- and wave-dominated estuaries can be recognized: (1) an outer zone dominated by marine processes (waves and tidal currents); (2) a low-energy central zone; and (3) an inner, river-dominated zone (Dalrymple et al., 1992; Boyd et al., 1992). The central zone, being an area of convergence and relatively lower energy regime, is usually characterized by the finest grained bedload sediment in the system. Dalrymple et al. (1992) argue that a

net landward transfer of sediment derived from outside the estuary mouth distinguishes estuaries from deltas (where net sediment transport is seaward). Deltaic deposition within the estuary is also common in the river-dominated zone as bay-head deltas, which are common in wave-dominated estuaries.

12.2.6 Wave-dominated estuaries

Wave-dominated estuaries are characterized by fronting barrier islands, bars or spits that are built by alongshore and onshore transport of sand at the estuary mouth (e.g. Miramachi, New Brunswick, Canada; Fig. 12.5). This barrier attenuates most or all of the offshore wave energy, protecting the estuary. However, wind-generated waves behind the barrier are an important physical process,

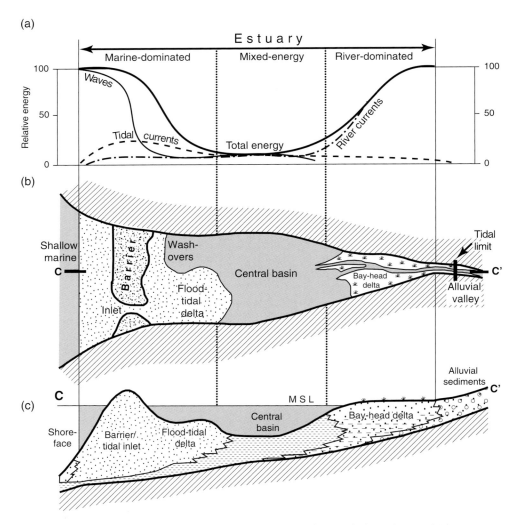

Fig. 12.5 Conceptual model of a wave-dominated estuary. (a) Energy regime. (b) Morphological units. (c) Facies association. MSL, mean sea level. (Source: Dalrymple et al. 1992.)

especially in large estuaries. Tidal currents maintain tidal inlets along the barrier shoreline, but most of the tidal energy is dissipated by friction within the inlet and flood-tidal delta systems. In contrast, fluvial energy decreases seaward, resulting in a low-energy central estuary bounded at the top and bottom by higher-energy regimes (Dalrymple et al., 1992). This zonation is reflected in the morphology of wave-dominated estuaries: a wave-generated sand or gravel body and associated tidal deltas and shoals fronting the estuary (Hayes, 1979), coarse-grained fluvial deposits at the estuary head form a bay-head delta, and a low-energy central basin that acts as the prodelta of the bay-head delta where fine-grained, organic-rich muds and in some shallower examples, salt-marsh sediment accumulates (Dalrymple et al., 1992).

Bay-head deltas develop at the head of wave-dominated estuaries where fluvial-derived sediment progrades into the estuary. These deltas may have a wave-, tide- or fluvial-dominated form (terminology of Galloway, 1975), and primarily encompass the tidally influenced freshwater zone of the estuary. Although these deltas are often progradational, within overall retrograding estuarine systems where sediment transport is dominantly in a landward direction (Dalrymple et al., 1992), increased rates of relative sea-level rise can reverse this trend, resulting in back-stepping (transgression) of the bay-head delta system (Rodriguez et al., 2010). Likewise, accelerating sea-level rise will cause barriers, flood-tidal deltas, and other tidal deposits to migrate landward into the estuary.

12.2.7 Tide-dominated estuaries

Tide-dominated estuaries occur where tidal-current energy exceeds wave energy at the mouth of the estuary. Elongate sandbars develop at the estuary mouth and dissipate wave energy (Hayes, 1975; Dalrymple et al., 1992). These estuaries are characterized by a funnel-shaped embayment that narrows in an up-estuary direction (Fig. 12.6). Because of this morphology and the strong tidal currents, the current velocity during flood tides may be amplified up-estuary due to convergence (Nichols and Biggs, 1985; Dalrymple et al., 1992). The zone where flood-tidal and fluvial energy are equal is situated just landward of the tidal energy maximum in most estuaries where measurements exist (Dalrymple et al., 1992). The zone of minimal energy that is so prevalent in the central basin of wave-dominated estuaries is not as pronounced in tide-dominated systems, and sedimentary response reflects this as sand bodies may extend along the length of the estuary (Woodroffe et al., 1989; Dalrymple et al., 1992; FitzGerald et al., 2000). An increase in the rate of sea-level rise will cause estuarine sand bodies to move landward up the estuary, and likewise peripheral marsh and tidal-flat environments to migrate onto uplands and up-estuary.

12.2.8 Effects of floods: New England estuaries

One of the chief attributes of estuaries is that they accumulate sediment that is derived from the inflowing river as well as sediment imported from outside the estuary mouth as a result of circulation set up by flow in the river, tides and density differences between the ocean and riverine waters. Although this is the normal sedimentation regime, some estuaries are dominated by high-energy discharge events during which time they export bedload sediment to the coastal ocean. For example, the bedrock-cut estuaries of New England (USA) discharge sand during major floods, which nourishes nearby barriers (FitzGerald et al., 2002). Typically, channel sediments of these estuaries are composed of medium- to coarse-grained sand and are texturally immature. These rivers may accumulate sand for tens of years until a spring freshet raises stage levels and increases freshwater flow by one to two orders of magnitude (Fenster and FitzGerald, 1996). During these events, sand is flushed from settling basins, eroded from estuarine point bars and from the channel bottom, and ultimately transported to the coastal ocean.

Spring freshets in New England occur during late winter or early spring, when an intensive rain event melts snow and the combined water flows overland into nearby streams and eventually into major rivers. During this period, freshwater discharge may supplant the entire estuarine saltwater tidal prism, despite tidal ranges exceeding 2.5 m, resulting in unidirectional seaward flow in the estuary that persists for one to several weeks (FitzGerald et al., 2002). High discharge events can also be caused by intense precipitation accompanying hurricanes or major storms. Major spring freshets and flood events not only control net coarse-grained sediment pathways in these estuaries, but also bedform hierarchy and orientation. For example, in the Kennebec River estuary on the central coast of Maine, USA, under normal freshwater discharge conditions, bedforms (megaripples and sandwaves) exhibit variable orientations depending on their location and tidal conditions. However, the largest bedforms in the estuary (transverse bars) remain seaward-oriented throughout the year, indicating the dominance of seaward sediment transport and the controlling nature of high-magnitude events (Fig. 12.7).

The importance of spring freshets in supplying coarse-grained sediment to estuaries and the offshore has application outside of New England. Meade (1972) suggested that during infrequent floods, the saltwater wedge is pushed entirely out of the estuary, and riverine sediment is carried out to sea. Milliman (1980) has also shown that 80% of the sediment moved through the Fraser Estuary in British Columbia, Canada, occurs during spring freshets, as indicated by ebb-oriented sandwaves and other hydrographic data. Along the Oregon coast, USA, sand that accumulates within the estuary during low-flow conditions in the

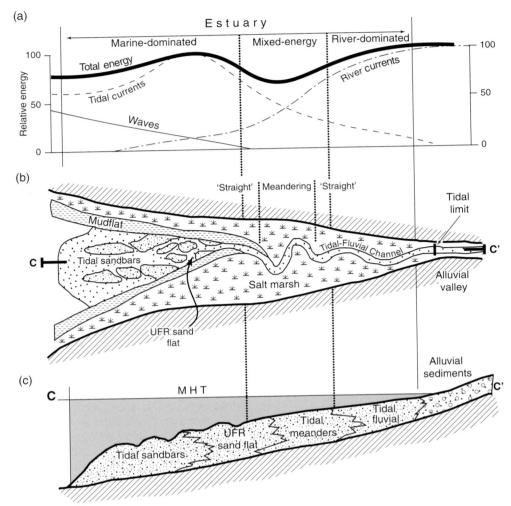

Fig. 12.6 Conceptual model of a tide-dominated estuary. (a) Energy regime. (b) Morphological units. (c) Facies association. MHT, mean high tide; UFR, upper flow region. (Source: Dalrymple et al. 1992.)

summer is flushed out of the estuary during winter high-discharge conditions (Boggs and Jones, 1976).

12.2.9 Future of estuaries

Predictive models of global warming, atmospheric circulation and rising sea level, based on carbon dioxide and other greenhouse gas emissions, indicate that estuaries may be impacted during this century in a number of ways: (1) inundation and destruction of wetlands occurring at a faster rate than their expansion and migration into uplands; (2) changes in freshwater influx and associated sediment load; (3) alteration of salinity gradients causing changes in estuarine circulation, exchange, mixing and sedimentation processes; and (4) a general redistribution of fauna and flora as salinity and habitats are altered.

Wetlands form a highly productive environment along the periphery of many estuaries. These marshes may reach a threshold and become threatened if they cannot accrete at a rate to keep pace with rising sea level. The loss of these marshes and wetlands due to drowning and edge erosion will not be compensated for by new marsh growth and expansion onto adjacent uplands. For instance, while low-salinity marshes can expand vertically as well as laterally through organic and mineral accumulation processes, their saline counterparts can only expand vertically, which threatens the future of estuaries and shifting habitats, especially those associated with large expanses of salt marsh. Marshlands unable to accrete vertically and keep pace with sea-level rise will create new accommodation space and will support a larger tidal prism. These changes will ultimately affect the dynamic exchange of fresh- and

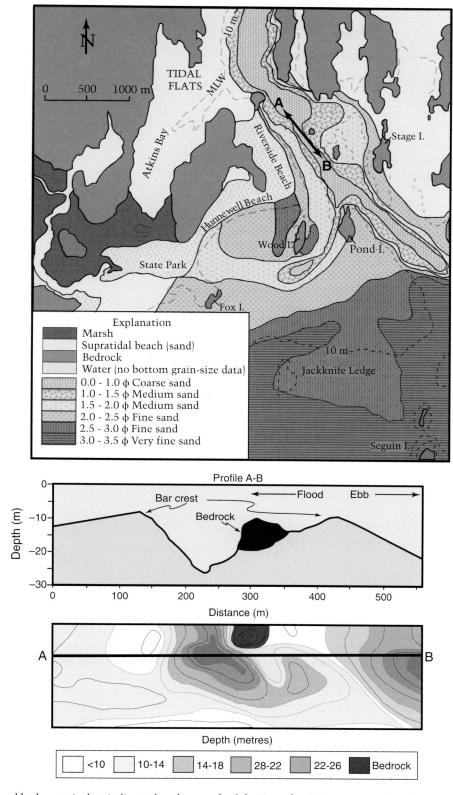

Fig. 12.7 Grain size and bathymetric data indicate that the mouth of the Kennebec River estuary along the central Maine coast, USA, is dominated by a seaward sediment transport. The channel is floored by coarse to medium sand that is sourced from the river. Two large seaward-oriented transverse bars demonstrate that sand is exported from this estuary. MLW, mean low water. For colour details, please see Plate 28.

saltwater masses and the resulting dispersal of sediment. As estuaries widen due to marsh-edge submergence, wind-waves will contribute to edge erosion and accelerate the destruction of marsh further.

One expected consequence of global warming will be changing weather patterns that will lead to greater precipitation levels in some drainage basins and lesser amounts in others. These changes will affect the freshwater discharge to estuaries and the volume of sediment that they deliver; the influx of water and sediment may be modified further by land-use practices within the watershed. Sediment deposition and land-building processes will be impacted, as will salinity gradients and the overall estuarine circulation. Because estuarine habitats are controlled by salinity, sediment influx and duration of tidal inundation, among many other factors, it is anticipated that faunal and floral communities will shift, migrate or be replaced completely, as conditions change.

Global warming and climatic shifts may also produce subtle changes to estuaries. For example, resistance of tidal flats to erosion is affected by biotic films, which may be expected to increase in many latitudes due to warming and greater productivity. Widdows et al. (2004) report that 50% of sediment accumulation on tidal flats in the Western Scheldt estuary, the Netherlands, is due to bio-stabilization, including biofilms, short-term stabilizers (e.g. burrowing clams) and longer-term stabilizers by persistent biota (e.g. mussels). Conversely, warming will reduce the volume of ice-rafted sediment that is contributed to marshes as the effects of ice diminish (Argow and FitzGerald, 2006). Climate change will affect storm magnitude and frequency, thereby influencing both sedimentation on estuarine wetlands as well as marsh destruction caused by storm wave erosion (Howes et al., 2010).

12.3 Tidal inlets

12.3.1 Morphodynamics

Diversity in the morphology, hydraulic signature, and sediment transport patterns of tidal inlets attests to the complexity of their processes. The variability in oceanographic, meteorological and geological parameters, such as tidal range, wave energy, sediment supply, storm magnitude and frequency, freshwater influx, and geological substrate, and the interactions of these factors, are responsible for this wide range in tidal-inlet settings.

12.3.1.1 What is a tidal inlet?

A tidal inlet is defined as an opening in the shoreline through which water penetrates the land, thereby providing a connection between the ocean and bays, lagoons, and marsh and tidal-creek systems. Tidal currents maintain the main channel of a tidal inlet.

The second half of this definition distinguishes tidal inlets from large, open embayments or passageways along rocky coasts. Tidal currents at inlets are responsible for the continual removal of sediment dumped into the main channel by wave action. Thus, according to this definition, tidal inlets occur along sandy or sand and gravel barrier coastlines, although one side may abut a bedrock headland. Some tidal inlets coincide with the mouths of rivers (estuaries), but in these cases inlet dimensions and sediment transport trends are still governed, to a large extent, by the volume of water exchanged at the inlet mouth and the reversing tidal currents, respectively.

At most inlets, over the long term, the volume of water entering the inlet during the flooding tide equals the volume of water leaving the inlet during the ebbing cycle. This volume is referred to as the tidal prism. The tidal prism is a function of the open water area and tidal range in the backbarrier complex as well as frictional factors, which govern the ease of flow through the inlet and lagoon.

12.3.1.2 Tidal-inlet morphology

Specifically, a tidal inlet is the area between two barriers or between the barrier and the adjacent bedrock or glacial headland. Commonly, the recurved ridges of spits, consisting of sand that was transported toward the backbarrier by refracted waves and flood-tidal currents, form the sides of the inlet. The deepest part of an inlet, the inlet throat, is located normally where spit accretion of one or both of the bordering barriers constricts the inlet channel to a minimum width and minimum cross-sectional area. Here, tidal currents normally reach their maximum velocity. Often the strength of the currents at the throat causes sand to be removed from the channel floor, leaving behind a lag deposit consisting of gravel or shells, or in some locations exposed bedrock or indurated sediments.

12.3.1.3 Tidal deltas

Closely associated with tidal inlets are sand shoals and tidal channels located on the landward and seaward sides of the inlets. Flood-tidal currents deposit sand landward of the inlet, forming flood-tidal deltas, and ebb-tidal currents deposit sand on the seaward side, forming an ebb-tidal delta.

The presence or absence, and size and development of flood-tidal deltas are related to a region's tidal range, wave energy, sediment supply, and backbarrier accommodation space. Tidal inlets that are backed by a system of tidal channels and salt marsh (mixed-energy coast) usually contain a single horseshoe-shaped flood-tidal delta (e.g. Essex River Inlet, Massachusetts, USA; Fig. 12.8). Contrastingly, inlets that are backed by large, shallow bays may contain multiple flood-tidal deltas. Along some microtidal coasts, such as Rhode Island, USA, flood deltas form at the end of narrow inlet channels cut through the barrier. Changes

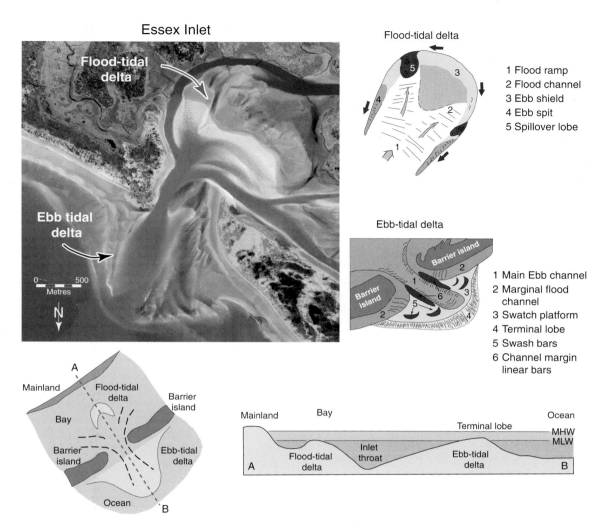

Fig. 12.8 Ebb- and flood-tidal models. Aerial photograph of Essex Inlet, Massachusetts, USA. MHW, mean high water; MLW, mean low water. (Source: Hayes 1979. Photograph: FitzGerald 1996.) For colour details, please see Plate 29.

in the locus of deposition at these deltas produce a multi-lobate morphology, resembling a lobate river delta (Boothroyd et al., 1985). Flood-delta size commonly increases as the amount of open water area in the backbarrier increases. In some regions, flood deltas have become colonized and altered by marsh growth, and are no longer recognizable as former flood-tidal deltas. At other sites, portions of flood-tidal deltas are dredged to provide navigable waterways, and thus are highly modified.

Flood-tidal deltas are best developed in areas with moderate to large tidal ranges (1.5–3.0 m) because in these regions they are well exposed at low tide. As tidal range decreases, flood deltas become largely sub-tidal shoals. Most flood-tidal deltas have similar morphologies, consisting of the following components (Hayes, 1975, 1979):

• **Flood ramp.** This is a landward shallowing channel that slopes upward toward the intertidal portion of the flood-tidal delta. Strong flood-tidal currents and landward sand transport in the form of flood-oriented sandwaves dominate the ramp.
• **Flood channels.** The flood ramp splits into two shallow flood channels. Like the flood ramp, these channels are dominated by flood-tidal currents and flood-oriented sandwaves. Sand is delivered through these channels onto the flood delta.
• **Ebb shield**. This defines the highest and landward-most part of the flood delta and may be partly covered by marsh vegetation. It shields the rest of the delta from the effects of the ebb-tidal currents.
• **Ebb spits**. These spits extend from the ebb shield toward the inlet. They form from sand that is eroded from the ebb

shield and transported back toward the inlet by ebb-tidal currents.

- **Spillover lobes**. These are lobes of sand that form where the ebb currents have breached through the ebb spits or ebb shield, depositing sand in the interior of the delta.

Through time, some flood-tidal deltas accrete vertically and/or expand in lateral extent. This is evidenced by an increase in areal extent of marsh grasses, which require a certain elevation above mean low water to survive. At laterally migrating inlets, new flood-tidal deltas are formed as the inlet moves along the coast and encounters new open water areas in the backbarrier. At most stable inlets, however, sand comprising the flood delta is simply recirculated. The transport of sand on flood deltas is controlled by the elevation of the tide, and the strength and direction of the tidal currents. During the rising tide, flood currents reach their strongest velocities near high tide, when the entire flood-tidal delta is covered by water. Hence, there is a net transport of sand up the flood ramp, through the flood channels and onto the ebb shield. Some of the sand is moved across the ebb shield and into the surrounding tidal channel. During the falling tide, the strongest ebb currents occur near mid to low water. At this time, the ebb shield is out of the water and diverts the currents around the flood-tidal delta. The ebb currents erode sand from the landward face of the ebb shield and transport it along the ebb spits and eventually into the inlet channel, where once again it will be moved onto the flood ramp, thus completing the sand gyre.

Ebb-tidal deltas are accumulations of sand that have been deposited by the ebb-tidal currents and that have been subsequently modified by waves and tidal currents. Ebb-tidal deltas exhibit a variety of forms, dependent on the relative magnitude of wave and tidal energy of the region, as well as geological controls. Along mixed-energy coasts, most ebb-tidal deltas contain the same general features, including:

- **Main ebb channel**. This is a seaward-shallowing channel that is scoured in the ebb-tidal delta sands and inlet throat. It is dominated by ebb-tidal currents.
- **Terminal lobe**. Sediment transported out the main ebb channel is deposited in a lobe of sand, forming the terminal lobe. The deposit slopes relatively steeply on its seaward side. The outline of the terminal lobe is well-defined by breaking waves during storms or periods of large wave swell at low tide.
- **Swash platform**. This is a broad, shallow sand platform located on both sides of the main ebb channel, defining the general extent of the ebb-tidal delta.
- **Channel margin linear bars**. These are bars that border the main ebb channel and sit atop the swash platform. These bars tend to confine the ebb flow and are exposed at low tide.

- **Swash bars.** Waves breaking over the terminal lobe and across the swash platform form arcuate-shaped swash bars that migrate onshore. The bars are usually 50–150 m long, 50 m wide and 1–2 m high.
- **Marginal-flood channels.** These are shallow channels (0–2 m deep at mean low water) located between the channel margin linear bars and the onshore beaches. The channels are dominated by flood-tidal currents.

12.3.1.4 Ebb-tidal delta morphology

The general shape of an ebb-tidal delta and the distribution of its sand bodies are determined by the relative magnitude of different sand transport processes operating at a tidal inlet. At tide-dominated inlets, ebb-tidal deltas are elongate, having a main ebb channel and channel margin linear bars that extend far offshore. Wave-generated sand transport plays a secondary role in modifying delta shape at these inlets. Because most sand movement is in onshore-offshore direction, the ebb-tidal delta overlaps a relatively small length of inlet shoreline. This has important implications concerning the extent to which the inlet shoreline undergoes erosional and depositional changes.

Wave-dominated inlet systems tend to be small relative to tide-dominated inlets. Their ebb-tidal deltas are driven onshore, close to the inlet mouth, by the dominant wave processes. Commonly, the terminal lobe and/or swash bars form a small arc outlying the periphery of the ebb-tidal delta. In many cases, the ebb-tidal delta of these inlets is entirely subtidal. In other instances, sand bodies clog the entrance to the inlet, leading to the formation of several major and minor tidal channels.

At mixed-energy tidal inlets, the shape of the delta is the result of tidal and wave processes. These deltas have a well-formed main ebb channel, which is a product of ebb-tidal currents. The swash platform and sand bodies of this inlet type substantially overlap the inlet shoreline many times the width of the inlet throat due to wave processes and flood-tidal currents.

Ebb-tidal deltas may also be highly asymmetric, such that the main ebb channel and its associated sand bodies are positioned primarily along one of the inlet shorelines. This configuration normally occurs when the major backbarrier channel approaches the inlet at an oblique angle, or when a preferential accumulation of sand on the updrift side of the ebb delta causes a deflection of the main ebb channel along the downdrift barrier shoreline.

12.3.2 Tidal-inlet formation and evolution

The formation of a tidal inlet requires the presence of an embayment and the development of barriers. In coastal plain settings, the embayment or backbarrier was often created through the construction of the barriers themselves, like much of the East Coast of the USA or the

Friesian Island coast along the North Sea. In other instances, the embayment was formed due to rising sea level inundating an irregular shoreline during the late Holocene. The embayed or indented shoreline may have been a rocky coast, such as that of northern New England and California, or it may have been an irregular unconsolidated sediment coast, such as that of Cape Cod in Massachusetts or parts of the Oregon coast. The flooding of former river valleys has also produced embayments (drowned river valley) associated with tidal-inlet development, such as Mobile Bay in the northern Gulf of Mexico.

12.3.2.1 Breaching of a barrier

Rising sea level, exhausted sediment supplies, and human influences have led to thin barriers that are vulnerable to breaching. The breaching process normally occurs during storms, after waves have destroyed the foredune ridge, and storm waves have overwashed the barrier, depositing sand aprons (washovers) along the backside of the barrier (see Chapter 9). Even though this process may produce a shallow overwash channel, barriers are rarely cut from their seaward side. In most instances, the breaching of a barrier is the result of the storm surge heightening waters in the backbarrier bay (FitzGerald and Pendleton, 2002). When the level of the ocean tide falls, the elevated bay waters flow across the barrier toward the ocean, gradually incising the barrier and cutting a channel. If subsequent tidal exchange between the ocean and bay is able to maintain the channel, a tidal inlet is established (Fig. 12.9). The breaching process is facilitated when offshore winds accompany the falling tide and enhance seaward flow through an overwash channel (Fisher, 1962). Many tidal inlets that are formed by this process are ephemeral and may exist for less than a year, especially if stable inlets are

located nearby. Along the southeast coast of Australia, open versus closed tidal inlets is closely related to drainage basin size of the associated river system and seasonal or event precipitation. Inlets that tend to stay open have large catchments with regular precipitation, whereas those that are infrequently open have small catchments and low precipitation. In some instances, a new breach may capture part of the bay tidal prism so that it remains open, decreasing tidal prism at other inlets. In the case where bay area is increasing over the long term, such as is happening in coastal Louisiana, USA, due to rapid relative sea-level rise and wetland erosion, breaches may remain open to accommodate the increasing bay tidal prism. Barriers most susceptible to breaching are long and thin and wave-dominated.

12.3.2.2 Spit building across a bay

The development of a tidal inlet by spit construction across an embayment usually occurs early in the evolution of a coast. The sediment to form these spits may have come from erosion of the nearby headlands, discharge from rivers, or from the landward movement of sand from inner-shelf deposits. Most barriers along the coast of the USA and elsewhere in the world are 3000 to 5000 years old, coinciding with a deceleration of Holocene sea-level rise. It was then that spits began enclosing portions of the irregular rocky coast of New England and the West Coast of the USA, parts of Australia, and many other regions of the world. As a spit builds across a bay, the opening to the bay gradually decreases in width and in cross-sectional area. It may also deepen. Coincident with the decrease in size of the opening is a corresponding increase in tidal current velocities. Assuming the tidal prism of the bay remains approximately constant, as the opening gets

Fig. 12.9 Photographs illustrating the formation of New Inlet. (Source: Photographs by Duncan FitzGerald.)

smaller, the current velocities must increase. A tidal inlet is formed as the spit reaches a stable configuration.

12.3.2.3 Drowned river valleys

Tidal inlets have also formed at the entrance to drowned river valleys due to the growth of spits and the development of barrier islands that have served to narrow the mouths of the estuaries. It has been shown through stratigraphic studies, particularly along the East and Gulf Coasts of the USA, that in addition to drowned river valleys, many tidal inlets are positioned in palaeo-river valleys in which there is no river leading to this site today (e.g. Siringan and Anderson, 1993). These incised valleys represent old river courses that were active during the Pleistocene when sea level was lower and they incised down into the exposed continental shelf in response to base level lowering. Modern tidal inlets become situated in these valleys because tidal currents more easily remove the unconsolidated sediment filling the valleys than the more indurated sediments comprising the interfluves.

12.3.2.4 Lateral inlet migration

Some tidal inlets have been positionally stable since their formation, whereas others have migrated long distances along the shore. In New England, USA, and along other glaciated coasts, stable inlets are commonly anchored next to bedrock outcrops or resistant glacial deposits. Along the California coast, most tidal inlets have formed by spit construction across an embayment, with the inlet becoming stabilized adjacent to a bedrock headland. In coastal plain settings, stable inlets are commonly positioned in former river valleys. One factor that appears to separate migrating inlets from stable inlets is the depth to which the inlet throat has eroded. Deeper inlets are often situated in consolidated sediments that resist erosion. The channels of shallow, migrating inlets are eroded into sand.

Although the vast majority of tidal inlets migrate in the direction of dominant longshore transport, there are some inlets that migrate updrift (Aubrey and Speer, 1984). In these cases, the drainage of backbarrier tidal creeks control flow through the inlet. When a major backbarrier tidal channel approaches the inlet at an oblique angle, the ebb-tidal currents coming from this channel are directed toward the margin of the inlet throat. If this is the updrift side of the main channel, then the inlet will migrate in that direction. This is similar to a river, where strong currents are focussed along the outside of a meander bend causing erosion and channel migration. Inlets that migrate updrift are usually small to moderately sized, and occur

along coasts with small to moderate net sand longshore transport rates.

12.3.2.5 Landward inlet migration

Similar to laterally migrating tidal inlets, a landward migrating inlet's position along the shoreline and future retreat path may be controlled by erodability of the substrate or backbarrier morphology governing tidal hydraulics. For example, results from regional stratigraphic studies have demonstrated that tidal inlets in Louisiana, USA, are often actively incising into former distributary channel deposits because the relatively coarse-grained fluvial channel deposits erode more readily than the semi-indurated clays in which they incise (Kulp et al., 2006).

Tidal inlets that occur along retrograding (transgressive) shorelines migrate landward with the adjacent barriers. As the shoreline retreats, the zone where processes governing tidal-inlet morphology are focussed at the inlet throat – tidal flushing of sand transported to the inlet by waves – is translated in a landward direction. Inlet-throat landward migration is facilitated by scouring in the landward portions of the channel, and deposition of sediment in the seaward portion of the channel; a process that is greatly accelerated during storms (Miner et al., 2009). During barrier retreat, a large component of landward sand transport occurs at flood-tidal deltas and recurved spits associated with landward migrating inlets (Armon, 1979; Kulp et al., 2006).

12.3.3 Tidal-inlet relationships

Tidal inlets throughout the world exhibit several consistent relationships that have allowed coastal engineers and marine geologists to formulate predictive models: (1) inlet throat cross-sectional area is closely related to tidal prism (O'Brien, 1931, 1969; Jarrett, 1976); and (2) ebb-tidal delta volume is a function of tidal prism (Walton and Adams, 1976).

The size of a tidal inlet is tied closely to the volume of water going through it (Fig. 12.10a; O'Brien, 1931, 1969). Although inlet size is primarily a function of tidal prism, to a lesser degree inlet cross-sectional area is also affected by the delivery of sand to the inlet channel. For example, at jettied inlets, tidal currents can more effectively scour sand from the inlet channel and therefore they maintain a larger throat cross-section than would be predicted by the O'Brien relationship. Similarly, for a given tidal prism, Gulf Coast inlets have larger throat cross-sections than do West Coast inlets in the USA. This is explained by the fact that wave energy is greater along the West Coast and therefore the delivery of sand to these inlets is higher than at Gulf Coast inlets. Jarrett (1976) has

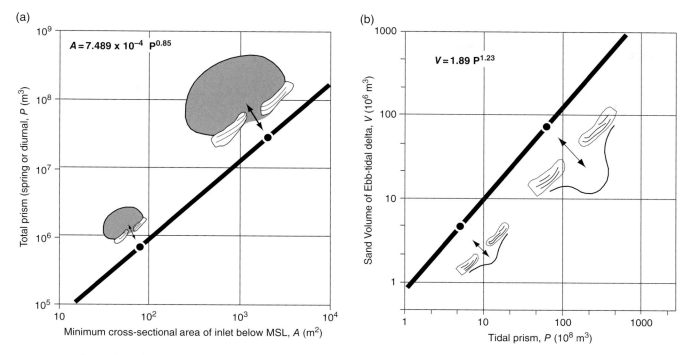

Fig. 12.10 Relationships demonstrating a correspondence between tidal prism and inlet cross-sectional area (a) and sand volume of the ebb-tidal delta (b). MSL, mean sea level. (Source: (a) Adapted from O'Brien 1931, 1969. (b) Adapted from Walton and Adams 1976.)

improved the tidal prism–inlet cross-sectional area regression equation for inlets in the USA by separating them into three classes the low-energy Gulf Coast inlets, moderate-energy East Coast inlets, and higher-energy West Coast inlets. Even better correlations are achieved when structured inlets are distinguished from natural inlets. Marchi (1990) and D'Alpaos et al. (2010) have confirmed the theoretical relationship between tidal inlet-size and tidal prism.

It is important to understand that the dimensions of the inlet channel are not static, but rather the inlet channel enlarges and contracts slightly over relatively short time periods (<1 year) in response to changes in tidal prism, variations in wave energy, effects of storms, seasonal changes in dominant wind direction, meteorologically driven tides, and other factors. For instance, the inlet tidal prism can vary by more than 30% from neap to spring tides due to increasing tidal ranges. Consequently, the size of the inlet varies as a function of tidal phases. Along the southern East Coast of the USA, water temperatures may fluctuate seasonally by as much as 20 °C. This causes the surface coastal waters to expand when they warm, raising mean sea level by 30 cm or more (Nummedal and Humphries, 1978). In the summer and early fall when mean sea level reaches its highest seasonal elevation, spring tides may flood backbarrier surfaces that

normally are above tidal inundation. This produces larger tidal prisms, stronger tidal currents, increased channel scour, and larger inlet cross-sectional areas. At some Virginia inlets, this condition increases the inlet throat by 5 to 15% (Byrne et al., 1975). Longer-term (>1 year) changes in the cross-section of inlets are related to inlet migration, sedimentation or erosion in the backbarrier, morphological changes of the ebb-tidal delta, and anthropogenic influences.

Walton and Adams (1976) showed that the volume of sand contained in the ebb-tidal delta is also closely related to the tidal prism (Fig. 12.10b). Moreover, they showed that the relationship was improved slightly when wave energy was taken into account in a manner similar to Jarrett's divisions. Waves are responsible for transporting ebb-tidal delta sand back onshore, thereby reducing the volume of the ebb-tidal delta. Therefore, for a given tidal prism, ebb-tidal deltas along the West Coast contain less sand than do equal-sized inlets along the Gulf or East Coasts of the USA.

The Walton and Adams relationship works well for inlets all over the world. However, field studies have shown that the volume of sand comprising ebb-tidal deltas changes through time due to the effects of storms (Miner et al. 2009), changes in tidal prism (FitzGerald et al. 2007), or processes of inlet sediment by-passing (FitzGerald, 1984). Sand by-passing at a tidal inlet is commonly achieved by large bar complexes migrating across the ebb delta and attaching to the landward inlet shoreline. These large bars may contain more than 300,000 m³ of sand and represent more than 10% of sediment volume of the ebb-tidal delta (FitzGerald, 1988).

12.3.4 Sand transport patterns

The movement of sand at a tidal inlet is complex due to reversing tidal currents, effects of storms, and interaction with the longshore transport system. Tidal inlets contain short-term and long-term reservoirs of sand, varying from the relatively small sandwaves flooring the inlet channel that migrate metres during each tidal cycle to the large flood-tidal delta shoals where some sand is recirculated but the entire deposit may remain stable for hundreds or even thousands of years. Sand dispersal at tidal inlets is complicated, because in addition to the onshore-offshore movement of sand produced by tidal and wave-generated currents, there is constant delivery of sand to the inlet and transport of sand away from the inlet by the longshore transport system. In the following discussion, the patterns of sand movement at inlets are described, including how sand is moved past a tidal inlet.

The ebb-tidal delta has segregated areas of landward versus seaward sediment transport that are controlled primarily by the way water enters and discharges from the inlet as well as the effects of wave-generated currents. During the ebbing cycle, the tidal flow leaving the back-barrier is constricted at the inlet throat, causing the currents to accelerate in a seaward direction. Once out of the confines of the inlet, the ebb flow expands laterally and the velocity slows. Sediment in the main ebb channel is transported in a net seaward direction and eventually deposited on the terminal lobe due to this decrease in current velocity. One response to this seaward movement of sand is the formation of ebb-oriented sandwaves with heights of 1–2 m.

In the beginning of the flood cycle, the ocean tide rises while water in the main ebb channel continues to flow seaward as a result of momentum. Due to this phenomenon, water initially enters the inlet through the marginal flood channels, which are the pathways of least resistance. The flood channels are dominated by shore-parallel and landward sediment transport and are floored by flood-oriented bedforms. On both sides of the main ebb channel, extending from the channel to the inlet shorelines, the swash platform is most affected by landward flow produced by the flood-tidal currents and breaking waves. As waves shoal and break, they generate landward flow, which augments the flood-tidal currents but retards the ebb-tidal currents. The interaction of these forces acts to transport sediment in a net landward direction across the swash platform. In summary, at many inlets there is a general trend of seaward sand transport in the main ebb channel, which is countered by landward sand transport in the marginal flood channels and across the swash platform.

Along most open coasts, particularly in coastal plain settings, shore-oblique dominant wave approach causes a net movement of sediment in an alongshore direction, which along the East Coast of the USA varies from 100,000 to 200,000 m³/yr (Komar, 1997). The manner whereby sand moves past tidal inlets and is transferred to the downdrift shoreline is called inlet sediment by-passing. The primary mechanisms of sand by-passing natural inlets include: (1) stable inlet processes; (2) ebb-tidal delta breaching; and (3) inlet migration and spit breaching (Fig. 12.11). One of the end products in all the different mechanisms is the landward migration and attachment of large bar complexes to the inlet shoreline. Discussion of bar attachment processes can be found in FitzGerald (1982) and FitzGerald et al. (2000).

Sediment by-passing by stable inlet processes occurs at inlets that do not migrate and whose main ebb channels remain approximately in the same position. Sand enters the inlet by wave action along the beach, flood-tidal and wave-generated currents through the marginal flood channel, and waves breaking across the channel margin linear bars. Most of the sand that is dumped into the main channel is transported seaward by the dominant ebb-tidal currents and deposited on the terminal lobe.

At lower tidal elevations, waves breaking on the terminal lobe transport sand along the periphery of the delta toward the landward beaches in much the same way as sand is moved in the surf and breaker zones along beaches. At higher tidal elevations, waves breaking over the terminal lobe create swash bars on both sides of the main ebb channel. The swash bars (50–150 m long, 50 m wide) migrate onshore due to the dominance of landward flow across the swash platform. Eventually, they attach to channel margin linear bars forming large bar complexes. Bar complexes tend to be parallel to the beach and may be more than a kilometre in length. They are fronted on their landward side by a slipface (25–33°), which may be up to 3 m in height.

The stacking and coalescence of swash bars to form a bar complex is the result of the bars slowing their onshore migration as they move up the nearshore ramp. As the bars gain a greater intertidal exposure, the wave bores that cause their migration onshore act over an increasingly shorter period of the tidal cycle. Thus, their rate of movement onshore decreases. Eventually, the entire amalgamated bar complex migrates onshore and welds to the upper beach. When a bar complex attaches to the downdrift inlet shoreline, some of this newly accreted sand is then gradually transported by wave action to the downdrift beaches, thus completing the inlet sediment

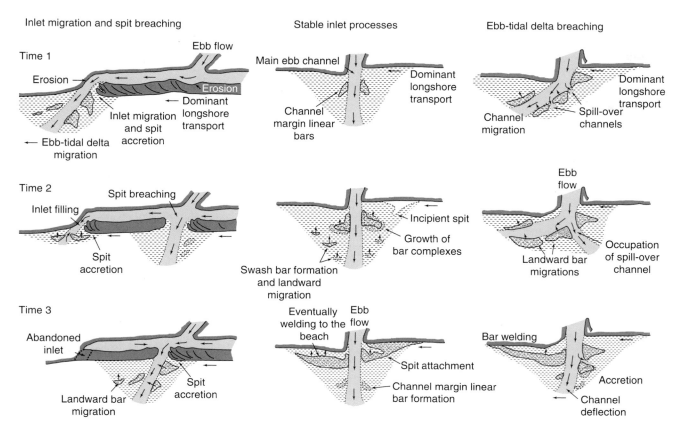

Fig. 12.11 Mechanisms of sediment by-passing tidal inlets. (Source: FitzGerald et al. 2001.)

by-passing process. It should be noted that some sand by-passes the inlet independent of the bar complex. In addition, some of the sand comprising the bar re-enters the inlet via the marginal flood channel and along the inlet shoreline.

Sediment by-passing by ebb-tidal delta breaching occurs at inlets with a stable throat position, but whose main ebb channels migrate across their ebb-tidal deltas like the wag of a dog's tail. Sand enters the inlet in the same manner as described earlier for stable inlet processes. However, at these inlets, the delivery of sediment by longshore transport produces a preferential accumulation of sand on the updrift side of the ebb-tidal delta. The deposition of this sand causes a deflection of the main ebb channel until it nearly parallels the downdrift inlet shoreline. This circuitous configuration of the main channel results in inefficient tidal flow through the inlet, ultimately leading to a breaching of a new channel through the ebb-tidal delta. The breaching process normally occurs during spring tides or periods of storm surge, when the tidal prism is very large. In this state, the ebb discharge piles up water at the entrance to the inlet where the channel bends toward the downdrift inlet shoreline. This causes some of the tidal waters to exit through the marginal flood channel or flow across low regions on the channel margin linear bar. Gradually over several weeks, or convulsively during a single large storm, this process cuts a new channel through the ebb delta, thereby providing a more direct pathway for tidal exchange through the inlet. As more and more of the tidal prism is diverted through the new main ebb channel, tidal discharge through the former channel decreases, causing it to fill with sand.

Sand that was once on the updrift side of the ebb-tidal delta, and which is now on the downdrift side of the new main channel, is moved onshore by wave-generated and flood-tidal currents. Initially, some of this sand aids in the filling of the former channel, while the rest forms a large bar complex that eventually migrates onshore and attaches to the downdrift inlet shoreline. The ebb-tidal breaching process results in a large packet of sand by-passing the inlet. Similar to the stable inlets discussed earlier, some sand by-passes these inlets in a less dramatic fashion, grain by grain, on a continual basis.

It is noteworthy that at some tidal inlets the entire main ebb channel is involved in the ebb-tidal delta breaching process, whereas at others just the outer portion of the main ebb channel is deflected. In both cases, the end product of the breaching process is a channel realignment to a configuration that more efficiently conveys water through the inlet, as well as sand being by-passed in the form of a bar.

A final method of inlet sediment by-passing, by inlet migration and spit breaching, occurs at laterally migrating inlets. In this situation, an abundant sand supply and a dominant longshore transport direction cause spit building at the end of the barrier. To accommodate constant tidal prism and spit construction, the inlet migrates by eroding the downdrift barrier shoreline as updrift spit platform deposits fill the channel. Along many coasts, as the inlet is displaced further along the downdrift shoreline, the inlet channel that extends into the backbarrier lengthens, retarding the exchange of water between the ocean and backbarrier. This condition leads to large water-level differences between the ocean and bay, making the barrier highly susceptible to breaching, particularly during storms. Ultimately, when the barrier spit is breached and a new inlet is formed in a hydraulically more favourable position, the tidal prism is diverted to the new inlet and the old inlet closes. When this happens, the sand comprising the ebb-tidal delta of the former inlet is transported onshore by wave action, commonly taking the form of a landward migrating bar complex. It should be noted that when the inlet shifts to a new position along the updrift shoreline, a large quantity of sand has effectively by-passed the inlet. The frequency of this inlet sediment by-passing process is dependent on inlet size, rate of migration, storm history and backbarrier dynamics.

Depending on the size of the inlet, the rate of sand delivery to the inlet, the effects of storms, and other factors, the entire process of bar formation, its landward migration, and its attachment to the downdrift shoreline may take from six to 10 years. The volume of sand by-passed can range from 100,000 to over 1,000,000 m³. The bulge in the shoreline that is formed by the attachment of a bar complex is gradually eroded and smoothed, as sand is dispersed to the downdrift shoreline and transported back toward the inlet. In some instances, a landward migrating bar complex forms a saltwater pond as the tips of the arcuate bar weld to the beach, stabilizing its onshore movement. Although the general shape of the bar and pond may be modified by overwash and dune building activity, the overall shoreline morphology is frequently preserved. Lenticular-shaped coastal ponds or marshy swales become diagnostic of bar migration processes and are common features at many inlets (Hayes and Kana, 1976).

12.3.5 Tidal-inlet effects on adjacent shorelines

In addition to the direct consequences of spit accretion and inlet migration are the effects of volume changes in the size of ebb-tidal deltas, sand losses to the backbarrier, processes of inlet sediment by-passing, increasing inlet cross-sectional area, and wave sheltering of the ebb-tidal delta shoals (FitzGerald, 1988).

The degree to which barrier shorelines are influenced by tidal-inlet processes is dependent on their size and number. As the O'Brien relationship demonstrates, the cross-sectional area of an inlet is governed by its tidal prism. This concept can be expanded to include an entire barrier chain in which the size and number of inlets along a chain are primarily dependent on the amount of open water area behind the barrier and the tidal range of the region. In turn, these parameters are a function of other geological and physical oceanographic factors. Wave-dominated, microtidal coasts tend to have long barrier islands and few tidal inlets, and mixed-energy coasts have short, stubby barriers and numerous tidal inlets (Hayes, 1975, 1979). Presumably, the mesotidal conditions produce larger tidal prisms than occur along microtidal coasts, which necessitate more holes in the barrier chain to let the water into and out of the backbarrier. Many coastlines follow this general trend but there are many exceptions due to the influence of sediment supply, large versus small bay areas, dominance of meteorologically driven tides, and other geological controls (Davis and Hayes, 1984).

Tidal deltas and associated backbarriers may be important sediment sinks. Sand is not only trapped temporarily on ebb-tidal deltas, but longer-term losses of sediment occur when sand is transported into the backbarrier. At inlets dominated by flood-tidal currents, sand is continuously transported landward, enlarging flood-tidal deltas and building bars in tidal creeks. Sand can also be transported into the backbarrier of ebb-dominated tidal inlets during severe storms. During these periods, increased wave energy produces greater sand transport to the inlet channel. At the same time, the accompanying storm surge increases the water surface slope at the inlet, resulting in stronger than normal flood-tidal currents. The strength of the flood currents, coupled with the high rate of sand delivery to the inlet, results in landward sediment transport into the backbarrier. Along the Malpeque barrier system in the Gulf of St Lawrence, New Brunswick, USA, it has been determined that over a 33-year period, 90% of the sand transferred to the backbarrier took place at tidal inlets and at former inlet locations along the barrier (Armon, 1979).

Sediment may also be lost at laterally migrating inlets when sand is deposited as channel fill. If the channel scours below the base of the barrier sands, then the beach sand, which fills this channel, will not be replaced entirely by the deposits excavated on the eroding portion of the channel. Because up to 40% of the length of barriers is underlain by tidal-inlet fill deposits ranging in thickness from 2 to 10 m (Moslow and Heron, 1978; Moslow and Tye, 1985), this volume represents a large, long-term loss of sand from the coastal sediment budget. Another major process producing sand loss at laterally migrating inlets is associated with the construction of recurved spits that build into the backbarrier. For example, along the East Friesian Islands, recurved spit development has caused the lengthening of barriers along this chain by 3–11 km since 1650 (FitzGerald et al., 1984). During this stage of barrier evolution, the large size of the tidal inlets permitted ocean waves to transport large quantities of sand around the end of the barrier, forming recurved ridges that extend far into the backbarrier. Due to the size of the recurves and the length of barrier extension, this process has been one of the chief natural mechanisms of bay infilling (FitzGerald and Penland, 1987).

Ebb-tidal deltas represent large reservoirs of sand that may be comparable in volume to that of the adjacent barrier islands along mixed-energy coasts (e.g. northern East and West Friesian Islands, Massachusetts, southern New Jersey, Virginia, South Carolina, Georgia and Louisiana, USA). For instance, the ebb-tidal delta volume of Stono and North Edisto Inlets in South Carolina is $197 \times 10^6 \, m^3$ and the intervening Seabrook-Kiawah Island barrier complex contains $252 \times 10^6 \, m^3$ of sand (Hayes et al., 1976). In this case, the deltas comprise 44% of the sand in the combined inlet-barrier system. The magnitude of sand contained in ebb-tidal deltas suggests that small changes in their volume dramatically affect the sand supply to the landward shorelines.

A transfer of sand from the barrier to the ebb-tidal delta takes place when a new tidal inlet is opened, such as the formation of Ocean City Inlet when Assateague Island, Maryland, USA, was breached during the 1933 hurricane. Initially, the inlet was only 3 m deep and 60 m across, but quickly widened to 335 m when it was stabilized with jetties in 1935. Since the inlet formed, more than 14 million cubic metres of sand has been deposited on the ebb-tidal delta (Stauble and Cialone, 1996). Trapping the southerly longshore movement of sand by the north jetty and growth of the ebb-tidal delta have led to serious erosion along the downdrift beaches. The northern end of Assateague Island has been retreating at an average rate of 11 m/yr. The rate of erosion lessened when the ebb-tidal delta reached an equilibrium volume and the inlet began to by-pass sand (Stauble and Cialone, 1996).

The shallow character of ebb-tidal deltas provides a natural breakwater for the landward shorelines. This is especially true during lower tidal elevations, when most of the wave energy is dissipated along the terminal lobe. During higher tidal stages, intertidal and subtidal bars cause waves to break offshore, expending much of their energy before reaching the beaches onshore. The sheltering effect is most pronounced along mixed-energy coasts where tidal inlets have well-developed ebb-tidal deltas.

The influence of ebb TIDAL DELTA is particularly well illustrated by the history of Morris Island, South Carolina, USA, which forms the southern border of Charleston Harbor. Before human modification, the entrance channel to the harbour paralleled Morris Island and was fronted by an extensive shoal system. In the late 1800s, jetties were constructed at the harbour entrance to straighten, deepen and stabilize the main channel. During the period prior to jetty construction (1849–1880), Morris Island had been eroding at an average rate of 3.5 m/yr. After the jetties were in place, the shoals eroded and gradually diminished in size, and so did the protection they afforded Morris Island, especially during storms. From 1900 to 1973, Morris Island receded 500 m at its northeast end, increasing to 1100 m at its southeast end, a rate three times what it had been prior to jetty construction (FitzGerald, 1988).

Tidal inlets interrupt the wave-induced longshore transport of sediment along the coast, affecting both the supply of sand to the downdrift beaches and the position and mechanisms whereby sand is transferred to the downdrift shorelines. The effects of these sediment by-passing processes are exhibited well along the Copper River Delta barriers in the Gulf of Alaska (Fig. 12.12). From east to west along the barrier chain, the width of the tidal inlets increases, as does the size of the ebb-tidal deltas (Hayes, 1979). In this case, the width of the inlet can be used as a proxy for the inlet's cross-sectional area. These trends reflect an increase in tidal prism along the chain, which is caused by an increase in bay area from east to west, while tidal range remains constant. Also quite noticeable along this coast is the greater downdrift offset of the inlet shoreline in a westerly direction. This morphology is coincident with an increase in the degree of overlap of the ebb-tidal delta along the downdrift inlet shoreline. The offset of the inlet shoreline and bulbous shape of the barriers are produced by sand being trapped at the eastern, updrift end of the barrier. The amount of shoreline progradation (build out) is a function of inlet size and extent of its ebb-tidal delta. What we learn from the sedimentation processes along the Copper River Delta barriers is that tidal inlets can impart a very important signature on the form of the barriers (FitzGerald, 1996).

In an investigation of barrier islands shorelines in mixed-energy settings throughout the world, Hayes (1979)

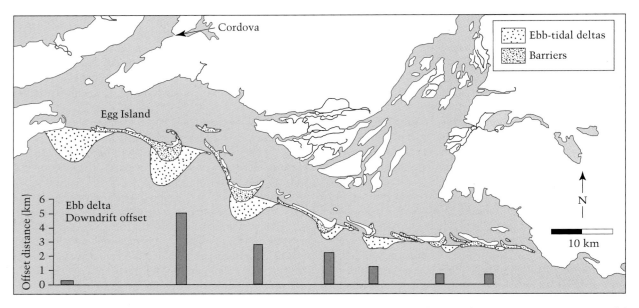

Fig. 12.12   Copper River Delta barrier chain, Gulf of Alaska, illustrating how increase in bay area from east to west increases tidal prism, producing larger tidal prisms and greater effects on barrier island morphology. (Source: Hayes 1979.)

noted that many barriers exhibit a 'chicken leg drumstick' barrier island shape. In this model, the meaty portion of the drumstick barrier is attributed to waves bending around the ebb-tidal delta producing a reversal in the longshore transport direction. This process reduces the rate at which sediment by-passes the inlet, resulting in a broad zone of sand accumulation along the updrift end of the barrier. The downdrift, or thin part of the drumstick, is formed through spit accretion. Later studies demonstrated that landward-migrating bar complexes from the ebb-tidal delta determine barrier island morphology and overall shoreline erosional-depositional trends, particularly in mixed-energy settings (Box 12.1; FitzGerald et al., 1984).

12.3.6   Human influences

Dramatic changes to inlet beaches can result from human influences, including the obvious consequences of jetty construction that reconfigures an inlet shoreline. By preventing or greatly reducing an inlet's ability to by-pass sand, the updrift beach progrades, while the downdrift beach, whose sand supply has been diminished or completely cut off, erodes. There can also be more-subtle human impacts, which can equally affect inlet shorelines, especially those associated with changes in inlet tidal prism, sediment supply, and the longshore transport system. Nowhere are these types of impacts better

demonstrated than along the central Gulf Coast of Florida, USA, where development has resulted in the construction of causeways, extensive backbarrier filling and dredging projects, and the building of numerous engineering structures along the coast. A detailed study of this region by Barnard and Davis (1999) has revealed that since the late-1980s, 17 inlets have closed along this coast and at least five closures can be traced to human influences caused primarily by changes in inlet tidal prism. For example, access to several barriers has been achieved through the construction of causeways that extend from the mainland across the shallow bays. Along most of their lengths, the causeways are dyke-like structures that partition the bays, thereby changing bay areas and inlet tidal prisms. In some instances, tidal prisms were reduced to a critical value, causing inlet closure. At these sites, the tidal currents were unable to remove the sand dumped into the inlet channel by wave action. Similarly, when the Intracoastal Waterway was constructed along the central Gulf Coast of Florida in the early 1960s, the dredged waterway served to connect adjacent backbarrier bays, thereby changing the volume of water that was exchanged through the connecting inlets. The Intracoastal Waterway lessened the flow going through some inlets, while at the same time increased the tidal discharge of others. This resulted in the closure of some inlets and the enlargement of others (Barnard and Davis, 1999).

CASE STUDY BOX 12.1   Friesian barrier island shape

Studies of the Friesian Islands demonstrate that inlet processes exert a strong influence on the dispersal of sand and in doing so dictate barrier form (FitzGerald et al., 1984). In addition to drumsticks, the East Friesian Islands exhibit many other shapes. Inlet sediment by-passing along this barrier chain occurs, in part, through the landward migration of large swash bars (>1 km in length) that deliver up to 300,000 m³ of sand when they weld to the beach. In fact, it is the position where the bar complexes attach to the shoreline that determines the form of the barrier along this coast (Fig. 12.13). If the ebb-tidal delta greatly overlaps the downdrift barrier, then the bar complexes may build up the barrier shoreline some distance from the tidal inlet, forming 'humpbacked barriers' (e.g. Norderney). If the downdrift barrier is short and the ebb-tidal delta fronts a large portion of the downdrift barrier, then bar complexes weld to the downdrift end of the barrier, forming 'downdrift bulbous barriers' (i.e. Baltrum).

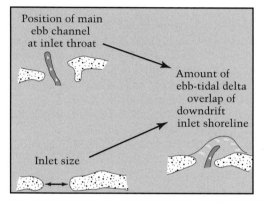

Barrier Island planform

controlled by:
1. Position of channel thalweg
2. Inlet size
3. Amount of delta overlap
4. Location of bar weldings

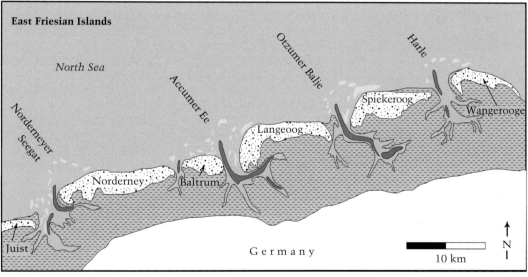

Fig. 12.13   The planform of barriers along the East Friesian Islands is a product of where landward migrating bars attach and weld to the barrier shoreline.

12.3.7   Tidal-inlet stratigraphy

Tidal-inlet fill sequences and ebb and flood delta stratigraphy have been imaged using high-resolution shallow seismic reflection profiling over water and ground-penetrating radar (GPR) on land, the latter offering an order of magnitude finer resolution than seismic reflection data. The reflectors produced by these systems coincide with large-scale erosional and accretionary surfaces, thereby providing a means of documenting the sedimentation history of tidal-inlet fill sequences and tidal deltas in great detail. Cores taken in conjunction with the geophysical data provide a means of ground-truthing the interpretation of the various reflectors and produce a detailed characterization of individual tidal facies. The results of several studies dealing with active tidal inlets and tidal delta deposits as well as palaeo-inlet locations are presented in this section to illustrate the types of facies architecture associated with tidal-inlet sequences, including their geophysical characterization, when available.

12.3.7.1 Inlet fill sequences

The size, geometry and facies characteristics of tidal-inlet fills are dependent on a number of factors that define the dimensions of the inlet system, migrational behaviour of the inlet channel, and conditions under which the inlet fills with sediment. The size of the inlet channel is controlled by tidal prism, and thus regions with large to moderate tidal ranges (>2 m) tend to have deep tidal inlets (>8 m deep; e.g. barrier coasts of southern Virginia, South Carolina, Georgia, East and West Friesian Islands, Copper River Delta barriers, some of the Algarve inlets in Portugal), whereas microtidal coasts produce small, shallow inlets (<6 m deep; e.g. northern South Carolina, southern North Carolina, Rhode Island, much of the east and west coasts of Florida, Nile River delta). As pointed out by Davis and Hayes (1984), large tidal inlets can also occur along microtidal, wave-dominated coasts, if the inlet connects to a large bay and accesses a large tidal prism (e.g. Barataria Pass, Louisiana, depth = 18 m; Pensacola Bay entrance, Florida, depth = 10 m; Beaufort Inlet, North Carolina, depth = 10 m, pre-dredging). Likewise, small tidal inlets occur in mesotidal regions where the inlet drains a small tidal basin (e.g. Captain Sam's Inlet, South Carolina, depth = 5 m). Thus, the dimensions of inlet deposits along any coast are variable and highly dependent on the tidal prism and how easily the backbarrier tidal prism is accessed through time.

The geometry and facies characteristics of inlet deposits are also dictated by the dynamics of the inlet channel. Three scenarios of inlet behaviour are presented here, building on the pioneering work of many previous authors (Bruun, 1966; Kumar and Sanders, 1974; Moslow and Heron, 1978; FitzGerald, 1984, 1988; Tye, 1984; Moslow and Tye, 1985; Tye and Moslow, 1993). These dynamic models provide a framework for developing stratigraphic models for inlet fills and identifying them in the sedimentological record.

(1) Migrating inlets: As the downdrift side of the inlet erodes, the channel fills with sand due to progradation of the updrift spit, and thus the strike section of the deposit is dependent on the distance that the inlet migrates along the coast and the depth of the channel. In GPR and shallow seismic reflection profiles, these deposits are characterized by large-scale, steep to moderately dipping, accretionary surfaces oriented in a downdrift direction. In dip sections, the deposit is thickest at the inlet throat and thins in both a seaward and landward direction.

(2) Breaches(called hurricane passes in the Gulf of Mexico and breaches elsewhere): Inlets that are cut during storms and later close, producing a fill sequence that is formed through spit accretion from one or both sides of the inlet. Filling may also involve the landward migration of a large bar that closes off the inlet mouth. Subsequent overwash activity fills the channel and/or tidal or wave-induced circulation in the lagoon contributes fine-grained sediment. Thus, the channel fill may consist predominantly of sand, or sand can be mixed or inter-layered with mud. In certain instances, a clay plug may comprise the inner abandoned channel sequence if the influx of sand is prevented due to barrier reconstruction in front of the breach. GPR and shallow seismic strike sections of these deposits exhibit a variety of channel geometries, including simply conformable, prograded, accretionary and complex (terminology after Mitchum et al, 1977).

(3) Channel reorientation: Tidal inlets of this type are characterized by a deflection of the main ebb channel due to the preferential accumulation of sand on the ebb-tidal delta caused by the dominant longshore system, such that the channel becomes skewed along the downdrift shoreline. The inlet channel can erode deeply into the adjacent barrier, as depicted in Fig. 12.14 at Capers and Price Inlets, South Carolina, USA. In this configuration, the low hydraulic gradient of the channel and inefficient tidal exchange eventually cause a breaching of the ebb-tidal delta to effect a more direct pathway for tidal flow (see section on ebb-tidal breaching). Initially, active tidal deposition characterizes the channel fill, until a collapse of the ebb-tidal delta and the ensuing landward migrating bar complex closes the former inlet mouth, resulting in lower-energy currents and finer-grained sedimentation (Tye, 1984; Moslow and Tye, 1985, Tye and Moslow, 1993). The inlet fill consists of a fining-upward sequence that can be dominated by mud if the former inlet channel closes completely and there is ample mud in the backbarrier system. This type of succession is commonly capped by marsh deposition (Tye, 1984).

In light of the above models and the results of numerous field investigations, discussed below, it has been shown that tidal-inlet deposits can be identified within barrier lithosomes, in backbarrier deposits, on continental shelves, and in the rock record by one or more of the following characteristics (see Moslow and Heron, 1978; Tye, 1984; Moslow and Tye, 1985; Siringan and Anderson, 1993; Tye and Moslow, 1993; FitzGerald et al., 2001):

(1) A sharp contact between the base of the inlet channel and underlying strata, which in deep inlets is commonly a scoured Pleistocene surface. In shallow inlets, the base of the inlet is cut into the barrier lithosome and into the shoreface in a seaward direction, or into lagoonal deposits in a landward direction.

(2) Strike sections of inlet fills imaged in shallow seismic and GPR profiles exhibit a range of reflector geometries that include various types of single and multiple channel cut-and-fills and/or repetitive sigmoidal-oblique reflectors dipping downdrift that are formed by a migrating inlet and spit system.

(3) A general fining-upwards sequence that is in contrast to the coarsening-upwards grain-size trend of most barrier lithosomes (due to their being underlain by either lagoon or nearshore fine-grained sediments). Considerable variability (fining and coarsening trends) can exist within units comprising inlet fills, including layers of mud.

(4) The base of the inlet sequence usually consists of a lag deposit and is composed of poorly sorted shell hash, whole and broken shells, heavy-mineral concentrations, rip-up clasts, coarse sand, and/or fine gravel.

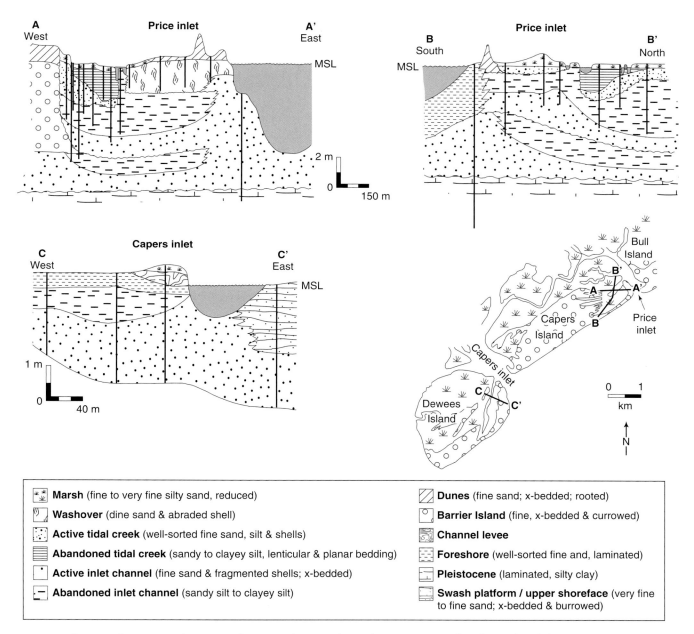

Fig. 12.14 Stratigraphic sections for Price and Capers Inlets, South Carolina, USA. Note that these inlets did not migrate, rather their main channels were deflected south due to the dominant southerly longshore transport and then breached back to straight channel course. MSL, mean sea level. (Source: Tye and Moslow 1993.)

(5) Tidal-inlet fills exhibit a wide variety of sedimentary structures, including planar and trough cross-bedding, graded beds, mud laminations and mud drapes, and shell hash layers. Bioturbation is rare in active sandy channel fill sequences but common in abandoned inlet fills.

12.3.7.2 Examples: North and South Carolinas

The stratigraphy of the Outer Banks of North Carolina is known from numerous coring studies (e.g. Moslow and Heron, 1978; Susman and Heron, 1979; Tye and Moslow, 1993). This body of work provides a basis for characterizing inlet deposits along wave-dominated coasts typified by tidal inlets that open during major storms and close during post-storm recovery periods or ones in which the inlet channel migrates several kilometres along the coast. For example, Shackleford Banks, located just west of Cape Lookout, is 14 km long but has tidal-inlet sediments that extend to depths of 9–20 m and underlie at least 11 km of the barrier (Susman and Heron, 1979). A core through the western end of the

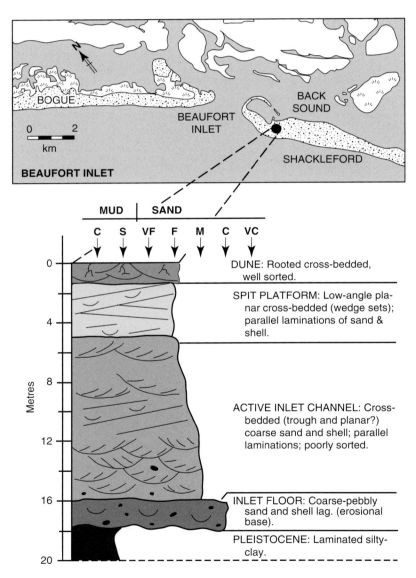

Fig. 12.15 Core log from the western end of Shackleford Banks, North Carolina, USA. A coarse lag deposit defines the base of the channel in an overall fining-upward inlet sequence. (Source: Moslow and Tye 1985. Reproduced with permission of Elsevier.)

island shows a typical tidal-inlet sequence, consisting of a coarse pebbly sand and shell lag sitting atop an erosional contact with the underlying Pleistocene units. The coarse sediment lag is overlain by an 11-m thick, planar to cross-bedded, poorly sorted, fining-upward (medium to fine sand) sequence that is topped by spit platform and dune sands (Moslow and Tye, 1985; Fig. 12.15). Tye and Moslow (1993) report that wave-dominated barrier island coasts are composed of 15 to 80% tidal-inlet sediments. In strike section, inlet deposits are lenticular to wedge-shaped, whereas dip-sections thin in both a landward and seaward inter-fingering with flood and ebb-tidal delta sediments, respectively. The size and dimension of inlet deposits are largely a product of tidal prism and geological factors that control channel deepening.

Central South Carolina is a mixed-energy coast, having relatively short, stubby barrier islands, numerous tidal inlets, well-developed ebb deltas and a backbarrier consisting of marsh and tidal creeks (Hayes, 1975, 1979). Although South Carolina's 'mixed-energy' inlets tend to be stable, some of them have had histories in which the seaward portion of the inlet channel is deflected downdrift and then breaches back to a more straight, hydraulically efficient, channel configuration (FitzGerald, 1988; see Fig. 12.11 on ebb-tidal delta breaching). Inlet sequences formed by these processes have been studied by Moslow and Tye (1985) and Tye and Moslow (1993) and are

(a)

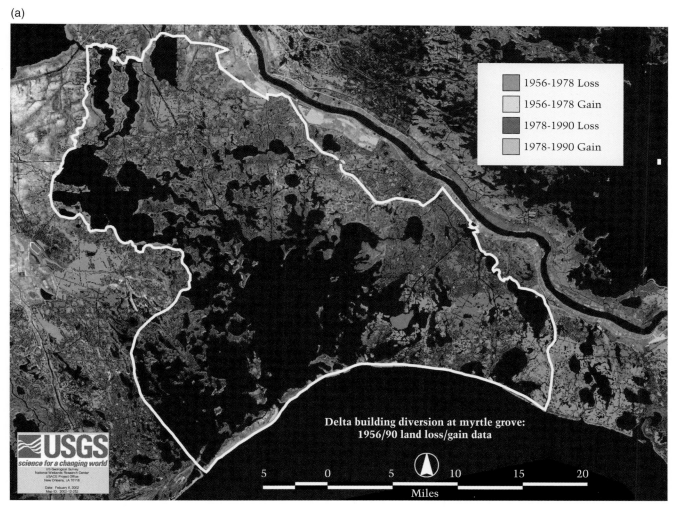

Fig. 12.16 Wetland loss. (a) Barataria Bay, located along the central Louisiana coast, has experienced extensive wetland loss during the past 60 years (Source: Barras 2006), (b) resulting in greater tidal exchange through the tidal inlets. This increasing tidal prism has enlarged the size of the tidal inlets, resulting in the movement of sediment offshore to the ebb deltas and the loss of sand along the barrier chain. For colour details, please see Plate 30.

presented for Price and Capers Inlets in Fig. 12.14. These types of inlet deposits are more complex than the wave-dominated examples, because in addition to active channel-fill sand, they contain inactive channel deposits, welded-bar and washover facies, and tidal-creek sediments. As depicted in stratigraphic sections in Fig.12.14, when the inlet channel migrates and erodes the downdrift barrier, active inlet fills are deposited from wave and tidal introduction of sand to the channel. However, after the breaching event occurs and the former inlet mouth is

closed by a landward-migrating bar complex (see Fig. 12.11), the strong reduction in wave and tidal energy leads to the deposition of muddy sediment (abandoned inlet channel fill). In the case of Price Inlet, this migration and breaching process has occurred twice, leaving behind two active and inactive fill sequences (Tye and Moslow, 1993). These units are topped by tidal-creek and marsh sediment. The most obvious difference between the North Carolina wave-dominated inlet deposits and the mixed-energy examples from the South Carolina coast is the

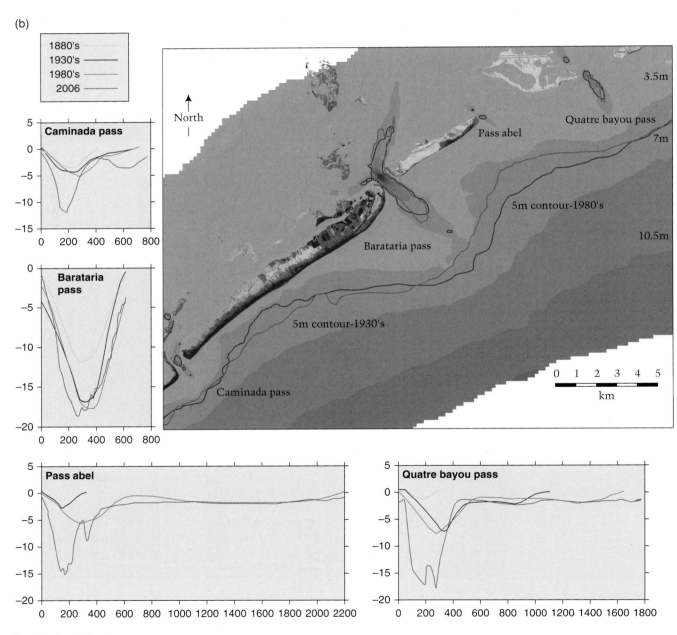

Fig. 12.16   (continued)

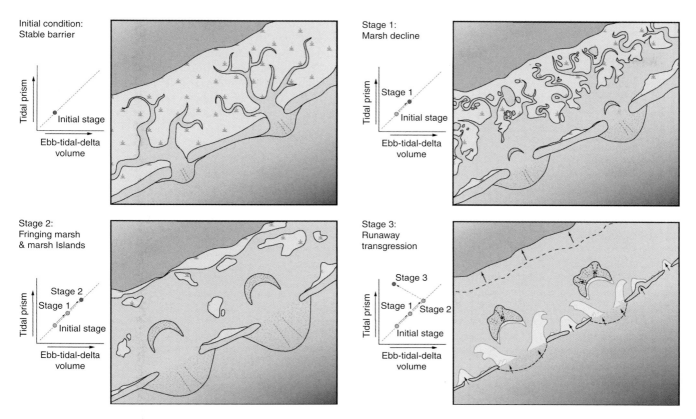

Fig. 12.17 Conceptual evolutionary model of a mixed-energy barrier coast to a regime of accelerated sea-level rise. (Source: FitzGerald et al. 2006.) For colour details, please see Plate 31.

presence of muddy, inactive channel sediment. It should also be recognized that some mixed-energy tidal inlets migrate and produce sedimentary sequences similar to those of North Carolina, such the East Friesian Island inlets (FitzGerald, 1996).

12.3.8 Tidal-inlet response to sea-level rise

Accelerated sea-level rise (SLR) may lead to the drowning of marshes, increased tidal prism, and a growth in the volume of sand contained in ebb- and flood-tidal deltas (Fig. 12.16). If salt marshes cannot maintain their areal extent through vertical accretion, then these wetlands will be transformed into intertidal and subtidal environments. In mixed-energy settings, backbarrier salt marshes are characterized by an almost flat topographic profile, generally covering a wide area (width = 2–10 km). This unique hypsometry will lead to a significant increase in the tidal prism if SLR causes frequent flooding of the supratidal marsh surface or the marshes are eroded and segmented. An increase in the tidal range in the backbarrier may also occur as marshlands convert to

open water. Increasing water levels and larger tidal inlets will reduce frictional resistance in tidal-wave propagation through the inlet, which will result in higher tidal elevations, increased tidal ranges, and enlarging tidal prism (Mota Oliveira, 1970). As the O'Brien (1931, 1969) relationship predicts, an increase in tidal prism will strengthen the tidal currents and enlarge the size of the tidal inlet (see Fig. 12.10). During the process of increasing tidal exchange between the backbarrier and ocean, the potential for dramatic changes to the inlet shoreline exist. It is well known that tidal prism is the primary factor governing the volume of the ebb-tidal delta (Walton and Adams, 1976). Thus, as salt marshes are replaced with open water, enlarged tidal prisms will enable the sequestration of an increasingly greater quantity of sand on ebb-tidal deltas. At the same time, the increase in extent of subtidal and intertidal areas landward of inlets will create accommodation space and promote the formation and/or enlargement of flood-tidal deltas (FitzGerald et al., 2004). As the backbarrier marsh converts to an open-water lagoon, the ensuing changes in inlet hydraulics from ebb to flood dominance will aid shoal development in the backbarrier (Mota Oliveira, 1970).

The capture of sand at inlets and within the backbarrier diminishes sand reservoirs of the barrier islands, leading to breaching and eventually retrogradational barriers. The rate at which barrier chains may evolve into a transgressive barrier system depends on the future rate of SLR, marsh accretion rates, and the volume of sand contained in the barrier system. FitzGerald et al. (2006) presented a conceptual model depicting the response of mixed-energy barrier island chains to a regime of accelerated SLR (Fig. 12.17). This evolutionary model is based on marsh conversion to open water causing an increase in tidal prism and growth of the ebb shoals. Sand is also lost from the littoral system as it is moved into the backbarrier to form flood shoals. In particular, flood-tidal deltas will form as the inlets change from ebb- to flood-dominated systems in concert with the transformation of marsh and tidal creeks to open bays. However, this change will not retard the growth of the ebb deltas because that volume depends on the tidal prism. The scenario presented in Fig. 12.17 suggests that barrier islands in mixed-energy settings will change drastically in a regime of accelerated SLR.

12.4   Summary

As discussed in this chapter, tidal inlets and estuaries occur in a variety of settings, but most commonly along coastal-plain depositional coasts. The hydrodynamics, sediment transport trends, gross facies relationships, and morphology of these systems are controlled by the interaction and relative magnitude of wave, tides, and in the case of estuaries, fluvial processes. The abundance and type of sediment contributed to these systems is an important factor in dictating morphology and facies assemblages. Some basics tenets of tidal inlets include:
- Tidal prism governs inlet size and volume of sand contained in the ebb-tidal delta.
- Sediment by-passes inlets through a number of mechanisms leading to bar welding to the inlet shoreline.
- The greatest extent of change to barrier shorelines occurs in the vicinity of tidal inlets.
- Tidal-inlet sequences consist of active and inactive deposits with a general fining-upward trend.

Estuaries can be classified according to their geological setting, salinity distribution, or sedimentary and morphological regime. Sedimentologically, there are two end members, which reflect the relative dominance of wave versus tidal energy. The wave-dominated estuary is characterized by a bay-head delta, a central low-energy basin, and an estuarine mouth partly closed off by a single or multiple barriers, and tidal inlets. The tide-dominated system is typically funnel-shaped and consists of extensive marsh and tidal-flats deposits with coarse-grained sediment arranged in elongated tidal sand bodies. A third system has been identified in rock-bound estuaries in New England, USA. These systems are unique in that they periodically discharge coarse sediment to the coast.

References

Argow, B. and FitzGerald, D.M., 2006, Winter processes on northern salt marshes: evaluating the impact of in-situ peat compaction due to ice loading, Wells, ME. *Estuarine, Coastal and Shelf Science*, **69**, 360–369.

Armon, J.W., 1979. Landward sediment transfers in a transgressive barrier island system, Canada. In: S.P. Leatherman (Ed.), *Barrier Islands: From the Gulf of St. Lawrence to the Gulf of Mexico*. Academic Press, New York, USA, pp. 65–80.

Aubrey, D.G. and Speer, P.E., 1984. Updrift migration of tidal inlets. *Journal of Geology*, **92**, 531–546.

Barnard, P.L. and Davis, R.A., 1999. Anthropogenic versus natural influences on inlet evolution: West-Central Florida. *Proceedings, Coastal Sediments '99, ASCE*, pp. 1489–1504.

Barras, J.A., 2006. *Land Area Change in Coastal Louisiana after the 2005 Hurricanes – A Series of Three Maps*. US Geological Survey Open-File Report 06–1274.

Biggs, R.B. and Howell, B.A., 1984. The estuary as a sediment trap: alternate approaches to estimating its filtering efficiency. In: V.S. Kennedy (Ed.), *The Estuary as a Filter*. Academic Press, New York, USA, pp. 107–129.

Boggs, S., Jr. and Jones, C.A., 1976. Seasonal reversal of floodtide dominant sediment transport in a small Oregon estuary. *Geological Society of America Bulletin*, **87**, 419–426.

Boothroyd, J.C., Friedrich, N.E. and McGinn, S.R., 1985. Geology of microtidal coastal lagoons, RI. In: G.F. Oertel and S.P. Leatherman (Eds.), Barrier Islands. *Marine Geology*, **63**, pp. 35–76.

Boyd, R., Dalrymple, R. and Zaitlin, B.A., 1992. Classification of clastic coastal depositional environments. *Sedimentary Geology*, **80**, 132–150.

Bruun, P. (1966). *Tidal Inlets and Littoral Drift*. Universitelsforlaget, Oslo, Norway.

Byrne, R.J., Bullock, P. and Taylor, D.G., 1975. Response characteristics of a tidal inlet: a case study. In: L.E. Cronin (Ed.), *Estuarine Research*, Vol. **2**. Academic Press, New York, USA, pp. 201–216.

Cameron, W.M. and Pritchard D.W., 1963. Estuaries. In: M.N. Hill (Ed.), *The Sea*, Vol. **2**. Wiley, New York, USA, pp. 306–324.

Coleman, J.M. and Wright, L.D., 1975. Modern river deltas: variability of processes and sand bodies. In M.L. Brousard (Ed.), *Deltas – Models for Exploration*. Houston Geological Society, Houston, TX, USA, pp. 99–149.

Cooper, J.A.G., 1993. Sedimentation in a river dominated estuary. *Sedimentology*, **40**, 979–1017.

D'Alpaos, A., Lanzoni, S., Marani, M. and Rinaldo, A., 2009. On the O'Brien-Jarrett-Marchi law. *Rendiconti Lincei*, **20**, 225–236, doi: 10.1007/ s12210-009-0052-x.

Dalrymple, R.W. and Choi, K.S., 2007. Morphologic and facies trends through the fluvial-marine transition in tide-dominated depositional systems: a systematic framework for environmental and sequence-stratigraphic interpretation. *Earth Science Reviews*, **81**, 135–174.

Dalrymple, R.W., Zaitlin, B.A. and Boyd, R., 1992. Estuarine facies models: conceptual basis and stratigraphic implications. *Journal of Sedimentary Petrology*, **62**, 1130–1146.

Davis R.A. Jr. and Hayes, M.O., 1984. What is a wave-dominated coast? *Marine Geology*, **60**, 313–329.

Fenster, M.S. and FitzGerald, D., 1996. Morphodynamics, stratigraphy, and sediment transport patterns in the Kennebec River estuary, Maine, USA. *Sedimentary Geology*, **107**, 99–120.

Fisher, J.J., 1962. Geomorphic expression of former inlets along the Outer Banks of North Carolina. Unpublished Masters Thesis, University of North Carolina, Chapel Hill, NC, USA.

FitzGerald, D.M., 1982. Sediment bypassing at mixed energy tidal inlets. *Proceedings of the 18th Coastal Engineering Conference, ASCE*, pp. 1094–1118.

FitzGerald, D.M., 1984. Interactions between the ebb-tidal delta and landward shoreline, Price Inlet, South Carolina. *Journal of Sedimentary Petrology*, **54**, 1301–1316.

FitzGerald, D.M., 1988. Shoreline erosional-depositional processes associated with tidal inlets. In: D.G. Aubrey and L. Weishar (Eds.), *Hydrodynamics and Sediment Dynamics of Tidal Inlets*. Springer, Berlin, Germany, pp. 186–225.

FitzGerald, D.M., 1996. Geomorphic variability and morphologic and sedimentological controls on tidal inlets. In: A.J. Mehta (Ed.), *Understanding Physical Processes at Tidal Inlets*. Journal of Coastal Research, Special Issue, No. 23, pp. 47–71.

FitzGerald, D.M. and Pendleton, E., 2002. Inlet formation and evolution of the sediment bypassing system: New Inlet, Cape Cod, Massachusetts. *Journal of Coastal Research*, **36**, 290–299.

FitzGerald D.M. and Penland S., 1987. Backbarrier dynamics of the East Friesian Islands. *Journal of Sedimentary Petrology*, **57**, 746–754.

FitzGerald, D.M., Penland, S. and Nummedal, D., 1984. Control of barrier island shape by inlet sediment bypassing: East Friesian Islands, West Germany. *Marine Geology*, **60**, 355–376.

FitzGerald, D.M., Buynevich, I.V., Fenster, M.S. and McKinlay, P.A., 2000. Sand dynamics at the mouth of a rock-bound, tide-dominated estuary. *Sedimentary Geology*, **131**, 25–49.

FitzGerald, D.M., Kraus, N.C. and Hands, E.B., 2001. *Natural Mechanisms of Sediment Bypassing at Tidal Inlets*. ERDC/CHL-IV-A, US Army Engineer Research and Development Center, Vicksburg, MS, USA (http://chl.wes.army.mil/library/publications/cetn).

FitzGerald, D.M., Buynevich, I.V., Davis, R.A., Jr. and Fenster, M.S., 2002. New England tidal inlets with special reference to riverine associated inlet systems. *Geomorphology*, **48**, 179–208.

FitzGerald, D.M., Kulp, M., Penland, P., Flocks, J. and Kindinger, J., 2004. Morphologic and stratigraphic evolution of ebb-tidal deltas along a subsiding coast: Barataria Bay, Mississippi River Delta. *Sedimentology*, **15**, 1125–1148.

FitzGerald, D.M., Buynevich, I.V. and Argow, B., 2006. Model of tidal inlet and barrier island dynamics in a regime of acceler-

ated sea-level rise. *Journal of Coastal Research, Special Issue*, **39**, 789–795.

FitzGerald, D.M., Howes, N., Kulp, M., Hughes, Z., Georgiou, I. and Penland, S., 2007. Impacts of rising sea level to backbarrier wetlands, tidal inlets, and barriers: Barataria Coast, Louisiana. *Coastal Sediments '07, Conference Proceedings*, CD-ROM13.

Galloway, W.E., 1975. Process framework for describing the morphologic and stratigraphic evolution of deltaic depositional systems. In: M.L. Broussard (Ed.), *Deltas – Models for Exploration*. Houston Geological Society, Houston, TX, USA, pp. 87–98.

Harris, P.T., Heap, A.D., Bryce, S.M., Porter-Smith, R., Ryan, D.A. and Heggie, D.T., 2002. Classification of Australian clastic coastal depositional environments based upon a quantitative analysis of wave, tidal, and river power. *Journal of Sedimentary Research*, **72**, 858–870.

Hayes, M.O., 1975. Morphology of sand accumulations in estuaries. In: L.E. Cronin (Ed.), *Estuarine Research*, Vol. **2**. Academic Press, New York, USA, pp. 3–22.

Hayes, M.O., 1979. Barrier island morphology as a function of tidal and wave regime. In: S.P. Leatherman (Ed.), *Barrier Islands: From the Gulf of St. Lawrence to the Gulf of Mexico*. Academic Press, New York, USA, pp. 1– 28.

Hayes, M.O. and Kana, T.W., 1976. *Terrigenous Clastic Depositional Environments*. Technical Report 11-CRD. Geology Department, University of South Carolina, Columbia, SC, USA.

Hayes, M. O., FitzGerald, D.M., Hulmes, L.J. and Wilson, S.J., 1976. Geomorphology of Kiawah Island. In: M.O. Hayes and T.W. Kana (Eds.), *Terrigenous Clastic Depositional Environments*, Technical Report 11-CRD. Geology Department, University of South Carolina, Columbia, SC, USA, pp. 80–100.

Howes, N.C., FitzGerald, D.M., Hughes, Z.J. et al., 2010. Hurricane-induced failure of low salinity wetlands. *Proceedings of the National Academy of Sciences*, **107**(32), 14014–14019, doi: 10.1073/pnas.0914582107.

Jarrett, J.T., 1976. *Tidal Prism-Inlet Area Relationships*. GITI Report No. 3, US Army Corps of Engineers, Waterways Experiment Station, Vicksburg, MS, USA.

Komar, P.D., 1997. *Beach Processes and Sedimentation*. Prentice Hall, New York, USA.

Kulp, M.A., FitzGerald, D., Penland, S. et al., 2006. Stratigraphic architecture of a transgressive tidal inlet-flood tidal delta system: Raccoon Pass, Louisiana. *Journal of Coastal Research, Special Issue*, **39**, 1731–1736.

Kumar R. and Sanders J.E., 1974. Inlet sequences: a vertical succession of sedimentary structures and textures created by the lateral migration of tidal inlets. *Sedimentology*, **21**, 291–323.

Manning, A.J., Martens, C., de Mulder, T. et al., 2007. Mud floc observations in the turbidity maximum zone of the Scheldt Estuary during neap tides. *Journal of Coastal Research, Special Issue*, **50**, 832–836.

Marchi, E., 1990. Sulla stabilità delle bocche lagunari a marea. *Rendiconti Lincei*, **9**, 137–150.

Masselink, G. and Hughes, M.G., 2003. *Introduction to Coastal Processes and Geomorphology*. Oxford University Press, London, UK.

Meade, R.H., 1972. Transport and deposition of sediments in estuaries. In: B.W. Nelson (Ed.), *Environmental Framework of*

Coastal Plain Estuaries. Geological Society of America Memoir **133**, pp. 91–120.

Mehta, A. and McAnally, W., 2008. Fine sediment transport. In: M.H. Garcia (Ed.), *Sedimentation Engineering: Processes, Measurements, Modeling, and Practice*. American Society of Civil Engineers, Reston, VA, USA, pp. 253–306.

Milliman, J.D., 1980. Sedimentation in Fraser River and its estuary, southwestern British Columbia, Canada. *Estuarine and Coastal Marine Science*, **10**, 609–633.

Miner, M.D., Kulp, M.A., FitzGerald, D.M. and Georgiou, I.Y., 2009. Hurricane-associated ebb-tidal delta sediment dynamics. *Geology*, **37**, 851–854.

Mitchum, R.M, Vail, P.R. and Sangree J.B., 1977. Seismic stratigraphy and global changes of sea level. Part 6: Stratigraphic interpretation of seismic reflection patterns in depositional sequences. In: C.E. Payton (Ed.), *Seismic Stratigraphy – Applications to Hydrocarbon Exploration*. AAPG Memoir 26, Tulsa, OK, USA, pp. 117–133.

Moslow, T.F. and Heron, S.D., Jr., 1978. Relict inlets: preservation and occurrence in the Holocene stratigraphy of southern Core Banks, North Carolina. *Journal of Sedimentary Petrology*, **48**, 1275–1286.

Moslow, T.F and Tye, R.S., 1985. Recognition and characterization of Holocene tidal inlet sequences. *Marine Geology*, **63**, 129–152.

Mota Oliveira, I.B., 1970. Natural flushing ability in tidal inlets. *American Society Civil Engineers Proceedings 12th Coastal Engineering Conference*, Washington, DC, USA, pp. 1827–1845.

Nichols, M.M. and Biggs, R.B., 1985. Estuaries. In: R.A. Davis Jr. (Ed.), *Coastal Sedimentary Environments* (2nd edn). Springer-Verlag, New York, USA, pp. 77–186.

Nummedal, D. and Humphries, S.M., 1978. *Hydraulics and Dynamics of North Inlet, South Carolina, 1975–1976*. GITI Report No. 16, CERC, USACE, Fort Belvoir, VA, USA.

O'Brien, M.P., 1931. Estuary tidal prisms related to entrance areas. *Civil Engineer*, **1**, 738–739.

O'Brien, M.P., 1969. Equilibrium flow areas of inlets on sandy coasts. *Journal of Waterways, Harbors, and Coastal Engineers, ASCE*, **95**, 43–55.

Prandle, D., Lane, A. and Manning, A.J., 2006. New typologies for estuarine morphology. *Geomorphology*, **81**, 309–315.

Pritchard, D.W., 1952. Estuarine hydrography. In: H.E. Landsberg (Ed.), *Advances in Geophysics*. Academic Press, New York, USA, pp. 243–280.

Pritchard, D.W., 1967. What is an estuary: physical viewpoint. In: G.H. Lauff (Ed.), *Estuaries*. American Association for the Advancement of Science, Washington, DC, USA, pp. 3–5.

Rodriguez, A.B., Simms, A.R. and Anderson, J.B., 2010. Bay-head deltas across the northern Gulf of Mexico back step in response to the 8.2 ka cooling event. *Quaternary Science Reviews*, **29**, 3983–3993.

Siringan F.P. and Anderson, J.B., 1993. Seismic facies, architecture, and evolution of the Bolivar Roads tidal inlet/delta complex, East Texas Gulf Coast. *Journal of Sedimentary Petrology*, **63**, 794–808.

Stauble, D.K. and Cialone, M.A., 1996. Ebb shoal evolution and sediment management techniques Ocean City Inlet, Maryland. *Proceedings 9th National Conference on Beach Nourishment*, St. Petersburg, FL, USA, pp. 209–224.

Susman, K.R. and Heron, S.D., 1979. Evolution of a barrier island-Shackleford Banks, Carteret County, North Carolina. *Geological Society of America Bulletin*, **90**, 205–215.

Tye, R.S., 1984. Geomorphic evolution and stratigraphy of Price and Capers Inlets, South Carolina. *Sedimentology*, **31**, 655–674.

Tye, R.S. and Moslow, T.F., 1993. Tidal inlet reservoirs: Insights from modern examples. In: E.G. Rhodes and T.F. Moslow (Eds), *Marine Clastic Reservoirs: Examples and Analogues*. Springer-Verlag, New York, USA, pp. 236–252.

Uncles, R.J. 2002. Estuarine physical processes research: some recent studies and progress. *Estuarine, Coastal and Shelf Science*, **55**, 829–856.

Valle-Levinson, A., 2010. Definition and classification of estuaries. In: A. Valle-Levinson (Ed.), *Contemporary Issues in Estuarine Physics*. Cambridge University Press, Cambridge, UK, pp. 1–11.

Walton, T.L. and Adams, W.D., 1976. Capacity of inlet outer bars to store sand. *Proceedings of the 15th Coastal Engineering Conference*, ASCE, Honolulu, Hawaii, pp.1919–1937.

Widdows, J., Blauw, A., Heip, C.H.R. et al., 2004. Role of physical and biological processes in sediment dynamics of a tidal flat in Westerschelde Estuary, SW Netherlands. *Marine Ecology Progress Series*, **274**, 41–56.

Woodroffe, C.D., Chappell, J., Thom, B.G. and Wallensky, E., 1989. Depositional model of a macrotidal estuary and floodplain, South Alligator River, northern Australia. *Sedimentology*, **36**, 737–756.

13 Deltas

Aix Marseille Université, Institut Universitaire de France, Europôle Méditerranéen
de l'Arbois, Aix en Provence Cedex, France

13.1 Deltas: definition, context and environment

13.1.1 A definition of deltas

Wright (1985) defined deltas as subaqueous and subaerial coastal accumulations of river-derived sediments adjacent to, or in close proximity to, the source stream, including the deposits that have been secondarily reworked by waves, currents or tides. Most river deltas are formed on the margins of marine basins, and the deltas of some of the biggest rivers are the largest coastal landforms in the world (Evans, 2012), but deltas exist in all sizes (Fig. 13.1). It is relevant to note at the outset that sediment transported from deltas is important in sourcing adjacent coasts in sediments, both relict and modern

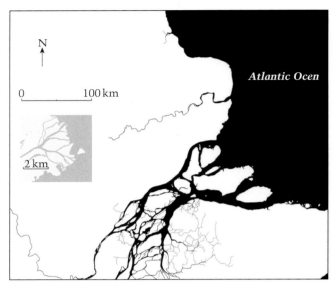

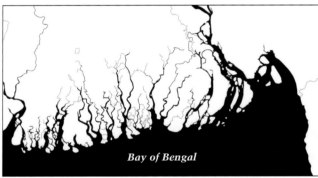

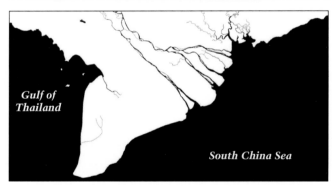

Fig. 13.1 The Amazon, Ganges-Brahmaputra and Mekong deltas, considered as the three largest deltas in the world. These megadeltas are also the largest coastal landforms. The inset in the Amazon (delta area: c. 465,000 km²) shows, in comparison, the Mossy river delta (area: c. 20 km²), which flows into Lake Cumberland, Saskatchewan, Canada, and is 23,000 times smaller than the Amazon.

(Fig. 13.2). The word delta, however, is used in a more general sense to describe any feature resulting from this type of marginal accumulation, including in lakes, lagoons, ponds and reservoirs. Deltas form where conditions in the receiving basin are not energetic enough to disperse all of the sediment brought down by rivers. The term 'delta' comes from Greek for the letter D, and is attributed to Herodotus, the 5th century BC Greek historian who first recognized the similarity of the shape of the subaerial Nile delta to this letter.

Modern deltas are a product of fluvial sediment sequestering on the present coasts (Fig. 13.3). Deltas also act as filters, sinks and reactors for continental materials, including carbon, on their way to the ocean. They are characterized by low topography, by commonly high productivity, rich and biodiverse ecosystems, and a wide range of related ecosystem services such as coastal defence, drinking-water supply, recreation, green tourism, and nature conservation. Many deltas support intense agriculture and fisheries, and are food baskets for many nations. Industry and transport in some deltas are also very important, leading to the development of major urban centres, ports and harbours. Deltas are home to nearly 600 million people. These low-lying areas are, however, vulnerable to a large range of hazards such as catastrophic river floods, tsunamis, cyclones, subsidence and global sea-level rise, and this vulnerability is increasing as a result of reduced sediment flux from rivers, caused by human interventions. Although deltas may develop dynamic resilience and adapt to changes in sediment supply and sea level, commonly through re-organization of their channels and their patterns of sedimentation, they have become economic and environmental hotspots, as human pressures have increasingly diminished these capacities of resilience, adaptation and hazard absorption.

13.1.2   The tectonic context of deltas

Deltas have generally formed on sites where subsidence has led to surface lowering, thus attracting rivers that have eroded their catchments and supplied large amounts of sediment over geological time. Subsidence in deltas appears to be generated by various mechanisms affecting the lithosphere such as temperature changes, changes in lithospheric thickness, loading and unloading (Evans, 2012). The mechanism of delta subsidence is self-reinforcing, since large sediment supplies imply significant loading that further contributes to subsidence, thus maintaining deltas in their locations. Over time, deltas are the sites of accumulation of thick sediment sequences that sometimes serve as reservoirs for hydrocarbons. In modern deltas, subsidence is also caused by compaction resulting from the dewatering of the superficial Holocene and underlying sediments, which are generally rich in water because of deposition in subaqueous and organic-rich environments.

The locations of the major deltas of the world vary, but correspond to clearly identified tectonic situations (Evans,

Fig. 13.2 A small river delta on the coast of Sierra Leone, West Africa. The Moa River has supplied sand to the coast, which is reworked by waves to form a series of bars that are built up into disjointed barriers in the vicinity of the mouth. The bars and built-up barriers serve as reservoirs that source longshore sand redistribution by littoral drift. The numerous beach ridges bordering the delta to the south were built over the last 5000 years, mainly from relict fluvial sand deposited by the Moa and other rivers on the inner continental shelf to the south during periods of lower sea level in the Pleistocene.

2012). Many deltas have formed at rifted continental margins, notable examples being the Amazon and the Niger. These deltas are built on faulted continental crust but extend onto oceanic crust, thus illustrating the important role of deltas in building out continental margins. Major rivers have generally been diverted by orogenic mountains towards accretionary plate margins, where large deltas and broad sediment-rich continental shelves have developed. Apart from the Amazon and Orinoco in South America, other examples include the Mississippi and Mackenzie in North America, and the Shatt-el-Arab (Tigris-Euphrates) and Indus in Asia. Other deltas occur at the seaward terminations of drainage basins located within such orogenic belts as in the cases of the Yukon in Alaska and the Magdalena in Columbia. In other cases, the deltas are located on the edges of marginal basins in which the river courses are controlled by faults. Examples include the Nile, Yangtze (Chang Jiang) and

Colorado. Still others form the terminations of rivers flowing parallel to such mountain belts, as in the case of the Irrawaddy. The Rhine is encased in a graben and entirely situated on a craton, building out on the shallow North Sea rather than on a continental margin. Evans (2012) noted that very few rivers manage to cross the orogenic fold-belts of the world to produce deltas on the oceanic coasts. Examples of deltas associated with rivers in this unusual situation include the Fraser River, which cuts across the Rocky Mountains, the Brahmaputra across the Himalaya, and the Ebro across the Cordillera Cantabrica in Spain.

13.1.3 Why do some rivers form deltas?

River-sediment input through geological time has been important in assuring the progradation of coasts, and in the growth of continental shelves and the ocean floor. Not

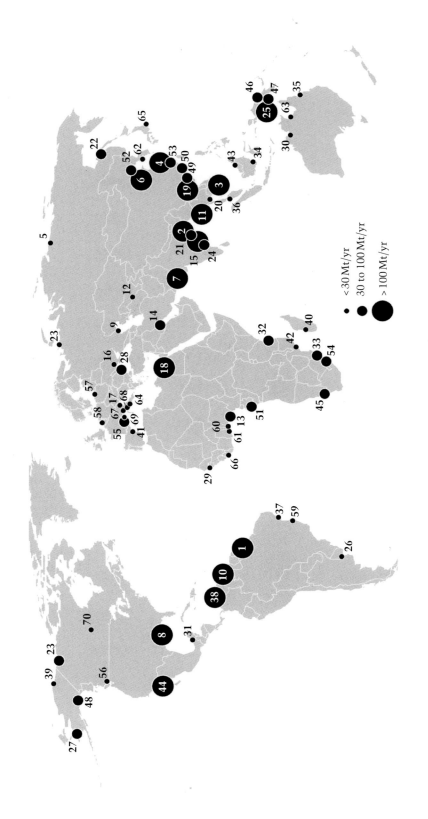

Fig. 13.3 Sediment input to the global ocean by a number of large rivers. Many of these rivers have constructed deltas. A large concentration of rivers drains the tectonically mobile zones associated with the Himalaya and the Tibetan plateau, as well as the steep, tectonically active island catchments of the western Pacific, such as Indonesia. Numbers refer to all the deltas cited in the text, based, from 1 to 43, on the areas shown in Fig. 13.4. Beyond 43, numbers are not representative of delta area. 1, Amazon; 2, Ganges-Brahmaputra; 3, Mekong; 4, Yangtze; 5, Lena; 6, Yellow; 7, Indus; 8, Mississippi; 9, Volga; 10, Orinoco; 11, Irrawaddy; 12, Amu Darya; 13, Niger; 14, Shatt-el-Arab; 15, Godavari; 16, Dneiper; 17, Po; 18, Nile; 19, Red; 20, Chao Phraya; 21, Mahanadi; 22, Pechora; 23, Mackenzie; 24, Krishna; 25, Fly; 26, Parana; 27, Yukon; 28, Danube; 29, Senegal; 30, Ord; 31, Grijalva; 32, Tana; 33, Zambezi; 34, Mahakam; 35, Burdekin; 36, Klang; 37, Sao Francisco; 38, Magdalena; 39, Colville; 40, Mangoky; 41, Ebro; 42, Pungoe; 43, Baram; 44, Colorado; 45, Orange; 46, Purari; 47, Sepik; 48, Copper; 49, Pearl; 50, Choshui; 51, Congo; 52, Liao He; 53, Kaoping; 54, Limpopo; 55, Rhône; 56, Fraser; 57, Vistula; 58, Rhine-Meuse; 59, Jequitinhonha; 60, Ouémé; 61, Mono; 62, Han; 63, McArthur; 64, Tiber; 65, Tone; 66, Moa; 67, Arno; 68, Ombrone; 69, Var; 70, Mossy. (Source: Adapted from Walsh and Nittrouer 2009. Reproduced with permission of Elsevier.)

all rivers form deltas, however. The development and size of deltas (Fig. 13.4) depend on a number of conditions that include sediment supply, grain size, the morphology of the coast and receiving basin, and the oceanographic conditions in the receiving basin. Delta formation requires trapping of sediment in the vicinity of the river-mouth. The largest supplies of sediments come from rivers draining tectonically mobile zones associated with mountain ranges, notably the Himalaya and the Tibetan plateau, and from the steep, tectonically active island catchments of the western Pacific such as Indonesia, where there is a particularly large concentration of deltas (see Fig. 13.3). The sediment collected by rivers represents the products of weathering and erosion of the continents on timescales ranging from the present to millions of years. These processes are also important in determining the mineralogy and grain sizes of sediments supplied to the world's deltas. Rivers carry sediment as dissolved load, suspended sediment, or bedload, depending on grain size and flow velocity. A few large rivers supply suspended and/or dissolved loads that account for 80–90% of the total global river load to oceans, together with significant organic and pollutant loads (Woodroffe and Saito, 2011). The amount and grain size of the sediment brought down by rivers to their mouths have a determining influence, depending on the hydrodynamic context, on whether a delta forms or not, as well as on delta morphology. Orton and Reading (1993) suggested that the available grain size has an influence on:

(1) the gradient and channel pattern of the fluvial system on the delta plain; (2) the mixing behaviour of sediment as it discharges into the receiving basin waters at the river-mouth; (3) the type of shoreline and its response to wave energy and tidal regime; and (4) the deformation and resedimentation processes on the subaqueous delta front. Suspended sediment may be completely dispersed offshore and alongshore where waves and currents are active. About 90% of the sediment load of the Amazon (Box 13.1), the world's largest river and delta, is clay, and to a lesser extent, silt. Materials eroded mainly from the Andes, and then transported, stored and remobilized over thousands of kilometres of Amazon River catchment, undergo intense weathering, resulting in the almost exclusively muddy sediment supplied to the coast by this river. Rivers supplying sand and gravel tend to have more chances of having deltas at their mouths, because these larger clast sizes, mainly transported as bedload, require more wave and current transport energy. Small rivers account for a dominant part of the total quantity of fluvial sediment supplied to the oceans (Milliman et al., 1999). They also have, however, less chances of having deltas formed at their mouths, because their lower sediment loads are more readily dispersed by waves and currents. Generally, large deltas are associated with large rivers, although exceptions exist, the most important of which is the Congo River, which lacks a well-developed delta, much of its sediment cascading down a 450-m deep canyon at its mouth.

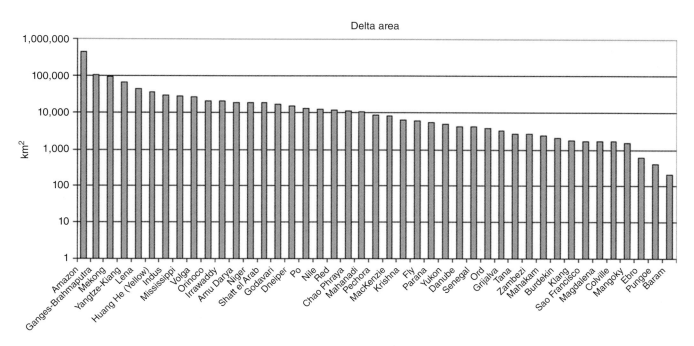

Fig. 13.4 The areas of some of the world's subaerial deltas. (Source: Adapted from Coleman and Huh 2004. Reproduced with permission from the author.)

CASE STUDY BOX 13.1   Morphology and processes of sedimentation at the mouth of the Amazon River

The Amazon, the world's biggest river, has an annual mud discharge of about 754–1000 × 10⁶ t. This large mud supply reflects sediment sourcing from the Andes orographic belt, weathering in a tropical-equatorial climate, and the extremely high water discharge (about 173,000 m³/s). The mouth of the river (Fig. 13.5a) shows a deltaic morphology typical of strong river and tidal influence. Wave activity is also important and about 15–20% of the mud supplied by the Amazon forms large migrating coastal banks that are driven by waves. As a result of the large river discharge and the muddy sediment load, much of the sediment is trapped on the shelf, to the benefit of a subaqueous delta (Fig. 13.5b, c), the uppermost portion of which is the present shoreline. The subaqueous delta is the most important part of a Holocene clinoform structure on this shelf built by the Amazon. Under the present sea-level conditions,

the sediment-trapping efficiency on the coast and shelf of the Amazon is close to 100%, such that sediments currently suplied by the Amazon do not reach the Amazon deep-sea fan. The fate of the mud supplied to the shelf by the Amazon can be considered, in simple terms, as involving four processes, schematically synthesized in Fig. 13.5(b, c): (1) hyperpycnal flows; (2) formation of a shelf-based estuarine-turbidity maximum (ETM); (3) offshore bottom gravity flows; and (4) onshore wind- and wave-driven lower-concentration muds driven towards a zone where mud-bank formation occurs. From there, the banks migrate along the coast from the mouth of the Amazon to the Orinoco delta in Venezuela, 1500 km to the northwest under wave-induced longshore drift. The Amazon is the world's largest delta by virtue of the important growth of a river- and tide-dominated delta plain at its mouth.

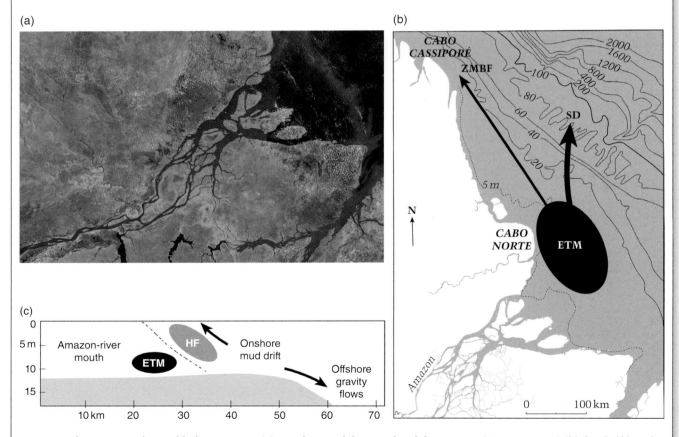

Fig. 13.5   The Amazon, the world's biggest river. (a) Aerial view of the mouths of the Amazon (Source: NASA); (b) the shelf-based estuarine-turbidity maximum (ETM), subaqueous delta (SD) and zone of mud-bank formation (ZMBF); (c) schematic cross-section of mud transport pathways, including hyperpycnal flows (HF).

13.2 Delta sub-environments

13.2.1 The delta plain

Deltas are relatively diversified landforms that can include various units (Fig. 13.6). They may also be linked to deep-sea fans that have built up over geological time from sediment supplied by rivers. The visible part of the delta is the subaerial delta plain. A typical delta plain comprises an upper part where river influence is generally dominant, and a lower part where estuarine and marine processes tend to be dominant (Fig. 13.7). The transition between the upper and lower plains corresponds to the inland limits of saltwater influence. This is generally marked by a change from mangroves (in tropical to sub-tropical deltas) or salt marshes (tropical to temperate deltas) in the lower delta plain, to brackish and then freshwater marshes, including marshy forests, in the upper delta plain.

The commonly high sediment supply and the morphodynamic response of the delta to this supply may

result in the presence of active and abandoned sections of the delta plain, the former flanking the active river channel(s), and the latter exhibiting palaeochannels that were former river courses abandoned as the delta evolves. River channels are commonly encased between river levees, which are natural embankments that have accreted over time, and the floodplain comprises low-lying basins between active or abandoned channels. The floodplain is subject to inundation during the periods of highest discharge, and thus gradually builds up from sediment deposited during such events, which are commonly seasonal in tropical deltas. The entire system of channels and floodplain progressively aggrades as sediment is delivered to the delta from upstream. Levees are best developed where fluvial processes act as the dominant force shaping the delta.

Natural levees can be breached during floods. The water flows channelized in these breaches result in the formation of crevasse splays (Fig. 13.7), which serve as spillways for sediment-charged water flows into the interdistributary basins. Crevasse splays become gradually infilled. On delta coasts where wave processes are important, beach ridges are commonly formed, their plan-view shapes illustrating the patterns of longshore drift, which may be quite complex, commonly involving bi-directional drift. Where tides exert the dominant influence, delta plains are commonly characterized by an intricate shoreline comprising numerous distributary mouths that broaden seaward, separating broad intertidal zones with salt-tolerant vegetation. The mouths of these distributaries commonly exhibit linear shoals aligned with the direction of tidal flux.

13.2.2 The subaqueous delta

Subaerial delta plains are generally fronted by a subaqueous delta (Fig. 13.6) consisting of a shallow delta front and a prodelta of finer muds in deeper water, often inter-bedded with laminae of silt or thin shell-beds. Where the water discharge is large and the sediment supply essentially consisting of mud, as in the Amazon (see Box 13.1), much of the fluvial sediment can be transported several kilometres offshore of the mouth, enabling significant development of the subaqueous delta. The angle of deposition of prograding deltaic slope deposits is generally very gentle, except on steep continental shelves such as on the French Riviera in southeastern France, where high slope angles characterize delta-front deposits, as in the case of the Var River delta (Anthony and Julian, 1997).

13.2.3 Deltas and deep-sea fans

The slopes and basins offshore of the world's major deltas are generally characterized by deep-sea fans cut by canyons at their heads (Fig. 13.6). Although deep-sea

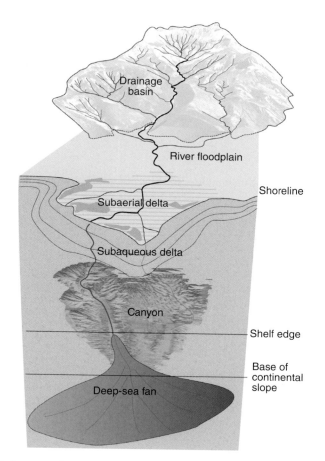

Fig. 13.6 Subaerial and subaqueous components of a typical large delta.

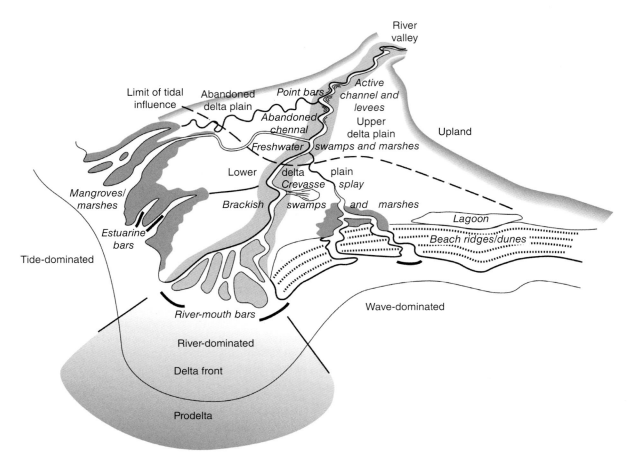

Fig. 13.7 Schematic synthesis of most features typical of subaerial deltas.

fans are often remote from the delta, these features and their canyons are an essential part of the deltaic system, and sometimes form gigantic accumulations that dwarf most other sedimentary accumulations (Evans, 2012). The Bengal fan, for instance, reaches as far south as the tip of the Indian subcontinent and covers an area of $3 \times 10^6 \, \text{km}^2$ and exceeds 4 km in thickness, and the fan of the Indus delta, which is more than 3 km thick, has an area of $1 \times 10^6 \, \text{km}^2$. Deltas are generally separated from their deep-sea fans during periods of high sea level by wide continental shelves, as at present, but some deltas, such as the Magdalena in Colombia, have built out almost to the shelf edge such that sediment goes through a canyon to the fan beyond. Low sea-level stands are particularly favourable to deep-sea fan growth, as most of the world's major deltas are able to prograde across the shelf to form shelf-edge deltas and supply the fans with vast quantities of sediment. Under the present high sea-level conditions, some deltas succeed in trapping sediment on the shelf, thus depriving their deep-sea fan of sediment, as in the case of the Amazon.

13.3 The morphodynamic classification of river deltas

Galloway (1975) classified deltas in terms of three end-point members: river-, wave- or tide-dominated, which he grouped into the now widely employed ternary diagram (Fig. 13.8). River-dominated deltas are typically elongate; wave-dominated deltas exhibit cuspate shorelines; and tide-dominated deltas have a plan-view morphology similar to that of estuaries with multiple tidal-influenced mouths. Orton and Reading (1993) added grain-size characteristics to this classification.

Individual river-mouths have generally been classified in the delta ternary diagram on the qualitative basis of the perceived influence of river flow, waves and tides. The classification of coastal depositional sequences proposed by Boyd et al. (1992) offers a framework of coastal evolution linking estuaries and deltas, with a propensity for the former to evolve into the latter as a function of time and adequate sediment supply (Fig. 13.9). This classification also includes the dichotomy between depositional wave-dominated systems,

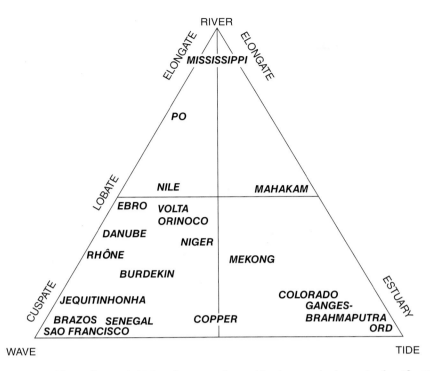

Fig. 13.8 Ternary diagram proposed by Galloway (1975) and commonly used in the morphodynamic classification of deltas on the basis of the relative balance between river, wave and tide processes.

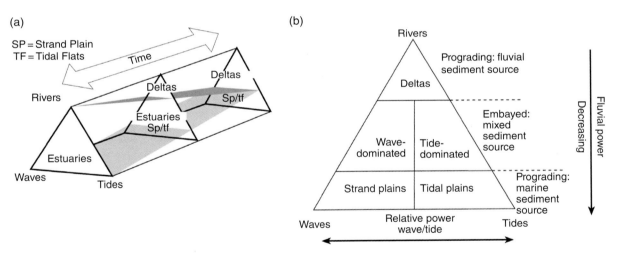

Fig. 13.9 (a) Prism showing a general coastal classification scheme based on inputs of rivers, waves and tidal processes, and their variation in time (sea-level changes), with many estuaries developing into deltas as their mouths become infilled; and (b) a cross-section through the prism showing deltas relative to other coastal landforms. (Source: Adapted from Boyd et al. 1992. Reproduced with permission of Elsevier.)

which result in strand plains comprising beach ridges, typical of wave-dominated deltas, and tide-dominated systems characterized by tidal flats that are well developed in tide-dominated deltas.

13.3.1 River-dominated deltas

River-dominated deltas are characterized by large fluvial discharge, but their development is best favoured where tidal and wave influences are weak, such as in lakes and inland seas, a fine example being the Volga delta on the shores of the Caspian Sea. The best-known example of a river-dominated delta is the Mississippi delta (Fig. 13.10), which commonly figures at the river-dominated apex of the ternary diagram (Fig. 13.8). In the low-energy settings where river-dominated deltas develop best, the currents generated by waves and tides are too weak to redistribute the fluvial sediment and waves do not rework this sediment into the usual shoreline forms such as beaches. Waves can sometimes produce thin linear and discontinuous beach ridges. Where these rest on mud, they are called cheniers (a term from Louisiana based on the French word *chêne*, or oak, because of the oak trees colonizing these shelly ridges in the Mississippi delta). As a result, river-dominated deltas commonly have an elongated shape determined by the downstream extension of distributaries, flanked by levees, into the receiving basin (Fig. 13.10). The modern Mississippi delta comprises several distributary channels that are building out into deep water in the Gulf of Mexico, where wave energy is low, except during occasional hurricanes, and where tides are extremely weak. River-dominated deltas are also commonly characterized by the lateral migration of meandering channels. As each distributary channel lengthens out into the receiving basin, its gradient diminishes, thus decreasing its hydraulic and sediment-transport competence. This commonly results in the process of avulsion, where channel switching occurs to take advantage of alternative, shorter, higher-gradient routes that are more efficient hydraulically. In certain situations, where rivers discharge into sheltered embayments, with attenuated wave and tidal activity, bay-head deltas can develop. A major example of a bay-head delta is the Shatt-El-Arab, the delta of the Tigris and Euphrates Rivers. Bay-head deltas are common in embayments closed off by wave-built barriers to form back-barrier lagoons, behind which fluvial sediment accumulates at the river-mouth as a delta (Fig. 13.11).

13.3.2 Wave-dominated deltas

Wave-dominated deltas prevail in settings where wave energy is efficient, leading to the reworking and redistribution of fluvial bedload that thus shapes the delta

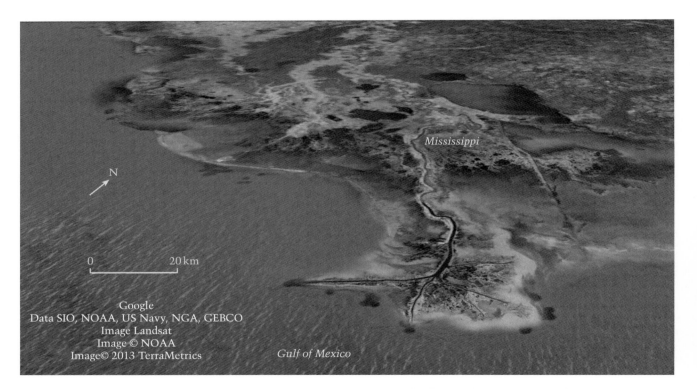

Fig. 13.10 A Google Earth image of the Mississippi River delta, an iconic example of a river-dominated delta. The Mississippi proper discharges into deep water near the continental shelf edge, following extreme, finger-like progradation of the levee-lined channels, whereas the Atchafalaya, a later branch of the Mississippi, discharges into shallow water. Parts of the delta no longer fed in sediment as a result of channel switching are subject to wave erosion, as in the western sector where barrier islands of sand have formed a lobate shoreline as the delta plain is reworked and the mud dispersed. (Source: Google. Data SIO, NOAA, US Navy, NGA, GEBCO. Image Landsat. Image © NOAA. Image © TerraMetrics.)

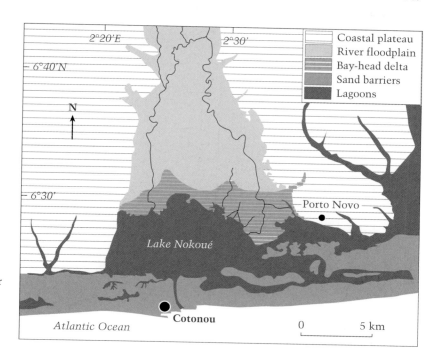

Fig. 13.11 A bay-head delta in Benin, West Africa, formed behind a beach-ridge barrier several kilometres wide that has enclosed a lagoon, forming Lake Nokoué. The Ouémé River has a catchment of about 50,000 km². The floodplain comprises two distributary channels, each of which is prograding into the low-energy lagoon.

Fig. 13.12 The Jequitinhonha River delta in Brazil is a fine example of a wave-dominated delta, showing the typical cuspate shoreline morphology.

morphology. Wave-dominated deltas are probably the most common of the three delta types, given the commonality of waves. The wave factor concerns both the direct build-up of beaches and the longshore transport of river bedload to form spits and barriers that may impound lagoons. Wave-dominated deltas therefore commonly exhibit sequences of shore-parallel beach ridges or dune-decorated ridges, recording progradational sequences of wave reworking of fluvial sediment. The shape of wave-dominated deltas is often cuspate, fine examples being the deltas of the Sao Francisco and the Jequitinhonha (Fig. 13.12) in Brazil, and the Po, Tiber and Ombrone in Italy. Where large rivers debouch on coasts subject to strong longshore drift, this commonly leads to strongly longshore-skewed deltas that may eventually cease to prograde, with sand integrally evacuated downdrift, giving a simple equilibrium drift-aligned delta shoreline, as in the example of the Senegal delta (Fig. 13.13). In extreme cases of wave influence involving strong unidirectional longshore drift, delta development can be stalted, as bedload is integrally transported downdrift of the river-mouth. A fine example is that of the Mono River in Benin (Anthony et al., 1996). Despite significant infill of its lower valley, the Mono, which supplies about 100,000 m³/yr of sand to the Bight of Benin coast, cannot prograde because the sand is swept

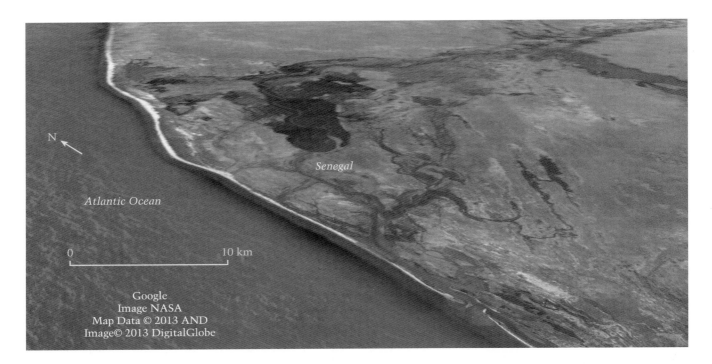

Fig. 13.13 Google Earth images of the Senegal River delta and the Mono River, in West Africa, two examples where strong wave-induced longshore drift impedes delta shoreline progradation, commonly generating downdrift river-mouth deflection and spit growth. (Source: Google. Image NASA. Map Data © 2013 AND, Image © 2013 DigitalGlobe.)

alongshore to the east by very strong wave-induced longshore drift of up to 1.5 million m³/yr.

13.3.3 Tide-dominated deltas

Tide-dominated deltas occur where large tides are the overarching control on delta morphology. The tidal influence is commonly expressed by the extensive development of tidal flats comprising a maze of meandering tidal creeks. The shoreline of tide-dominated deltas is generally extremely irregular, comprising multiple broad, trumpet- to funnel-shaped estuary mouths (Fig. 13.14). The bedload in these channels is deposited as elongate bars and islands perpendicular to the coastline and aligned with bi-directional tidal currents, and the channels taper upstream. The tidal intrusion upstream in the channels is generally important, inducing water-level changes, sometimes considerable distances upstream, as in the case of the Amazon. In certain tide-dominated deltas, sand ridges and cheniers can form discrete barriers built up under the influence of occasional episodes of strong waves. Walker (1992) considered tide-dominated deltas as estuarine systems rather than true deltas, a point of view that has become untenable as better scientific studies, especially over the last decade, of many large deltaic systems in Asia subject to dominant tidal influence, have shown (Woodroffe and Saito, 2011). In fact, five of the seven largest modern deltas (see Fig. 13.4) are largely tide-dominated, including the two largest, the Amazon and the Ganges-Brahmaputra (see Fig. 13.1). The propensity for large tidal ranges to be associated with such deltas is probably related in part to important upbuilding and outbuilding of large, shallow continental shelves over geological time by the sediments brought down by these rivers. Although resembling tide-dominated estuaries, tide-dominated deltas are characterized, like other deltaic systems, by net terrestrial sediment supply to the coast.

13.3.4 Quantifying river, wave and tide controls

Various attempts have been made to quantify delta morphodynamic controls using ratios combining river discharge, wave height and tidal range. Bhattacharya and Giosan (2003) highlighted the potential for asymmetric development of deltas under wave influence and proposed an asymmetry index that attempts to express the importance of wave influence relative to riverine processes (Fig. 13.15). Wave-induced asymmetry has been further analysed by Ashton and Giosan (2011), who showed from a numerical model that the angles from which waves approach a delta can have a first-order influence on its plan-view morphologic evolution and sedimentary architecture. Hori and Saito (2008) have proposed quantitative indices based on tidal range, wave height and suspended sediment load, to distinguish

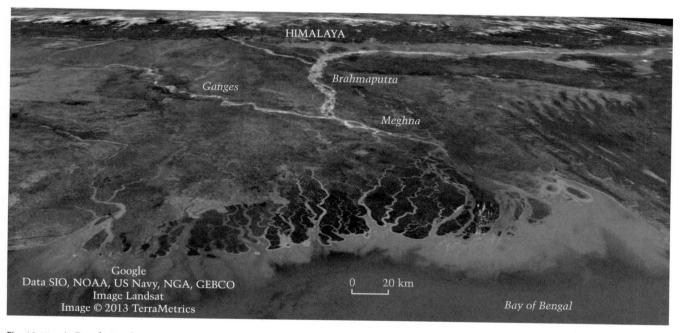

Fig. 13.14 A Google Earth image of the Ganges-Brahmaputra delta. The Ganges and Brahmaputra now flow in the Meghna main stem delta channel. The strongly tide-dominated Sundarbans in the west represent the abandoned delta plain of the Ganges River. (Source: Google. Data SIO, NOAA, US Navy, NGA, GEBCO. Image Landsat. Image © TerraMetrics.)

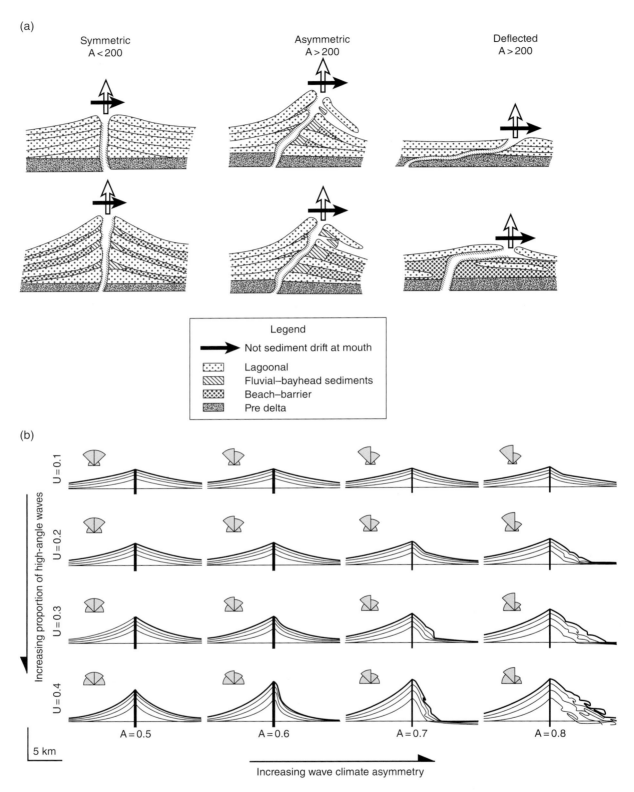

Fig. 13.15 Delta plan-form shoreline asymmetry and waves: (a) an asymmetry index of generalized delta morphologies; (b) plan view time series of simulated delta shorelines in the vicinity of a wave-dominated river-mouth for different high-angle waves, with insets of area-normalized wave energy rose type plots. The simulation is based on shorelines plotted at intervals of 65.4 model years, with a final shoreline at 327 model years. U is the proportion of high angle waves and A is the net asymmetry. (Source: (a) Bhattacharya and Giosan 2003. Reproduced with permission of John Wiley & Sons. (b) Ashton and Giosan 2011. Reproduced with permission of John Wiley & Sons.)

between wave-influenced, mixed tide- and wave-influenced, and tide-influenced deltas, retaining the valid assumption that all deltas are strongly influenced by fluvial discharge and sediment load (Fig. 13.16). Although such attempts are a useful basis for a better understanding of deltas, these coastal forms are extremely diverse, and many suffer from a lack of basic environmental data, which is necessary in determining such indices. Many deltas in the world, especially in tropical countries, are still rather poorly studied, and have difficult field access. High-resolution remote-sensing (Box 13.2) is a modern tool in the study of the morphology of many deltas and their floodplains (Syvitski et al., 2012).

(a)

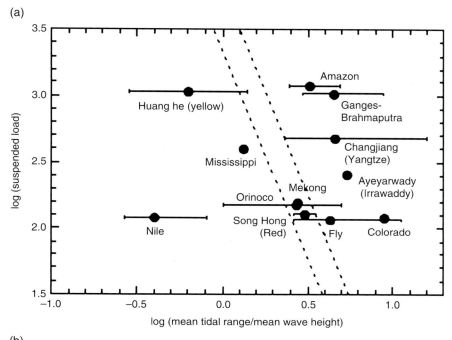

(b)

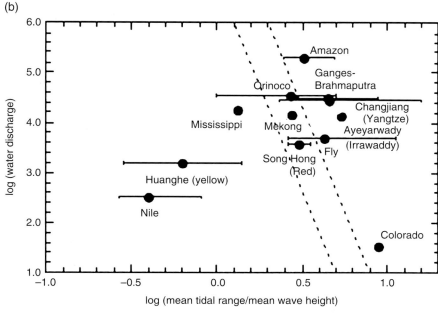

Fig. 13.16 Semi-quantitative classification of major river deltas based on a plot of log of mean tidal range/mean wave height versus: (a) log of suspended load; and (b) log of water discharge. The scheme assumes that all deltas are strongly influenced by fluvial discharge and sediment load. The dashed lines dividing the three groups (wave-influenced, mixed tide- and wave-influenced, and tide-influenced deltas) were drawn by eye. (Source: Hori and Saito 2008. Reproduced with permission of John Wiley & Sons.)

METHODS BOX 13.2   Modern approaches in delta studies

Much of the early research on deltas was conducted in the Mississippi and benefited greatly from exploratory oil and gas studies. These efforts led to early definitions of deltaic sedimentation based on stratigraphy. At the same time, the use of aerial photography enabled early observations and characterization of sediment plumes from the Mississippi into the Gulf of Mexico. The combination of these approaches and field observations led to the erection of conceptual models of delta development that gave rise,

for instance, to the ternary diagram of delta morphodynamic classification (see Fig. 13.8). Stratigraphic analysis was considerably modernized with the advent of sequence stratigraphy, considered as the most recent and revolutionary paradigm in the field of sedimentary geology. Recent studies on processes involved in subaqueous deltaic sediment transport have also improved our understanding of shelf sedimentation. Figure 13.17(a) shows an isopach map, based on digital X-radiography, of observed deposition

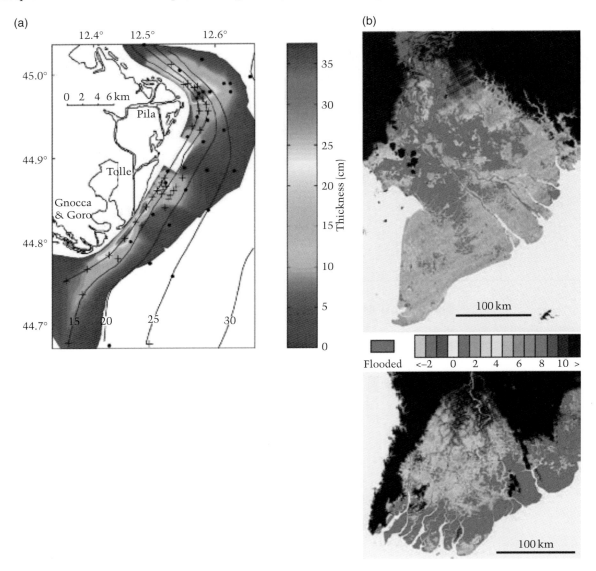

Fig. 13.17   A selection of methods used in delta studies. (a) Digital X-radiography of Po River flood deposition on the shoreface; (b) SRTM (Shuttle Radar Topography Mission) and MODIS (Moderate Resolution Imaging Spectroradiometer) imaging of the Mekong (top) and Irrawaddy (bottom) deltas; (c) time series imagery of deltas. (Source: (a) Wheatcroft et al. 2006. Reproduced with permission from Elsevier. (b) Adapted from Syvitski et al. 2009. Reproduced with permission of Nature Publishing Group. (c) Wolinsky et al. 2010. Reproduced with permission from John Wiley & Sons.) For colour details, please see Plate 32.

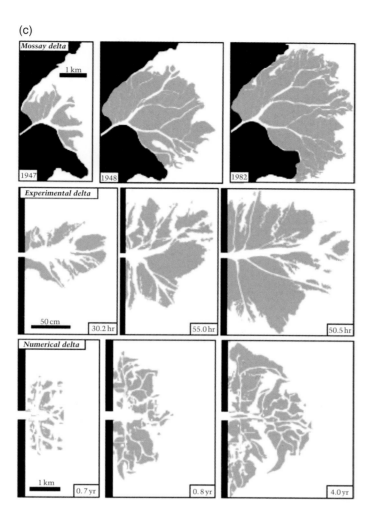

(c)

Mossay delta 1 km 1947 1948 1982

Experimental delta 50 cm 30.2 hr 55.0 hr 50.5 hr

Numerical delta 1 km 0.7 yr 0.8 yr 4.0 yr

Fig. 13.17 (Continued)

associated with the autumn 2000 Po River flood. Deltas are commonly relatively inaccessible and this, together with their large size, has been an obstacle in their study. In this regard, remote-sensing methods, the performance of which has steadily improved over the years, constitute an important breakthrough in delta studies. Over the last decade, data from remote-sensing systems such as AMSR-E (advanced microwave scanning radiometer), SRTM and MODIS are providing new quantitative information that considerably improves the characterization of floodplain and deltaic landscapes and their susceptibility to flooding (Syvitski et al., 2012). Figure 13.17(b) shows examples of the use of SRTM and MODIS imagery in highlighting patterns of flooding in the Mekong and Irrawaddy deltas. Another important development in research on deltas is that of numerical modelling, which is rapidly permeating all aspects of delta studies, from shoreline shape through avulsion and backwater dynamics to subsidence. Figure 13.17(c) shows time series imagery of experimental, numerical and field-scale deltas used by Wolinsky et al. (2010) to derive laws that govern the growth of river-dominated deltas.

13.3.5 Spatial and temporal morphodynamic variability

The morphology of the delta plain and the variety of the subenvironments it contains are closely related to the dominant type of process affecting the delta. Processes generated by river flow, waves and tides affect most river-mouths to varying extents. Whereas each of these three agents may clearly dominate the process regime of deltas, they may also show considerable variation and overlap in space and time. Variations in space affect large deltas, and arise as a result of alongshore gradients in the three forces, notably waves and tides. Morphologically, these are also the most diverse deltas.

The Orinoco delta shows a variety of features that reflect differences alongshore in river-, tide- and wave-domination (Fig. 13.18). The deltas of some large Asian rivers have also been shown to exhibit alongshore morphodynamic variability, especially the Red River delta in Vietnam (Woodroffe and Saito, 2011). Such alongshore variations complicate the classification of delta types. Temporal variations commonly reflect shifts in fluvial sediment supply that affect the balance between fluvial and marine (especially wave) processes, and may occur over all timescales, ranging from multi-millenial

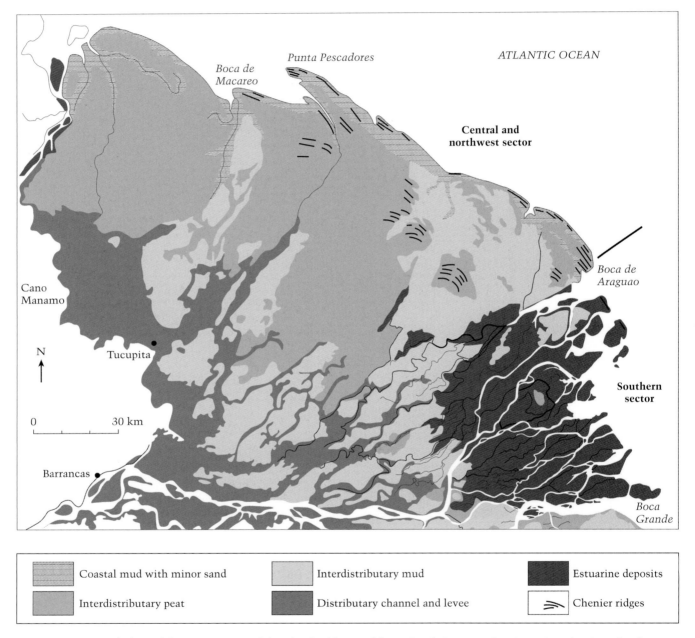

Fig. 13.18 Geomorphology of the Orinoco River delta, the third largest delta in South America, showing a river-dominated and tidally influenced southern sector and a wave-dominated central and northwestern sector. Progradation in the late Holocene has essentially been dominated by the southern distributary in the southern delta (Boca Grande delta mouth), resulting in much of the central and northwestern delta plain becoming sediment-starved and subject to widespread accumulation of peat deposits. This also appears to explain the wave-dominated morphology of these parts of the delta, characterized by numerous sandy cheniers between Boca de Araguao and Boca de Macareo channels. (Source: Adapted from Warne et al. 2002a.)

to seasonal variations in flow. Temporal variations are generally associated with swings, commonly from river- to tide-domination, where seasonal variations in river discharge are important, as in the monsoonal deltas of Southeast and East Asia (Woodroffe and Saito, 2011). In tide-dominated deltas, in particular, there is a geographical separation between areas that remain river-dominated and the abandoned parts of the delta, which become dominated by other processes, a fine example being the Ganges-Brahmaputra delta (Woodroffe and Saito, 2011). The active main stem channel

of this delta, called the Meghna, is river-dominated, whereas the Sundarbans, which include the moribund delta of the Ganges to the west, is tide-dominated.

Deltas commonly considered as wave-dominated, such as the Danube, the Rhône and the Ebro in Europe, and the Volta in West Africa (Fig. 13.19), show, more commonly, a lobate rather than a cuspate morphology, which clearly expresses a compromise between river and wave influence, more favourable to the retention of coarse sediment within the confines of the delta, as well as deltaic

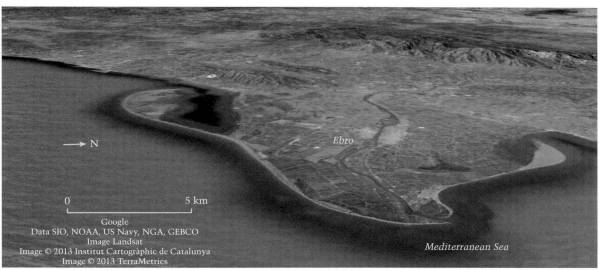

Fig. 13.19 The Volta River delta in Ghana and the Ebro River delta in Spain, two examples of deltas showing a lobate morphology representing a compromise between strong wave influence and river discharge. (Source: Google. Data SIO, NOAA, US Navy, NGA, GEBCO. Image Landsat. Image © Institut Cartographic de Catalunya. Image © TerraMetrics.)

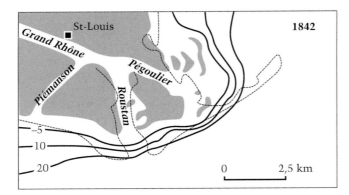
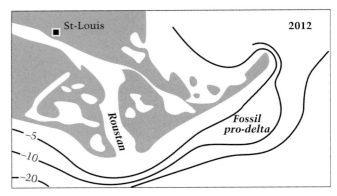

Fig. 13.20   The main delta mouth sector of the Rhône River showing morphological changes over a 70-year period associated with the closure of the Piémanson and Pégoulier mouths by spits. These changes are indicative of increasing wave domination following engineering works on the Rhône that have modified the river flow. (Adapted from Sabatier et al. 2006.)

self-organization. Even where wave influence is strong, large-scale delta morphodynamic adjustments can command the longshore drift cell structure, preventing sand leakage from the system. This occurs notably through what may be termed as the 'hydraulic groyne' effect of river (and tidal) discharge, wherein the strong outflowing river-mouth jet refracts waves. Wave-angle control has been advocated as a cause of drift reversal downdrift of wave-influenced delta mouths (Ashton and Giosan, 2011). Similar drift reversal also occurs, however, well downdrift of delta mouths where shoreline orientation changes rapidly due to more pronounced progradation in the mouth(s) sector as a result of the hydraulic groyne effect on waves, resulting in conservation of a large share of the bedload for progradation. The downdrift delta termini are commonly prominent sand spits, as on the Volta, Ebro (Fig. 13.19) and Rhône deltas. Where several distributary mouths occur, multiplying the hydraulic groyne effect, pronounced longshore variability in wave-induced sand transport ensues, resulting in multiple drift cells that assure sand retention, as in the Mekong and Niger deltas, signalling a clear balance between river, wave and tidal domination. By affecting the way waves redistribute delta sediments, these controls express fluvial (and tidal) determinants on delta development even when the ambient wave influence is strong, as shown by the wave-formed beach-ridge sets common in these deltas. From work on the Danube delta, Giosan et al. (2005) have further suggested that the river-mouth morphodynamics are highly non-linear, involving multiple feedbacks among subaerial deltaic progradation, deposition on the subaqueous delta, current and wave hydrodynamics, and wave-current interactions. Despite the complexity of these sedimentation processes, the resulting morphology at the mouth exhibits a tendency to self-organize.

Over time, subtle temporal shifts from mixed river and wave domination to stronger wave domination may appear where river discharge has been affected by either competition between channels or by engineering modifications to the detriment of a channel, as in the Rhône delta (Fig. 13.20). In contrast, the former delta lobes of the Mississippi had a highly lobate shape that indicates more important past wave processes than in the case of the modern strongly river-dominated delta.

13.4   Sediment trapping processes in deltas and coastal sediment redistribution

13.4.1   Delta-plain deposition

Deltas grow by upward accumulation and offshore progradation. The processes involved in these two mechanisms and in their balance are relatively complex. During periods of high discharge, sediment entrainment within the river channels becomes very important. Where a wide range of sediment sizes is available, the coarsest material is transported as bedload in the channels. Channel banks may be submerged, resulting in inundation of the adjacent floodplains. Fine sand and mud transported as suspension load during these floods are deposited along the banks of the channels where flow competence diminishes, thus forming levees. The breaching of levees through crevasse splays enables the rapid deposition of fans of coarse sediments over the fine-grained floodplain deposits with the sudden drop in flow competence. Floodplains are generally colonized by vegetation comprising woodlands or swamps with lakes and ponds in which peat can form. The sedimentary succession in many deltas can include significant amounts of peat, which is highly compressible compared to clay, silt and sand. The type of clay influences such organic matter accumulation and subsequent peat development by restricting or allowing for pore-water flow in the tidal and overbank deposits that comprise the delta plain.

Restricted pore-water flow encourages wetland conditions that further promote the accumulation of organic matter. Edmonds and Slingerland (2010) have used a numerical flow and transport model to show that sediment cohesiveness strongly influences the morphology of deltas. Their model showed that highly cohesive sediments form river-dominated deltas with rugose shorelines and highly complex floodplains, whereas less cohesive sediments result in fan-like deltas with smooth shorelines and flat floodplains. Their simulations also suggested that sediment cohesiveness also controls the number of channels that form within the deltas, and the average angle of bifurcation of those channels.

River channels may meander or have an anastomosed pattern, the implications of which are discussed later. Anastomosed channels are characterized by frequent avulsions. This occurs during extreme floods, but abandonment can also be the outcome of a progressive drop in competence as the river channel lengthens and its gradient diminishes. Repeated river avulsions are very important in the building of deltas. Another cause of avulsions is engineering works or modifications. Avulsions may also be caused by tectonic events. Sudden avulsions can have dramatic consequences where settlements are affected by the new channel course. Major avulsions may take decades to be completed, because of the necessity of significant incision into the floodplain in order to carve out a stable channel capable of carrying the river discharge during floods (Syvitski et al., 2012). The abandoned channel and its levees are left as upstanding ridges on the deltaic plain. As in river floodplains, such abandoned channels become static water bodies in which accumulate floodplain muds rich in organic matter.

13.4.2   Estuarine processes in deltas

The majority of the world's deltas are characterized by dynamic processes involving interactions between fresh- and saltwater, such that estuarine dynamics are an integral part of the functional mechanisms and sediment transport processes in these deltas. Estuarine processes may occur within the confines of the river-mouth or in an unconfined offshore setting, depending on river flux and grain size. Classical river-mouth estuarine processes may be completely excluded from small coarse-grained so-called Gilbert-type or fan-deltas fed by short streams debouching from steep mountainous hinterlands, a possibility in parts of the western Mediterranean and on the steep Pacific coast of South America. More commonly, seawater penetrates up the river channels, either mixed to varying degrees with the river freshwater discharge, or as a classical salt wedge beneath the overlying river discharge. Water and bedload generally move seaward in the river-dominated portion of the fluvial-marine transition, whereas in the tide-dominated portion, the net movement integrated over the tidal cycle (comprising the upflowing flood tide and the downflowing ebb tide and river flow) may be either seaward or landward, depending on river discharge and on the neap-spring tidal stage. The flows in the estuarine reaches of deltas are, thus, often characterzsed by mutually 'evasive' flood- and ebb-dominated transport pathways around elongate tidal banks in the delta channels and major tidal channels.

At the interface between downflowing freshwater and the upflowing salt wedge, tidal and density current activity in delta channels can generate alternating deposition and resuspension of fine material in an estuarine turbidity maximum (ETM), a zone exhibiting generally very high suspended sediment concentrations (SSC). Salt-wedge intrusion, a fundamental mechanism in the infill of estuaries, can lead to the re-introduction into delta channels of sediment deposited by a river in the nearshore or offshore zone.

The sedimentary processes that lead to the deposition of mud, notably in open-mouthed tide-dominated deltas, do not simply involve monotonous and slow steady-state accumulation. Sediment accumulating in the water column within the ETM is usually derived from gravitational circulation and tidal pumping of bottom sediments, because the ETM is a mobile pool of sediment maintained by the freshwater and saltwater convergence processes. SSC in the ETM may be so high as to generate fluid muds, especially following strong resuspension by spring tidal currents and waves. Fluid-mud bodies beneath the turbidity maximum can generate very thick mud layers in deltas where tides are strong, such as the Fly and the Amazon. Sediments commonly undergo exchange with the overlying water column, and these processes are further subjected to vertical and landward or seaward translation as a function of the neap-spring tidal cycle and seasonal or shorter-term variations in river discharge and sea level. Changes in river discharge, especially in tropical, seasonal settings, and neap-spring tidal variations often result in large variations in the concentration of suspended sediments by influencing resuspension processes, but significant sediment export can also result during periods of strong freshwater outflow. River mud may, thus, be delivered to adjacent shores by coastal currents generated by waves, tides and wind stress.

13.4.3   River-mouth plumes

River sediment that is not deposited on delta floodplains attains the coast. The fine load may be deposited offshore on the subaqueous part of the delta or transported alongshore. In large river systems such as the Amazon, the momentum of the freshwater outflow may even preclude saltwater intrusion, and the freshwater column

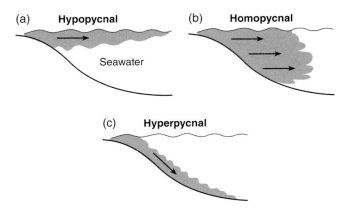

Fig. 13.21 Flow types of sediment-laden river waters in a receiving marine basin: (a) hypopycnal flows are characterized by bouyant plumes above the denser seawater; (b) homopycnal flows are characterized by river plumes of about the same density as seawater; (c) hyperpycnal flows are characterized by highly sediment-charged plumes of density greater than that of seawater, resulting in them sinking to the sea floor. (Source: Adapted from Bates 1953.)

directly moves several kilometres out into the ocean, generating physical and biogeochemical impacts in coastal and deeper waters far beyond the region of the easily identifiable turbid water plume of river-borne sediment (Bianchi and Allison, 2009). Bianchi and Allison (2009) termed these large-river delta-front estuaries (LDE), and have considered that they have a global impact on marine biogeochemistry by acting as both 'drivers' and 'recorders' of natural and anthropogenic environmental change.

Where river flow meets the sea, which is the case in most deltas, the marked differences in density resulting from contrasts in salinity between freshwater and saltwater, and/or variation in sediment concentration through the water column, determine a number of processes that are fundamental to fine-grained marine deltaic sedimentation. Density differences between water bodies were first identified and used as a basis for classifying deltas in terms of process by Bates (1953), who distinguished between hypopycnal, homopycnal and hyperpycnal flows (Fig. 13.21). The first consists of flows wherein the incoming river water and its sediment load are less dense than seawater. The river flow commonly forms a thin bouyant plume that contrasts in colour with the adjacent seawater. The plume is gradually broken up by surface waves along its seaward edges and by internal waves in the freshwater-saltwater junction, allowing the sediment to settle to the seafloor or be dispersed. Homopycnal flows occur where there are no density contrasts between the inflowing river water and the receiving water body. This is commonly the case where rivers flow into lakes, both water bodies being fresh.

Where the river flow is rendered denser than the receiving basin water by a high load of suspended sediment, hyperpycnal flow results, with rapid sinking of the sediment-charged plume.

Hyperpycnal plumes commonly occur where river discharge is extremely turbid, as in the case of the Amazon and the Yellow (Huang He) River. These flows also occur more episodically on other delta fronts during particularly high spates of turbid discharge, as on the small Var River delta in France. Hyperpycnal flows can also occur occasionally on many rivers after a period of very low river discharge under the influence of significant salt-wedge activity, a fine example having been described from the Mississippi (Woodroffe and Saito, 2011). During a period of particularly low flow, a salt wedge extended up the Mississippi river almost to the point of the water intake for the city of New Orleans. The subsequent increase in river discharge flushed out the ETM formed at the tip of this salt wedge onto the delta-slope as a dense hyperpycnal flow. On the Amazon shelf, these high-density plumes (up to 10 g/l) source the shelf-based ETM formed off the mouth of the river (see Box 13.1).

Wright (1985) provided a more detailed description of river-mouth processes and morphology as a function of the nature of the flow, the geometry of the river-mouth, the relative depths of the river and the basin into which it empties, the sediment load, and the grain size and density. Where the basin into which the river discharges is shallow, or where shallow bars occur in the river-mouth, the river flow can be slowed down by friction with the sea floor, leading to bifurcated distributaries, as in the case of the Mississippi river. The hydrographic structure and dynamics of the plumes emitted from the two mouths of the Mississippi delta, the Mississippi proper and the Atchafalaya, differ because of differences in the depth of the continental shelf (Bianchi and Allison, 2009). The former discharges into deep water near the continental shelf edge, whereas the latter discharges into shallow water on a 150-km wide shelf.

River outflow is generally well mixed, due to the relatively shallow receiving basin and the common mixing influence of strong tide- or wind-driven currents, and waves. Well-mixed plumes due to strong wind forcing in the presence of weak tides occur off the Mississippi shoreface. The tide- and wind-driven currents, together with turbulence generated by variations in wave energy and high river discharge, lead to intense sediment resuspension, whereas the absence of stratification allows turbid sediment to mix through the water column. The prodelta is often muddy because of the offshore settling of mud, either from initial hypopycnal plumes or from hyperpycnal plumes sometimes associated with an offshore-displaced ETM, as off the Amazon (see Box 13.1). Sediment concentrations in river plumes may, however, settle out very rapidly near the delta

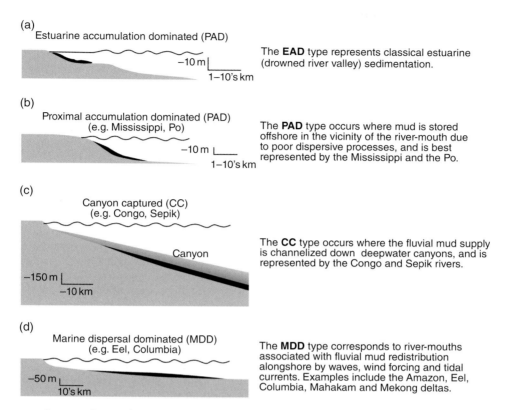

(a) Estuarine accumulation dominated (PAD)

−10 m

1–10's km

The **EAD** type represents classical estuarine (drowned river valley) sedimentation.

(b) Proximal accumulation dominated (PAD)
(e.g. Mississippi, Po)

−10 m

1–10's km

The **PAD** type occurs where mud is stored offshore in the vicinity of the river-mouth due to poor dispersive processes, and is best represented by the Mississippi and the Po.

(c) Canyon captured (CC)
(e.g. Congo, Sepik)

Canyon

−150 m

−10 km

The **CC** type occurs where the fluvial mud supply is channelized down deepwater canyons, and is represented by the Congo and Sepik rivers.

(d) Marine dispersal dominated (MDD)
(e.g. Eel, Columbia)

−50 m

10's km

The **MDD** type corresponds to river-mouths associated with fluvial mud redistribution alongshore by waves, wind forcing and tidal currents. Examples include the Amazon, Eel, Columbia, Mahakam and Mekong deltas.

Fig. 13.22 Major types of marine dispersal of river-supplied mud. (Source: Adapted from Walsh and Nittrouer 2009. Reproduced with permission of Elsevier.)

mouth in certain high-discharge rivers such as the Po and the Red, thus, sometimes depriving the prodelta of mud.

Critical to these aspects of sedimentation are the processes across the transition from river-mouth to shoreline. Although this transition is important in deltaic sedimentation, much still remains to be understood of the morphodynamics involved. The behaviour of rivers near their mouths is fundamentally different from that farther upstream. The area near the mouth is known as the 'backwater', and it can be expansive for large, low-sloping rivers, thus creating non-uniform flow that decelerates towards the river-mouth (Lamb et al., 2012). The Mississippi backwater zone during low flow extends, for instance, 500 km upstream of the river-mouth. This backwater zone is dynamic, and its upstream extent is sensitive to both river discharge and water-surface elevation at the river-mouth. The water-surface elevation at the river-mouth can be affected, in turn, by tidal variation, storm surges, sea level and river plume dynamics. Large floods push the upstream boundary of backwater toward the shoreline, and this affects patterns of sediment convergence that control erosion and deposition (Lamb et al., 2012). Modelling by Lamb et al. (2012) shows that during

such high-discharge phases, the normal flow depth can become larger than the water depth at the river-mouth, and this results in a drawdown of the water surface, spatial acceleration of flow, and erosion of the riverbed, the elevation of the water surface at the river-mouth being relatively insensitive to discharge, due to lateral spreading of the river plume.

13.4.4 Marine mud storage and dispersal off river-mouths

Walsh and Nittrouer (2009) have attempted to classify large river-mouths in terms of the way fluvial mud is stored or redistributed (Fig. 13.22). The most remarkable dispersal system is that of the Amazon (see Box 13.1). About 15–20% of the mud supplied by this river becomes concentrated by various processes into a series of coastal banks that migrate under wave influence towards the mouth of the Orinoco (Anthony et al., 2010). The rest contributes to the growth of a clinoform structure (the subaqueous delta (SD) in Fig. 13.5), which is an important accumulation found on the continental shelf seaward of the mouths of some large rivers and characterized by foreset beds with high rates of sediment accumulation, commonly exceeding several centimetres per year. Other examples of this type of system include the Fly, Yangtze, Ganges-Brahmaputra and Po Rivers. Hyperpycnal flows may serve as sediment feeders of offshore gravity flows by wave action, especially during cyclones.

13.4.5 Bedload deposition in deltaic river-mouths

Bedload supply to the coast is particularly important during important river-flood events. Under these conditions of strong river-domination of the delta channel dynamics, estuarine processes involved in the formation of mutually evasive bedload transport pathways within the main delta channel(s) can be effaced, and the bedload transported directly to the mouths of these channels. Edmonds and Slingerland (2007) found from modelling work that the distance between the river-mouth and the river-mouth bar was proportional to the river jet momentum flux and inversely proportional to grain size. The larger the momentum flux and finer the grain size, the larger this distance. River-mouth bars are commonly sandy shallow-water deposits where they are subject to wave action, which inhibits mud deposition or resuspends mud deposited during preceding phases of lower energy. Bars deficient in mud also occur as a result of rapid muddy sedimentation between the river-mouth and the bar, especially when large, rapidly sinking flocs are an important part of the suspended load.

River-mouth bars are important in delta development as shown in the next section. They are generally subject to two modes of development: (1) they are built up by wave action to form sand barriers that provide shelter for rapid fine-grained sedimentation in back-barrier delta plains and lagoons; (2) they serve as sources and longshore transport pathways for sand that contributes to the development of adjacent beaches and barriers (see Fig. 13.2). On some open ocean coasts, especially where constant swell waves occur, and where bedrock headlands are absent, sand supplied in large quantities by rivers, especially those draining granitic and sandstone catchments, can be transported alongshore for hundreds of kilometres away from river-mouth deltas, forming massive accumulations of successive beach ridges, as on parts of the coast of West Africa (Anthony and Blivi, 1999) and the Pacific coast of Mexico.

13.5 Delta initiation, development and destruction

13.5.1 Deltas and sea level

Because deltas are low-lying coastal landforms, they are highly sensitive to changes in relative land and sea levels, which determine the base level to which the mouths of

their channels adjust. The notion of eustatic sea-level change and its control of base level as the primary factor governing delta development is supported by various studies and experiments. It is now recognized that the modern deltas of the world started developing following the slowing down of the postglacial rise of sea level between about 9000 and 6000 years BP, and this occurred more or less simultaneously for deltas around the world (Stanley and Warne, 1994). A rising sea level constantly creates space available for sediments to accumulate. This space is called 'accommodation space', and in the case of deltas, it needs to be filled with sediments brought down by rivers. As sea-level rise commonly outpaced sediment supply before 9000 years BP, this favoured the development of estuaries, rather than deltas. As sea level stabilized, there was a need for a lesser quantity of river-borne sediments, since sediment accommodation space was no longer being created by sea level and, over the last 5000 years or so, this has favoured the massive development of deltas associated with rivers that provide abundant sediment. Where sea level remains fairly constant and there is a continuous supply of sediment, the delta will grow seaward (a process called progradation), and the boundaries between the major sedimentary formations or facies (i.e. sedimentary units with a particular lithological and palaeontological character) such as the distal delta, prodelta, nearshore delta-front sands and delta plain sediments, will be approximately horizontal (Evans, 2012). Where sediment supply and sea-level rise are balanced (a rare combination), the delta will grow upward (aggradation), with the coastline remaining 'in situ' as the delta plain builds up to match the rise in sea level. In this case, the various sedimentary units or facies will accumulate side by side rather than overlying one another (Evans, 2012). Alternatively, where sea level is rising and sedimentation cannot keep pace with the rise, the delta will retreat landwards (retrogradation), although sediment is still accumulating, and sediments offshore will overlie nearshore sediments, which will in turn overlie the coastal plain sediments, thus giving a retrograding delta that eventually evolves back into an estuary (see Fig. 13.9). All the three scenarios of progradation, aggradation and retrogradation have occurred on Holocene deltas.

13.5.2 Progradation versus aggradation

Deltas, as shown in the preceding section, grow through seaward advance, or progradation, and upward build-up, or aggradation. Delta distributary networks play an important role in both processes, supplying sediment for both (Jerolmack, 2009). By leading to seaward advance of the delta, progradation creates deltaic space that needs to be filled by vertical accumulation represented by aggradation (Box 13.3). In delta studies, the space created by

CONCEPTS BOX 13.3 Deltaic channel growth and delta development

This box illustrates some of the processes driving river delta growth and channel development. These processes also create new land through delta lobe extension, avulsion and channel abandonment. The relative rate of progradation relative to aggradation is a fundamental parameter controlling the development of the channel network, and progress in recent years in understanding the way the morphology and dynamics of such channel networks respond to the 'accommodation space' created by progradation has been reviewed by Jerolmack (2009). Some of these processes are shown in Fig. 13.23. They also highlight the importance of river-mouth bars in generating subsequent channel bifurcations.

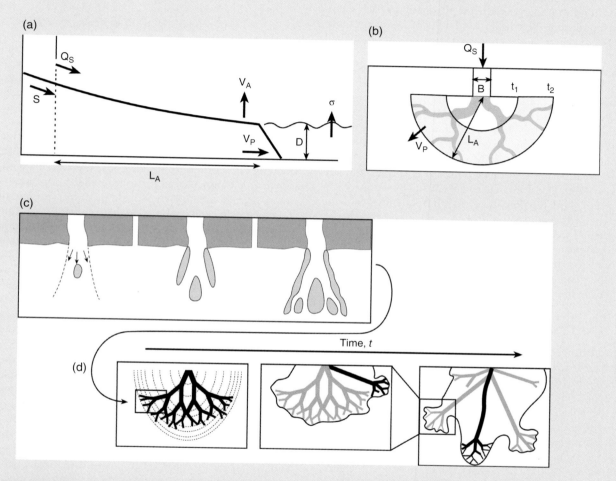

Fig. 13.23 Some processes of river delta growth and channel development. (a) and (b) are, respectively, a cross-section and a plan-view of a river delta prograding into the sea. The delta progrades radially from time t_1 to t_2 by sequential channel bifurcations. L_A = delta length; S = river slope upstream of the delta; Q_s = sediment supply rate; V_P = progradation rate; V_A = aggradation rate; D = water depth of receiving basin; σ = subsidence + sea-level rise; B = delta fan width, an arc that increases as a function of the radial distance of delta growth. (c) Shows the growth of a branching distributary network under rapid progradation. In this diagram, the river emptying into the sea deposits an initial mouth bar that progrades, and towards which levees also prograde. This prograding system stagnates at some distance seaward of the mouth, forcing a flow bifurcation. The process repeats, with channels reducing their width and length to create an idealized fractal network commonly associated with some delta networks. (d) Under steady sea-level rise, the rate of progradation slows down as a result of increasing delta size and water depth, whereas aggradation driven by this sea-level rise becomes important. Channel aggradation results in avulsion and a new channel path to the sea is created (black channel in centre); progradation at the tip of this new channel is rapid, whereas the old network (in grey) becomes abandoned. Repeated avulsions over time (right) create new distributaries with branching tips. Abandoned channels (in grey) on the delta plain are progressively infilled and effaced by overbank sedimentation, whereas the abandoned channel tips are reworked by waves or inundated by flooding resulting from relative sea-level rise. (Source: Jerolmack 2009. Reproduced with permission of Elsevier.)

progradation is sometimes also called 'accommodation space', which is not quite the same as that created by sea-level rise, and to which the term is commonly applied. Depending on variations in the rate of deposition and bank erosion, more or less influenced by sea-level changes, and in delta slope and grain size of sediment delivered to the delta, river channels show two morphodynamic end members, meandering and anastomosed. Meandering creates a large area of channel reworking, whereas anastomosed channels create narrow, vertically stacked channel deposits, and tend to undergo avulsion, resulting in river channel abandonment (Box 13.3). Avulsions often occur about a persistent spatial node, creating a fan-like morphology of delta lobes.

As sea-level rise slowed down about 6000 years ago, the progradation of many deltas was accompanied by the development of branching distributaries. Over the last 6000 years of relative sea-level stabilization, the development of many deltas has been characterized by major changes in distributary networks, as Törnqvist (1994) has demonstrated for the Rhine-Meuse delta, which switched from a meandering to an anastomosed and then a meandering system. These changes were caused by variations in deposition, bank erosion and avulsion frequency, resulting from eustatic sea-level control as well as in changes in fine-grained sediment delivery to the delta. A comparable development to that of the Rhine-Meuse system has been proposed for the Mississippi, where sea-level rise in the mid-Holocene was associated with multiple, anastomosed large-scale distributary branches (Gouw and Autin, 2008).

13.5.3 Sediment retention and delta growth

Where sediment retention has been efficient under the relatively stable sea-level conditions of the last 6000 years, deltas have grown very rapidly, as in the case of the Mekong (Box 13.4). The rate of advance of some deltaic shorelines into the adjacent seas can attain several metres a year. Sediment retention depends on the grain size of the river-borne sediment, the morphology and hydrodynamics of the receiving basin, and the interactions between the river outflow and the basin. The coarser the grain size the greater the retention tendency. Where sheltering is provided from wave and current activity, as in embayed and gulf settings, sediment retention can be high, leading to rapid and sustained deltaic progradation, especially if the setting is shallow. Sediment retention is, however, only one of many parameters that determine patterns of delta evolution. Other factors, such as the development of peat, compaction and subsidence may also affect deltaic evolution, generally leading to lowering of the surface of the delta plain as a result of dewatering of organic-rich and rapidly deposited sediments under the increasing weight of fresh sediment inputs over the frequently inundated floodplains. Peat compaction, in particular, may contribute considerably to total

delta subsidence, although with marked variations in time and space (Fig. 13.25). Local rates of up to 15 mm/yr can occur in some deltas, depending on the sedimentary sequence (van Asselen et al., 2011). Subsidence is important as it creates further 'accommodation space' for fresh sediments. Where rates are high, they may lead to the formation of ponds in delta plain marshes. Where subsidence due to peat compaction exceeds relative sea-level rise, it poses a serious risk of delta vulnerability to flooding and eventual drowning.

13.5.4 Delta destruction

Because of the instability that characterizes many deltaic river courses, parts of the delta plain may become abandoned where channel switches occur. The delta shoreline in such areas thus becomes deprived of sediment. Since wave and tidal processes continue to operate, erosion and marine flooding of the delta plain are two common outcomes of abandonment, resulting in coastline retreat as sediment is dispersed alongshore or offshore. Marine flooding is favoured by the fact that continuous compaction is not compensated for by the addition of new sediment. Ponds created by subsidence can progressively increase in size and coalesce. In deltas with a large amount of fine-grained sediment, the rate of compaction is slow, allowing minimal marine reworking of the upper part of the abandoned deltaic progradational sequence. Surface winnowing of mud commonly results in a thin sand cover, derived from eroded channels and river-mouth bars, and that may then be built up into barrier deposits by waves. These may form isolated cheniers, or a series of retreating barrier islands, such as the Chandeleur Islands east of the present Mississippi, behind which fine sediment may temporarily accumulate, to be subsequently reworked as the barriers migrate landward. The barrier islands may end up being submerged, to form shoals such as Ship shoal in the Mississippi delta. The abandoned portions of many of the large tide-dominated deltas of southern and Southeast Asia tend to become even more tide-dominated, with former distributaries and their tributary channels adjusting their morphology to greater tidal influence, and with mangrove forests commonly becoming widespread over much of the lower abandoned delta plain (Woodroffe and Saito, 2011). Even in areas of low tidal range, this reduction of river flow after abandonment of a distributary can result in relative tide domination of the delta, as in the case of the McArthur Delta in Australia (Woodroffe and Saito 2011).

In large deltas, the former prograded parts of the abandoned delta commonly form retreating lobes identifiable on the shoreface (Fig. 13.26). In the Mississippi, delta lobes have an area of approximately 30,000 km^2, a thickness of about 35 m, an average life cycle of about 1500 years, and generally undergo a cycle involving initial

CASE STUDY BOX 13.4   Changes in the Mekong delta

The Mekong delta (Fig. 13.24a) is considered as having the world's third largest delta plain, with an area estimated at 93,781 km². A high sediment supply and a suitable geological context, comprising a relatively wave-sheltered and shallow substrate, favoured very rapid growth of the Mekong delta over the last 6000 years, the delta advancing more than 200 km out into the South China Sea between 5300 and 3500 years ago, at rates of up to 16 m/yr, and then at lower average rates of less than 10 m/yr as the delta shoreline became increasing exposed to wave influence

following the rapid progradation (Ta et al., 2002; Xue et al., 2010). This phase was marked by wave-induced longshore drift of essentially fine-grained sediment towards the more sheltered western part of the delta, where shoreline advance rates over the last 3500 years, including into the Gulf of Thailand, were up to 26 m/yr. Figure 13.24(b) is a simplified reproduction of the latter stages of progradation in the area of the delta mouths, characterized by the construction of numerous beach ridges. Over 70% of the muddy mangrove shoreline in the western sector, where

(a)

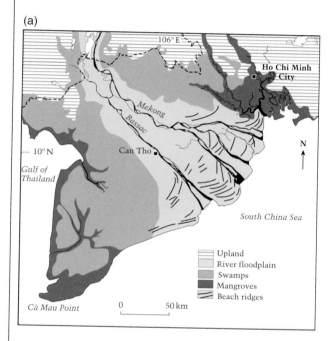

(b)

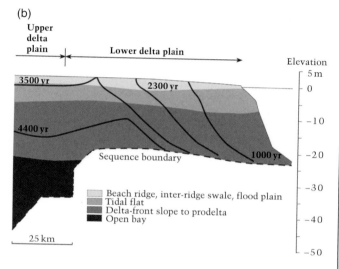

(c)

(d)

Fig. 13.24 Morphology of the Mekong River delta (a); synthetic cross-sectional view of growth patterns of the delta during the Holocene (b) (Source: Adapted from Tamura et al. 2012); examples of increasing human pressures on the delta (c,d).

the late Holocene advance rates were highest, has been eroding over the last decade, at a mean rate exceeding 10 m/yr. The shoreline sector of the delta channel mouths exhibits strong spatial contrasts, wherein tracts of rereating shoreline alternate with advancing and stable tracts. Figures 13.24(c and d) show some of the effects of strong human pressures from the growing needs of economic development on the delta. Figure 13.24(c) shows sand, extracted from the bed of one of the main deltaic channels, being transferred from a barge to a storage depot near Can Tho (February 2012), the largest city in the delta (2011 population: 1.2 million). Sand is massively extracted from the beds of the various channels to meet the growing needs of the construction and civil industries in Cambodia and Vietnam, and for export. Figure 13.24(d) shows in the foreground shrimp ponds constructed at the expense of wetlands behind a beach ridge that is now being eroded, thus requiring recourse to rock armouring for shoreline protection, which is being prepared in the background (February 2012). Erosion of the Mekong delta will be aggravated in the future as sand extraction continues, and especially as sediment is retained in reservoirs associated with a cascade of large hydropower dams planned on the river over the next 20 years. The Mekong delta has been identified as one of the most vulnerable to sea-level rise (Syvitski et al. 2009).

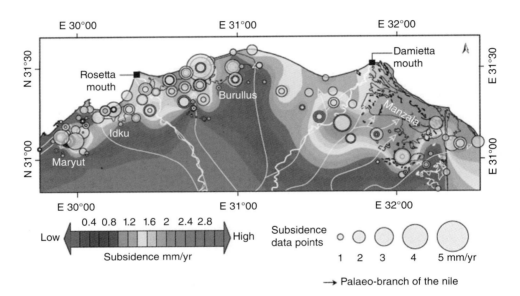

Fig. 13.25   Subsidence in the Nile delta in the course of the Holocene. (Source: Adapted from Marriner et al. 2012. Reproduced with permission from the Geological Society of America.) For colour details, please see Plate 33.

progradation, enlargement, distributary abandonment, and transgression and erosion (Coleman et al., 1998). Sand-rich delta lobes on the shoreface can release sediment that is transported seaward during storms, or shoreward during fair-weather phases. Abandoned lobes can also generate changing patterns of wave refraction and directional approach to the shore, which can affect delta shoreline patterns.

Deltas can thus undergo major changes in sedimentation, morphology and hydrodynamics, at various timescales. The net result of variations in river discharge, delta channel switching and delta plain abandonment, storm erosion, tectonics, human interventions (see later), and over longer timescales, sea-level changes, is a complex stratigraphy that may involve important imbrication of muddy and sandy deposits. As new delta branches develop with their portions of plain, their deposits are likely to overlap those of the abandoned part of the delta. Repetition of this results in deposits of successive deltaic lobes overlapping one another to produce a complicated stratigraphic scenario. In sand-rich deltas subject to significant wave influence, some of these processes are expressed over the subaerial delta plain by variable beach-ridge sets, commonly including series of truncated beach ridges (Fig.13.27).

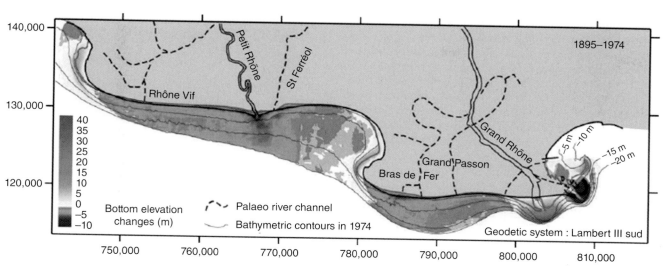

Fig. 13.26 Reduced sediment supply and/or changes in river-mouth locations on deltas can result in sediment lobes in the nearshore zone that are subject to erosion. Here, secular bathymetric changes off the Rhône delta show the long-term erosion of the lobe off the Petit Rhône channel, where water and sediment discharge became reduced by engineering works, relative to the Grand Rhône channel, less affected by these transformations, and the mouth of which has also shifted westwards, resulting in the formation of an abandoned sediment lobe that is also being eroded. (Source: Sabatier et al. 2009. Reproduced with permission of Elsevier.)

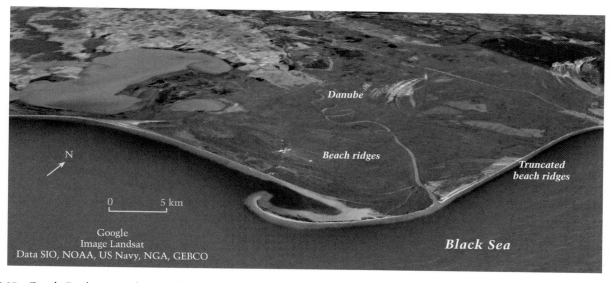

Fig. 13.27 Google Earth image of part of the Danube river delta, showing shoreline reorientations involving the truncation of beach-ridge sets. (Source: Google. Image Landsat. Data SIO, NOAA, US Navy, NGA, GEBCO.)

13.6 Syn-sedimentary deformation in deltas and ancient deltaic deposits

13.6.1 Syn-sedimentary deformation

Many megadeltas store considerable amounts of sediment, which may lead to flexing and fracturing of the continental shelf. Subsidence caused by tectonics, isostasy, sediment compaction and anthropogenic processes, combined with eustatic sea-level rise, results in drowning and increased flood risk within densely populated deltas. In some fault-bounded settings, the deltaic sediments may be disturbed by tectonic events. Prograding muddy deltaic sediments commonly have very high water content and are intrinsically unstable, especially where overlain by coarser nearshore and delta-plain sediments (Evans, 2012). The sediments are especially susceptible to considerable sliding and gravity-induced spreading on the gentle delta slopes, resulting in the development of slide-plains and faults, along which sediment movements are further favoured by the high pressure induced in the pore-fluids by sedimentary loading. Gas may also be generated in the muddy sediments, disorganizing the stratification. Sediment pressures can generate various forms of diapirism, which is the upward injection of underlying sediment, such as mud domes and salt domes. The latter occur where the deltaic sediments have accumulated over previously deposited evaporites. In the Mississippi, following important flood deposition of sediment, mud domes can protrude through to the surface of the delta plain, producing 'mud lumps' that block navigation channels and that have to be removed by dredging.

13.6.2   Ancient deltaic deposits

Ancient deltaic deposits are common in the geological record in many areas around the world, such as the Tertiary of the USA Gulf coast and the Jurassic of the Interior Seaway of the eastern USA, and the Carboniferous deposits of Europe, notably the North Sea, and of the USA and Nigeria (Evans, 2012). Sand bodies in these deposits are the former channels, beaches, dunes and barriers of deltaic shorelines. Delta plain channel sands form long, thin sand bodies, hence their name of 'shoestring sands', which are encased in the fine-grained deposits of the delta plain. Channel sand bodies in large river systems may be several kilometres wide and hundreds of metres thick. In ancient, wave-dominated deltas, successive beach ridges and dune sand bodies can form sand sheets tens of kilometres wide and tens of metres thick. These wave-dominated deposits are much more commonly recognized in the ancient record. Despite the numerous modern and Holocene examples of tide-dominated deltas, such deltas have not been widely recognized in the ancient record, although modern methods of fine-tuned geological reconnaissance and monitoring are contributing to their better recognition, as Plink-Björklund (2012) has shown. In ancient tide-dominated deltas, the sands of tidal bars, which often show bi-modal cross-stratification, form elongated sand bodies, sometimes several kilometres long, hundreds of metres wide and tens of metres thick. Over time, burial of the marshes, swamps and other organic matter that accumulated on delta plains and in the shallow offshore areas of deltas has resulted in the constitution of gas, oil and, to a lesser extent, coal deposits, whereas some of the sand bodies have proven to be valuable reservoir rocks.

13.7   Deltas, human impacts, climate change and sea-level rise

13.7.1   Humans and deltas

Deltas are generally rich in biodiversity, and also commonly constitute the agricultural and/or industrial backbone of a country, offering disproportionately large ecosystem services and other advantages relative to their size. Many large deltas have not only acted as cradles of civilization by providing rich agricultural lands, but have continued, since Antiquity, to be zones of strategic importance for a nation, as in the case of the Nile delta and Egypt, considered as 'a gift of the Nile'. Deltas often contribute to a large percentage of a country's gross domestic product (GDP). The Pearl River delta covers only 0.5% of the Chinese territory and 4.5% of China's population, but accounts for 10% of the country's GDP, whereas the Yangtze delta provides more than 20% of the national GDP of China (Kuenzer and Renaud, 2012). The Mekong delta hosts a population of nearly 18 million. Considered as one of Asia's main food baskets, it provides 50% of Vietnam's food, 90% of Vietnam's rice production (Vietnam is the world's second most important rice exporter) and 60% of its seafood, both with export values of several billion US dollars, while being an extremely active area in overall agriculture and animal husbandry.

13.7.2   The difficulty of disentangling climate change from human impacts

Deltas are considered as being particularly vulnerable to sea-level rise, to changes in coastal oceanographic conditions such as storminess and wave energy, and to changes in catchment flow conditions, all of which are liable to be caused by climate change. Changes in catchment conditions will likely affect rates of landscape erosion, river flow and storm intensity, all of which contribute to the release and transport of sediments. Changes in temperature and rainfall regimes are deemed to affect both glacier and torrential stream activity, and enhance the release of sediment from physically weathered mountain catchments, as well as from agricultural lands exposed to increased erosion.

Climate change is an exceedingly complex issue, involving teleconnections at the global scale, and with potentially variable regional and local impacts. As a result, it is extremely difficult to isolate climate change effects from natural and anthropogenic contributions to rapidly changing sediment fluxes from rivers to the ocean (e.g. Wang et al., 2011). Yet distinguishing between the two is important if we are to accurately predict future river discharge trends, which is critical to deltas. There are abundant reports in the literature of evidence from various areas of the globe of the regional impact of climate change

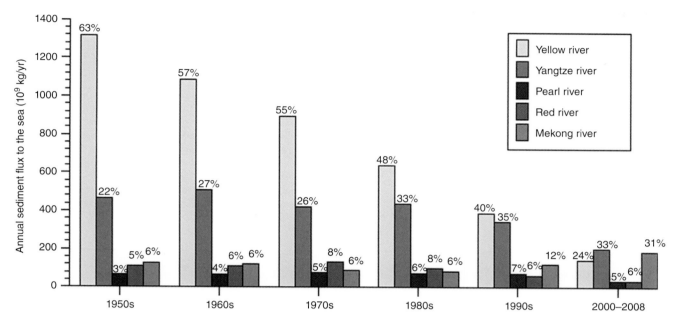

Fig. 13.28 Decadal variations in sediment flux to the western Pacific Ocean from five major rivers in East and Southeast Asia. (Source: Wang et al. 2011. Reproduced with permission of Elsevier.) For colour details, please see Plate 34.

on river discharge. Wang et al. (2011) have suggested that El Niño Southern Oscillation (ENSO) events affect the precipitation and sediment yield of major rivers in Asia.

There is much more certainty regarding the effects of humans on deltas. Although the delivery of more than 70% of river sediment to the world's oceans by rivers in Asia and the Pacific islands results from rapid tectonic uplift and erosional processes, the sediment loads of many of these rivers have been substantially increased as a result of human activities (Woodroffe and Saito, 2011). At the same time, human activities in five of the largest river catchments in Asia have also sharply reduced the sediment they deliver to the ocean (Fig. 13.28), with negative effects on the five megadeltas of these Asian rivers (Wang et al., 2011). The advantages offered by deltas in terms of economic development are being eroded away by large-scale human interventions that are directly implemented in river basins upstream of the deltas, as well as in the deltas themselves (Table 13.1). Chief among these human interventions are flow regulation by dams and sediment entrapment by reservoirs.

Dams are, however, relatively recent in the history of deltas, whereas human alteration of landscapes that directly impacts on deltas has been ongoing since the advent and expansion of agriculture, and the formation of many large deltas such as the Danube in Europe, and the Indus, Krishna, and Godavari deltas in Asia, has been significantly influenced by humans (Giosan, 2011). The development of many of the deltas in the Mediterranean

has also fluctuated as a function of human interventions in catchments. The rapid growth and decline of several of the smaller Mediterranean deltas has been due essentially to changes in human population dynamics and interventions on catchments, notably deforestation, population decline and reforestation, fine examples being the Po, Arno and Ombrone deltas in Italy. The Arno and Ombrone deltas, formed only over the last 2500 years, following large-scale deforestation of the catchment hillslopes, have prograded by up to 7 km over this time span (Pranzini, 2001). The Po delta, the largest in the Mediterranean, is even younger. The onset of active formation of the Po delta (Fig. 13.29), strongly influenced by fluctuations in population during Roman and Renaissance times, has been set at about 2000 years BP (Simeoni and Corbau, 2009).

Sediment retention by dams and other hydrological modifications on rivers restrict the supply of river sediment to deltas. This is exacerbated in certain deltas by accelerated subsidence due to groundwater and hydrocarbon extraction (Ericson et al., 2006; Syvitski et al., 2009). In many deltas affected by these activities, this has led to potentially catastrophic imbalances in delta plain sedimentation at the same time that eustatic sea level is rising. The primacy of human destabilization of deltas, relative to eustatic sea-level rise, is borne out, however, in these studies. Based on a sample of 40 deltas worldwide, representing river basins draining 30% of the Earth's landmass, and 42% of global terrestrial run-off, and hosting

Table 13.1 A range of anthropogenic effects on river catchments and deltas liable to affect the sediment balance of deltas and their susceptibility to sinking and erosion. The construction of dams and reservoirs stands out as a primary cause of decreased sediment supply to deltas.

Activities leading to reduction of fluvial sediment supply	Activities leading to increase in fluvial sediment supply	Direct interventions on deltas
• Construction of dams and reservoirs • Sand and gravel extraction in river channels • Abandonment of cultivated land • Reforestation	• Deforestation • Agricultural land-use and grazing • Waste from quarries, mines and sewage disposal • Construction of settlements with roads and buildings	• Construction of artificial levees to control water flow and prevent delta plain flooding → drop in delta plain sedimentation, with effects on balance between aggradation and sinking • Extraction of water, oil and gas → accelerated sinking • Clay mining → net delta plain sediment loss • Sand and gravel extraction in delta channels → channel deepening, modification of channel dynamics and sedimentation (bank erosion, shoreline erosion, saltwater intrusion during low flow, fine-grained sediment dispersal farther offshore during high flow) • Mangrove removal for agriculture and shrimp farming → lesser coastal protection, modification of sedimentation patterns • Coastal defence works → modification of alongshore sediment transport

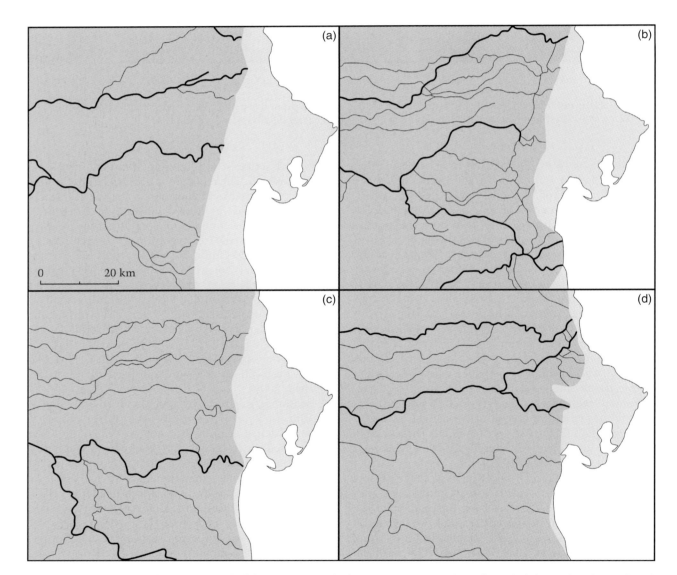

Fig. 13.29 The Po delta in Italy: (a) at the end of the Bronze Age; (b) during Roman times; (c) at the end of the Medieval period; and (d) at the beginning of the 17th century. (Source: Simeoni and Corbau 2009. Reproduced with permission of Elsevier.)

nearly 300 million people, Ericson et al. (2006) carried out an assessment of contemporary sea-level rise and concluded that decreased accretion of fluvial sediment resulting from upstream siltation of artificial impoundments and consumptive losses of run-off from irrigation were the primary determinants of delta sinking in nearly 70% of the deltas. Their study showed that direct anthropogenic effects determine relative sea-level rise in the majority of deltas, with a relatively less important role for eustatic sea-level rise. Approximately 20% of the deltas showed accelerated subsidence, whereas only 12% showed eustatic sea-level rise as the predominant effect. The authors came to the conclusion that serious challenges to human occupancy of deltaic regions worldwide are being conveyed by factors that to date have been studied less comprehensively than the question of climate change and sea-level rise. Syvitski et al. (2009) determined that since about 2000, 85% of 33 deltas they studied worldwide experienced severe flooding, resulting in the temporary submergence of 260,000 km². They conservatively estimated that the delta surface area vulnerable to flooding could increase by 50% under the current projected values for sea-level rise in the 21st century, a figure liable to increase if deltas continue to be deprived of sediment by dams and reservoirs.

13.7.3 Delta vulnerability

In many deltas, the drop in sediment load in the face of relative sea-level rise induced by subsidence and eustasy is resulting in greater erosion of their shores and in sinking of their delta plains (Fig. 13.30), resulting in a loss of wetlands, in large-scale growth of areas of open water within the delta plains, and consequently in the loss of valuable infrastructure built over decades. One of the highest rates of land loss over the last century of any system on Earth has been that of the Mississippi delta, terminus of one of the most managed rivers in the world, and the land area of which has decreased by about 20%.

Delta vulnerability also implies that people living in deltas are exposed to both marine and riverine catastrophic flooding. Syvitski et al. (2009) identified many deltas (Table 13.2) that are now sinking at rates that are much faster than global sea-level rise. They identified three categories of deltas in order of increasing risk: (1) reduced aggradation that can no longer keep up with local sea-level rise (Brahmaputra, Godavari, Indus, Mahanadi, Parana and Vistula); (2) reduced aggradation plus accelerated compaction largely exceeding the rates of global sea-level rise (Ganges, Irrawaddy, Magdalena, Mekong, Mississippi, Niger and Shatt-el-Arab); and (3) virtually no aggradation and/or very high accelerated compaction (among which are the Chao Phraya, Colorado, Krishna, Nile, Pearl, Po, Rhône, Sao Francisco, Yangtze and Yellow). The Pearl and

Mekong deltas are particularly vulnerable because of their high population densities.

Apart from the delta-weakening effects of sediment retention upstream, the buffering capacity of deltas to subsidence and sea-level rise is also impacted by direct interventions often carried out on a large-scale within the deltas themselves. These include the large-scale conversion of mangroves to shrimp farms, thus reducing both the storm-buffering and sediment-trapping capacities assured by mangroves (see Box 13.4). This further reduces the rich biodiversity assured in part by mangroves. Delta channels such as those of the Mississippi have had their levees reinforced and heightened to prevent overbank floods. Deltas are increasingly armoured with engineering works and dykes to protect populations from storms, shrimp farms from saltwater intrusion, and transport infrastructure from shoreline erosion, as in the case of the heavily armoured Var delta on the French Riviera (Fig. 13.31). These works affect overbank sedimentation important in the aggradation of delta plains, as well as longshore sediment transport. In many deltas, the natural channels have been strongly modified by engineering works. Another important activity in some deltas is aggregate extraction from river channels upstream, but also sometimes from the deltaic river channels, to meet the growing needs of the civil industry, housing and transport. All of these activities are practised on a large-scale in the Mekong delta, resulting in changes in channel depths and modifications in the distribution of fine-grained sediment that may be involved in the rampant erosion affecting the western muddy part of this megadelta (see Box 13.4). Large-scale aggregate extraction in the Pearl River in China has also led to dramatic changes in the Pearl megadelta (Wang et al., 2011).

13.7.4 Mitigating delta vulnerability

It can be claimed that the scientific knowledge for mitigating delta vulnerability largely exists because of the significant progress in understanding delta systems and their morphodynamics over the last decades. However, deltaic processes, like other coastal processes, may be characterized by non-linear dynamics and uncertainty. Even in the world's best-studied delta, which is no doubt the Mississippi delta, there are uncertainties regarding the outcomes of mitigation attempts. Plans to restore this delta, for instance, include partial diversion of the Mississippi river, replicating the effect of avulsion and reinitiating natural deltaic land-building processes. Kolker et al. (2012) have shown, however, that although the basic physical principles underpinning river diversions are relatively straightforward, there is considerable controversy over whether diversions can and do deliver enough sediment to the coastal zone to build subaerial land

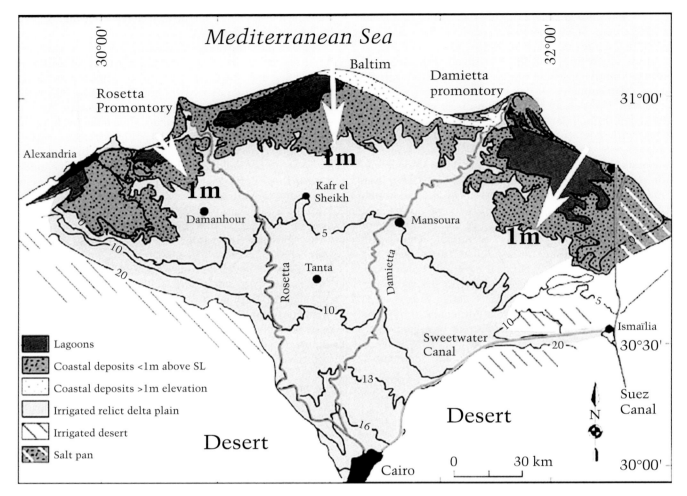

Fig. 13.30 Map showing the elevation contours in the Nile delta between the present coastline and the delta apex at Cairo. Projections of relative sea-level rise and compaction rates indicate encroachment of seawater to inland extents of 10–45 km up to the +1-m contour (arrows) in an average of about 130 years from present time. SL, sea level. (Source: Stanley and Corwin 2013. Reproduced with permission of The Coastal Education & Research Foundation, Inc.)

Table 13.2 Risk status of representative deltas in the world. (Source: Syvitski et al. 2009. Reproduced with permission of Nature Publishing Group.)

Deltas not at risk	Deltas at risk	Deltas at great risk	Deltas in peril	Deltas in great peril
Amazon (Brazil)	Amur (Russia)	Brahmaputra (India)	Ganges (Bangladesh)	Chao Phraya (Thailand)
Congo (Congo DR)	Danube (Romania)	Godavari (India)	Irrawaddy (Myanmar)	Colorado (Mexico)
Fly (Papua New Guinea)	Han (Korea)	Indus (Pakistan)	Magdalena (Colombia)	Krishna (India)
Orinoco (Venezuela)	Limpopo (Mozambique)	Mahanadi (India)	Mekong (Vietnam)	Nile (Egypt)
Mahakam (Indonesia)		Parana (Argentina)	Mississippi (USA)	Pearl (China)
		Vistula (Poland)	Niger (Nigeria)	Po (Italy)
			Shatt-El-Arab (Iraq)	Rhône (France)
				Sao Francisco (Brazil)
				Tone (Japan)
				Yangtze (China)
				Yellow (China)

Fig. 13.31 Google Earth image of the Var river delta on the French Riviera. This is a fine example of radical human transformation of a delta. Changes in the course of the 20th century included the construction of several barrages in the lower course of the river a few kilometres upstream of the delta, reclamation and extension of the delta plain through infill, and complete armouring of the shoreline for the construction of the international airport of Nice. Up to 3.5 km² of additional space was gained by reclamation of part of the steep, muddy delta front through an impressive series of groynes, embankments and infill structures. Part of this reclamation fill, including a harbour breakwater, collapsed on 16 October 1979, causing numerous casualties, and a small tsunami that led to further damage of sea-front property. The natural supply of fluvial gravel to the adjacent beaches, especially that of Nice, has been completely cut-off by the delta armouring, leaving present beach sediment budgets with zero natural inputs, resulting in chronic erosion that constantly needs to be contained by frequent nourishment. (Source: Google. Data SIO, NOAA, US Navy, NGA, GEBCO. Image Landsat. Image © DigitalGlobe.)

on restoration-dependent timescales. In a case study of crevasse-splay dynamics at the West Bay Mississippi River Diversion project, the largest diversion in the Mississippi river that was specifically constructed for coastal restoration, these authors found that most sediments were distributed downstream of the initially defined project boundaries, in contrast to simple sedimentary models, which predicted maximum sediment deposition closest to the riverbank. There is a need for further work on these aspects of delta sedimentation, termed 'restoration sedimentology' by Edmonds (2012), and which are a necessary forerunner of successful ecological restoration.

The vulnerability of river deltas has become a major issue, with a call by leading delta specialists for the year 2013–2014 as International Year of Deltas (IYD) (Foufoula-Georgiou et al., 2011). As noted by these scientists, despite the many meetings and reports calling for action on deltas worldwide, and the urgency of scientific and policy-making communities to understand these threatened systems and to provide support for global, regional and local plans for protection, mitigation, and response to change, progress has been slow, focus lacking, and synergies not

fully realized. Although our concern in this chapter is essentially with the geomorphology and processes shaping deltas, it is important to recall that deltas span a wide range of earth-science, biological, social and economic domains (Kuenzer and Renaud, 2012), and their study and rational management require a trans-disciplinary approach (Box 13.5). Deltas are probably the most iconic of Earth systems in terms of the difficulties of reconciling environmental well-being and economic growth. The increasing vulnerability of individual deltas throughout the world is characteristic of countries where economic growth has been rapid over the last three decades, such as China, or where growth rates are rising sharply, as in many densely populated tropical countries, such that mitigating delta vulnerability is sometimes viewed as being tantamount to staving off economic growth and affecting the livelihood of millions of people. Coordinated plans at a global scale are the most desirable approach in mitigating delta vulnerability, but there are chances that success will be doubtful, given the failure of many global initiatives to regulate environmental change. One way forward could be the multi-disciplinary development of a universal framework

CONCEPTS BOX 13.5   Initiatives of awareness and actions on the perils faced by deltas, and coordination efforts aimed at saving them

The International Year of Deltas (IYD) proposal, depicted in Fig. 13.32(a), comprises a number of initiatives aimed at consolidating efforts of awareness of the value and vulnerability of deltas and cooperation at various levels. It is a vibrant plea for data-sharing on deltas, and for a better understanding of the human and environmental dynamics of deltas. Global initiatives, alliances and networks on deltas, some of which are shown in Fig. 13.32(b), are aimed, among other things, at improving cooperation and enhancing delta resilience. Other initiatives involving deltas include the Biosphere Reserves for Environmental and Economic Security (BREES) of UNESCO, a climate change and poverty alleviation programme in the Asia-Pacific Region. There are also numerous initiatives targeting individual deltas on all continents.

(a)

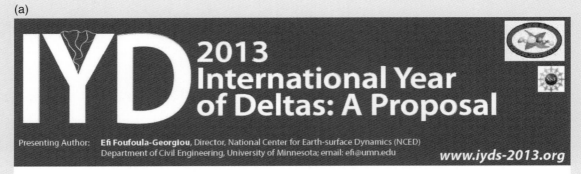

We propose that 2013-2014 be designated the International Year of Deltas to:
(1) increase awareness and attention to the value and vulnerability of deltas worldwide,
(2) promote and enhance international and regional cooperation at the scientific, policy, and stakeholder levesl, and
(3) launch a 10-year committed initiative towards understanding and modeling these complex socio-ecological systems as the cornerstone of ensuring preparedness in protecting or restoringthem in a rapidly changing environment.

Four broad challenges can provide a framework for organizing IDY activities:
(1) Sharing data, knowledge, and culture in deltas around the world and raising local and global awareness
(2) Understanding human-environmental dynamics in regions where the connections are intimate, and the impacts of catastrophic change
(3) Creating the capacity to model and evaluate trade-offs for decision making and adaptive management
(4) Creating new connections between scientists, decision makers, and the public as stakeholders in a sustainable future

(b)

Current global delta initiatives, alliances and networks				
Initiative	World estuary alliance www.estuary-alliance.org	Delta alliance www.delta-alliance.org	Connecting delta cities www.deltacities.com	USGS Dragon initiative http://deltas.usgs.gov/	WWF Ecoregion Initiative http://wwf.panda.org/about_our_earth/ecoregions/ecoregion_list/
Goals	Creating awareness, developing and integrating knowledge, promoting knowledge exchange, exchange of best practise, initiating joint projects, network building, connecting individuals and organizations across all sectors (individuals, science institutes, NGOs, policy makers); international events, conferences, and communication			← + analysing river deltas in a comparable manner, especially utilizing earth observation data	Regional WWF projects in respective countries to protect biodiversity and the environment
Deltas concerned	Indus (Pakistan); Ganges-Brahmaputra (India, Bangladesh); Guadalquivir (Spain); Mekong (Vietnam); Rhine (Netherlands); Zambezi (Mozambique).	Ciliwung (Indonesia); Mahakam (Indonesia); Ganges-Brahmaputra (India, Bangladesh); Mekong (Vietnam); Yangtze (China); Nile (Egypt); Rhine (Netherlands); Mississippi (USA).	Ciliwung (Indonesia); Mekong (Vietnam); Pearl River (China); Rotterdam/Rhine (Netherlands); Mississippi (USA).	Irrawaddy (Myanmar); Chao Phraya (Thailand); Amazon (Brazil); Danube (Rumania); Ganges (India); Yellow River (China); Lena (Russia); Mekong (Vietnam); Mississippi (USA); Nile (Egypt); Okavango (Botswana); Rhine (Netherlands); Selenga (Russia); Volga (Russia); Yangtze (China).	Niger (Nigeria); Indus (Pakistan); Volga (Russia); Danube (Rumania); Lena (Russia); Lower Mekong (Vietnam); Lower Congo (Congo); Lower Mississippi (USA); Lower Amazon (Brazil); Lower Orinoco (Venezuela)

Fig. 13.32   Inititatives aimed at saving deltas. (a) Elements of a poster calling for the year 2013–2014 as International Year of Deltas (IYD); (b) a number of current global initiatives, alliances and networks on deltas. (Source: (a) Adapted from Foufoula-Georgiou et al. 2011; (b) Adapted from Kuenzer and Renaud 2012.)

of vulnerability assessment and resilience modelling applicable to any delta. This will require achieving the objectives set out in an initiative such as the IYD (Box 13.5).

13.8 Summary

Deltas are subaerial and subaqueous coastal accumulations of river-derived sediments, and are formed where conditions in the receiving basin are not energetic enough to disperse all of the sediment brought down by rivers. The biggest deltas are the largest coastal landforms in the world. Deltas are important in supplying sediment to adjacent coasts. Deltas are characterized by low topography, usually high productivity, rich and biodiverse ecosystems, and a wide range of related ecosystem services. Many deltas support large rural populations that live on agriculture and fisheries, but deltas can also host important urban and industrial centres, while supporting a range of transport activities.

Deltas are relatively diversified landforms, which can include various units including a subaerial plain, which represents the area above sea level. They are commonly classified in terms of river-, wave- or tide-domination, which are considered as the dominant processes that shape the delta shoreline and morphology. River-dominated deltas are typically elongate; wave-dominated deltas exhibit cuspate shorelines; and tide-dominated deltas have irregular shorelines with numerous tide-influenced channels. Delta morphology is, however, extremely diverse, and only a few deltas readily fall clearly into these three categories.

Deltas are highly sensitive to changes in relative land and sea levels, which determine the base level to which the mouths of their channels adjust. The modern deltas of the world started developing following the slowing down of the postglacial rise of sea level between about 9000 and 6000 years BP. Sea-level changes also affect patterns of delta development, by influencing the rates of progradation and aggradation. Both of these processes involve morphodynamic changes in delta channel type and growth. Deltaic channels are commonly meandering or anastomosed, the latter being particularly prone to channel switches, which result in abandonment and erosion of parts of the delta that no longer receive fluvial sediment.

The study of deltas spans a wide range of earth-science, biological, social and economic domains, and their rational management requires a trans-disciplinary approach. The development and dynamics of many deltas in the world have been strongly influenced by humans over the last two to three millenia. Because of the commonly dense population densities in deltas, especially in developing countries, many deltas are today caught in the cross-fire between environmental and economic preoccupations. As

low-elevation coastal landforms, they are particularly vulnerable to a large range of hazards such as catastrophic river floods, tsunamis, cyclones, subsidence and global sea-level rise, and this vulnerability is increasing as a result of reduced sediment flux from rivers, caused by human interventions. Although deltas may develop dynamic resilience and adapt to changes in sediment supply and sea level, commonly through re-organization of their channels and river-mouths, they have become economic and environmental hotspots, as human pressures have increasingly diminished these capacities of resilience, adaptation and hazard absorption.

Key publications

Evans, G., 2012. Deltas: the fertile dustbins of the world. *Proceedings of the Geologists' Association*, **123**, 397–418.
[*A very recent review of deltas, including their geological aspects*]
Kuenzer, C. and Renaud, F.G., 2012. Chapter 2. Climate and environmental change in river deltas globally: expected impacts, resilience, and adaptation. In: F.G. Renaud and C. Kuenzer (Eds.), *The Mekong Delta System: Interdisciplinary Analyses of a River Delta*. Springer, Dordrecht, the Netherlands, pp. 7–46.
[*A useful analysis of the human dimension of deltas and their vulnerability to climate change*]
Woodroffe, C.D. and Saito, Y., 2011. River-dominated coasts. In: E. Wolanski and D.S. McLusky (Eds), *Treatise on Estuarine and Coastal Science*, Vol **3**. Academic Press, Waltham, UK, pp. 117–135.
[*A fine and accessible presentation of deltas*]
Wright L.D., 1985. River deltas. In R.A. Davis (Ed.), *Coastal Sedimentary Environments*. Springer-Verlag, New York, USA, pp. 1–76.
[*Summarizes many of the basic concepts behind deltas and their development*]

References

Anthony, E.J. and Blivi, A.B., 1999. Morphosedimentary evolution of a delta-sourced, drift-aligned sand barrier-lagoon complex, western Bight of Benin. *Marine Geology*, **158**, 161–176.

Anthony, E.J. and Julian, M., 1997. The 1979 Var Delta landslide on the French Riviera: a retrospective analysis. *Journal of Coastal Research*, **13**, 27–35.

Anthony, E.J., Lang, J. and Oyédé, L.M., 1996. Sedimentation in a tropical, microtidal, wave-dominated coastal-plain estuary. *Sedimentology*, **43**, 665–675.

Anthony, E.J., Gardel, A., Gratiot, N. et al., 2010. The Amazon-influenced muddy coast of South America: a review of mud bank-shoreline interactions. *Earth-Science Reviews*, **103**, 99–129.

Ashton, A.D. and Giosan, L., 2011. Wave-angle control of delta evolution. *Geophysical Research Letters*, **38**, L13405, doi: 10.1029/2011GL047630.

Bates, C.C., 1953. Rational theory of delta formation. *Bulletin of the American Association of Petroleum Geologists*, **37**, 2119–2162.

Bhattacharya, J.P. and Giosan, L., 2003. Wave-influenced deltas: geomorphological implications for facies reconstruction. *Sedimentology*, **50**, 187–210.

Bianchi, T.S. and Allison, M.A., 2009. Large-river delta-front estuaries as natural 'recorders' of global environmental change. *Proceedings of the National Academy of Science of the United States of America*, **106**, 8085–8092.

Boyd, R., Dalrymple, R.W. and Zaitlin, B.A., 1992. Classification of clastic coastal depositional environments. *Sedimentary Geology*, **80**, 139–150.

Coleman, M. and Huh, O.K., 2004. *Major Deltas of the World: A Perspective from Space.* Coastal Studies Institute, Louisiana State University, Baton Rouge, LA, USA, www.geol.lsu.edu/WDD/PUBLICATIONS/C&Hnasa04/C&Hfinal04.htm.

Coleman, J.M., Roberts, H.H. and Stone, G.W., 1998. Mississippi River Delta: an overview. *Journal of Coastal Research*, **14**, 698–716.

Edmonds, D.A., 2012. Restoration sedimentology. *Nature Geoscience*, **5**, 758–759.

Edmonds, D.A. and Slingerland, R.L., 2007. Mechanics of river mouth bar formation: implications for the morphodynamics of delta distributary networks. *Journal of Geophysical Research*, **112**, F02034.

Edmonds, D.A. and Slingerland, R.L., 2010. Significant effect of sediment cohesion on delta morphology. *Nature Geoscience*, **3**, 105–109.

Ericson, J.P., Vörösmarty, C.J., Dingman, S.L., Ward, L.G. and Meybeck, M., 2006. Effective sea-level rise and deltas: causes of change and human dimension implications. *Global and Planetary Change*, **50**, 63–82.

Evans, G., 2012. Deltas: the fertile dustbins of the world. *Proceedings of the Geologists' Association*, **123**, 397–418.

Foufoula-Georgiou, E., Syvitski, J.P.M., Paola, C. et al., 2011. International Year of Deltas 2013: A proposal. *EOS*, **92**, 340–341.

Galloway, W.E., 1975. Process framework for describing the morphologic and stratigraphic evolution of delta depositional systems. In: M.L. Broussard (Ed.), *Deltas: Models for Exploration.* Texas Geological Society, Houston, TX, USA, pp. 87–98.

Giosan, L., 2011. Are Old World deltas human constructs? *Geophysical Research Abstracts*, **13**, EGU2011–4564.

Giosan, L., Donnelly, J.P., Vespremeanu, E., Bhattacharya, J.P., Olariu, C. and Buonaiuto, F.S., 2005. River delta morphodynamics: examples from the Danube delta. In: L. Giosan and J.P. Bhattacharya (Eds.), *River Deltas – Concepts, Models and Examples.* SEPM Special Publication 83, pp. 393–411.

Gouw, M.J.P. and Autin, W.J., 2008. Alluvial architecture of the Holocene Lower Mississipi Valley (USA) and a comparison with the Rhine-Meuse delta (The Netherlands). *Sedimentary Geology*, **204**, 106–121.

Hori, K. and Saito, Y., 2008. Classification, architecture and evolution of large river deltas. In: A. Gupta (Ed.), *Large Rivers: Geomorphology and Management.* John Wiley & Sons, pp. 214–231.

Jerolmack, D.J., 2009. Conceptual framework for assessing the response of delta channel networks to Holocene sea level rise. *Quaternary Science Reviews*, **28**, 1786–1800.

Kolker, A.S., Miner, M.D. and Weathers, H.D., 2012. Depositional dynamics in a river diversion receiving basin: the case of the West Bay Mississippi River Diversion. *Estuarine, Coastal and Shelf Science*, **106**, 1–12.

Kuenzer, C. and Renaud, F.G., 2012. Chapter 2. Climate and environmental change in river deltas globally: expected impacts, resilience, and adaptation. In: F.G. Renaud and C. Kuenzer (Eds.), *The Mekong Delta System: Interdisciplinary Analyses of a River Delta.* Springer, Dordrecht, the Netherlands, pp. 7–46.

Lamb, M.P., Nittrouer, J.A., Mohrig, D. and Shaw, J., 2012. Backwater and river plume controls on scour upstream of river mouths: implications for fluvio-deltaic morphodynamics. *Journal of Geophysical Research*, **117**, F01002, doi: 10.1029/2011JF002079.

Marriner, N., Flaux, C., Morhange, C. and Kaniewski, D., 2012. Nile Delta's sinking past: quantifiable links with Holocene compaction and climate-driven changes in sediment supply? *Geology*, **40**, 1083–1086.

Milliman, J., Farnsworth, K.L. and Albertin, C.S., 1999. Flux and fate of fluvial sediments leaving large islands in the East Indies. *Journal of Sea Research*, **41**, 97–107.

Orton, G.J. and Reading, H.G., 1993. Variability of deltaic processes in terms of sediment supply, with particular emphasis on grain size. *Sedimentology*, **40**, 475–512.

Plink-Björklund, P., 2012. Effects of tides on deltaic deposition: causes and responses. *Sedimentary Geology*, **279**, 107–133.

Pranzini, E., 2001. Updrift river mouth migration on cuspate deltas: two examples from the coast of Tuscany, Italy. *Geomorphology*, **38**, 125–132.

Sabatier, F., Maillet, G., Fleury, J. et al., 2006. Sediment budget of the Rhône delta shoreface since the middle of the 19th century. *Marine Geology*, **234**, 143–157.

Simeoni, U. and Corbau, C., 2009. A review of the Delta Po evolution (Italy) related to climatic changes and human impacts. *Geomorphology*, **107**, 64–71.

Stanley, D.J. and Corwin, K.A., 2013. Measuring strata thicknesses in cores to assess recent sediment compaction and subsidence of Egypt's Nile delta coastal margin. *Journal of Coastal Research*, **29**, 657–670.

Stanley, D.J. and Warne, A.G., 1994. Worldwide initiation of Holocene marine deltas by deceleration of sea-level rise. *Science*, **265**, 228–231.

Syvitski, J.P.M., Kettner, A.J., Overeem, I. et al., 2009. Sinking deltas due to human activities. *Nature Geoscience*, **2**, 681–689.

Syvitski, J.P.M., Overeem, I., Brakenbridge, G.R. and Hamon, M., 2012. Floods, floodplains, delta plains – a satellite imaging approach. *Sedimentary Geology*, **267–268**, 1–14.

Ta, T.K.O., Nguyen, V.L., Tateishi, M., Kobayashi, I., Tanabe, S. and Saito, Y., 2002. Holocene delta evolution and sediment discharge of the Mekong River, southern Vietnam. *Quaternary Science Reviews*, **21**, 1807–1819.

Tamura, T., Saito, Y., Bateman, M.D., Nguyen, V.L., Ta, T.K.O and Matsumoto, D., 2012. Luminescence dating of beach ridges for characterizing multi-decadal to centennial deltaic shoreline changes during Late Holocene, Mekong River delta. *Marine Geology*, **326–328**, 140–153.

Törnqvist, T.E., 1994. Middle and late Holocene avulsion history of the River Rhine (Rhine-Meuse delta, Netherlands). *Geology*, **22**, 711–714.

van Asselen, S., Karssenberg, D. and Stouthamer, E. 2011. Contribution of peat compaction to relative sea-level rise within Holocene deltas. *Geophysical Research Letters*, **38**, L24401, doi: 10.1029/2011GL049835.

Walker, R.G., 1992. Facies, facies models and modern stratigraphic concepts. In: R.G. Walker and N.P. James (Eds.), *Facies Models: Response to Sea Level Change*. Geological Association of Canada, St John's, NL, Canada, pp. 1–14.

Walsh, J.P. and Nittrouer, C.A., 2009. Understanding fine-grained river-sediment dispersal on continental margins. *Marine Geology*, **263**, 34–45.

Wang, H., Saito, Y., Zhang, Y., Bi, N., Sun, X. and Yang, Z., 2011. Recent changes of sediment flux to the western Pacific Ocean from major rivers in East and Southeast Asia. *Earth-Science Reviews*, **108**, 80–100.

Warne, A.G., Meade, R.H., White, W.A. et al., 2002. Regional controls on geomorphology, hydrology and ecosystem integrity in the Orinoco Delta, Venezuela. *Geomorphology*, **44**, 273–307.

Wheatcroft, R.A., Stevens, A.W., Hunt, L.M. and Milligan, T.G., 2006. The large-scale distribution and internal geometry of the fall 2000 Po River flood deposit: evidence from digital X radiography. *Continental Shelf Research*, **26**, 499–516.

Wolinsky, M.A., Edmonds, D.A., Martin, M. and Paola, C., 2010. Delta allometry: growth laws for river deltas. *Geophysical Research Letters*, **37**, L21403, doi: 10.1029/2010GL044592.

Woodroffe, C.D. and Saito, Y., 2011. River-dominated coasts. In: E. Wolanski and D.S. McLusky (Eds.), *Treatise on Estuarine and Coastal Science*, Vol **3**. Academic Press, Waltham, UK, pp. 117–135.

Wright L.D., 1985. River deltas. In R.A. Davis (Ed.), *Coastal Sedimentary Environments*. Springer-Verlag, New York, USA, pp. 1–76.

Xue, Z., Liu, J.P., DeMaster, D., Nguyen, V.L. and Ta, T.K.O., 2010. Late Holocene evolution of the Mekong subaqueous delta, southern Vietnam. *Marine Geology*, **269**, 46.

14 High-Latitude Coasts

AART KROON

Center for Permafrost (CENPERM), Department of Geosciences and Natural Resource Management,
University of Copenhagen, Copenhagen, Denmark

14.1 Introduction to high-latitude coasts

This chapter deals with the description of processes and morphologies of coastal environments in a specific area – the high latitudes. High-latitude coasts are all located in polar zones: northward of the Polar or Arctic Circle or southward of the Antarctic Circle. The Arctic coastlines represent approximately one-third of the world's coastlines (Lantuit et al., 2011) and cover a broad spectrum of geological and oceanographic settings, resulting in a wide variety of coastal geomorphologies (Nikiforov et al., 2005; Lantuit et al., 2011). It includes the northern shores of Norway and Svalbard, Russia and Siberia, Alaska, northern Canada and Greenland (Fig. 14.1). The Antarctic coastlines are very often covered by ice and snow; only 2% of the Antarctic continent is not permanently covered in ice and snow (Anisimov et al., 2007). The distinctive high-latitude coastal conditions, processes and landforms are caused by common factors such as strong seasonality, cold temperatures, (dis)continuous permafrost, and the presence of seasonal or perennial sea-ice cover (Forbes, 2011). Thus, the freezing temperatures and the impact of ice and snow make the polar coastal environments unique and different from the coastal environments in temperate zones and in the tropics.

High-latitude coasts include a wide variety of coastal geomorphologies. All typical major coastal features are present, like sandy and gravel beaches, barrier islands, spits, deltas, salt marshes, glacial bluffs and rocky cliffs. Most of the sedimentary landforms are associated with coastal plains, for example in the low-relief areas along the northern slope of Alaska (Reimnitz et al., 1990; Mars and Houseknecht, 2007; Jones et al., 2008, 2009), the Canadian Beaufort Sea (Harper, 1990; Hequette and Barnes, 1990; Hill et al., 1994; Ruz et al., 1994; Lajeunesse and Hanson, 2008; Lantuit and Pollard, 2008), the Siberian and Russian north shore (Nikiforov et al., 2005; Streletskaya et al., 2009) or along the fringes of major deltas like the MacKenzie delta in Canada (Jenner and Hill, 1998; Solomon, 2005) or the Lena delta in Siberia (Box 14.1). Other sedimentary features develop in high-relief areas, along fjords and hard-rock cliffs, as observed along the coasts of Greenland (Kroon et al., 2010; Funder et al., 2011) or Antarctica (Butler, 1999). The sources for the sediments at the shore are often rivers that are fed by the melting glaciers and drain the pro-glacial and fluvial valleys (Rachold et al., 2005; Gordeev, 2006).

Coastal Environments and Global Change, First Edition. Edited by Gerd Masselink and Roland Gehrels.
© 2014 John Wiley & Sons, Ltd. Published 2014 by John Wiley & Sons, Ltd. Companion Website: www.wiley.com/go/masselink/coastal

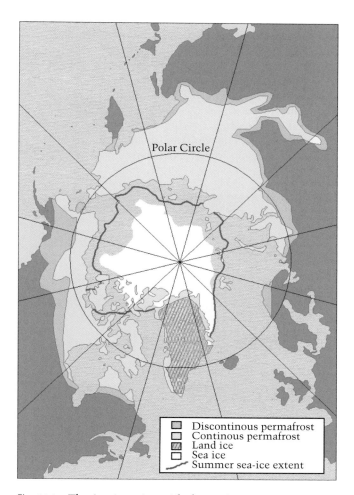

Fig. 14.1 The Arctic region with the sea-ice coverage in summer (sea-ice extent in September 2011; National Snow and Ice Data Center, Boulder, USA) and with continuous and discontinuous permafrost zones. (Source: Adapted from Romanovsky et al. 2002. Reproduced with permission of the American Geophysical Union.)

Legend:
☐ Discontinous permafrost
☐ Continous permafrost
▨ Land ice
☐ Sea ice
— Summer sea-ice extent

Polar Circle

Sediments are derived from the erosion of glacial and periglacial deposits that were formed during the latest glaciations in the present coastal plain. The high-latitude coasts in this chapter do not include the shorelines of, for example, southern Canada or islands where glacial meltwater with sediments feed the coastal environments, but where the typical ice-related coastal processes are less dominant. A nice review of these paraglacial coasts is presented in Forbes and Syvitski (1994).

Global changes in climate induce many changes along the high-latitude coasts. Sea levels are rising due to an increased freshwater flux from the glaciers and land-ice masses (see Anisimov et al., 2007). At the same time, the ice coverage of the coastal waters decreases and the open water periods in summer extend. This causes extra wave activity with higher erosion rates along many of the shorelines (e.g. Hequette and Barnes, 1990; Mars and Houseknecht, 2007; Lajeunesse and Hanson, 2008; Lantuit and Pollard, 2008; Jones et al., 2009). High-latitude coasts have a special sea-level history. For most of the Holocene, the high-latitude coasts observe a relative sea-level fall (e.g. Colhorn et al., 1992; Hjort et al., 1997; Whitehouse et al., 2007; Shennan, 2009; Woodroffe and Long, 2009; Pedersen et al., 2011), mainly due to isostatic uplift after the retreat of the glaciers.

Environmental forcing induces processes in coastal environments that cause coastal responses. Rachold et al. (2005) sketched these interrelations for a high-arctic environment (Fig. 14.3). In general, the coastal dynamics work through a complex system of processes, responses and feedback mechanisms forced by internal, as well as external, environmental factors. Changing environmental forcing, for instance global climate change, can force changes in the coastal processes, with subsequent coastal responses. The resulting coastal change has possible implications on natural and human systems, and can further influence the ongoing changes in environmental forcing through various feedback mechanisms.

CASE STUDY BOX 14.1   Shoreline evolution patterns along a major Arctic delta: Lena delta, Siberia

Arctic deltas occur in many sizes along the high-latitude coasts. Large deltas are located along the coastal plain of Siberia (e.g. the Lena delta; Fig. 14.2; Gordeev, 2006), Canada (e.g. the MacKenzie delta; Jenner and Hill, 1998) and Alaska (e.g. Colville River delta; Walker and Hudson, 2003). Smaller deltas are located in the coastal zone of many fjords and open seas in the high-arctic region of, for example, Greenland. The deltas form the transition between the terrestrial drainage basin and the sea, and act as temporal sediment traps for terrestrial materials. Meltwater from glaciers and snow-covered land are the major input in terms of river discharge during the spring and summer seasons. The main sediment influx towards

the coastal areas comes from these rivers. Minor sources of sediment transport towards the delta are through cross-shore reworking of sediments on the delta slope and through lateral transport from the adjacent shores by coastal processes due to ice, waves and tides. Even smaller amounts of sediment can be added by stranded sediment-loaded ice from the sea.

Changes in fluvial channel patterns on deltas have a significant impact on the coastal morphology along its fringes. Lateral channel migration can locally cause cliff erosion and introduce an extra sediment source into the local budget of an active delta plain. Stabilization of channels, or even channel lobe switching, reduces the fluvial impact on the delta and introduce the formation of beach ridges and spits along the (former) delta edge (Jerolmack, 2009). These accumulative features are formed in the ice-free summer periods, and are fed by alongshore sediment input from adjacent shores due to wave-driven alongshore currents, and by the reworking of the sediments on the delta plain by wave-driven cross-shore processes. Sandy spits and small barriers often fringe the shoreline of a delta. These features are typically formed and active in the ice-free periods, when coastal processes by waves and drifting ice rework the delta front and adjacent coastal cliffs. Local sources of sediment on the delta are former glacial deposits or former beach ridges close to the active channel.

The Lena delta in Siberia is built up in different phases. The oldest delta lobes are located in the western part of the delta (Fig. 14.2). Thermokarst lakes are well developed in this area. The western facing shoreline has barrier islands that are typically formed by the alongshore and cross-shore wave-driven processes in the ice-free season. The shoreline is typically in the barrier island formation or barrier retreat stages in the schematic shoreline evolution model of Ruz et al. (1992; see Fig. 14.8), where headlands between the eroded thermokarst lakes deliver the sediment. The coastline is erosive: location 1 in Fig. 14.2 is an erosive coast with sandy cliffs that are highly erosive, while locations 2 and 3 are thermokarst coasts with small sandy cliffs undergoing moderate erosion of c. 2 m/yr (Lantuit et al., 2009). The northern-facing shoreline of these western parts lacks the well-developed barrier islands and spits, probably because of the lack of wave-driven longshore currents. The most recent delta lobes are located in the eastern part. Here, the thermokarst lakes are still less developed

and the coast receives a lot of sediments from the Lena. The coast is characterized by frontal delta lobes, which are accumulative to stable (locations 4 and 5 in Fig. 14.2; Lantuit et al., 2009). The northeastern part of the Lena delta is typically between the oldest delta lobes in the west and the present delta lobes in the east. There are still some tributaries flowing through this area, but thermokarst lakes develop and the shoreline is already straightened due to wave-driven coastal processes.

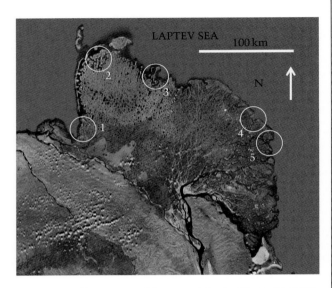

Fig. 14.2 Satellite image of the Lena delta in Siberia (72°50′N, 126°50′E). Locations: 1, western Arga Island; 2, northeastern Arga Island; 3, northern Arga Island; 4, Tumatskaya channel; 5, Trofimovskaya channel. (Source: 2011 TerraMetrics, Google Earth.)

14.2 Ice-related coastal processes

Ice is an essential element in the coastal areas of high-latitude coasts. The presence of ice in coastal waters dampens or even stops the wave activity, and thus reduces the stresses on the substrate. Forbes and Taylor (1994) present a summary of different ice types and their associated coastal processes and products (Table 14.1).

Freezing temperatures create sea ice in the ocean and open seas. The sea ice covers the relatively warm ocean water and is typically about 3–5 m thick. Normally, this sea ice masses together into pack ice and drifts on ocean currents along the shores in spring and summer with velocities that can reach over 100 m per day (Wadhams, 2000). The sea ice can be seasonal or perennial; seasonal sea ice is formed in the winter season and completely melts in the next summer season, while perennial sea ice is older than one season. Sea ice directly influences the

driving forces of the waves and tidal currents during the summer season. Pack ice or individual icebergs dissipate the energy of the waves in their surroundings, and the presence of ice reduces the fetch of open water areas in summer, and thus the intensity of wave processes (Forbes and Taylor, 1994; Trenhaile, 1997).

In the nearshore coastal waters, the impact of regular hydrodynamic processes driven by wind, waves and tides is strongly limited by ice (Fig. 14.4). The length of the ice-free period and the size of the open water area are the most important factors in governing the overall activity of waves at the shorelines (Wiseman et al., 1974; Reimnitz et al., 1987; Trenhaile, 1997). In winter, most of the coastal waters are frozen and there is no sediment transport; morphological changes do not occur. However, this situation changes when air temperatures rise to values above 0°C in spring and early summer. At this stage, the snow melts on land and this causes an activation of the river discharge. This

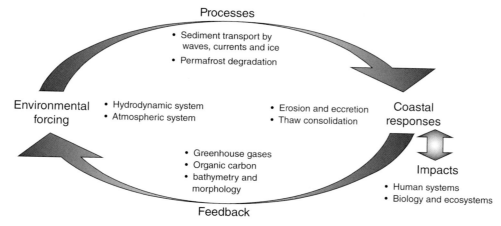

Fig. 14.3 Environmental forcing, coastal processes and responses, impacts and feedback. (Source: Rachold et al. 2005. Reproduced with permission of Springer Science+Business Media.)

Table 14.1 Summary of ice types in coastal waters and their associated processes.

Ice type	Coastal processes
Glacial ice, including ice shelves	Tidewater calving and ice-contact deposition
	Iceberg drifting and grounding
Permafrost and ground ice	Erosion by mechanical and thaw failure
Snow and ice on beaches	Co-deposition of sediments and snow or ice
Surface ice cover on seas and lakes, pressure ridges, ice floes and fragments	Ice as a barrier to surface wave motion
	Enhanced hydrodynamic scour
	Ice scour
	Ice ride-up and pile-up
	Ice rafting
Frazil-, slush- and anchor ice	Sediment entrainment by ice

fresh river discharge transports its sediments over the still frozen and slowly thawing coastal waters (Fig. 14.5c). Quite often, the sediment bypasses the coastal area (delta). During this time, marine influence remains minimal as a result of protection afforded by sea-ice cover (Short and Wiseman, 1975). Hereafter, the ice in the coastal waters further melts in early summer. This is the result of increased melting of the ice surface by rising air temperatures, freshwater influx at delta mouths, and influx of relatively warm ocean water. At this stage, a series of ice-related processes occur in the coastal waters: ice-souring or ice-grounding, ice-rafting and ice pile-up and ice ride-up (Fig. 14.4; Forbes and Taylor, 1994; Are et al., 2008). Wind, waves and tidal currents induce a transport of ice and associated sediments. Ice-scouring or ice-grounding is caused by grounded ice on the shoreface. Quite often, icebergs that are calved from a tidal glacier in a fjord may induce this ice-scouring at the bottom. The morphological product is a trench with an associated ridge on each side. These trenches are easily re-worked by wave action during the succeeding summer season,

especially during the passage of a storm. Ice-rafting is the transport of ice or ice fragments, like slush ice, towards the shore. This ice-rafting is an important process in high-latitude tidal flats and salt marshes for delivering and redistributing sediment (Dionne, 1993; Argow et al., 2011). The ice on the coastal waters can finally pile-up (involving broken ice fragments) or ride-up (involving larger ice plates) at the shore, creating ice stacks of several metres in height. The melted ice fragments often produce hummocky holes on the beach near high-tide levels and are called kettle holes (Short and Wiseman, 1974) or melt pits (Butler, 1999). These features are quite small and often disappear through reworking by waves later on in the season. The summer period will lead to increased wave action along the shores as the fetch over free water increases. However, the presence of pack-ice by the drift upon the ocean currents along the open waters (e.g. the east coast of Greenland) may still significantly reduce the fetch and dampen wave activity. Short and Wiseman (1974) distinguish a sequence of freezing processes on a beach and adjacent lagoon in northern Alaska. They describe seven phases: (1) solidification of the beach face; (2) formation of snow cover; (3) deposition of ice cakes on the beach (these are single pieces of sea ice smaller than 10 m); (4) formation of ice slush in the lagoons; (5) deposition of ice slush berms on the beach face; (6) formation of inter-bedded layers of sediment and ice; and (7) formation of icefoot features.

There are three major sources for sediments that contribute to the coastal zone. Most of the sediments are delivered by meltwater from the glaciers. These glaciers yield fluvial discharge over land and deliver sediments in the pro-glacial valley and the coastal area. Two other mechanisms deal with the direct transport of sediments by ice. Ice can be calved from tidewater glaciers, whereby ice with a high sediment load drifts directly into the local coastal waters. In this way, ice can directly transport

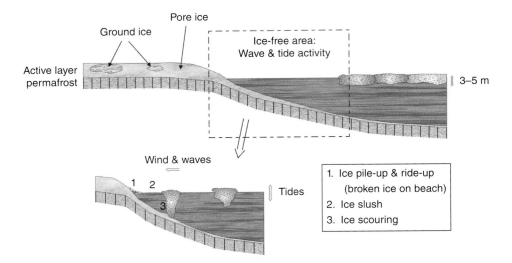

Fig. 14.4 Ice and coastal processes.

glacial sediments onto the beaches and indirectly shovel sediments from the sea-bottom onshore (Fig. 14.5a). In addition, ice can enter the coastal environment through sea ice, either as pack ice or slush (Fig. 14.5b). The latter will also dissipate the energy of waves in open water and will reduce the wave impacts on the shore.

14.3 Terrestrial ice in coastal environments

The presence of ice in the sediment matrix of sedimentary cliffs strengthens the material. On land, the ground has typically two layers: an active layer at the surface, and a permafrost layer beneath the active layer (Fig. 14.4). The active layer is thus the upper layer, where the ground temperature is above 0°C in summer. The thickness of the active layer ranges from several centimetres to several metres (Romanovsky et al., 2002; Grosse et al., 2011). Many factors determine this thickness of the active layer, including atmospheric temperature and solar insulation, length of the snow coverage period on the surface, vegetation coverage, soil type, etc. Normally, the ice is stored in the matrix of the loose sediments and forms pore-ice between the grains or larger ground-ice bodies like veins, lenses or massive ice blocks (Forbes and Taylor, 1994; Fig. 14.4). In summer, melting of ice in the active layer and snow create poorly drained areas with a lot of water, and soils become unstable. The process of solifluction is widespread, and destabilizes the upper active layer and decreases the stability of glacial bluffs and cliffs along the shorelines (Matsuaka, 2001). The processes of freeze-and-thaw of water in the active layer also produce a number of typical periglacial morphologies in the coastal zone, such as polygon-soils, stone-circles, and open and closed

system pingos. Thermokarst is an irregular topography that is widespread and is caused by spatial dissimilarity of these processes in the active layer due to differences in grain sizes, water content and micro-climate (Fig. 14.6).

The permafrost layer below the active layer is frozen all year round. Permafrost is defined as the ground that is frozen, below 0°C, for more than two successive years. The permafrost is often spatially continuous under high-Arctic conditions, where the mean air temperature in summer stays below 6°C, and may be patchy in warmer areas like those with low-Arctic conditions (see Fig. 14.1). Discontinuous permafrost is often observed outside the Arctic and Antarctic zones. The thickness of the permafrost layer can be up to 1500 m, and depends on the atmospheric temperature at its top and the geothermal heat at its bottom. The ice-content in permafrost soils may reach volumetric values over 60% (Romanovski et al., 2002); it is normally manifested as frozen water in the pores or as separate ice blocks. The strength of the deeply frozen permafrost layer is large, and the layer is almost impermeable. The permafrost layer hinders the meltwater within the active layer to infiltrate or drain. This means that the water stays in the active layer. This enhances the extensive formation of peat in low-relief areas and triggers solifluction processes in high-relief areas. At the same time, the daily and seasonal freeze-thaw cycles enhance the thermokarst processes in the active layer. In the case of the discontinuous permafrost, however, the meltwater in the active layer can drain. A permafrost layer often continues under the ocean as offshore permafrost when seawater temperatures are close to 0°C. Most of the de-glaciated areas in the coastal environment consist of soils with permafrost. This permafrost fills up the matrix and gives extra stability, as the strength of the material is increased.

Fig. 14.5 (a) Sediment transport through drifting ice (icebergs) in the Sermilik fjord, southeast Greenland. (b) Slush ice on a beach and icebergs in a fjord. (c) Fluvial discharge and sediment delivery into a fjord during the break-up of the ice in early summer, Young Sound, northeast Greenland (26 June 2010). (Source: Day 178, 2010; GeoBasis Zackenberg.)

Fig. 14.6 Thermokarst surface on the coastal plain of Alaska, close to Prudhoe Bay.

14.4 Coastal geomorphology and coastal responses

The coastal environments of high-latitude coasts include a wide variety of landforms. Sand and gravel beaches, barrier islands, spits, deltas, salt marshes, glacial bluffs and rocky cliffs all occur, and they show similar behaviour as those in temporal and tropical latitudes. However, the frequency and intensity of coastal processes by waves and tides are limited. The wave processes are spatially restricted by sea ice and drifting ice, and are temporally restricted to the thawing and ice-free seasons. The evolution of features is thus slower and ice-related processes (see Fig. 14.4) reshape the classic wave and current related shapes. Typical features on beaches in summer are ice-pushed ridges, pitted kettle holes, ice rafted fragments and beach ridges (Short and Wiseman, 1974; Hill et al., 1994; Butler, 1999). The high-latitude coasts can be broadly classified

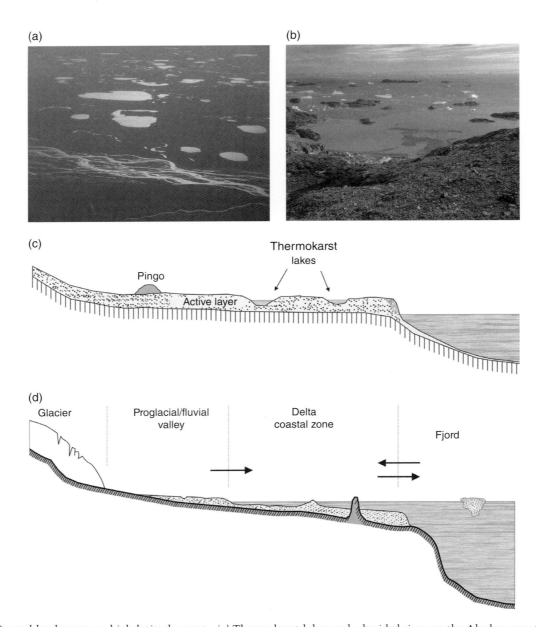

Fig. 14.7 Coastal landscapes on high-latitude coasts. (a) Thermokarst lakes and a braided river on the Alaskan coastal plain. (b) Delta at high tide in a high-relief area, Sermilik, southeast Greenland. The distance from the delta mouth towards the glacier is c. 2 km. Note the icebergs and plume of sediment in the fjord. Schematic coastal landscapes for: (c) a low-relief area; and (d) a high-relief area (Source: Data from Nielsen 1994). Transects in a low-relief area are typically in the order of 10–1000 km, like those over tundras with major rivers on coastal plains in Siberia and Canada. The transects in a high-relief area are typically small and in the order of kilometres, like in eastern Greenland where glaciers are close to open water. For colour details, please see Plate 35.

into two types of coastal landscapes: (1) coastal landscapes in low-relief areas; and (2) coastal landscapes in high-relief areas (Fig. 14.7). Both landscapes have of a wide variety of distinctive geomorphological features as discussed next.

14.4.1 Low-relief areas

Coastal landscapes of low-relief areas are typically characterized by a wide and flat sloping coastal plain over several hundreds of kilometres across (Fig. 14.7c). The tundra is the main unit in this type of coastal environment and

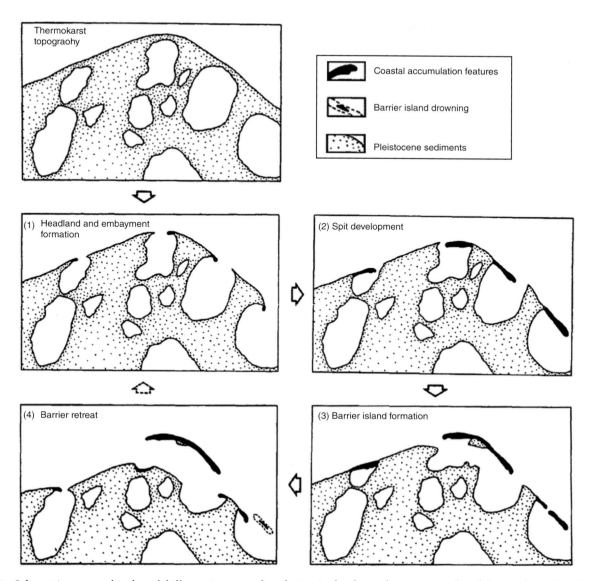

Fig. 14.8 Schematic geomorphical model illustrating coastal evolution in the thermokarst topography of the southern Canadian Beaufort Sea. (Source: Ruz et al. 1992. Reproduced with permission of Elsevier.)

glaciers are not in the immediate vicinity of the actual shoreline. The active layer of the tundra on top of the continuous permafrost layer creates several specific morphologies, such as pingos. Thermokarst is widely observed at different scales; from ice wedges several metres wide up to thermokarst lakes with diameters of several tens of kilometres (Fig. 14.7a). Large rivers drain the hinterland and tundra area, and build extensive deltas near the shore. The sediments at the shore are typically reworked by drifting ice, waves and tides. Examples of those coastal plains with large river deltas are found in Siberia, Alaska and northern Canada (e.g. the MacKenzie delta).

The evolution of sandy and gravel barriers in areas with abundant thermokarst lakes in Canada has been studied by Hill (1990; Hill et al., 1994) and Ruz et al. (1992), with the latter showing an example of the formation of barrier islands and spits on a transgressive shore in a low-relief area in northern Canada (Fig. 14.8). The accumulative features were located at the seaward side of a coastal embayment. This coastal embayment was formed in an earlier phase, through coastal erosion of a thermokarst lake embankment. Hequette and Barnes (1990) studied the shoreface evolution in an area with unconsolidated bluffs along the Canadian Beaufort Sea. They concluded that factors that influence the retreat of an unconsolidated bluff are sediment texture, ground-ice content, cliff height, wave energy and shoreface gradient. They described an erosion of the lower shoreface between 12 and 15 m water depth of up to 1 m, probably caused by sea-ice gouging. The middle shoreface between 5 and 9 m accreted due to ice-push processes, and the upper shoreface eroded due to wave and current processes. They stated that the retreat of the coastal bluffs at the shore was indirectly caused by the sea-ice processes, because they steepened the shoreface profiles, which should thereafter adjust towards a new equilibrium profile. Kobayashi et al. (1999) estimated the erosion rates of the glacial bluffs along the Alaskan shores with a simple retreat model.

The sediment flux of the major rivers that enters high-latitude, low-relief coastal zones is generally less than those in temperate or tropical latitudes. This is due to a range of factors. The major rivers are only seasonally active (Short and Wiseman, 1975; Gordeev, 2006). Most of the time, the supply of sediments is limited due to low rates of weathering. In addition, many large rivers in the coastal plains of Canada and Siberia flow through extensive tundra areas, where the amount of mineralogical sediments is scarce and (dis)continuous permafrost zones are widespread (Rachold et al., 2005; Gordeev, 2006).

14.4.2   High-relief areas

Coastal landscapes of high-relief areas are typically characterized by a relatively short lateral distance, often less than 50 km, between a glacier/land-ice mass and the coastal waters (Fig. 14.7d). The landscape elements are glacier, proglacial and fluvial valley, delta and fjord or open coastal water. The melting ice of glaciers is delivered to the icefront by supra-, en- and sub-glacial streams. The sediments in these streams are flowing in the pro-glacial valley, and the water may form a concentrated braided drainage pattern that results in a small river. The sediments are further transported through the fluvial valley towards the coastal areas. Quite often the sediment delivery is large enough to build an alluvial fan or a delta. The sediments at the delta front or near the shoreline of the alluvial fan can be further transported to the fjords or open coastal waters, or reworked and redistributed along the shores by drifting ice, waves and tides. Examples of these landscapes in high-relief areas can be found in Greenland and Antarctica. The scale of the Arctic landscape model is variable, and ranges from systems with a distance from the glacier towards the open water of some kilometres, like Sermilik in southeast Greenland (Fig. 14.7b), to tens of kilometres.

The characteristics of beach-ridge plains and their formation have been studied by Butler (1999) in the Antarctic, and Sanjaume and Tolgensbakk (2009) along the Arctic Norwegian shore. Most of the beach ridges are formed in periods of ice-free coastal waters, and the material in the sandy and gravel ridges is often of local origin, from eroding unconsolidated bluffs or small rivers. However, there exists a large variability in beach-ridge crest levels between various bays and parts of the same bay, which invalidates the explanations of global processes like sea-level change being responsible for beach-ridge development (Sanjaume and Tolgensbakk, 2009). Kroon et al. (2011) showed an example of coastal reworking on a small delta in northeast Greenland. Here, spits developed on the delta platform in front of an abandoned delta lobe, and slowly migrated along the shore due to wave-driven sediment transport and landward due to overwash processes (Fig. 14.9 and Fig. 14.10).

14.4.3   Coastal change

Coastal responses over larger areas in the Arctic are often studied using remote-sensing observations, such as satellite images (Harper, 1990; Mars and Houseknecht, 2007; Lajeunesse and Hanson, 2008; Lantuit and Pollard, 2008; Jones et al., 2009, Lantuit et al., 2011). Most of these studies show a general retreat of shoreline position. However, there are differences on a local or regional scale. Mars and Housknecht (2007) looked to the Arctic coast of Alaska and observed accumulation in terms of spit and barrier island growth, deltaic fillings of lakes and bays, and areas of thermokarst lakes that are drained. However, they detected an increase of erosion rates over the last 50 years, with rates of coastal erosion being greatest at open ocean shorelines compared to bays and lagoons. Jones et al. (2009) attribute the shift in rate and pattern of land losses along the Alaska Beaufort Sea to a decline in sea-ice content, an increasing summer sea-surface temperature, a rising sea level, and an increase in storm power and wave action. Lantuit et al. (2011) give an overview of rates of shoreline changes in the Arctic. Most of the coastlines are retreating, with maximum retreat rates along the Alaskan and Canadian Beaufort Sea (where it is exceeding 1.1 m/yr). The coastlines of the Canadian Archipelago and Svalbard are stable. The persistence of sea ice during the summer season, the rocky character of the coast, and the vertical isostatic movements are probably the main reasons for this. In addition, the erosion rates are higher for unlithified coasts, and are very weakly related with ground-ice contents.

Coastal change exhibits great spatial and temporal variability, and can be stable or extremely dynamic, depending on factors like coastal geomorphology, lithology and cryology, as well as relative sea-level trends, storm frequencies, thermal conditions and sea-ice conditions (Manson et al., 2005; Lantuit et al., 2011). Spatial and temporal variability

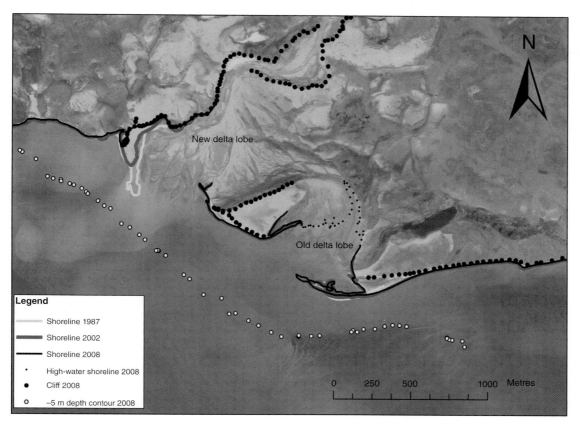

Fig. 14.9 Coastal reworking by waves and wave-driven currents along the fringes of an abandoned delta lobe in Zackenberg, northeast Greenland. (Source: Adapted from Kroon et al. 2011.)

Fig. 14.10 Erosion of c. 20-m high glacial bluffs at an active delta mouth in the Zackenberg area, northeast Greenland (see Fig. 14.9; cliff end is c. 50 m to the south). The eroded blocks and the scarps in the bluff contain large blocks of ground ice.

in factors affecting coastal change can greatly affect the shoreline erosion potential, for example, when the timing of the peak frequency of storms and maximum open water coincide with deep thaw of ice-rich cliffs and rapid relative sea-level rise, the erosion potential is substantially enhanced (Manson et al., 2005). Hence, climate change can significantly alter the erosion potential in some coastal areas by influencing the timing of events. Coastal responses are often related to distinct temporal and spatial scales (De Boer, 1992). The hierarchy of coastal features and associated dynamics in high-latitude coastal environments is shown in Fig. 14.11. Events may cause quick responses on a local scale. An extreme river discharge due to the breach of ice-dammed lakes can induce bank erosion in channel bands on a fluvial plain or delta (see Fig. 14.9 and Fig. 14.10), and summer storms with high waves can erode sandy beaches and cause overwash deposits on spits and barriers (see Fig. 14.9). These local changes typically occur on a spatial scale of up to a few kilometres. Coastal responses on a larger scale, with spatial dimensions between 1 and 100 km, typically include glacial bluff erosion by waves and

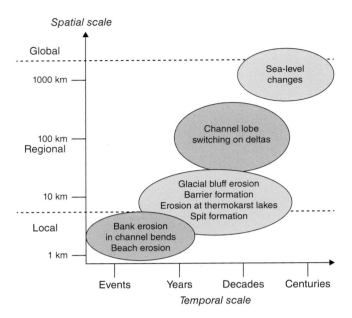

Fig. 14.11 Spatial and temporal scales for morphological features in Arctic coastal environments.

coastal thermal processes, or the formation of a spit or barrier at the entrance of a former thermokarst lake. Coastal responses on a regional scale have alongshore spatial dimensions of over 100 km. An example of this type of response is the shift of a channel lobe on a delta (Jerolmack, 2009), such as along the northern shores of Alaska, Canada or Siberia, which typically occurs on a timescale of decades or centuries. Even larger timescales involve the development of a coastal plain over centuries to millennia. At the local scales, the detailed description of coastal processes by wind, ice, waves and tides are important in order to understand the coastal responses. However, these detailed coastal processes are less useful studies at the larger scales. Coastal responses on the largest scales are often schematized and expressed by simple process variables such as the relative sea-level rise. The slowly responding geological processes of, for example, isostasy are not important at the local scales, but certainly influence the long-term evolution of the morphology at the larger scales. Rising global sea levels will affect the shape of the shorelines over the coming decades. However, most severe and catastrophic shoreline changes occur as a consequence of local and regional scale problems (Gratiot et al., 2008).

14.5 Relative sea-level change

As outlined in Chapter 2, relative sea-level change is a combined effect of volumetric changes of ocean water (eustasy) and vertical displacements of land (isostasy).

The eustatic changes are of a global nature, and originate from temporal differences in the ocean basin geometry of the Earth and in the volume of entrapped water in glaciers and on land, and thermally induced volumetric changes of seawater. The periodic changes in eustasy are associated with climatically forced ice ages and interglacials during the Quaternary; ice ages correspond to relative sea-level lowstands and interglacials to relative sea-level high-stands. The isostatic displacements are of a more regional or local nature, and are the result of loading and unloading of mass onto the Earth's crust, as well as, on a longer time-scale, tectonic movements. Isostatic changes have mainly been the result of adjustment of the crust to expansion, and melting of glaciers and ice caps during the Quaternary. The weight of the ice causes the crust beneath it to sub-side as glaciers expand laterally and grow in height. A glacial isostatic rebound occurs when the load is reduced during interglacials and the crust slowly responds by raising the surface. The isostatic movements can be modelled as a function of glacier growth if the geophysical properties of the crust are known (lithosphere thickness and upper and lower mantle viscosity). The glacial history can be described when isostatic displacements are measured. The relative sea-level changes differ in space and time, and reflect a variety of geophysically and climatologically driven processes. The relative sea-level changes in the past can be used to predict future relative sea-level rise rates due to global warming.

Glacio-isostatic adjustment varies considerably around the Arctic, some regions experiencing uplift, such as the Canadian Archipelago (Lajeunesse and Hanson, 2008), Alaska (Shennan, 2009) and Antarctica (Colhorn et al., 1992; Hjort et al., 1997), and some experiencing subsidence, such as some regions along the Siberian coasts (ACIA, 2005; Whitehouse et al., 2007). In Greenland, the relative sea level fell rapidly during the early Holocene due to isostatic rebound following deglaciation, whereas the sea transgressed low-lying coastal areas during the late Holocene (e.g. Hjort, 1997; Woodroffe and Long, 2009; Funder et al., 2011; Pedersen et al., 2011).

Beach ridges (e.g. Colhorn et al., 1992; Hjort et al., 1997; Funder et al., 2011; Pedersen et al., 2011) or salt marshes (e.g. Shennan, 2009; Woodroffe and Long, 2009) are often used to reconstruct relative sea-level curves in high-latitude coastal environments. Beach ridges are often present on sedimentary Arctic environments when the glacio-isostatic adjustment exceeds the absolute water-level change. Active beach ridges or storm ridges are generally formed at a position well above mean sea level at the upper limit of the wave run-up. However, many studies are not explicit about the coupling between water levels, wave energy levels (wave run-up) and tides, and the height of the ridge during their formation. Sanjaume and Tolgensbakk (2009) even doubt a direct coupling between global processes like sea-level change and beach-ridge development. In fjords, the crest of active fair-weather beach ridges is often related to the high-water mark at spring tides (Pedersen et al., 2011). However, on open ocean coasts, the beach-ridge height is probably more related to extreme events, like storms, with associated water-level surges (e.g. Butler, 1999). The local morphology of the coastal stretch and the availability of sediment also influence the shape and location of beach ridges on a plain. A number of successive beach ridges on a coastal plain in the high Arctic (Fig. 14.12a, b) can indicate former shorelines for different phases of, for example, the Holocene, and might indicate the local relative sea-level change. An example of a relative sea-level curve for the Holocene in northeast Greenland is presented in Fig. 14.12c. This curve is partly based on optically stimulated luminescence (OSL) dating of sandy material in specific beach ridges at different heights above the present mean sea level. This reconstruction is made under the assumptions that: (1) the local wave conditions, surge levels and coastal profile were constant in the Holocene; and (2) the wave set-up and wave run-up in the fjord areas were negligible (because the wave field in the short and ice-free summer season is low-energetic and fetch limited). This implies that the measured levels of the specific ridges are measures of the former high-tide marks of palaeo sea level at the time of beach-ridge formation. The relative sea-level curve clearly presents a relative sea-level fall after the deglaciation of the area during the Holocene, followed by a more or less stable sea level in the last 2000 years. Woodroffe and Long (2009) and Shennan (2009) dated cores from salt marshes to establish the relative sea-level curves for, respectively, west Greenland and Alaska. The high-latitude salt marshes often have a limited bioturbation, but may experience disturbance from grazing or problems with ice rafting (Dionne, 1993; Argow et al., 2011), or kettle holes or willows (Forbes and Taylor, 1994), that influence the sedimentary record. However, despite these problems, salt marshes are still well suited to study the local sedimentation record and associated related relative sea-level curves.

14.6 Climate change predictions and impacts for high-latitude coasts

The Arctic is both an active player in, and a sensitive responder to, climate change, and is thus a key region in the global climate system. The last decade's intensified research in the Arctic has revealed that these high latitudes are experiencing the strongest air and sea-surface temperature increases on Earth (ACIA, 2005; Anisimov et al., 2007; Miller et al., 2010). Solar radiation is the primary driver of temperature in the Arctic. Strong seasonality, where midnight sun and polar nights prevail for long periods, is a prominent feature of high-latitude environments, and

exerts a major influence on almost every aspect of the high-latitude coasts (Forbes, 2011). Absolutely speaking, solar insolation is, however, low because a considerable amount of energy is reflected as a result of the low angle of insolation and the snow and ice cover. Furthermore, temperatures in the coastal regions are influenced by their proximity to the ocean. Most of the coastal locations are cooler in summer than their interior counterparts, and correspondingly warmer in winter than their interior counterparts. Sea-surface temperatures are strongly influenced by factors like ocean circulation and sea-ice cover (Forbes, 2011). Cloud cover also has an important moderating effect on the coastal environment, as it controls the summer-radiation balance and thereby the loss of ice in the terrestrial and marine environments, for example, thawing of ice-rich materials and the resulting coastal erosion response (Forbes, 2011).

Arctic sub-regions have experienced some of the greatest warming rates on Earth in recent decades (Anisimov et al., 2007). Average annual temperatures have risen by about 2–3°C since the 1950s, and in winter by up to 4°C (ACIA, 2005). In general, average Arctic temperatures have increased at almost twice the global average rate in the past 100 years. However, this general warming trend is superimposed on a pattern of high decadal variability and large regional differences (Anisimov et al., 2007).

The extent of the sea-ice cover in the high latitudes has been measured with high frequency by satellite-borne instruments since the 1970s. There is an overall decline in the extent and thickness of sea-ice cover in the Arctic region (e.g. Parkinson et al., 1999; Johannesen et al., 2004; Serreze et al., 2007), and the ice-free season is increasing (Belchansky et al., 2004). Loss of the ice cover will alter the surface energy budget and a significantly warmer Arctic in autumn and winter is expected as a result of larger heat fluxes from the ocean to the atmosphere (Serreze et al., 2007). Ice and snow acts as important agents in positive feedback mechanisms, such as the ice/snow albedo feedback contributing to Arctic amplification of the climate forcing (Miller et al., 2010).

Sea-level rise is one of the most important consequences of climate change and has the potential to cause significant global impacts on ecosystems and societies. Changes in relative sea level cause coastal changes by either inundating low-lying areas of land (transgression) or subaerially exposing areas of land previously flooded (regression), directly affecting shoreline erosion and accretion, and coastal stability in general. Hence, increases in sea level are likely to increase coastal erosion rates in low-elevation areas and affect sediment transport in coastal regions (ACIA, 2005).

The impacts of climate change due to increasing atmospheric and ocean temperatures have already led to a number of changes in all high-latitude coastal environments

(a)

(b)

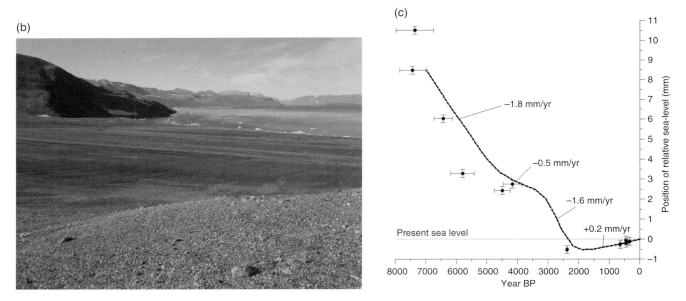

Fig. 14.12   (a, b) Beach-ridge plain at Grønnedal, northeast Grønland, with Holocene beach ridges from the present water level up to 36 m above mean sea level. (c) Typical sea-level curve at the same location in northeast Greenland. The crosses are optically stimulated luminescence (OSL) dating on the lower beach ridges. (Source: Adapted from Pedersen et al. 2011. Reproduced with permission of John Wiley & Sons.) For colour details, please see Plate 36.

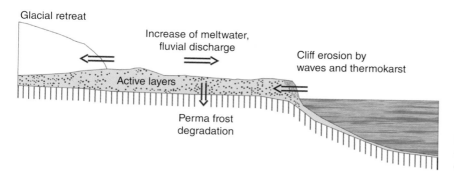

Fig. 14.13 Changes in the high-latitude coastal environments as a result of a warmer climate.

(Fig. 14.13): glaciers have retreated on land and in the fjords, the discharge of water and sediment from meltwater rivers has increased (Anisimov et al., 2007), permafrost has degraded and become more discontinuous (Grosse et al., 2011), and coastal cliffs in glacial bluffs and along coastal plains have shown amplified erosion due to wave processes (e.g. Jones et al., 2009; Lantuit et al., 2011) and thermokarst (Burn and Zhang, 2009; Lantuit et al., 2009; Grosse et al., 2011). The extra sediment discharge towards the coast locally causes more sedimentation in pro-glacial and fluvial valleys, and may lead towards an extension and increase in the volumes of alluvial fans and deltas (positive feedback; e.g. Anisimov et al., 2007). The increase in air and seawater temperatures in the high latitudes in the near future will induce a reduction of the permafrost zones and a shift of the discontinuous permafrost towards higher latitudes (Grosse et al., 2011). This will probably lead to a number of changes in the coastal environments (negative feedbacks; Lantuit et al., 2009):

(1) Wave-driven coastal processes on shorelines may become more dominant and induce erosion of cliffs and beaches (increase of the stress). This is already happening along the northern shores of Alaska and the Bering Strait.

(2) The exposure of the eroding cliffs will locally enhance the thawing of the permafrost in the cliff. This reduces the strength of the cliffs in sedimentary environments (e.g. glacial bluff).

(3) The coastal areas around the shoreline cliffs may experience more thermokarst, which also decreases the strength of the material.

The net effect of all these changes will probably be an enhanced erosion of the shoreline, and a shift towards more wave-related coastal features where eroded sediments accumulate. The erosion of shorelines also causes additional fluxes of carbon and nitrogen away from source areas (Grosse et al., 2011; Lantuit et al., 2011). The number and size of lakes has increased over the last decades on several coastal plains in high-latitude environments, for example, on the northern slope of Alaska and in the MacKenzie delta in Canada (Jones et al., 2009). The amplified erosion of the coastal cliffs along the shores is caused by the longer ice-free seasons and the decreased size of the sea-ice coverage in the coastal waters. These enhance the duration and the intensity of the wave processes, respectively. At the same time, the active layer in the cliffs increases, and the stability of the cliffs decrease, due to thermokarst. Quantitative rates of coastal shoreline responses are presented in Lantuit et al. (2011) for the Arctic coastal environments. However, it is still impossible to give an idea of the acceleration of shoreline retreat due to climate change.

14.7 Future perspectives

The effects of climate change are one of the most serious environmental issues threatening the Arctic environment. Arctic sub-regions have shown some of the most rapid rates of warming in recent years, and substantial environmental impacts of great complexity and regional differences are expected (Anisimov et al., 2007). Evidence of a warming climate is already widespread across the Arctic, with potentially dramatic impacts on Arctic ecosystems, including sea levels, sea ice, waves, permafrost, plant and animal species, and human use of the coastal zone (ACIA, 2005; Forbes, 2011; Box 14.2). Arctic amplification is expected to cause an Arctic hemisphere temperature increase, maybe several times the global mean, but the exact magnitude is still uncertain as it depends on multiple factors and complex mechanisms (Miller et al., 2010)

Several studies indicate an accelerating decrease in the Arctic sea-ice extent, and most current climate models significantly underestimate the Arctic sea-ice decline compared to the observed trend (Stroeve et al., 2007). As sea ice plays an important role in the climate system, influencing ocean-to-atmosphere fluxes, surface albedo and ocean buoyancy, the reduction in sea-ice extent and cover will have numerous impacts. Loss of the ice cover will alter the surface energy budget, and a significantly warmer Arctic in autumn and winter is expected as a result of larger heat fluxes from the ocean to the atmosphere (Serreze et al., 2007). Ice and snow act as important

CASE STUDY BOX 14.2   Coastal inhabitants and impacts of global change

The coastal areas of the Arctic and Antarctic regions have always been sparsely inhabited. The climate is harsh and the faunal resources are relatively scarce. In addition, the harsh conditions hinder exchange of people and goods. Most of the open waters in these high-latitude areas are not intensively used for shipping, because the ice-free periods are limited and navigation is problematic.

The original inhabitants in Arctic coastal environments were Inuits. They are Arctic hunters and use the coastal areas for housing and as places to store their food. The partly ice-covered coastal waters are used as hunting grounds for fish and all types of whales. The impact of the Inuit culture on the Arctic environment was small: they did not develop large towns, their transport ways were quite often over ice, and they did not build large infrastructures. However, non-native people started to settle in the Arctic at the end of the 19th century. Sometimes they came as hunters, as in eastern Greenland. Major activities developed around commercial fisheries, mining and oil exploration/exploitation. The latter activity is especially developed in Alaska along the northern shore and has led to a more sophisticated infrastructure: pipelines have been built, mines and oil infrastructure constructed, and shipping harbours developed.

The pollution in high-latitude coastal areas created by, for example, shipping accidents or oil and gas exploration are also major threats to the environment. The regeneration capacity of nature is very small in these environments, compared with other areas. Temperatures are often very low and chemical processes are slow, preventing a quick recovery. Moreover, the vegetation coverage in coastal high-latitude areas is sparse and it takes a long time before damaged vegetation recovers. Coastal zone management is not widely applied in the high-latitude areas. Many activities are focussed on fisheries, mines, oil industry and hunting activities. The coastal zone management is in general driven by sector planning. It often focusses on a limited number of activities and is affected by the dominant economic sector.

The present-day erosion rates of many shorelines in the Arctic due to the increased wave impact over longer ice-free seasons have a significant impact on anthropogenic activities. Smaller coastal villages in the low-relief areas suffer from erosion and have to protect their infrastructures (e.g. Forbes, 2011; Fig. 14.14). In addition, the coastal erosion leads to extra losses of cultural heritage and archaeological sites (Jones et al., 2008; Kroon et al., 2010; Westley et al., 2011). For human communities in the Arctic, the impacts of climate change will most likely be mixed, and to some extent will be controlled by local conditions. Traditionally, Arctic communities, especially indigenous people, have been living in close relationship with the local environment. Seasonal changes in access to local resources and the changing ecological conditions have fostered a natural resilience and cultural ability to adapt in many Arctic communities. But it seems that the tremendous ongoing changes of the Arctic coastal environments now threatens the subsistence activities of some indigenous people, for example traditional use of the coast by Inuit communities in Canada, Alaska and northwestern Greenland is threatened by the disappearance of sea ice (ACIA, 2005; Anisimov et al., 2007; Forbes, 2011).

Fig. 14.14   Oil-pipelines and additional infrastructure required for the oil industry, northern shore Alaska, near Prudhoe Bay.

agents in positive feedback mechanisms, like the ice/snow albedo feedback contributing to Arctic amplification of the climate forcing (Miller et al., 2010). Sea-ice decline will affect vulnerable ecosystems, cause changes in coastal ecology and bioproductivity, and furthermore result in critical habitat loss and possible harmful effects on ice-dependent marine wildlife (Anisimov et al., 2007). The projected reductions in sea ice and breakup of thick, old multi-year landfast sea ice along the shoreline is also likely to exacerbate the collapse of ice shelves and decline of tidewater and outlet glaciers in the Arctic region. Furthermore, melting of ice is expected to affect the

Arctic's freshwater system, which could in turn affect ocean circulation patterns (Serreze et al., 2007). Rising global sea levels will affect the shape of the shorelines over the coming decades. Most severe and catastrophic shoreline changes occur as a consequence of local and regional scale problems (Gratiot et al., 2008). The increased coastal wave action due to sea-ice reduction and the permafrost reduction results in a potential increase in wave-induced coastal erosion (ACIA, 2005).

14.8 Summary

This chapter on high-latitude coast provides information about the coastal evolution and coastal morphodynamics in polar environments. The distinctive high-latitude coastal conditions, processes and landforms are caused by common factors such as strong seasonality, cold temperatures, (dis)continuous permafrost, and the presence of seasonal or perennial sea-ice cover. In this way, these coasts substantially differ from many other coastal environments discussed in this book.

• Ice-related coastal processes include ice pile-up and ice ride-up on the beach, ice slush and ice scouring by icebergs. The impact of wave- and tide-related coastal processes is often limited by the limited ice-free season, and by the limited fetches due to the presence of pack ice in deeper water.

• Terrestrial ice in coastal environments is often stored in a permafrost layer. This permafrost layer is strong and stabilizes coastal cliffs. However, the active layer on top of this permafrost layer often weakens the strength of the material and solifluction processes are widely observed on many slopes in coastal areas.

• There is a wide variety of landforms. Sand and gravel beaches, barrier islands, spits, deltas, salt marshes, glacial bluffs and rocky cliffs all occur and they show similar behaviour as those in temporal and tropical latitudes. However, the frequency and intensity of coastal processes by waves and tides are limited.

• The coastal responses show a large variation in space, especially along the fringes of major Arctic delta systems. Large erosion rates (>1 m/yr) are observed on former delta lobes, while accretion typically occurs along the shores in the vicinity of present delta distributaries. The spatial and temporal scales for different morphological features in Arctic coastal environments are strongly related.

• The relative sea-level change over the Holocene is often characterized by a fall, due to disappearance of ice, followed by a stabilization or small increase. Data from beach ridges and salt marshes are often dated and used to determine the local sea-level curve for a specific location.

• The environmental forcing induced by climate warming and increased storm-event frequencies is expected to trigger landscape instability, increased sediment and nutrient supply from land, intensified shoreline erosion and increased sediment transport due to increasing open-water conditions, as well as rapid responses to change and increased hazard exposure. The coastal response to global and regional climate changes can result in land and habitat loss, and thus affect vulnerable biological and human systems.

Key publications

Forbes, D.L. and Taylor, R.B., 1994. Ice in the shore zone and the geomorphology of cold coasts. *Progress in Physical Geography*, **18**, 59–89.
[*Very nice overview of processes related to ice and their impact on coastal morphologies*]

Forbes, D.L. (Ed.), 2011. *State of the Arctic Coast 2010 – Scientific Review and Outlook*. International Arctic Science Committee, Land-Ocean Interactions in the Coastal Zone, Arctic Monitoring and Assessment Programme, International Permafrost Association. Helmholtz-Zentrum, Geesthacht, Germany. (http://arcticcoasts.org)
[*Comprehensive and up-to-date account of the impact of climate change on Arctic coastal systems*]

Hill, P.R., Barnes, P.W., Héquette, A. and Ruz, M.–H., 1994. Arctic coastal plain shorelines. In: R.W.G. Carter and C.D. Woodroffe (Eds), *Coastal Evolution: Late Quaternary Shoreline Morphodynamics*. Cambridge University Press, Cambridge, UK, pp. 341–372.
[*Nice summary of Arctic coastal features and their morphodynamics*]

Trenhaile, A.S., 1997. *Coasts in cold environments*. In: *Coastal Dynamics and Landforms*. Clarendon Press, Oxford, UK, pp. 290–309.
[*Chapter with a comprehensive account of coastal processes and their associated Arctic forms*]

References

ACIA, 2005. *Arctic Climate Impact Assessment*. Cambridge University Press, Cambridge, UK.

Anisimov, O.A., Vaughan, D.G., Callaghan, T.V. et al., 2007. Polar regions (Arctic and Antarctic). In: M.L. Parry, O.F. Canziani, J.P. Palutikof, P.J. Van der Linden and C.E. Hanson (Eds), *Climate Change 2007: Impacts, Adaptation and Vulnerability*. Contribution of Working Group II to the Fourth Assessment Report on Intergovernmental Panel on Climate Change. Cambridge University Press, Cambridge, UK, pp. 653–685.

Are, F., Reimnitz, E., Grigoriev, M., Hubberten, H.-W. and Rachold, V., 2008. The influence of cryogenic processes on the erosional Arctic shoreface. *Journal of Coastal Research*, **24**, 110–121.

Argow, B.A., Hughes, Z.J. and FitzGerald, D.M., 2011. Ice raft formation, sediment load, and theoretical potential for ice-rafted sediment influx on northern coastal wetlands. *Continental Shelf Research*, **31**, 1294–1305.

Belchansky, G.I., Douglas, D.C. and Platonov, N.G., 2004. Duration of the Arctic Sea ice melt season: regional and interannual variability, 1979–2001. *Journal of Climate*, **17**, 67–80.

Butler, E.R.T., 1999. Process environments on modern and raised beaches in McMurdo Sound, Antarctica. *Marine Geology*, **162**, 105–120.

Burn, C.R. and Zhang, Y., 2009. Permafrost and climate change at Herschel Island (Qikiqtaruq), Yukon Territory, Canada. *Journal of Geophysical Research*, **114**, F02001, doi: 10.1029/2008JF001087.

Colhoun, E.A., Mabin, M.C.G., Adamson, D.A. and Kirk, R.M., 1992. Antarctic ice volume and contribution to sea-level fall at 20,000 yr BP from raised beaches. *Nature*, **358**, 316–319.

De Boer, D.H., 1992. Hierarchies and spatial scale in process geomorphology: a review. *Geomorphology*, **4**, 303–318.

Dionne, J.C., 1993. Sediment load of shore ice and ice rafting potential, upper St. Lawrence Estuary, Quebec, Canada. *Journal of Coastal Research*, **9**, 628–646.

Forbes, D.L. (Ed.), 2011. *State of the Arctic Coast 2010 – Scientific Review and Outlook*. International Arctic Science Committee, Land-Ocean Interactions in the Coastal Zone, Arctic Monitoring and Assessment Programme, International Permafrost Association. Helmholtz-Zentrum, Geesthacht, Germany. (http://arcticcoasts.org)

Forbes, D.L. and Syvitski, J.P.M., 1994. Paraglacial coasts. In: R.W.G. Carter and C.D. Woodroffe (Eds), *Coastal Evolution: Late Quaternary Shoreline Morphodynamics*. Cambridge University Press, Cambridge, UK, pp. 373–424.

Forbes, D.L. and Taylor, R.B., 1994. Ice in the shore zone and the geomorphology of cold coasts. *Progress in Physical Geography*, **18**, 59–89.

Funder, S., Goosse, H., Jepsen, H. et al., 2011. A 10,000-year record of Arctic Ocean sea-ice variability – view from the beach. *Science*, **333**, 747–750.

Gordeev, V.V., 2006. Fluvial sediment flux to the Arctic Ocean. *Geomorphology*, **80**, 94–104.

Gratiot, N., Anthony, E.J., Gardel, A., Gaucherel, C., Proisy, C. and Wells, J.T., 2008. Significant contribution of the 18.6 year tidal cycle to regional coastal changes. *Nature Geoscience*, **1**, 169–172.

Grosse, G., Romanovsky, V., Jorgenson, T., Anthony, K.W., Brown, J. and Overduin, P.P., 2011. Vulnerability and feedbacks of permafrost to climate change. *EOS Transactions*, **92**, 73–74.

Harper, J.R., 1990. Morphology of the Canadian Beaufort Sea coast. *Marine Geology*, **91**, 75–91.

Hequette, A. and Barnes, P.W., 1990. Coastal retreat and shoreface profile variations in the Canadian Beaufort Sea. *Marine Geology*, **91**, 113–132.

Hill, P.R., 1990. Coastal geology of the King Point Area, Yukon Territory, Canada. *Marine Geology*, **91**, 93–111.

Hill, P.R., Barnes, P.W., Héquette, A. and Ruz, M.-H., 1994. Arctic coastal plain shorelines. In: R.W.G. Carter and C.D. Woodroffe (Eds), *Coastal Evolution: Late Quaternary Shoreline Morphodynamics*. Cambridge University Press, Cambridge, UK, pp. 341–372.

Hjort, C., 1997. Glaciation, climate history, changing marine levels and the evolution of the northeast water polynia. *Journal of Marine Systems*, **10**, 23–33.

Hjort, C., Ingolfsson, O, Moller, P. and Lirio, J.M., 1997. Holocene glacial history and sea-level changes on James Ross Island, Antarctica Peninsula. *Journal of Quaternary Science*, **12**, 259–273.

Jenner, K.A. and Hill, P.R., 1998. Recent, arctic deltaic sedimentation: Olivier Islands, Mackenzie Delta, North-west Territories, Canada. *Sedimentology*, **45**, 987–1004.

Jerolmack, D.J., 2009. Conceptual framework for assessing the response of delta channel networks to Holocene sea level rise. *Quaternary Science Reviews*, **28**, 1786–1800.

Johannesen, O.M., Bengtson, L., Miles, M.W. et al., 2004. Arctic climate change: observed and modelled temperature and sea-ice variability. *Tellus*, **56A**, 328–341.

Jones, B.M., Hinkel, K.M., Arp, C.D. and Eisner, W.R., 2008. Modern erosion rates and loss of coastal features and sites, Beaufort Sea coastline, Alaska. *Arctic*, **61**, 361–372.

Jones, B.M., Arp, C.D., Jorgensen, M.T., Hinkel, K.M., Schmutz, J.A. and Flint, P.L., 2009. Increase in the rate and uniformity of coastline erosion in Arctic Alaska. *Geophysical Research Letters*, **36**, L03503, doi: 10.1029/2008GL036205.

Kobayashi, N., Vidrine, J.C., Nairn, R.B. and Soloman, S.M., 1999. Erosion of frozen cliffs due to storm surge on Beaufort Sea Coast. *Journal of Coastal Research*, **15**, 332–344.

Kroon, A., Jakobsen, B.H. and Pedersen, J.B.T., 2010. Coastal environments around Thule settlements in Northeast Greenland. *Danish Journal of Geography*, **110**, 143–154.

Kroon, A., Pedersen, J.B.T. and Sigsgaard, C., 2011. Morphodynamic evolution of two deltas in arctic environments, east coast of Greenland. *The Proceedings of the Coastal Sediments 2011*. World Scientific, Singapore, pp. 2299–2310.

Lajeunesse, P. and Hanson, M.A., 2008. Field observations of recent transgression on northern and eastern Melville Island, western Canadian Arctic Archipelago. *Geomorphology*, **101**, 618–630.

Lantuit, H. and Pollard, W.H., 2008. Fifty years of coastal erosion and retrogressive thaw slump activity on Herschel Island, southern Beaufort Sea, Yukon Territory, Canada. *Geomorphology*, **95**, 84–102.

Lantuit, H., Rachold, V., Pollard, W.H., Steenhuisen, F., Ødegård, R. and Hubberten, H.-W., 2009. Towards a calculation of organic carbon release from erosion of Arctic coasts using non-fractal coastline datasets. *Marine Geology*, **257**, 1–10.

Lantuit, H., Overduin, P.P., Couture, N. et al., 2011. The Arctic Coastal Dynamics database: a new classification scheme and statistics on Arctic permafrost coastlines. *Estuaries and Coasts*, **35**(2), 383–400, doi: 10.1007/s12237-010-9362-6.

Manson, G.K., Solomon, S.M., Forbes, D.L. et al., 2005. Spatial variability of factors influencing coastal change in the western Canadian Arctic. *Geo-Marine Letters*, **25**, 138–145.

Mars, J.C. and Houseknecht, D.W., 2007. Quantitative remote sensing study indicates doubling of coastal erosion rate in past 50 yr along a segment of the Arctic coast of Alaska. *Geology*, **35**, 583–586.

Matsuaka, N., 2001. Solifluction rates, processes and landforms: a global review. *Earth-Science Reviews*, **55**, 107–134.

Miller, G.H., Alley, R.B., Brigham-Grette, J. et al., 2010. Arctic amplification: can the past constrain the future? *Quaternary Science Reviews*, **29**, 1779–1790.

Nielsen, N., 1994. Geomorphology of a degrading arctic delta, Sermilik, Southeast Greenland. *Geografisk Tidsskrift (Danish Journal of Geography)*, **94**, 46–57.

Nikiforov, S.L., Pavlidis, Y.A., Rachold, V. et al., 2005. Morphogenetic classification of the Arctic coastal zone. *Geo-Marine Letters*, **25**, 89–97.

Parkinson, C.L., Cavalieri, D.J., Gloersen, P., Zwally, H.J. and Comiso, J.C., 1999. Arctic sea ice extents, areas and tends, 1978–1996. *Journal of Geophysical Research*, **104**(C9), 20837–20856.

Pedersen, J.B.T., Kroon, A. and Jakobsen, B.H., 2011. Holocene sea-level reconstruction in the Young Sound region, Northeast Greenland. *Journal of Quaternary Science*, **26**, 219–226.

Rachold V., Are, F., Atkinson D., Cherkashov, G. and Solomon S., 2005. Arctic Coastal Dynamics (ACD): an introduction. *Geo-Marine Letters*, **25**, 63–68.

Reimnitz, E., Kempema, E.W. and Barnes, P.W., 1987. Anchor ice, seabed freezing, and sediment dynamics in shallow arctic seas. *Journal of Geophysical Research*, **92**, 14671–14678.

Reimnitz, E., Barnes, P.W. and Harper, J.R., 1990. A review of beach nourishment from ice transport of shoreface materials, Beaufort Sea, Alaska. *Journal of Coastal Research*, **6**, 439–469.

Romanovsky, V.E., Burgess, M., Smith, J., Yoshikawa, K. and Brown, J., 2002. Permafrost temperature records: indicators of climate change. *EOS Transactions*, **83**, 589–594.

Ruz, M.-H., Héquette, A. and Hill, P.R., 1992. A model of coastal evolution in a transgressed thermokarst topography, Canadian Beaufort Sea. *Marine Geology*, **106**, 251–278.

Sanjaume, E. and Tolgensbakk, J., 2009. Beach ridges from the Varanger Peninsula (Arctic Norwegian coast): characteristics and significance. *Geomorphology*, **104**, 82–92.

Serreze, M.C., Holland, M.M. and Stroeve, J., 2007. Perspectives on the Arctic's shrinking sea-ice cover. *Science*, **315**, 1533–1536.

Shennan, I., 2009. Late Quaternary sea-level changes and palaeo-seismology of the Bering Glacier region, Alaska. *Quaternary Science Reviews*, **28**, 1762–1773.

Short, A.D. and Wiseman, W.M.J., 1974. Freezeup processes on Arctic beaches. *Arctic*, **27**, 215–224.

Short, A.D. and Wiseman, W.M.J., 1975. Coastal breakup in the Alaskan Arctic. *Geological Society of American Bulletin*, **86**, 199–202.

Solomon, S.M., 2005. Spatial and temporal variability of shoreline change in the Beaufort Mackenzie region, Northwest Territories, Canada. *Geo-Marine Letters*, **25**, 127–137.

Streletskaya, I.D., Vasiliev, A.A. and Vanstein, B.G., 2009. Erosion of sediment and organic carbon from the Kara Sea Coast. *Arctic, Antarctic, and Alpine Research*, **41**, 79–87.

Stroeve, J., Holland, M.M., Meier, W., Scambos, T. and Serreze, M., 2007. Arctic sea ice decline: faster than forecast. *Geophysical Research Letters*, **34**, L09501.

Trenhaile, A.S., 1997. *Coasts in cold environments*. In: *Coastal Dynamics and Landforms*. Clarendon Press, Oxford, UK, pp. 290–309.

Wadhams, P., 2000. *Ice in the Ocean*. Gordon and Breach Scientific Publishers, London, UK.

Walker, H.J. and Hudson, P.F., 2003. Hydologic and geomorphic processes in the Colville River delta, Alaska. *Geomorphology*, **56**, 291–303.

Westley, K., Bell, T., Renouf, M.A.P. and Tarasov, L., 2011. Impact assessment of current and future sea-level change on coastal archaeological resources-illustrated examples from northern Newfoundland. *The Journal of Island and Coastal Archaeology*, **6**, 351–374.

Whitehouse, P.L., Allen, M.B. and Milne, G.A., 2007. Glacial isostatic adjustment as a control on coastal processes: an example from the Siberian Arctic. *Geology*, **35**, 747–750.

Wiseman, W.M.J., Suhayda, J.N., Hsu, S.A. and Walters, C.D., 1974. Characteristics of nearshore oceanographic environment of Arctic Alaska. In: J.C. Reed and J.E. Slater (Eds), *The Coast and Shelf of the Beaufort Sea*. Arctic Institute of North America, Arlington, VA, USA, pp. 49–64.

Woodroffe, S.A. and Long, A.J., 2009. Salt marshes as archives of recent relative sea level change in West Greenland. *Quaternary Science Reviews*, **28**, 1750–1761.

15 Rock Coasts

WAYNE STEPHENSON

Department of Geography, University of Otago, Dunedin, New Zealand

15.1 Introduction

Rock coasts are the most common coastal type, making up around 80% of the world's coastline (Emery and Kuhn, 1982), and are found at all latitudes and in all morphogenetic environments (Trenhaile, 1987), including lakes and estuaries. One of the most important management challenges that rock coasts present is that rock coasts are erosional in origin, thus the management of rock coasts may require preservation of erosional processes (and landforms) that are often seen as destructive or are unwanted in other coastal landscapes. This is particularly the case with sea cliffs, where management is commonly applied to stop cliff erosion, resulting in a modification of the cliff form. This can cause erosion of adjacent beaches where the eroding cliff is an important source of sediment for the beach. Such management problems highlight the interconnectedness of rock coasts with other coastal systems and the importance of an integrated approach to rock coast management.

The erosional origin of rock coasts and rock coasts landforms also makes reconstructing the evolutionary history particularly difficult. Understanding how rock coasts have evolved remains unclear, particularly as the influence of changing sea levels over the Quaternary, the role of inheritance (where multiple sea levels have shaped a shoreline), and the relative contributions of a multitude of processes are extremely complex. In addition, geology and lithology cause enormous variability in morphology from site to site. It is perhaps not surprising then that rock coast scientists are yet to develop a coherent morphodynamic framework for the study of rock coasts in the way the beach morphodynamic model has shaped beach research over the last 35 years (see Chapter 7). The consequence of current uncertainty is that predicting and understanding climate change impacts on rock coasts is still in its infancy. It has been argued recently that rock coasts are not as resilient to climate change and sea-level rise as has traditionally been thought (Naylor et al., 2010). This is because of the erosional

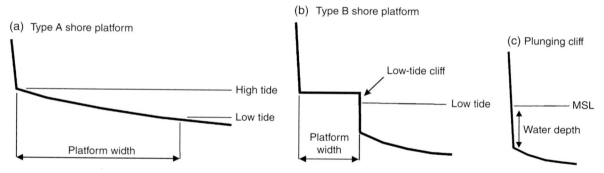

Fig. 15.1 Three major rock coast morphologies: (a) sloping shore platform; (b) near-horizontal shore platform; and (c) plunging cliff. MSL, mean sea level. (Source: Adapted from Sunamura 1992. Reproduced with permission of John Wiley & Sons.)

origin of rock coasts and the lack of a dynamic response (e.g. accretion) to sea-level rise that other systems such as reefs, beaches and salt marshes are capable of exhibiting. Rock coasts have no capacity to recover from storm events in the way beaches can rebuild as sediment is transported onshore.

Rock coast morphology can be broadly categorized into three types (Fig. 15.1): (1) sloping shore platforms backed by a cliff (Type A shore platform); (2) near-horizontal platform with a landward cliff and characterized by a seaward drop, sometimes referred to as a low-tide cliff as the upper part of the cliff is seen at low tide (Type B shore platform); and (3) plunging cliffs where no platform is present. Water depths in front of a plunging cliff are deep enough that waves do not break against the cliff face, but reflect off it. Despite this simple tripartite classification, rock coast morphology is far more complex than Fig. 15.1 suggests, as highlighted in Fig. 15.2. For example, many cliffs are fronted by beaches. Whether or not shore platforms are backed by an active cliff has much to do with recent changes in sea level and/or tectonic processes. Numerous shore platforms can be found that are backed by beaches resting in front of inactive and abandoned cliffs.

15.2 Geology and lithology

Rock coasts and associated landforms develop in a bewildering array of geological and lithological settings, from hard and highly resistant granite to relatively weak consolidated gravels and loess, and with variations in the density of bedding and joints (Fig. 15.3). The overall similarity of coastal rock landforms (cliffs and platforms) belies that complexity, and is often expressed in differing rates of development and a variety of resulting morphologies. Geology and lithology have a significant bearing on the efficiency of erosional processes and consequently are a primary control on the morphology of rock coasts. The height and steepness of cliffs are largely a function of rock strength. Geological structure also plays an important role

in determining how shore platforms and cliffs evolve, and the scale at which they erode. This is because the spacing and number of joints determine the size and scale at which erosion occurs. Rock coasts eroded by wave action do so at a scale similar to the block sizes yielded by joint sets and bedding planes. Consequently, determining the amount of wave energy needed to erode blocks from rock coasts can be estimated by assessing geological structure. Susceptibility to weathering is also a function of rock structure, as joints and bedding planes permit the intrusion of water, salts and biological agents to weaken discontinuities (Fig. 15.3). Lithology will also determine the effectiveness of weathering processes because fine-grained sedimentary rocks, such as mud or silt stones, are far more susceptible to wetting and drying or salt weathering than are igneous or well-metamorphosed rocks.

The most important consequence of geology and lithology is rock resistance, that is, the resisting force that rock has to erosive processes, particularly waves. Many attempts to understand rock coast development have used the relationship between the erosive force of waves and rock resistance as a guiding paradigm. Work particularly by Japanese geomorphologists (Sunamura, 1992) has focussed on quantifying both sides of the equation:

$$F_W \geq F_R \qquad (15.1)$$

where F_W is the wave erosive force and F_R is the rock resistance (Box 15.1). Eqn. 15.1 shows that rock coast development begins when the erosive force of waves exceeds the resisting force of rock. More effort is needed to better quantify each side of eqn. 15.1, but this is problematic because of the inherent difficulties of measuring those processes and factors that determine the value of each side of the equation (Fig. 15.4). Rock strength is initially a function of geology and lithology, but that strength is almost always reduced by weathering and biological activity to the point where erosion can occur. While this equation provides an important conceptual tool and a starting

Fig. 15.2 Variety of rock coast morphologies in differing geological settings and rock types: (a) granitic coast, Putuo Island, East China; (b) cliff in greywacke sandstone, Catlins Coast, southeast coast South Island, New Zealand; (c) sandstone plunging cliff, Sydney Harbour, Australia; (d) sloping shore platform developed in Blue Lias limestone, Glamorgan Heritage Coast, South Wales, UK; (e) horizontal shore platform developed in Jurassic sandstone, Curio Bay, Southland, New Zealand; (f) horizontal platform developed in greywacke sandstone, Victoria, southeastern Australia; (g) sloping platform developed in mudstone with inactive sea cliff behind, Kaikoura Peninsula, South Island, New Zealand; and (h) cliff developed in consolidated glacial till, Lake Pukaki, South Island, New Zealand. (Source: Photographs by Wayne Stephenson.) For colour details, please see Plate 37.

(g)

(h)

Fig. 15.2 (continued)

Fig. 15.3 Weathering along rock joints in limestone on shore platforms on the Glamorgan coast of South Wales, UK. Runnels no longer hold water after the tide has receded because the layer of limestone has been detached from the layer below, allowing water to drain through the joint. Compare those with runnels holding water with active biology present. Schmidt hammer case is 32 cm long. (Source: Photograph by Wayne Stephenson.)

point for examining rock coasts, it does not tell us what happens to a rock coast systems once F_W exceeds F_R. Deterministic (mathematical) models are required once the threshold is passed, in order to understand rock coast development (e.g. Trenhaile, 2000).

Rock resistance has important consequences for long-term coastal landform and landscape evolution. Harder rocks types are able to persist in the landscape if abandoned by a falling sea level, or when cut on a lower sea level and subsequently raised by tectonic forces. This means that

many rock coasts have inherited morphology from previous glacial or interglacial sea levels. Many sea cliffs and shore platforms have undoubtedly undergone at least two episodes of erosion; one at the current sea level and the other one during the last interglacial.

The large-scale (kilometres) configuration of coastlines is mainly a function of geology, with the geology providing the framework for determining the plan shape of coasts and the types of landforms. Even sedimentary landforms, such as beaches and estuaries, accumulate on basement rock, and their size and shape are also significantly affected by the underlying geological structure. An additional role is played by the characteristics of the coastal catchment. For example, along the New South Wales coast, Australia, with its many rock headlands and embayments, the width and depth of those embayments have been linked to the drainage basin area, with wider bay entrances associated with large basin areas, rather than lithology at the coast (Bishop and Cowell, 1997).

15.3 Processes acting on rock coasts

Rock coasts are exposed to a myriad of geomorphic processes that reflect geographical distributions, latitudinal gradients and climate. Ocean waves remain the single most important driver of change on rock coasts, but the efficiency of wave erosion is affected by geological setting and subaerial processes, particularly weathering, which acts to reduce rock strength (Fig. 15.4). In addition to marine and subaerial processes, biological activity can also be important as an erosive agent, but also as a protective one, reducing the efficiency of other processes. In some cases, biological activity can construct rock coasts.

METHODS BOX 15.1   Determining rock resistance

In order to understand and predict how rock coasts erode and develop, it is necessary to be able to represent force and resistance as expressed in eqn. 15.1. While wave forces can be measured and represented quantitatively through direct measurement, rock resistance or strength presents far more difficulties and complications. This is because a large number of factors contribute to rock strength or, alternatively, contribute to rock strength reduction (Fig. 15.4). Rock resistance has been of interest to engineering, geology and geomorphology, and a large

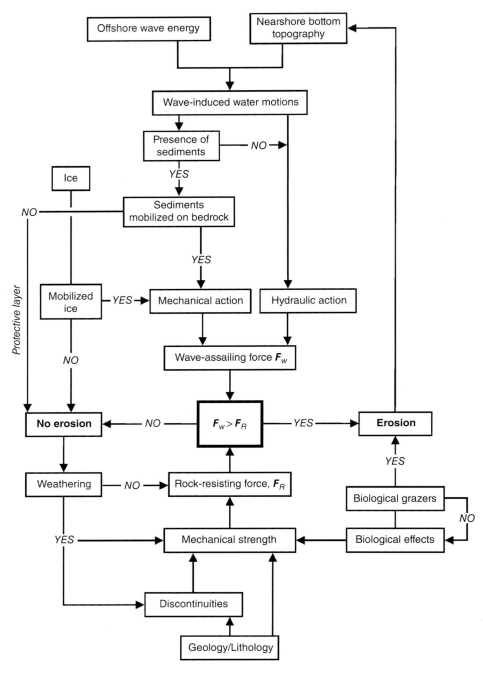

Fig. 15.4 Interaction of factors that determine the strength of rock and erosive force of waves on rock coasts. (Source: Adapted from Sunamura 1992. Reproduced with permission of John Wiley & Sons.)

body of literature exists detailing various means of measuring rock strength. Any consideration of rock strength must distinguish between intact rock strength and rock mass strength, the latter accounts for fractures, bedding planes and joints.

The most challenging aspect of representing rock strength is dealing with fractures, and this introduces a scale problem. At the scale greater than joint spacing, fracture spacing or bedding thickness, rock resistance can really only be represented qualitatively or semi-empirically. Intact rock strength can be represented in at least three ways: compressive strength, tensile strength and sheer strength. Which type of strength is used depends on how erosion is achieved in the field. Measuring intact rock strength can be relatively easy, using unconfined compressive strength testing in the laboratory through the point load test, triaxial test or Brazilian test, all of which are detailed by the American Society for Testing Materials. The Schmidt hammer (Fig. 15.5) is an easy and cheap method for use in the field for measuring compressive strength, and is widely used by geomorphologists. However, intact rock strength becomes increasing difficult to measure where weathering processes are relevant. The Schmidt hammer does permit assessment of rock strength of weathered and unweathered rock (Table 15.1). Intact rock strength becomes relatively useless once a rock mass is fractured, jointed or contains bedding planes.

A variety of semi-empirical or qualitative classification schemes exist to represent rock mass strength, but because these are not quantitative, they cannot be utilized in eqn. 15.1. Attempts to quantify F_R have used intact rock strength represented by compressive strength. Sonic wave velocity through the rock mass can be used to account for discontinuities, but it remains to be demonstrated that this approach adequately represents rock mass strength. Discontinuities in a rock mass subjected to wave action will determine the size of erosion products, and rock mass strength is more important than intact rock strength on most coasts. In some cases, intact rock strength can be very high, but the presence of discontinuities results in a relatively low rock mass strength. In rock coast geomorphology, little attention has been given to the significant efforts made by engineers and engineering geologists to determine rock mass strength, and much learnt by these disciplines could be applied to the problem of rock resistance in coastal geomorphology.

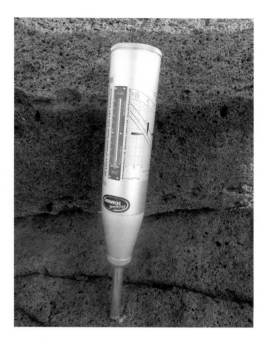

Fig. 15.5 The Schmidt hammer. (Source: Photograph by Wayne Stephenson.)

Table 15.1 Schmidt hammer rebound values comparing weathered and unweathered rocks on shore platforms.

Location	Rock type	Unweathered rebound number	Weathered rebound number	% Weathered
Kaikoura Peninsula, NZ	Calcareous mudstone	35	28	80
Kaikoura Peninsula, NZ	Mudstone	32	24	75
Kaikoura Peninsula, NZ	Limestone	50	31	62
Glamorgan, Wales, UK	Limestone	47	42	89
The Gower, Wales, UK	Limestone	53	41	77
Start-Prawle, Devon, UK	Mica Shist	46	31	67
Marengo, Victoria, Australia	Greywacke	47	36	77

15.3.1 Waves

Waves impacting on rock coasts impart a variety of forces, but understanding the erosional efficiency of these forces still requires a great deal of research, with there being relatively few field studies. Waves erode rock coasts and transport weathered debris, but the precise relationship between wave forces and rock resistance as expressed by Sunamura (1992) remains unclear. Waves can erode by directly expending force and quarrying blocks, or by abrasion when waves mobilize sediment against bedrock. Only two attempts have been made to assess the forces exerted by waves on rock coasts, with mixed success (Stephenson and Kirk, 2000a; Trenhaile and Kanyaya, 2007).

Wave quarrying is the breaking free and removal of rock fragments by shock pressures from breaking waves (especially plunging breakers), water hammer and the compression of air in rock joints. Shock pressures and water hammer occur when waves break directly against rock surfaces and both breaking and broken waves cause air compression in joints and bedding planes. Water then rushes into voids and this is followed by a sudden explosive release as the water recedes. It is this sudden release of pressure that causes stress on the rock (Trenhaile, 1987). Wave quarrying is limited in extent over a shore platform, since maximum pressures occur at or close to still-water level and the position of still water is controlled by the tide. Measured waves on platforms in the Bay of Fundy, Canada, have been shown to generate forces large enough to remove blocks from the platform (Trenhaile and Kanyaya, 2007). Recent studies have demonstrated the effectiveness of wave quarrying on platforms and cliffs (e.g. Hall et al., 2008). While these studies demonstrate wave quarrying in operation, it remains unclear whether wave forces are capable of quarrying unweathered rock. It is likely that in all of these cases, weathering along joints and bedding planes has occurred so that the resistance of the rock mass has been reduced by weathering (Fig. 15.4).

Abrasion has been identified as an important erosive process on shore platforms by some authors (Robinson, 1977; Blanco-Chao et al., 2007). Abrasion relies on the presence of sediment and sufficient wave energy to move it, and generally occurs in a narrow zone and at the back of shore platforms. Abrasion works best where accumulated sediments are as hard as, or harder than, the underlying bedrock and can be washed back and forth by wave action so that the sediment works as an abrasive. If sediment accumulations are too deep to be moved by waves, then abrasion ceases and a protective function is provided by the sediment. Robinson (1977) measured abrasion rates of up to 4.5 cm/yr, with two-thirds occurring in winter when storms were more prevalent. Blanco-Chao et al. (2007) demonstrated how abrasion was effective on shore platforms with erosion rates between 0.13 and 1.8 mm/yr, which are comparable to erosion rates reported from platforms where abrasion was absent (Stephenson and Finlayson, 2009). The significance of abrasion to shore platform development is unclear and it has not been demonstrated to be a primary formative mechanism. Given that abrasion is a spatially and temporally discrete process and works most effectively only where abrasive sediments

Fig. 15.6 Cliff top boulder, Little Beecroft Head, Jervis Bay, New South Wales, Australia. This boulder measures 3.5 × 2.1 × 1.3 m, weighs approximately 23 tonnes and sits 30 m above sea level (Switzer and Burston, 2010). (Source: Photograph by Wayne Stephenson.)

accumulate (to a limited depth), typically at the landward margin of shore platforms, large parts of shore platforms, and indeed whole platforms, are currently eroded by other processes, therefore abrasion is probably not a significant process for the overall development of shore platforms.

A key debate in coastal science is whether coastal evolution is mainly driven by infrequent but high-magnitude events, such as tsunamis or very powerful storms, or by frequent but moderate-magnitude events (see Chapter 5). Rock coasts are excellent places to test this debate as very large boulders and megaclasts (boulders are 256–4096 mm across the intermediate axis, while megaclasts have an intermediate axis diameter greater than 4096 mm) are often found along rock coasts, on shore platforms and cliff tops. Such large boulders seem only moveable by the very largest of waves (Fig. 15.6). Cliff top boulders and blocks situated apparently above or beyond the reach of storm waves are frequently cited as evidence for the occurrence of tsunamis (Noormets et al., 2002). It is rare for very large storms to be observed moving large boulders. While tsunamis have been a 'popular' explanation for cliff top boulders, recent work calls the tsunami hypothesis into serious question (Switzer and Burton, 2010), and has documented the ability of storm waves to deposit boulders on cliff tops tens of metres above sea level (Hall et al., 2008). However, where massive boulders (>250 tonnes) are positioned above the reach of storm waves and seem too large to have been transported by storm waves, then tsunamis seem a plausible explanation. While very large storm waves and tsunamis are responsible for wave quarrying, the significance of these events to long-term rock coast evolution is not known (see Chapter 5).

(a)

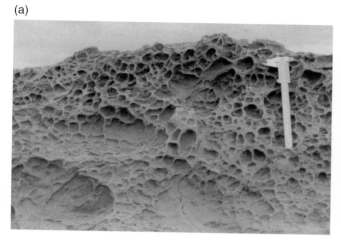

(b)

Fig. 15.7 (a) Tafoni weathering of supratidal mudstone, Kaikoura Peninsula, South Island, New Zealand. Vernier callipers are 23 cm long. (b) Water-layer weathering on mudstone shore platform, Kaikoura Peninsula, South Island, New Zealand. Note that raised rims form along joints where water is less likely to evaporate and dry the joint. Consequently, the joints undergo fewer wetting and drying cycles than the flat pans, enhancing the micro-relief. (Source: Photographs by Wayne Stephenson.)

15.3.2 Weathering

Coastal environments are very prone to weathering, due to an abundance of salts, water and temperature extremes, and a large exposure of rock to be broken down. Apart from causing an overall downwearing of shore platforms and weakening of rock, weathering also gives rise to distinctive weathering features (Fig. 15.7). While both physical and chemical weathering processes generally operate in concert, consideration of weathering usually focusses on salt weathering, wetting and drying, and solution, which combine to produce a process known as water-layer weathering. In colder high-latitude climates, freeze thaw can also be important.

- **Salt weathering.** Salts cause rock weathering by exerting pressures within rock as crystals grow from solution, by expanding when salt crystals are heated, and from volume changes induced by hydration. The effectiveness of salt weathering is determined by the type of the salts, the nature of the rock (e.g. degree of porosity, tensile strength), and the type of environment. Generally, temperate to hotter climates promote salt weathering. The degree of saturation of the solution is also important, as is the duration of exposure to super-saturated conditions. The significance of salt weathering in coastal environments is difficult to determine, since it occurs in unison with wetting and drying.

- **Wetting and drying.** Repeated wetting and drying causes internal stress to many rocks, especially where it results in expansion and contraction of the rock. Wetting and drying is most common in the intertidal zone, but can also occur in the supratidal zone, where wave splash delivers water to higher elevation, or further inland. Shales and

mudstones are most susceptible, because they can contain clay minerals that expand when wet and shrink when dry. Measurements of erosion made with micro-erosion meters (Box 15.2) have revealed that rock surfaces can also expand, or swell, and this is most likely the result of wetting and drying (Stephenson and Kirk, 2001).

- **Solution.** Minerals within rock can be dissolved by solution. The dissolution of limestone (calcium carbonate) by water containing carbon dioxide (acidic water) is perhaps the best-known example and is important because of the significant amount of rock coast composed of limestone around the world. Rocks containing feldspars are also susceptible to solution to varying degrees. Solubility of rocks, in addition to the mineral composition, is determined by water temperature, pH and the amount of carbon dioxide in the water. Paradoxically, seawater is super-saturated with calcium carbonate and so should not be able to achieve solution, but variations in water temperature and biological activity alter the concentrations and pH from alkaline to acidic so that solution can occur. Acidic groundwater can also occur at the coast and cause solution.

- **Water-layer weathering.** On shore platforms, where the generally flat surface allows water to pool and subsequently dry out, a process termed water-layer weathering has been identified. This process involves wetting and drying, salt weathering, and, depending on the lithology of the rock, also chemical weathering. The relative contribution of these individual processes is unknown. The process of water-layer weathering produces a distinct morphology where shallow pools are surrounded by raised rims associated with joints (Fig. 15.7b). The raised rims retain water in the joint so that they undergo fewer wetting

METHODS BOX 15.2   Measuring surface lowering on shore platforms

The micro-erosion meter (MEM; Fig. 15.8) was developed to measure surface erosion of bedrock in caves, but was later adapted for use on shore platforms. The instrument works by relocating precisely positions on a rock surface, using three bolts installed in the rock as reference points. An engineer's dial gauge shows the elevation of the rock surface in relation to the three bolts, and subsequent readings reveal the amount of surface lowering. The original MEM allowed for three readings to taken. A modified version, called the traversing micro-erosion meter, allowed for many more positions,

and depending on the size and specific design, in excess of 200 points can be repeatedly measured. A further modification was the addition of a digital dial gauge so that recordings could be logged directly to a computer in the field.

The MEM has been used to measure rates of erosion over a variety of timescales, from one or two years, to 30 years in the case of Stephenson et al. (2010). Rates reported vary greatly, from a few microns per year to 70 mm/yr on chalk. Stephenson and Finlayson (2009) averaged all published erosion rates from MEM studies

(a)

(b)

Fig. 15.8   The micro-erosion meter (MEM) used for measuring down-wearing rates on shore platforms. (a) The original design that allowed three measurements to taken. (b) The traversing meter, allowing a large number of measurements, in this example 180, to be taken. The electronic dial gauge can be connected to a laptop computer in the field to facilitate rapid data collection. (Source: Photographs by Wayne Stephenson.)

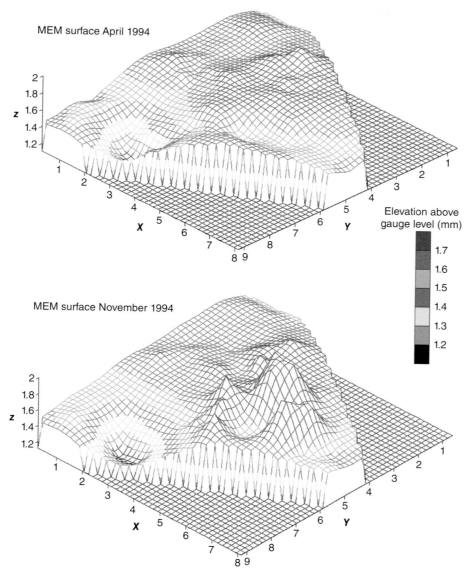

Fig. 15.9 Wire frame surface plots from a traversing micro-erosion meter (MEM) site. The upper plot is the surface in April 1994 and the lower plot is the same surface in November 1994. Note surface swelling evident at the lower right corner of the surface indicated by the dark blue peaks that have risen up. Elevations are in millimetres and relative to the MEM gauge zero level. For colour details, please see Plate 38.

on shore platforms and found that the global average rate of surface lowering on shore platforms was 1.486 mm/yr (n = 50; sd = 1.900 mm/yr). A key feature of published lowering rates is that they exhibit a wide range of values and often large standard deviations, so that mean rates must be used cautiously. The traversing MEM generates enough data points to permit construction of a digital terrain model of a bolt site (Fig. 15.9). Stephenson et al. (2010) used this technique to calculate volumes of material

eroded from bolt sites and determined the sediment budget for shore platforms on Kaikoura Peninsula, South Island, New Zealand. The intertidal shore platforms on Kaikoura Peninsula have an area of 740,579 m^2 and yield 1032 m^3/yr at the micro-scale measured with the MEM. This calculation does not include larger-scale erosion products and does not represent a complete sediment budget.

While intended to measure vertical erosion of rock surfaces, the MEM has also revealed that rock surfaces are dynamic, undergoing surface lowering (not necessarily erosion) and swelling where rock surfaces actually rise (Fig. 15.9). Swelling has been explained as expansion resulting from absorption of moisture or the growth of salt crystals. Stephenson and Kirk (2001) presented the first attempt to analyse this phenomenon, and subsequently a number of authors have attempted to better understand rock swelling (e.g. Porter and Trenhaile, 2007). Detailed explanations for swelling remain vague, other than to say that wetting and drying, salt crystallization and endolithic algae play some role. The magnitude of surface expansion ranges from microns to millimetres (up to 8 mm was reported by Stephenson and Kirk, 2001), and over a range of temporal scales of a few hours to months. Swelling is generally regarded as an important weathering process that is a precursor to erosion, because it is thought to generate stresses that weaken the rock surface (reduce F_R).

and drying cycles than the pools, which lower more quickly. Therefore, the micro-morphology is enhanced through positive feedback, until the relief of the rims becomes too pronounced to withstand wave action (see Chapter 1, section 1.2.3).

• **Freeze-thaw.** At high latitudes, water freezing and expanding in rock mass and joints causes stresses that fracture rock. The frequency of these cycles is dependent on seasonal, synoptic and daily temperature variations, with sub-polar regions being the most susceptible. The effectiveness of freeze-thaw as a weathering agent is dependent on the number of temperature cycles through the freezing point of the water in the rock mass (approximately –2.2 °C for seawater). A well-known illustration of freeze-thaw impacts on rock coasts is the case of the chalk shore platforms along the English Channel coast in Sussex, when in 1985 a particularly cold winter produced excessive frost shattering, enhancing erosion of the platforms' surface (Robinson and Jerwood, 1987).

The relative contribution of each type of weathering described above at any one location will be dependent on environmental conditions, so that the importance of each will be different both spatially and temporally. For example, seasonal variations are likely to be important in temperate environments for wetting and drying. The suite of weathering processes is responsible for a number of micro- to meso-scale landforms, such as tafoni (Fig. 15.7a), pits and lapies (on limestone).

15.3.3   Biology

Bio-erosion is the removal of rock substrate by the direct action of organisms, usually by boring or grazing. On rock shores, micro-boring occurs as a result of cyanobacteria and chlorophytes. Boring invertebrates include sponges and molluscs, such as chiton, gastropods and bivalves. Rates of bio-erosion vary enormously, from microns to metres per year, depending on the organisms responsible and their population densities. Erosion is achieved through grazing, chemical reaction to secretions, and boring. Limpets feed by scraping the rock surface with their radula to ingest the rock and digest the endolithic algae contained in the rock. They leave behind a clearly visible trace (Fig. 15.10a). Trenhaile (1987) presented a summary of 42 investigations of bio-erosion rates, which range from 30 to 50 microns in three to four weeks by algae on carbonates, to 1 m/yr on limestone by boring *Paracentrotus lividus*.

Biology on rock coasts also serves to provide protection from other erosive processes. Measured erosion rates on shore platforms are sometimes lower in winter compared with summer because of extensive algae coverage during winter but not in summer. The main reason for this pattern is that cooler winter temperatures reduce desiccation and allow mats of algae to persist over platform surfaces, thus trapping sediment, providing protection from wave erosion, and preventing drying (no wetting and drying cycles). There has been very little research to determine the relative importance of bioprotection compared to bio-erosion and other erosion processes.

Bioconstruction on rock coasts has been acknowledged by a very few studies, and the geomorphic importance of it is not known. Spencer and Viles (2002) noted the protective role bioconstruction by vermetid molluscs and red algae plays on temperate rock coasts. Naylor and Viles (2000) documented reef building on the seaward edge of shore platforms in South Wales, UK, by *Sabellaria alveolata*. Reefs several metres in dimension develop as individual worms build tubes by cementing sand grains together (Fig. 15.10b). These reefs dissipate wave energy by reducing the water depth and forcing wave breaking, and by covering and protecting the rocky substrate.

15.3.4   Mass movement

In addition to marine and weathering processes, sea cliffs are subject to mass movement and processes that drive slope evolution. Thus, mass movement types, such as topple, fall, slide, avalanche and flow, can all be found occurring on rock coasts, depending on the climate, groundwater hydrology, and the geological and lithological framework of the coast in question (Lee and Clark, 2002). Mass movement modifies the cliff form, and work by Young et al. (2009) showed how falling debris can cause the base of the cliff to temporally prograde, at least until the material is transported away by waves, and also that mass movement is spatially discrete and episodic. Mass movement is a

(a)

(b)

Fig. 15.10 (a) Grazing scar left by a limpet on a shore platform surface. (b) *Sabellaria alveolata* constructed reef on a shore platform, Glamorgan coast, south Wales, UK. (Source: Photographs by (a) Wayne Stephenson and (b) Larissa Naylor.)

function of the volume, mobility and morphology of the sediment deposited at the base of the cliff, which in turn determines the susceptibility of the cliff base to wave attack. Surface- and groundwater also play important roles in cliff failure. Surface water washing down cliff faces transports sediment downslope, and on cliffs composed of unconsolidated or weakly lithified material this can be significant. Groundwater increases loading and can lubricate discontinuities (if present), reducing the strength of rock materials (Sunamura, 1992). Seepage of groundwater at the cliff face can erode material and weaken discontinuities. Water can also cause weathering and dissolution, further reducing rock strength. The effectiveness of groundwater and surface water in driving and aiding mass movement is controlled by rainfall (and evaporation), and mass movement events have been associated with periods of more intensive rainfall (see later).

15.4 Rock coast landforms

15.4.1 Sea cliffs

Sea cliffs can be defined in a variety of ways, based on height and slope, although there is little consensus as to what the limiting values of either component are. Bird (2004) suggested that cliffs have slopes greater than 40°, are often vertical, and sometime overhang and cut into rock. Cliffs less than a metre have been termed clifflets or micro-cliffs; cliffs between 1 and 100 m are simply 'cliffs'; cliffs rising 100–500 m are termed 'high cliffs'; and cliffs higher than 500 m are 'mega-cliffs' (Bird, 2004). Many sea cliffs are also developed in consolidated or semi-lithified sediments, so cliffs occur in sedimentary

deposits and rock with a wide range of geomechanical strengths (see Fig. 15.2).

Since sea cliffs are a type of hillslope, knowledge of hillslope processes and development is directly relevant. The one process that makes sea cliffs distinct is marine erosion, but even this is not too different from basal removal from hillslopes connected to rivers. Sea cliffs are not only subjected to marine processes, but also experience a suite of subaerial processes, including groundwater seepage, weathering and mass movement. The importance of non-coastal processes has been highlighted in a study by Greenwood and Orford (2008), who found that higher erosion rates were associated with higher rainfall events on coastal cliffs developed in drumlins in Northern Ireland. During periods when marine processes are inactive at the base of the cliff, subaerial processes dominate, and the slope of the cliff declines through landward rotation. In this case, the slope is said to be transport-limited. When marine processes regain dominance or are dominant, erosion results in a steeper profile, and the profile is considered weathering-limited. Cliff erosion is often viewed as a result of marine process, where waves undercut cliffs and reduce cliff support until collapse occurs, and then the fallen debris is removed by waves to expose the cliff and to repeat the cycle. However, sea cliffs can continue to erode even when abandoned by the sea or in the absence of undercutting (Lee and Clark, 2002). Cliff erosion is driven by a complex array of processes, including sediment supply to the base of the cliff, which temporarily protects against further marine erosion.

Cliff form is far more complex than suggested by Fig. 15.1. Simple plunging profiles are common, but the cliff form is dependent on tectonic setting, geology and

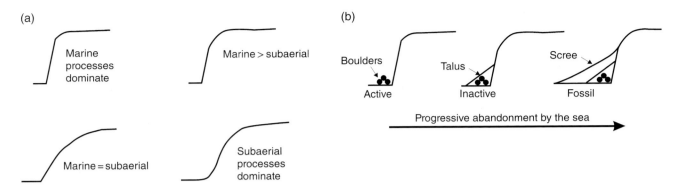

Fig. 15.11 Sea-cliff classification. (Source: Adapted from Emery and Kuhn 1982. Reproduced with permission of Geological Society of America.)

Fig. 15.12 Slope over wall profile in Late Miocene calcareous mudstone, Whakatakahe Head, Mahia Peninsula, North Island, New Zealand. (Source: David Kennedy.)

lithology, history of sea-level change, the degree of inheritance, and the relative contribution of marine and subaerial processes. Emery and Kuhn (1982) classified cliffs based on the relative importance of marine and subaerial processes (Fig. 15.11). Vertical cliffs result when marine processes dominate and the slope is reduced under the dominance of subaerial processes. Over time, as sea level falls or land uplift occurs, or as wave action is displaced from the base of the cliff by an oversupply of sediment, the cliff becomes inactive and eventually fossilizes as it is buried under accumulating debris (Fig. 15.11b). Cliffs also present profiles that result from the development of a sea cliff into an existing hillslope. This has occurred when sea-level rise since the last glaciation has initiated cliff development on flooded hillslopes. Such profiles are termed 'slope over wall', where a near-vertical cliff truncates a hillslope (Fig. 15.12).

The geological setting and lithology of cliff material have significant impacts on erosion rates, the morphology of cliffs, and the efficiency of the processes acting on them. Unlike beaches, cliffs are erosional and do not 'recover' following storms. Rates of cliff erosion vary enormously, from rates almost imperceptible on human timescales to tens of metres per year (Sunamura, 1992). Unsurprisingly, rates of erosion relate to rock resistance governed by those factors illustrated in Fig. 15.4. Cliff erosion should not always be viewed as a negative outcome of coastal processes, since eroded material is liberated into the coastal system and often supplies sediment to beaches, where locally this can represent a significant source of sediment. Dawson et al. (2009) found that cliff recession rates fell in some locations along the East Anglia coast of the UK when modelling the impacts of rising sea levels, because beaches in front of cliffs received more sediment from updrift erosion of cliffs. How fast cliffs erode is often a basic question asked by scientists and coastal managers. A variety of methods exist to measure erosion rates, and these typically involve historical maps, aerial photographs and geographic information systems (GIS). Advanced techniques, including orthorectification and geo-referencing of aerial photographs, have improved the accuracy of such methods (e.g. Lee and Clark, 2002). The advent of LIDAR (light detection and ranging) and terrestrial laser-scanning has enabled far greater detail and coverage of cliff erosion (Rosser et al., 2005), although the timescales over which these data are derived are still relatively short compared to long-term cliff recession over hundreds to thousands of years.

Where cliffs are fronted by beaches, there is usually a balance between beach volume and cliff erosion. If the beach holds a high enough volume of sand, the cliff is protected from direct wave attack (except perhaps under the most violent storms and elevated water levels). Under these conditions, sediment eroded subaerially from the cliff face accumulates at the base as a talus cone, causing the cliff profile to rotate to a lower slope and become more stable, especially if vegetation colonizes the declining slope. If the beach is depleted of sediment, erosion of talus

CASE STUDY BOX 15.3   Managing cliff erosion, Sandringham Beach, Melbourne, Australia

In the case of Sandringham Beach, a Melbourne metropolitan beach in Port Phillip Bay, Australia, a perceived threat of hotspot cliff erosion resulted in a response that simply transferred cliff erosion alongshore to a cliff that had previously been stable. Figure 15.13a illustrates the impact of a groyne built in 1990 to secure a nourished beach and intended to stabilize cliffs. Subsequently, cliff erosion occurred on the downdrift side of the groyne, necessitating further attempts to stop erosion and minimize the risk to beach users (fencing the cliff off from public access), none of which were successful. A decision was made to nourish the beach to protect the cliff from wave erosion and to hold this beach by building a second groyne (Fig. 15.13b), despite the obvious impact of the first groyne. Surprisingly, no attempt was made to quantify the rate of erosion or the likely impact of that erosion on infrastructure before this decision was taken. The main driver of the response was government concern over public liability from falling cliff material injuring or killing beach users. The second groyne was constructed in December 2006 and predictably caused erosion where there had previously been none in historical times. Further beach nourishment between the two groynes occurred in November 2008 (Fig. 15.13c) and additional sand was placed on the downdrift side, but this quickly dissipated and cliff erosion continued (Fig. 15.13c). The consequence of this approach has been to transfer cliff erosion to a third location along the shoreline and encase the beach with a rock groyne that detracts from the aesthetic qualities of the beach.

(a)

(b)

(c)

(d)

Fig. 15.13   Cliff erosion in Port Philip Bay, Victoria, Australia (see text for explanation). (Source: Photographs by (a, b, d) John Amiet and (c) Wayne Stephenson.)

debris yields sediment to re-supply the beach. If erosion exceeds the sediment supply in the talus cone, the cliff begins to be undercut by direct wave attack, and the cliff profile will become more vertical as erosion progresses. As a result, cliffs fronted by beaches undergo episodes of erosion and stability, but usually a beach with varying sediment volumes is maintained as the cliff retreats inland. It must also be remembered that erosion occurs episodically, both in time and space, so that erosion 'hotspots' can exist alongside relatively stable cliffs (Young et al., 2009). Attempts to address hotspot erosion by building walls, revetments or groynes can inadvertently initiate erosion on adjacent sections of cliffs that were previously relatively stable (Box 15.3).

Cliff erosion can be managed in a variety of ways, from highly engineered solutions involving rock walls and revetments, sculpting cliff profiles and re-vegetating, to hazard zoning and even abandonment of the coastline (Lee and Clark, 2002). Beaches can be engineered through nourishment projects to protect cliffs from direct wave attack (Fig. 15.13c). With time, the cliff is expected to stabilize and the slope to decline, resulting in a more stable form. Such an outcome is similar to sea level abandoning a cliff. From a management perspective, the challenge is to prevent the sea from removing basal debris so that the cliff face can stabilize; however, such an approach does not prevent subaerial or mass movement processes. An alternative approach is planning and zoning restrictions that prevent cliff-top development or development too close to cliff tops. Preventing cliff erosion can in fact cause beach erosion, by denying the source of sand to the beach, and make any defensive structure more vulnerable to wave attack as beach volume declines. This was a contributing factor in the case of Sandringham Beach and other beaches along Melbourne's bayside (Box 15.3). Thus, wherever possible, cliff retreat should be accommodated and space be preserved to allow retreat. Careful assessment is required to determine whether the rates of erosion are fast enough to threaten assets. If the rate of cliff retreat is calculated, then the amount of space required behind the cliff can be also be calculated to accommodate the retreat. Problems arise when assets are placed too close to cliff lines and are subsequently threatened. In some cases, it may be appropriate to abandon cliff-top assets where the value of those assets is less than the cost of stabilization.

15.4.2 Shore platforms

Shore platforms are sloping or near-horizontal rock surfaces that generally, but not always, occur in the intertidal zone (see Fig. 15.1). Platforms can be characterized as either sloping or sub-horizontal. Sunamura's (1992) terminology of Type A for sloping platforms and Type B for sub-horizontal platforms is also widely used (Fig. 15.14).

The key characteristic of a Type B platform is not only that the surface lies close to horizontal, but also that it is fronted by a cliff exposed at low tide. It is important to note that not all Type B platforms are absolutely horizontal; many, if not most, have some degree of slope. The low-tide cliff is an important feature because it controls wave transformation and wave energy dissipation across the platform. At low tide, and depending on the water depth in front of the low-tide cliff, significant amounts of wave energy are reflected. If the water is sufficiently shallow, waves will break before the low-tide cliff. At high tide, the low-tide cliff is submerged, but presents a sudden change in water depth that causes waves to break and dissipate across the platform, reducing the amount of wave energy that reaches the landward cliff. The distinction between Type A and B platforms made by Sunamura (1992) was based on the relative strength of wave forces and rock resistance, with Type A platforms occurring in softer rocks and exposed to less energy than Type B platforms. Sloping and sub-horizontal platforms referred to by Trenhaile (1987) were distinguished based on tidal range, with steeper platforms generally found in meso- to macro-tidal settings. Field evidence suggests that platform morphology is governed more by rock properties in micro-tidal environments, but that this is overridden by tidal range in locations where meso- and macro-tidal ranges occur.

Shore platforms can also be grouped according to elevation relative to sea level: (1) high-level platforms with average heights of at least 1 m above sea level; (2) normal platforms with an average elevation between mean low tide and mean high tide; and (3) platforms that are situated mainly below low water. However, elevation has proved to be an unreliable indicator for categorizing platforms, and field evidence shows that platforms occur through a variety of elevations, depending to some degree on rock hardness, exposure to wave energy, sea-level history and tectonic setting.

Platform width is another key morphological feature that determines the extent of a platform. Width can be defined as the horizontal distance from the low-tide level to the cliff platform junction. A relationship between wave energy and width might be expected, with wider platforms developing in response to higher wave energy, but this is not the case. Sunamura (1992) considered width to be controlled by wave energy and rock hardness. One of the most significant challenges to understanding platform width, and indeed platform development, is understanding whether the seaward edge of a platform retreats or is stable. Evidence is contradictory, and theoretical consideration by Sunamura (1992) suggested that under stable sea level it does not retreat. Stephenson (2001) attempted to measure low-tide cliff retreat from aerial photographs and was unable to detect any change in the position of

(a)

(b)

Fig. 15.14   (a) The seaward edge of a Type B platform at low tide, showing the low-tide cliff at low tide. (b) A Type A shore platform that dips gently into the sea, Kaikoura Peninsula, South Island, New Zealand. (Source: Photographs by Wayne Stephenson.)

low-tide cliffs on the platforms of Kaikoura Peninsula, South Island, New Zealand. The advent of more accurate measurements using GIS and orthorectified aerial photographs lends itself to revisiting this particular problem, and only then might we better understand what controls platform width. Understanding platform width is further complicated when inheritance is considered. Some platforms, and perhaps most, were probably initiated during the last interglacial high sea level, so that platforms today have inherited morphological characteristics related to at least one, and perhaps more than one, episode of development (Trenhaile, 2002).

De Lange and Moon (2005) proposed that in cases where the seaward edge of platforms had been static over the Holocene, the width of the platform is a measure of total cliff retreat, and the annual rate of retreat can then be calculated by dividing platform width by the period of time sea level had been at the current level (≈6500 years). Stephenson (2008) argued that the assumption that the seaward edge does not retreat is highly problematic, and that this method of calculating cliff retreat is not valid where a platform is inherited. This disagreement might seem simply academic, except that the technique proposed by De Lange and Moon (2005) has since been used in Australia for hazard planning purposes. The Shoalhaven City Council on the southern New South Wales coast commissioned a study of coastal slope and cliff erosion hazards (Rheinberger and Malorey, 2008). In this study, cliff recession was calculated based on the assumption that platforms along this coast are not inherited and that the seaward edge marks the original position of the shoreline 6500 years ago. The fact that shore platforms along the Shoalhaven coast are known to be inherited from at least one, if not two, interglacial sea

levels during the Quaternary was ignored. The approach also failed to consider higher mid-Holocene sea levels at 3000–2000 years BP. In this case, the cliff recession rates are mostly likely to have been over-estimated; consequently, land-use planning has been based on inaccurate assessments of cliff recession rates and are likely to be over-conservative.

The classic debate on how shore platforms develop has continued in recent times. Early arguments were concerned with the relative roles of marine processes and subaerial weathering. More recently, workers have attempted to measure erosional forces exerted by waves, but these efforts remain semi-empirical and are hampered by the difficulties of quantifying rock resistance. Stephenson and Kirk (2000a) argued that wave forces on platforms on the Kaikoura Peninsula, South Island, New Zealand, were too small to erode unweathered rock. Even large storms ($H_s \approx 6\,\text{m}$) failed to deliver sufficient energy to the cliff platform junction where platform extension occurs, because the larger waves break offshore due to the limited water depths across the shore platform, and significant wave energy dissipation occurs as the broken waves propagate across the platform. Stephenson and Kirk (2000b) further argued that weathering is necessary to reduce rock resistance to a level where waves can remove weathered material. Trenhaile and Kanyaya (2007) countered this by calculating wave forces in joints and bedding planes, and suggested that waves could accomplish block removal, because waves generate pressures in crevices capable of eroding blocks. This argument, however, depends on the assumption that blocks were already detached from neighbouring blocks and begs the question: what leads to block detachment prior to wave erosion? Undoubtedly, weathering (physical, chemical and biological) must play a significant role in preparing blocks for removal by waves. In this context, the scenario depicted by Trenhaile and Kanyaya (2007) is no different to that proposed by Stephenson and Kirk (2000a,b), where waves remove weathered debris.

The waves versus weathering debate continues because the role of waves has been considered at different scales by different researchers. Wave quarrying operates at the scale of centimetres to metres (i.e. the scale of blocks of rock determined by discontinuities) and, clearly, large waves are required to remove boulder-size blocks of rock. A key question then is: at what scale should rock coast erosion be investigated? Some clues as to the answer to this question are provided by Naylor and Stephenson (2010), who demonstrated how the scale of erosion is contingent on the geological setting, particularly the spacing of discontinuities (joints and bedding planes). On some rock coasts, block erosion dominates, because of a prevalence of densely spaced joints and bedding planes, while on other coasts, particularly sedimentary rock coasts, erosion is at the scale of sand and/or silt particles when there are no discontinuities in the rock mass. Measuring erosion across these widely varying scales on shore platforms is problematic. Traditionally, shore platform erosion has been measured using the micro-erosion meter (MEM) and its variant the traversing micro-erosion meter (see Box 15.2), but this technique does not capture larger-scale block erosion. Furthermore, shore platform development is achieved not only through gradual lowering of the platform surface, but also through erosion of the landward cliff. Erosion of the seaward edge may also occur through surface lowering or backwearing, but attempts to measure it have proven difficult. Consequently, calculating erosion rates on shore platforms with active cliffs requires a multi-method approach across differing scales of erosion. On their own, MEM-derived rates of platform lowering are not sufficient for predicting or modelling the long-term evolution of rock coasts.

A consideration of the relative roles of waves and weathering is relevant for coastal management. Failure to understand how rock coasts erode, and the relative contributions of the different processes to rock coast development, makes prediction of the impact of climate change and formulation of mitigating management strategies extremely difficult. There would be little point or value in mitigating against wave erosion on a rock coast if the primary driver of erosion is weathering or biological activity. Conversely, if platform development is primarily driven by wave erosion, changes in weathering regimes and groundwater hydrology could be ignored. Understanding the impacts of climate change on rock coasts gives new impetus to determining the relative roles of marine and subaerial processes in rock coast evolution.

15.4.3 Stacks and arches

Stacks and arches are often considered minor coastal landforms, and when one considers the scale at which they occur this is perhaps not surprising. However, such landforms are often important components of coastal landscapes valued by visitors to the coast and popular destinations for sightseeing. The economic value of coastal landscapes due to tourism is difficult to estimate, and in the case of rock coasts no attempt has been made yet to assess that value. However, it is likely to be significant and important to local economies that service coastal tourism. Examples include the Great Ocean Road tourist drive in Victoria, southeastern Australia, and the Pacific Coast Highway scenic drive in California, USA.

The loss of one of the sea stacks that makes up the Twelve Apostles and sea arch along the Port Campbell coast in Victoria, Australia, provides an interesting example. On 3 July 2005 at 9:18 am, one of the sea stacks collapsed (Fig. 15.15a,b). The collapse was witnessed by tourists viewing the Twelve Apostles (about 1.5 million annual visitors). Following this collapse, calls were made by a local government official for the remaining stacks to be protected from the sea. However, the stack fell on a day when the sea was relatively calm and the collapse was probably caused by an internal failure associated with the structure of the limestone. A few kilometres further west, the Island Archway sea arch collapsed on either 9 or 10 June 2009 (Fig. 15.15c,d) and, again, concerns were raised amongst local communities about the loss of an important tourist vista. However, the obvious point that two new stacks had been created from the arch was ignored, and that the processes causing the erosion were responsible for the scenery in the first place. Such events help to illustrate that rock coasts are actually systems capable of instantaneous change, and this fact has now been incorporated into visitor information boards at the sites. Arguably, such events add value to coastal tourist destinations by creating a sense of excitement and expectation that a similar event might occur again while being viewed.

15.5 Towards a morphodynamic model for rock coasts

Mathematical modelling has been applied in the study of rock coast evolution because the timescales over which this evolution occurs is too long to be observed. Attempts to understand the development of rock coasts over long timescales (>1000 years) are further hindered by the problem of inheritance, where coasts have undergone more than one episode of development as sea level has risen and fallen, or where tectonic uplift brings a submerged shoreline up to the position of an existing sea-level high stand.

Fig. 15.15 (a, b) Collapse of a sea stack at the Twelve Apostles on the Great Ocean Road in Victoria, Australia. (Source: Courtesy of Parks Victoria.) (c, d) Collapse of a sea arch at Lochard Gorge on the Great Ocean Road, Victoria, Australia. (Source: Photographs by Wayne Stephenson.)

Consequently, numerical models have been used to investigate the role of inheritance (Trenhaile, 2002). Modelling has also been used to better understand the relationship between morphology and processes. Trenhaile (2000) showed, with a model driven by mechanical wave erosion and a stable sea level, how shore platform width increased and gradient decreased until static equilibrium is attained. This same model has subsequently been modified to examine erosional continental shelves and volcanic island shelves, elevated marine terraces, the effect of surface downwearing by weathering on platform development, and the influence of Holocene sea-level variations on shore platform development. While substantial insights have been made using such models, it remains difficult to model the variability of tides, sea-level history, geology,

process regimes and time, consequently empirical field-based studies will continue to be required in order to advance modelling efforts in the future. However, empirical field-based studies are limited by the absence of a formative theoretical framework. Sunamura's (1992) bedrock erosion model, governed by the relationship between F_R and F_W is useful for understanding the interplay of factors that govern rock resistance and erosive forces, but as noted earlier in this chapter, it does not explain what happens once erosion begins (F_W exceeds F_R).

At the beginning of this chapter, the point was made that rock coasts have not been studied through a particular model, such as a morphodynamic framework, in the same way beaches have (see Chapter 7). While the morphodynamic model of Wright and Short (1984) has shaped

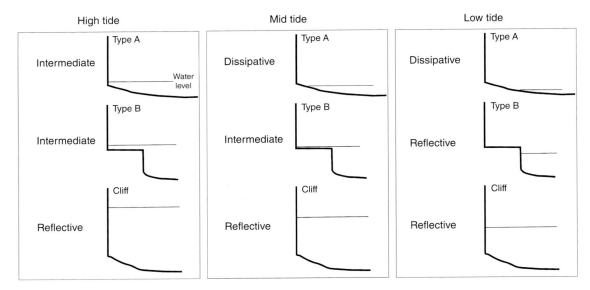

Fig. 15.16 Proposed morphodynamic model for rock coasts.

beach research over the last 35 years, rock coast geomorphologists have continued to struggle with basic questions about formative processes and how rock coasts have evolved. A contributing factor has probably been the relatively small number of researchers working on rock coasts and limited resources (compared to beach research). Consequently, there are relatively few process studies of rock coasts, although this has improved in the last decade, with a greater number of active researchers and published papers (see Naylor et al., 2010). There is still a relative paucity of data sets of rock coast processes and associated morphodynamic responses. Another significant barrier to the development of a conceptual morphodynamic model for rock coasts is that the response component of such a model requires very long timescales to become evident (or perhaps has already occurred due to inheritance), unlike beaches, which exhibit adjustments over observable and measureable timescales (minutes and hours to years).

One approach to a morphodynamic model for rock coasts might be to consider how wave energy is expended on different rock coast morphologies, enabling discussion of dissipative, intermediate or reflective wave energy conditions, in much the same way as on beaches. As outlined in Fig. 15.16, three types of rock coast morphologies (dissipative, intermediate, reflective) are taken as being broadly representative. How each type dissipates or reflects wave energy is a function of both the cross-shore profile and the water depth across the profile, which changes with the tide. Type A platforms are generally dissipative, similar to wide, low-gradient beaches, but wave reflection can occur at high tide when waves impact the cliff at the back of the platform (if present). At high tide,

therefore, the platform might be considered akin to an intermediate beach, since the upper part of the profile reflects some energy while the lower part can dissipate wave energy. Type B platforms in this model are considered intermediate, because they possess elements of both reflection and dissipation, depending on tide level. At low tide, the low-tide cliff presents a vertical face that reflects wave energy, but, depending on the water depth in front of the cliff, there may be dissipation when waves break seaward of the low-tide cliff (Fig. 15.17a). With rising tide level, wave energy dissipation occurs across the flooded platform, while wave reflection at the cliff at the back of the platform may take place at high tide (Fig. 15.17b). Cliffs are for the most part reflective of wave energy, unless the water depth in front of the cliff is shallow enough to cause waves to break.

The question that the above model begs is: what are the morphological responses? Since morphodynamics is concerned with the mutual co-adjustment of process and form, for a morphodynamic model of rock coasts to be useful, it must be able to predict how rock coasts respond to variations in wave energy. There are two possible scenarios: (1) we are unlikely to be able to define the adjustment because we are incapable of observing a response that occurs at a timescale beyond human measurement (and so remain reliant on deterministic modelling); and (2) contemporary rock coast morphology already represents the equilibrium form of the profile because sea level has occupied current rock coasts for approximately 6000–7000 years. Substantial challenges remain before a comprehensive morphodynamic model for rock coasts can be developed. However, such a model may be helpful in unifying disparate views of how rock coasts evolve, how they can

Fig. 15.17 Type B shore platform: (a) at low tide, showing wave dissipation and reflection on the seaward edge; and (b) at high tide, showing wave reflection from the cliff and a dissipative surface zone across the platform. (c) Profile across the platform. AHD, Australian Height Datum equivalent to mean sea level). (Source: Photographs by Wayne Stephenson.)

be managed, and for understanding how rock coasts will respond to climate change in the future.

15.6 Impacts of climate change on rock coasts

Predicting and understanding the impacts of climate change on rock coasts is still very much in its infancy, since much more attention has been focussed on coastal systems such as barriers, salt marshes and coral reefs. The irony is that these latter systems are capable of a dynamic response in the form of positive feedback to rising sea level, depending on local sediment budgets and biogenic growth. Reefs can increase in elevation by keeping up with sea-level rise (Chapter 16). Salt marshes can increase elevation through sedimentation (Chapter 10), and barriers can prograde where there is an oversupply of sediment (Chapter

9). In the case of rock coasts, there is no dynamic response in a system where erosion is the only response. Erosion may be slowed or accelerated depending on the interplay of processes outlined in Fig. 15.4. Modelling attempts to date have highlighted increased rates of erosion in response to sea-level rise, but this is hardly surprising given the erosional nature of rock coasts. In those cases where cliff erosion delivers a significant volume of sediment to the base of a cliff, erosion may slow temporarily until that sediment is redistributed alongshore, as demonstrated by Dawson et al. (2009).

15.6.1 Modelling climate change impacts

We are very much reliant on modelling to understand the impact of climate change on rock coasts. The likely impacts of climate change on rock coasts have been investigated with a number of different models and in a variety of geological settings (Dickson et al., 2007; Walkden and Dickson, 2008; Trenhaile, 2010). Trenhaile (2011) considered the impact of sea-level rise and increased storminess. Rising sea level was found to cause increased rates of cliff erosion, but higher waves had little impact. In one of the most comprehensive modelling attempts to understand climate change impacts on eroding coasts, Dawson et al. (2009) coupled a morphodynamic, a hydrodynamic, a reliability and a socio-economic model to predict climate change impacts on the Norfolk coast in the UK. This 72-km stretch of coast consists of eroding soft-rock cliffs and low-lying beach systems prone to flooding. A number of different scenarios were explored with the model, including variations in management, alongside different climate change scenarios and socio-economic change. Two key findings were: (1) sediment released from cliff erosion plays a significant role in protecting neighbouring low-lying land from flooding; and (2) flood risk in the area studied is an order of magnitude greater than erosion risk. A significant finding of this research is that it can make economic sense to allow coastal erosion to take place, sacrificing properties and land, because the costs associated with the coastal erosion can be offset by the benefits of achieving improved defence against coastal flooding. The modelling also demonstrated that, provided the coastal system is reasonably well understood, some confidence can be obtained in predicting the effects of climate change, both physically (flooding and erosion) and in terms of socio-economic impacts.

The impacts of climate change on rock coasts can be summarized by coupling climate change impacts with the bedrock erosion model (see Fig. 15.4) of Sunamura (1992). In Fig. 15.18, climate change impacts sit alongside a modified bedrock erosion model, and are linked to the bedrock model either by changing rock resistance (F_R) or changing erosive force (F_W). Not surprisingly, the associated impacts of climate change can work to both enhance and reduce the effectiveness of the various processes that govern F_W and F_R.

15.6.2 Sea-level rise

Sea-level rise has the potential to increase wave energy arriving on rock coasts, since increasing water depths reduce wave attenuation. As sea level rises, the water depth in front of rock coasts where waves break will progressively move shoreward. On shorelines with shore platforms fronting a cliff, this may mean the cliff is more susceptible to increased wave attack, at least until the cliff retreats enough to widen the platform, which mitigates against further wave erosion. On some cliffed coasts without a platform, increasing water depths may result in less wave energy and erosion, as high-impact waves breaking against the cliff (by plunging or spilling) may be replaced by wave reflection and standing wave motion, which is

significantly less energetic and destructive. For Type B shore platforms with a low-tide cliff, that cliff may be protected from further retreat as water depth increases. Consequently, landward cliff retreat will widen the platform and increase wave dissipation, so that increased cliff retreat may be short-lived.

The exact response of a specific coast will need to be explored on a case-by-case basis, as the nearshore profile and water depths will have a large bearing on the exact response. Furthermore, feedback mechanisms need to be considered; for example, will increased erosion deliver more sand to shallow water depths and reduce wave energy? Under predicted rates of sea-level rise over the next 100 years, water depths might be expected to increase by 0.18–0.58 m (Meehl et al., 2007). Exactly how much more erosion such a rise in sea level will cause can be estimated by modelling. Such modelling, however, must be designed for site-specific wave energies and geology, making broad generalizations very difficult (e.g. Dickson et al., 2007; Dawson et al., 2009).

15.6.3 Wave energy

Under future climate change scenarios, wave climates are expected to change. Wave heights have increased in recent decades in the North Atlantic (Komar and Allan, 2009) and in the Northwest Pacific (Sasaki et al., 2005). Meehl et al. (2007) reported that tropical storms will occur less frequently, but those storms will be more intense. Fewer mid-latitude storms are expected, as these systems shift to higher latitudes where larger waves are expected to be generated. Consequently, there will be an uneven geographical distribution of changes in wave heights, making prediction of climate change impacts even more difficult. It is not reasonable to expect that all rock coasts will simply be exposed to increased wave energy as a result of climate change. Changing climates may also reduce shore-bound ice on high-latitude rock coasts, opening these coasts up to further wave impacts, rather than being ice bound and protected from waves during winter (Chapter 14). Another consequence of this response is that mechanical erosion by ice pushed against the shore may be reduced.

15.6.4 Weathering, biology and mass movements

Climate change has the potential to alter weathering regimes, which may either enhance or reduce the efficiency of erosion. In colder latitudes, where frost shattering can contribute to rock coast erosion (e.g. Robinson and Jerwood, 1987), warming may reduce the number of freeze-thaw cycles, thus slowing erosion rates. Conversely, in temperate climates, warmer temperatures may mean a greater number of wetting and drying cycles, increased salt weathering, and more efficient chemical weathering.

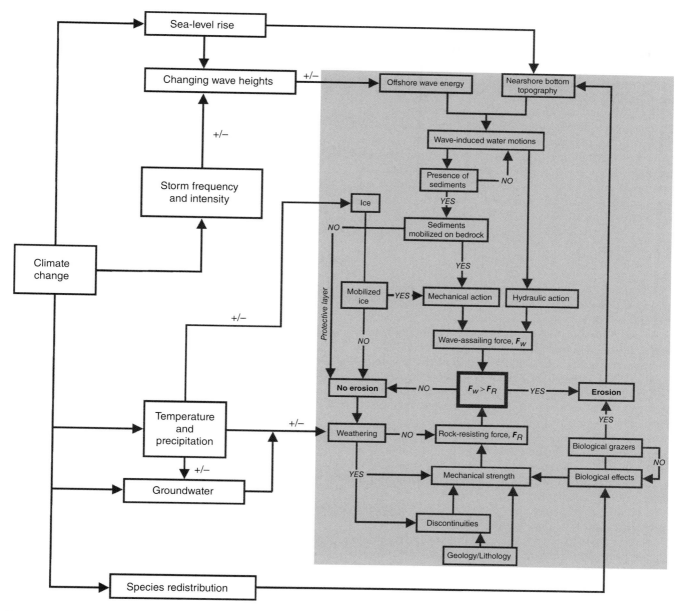

Fig. 15.18 Conceptual model of climate change impacts on rock coasts. Climate change drivers are coupled to a modified bedrock erosion model based on Sunamura (1992). Climate change can increase the erosive forces that constitute the wave-assailing force (F_W) and can induce changes to relevant processes (e.g. weathering and biology) that may increase or decrease the resisting force of rock (F_R). (Source: Adapted from Sunamura 1992.)

Increase in ocean acidity has the potential to enhance solution processes on limestone coasts or rock types containing significant amounts of calcium carbonate.

Climate change impacts on intertidal biology is an area of uncertainty in the overall question of how climate change will alter the geomorphology of rock coasts. Biologists have considered the impacts of climate change on rocky intertidal biota, with topics including the geographical redistribution of species as a consequence of

warmer climates (Hawkins et al., 2009). Warmer climates might cause increased desiccation of algae that provide protection against erosion and weathering, so that erosion rates may increase as biological protection is reduced. Ocean acidification may also reduce bioconstruction, as vermetid molluscs and red algae struggle to construct shells. The complexity of biological response to climate change has been acknowledged by Helmuth (2009), and simple predictions of poleward migration of warmer-water species cannot always be expected. Consequently, the subsequent geomorphic responses that arise from the inter-relationships between biota and geomorphology is extremely difficult, if not impossible, to predict, and is very much a black box in the model depicted in Fig. 15.18.

Finally, mass movement can be expected to increase under climate change scenarios, where increased rainfall results in more groundwater loading, which in turn reduces slope stability. Conversely, where groundwater is reduced, slopes can be expected to be less susceptible to mass movement. Thus, mass movement magnitude and frequency will be altered mainly depending on variability and changes in groundwater hydrology.

15.7 Summary

Of all the coastal systems covered in this book, rock coasts are perhaps the least well understood in terms of how they respond to climate change. Like many other coastal systems, rock coasts continue to be subjected to poor management decisions and inappropriate development. Future climate change impacts will serve to exacerbate the challenges of managing these coastal systems. Compounding such problems is the relatively poor understanding of how rock coasts evolve and how different processes operate in these systems. Significant progress has been made in recent years in terms of understanding contributing processes and rates of erosion, but there is still much to be learnt about the evolution of rock coasts.

• Rock coasts occupy 80% of the world's shoreline and are wholly erosive in origin.

• A detailed understanding of how processes operate on rock coasts and how they interact with geological setting is yet to be achieved, and this lack of understanding stands in the way of predicting with any confidence future impacts of climate change on rock coasts.

• Numerical modelling will be the principal tool for predicting and understanding climate change impacts on rock coasts.

Key publications

Dawson, R., Dickson, M., Nicholls, R. et al., 2009. Integrated analysis of risks of coastal flooding and cliff erosion under scenarios of long term change. *Climatic Change*, **95**, 249–288.
[*Excellent paper documenting fully integrated modelling of climate change impacts on soft cliff coasts with specific emphasis on coastal flooding versus coastal erosion*]
Dickson, M., Walkden, M. and Hall, J., 2007. Systemic impacts of climate change on an eroding coastal region

over the twenty-first century. *Climatic Change*, **84**, 141–166.
[*Excellent case study of climate change impacts on rock coasts*]
Lee, M. and Clark, M., 2002. *Investigation and Management of Soft Rock Cliffs*. Thomas Telford, London, UK.
[*Very useful review of soft cliff coasts*]
Naylor, L.A., Stephenson, W.J. and Trenhaile, A.S., 2010. Rock coast geomorphology: recent advances and future research directions. *Geomorphology*, **114**, 3–11.
[*Very accessible paper highlighting the way forward in rock coast research*]
Trenhaile, A.S., 2002. Rock coasts, with particular emphasis on shore platforms. *Geomorphology*, **48**, 7–22.
[*Excellent paper providing thorough review of shore platforms*]

References

Bird, E.C.F., 2004. Cliffs. In A. Goudie (Ed.), *Encyclopaedia of Geomorphology*. Routledge, London, UK, pp. 159–164.

Bishop, P. and Cowell, P., 1997. Lithological and drainage network determinants of the character of drowned, embayed coastlines. *The Journal of Geology*, **105**, 685–699.

Blanco-Chao, R., Perez-Alberti, A., Trenhaile, A.S., Costa-Casais, M. and Valcarcel-Diaz, M., 2007. Shore platform abrasion in a para-periglacial environment, Galicia, northwestern Spain. *Geomorphology*, **83**, 136–151.

Dawson, R., Dickson, M., Nicholls, R. et al., 2009. Integrated analysis of risks of coastal flooding and cliff erosion under scenarios of long term change. *Climatic Change*, **95**, 249–288.

De Lange, W.P. and Moon, V.G., 2005. Estimating long-term cliff recession rates from shore platform widths. *Engineering Geology*, **80**, 292–301.

Dickson, M., Walkden, M. and Hall, J., 2007. Systemic impacts of climate change on an eroding coastal region over the twenty-first century. *Climatic Change*, **84**, 141–166.

Emery, K.O. and Kuhn, G.G., 1982. Sea cliffs: their processes, profiles, and classification. *Geological Society of America Bulletin*, **93**, 644–654.

Greenwood, R.O. and Orford, J.D., 2008. Temporal patterns and processes of retreat of drumlin coastal cliffs – Strangford Lough, Northern Ireland. *Geomorphology*, **94**, 153–169.

Hall, A.M., Hansom, J.D. and Jarvis, J., 2008. Patterns and rates of erosion produced by high energy wave processes on hard rock headlands: the Grind of the Navir, Shetland, Scotland. *Marine Geology*, **248**, 28–46.

Hawkins, S.J., Sugden, H.E., Mieszkowska, N. et al., 2009. Consequences of climate-driven biodiversity changes for ecosystem functioning of North European rocky shores. *Marine Ecology, Progress Series*, **396**, 245–259.

Helmuth, B., 2009. From cells to coastlines: how can we use physiology to forecast the impacts of climate change? *Journal of Experimental Biology*, **212**, 753–760.

Komar, P.D. and Allan, J.C., 2009. Increasing hurricane-generated wave heights along the U.S. East Coast and their climate controls. *Journal of Coastal Research*, **24**, 479–488.

Lee, E.M. and Clark, A.R. 2002. *Investigation and Management of Soft Rock Cliffs*. Thomas Telford, London, UK.

Meehl, G.A., Stocker, T.F., Collins, W.D. et al., 2007. Global climate projections. In: S. Solomon, D. Qin, M. Manning et al. (Eds), *Climate Change 2007: The Physical Science Basis*. Contribution of Working Group I to the Fourth Assessment Report of the Intergovernmental Panel on Climate Change. Cambridge University Press, Cambridge, UK, pp. 747–845.

Naylor, L.A. and Stephenson, W.J., 2010. On the role of discontinuities in mediating shore platform erosion. *Geomorphology*, **114**, 89–100.

Naylor, L.A. and Viles, H.A., 2000. A temperate reef builder: an evaluation of the growth, morphology and composition of *Sabellaria alveolata* (L.) colonies on carbonate platforms in South Wales. *Geological Society Special Publication*, **178**, 9–19.

Naylor, L.A., Stephenson, W.J. and Trenhaile, A.S., 2010. Rock coast geomorphology: recent advances and future research directions. *Geomorphology*, **114**, 3–11.

Noormets, R., Felton, E.A. and Crook, K.A.W., 2002. Sedimentology of rocky shorelines. 2: Shoreline megaclasts on the north shore of Oahu, Hawaii – origins and history. *Sedimentary Geology*, **150**, 31–45.

Porter, N.J. and Trenhaile, A.S., 2007. Short-term rock surface expansion and contraction in the intertidal zone. *Earth Surface Processes and Landforms*, **32**, 1379–1397.

Rheinberger, T. and Malorey, D., 2008. *Coastal Slope Instability Study – Draft Report, Shoalhaven City Council Coastal Zone Management Study and Plan*. Snowy Mountain Engineering Corporation. Sydney, Australia.

Robinson, D.A. and Jerwood, L.C., 1987. Sub-aerial weathering of chalk shore platforms during harsh winters in southeast England. *Marine Geology*, **77**, 1–15.

Robinson, L.A., 1977. Marine erosive processes at the cliff foot. *Marine Geology*, **23**, 257–271.

Rosser, N.J., Petley, D.N., Lim, M., Dunning, S.A. and Allison, R.J., 2005. Terrestrial laser scanning for monitoring the process of hard rock coastal cliff erosion. *Quarterly Journal of Engineering Geology and Hydrogeology*, **38**, 363–375.

Sasaki, W., Iwasaki, S.I., Matsuura, T. and Iizuka, S., 2005. Recent increase in summertime extreme wave heights in the western North Pacific. *Geophysical Research Letters*, **32**, L15607, doi: 10.1029/2005GL023722.

Spencer, T. and Viles, H., 2002. Bioconstruction, bioerosion and disturbance on tropical coasts: coral reefs and rocky limestone shores. *Geomorphology*, **48**, 23–50.

Stephenson, W.J., 2001. Shore platform width – a fundamental problem. *Zeitschrift für Geomorphologie*, **45**, 511–527.

Stephenson, W.J., 2008. Discussion of De Lange, W.P. and Moon V.G., 2005. Estimating long-term cliff recession rates from shore platform widths. *Engineering Geology*, **101**, 288–291.

Stephenson, W.J. and Finlayson, B.L., 2009. Measuring erosion with the micro-erosion meter –contributions to understanding landform evolution. *Earth-Science Reviews*, **95**, 53–62.

Stephenson, W.J. and Kirk, R.M., 2000a. Development of shore platforms on Kaikoura Peninsula, South Island, New Zealand, Part I the role of waves. *Geomorphology*, **32**, 21–41.

Stephenson, W.J. and Kirk, R.M., 2000b. Development of shore platforms on Kaikoura Peninsula, South Island, New Zealand, Part II the role of subaerial weathering. *Geomorphology*, **32**, 43–56.

Stephenson, W.J. and Kirk, R.M., 2001. Surface swelling of coastal bedrock on inter-tidal shore platforms, Kaikoura Peninsula, South Island, New Zealand. *Geomorphology*, **41**, 5–21.

Stephenson, W.J., Kirk, R.M., Hemmingsen, S.A. and Hemmingsen, M.A., 2010. Decadal scale micro-erosion rates on shore platforms. *Geomorphology*, **114**, 22–29.

Sunamura, T., 1992. *The Geomorphology of Rocky Coasts*. Wiley, New York, USA.

Switzer, A. and Burston, J., 2010. Competing mechanisms for boulder deposition on the southeast Australian coast. *Geomorphology*, **114**, 42–54.

Trenhaile, A.S., 1987. *The Geomorphology of Rock Coast*. Oxford University Press, Oxford, UK.

Trenhaile, A.S., 2000. Modeling the development of wave-cut shore platforms. *Marine Geology*, **166**, 163–178.

Trenhaile, A.S., 2002. Rock coasts, with particular emphasis on shore platforms. *Geomorphology*, **48**, 7–22.

Trenhaile, A.S., 2010. Modeling cohesive clay coast evolution and response to climate change. *Marine Geology*, **277**, 11–20.

Trenhaile, A.S., 2011. Predicting the response of hard and soft rock coasts to changes in sea level and wave height. *Climatic Change*, **109**, 599–615.

Trenhaile, A.S. and Kanyaya, J.I., 2007. The role of wave erosion on sloping and horizontal shore platforms in macro- and mesotidal environments. *Journal of Coastal Research*, **23**, 298–309.

Walkden, M. and Dickson, M., 2008. Equilibrium erosion of soft rock shores with a shallow or absent beach under increased sea level rise. *Marine Geology*, **251**, 75–84.

Wright, L.D. and Short, A.D., 1984. Morphodynamic variability of surf zones and beaches: a synthesis. *Marine Geology*, **56**, 93–118.

Young, A.P., Flick, R.E., Gutierrez, R. and Guza, R.T., 2009. Comparison of short-term sea cliff retreat measurement methods in Del Mar, California. *Geomorphology*, **112**, 318–323.

16 Coral Reefs

PAUL KENCH

School of Environment, The University of Auckland, Auckland, New Zealand

16.1 Coral reefs in context

Distributed throughout the tropical oceans, coral reefs extend over approximately 300,000 km^2 of the Earth's surface. While relatively modest in spatial extent compared with other coastal systems considered in this book, coral reefs are considered among the most valuable ecosystems on Earth. Their ecological value is well recognized as they are zones of high biological diversity and habitat to 25% of known marine species. Furthermore, it is estimated that coral reefs provide goods and services in the order of $US375 billion to coastal communities on an annual basis (Best and Bornbusch, 2005). Less well recognized is the 'geomorphic value' provided by reefs as a consequence of the ecosystem services they provide. For example, coral reefs provide the physical foundation of a number of mid-ocean atoll nations (e.g. Kiribati, Tuvalu, Maldives and Marshall Islands). As well as providing habitat, coral reef structure regulates oceanographic processes that control reef and lagoon circulation, and wave energy levels that impact coastlines.

Coral reefs are unique coastal environments as they represent a delicate balance between ecological processes (organisms) that produce calcium carbonate ($CaCO_3$) material, and ecological and physical processes that act to breakdown and

Coastal Environments and Global Change, First Edition. Edited by Gerd Masselink and Roland Gehrels.
© 2014 John Wiley & Sons, Ltd. Published 2014 by John Wiley & Sons, Ltd. Companion Website: www.wiley.com/go/masselink/coastal

redistribute $CaCO_3$ to control the development and morphology of reef landforms. Consequently, without the presence of these organisms, coral reefs would not exist.

The geomorphic investigation of coral reefs had its inception in the earliest modern voyages of discovery to the tropical oceans. Navigators and naturalists (e.g. Cook, Banks and Darwin) in the 18th and 19th centuries provided the first accounts of the extent, distribution and shape of coral reefs. These observations spawned the great 'Coral Reef Problem', which sought to explain the structure and distribution of reef systems and which dominated global scientific debate for more than a century. However, developments in field sampling, instrumentation and analytical techniques over the past 100 years have seen significant advances in understanding the geomorphic complexities of reef systems and the process controls on reef landform development.

Contemporary global assessments of the ecological condition of coral reefs indicate they are under threat of collapse due to a range of stressors that include climate change, anthropogenic impacts and natural environmental variability (Hughes, et al., 2003). Understanding the geomorphic implications of global environmental change is of paramount importance in the ongoing, sustainable management of reef systems.

Despite the fundamental importance of ecological processes on coral reefs, this chapter focusses on the geomorphology of reef systems. The chapter begins by introducing the geomorphic complexity of coral reefs. It places reef systems within a morphodynamic framework that highlights the relationship between ecological and physical processes. The chapter then examines the formation and controls on development of coral reef platforms and reef islands, which are conspicuous features of reef systems. The challenges for managing coral reef landforms are examined, and the chapter concludes with a discussion of the geomorphic implications of global environmental change on coral reefs.

16.2 Coral reefs and their geomorphic complexity

Coral reefs possess a diverse range of geomorphic features that have evolved at a range of space and time scales and which show varying levels of persistence in the geological record. A major division of geomorphic units can be made between **coral reefs** and **reef sedimentary landforms**.

16.2.1 Coral reefs

To geomorphologists, coral reefs are three-dimensional structures consisting of veneers of living coral and reef-associated organisms that overlie vast sequences of previously deposited $CaCO_3$ that can extend several thousand

metres deep in mid-ocean atolls (Fig. 16.1). These structures evolve over geological (millennial) timescales. Geomorphic classifications of reefs are typically based on the size and shape of reefs, the presence of lagoons, and their proximity to continental landmasses (Table 16.1; Fig. 16.2). Atolls, fringing reefs, barrier reefs and reef platforms comprise the primary reef types. However, a diverse range of secondary reef features have also been described, but are generally smaller in size and are associated with the major reef morphologies (Table 16.1). Coral reefs range in area from less than 1 km^2, in the case of smaller patch reefs, to more than 100 km^2 in extent. Networks of reefs can form barrier complexes up to 2400 km in length, such as the Great Barrier Reef, which is the largest biological construction on Earth.

Beneath this macro-scale classification of reef types is a suite of meso-scale geomorphic units. Each reef possesses distinct morphological zones developed in response to water depth, exposure to hydrodynamic energy and ecological characteristics. There are four principal geomorphic zones: forereef; reef crest; reef flat; and backreef (characteristics of each zone are summarized in Table 16.2).

16.2.2 Reef sedimentary landforms

Reef sedimentary landforms are surficial accumulations of unconsolidated carbonate sediment deposited by wave and current processes on, or adjacent to, a coral reef structure. On geological timescales, they represent ephemeral stores of detrital material in the carbonate sediment budget. Geographic variations in processes produce a suite of reef sedimentary landform units that can be differentiated based on the location of sediments relative to the reef surface, reef type, presence of non-carbonate substrate, and elevation of landforms with respect to sea level.

In fringing reefs, sedimentation typically occurs toward the leeward edge of reefs at the interface between the reef and terrestrial environment. Typically, subaerial coastal plains, beaches, spits and barriers form through progradation of sediment across the backreef zones (Fig. 16.2a). In these settings, terrigenous sediments delivered to the coast mix with biogenic sediments and contribute to landform development.

In barrier reefs and atolls, the development of sedimentary landforms occurs in two major areas. First, sediment generated from reefs is transported to the lagoon where large subtidal sand aprons prograde into and contribute to the infilling of lagoons. Second, sediments can accumulate directly on reef platforms where they build subaerial deposits such as gravel ridges, reef islands and beaches (Fig. 16.2d–f). Reef islands and coastal plains are geomorphically important at the human timescale, as they

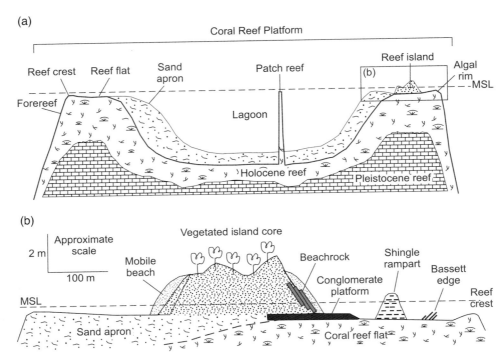

Fig. 16.1 Example of range of coral reef landforms: (a) cross-section of an atoll representing major elements of coral reef structure and large-scale reef geomorphic units typical of many reef settings; and (b) range of sedimentary landforms deposited on or adjacent to coral reefs. MSL, mean sea level. (Source: Reproduced from Kench et al. 2009a. Reproduced with permission of Cambridge University Press.)

Table 16.1 Major reef types and physical characteristics. (Source: Ladd 1977. Reproduced with permission of Elsevier.)

Reef type	Description	Location	Dimension/scale
Primary reef types			
Atolls	Annular reef rim enclosing a lagoon	(a) Oceanic	10^0–10^1 km diameter
		(b) Continental shelves, barrier lagoons	10^1–10^3 km² area
Barrier reef	Linear reefs separated from land by a lagoon	Continental shelves or platforms, or high islands	10^1–10^3 km long
Fringing reef	Reefs attached to shoreline without lagoon	Continental and island shorelines	10^{-1}–10^1 km long
Table/platform reef	Isolated intertidal reefs without lagoon	(a) Oceanic	10^{-1}–10^1 km diameter
		(b) Continental shelves, barrier lagoons	
Secondary reef types			
Faro	Ring-shaped reef with lagoon	Atoll lagoons or barrier lagoons	10^{-1}–10^1 km diameter
Knoll	Isolated mounds covered by living coral	Atoll and barrier lagoons	10^{-3}–10^1 km diameter
Patch reef	Small sea-level reefs	Submarine shelves or lagoons	10^{-1}–10^1 km diameter
Ribbon reef	Long, narrow, sinuous reef	Section of barrier or atoll reef system	10^{-1}–10^1 km long
Bank reef	Linear or semi-circular reef	Continental shelf or barrier lagoon	10^1 km long

form the foundation of a number of mid-ocean atoll nations, such as Tuvalu, Kiribati and the Maldives.

16.2.3 Coral reefs as eco-morphodynamic systems

Morphodynamics (see Chapter 1) provides a powerful framework to conceptualize the process controls on the formation and dynamics of coral reef landforms. Geomorphic development of coral reefs and associated sedimentary structures reflects a dynamic balance between constructive and destructive processes. Variability in that balance (e.g. in extent, rates, duration, timing) produces the diversity of macro- to micro-scale geomorphic units within reef systems (Fig. 16.1 and Fig. 16.2).

The term 'eco-morphodynamics' was recently introduced as a conceptual framework to provide more detailed exploration of the complex relationships that exist between ecological and physical processes and reef geomorphology

Fig. 16.2 Examples of coral reef landforms: (a) fringing reef, Great Barrier Reef (GBR), Australia; (b) fringing and barrier reefs around a high island in southern Japan; (c) mid-Pacific atoll; (d) reef platforms with platform islands, Maldives; (e) linear atoll island, Majuro atoll, Marshall Islands; (f) reef platform island, GBR, Australia; (g) productive forereef zone, Nadi Bay, Fiji; and (h) emergent reef flat, Jeh Island, Republic of Marshall Islands. (Source: (a, f, g, h) Paul Kench. (b–e) Kench 2013.)

(g)

(h)

Fig. 16.2   (Continued)

Table 16.2   Summary of principal geomorphic zones of a coral reef.

Reef zone and geomorphic description	Hydrodynamic environment	Length scale (m)	Depth range (metres below sea level)	Productivity (calcification)
Forereef The outer seaward margin of a reef, characterized by steep slopes (30–40°). The upper 30 m has living coral cover	High energy Exposed to incident swell, wave shoaling and breaking	5 in fringing reefs to 3000 in atolls	>0 to 3000	High (10 kg/m²/yr) Optimal coral growth occurs
Reef crest (algal ridge) Narrow morphological break in slope between forereef and reef flat. Comprises encrusting coralline algae	Highest energy Wave breaking and surf zone processes	10 to 40	1.0 to +1.0 Intertidal	High
Reef flat Directly leeward of the reef crest. A sub-horizontal morphology comprised of cemented coral/algal pavement. Depending on exposure, the reef flat is also host to isolated coral colonies, stands of living coral, mobile sand sheets, gravel deposits and storm blocks	Moderate to low-energy reformed waves Exposure to light and higher temperatures at low water	100 to 3000	3.0 to +1.0 Shallow–intertidal	Medium (4 kg/m²/yr)
Backreef Leeward of reef flat. In atolls and barrier reefs, comprises a lagoon. Characterized by sediment deposition and isolated coral patches (bommies). In fringing reefs, backreef does not commonly have a lagoon but merges with the coastline	Low energy – conducive to sedimentation	100 to 80,000	5 to 100	Low (0.4 kg/m²/yr)

(Kench et al. 2009a; Fig. 16.3). The eco-morphodynamic framework highlights a number of key aspects of the nature of co-adjustment of physical processes, reef ecological communities and geomorphic products:

• The cycling of $CaCO_3$ is of fundamental importance for supplying the building blocks (corals, sands and gravels) for landform construction. As outlined in Chapter 1, sediment is a time-dependent coupling mechanism of morphological change in coastal morphodynamics. Unique to coral reef systems, sediment is primarily produced by eco-

logical processes that produce $CaCO_3$. Furthermore, the rate of carbonate sediment generation varies greatly between coral reef settings (locations and geomorphic zones). Consequently, the 'carbonate sediment factory' is a highly space- and time-dependent coupling mechanism. This time-dependency emerges not only from the time lags in redistribution of material (as occurs in other coastal settings), but also from the timeframes associated with organism growth, mortality and conversion to detrital sediment (Perry et al., 2008).

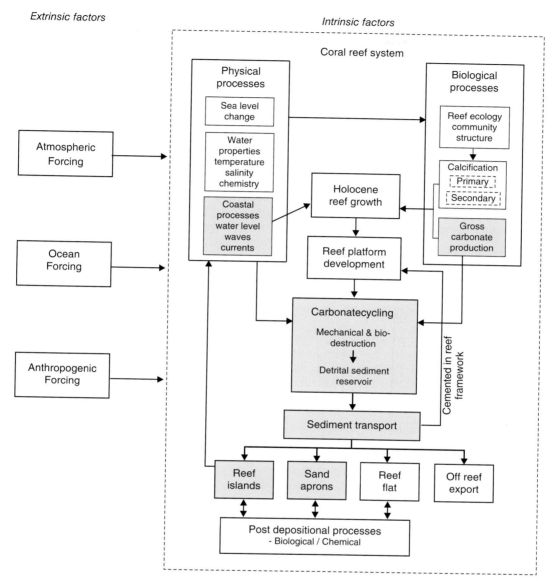

Fig. 16.3 Structure and function of the eco-morphodynamic model for coral reefs. The model shows co-adjustment of biological and physical processes, and coral reef morphology and reef sedimentary landforms that operate at a range of timescales. Grey shaded boxes and arrows highlight linkages between the contemporary eco-morphodynamic system at event to centennial timescales, which are embedded in the broader system that controls reef development at centennial to millennial timescales. (Source: Adapted from Kench et al. 2009a. Reproduced with permission of Cambridge University Press.)

- Alterations in boundary controls will drive change in the state of the coral reef system. Changes in boundary conditions can occur as a consequence of external alterations in the atmosphere–ocean system, such as sea-level rise, ocean temperature and chemistry variations, or internal alterations in boundary conditions as a result of anthropogenic impacts, such as coastal construction or resource exploitation.
- The geomorphic responsiveness (magnitude, style and timescales of geomorphic change) of reef landforms varies between different geomorphic units. For example, the

formation of coral reef platforms is modulated by sea-level oscillations at millennial timescales (Montaggioni, 2005). In contrast, shoreline dynamics of reef islands occur at event to decadal timescales in response to alterations in wave energy input (Kench and Brander, 2006a).
- Feedbacks exist that can be temporally specific or cascade across timescales. For example, while sea-level oscillations govern the pattern of reef growth at millennial timescales, at shorter timescales the reef structure modulates wave and current processes (Kench and Brander, 2006b). The characteristics of incident energy in turn control the structure of

ecological communities (Chappell, 1980), reef morphology, sedimentation processes and short-term geomorphic change of beach and island shorelines.

• Feedbacks are non-linear and have significant time lags. For instance, changes in ecological condition of a reef may occur in response to short-term stresses, such as storms, human impact or disease. However, depending on the magnitude and temporal scale of ecological change (severity, persistence or ephemeral transition), alterations in the carbonate budget may, or may not, propagate through the system to yield detectable changes in the geomorphic system at decadal to centennial timeframes (Kench et al., 2009a).

16.2.4   The carbonate factory: a critical control on geomorphic development

Central to the development of reef structure and sedimentary landforms is the production of calcium carbonate ($CaCO_3$) sediment, which in coral reefs is almost exclusively produced through ecological processes. The major producers of sediment in this 'carbonate factory' are commonly divided into a three groups:

(1) Corals are considered the major constructional components that contribute to framework development of reef structure (also referred to as primary producers). Hermatypic corals are characterized by a symbiotic relationship between a coral animal and single-celled algae, zooxanthellae, which live within coral tissue. In return for protection and essential nutrients, the algae play a role in light-enhanced calcification and provide energy to the coral host, allowing it to secrete a rigid skeleton of $CaCO_3$. Calcification is the biologically mediated mechanism by which calcium (Ca^{2+}) and carbonate (CO_3^{2-}) ions derived from supersaturated seawater are converted to $CaCO_3$ (Kinzie and Buddemeier, 1996).

(2) Secondary producers are organisms that encrust the reef surface. Calcareous algae are the major group of secondary producers and have two major roles in reef development. Firstly, they contribute directly to the growth of the reef structure. However, their growth rates are typically an order of magnitude less than corals. Secondly, coralline algae play a major role in cementing detrital sediment into the reef framework.

(3) A suite of additional organisms that dwell on and within the reef also secrete $CaCO_3$. Such benthic producers include molluscs, calcareous algae, foraminifera, bryozoans and echinoderms. The major distinction between these organisms and the primary and secondary producers is the fact that benthic producers do not contribute directly to reef growth. Once they die, their skeletal remains contribute directly to the detrital sediment reservoir.

From a geomorphic perspective, the most important consequence of reef metabolic processes is calcification. Rates of calcification on coral reefs are spatially variable and are dependent on the range, density and growth rates of organisms across reefs. Field studies have identified typical rates of carbonate production of 10 kg/m²/yr on productive (coral-rich) forereef zones (Fig. 16.2g), 4 kg/m²/yr on reef flats (Fig. 16.2e), and lower rates, in the order of 0.8 kg/m²/yr in lagoons and rubble substrates (Kinsey, 1985).

The geomorphic development of a reef and its associated sedimentary structures is consequently reliant upon the growth of corals and other calcifying organisms. However, reducing the dynamics of the carbonate factory to one driven by carbonate production alone oversimplifies the geomorphic development of reefs and associated sedimentary deposits, which is dependent upon a range of additional ecological, chemical and physical processes that cycle carbonate sediment within and through reef systems. Some of these processes can aid the construction of reef landforms, some convert framework to detrital carbonate sediment, and others can remove reef framework. The set of constructive processes include sediment production by $CaCO_3$ secreting organisms and precipitation of cements that bind and stabilize sediments. Destructive processes include bioerosion, which is the action of organisms in destroying reef framework through mechanical boring, etching and chemical dissolution, and physical processes, in which waves mechanically break the skeletal structure of carbonate material (Scoffin, 1992). Physical processes that mechanically erode, transport and deposit $CaCO_3$ are of paramount importance in controlling the distribution, structure and morphology of reefs and sedimentary landforms.

16.2.5   Environmental controls on the carbonate factory

Cold-water corals have a broad distribution, but are predominantly located at higher latitudes and at depths ranging from 40 to 1000 m (Fig. 16.4). While cold-water corals do secrete $CaCO_3$, they differ biologically from their tropical counterparts as they do not photosynthesize, and therefore do not have zooxanthellae. Rather, they gain their energy from zooplankton that drift past the coral framework on currents. Shallow tropical and sub-tropical coral reefs, which are the focus of this chapter, are found between the latitudinal limits of 28°N and 28°S (Fig. 16.4). Within this broad zone, reefs occur in two biogeographic regions. The Indo-Pacific province is centred on Indonesia and the Philippines, and is characterized by high coral biodiversity (80 genera and 500 species). The Atlantic province is centred on the Caribbean, and is characterized by lower biodiversity (20 genera, 65 species). The broad latitudinal limits and the complex distribution of reefs within this region are controlled by variations in a number of key environmental parameters that directly affect zooxanthellae and, due to their symbiotic relationship with corals, influence coral survivorship and growth (Kleypas et al., 1999).

Sea-surface temperature (SST) exerts a major control on coral survivorship. The SSTs within which hermatypic corals thrive range from 17°C to 34°C, and coral growth is

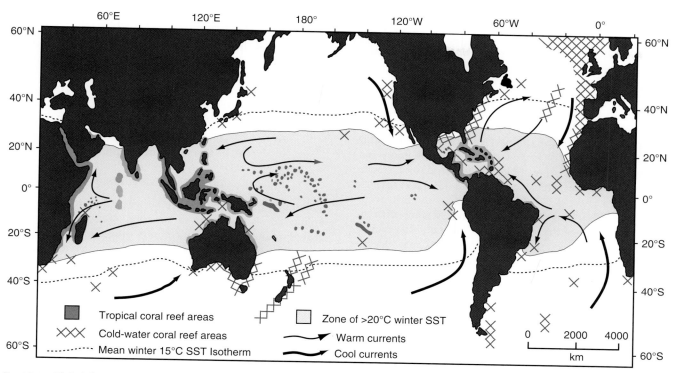

Fig. 16.4 Global distribution of coral reefs. Note the cool-water zones on the western margin of South America and Africa, which precludes coral reef development. SST, sea-surface temperature.

limited when temperatures lie outside this range for considerable periods of time. The effect of SST on coral distribution is clearly expressed through a decline in species richness with increasing latitude (decreasing water temperature). Furthermore, corals are absent along the eastern margins of Africa and South America where cool-water upwelling prevails. Short-term elevation in water temperature can severely stress corals, causing them to expel their zooxanthellae. The loss of zooxanthellae causes corals to lose their pigment and the coral skeleton takes on a white appearance, a process known as bleaching (Box 16.1).

Light is also a critical control on coral growth because the symbiotic zooxanthellae rely on light for photosynthesis. Consequently, reef-building corals are constrained to the shallow photic zone (the depth of water at which surface light levels are reduced to 1%). The depth of the photic zone varies considerably, from >90 m in non-turbid waters to <5 m in highly turbid environments. The decrease in light level with depth reduces coral diversity. Although individual reef-building corals can be found up to depths of 100 m, generally reef development occurs at depths less than 50 m.

Salinity, turbidity and nutrient levels are further limits on coral growth. While corals can endure a large range in salinity (23–42 ppt), reefs are generally absent from rivermouths, which have characteristically low salinity levels and high sediment yields. Turbidity refers to the presence of fine sediment suspended in the water column and affects corals in two ways. Firstly, through impacting light penetration, thereby reducing the water depth within which coral can thrive. Secondly, sedimentation smothers coral. Variations in the nutrient load of reef waters can also regulate coral growth. Corals thrive in low-nutrient waters, whereas high-nutrient levels can severely inhibit coral production, and sometimes lead to the replacement of coral communities with macro-algal communities.

Water level provides a further constraint on the vertical limit of coral and reef development (approximately mean low-water spring tide level). Above this limiting water level, corals are exposed to direct solar radiation, and the combination of elevated temperature and desiccation promote the death of exposed coral tissue. The effect of limiting water level at the colony level is demonstrated by the presence of micro-atolls on reef flats. Micro-atolls are discoidal coral colonies up to 6 m in diameter, which have living coral tissue around the edge, but are dead on the surface due to subaerial exposure. The surface elevation of micro-atolls is directly related to water level and they have been used to reconstruct past sea levels (Woodroffe and McLean, 1990).

At the platform scale, exposure to wave energy also modulates the distribution of corals and their growth morphology and growth rate (Chappell, 1980). Corals have adapted their morphology so that robust encrusting forms

CONCEPTS BOX 16.1   Bleaching and coral reefs

Coral bleaching is the process by which corals lose their pigmentation and take on a white or 'bleached' appearance. Bleaching is a stress response of corals to unusually cold or warm water temperatures, low salinity or pollution. Under these stresses, corals can expel zooxanthellae, which provide the photosynthetic pigment and colour to corals. Over the past two decades, bleaching has been most closely linked to periods of elevated water temperatures. Widespread bleaching episodes have been associated with El Niño events in 1982–1983, 1987–1988, 1994–1995 and 1997–1998. In particular, the last event promoted bleaching in every coral reef region, with an estimated 16% of the world's reefs affected. However, it is important to note that El Niño does not cause bleaching, rather it elevates the probability that bleaching events (sustained warm-water periods) may occur.

It is commonly perceived that bleaching results in the death of corals. However, while corals can die from bleaching, they can also partially or fully recover. From a geomorphic perspective, it is important to consider the spatial extent of bleaching. Within any one reef system, the intensity and occurrence of bleaching is spatially variable and dependent upon climatic conditions, water depth, exposure to wave energy and current flow. For example, shallow depths, low cloud cover and low water motion are all conducive to more intense bleaching. The susceptibility to bleaching also varies among coral taxa, with branching forms more susceptible than massive forms to bleaching episodes. Bleaching can also induce a range of impacts on coral communities that may propagate through the eco-morphodynamic system. Such impacts may include a decline in abundance of temperature-sensitive coral and reef-associated species, reduced coral growth rates, increased susceptibility of corals to disease, enhanced bio-erosion, and impaired reproduction and recruitment.

As a consequence of the heightened awareness of coral bleaching and potential impacts on reef systems, there is great concern over the future survival of coral reefs with global warming. A key question for the future trajectory of living corals and geomorphology of reefs is whether corals can change their tolerance to an increase in frequency of thermal stress, associated with rising sea-surface temperatures. There is some evidence to suggest that corals can acclimate to higher temperatures through switching to a more thermally tolerant type of symbiont algae. Nevertheless, increased bleaching poses a serious threat to coral reefs, and it is likely that reef response will be patchy. Some reefs may experience a decline in living coral cover, some reefs may experience inhibited growth, and some reefs may undergo ecological shifts to more temperature-tolerant massive species (for a review, see Lough, 2011).

can survive in high-energy environments, whereas more delicate branching forms thrive in moderate- and low-energy zones. In higher-energy settings, vigorous mixing delivers nutrients and flushes metabolic wastes from corals, and promotes optimal growth rates. As a result, coral reefs in exposed settings and on windward sides of reefs are generally better-developed and more robust than in sheltered settings.

16.3   Coral reef development

16.3.1   Coral growth versus reef growth

There is a marked difference between the growth of individual corals and growth of coral reef platforms. The growth of individual coral colonies ranges from 1 cm/yr for some slow-growing massive corals to as much as 30 cm/yr for rapid-growing branching forms. Coral growth also varies spatially in response to the ecological zonation of a reef. It is a common perception that coral reefs are principally composed of interlocking coral skeletons, in growth position, creating a rigid wave-resistant framework and which provide the structural stability of a reef (Hubbard et al., 1998). This perception was largely based on extrapolation of observations of live coral reef assemblages on the surface of modern reefs and, consequently, it was assumed that biological construction was the primary process in reef framework development. However, detailed analysis of reef cores reveals that secondary producers (coralline algae), detrital rubble, sand and void space can comprise up to 70% of the volume of reef framework (Hubbard et al., 2001). Such findings question the role of in situ coral in reef framework development and highlight the fact that bio-erosion, transportation, encrustation and cementation are equally important in reef construction as calcification. Therefore, from a geomorphological perspective, the growth of a coral reef reflects the net processes of biological construction and mechanical and chemical deposition (e.g. detrital sediment and cementation), as well as bio-erosion and mechanical and chemical erosion (transport and solution).

16.3.2   The geological formation and distribution of coral reefs

Charles Darwin provided the first integrated process model of the formation of coral reefs, based on observations during the voyage of the HMS *Beagle* (Darwin, 1842). Recognizing that reef-building corals thrive in the shallow photic zone, Darwin proposed subsidence of volcanic substrate as the most likely mechanism that produces vast sequences of reefal limestone in mid-ocean basins. Darwin's model provides a genetic sequence of oceanic reef types, which evolve as a central volcanic island subsides beneath the sea (Fig. 16.5). These reef types include: fringing reefs that adjoin and

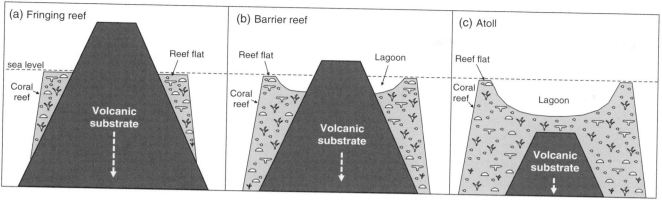

Fig. 16.5 Darwin's subsidence model of coral reef development, accounting for the formation of: (a) fringing; (b) barrier; and (c) atoll reefs. Note the gradual subsidence of the volcanic core and maintenance of reef growth near sea level. See Fig. 16.2 for examples of each reef type. (Source: Kench 2013.)

surround a volcanic high island; barrier reefs with a well-developed lagoon between the reef edge and much-reduced volcanic island; and atolls characterized by an annular reef rim enclosing a central lagoon and where the volcanic cone has subsided beneath the sea surface (Fig. 16.5 and Fig. 16.2b, c).

Confirmation of the subsidence theory was sought from a series of deep-drilling programmes in the central Pacific. Attempts were first made on Funafuti atoll in Tuvalu (central Pacific) in the years 1896–1898, where drilling reached a maximum depth of 340 m and terminated in limestone rock. However, the final confirmation of Darwin's model followed deep drilling in Eniwetok atoll, Marshall Islands, in 1951, where two drill holes penetrated over 1200 m of shallow-water coral reef limestones and reached basaltic rocks at depths of 1287 m and 1411 m (Ladd et al., 1953). Based on these drill cores, rates of subsidence of the volcanic basement were estimated at 0.025 mm/yr.

Development of plate tectonic theory further provided the framework to account for the regional distribution of reef systems, which are commonly found in linear chains in mid-ocean settings. The movement and thermodynamics of ocean plates, which migrate away from mid-ocean spreading centres, cool and subside, and the behaviour of hotspots at intraplate locations provides a large-scale geophysical explanation for the Darwinian subsidence model and a modern explanation for the evolution, orientation and age structure of island archipelagos (Box 16.2).

16.3.3 Controls on the morphology of modern coral reefs

Darwin's model and plate tectonics provide a robust explanation for the structural development of mid-ocean atolls at very long timescales (millions of years). Of note, the typical rate of subsidence of atoll foundations is 0.01–0.1 m/yr. However, the contemporary morphology of coral reefs is controlled by a number of other factors that influence reef development at shorter timescales than subsidence (for a synthesis, see McLean and Woodroffe, 1994). Onto the geophysical template of subsidence and plate tectonics must be superimposed the effects of fluctuations in sea level of 120–150 m throughout the Quaternary. These oscillations in sea level have occurred at rates of 1–10 mm/yr, exerting a strong control on coral and reef growth. Modern coral reefs have occupied the accommodation space between sea level and the underlying substrate. During sea-level lowstands, coral substrate is exposed to subaerial weathering, which denudes the platform surface. As the platform is composed of $CaCO_3$, the exposed surface develops characteristic karstic (limestone) features, with preferential solution of lagoons as opposed to reef rims. As sea level rises and inundates the denuded surface, reef growth can resume in the newly created accommodation space (the result of solution and background subsidence). Upward reef growth continues until it reaches sea level, which is the vertical growth limit of reefs. Consequently, the morphology of the reef is controlled by the antecedent karst surface. Multiple sea-level oscillations throughout the Tertiary and Quaternary have produced alternating periods of subaerial erosion during sea-level lowstands and reef growth during highstands. The morphology of reefs, therefore, is inherited from multiple alternating episodes of growth and solution modulated by sea level. Consequently, modern reefs reflect reef growth in response to the Holocene marine transgression and form a relatively thin veneer over older reef limestones (see Fig. 16.1).

16.3.4 Styles of reef growth

Vertical reef growth in response to sea-level change has been a major research focus of reef geomorphology in recent decades. In general, there are two dominant

CASE STUDY BOX 16.2   Evolution of reef archipelagos in mid-ocean settings

The Hawaiian-Emperor chain of islands located in the Pacific Ocean is the best-known example of the movement and thermodynamics of ocean plates and behaviour of hotspots that control the evolution of linear island archipelagos. This chain of islands extends more than 3000 km and exhibits the entire spectrum of reef types identified by Darwin. The Pacific Plate migrates in a northwest direction away from the mid-oceanic ridge at a rate of 80–150 mm/yr, whilst at the same time subsiding at a rate of 0.02–0.03 mm/yr. Periodic volcanic activity associated with a fixed hotspot (thermal anomaly in the oceanic crust and asthenosphere) provides a mechanism for island formation. Once an island is formed, it is carried along by the underlying oceanic crust, away from the hotspot. Repetition of this process over geological timescales has produced long chains of islands, which increase in age with distance from the hotspot location (Fig. 16.6). As islands move along the conveyor of the oceanic plate, they subside and, where this occurs in the tropical seas, the islands support coral reef growth that follow Darwin's genetic sequence of reef types (fringing reefs to atolls). As islands migrate to higher latitudes, reduction in water temperature limits coral growth. Eventually, vertical reef growth can no longer keep up with subsidence (or ceases) and the coral atoll becomes submerged, forming a guyot (Fig. 16.6; Scott and Rotondo, 1983).

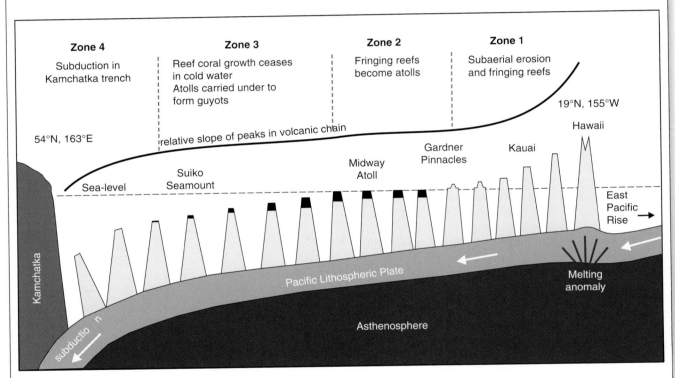

Fig. 16.6   Model of the geological formation and age structure of mid-ocean archipelagos. The example shows the development sequence of the Hawaiian-Emperor seamount on the Pacific Plate. (Source: Adapted from Scott and Rotondo 1983. Reproduced with permission from Springer Science + Business Media.)

growth responses to sea-level change. Firstly, vertical development in response to increased water depth over reefs as sea level rises. Secondly, lateral progradation as reefs reach their vertical growth limit or accommodation space. Based on radiometric dating and facies analysis of more than 1000 cores drilled through Holocene reefs in the Indo-Pacific and Caribbean reef systems, a number of vertical reef growth responses to sea level have been identified (Montaggioni, 2005). Initially, three styles of reef accretion were identified, which included 'keep-up', 'catch-up' and 'give-up' reefs (Fig. 16.7; Neumann and Macintyre, 1985).

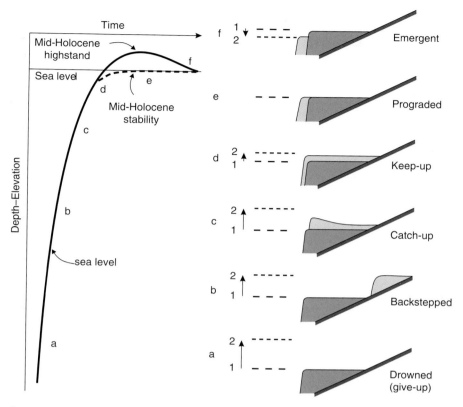

Fig. 16.7 Modes of reef growth in response to sea-level change. The sea-level curve is presented on the left, with reef growth modes on the right. (Source: Adapted from Woodroffe 2002. Data from Neumann & Macintyre 1985 and Hubbard 1997.)

Keep-up reefs track sea level as it rises, maintaining shallow frame-building communities and a reef crest at sea level. Catch-up reefs initially lag behind sea-level rise, either due to time delays for reef-building communities to re-colonize the platform surface or due to accelerations in sea-level rise above reef-building capacity. However, these reefs later exhibit rapid vertical growth that catch up with sea level as it slows or stabilizes. Give-up reefs fall behind rising sea level and, finding themselves in rapidly increasing water depths, shut down as failed reefs. Recently, this tripartite classification has been extended to incorporate a number of alternate development responses. Reef backstepping occurs where sites of reef growth are abandoned on sloping shelves and are re-established at new upslope locations. Triggers for backstepping may include periodic accelerations in sea-level rise or changes in water quality conditions. Lateral reef progradation becomes the dominant trend of reef development once reefs fill the vertical accommodation space as they reach sea level. This growth mode is typical of reefs under sea-level still-stand conditions, where accommodation space becomes constrained by substrate slope. Reef-flat emergence is common in settings where reefs have been affected by relative sea-level

fall. In such settings, keep-up and catch-up reefs are exposed by sea-level fall and reef-building communities die (Fig. 16.7; Woodroffe, 2002).

16.3.5 Regional differences in reef morphology

It is important to note that differences in Holocene sea-level dynamics between reef regions (Chapter 2) has modulated reef growth and the morphology of modern reefs and reef flats. In the Indo-Pacific, sea level first reached present level approximately 6000 years ago, rose to 1.0 m above present in the mid-Holocene, before falling to present. This pattern has imparted distinct characteristics on reef growth and morphology. Firstly, reef development was dominated by keep-up or catch-up growth in the early to mid-Holocene, until reefs filled the accommodation space. Secondly, Indo-Pacific reefs have been at sea level for at least 1000–6000 years. This period has been dominated by lateral reef growth, resulting in broad reef-flat surfaces. Thirdly, reef flats have become emergent following sea-level fall. In contrast, in the Caribbean, sea level has continued to rise throughout the Holocene, although the rate of rise has decreased over the past 5000 years. In response, Caribbean reefs have undergone catch-up in the early Holocene and have maintained a keep-up growth mode throughout the remainder of the Holocene.

16.3.6   Rates of reef growth

Vertical growth rates of framework reefs range from 1 to 30 mm/yr with a modal rate of 6–7 mm/yr (Montaggioni, 2005; Hopley et al., 2007). Highest rates of growth are found in highly porous frameworks of tabular and branching corals, while the lowest rates occur in foliose and encrusting coral and coralline algal reefs. Detrital dominated reefs (those where the reef fabric is composed of detrital sediment) exhibit vertical accretion rates that range from 0.2 to 40 mm/yr and can be divided into three growth rates that reflect differences in hydrodynamic energy regime. Lower modal rates of accretion of 1–3 mm/yr occur in low-energy lagoonal settings with mud-dominated materials. Rates of detrital reef accumulation in the order of 4–8 mm/yr characterize sand and rubble facies on moderate-energy reef-flat and backreef settings. Highest detrital rates of framework accumulation (>10 mm/yr) occur in storm settings in which sand and gravel sheets are able to be deposited episodically (Montaggioni, 2005).

Lateral reef accretion is the dominant growth mode once reefs reach sea level, as progradation across substrate provides the only additional accommodation space. In Indo-Pacific reefs, rates of lateral reef accretion range from 15 to 84 mm/yr (modal rate of 50 mm/yr) for reefs in semi-exposed to sheltered settings. Highest rates of lateral progradation occur in high-energy reef margins and range from 24 to 300 mm/yr (modal rate of 90 mm/yr) (Montaggioni, 2005).

It is also important to note that rates of reef growth can vary spatially within individual reef systems, according to ecological zonation and energy exposure. For instance, in the barrier reefs of Tahiti and Palau, windward outer reef margins were characterized by keep-up reef growth at rates of approximately 6 mm/yr, whereas leeward reef growth followed a catch-up mode at rates of 3–4 mm/yr. Net rates of reef accumulation also mask temporal variations in reef growth rate as Holocene reefs accreted. For example, as growth switched from catch-up to keep-up modes, reefs recorded a reduction in growth rates as they approached their vertical growth limit (Montaggioni, 2005; Hopley et al., 2007).

16.4   Reef island formation and morphodynamics

The generation of detrital sediment on reef platforms and its transfer by physical wave and current processes controls the formation, maintenance and ongoing development of a suite of geomorphic features in reef systems (see Fig. 16.1). Detrital sediment can be: (1) transported off-reef where it is lost to the geomorphic system; (2) reincorporated into the reef framework to contribute to reef development; (3) transported to lagoons where sedimentation can contribute to lagoon infill; or (4) deposited as subaerial accumulations of sediment, forming progradational plains and islands. The focus of this section is on the formation and change of reef islands, as they are conspicuous features of coral reefs and, for obvious reasons, preferred sites for human occupation.

Reef islands are accumulations of sand or gravel, deposited directly on reef platforms or over sedimentary deposits that infill lagoons (Stoddart and Steers, 1977). They are found on the peripheral rim of atolls, on barrier reefs and on isolated reef platforms (see Fig. 16.2). In general, reef islands are typically low-lying, rarely reaching more than 3.0–4.0 m above sea level. However, they do exhibit significant variation in morphology (size, elevation) and sediment composition. Four principle criteria have been employed to classify island types: (1) the location and number of islands on reef platforms, which reflects exposure to incident wave energy; (2) island planform shape and morphological complexity; (3) the calibre of sediments (sand cays, gravel islands or mixed sand and gravel motu); and (4) the presence, extent and type of vegetation (Hopley et al., 2007).

Based on morphological differences, early attempts were made to place islands in an evolutionary sequence. However, it is now understood that islands form a spectrum of different morphological states, controlled by site-specific interactions between wave energy exposure, sediment calibre and reef platform configuration.

16.4.1   Controls on reef island evolution

The evolution and ongoing development of reef islands is dependent on the interaction of five factors:

(1) Sea-level change, which, as outlined in section 16.3, controls reef platform development and modulates the character of incident wave energy to reef platforms.

(2) Substrate characteristics, which in coral reefs include the surface area, configuration, elevation and slope of the platform.

(3) Accommodation space, which defines the available volume for sediment accumulation as controlled by substrate elevation and sea level (Cowell and Thom, 1994). The upper limit of land building is controlled by storm wave run-up processes, which are modulated by relative sea level. For reef islands, the lower boundary defining accommodation space is the reef margin, reef elevation and lagoon depth.

(4) Process regime, which includes the incident wave inputs to a reef platform and the transformation of energy that governs reef-flat wave regimes, circulation and sediment transport patterns.

(5) Sediment supply, which is controlled by reef productivity and sediment generation processes.

Coral reefs are unique in that substrate adjustment (reef growth) occurs in response to sea-level change at centennial timescales. Therefore, one of the boundary conditions defining accommodation space exhibits morphodynamic feedback at timescales that overlap with the processes responsible for island evolution and dynamics. This co-adjustment has a number of important implications for the timing and conditions under which sedimentary landforms have evolved.

16.4.2 Models of reef island evolution

Sea level and its control on reef growth are commonly identified as a major control on the formation and future change of reef islands. In the eastern Indian and Pacific Oceans, studies have identified a sequential model of island formation in which sea-level stabilization, completion of vertical reef growth, and reef-flat development at sea level were precursors to island accumulation (Fig. 16.8a). In this region, sea level has been at, or slightly higher than, present for the past 6000 years. Vertical reef growth was rapidly constrained and replaced by lateral reef development, which has produced broad reef flats. These reef flats later became emergent as a consequence of late Holocene sea-level fall, providing the foundation for island accumulation in the mid to late Holocene. This model is supported by radiocarbon-dating evidence that shows islands began forming 4200 years ago in the Cocos (Keeling) Islands (Woodroffe et al., 1999), 2600 years ago on Makin Island, Kiribati (Woodroffe and Morrison, 2001), and 2000 years ago in Tuvalu (McLean and Hosking, 1991). Differences in the onset of island formation reflect differences in the time at which reefs reached sea level. Of further interest, the ages of reef islands in the Pacific indicate they formed following late Holocene sea-level fall. This synchronization of sea-level change followed by island formation has lead to the assertion that sea-level fall is a necessary pre-condition for island formation.

A different model of reef island formation has been identified in the Maldives, which are located in the central Indian Ocean. Dating evidence shows that islands in the Maldives formed in the mid-Holocene (5000 years ago), which is earlier than in the Pacific. Islands formed across submerged reefs and over infilled shallow lagoons prior to reefs achieving their vertical growth limit (Fig. 16.8b). Furthermore, radiocarbon dates indicate the islands formed very rapidly, in a 1500-year window, during a period when water level was deeper across reefs than present. Subsequent to island formation, the outer reef

platform continued to grow vertically and shut down the hydrodynamic process regime affecting island shorelines. The closing down of the reef-top process regime has contributed to relative stability of reef islands over the past 3500 years (Kench et al., 2005). Recent reconstruction of the sea-level history in the Maldives identified a late Holocene sea-level highstand 3500–2000 years ago of at least 0.5 m above present (Kench et al., 2009b). This episode of higher sea level post-dates reef island formation and presents evidence that islands can withstand higher sea levels of the order of those projected over the next 100 years. This alternative model of island evolution, in which island formation can occur prior to sea-level stabilization and reef-flat development, is likely to have similarities to the Caribbean, where landforms have developed under continual rising sea level throughout the late Holocene (Toscano and Macintyre, 2003).

Differences in the models of island development provide key insights into the role of sea level and reef development in island evolution. Firstly, it is apparent that island building has occurred under contrasting sea-level change histories, including sea-level rise (Fig. 16.8). Secondly, sea-level fall, while apparent in some examples, is not necessary for the onset of island building. Thirdly, islands form during latter stages of reef platform evolution, but reef-flat development at sea level is not necessary for island formation. Fourthly, analogues exist that suggest islands can withstand increases in sea level in the order of 0.5 m, which occur at millennial timescales.

16.4.3 The importance of sediment supply on island building

Reef island building requires the supply of sediment to fill the accommodation space. In general, reef island formation was initiated in the mid- to late-Holocene, but studies suggest that the subsequent accumulation history can follow different patterns. Maldivian reef islands appear to have formed in a narrow 1500-year window in the mid-Holocene and effectively ceased building 3500 years ago (Kench et al., 2005). In contrast, Warraber Island in the Torres Strait has formed incrementally over the past 3000 years in response to continued supply of sediment (Woodroffe et al., 2007). Furthermore, in storm-dominated settings, island development can be controlled by episodic deposition of storm-generated rubble ridges (Hayne and Chappell, 2001).

Variations in accumulation history of reef islands (and other sedimentary landforms) may be explained through temporal variations in sediment supply during the Holocene, and recognition that the relationship between reef carbonate productivity and sediment generation is non-linear. Variations in sediment availability are likely to reflect temporal alterations in the balance between reef

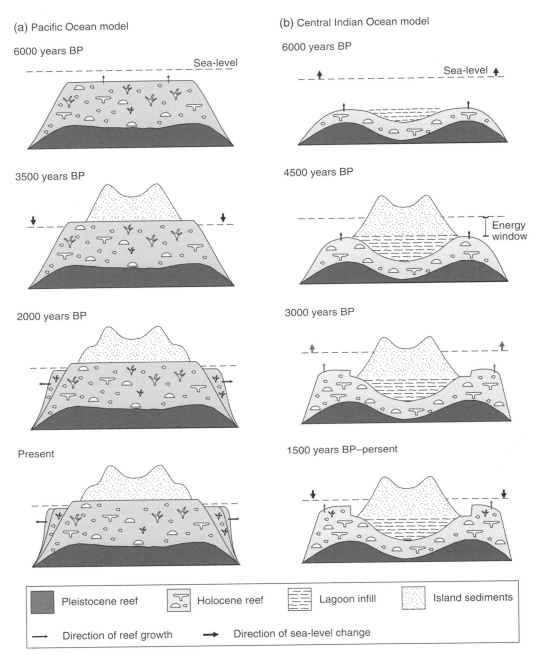

Fig. 16.8 Alternative models of reef island formation in: (a) the Pacific; and (b) in the central Indian Ocean. Note the relationships between sea level, reef growth and island development. (Source: Adapted from Kench et al. 2005.)

growth, productivity and sediment cycling processes. For example, onset of island building in the mid-Holocene may coincide with reefs either reaching sea level or reaching wave base and releasing a pulse of excess sediment. Episodes of sediment generation also occurred as a result of reef-flat emergence, which promoted shifts in reef-flat ecology, and thus carbonate production. Such sea-level forced changes in reef-flat ecology are likely to have altered the dominant skeletal components available for land building, from sediments generated from frame builders (coral and coralline algae) to those derived from other sediment producers (e.g. foraminifera, calcareous green algae).

16.4.4 *Process controls on island development*

The interaction of ocean swell with coral reefs is widely known to modulate oceanographic, ecological and geological processes in coral reef systems (Roberts et al., 1992). In a geomorphic context, wave interaction with coral reef platforms is recognized as the principal mechanism activating geomorphic processes on reefs and controlling the formation of reef islands. The geomorphic importance of wave-reef interactions are manifest in two ways. Firstly, reefs act as a filter to wave energy, controlling energy available for geomorphic work. Secondly, the planform configuration of reefs both refracts and diffracts wave energy, and controls the nodal location of sediment deposition on reef surfaces.

The filtering of wave energy is achieved through the direct physical interaction of waves with reef structure, which is modulated by water depth across the reef surface. Numerous studies have documented large reductions in wave energy of up to 97% as incident waves are transformed and break at the reef edge (e.g. Roberts and Suhayda, 1983). Despite the large reduction in energy at the reef edge, residual energy leaks onto reef surfaces. The magnitude of this energy is critical in controlling sediment entrainment and transport on the reef surface. At low-water stages, when the reef flat may be emergent, incident waves break at the reef crest and little energy is transmitted across the reef surface. However, at higher tidal levels, and during storms due to storm surge and wave set-up, water depth across the reef may exceed 2 m. Under such high-water conditions, wave breaking at the reef edge is less effective and greater amounts of wave energy can propagate onto the reef surface (Brander et al., 2004). Relative water depth across reef surfaces and reef width are critical controls on the presence and delivery of wave energy to island shorelines. Hardy and Young (1996) found that the maximum wave height on reef surfaces is depth-limited and never exceeds 0.6 of the water depth h, whereas the significant wave height does not exceed $0.4h$. Furthermore, Kench and Brander (2006b) show that wave energy diminishes with increasing reef width due to frictional dissipation of energy. The 'reef energy window index' was proposed to capture the influence of these physical properties on the level of geomorphic activity of reef surfaces (Kench and Brander, 2006b). In general, wave energy (and geomorphic work) on reefs and island shorelines increases as relative water depth H/h increases and reef width becomes narrower. In contrast, the magnitude of wave energy diminishes with increasing reef width and, therefore, the ability to stimulate sediment transport also declines.

The geomorphic importance of water depth control on wave height across reef platforms is evident at instantaneous through to geological timeframes. At instantaneous timescales, tidal variations modulate the magnitude of wave energy that propagates onto reef surfaces. At geological timescales, the relationship between sea level and reef growth modulates the ability of waves to interact with the reef surface, and controls the gross magnitude of wave energy propagating over reefs. During the mid-Holocene, reefs in catch-up growth mode were not constrained by sea level, and water depth over reefs was greater than present, allowing higher ocean wave energy to propagate onto reefs (Fig. 16.8b). As reefs caught up with sea level, this energy window closed. This period in the development of reefs has been characterized by the mid-Holocene 'high energy window' and has been implicated as an important phase of construction of a range of sedimentary landforms in the Great Barrier Reef (Hopley, 1984) and Maldivian reef islands (Kench et al., 2005).

Wave breaking at the reef edge and breaking of waves at island shorelines also induce the development of wave set-up and the potential for wave set-up gradients to control across-reef and alongshore flows on reefs and island shorelines (Fig. 16.9; Gourlay and Colleter, 2005). The magnitude and pattern of currents is therefore controlled by the incident wave field and have been shown to be sensitive to changes in incident wave characteristics. For example, in the Maldives, seasonal variations in wind direction are known to promote reversals in reef platform wave-induced current patterns, which in turn modulates shoreline change (Kench et al., 2009c).

Observations of nearshore hydrodynamics have also shown that reef platform shape can impart significant control on hydrodynamics of reef surfaces (Fig. 16.9). On small reef platforms it is possible that incident swell is able to refract completely around the reef edge and control depositional nodes on reefs (Gourlay, 1988; Fig. 16.9a). Under such circumstances of complete refraction around platforms, field measurements have shown that greater energy levels can occur on leeward shorelines (Kench et al., 2009a). As both wave and current energy generally decline in magnitude from reef edge across reef platforms, the intersection of energy gradients across small platforms promotes deceleration of currents, creating a nodal point for sediment accumulation (Fig. 16.9c). As islands build and prograde across reef surfaces, shorelines are placed in more energetic conditions. Consequently, shoreline run-up processes are more energetic and this mechanism allows seaward ridges to build higher with respect to sea level. This energy-sedimentation relationship provides a process mechanism for the concave morphology of island surfaces and the progressive increase in the seaward island margin closer to the reef edge (Fig. 16.9c).

On large atolls with linear atoll rims (see Fig. 16.2e), the refraction of waves around reef platforms and islands

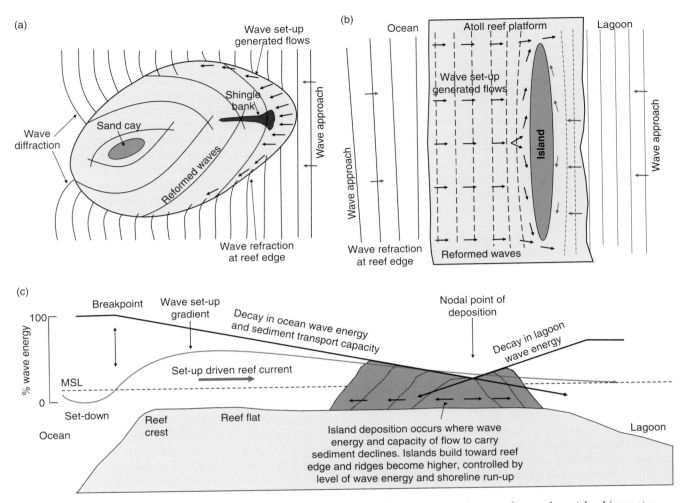

Fig. 16.9 Conceptual models of wave interaction with reef platforms, resultant current gradients and controls on island formation. (a) Isolated reef platforms where wave energy can refract around the platform (Source: Adapted from Gourlay 1988). (b) Wave interaction with a linear atoll reef rim. (c) Idealized ocean-to-lagoon cross-section showing the decay in wave energy and generation of wave set-up driven currents across reefs. The decay in wave energy and current energy reduces the capacity to transport sediment, providing an envelope for the accumulation of sediments at nodal points on reefs (islands). As islands build toward reef edge they accrete vertically in response to increased wave energy and run-up processes. (Source: Kench 2013.)

is prevented by the extent of the reef structure (Fig. 16.9b). In these settings, wave energy, current and sediment transport gradients are unidirectional from the reef edge to lagoon (Kench and McLean, 2004). Consequently, the limit of sediment accumulation is controlled by the competency of flow to transport sediment of varied grain size. Under such conditions, islands can accumulate through seaward deposition across reefs, and the seaward margin increases in height toward the reef edge (Fig. 16.9c). In large atolls, waves generated in the lagoon can influence the lagoon shoreline of islands, although this energy gradient is typically lower in magnitude and promotes the formation of lower-elevation lagoon ridges.

16.4.5 Morphodynamics of reef islands

Reef islands are dynamic landforms that are in continual adjustment to changes in process regime forced by alterations in climate, oceanographic boundary conditions and sediment supply. These boundary conditions force morphological change at a range of temporal scales.

Extreme events (cyclones and tsunamis) can promote both erosion and accretion responses on reef islands. Bayliss-Smith (1988) proposed one of the earliest morphodynamic models, in which the mode of morphological change was dependent upon variations in the interplay between the sedimentary character of islands and the

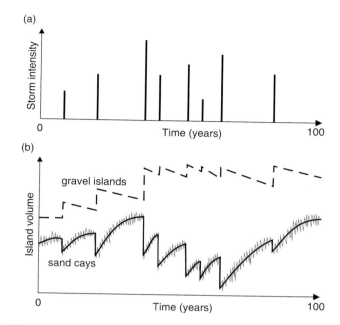

Fig. 16.10 Conceptual model of reef island morphodynamics in response to storm events. (a) Storm frequency and intensity. (b) Morphological response of sand cays and shingle islands. The grey line in (b) depicts short-term shoreline morphodynamics. (Source: Adapted from Bayliss-Smith 1988. Reproduced with permission of John Wiley & Sons.)

frequency and intensity of storms (Fig. 16.10). Sand cays found in environments with low storm frequency are susceptible to erosion during extreme wave events. For example, sand cays on the Belize barrier reef suffered extensive erosion in response to Hurricane Hattie (Stoddart, 1963). In contrast, in storm-dominated reef systems, islands are commonly composed of coarse rubble on their exposed reef margins, while islands on leeward reefs are typically composed of sand. Storm events can generate large volumes of material, by mechanically removing living coral communities and depositing this material on reef flats and on islands. For example, Hurricane Bebe in 1972 deposited 1.4×10^6 m³ of storm rubble on the windward reef flat of Funafuti atoll, Tuvalu (Maragos et al., 1973). Subsequent reworking of this storm rubble onto island shorelines increased island area by approximately 10%. Recent tsunami events provide analogues of the impact of extreme wave inundation on reef islands. The first quantitative observations of tsunami impacts were undertaken in the Maldives, where detailed mapping of island morphology and sediments before and after the Indian Ocean tsunami (2004) identified both erosion scarping of island shorelines and vertical building of islands through washover of beach sediments to island surfaces (Kench et al., 2006).

At seasonal timescales, oscillations in wind and wave climate have been correlated with predictable changes in

island shorelines. For example, Kench and Brander (2006a) examined the morphological sensitivity of 13 islands in the Maldives to predictable changes in wind and wave conditions controlled by the reversing monsoons. Results showed large (up to 53 m of beach change) and rapid excursions of beach material around island shorelines in response to monsoonally forced changes in nearshore circulation (Fig. 16.11). Of note, beach changes involved the alongshore reorganization of beach material, rather than the off/onshore exchange common on continental shorelines. However, seasonal oscillations in morphology were balanced at the annual scale. The magnitude of seasonal morphological change, and sensitivity of islands to change, was found to vary between islands as a function of reef platform shape, which controls wave refraction patterns. Kench and Brander (2006a) proposed the 'island oscillation index' to predict the sensitivity of islands to changes in wave climate. Analyses showed circular islands were most sensitive to changes in incident wave processes.

At decadal timescales, changes in wind regime (e.g. El Niño Southern Oscillation variations) and their impact on reef processes (waves, currents and sediment transport) have been implicated in the physical adjustment (migration) of islands on reef surfaces (Flood, 1986).

16.5 Management in reef environments

Practical management challenges in coral reefs and reef islands are primarily related to a multitude of human pressures, which include exploitation of reef resources (e.g. fishing and aggregate mining), construction of transport infrastructure (navigation channels through reefs, boat harbours and causeways that link islands), land reclamation (as a consequence of high population densities), waste disposal and pollution, and management of shoreline erosion or island instability (e.g. Fig. 16.12a, b). These pressures occur at a range of spatial scales. Anthropogenic impacts on reefs can be divided into three categories based on the geographic relationship between human activity and the reef system. Some impacts are local and direct, such as dredging or coral mining. In the Maldives, the selective removal of larger corals from reef flats has promoted ecological change (Brown and Dunne, 1988) and lowered the reef surface. Consequently, mining has produced an anthropogenically forced higher-energy window. Another set of impacts are proximal, such as harbour construction or resort development, and may have an alongshore influence on reef processes and shoreline stability. Other impacts are distant, but the effects are translated to reef systems. For example, logging of rainforests and agricultural practices in Southeast Asia have released sediment plumes to the coast, threatening the existence of over 20% of reefs (Burke et al., 2002). Understanding

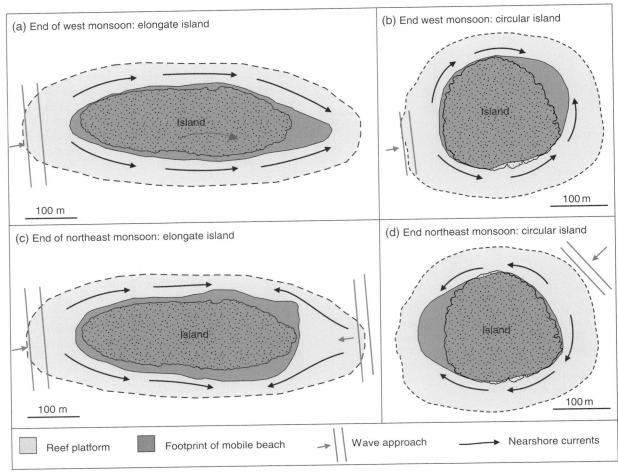

Fig. 16.11 Summary of seasonal beach change on reef platform islands in the Maldives. Note changes in nearshore currents in response to changing wave approach between seasons. Currents control the alongshore reorganization of beach material, promoting large excursions of beach position around the circular island (±40 m in (b) and (d)). In contrast, beach change on the elongate island is constrained to the eastern tip (±60 m in (a) and (c)). (Source: Adapted from Kench and Brander 2006a; Kench et al. 2009c.)

differences in the type and scale of impacts affecting reefs is essential for a number of reasons. Firstly, to identify key factors interfering with the natural dynamics of reefs systems; secondly, to recognize the baseline conditions confronting reefs in any future adjustment; and thirdly, to allow management strategies to focus at the appropriate space and timescale.

16.5.1 Management of shoreline instability

Historically, approaches to managing human activities and impacts have imparted a legacy of adverse environmental effects, as actions have directly compromised the integrity of the eco-morphodynamic system (Maragos, 1993; summarized in Table 16.3). The management of shoreline stability provides a useful example of the lack of

understanding of coastal processes in informing management decisions. Management of shoreline instability (erosion) is a prevalent issue in reef islands, and one that is expected to become more pervasive with future sea-level change. Management responses have typically included the standard suite of hard-engineering solutions to stabilize shorelines, such as seawalls, revetments and groynes (Fig. 16.12b–d). However, the use of hard-engineering structures is known to have a high failure rate in reef islands (Fig. 16.12e, f). There are a number of reasons for their poor performance. Firstly, structural techniques are commonly transposed from use on continental coastlines and are incompatible with the natural processes and dynamics of reef island shorelines. In general, reef islands are inherently dynamic and are in a constant state of adjustment on reef platforms, related to variations in

Fig. 16.12 Examples of anthropogenic transformation of reef islands and approaches to shoreline stabilization in developing countries: (a) densely populated and modified capital island of Malé, Maldives; (b) modified tourist island shoreline and reef, Maldives; (c) sandbag revetment, Tarawa atoll, Kiribati; (d) coral cobble seawall, South Tarawa atoll, Kiribati; (e) failed sandbag revetment, Bairiki island shoreline, Tarawa atoll, Kiribati; and (f) failed sandbag revetment, South Tarawa atoll, Kiribati. (Source: Photographs by Paul Kench.) For colour details, please see Plate 39.

Table 16.3 Summary of impacts of anthropogenic activities and natural processes on coral reef ecological and geomorphic processes. ENSO, El Niño Southern Oscillation; IPO, Interdecadal Pacific Oscillation. (Source: Kench et al. 2009a. Reproduced with permission of Cambridge University Press.)

Alterations to controlling factors of landform stability	Anthropogenic causes	Natural causes
Impacts on reef structure (a) Changes in reef ecology that shift the carbonate budget of reef platforms	• Increased nutrient loads • Increased sedimentation • Overfishing, dynamite fishing • Reef mining • Tourism impacts	• Global sea-level change • Decadal–interannual variations in sea level (IPO, ENSO) • Frequency, intensity and location of storms • Patterns of reef growth • Disease outbreaks • Storms, tsunamis
(b) Direct physical impacts on reef structure	• Reef blasting • Channel construction and dredging • Reef mining	
Impacts on reef sedimentary landforms *Change in hydrodynamic processes* (a) Alteration in wave processes (energy and direction): manifest as changes in water depth across reef surface that influence wave energy on reef platform (b) Change in reef platform current patterns (circulation, flushing)	• Channel dredging • Reef blasting • Coral mining • Engineering structures at shoreline or on reef (groynes, breakwaters) • Boat wakes • Causeway construction • Reclamation • Harbour construction	• Global sea-level change • Decadal–interannual variations in sea level (IPO, ENSO) • Climate-driven changes in angle of wave propagation • Frequency, intensity and location of storms • Variations in wave processes • Modifications to reef bathymetry
Change in sediment budget (a) Alterations in sediment generation (b) Change in littoral-reef sediment transport (c) Adjustment in sediment volume at shoreline	• Pollution of reef – decreasing productivity • Overfishing • Insertion of shore-perpendicular structures • Scouring of shoreline due to stormwater discharge • Removal of vegetation • Coral mining • Sand extraction or dredging	• Changes in reef ecology/productivity • Ecological perturbations (predator explosions, bleaching) • Storm magnitude and frequency • Formation of littoral barriers (e.g. beachrock development) • Altered current processes • Redistribution of sediment through oceanographic processes, overwash

wave, current and sediment transport processes. The nature of shoreline dynamics also varies between islands of varying shape. Unlike continental shorelines, reef islands have a 360° perimeter. Furthermore, short-term reef island shoreline dynamics are dominated by along-shore processes rather than by on/offshore exchange of sediments (see Fig. 16.11). Consequently, use of groynes can have extreme impacts on the seasonal alongshore reor-ganization of beach sediments and exacerbate island erosion (Fig. 16.13). Secondly, there are very few measurements of shoreline wave and current processes with which to evaluate maximum wave height and run-up levels in the design stage. Thirdly, materials used for construction of structures in many small-island settings rely on local aggregates and construction techniques, which contravene standard measures of 'best engineering practice' (Kench et al., 2003). In Kiribati, for example, revetments were tra-ditionally made of coral aggregate or more recently from sand bags filled with a mix of cement and carbonate sand, stacked against the coast (Fig. 16.12). Sand bags are small engineering units and are generally an order of magnitude smaller than the required size to withstand wave energy on moderate to high-energy shorelines. Consequently, they are commonly overtopped, leading to failure and sed-iment loss landward of structures.

16.5.2 Coastal management and planning

From the geomorphic perspective, approaches to manage-ment of reef systems must focus on maintaining the integ-rity of the process linkages in the eco-morphodynamic system. In reef systems, this requires maintenance of eco-logical and physical processes that modulate the carbonate factory and landform change.

Management and planning systems in reef environ-ments can be divided into two broad groups, which relate

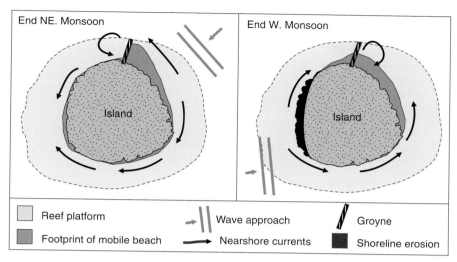

Fig. 16.13 Example of the impact of a groyne on nearshore circulation, seasonal beach dynamics and shoreline erosion on a circular reef platform island. Note the groyne disrupts the alongshore transfer of beach sediment (cf. Fig. 16.11b, d). At the end of the northeast monsoon, beach material is prevented from occupying the western shoreline (cf. Fig. 16.11d). Consequently, the island shoreline is exposed to erosion in the subsequent westerly monsoon season

to the level of development in each country. In developing countries, attempts to develop robust coastal management and planning processes have been constrained due to a lack of human capacity and weak governance structures. In such settings, there is little institutional recognition of the integrated nature of coral reef systems, with priorities on resource use and development fragmented among a number of government agencies (Aston, 1999). Furthermore, planning controls on coastal development are typically poorly designed and implemented, and practical management of shoreline stability is approached with the use of a limited number of structural solutions.

In contrast, in developed countries, management and planning approaches more commonly do embrace a more integrated approach to resolving impacts on coral reefs, although these responses are primarily focussed on the ecological condition and preservation of reefs. A well-known example is the management system adopted in Great Barrier Reef, which extends more than 2600 km along Australia's northeast coast. In 1975, the Australian Government created the Great Barrier Reef Marine Park, the largest marine managed area in the world. Underpinning management of the reef system is a multi-tiered zoning system that provides for differing levels of access and resource use, which is managed through a permitting process. The primary zones are general use, conservation park, habitat protection, and marine national park. In 2004, a revised zoning plan saw the most-protected zones increase from 4.4% to 33.3% of the area of the reef system. While the zoning system provides an ability to manage local and proximal activities within the reef system,

the management of external impacts is a more difficult challenge. In particular, declining water quality on the reef, as a consequence of human activities in the adjacent land-based catchments, has been identified as a major threat to the health of the reef.

16.6 Future trajectories of coral reef landforms

Anthropogenic activities will continue to exert a major impact on the ecological and geomorphic functioning of reefs through changes to the boundary conditions of the morphodynamic system and the carbonate factory. In some cases, such impacts will be extreme (e.g. hard-engineering structures and reef blasting) and will entirely compromise the eco-morphodynamic system. In addition, coral reefs are entering a phase of geomorphic alteration in response to global environmental change, which poses additional stressors to coral reefs. Global assessments of reef ecological condition suggest that 20% of the world's reefs have been destroyed, with a further 25% under imminent risk of collapse (Wilkinson, 2004). On the basis of reef ecological decline, it is widely assumed that reef landforms face a similarly bleak future. Entire loss of reefs and reef islands is widely projected, rendering inhabitants of small mid-ocean atoll nations the first environmental refugees of climatic change. Consequently, the fate of coral reefs and reef islands has captured international concern, with small-island states seen as one of the most sensitive indicators of the environmental consequences of global climatic change. However, such projections are based on the assumption that short-term changes in the

Table 16.4 Summary of proposed changes in boundary controls on the reef eco-morphodynamic system and geomorphic consequences.

Environmental property	Characteristic of change	Negative effects on coral system	Positive effects on coral system
Sea-surface temperature	• Increase in temperature • Increased intensity and frequency of warming events	• Excessive temperatures promote expulsion of zooxanthellae, inducing bleaching. Increased frequency of bleaching may lead to loss of reefs	• Potential expansion of latitudinal limits of coral survivorship and reef growth • Increased rate of calcification of corals
Ocean acidification	• Decrease in ocean pH due to ocean uptake of carbon dioxide	• Decrease calcification by coral • Weaker coral skeletons • Increase dissolution of carbonate	• Increased rate of sediment generation
Sea level	• Sea-level rise	• Potential to drown some reefs • Increases process window across reef surfaces (wave energy and sediment transport) • Increase geomorphic instability of reef islands (erosion, migration) • Inundation of islands	• Reactivate coral growth • Increase carbonate production on reefs and sediment reservoir • Islands physically adjust to changing process regime and sediment supply
Storms	• Increase in frequency and intensity	• Destroy living coral reef • Reduce interval for ecological recovery of reefs • Three-dimensional loss of reef structure • Increase wave attack on reef islands, promoting erosion	• Increase supply of detrital sediment on reefs

ecological condition of reefs have direct and immediate consequences on geomorphic processes. This assumption is invalid as it ignores non-linearities in the geomorphic system (see Fig. 16.3). This section provides an overview of the geomorphic implications of global climate change on coral reefs (see Smithers et al. (2007) and Kench et al. (2009a) for detailed reviews).

16.6.1 Changes in boundary conditions

Future environmental changes, such as ocean acidification, increases in SST (which can promote bleaching), changes in storm frequency and intensity, and sea-level rise, are all expected over the next century (IPCC, 2007), and will alter the boundary controls of the reef eco-morphodynamic system. These changes in boundary conditions are expected to impact on coral survivorship and living reef-building communities in a variety of ways, and are summarized in Table 16.4 (also see Box 16.1). While the direct impacts on coral physiology have been widely explored by marine ecologists (e.g. Lough, 2011), the geomorphic consequences of such changes are very poorly understood. It is clear that the degree of modification of each parameter and their impacts will be spatially variable. However, it is unclear whether the combined effects of changes in each variable will compound or negate impacts on reefs. For example, ocean

acidification is expected to inhibit calcification in corals and promote the direct dissolution of carbonate surfaces. Both effects are likely to promote degradation of reef platform surfaces. However, weaker skeletons may increase the rate of sediment production on reefs, which may benefit construction of sedimentary landforms. Furthermore, increases in SST have been shown to increase the calcification rate in corals, therefore, offsetting the effect of acidification.

16.6.2 Future geomorphic responses of coral reef platforms

Geological analogues provide insights into the future geomorphic response of reef platforms to sea-level rise. Reconstructions of the Holocene marine transgression show that the rise in sea level of 125 m was not smooth but may have been characterized by periods of fast sea-level rise separated by periods of relatively stability. In the Caribbean, three periods of increased sea-level rise have been identified. However, evidence for two of these periods is not unanimously accepted in the Pacific and Indian Oceans and is the subject of considerable debate (Blanchon, 2011). The best defined and most widely established episode of accelerated sea-level rise is Meltwater Pulse 1A, in which sea level rose approximately 20 m over a 500-year period (14,200 to 13,800 years BP). During this period, the rate of sea-level rise peaked at 43 mm/yr (Stanford et al., 2006). This episode has been linked to the drowning of some reefs. For example, in Hawaii a reef surface currently at −150 m below sea level has been shown to have shut down in this period. In the Caribbean, reefs at depths of 75–90 m, and 40–50 m around Barbados, provide further evidence of reef drowning in response to other periods of increased sea-level change. However, despite some examples of reef drowning in response to rapid rates of sea-level change, reef termination was not widespread and many reefs coped with these rapid rates of sea-level rise.

In a modern context, these historic rates of sea-level change (40–50 mm/yr) are five times larger than current IPCC projections (IPCC, 2007), and still twice the rate of recent higher projections that account for ice melt (Vermeer and Rahmstorf, 2009). In addition, past reef growth rates range from 4 to 30 mm/yr, indicating that reef platforms have the capacity to keep pace with current projections of sea-level rise. This evidence would suggest that coral reefs may not drown in the future. Such a conclusion is contrary to the popular perception that reefs will drown, which is promoted at major climate change conferences and by governments of small-island states. Consequently, the future trajectory of reef growth is the subject of considerable scientific debate.

Despite high-profile assertions that reef drowning will be widespread in the future, synthesis of scientific understanding of reef development suggests the future reef growth response is likely to be highly variable and dependent on several site-specific factors, including the environmental start-up conditions affecting coral communities, the current reef-building capacity of reefs, and the level of anthropogenic impact on reefs. There are some anthropogenically stressed reefs that are severely compromised in terms of the ecological processes and integrity of the geomorphic system (see Fig. 16.12a). Vertical reef growth is unlikely in these settings and will result in submergence of reef platforms as sea level rises. At less-impacted sites, or where the $CaCO_3$ budget is unaltered, reefs are likely to exhibit catch-up or keep-up growth responses. Catch-up reef growth is likely in situations where currently emergent or intertidal reef surfaces are initially submerged by rising sea level. Such reefs will require time to recolonize reef surfaces, and increase productivity, in order to reactivate vertical growth. The time lag for reactivation of reef growth will determine the degree of submergence and the temporal window before reefs catch up with sea level.

The effects of ocean temperature and acidification changes are expected to produce a range of second-order geomorphic responses on coral reefs, which include reduced growth rates, shifts in carbonate budget status, an increase in rocky and rubble surfaces, and reduced structural integrity of reefs (loss of three-dimensional structure).

Collectively, these scenarios indicate that the geomorphic response of reefs will be geographically variable and will be contingent upon the antecedent geomorphic condition of reefs and the environmental start-up conditions that affect individual reefs (Kench et al., 2009a).

16.6.3 Future geomorphic responses of reef islands

Reef islands are expected to undergo morphological adjustment in response to variations in climate, sea-level rise, changes in the sediment supply, and anthropogenic factors. In particular, sea-level rise and reef submergence are expected to raise water depths over reef platforms, which will increase the energy regime operating on reef surfaces (based on the H/h ratio, section 16.4.4). Therefore, reef islands are expected to enter a period of enhanced sediment transport and morphological change.

Kench and Cowell (2002a) adapted the Shoreface Translation Model for application to reef islands to take explicit account of non-erodable reef substrates and adopted perched beach principles. Model simulations undertaken on islands in Kiribati and the Maldives indicated the following:

• Islands will undergo a range of morphological responses, which include erosion and island narrowing, island migration on reef platforms, and vertical island building through overwash sedimentation (Fig. 16.14a, b).
• The magnitude and mode of island response will be highly variable and will be dependent on initial morphology (elevation, sediment volume) and accommodation space. For instance, higher-elevation islands with larger sediment volume are more resilient to change than lower-elevation islands with limited sediment volume.
• Overwash sedimentation of island ridges provides a mechanism for vertical building of island surfaces to keep pace with sea-level rise. Field evidence for overwash sedimentation as a process of island geomorphic adjustment has been demonstrated in a number of studies of storm and tsunami impacts on reef islands (Bayliss-Smith, 1988; Kench et al., 2006; Fig. 16.14c–f). For example, as a consequence of wave overwash during the Indian Ocean tsunami in 2004 and a surge event in 2007, some island ridges on Maldivian reef islands increased in elevation by 0.5 m (Fig. 16.14f).
• Overwash of entire island surfaces on small islands and inlet by-passing promotes migration of islands on reef platforms, while conserving or building the sediment volume.

Application of the Shoreface Translation Model to explore the morphological sensitivity of reef landforms to changes in sediment supply further indicated that shifts in the sediment budget may be of greater importance than sea-level change in controlling future landform adjustment (Kench and Cowell, 2002b).

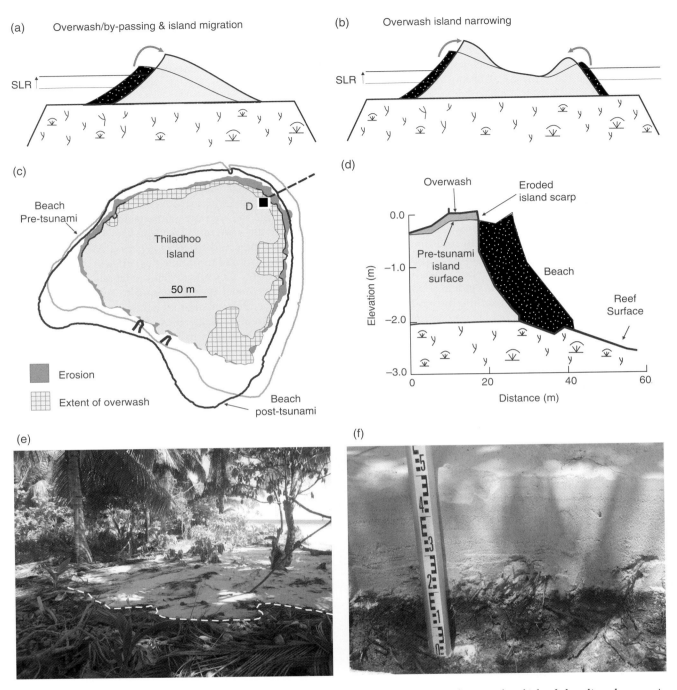

Fig. 16.14 Physical response of reef islands to sea-level rise. (a, b) Summary of model simulations of reef island shoreline change using the modified Shoreface Translation model (Source: Adapted from Kench and Cowell 2002a). SLR, sea-level rise. (c, d) Physical response of Thiladhoo reef island, Maldives, to the Indian Ocean tsunami, showing both shoreline erosion and overwash sedimentation (Source: Adapted from Kench et al. 2006). (e) Example of the extent of island overwash following the 2007 surge event in Gaaf Dhaal atoll, Maldives. (f) Excavated overwash deposit showing recently deposited sand (0.35 m thick) overlying the island soil (dark horizon). (Source: Photographs by Paul Kench.)

CASE STUDY BOX 16.3   Reef island response to sea-level rise

Low-lying atoll islands are widely perceived to erode in response to measured and future sea-level rise. Using historical aerial photography and satellite images, Webb and Kench (2010) undertook the first quantitative analysis of physical changes in 27 atoll islands in four atolls in Tuvalu, Kiribati and the Federated States of Micronesia in the central Pacific over a 19–61-year period of analysis.

This period of analysis corresponds with instrumental records that show a rate of sea-level rise of 2.0 mm/yr in the Pacific.

Results show that 86% of islands had either not changed (43% of islands) or had increased in area (43% of islands) over the timeframe of analysis. The largest decadal rates of increase in island area (~5% of islands) range

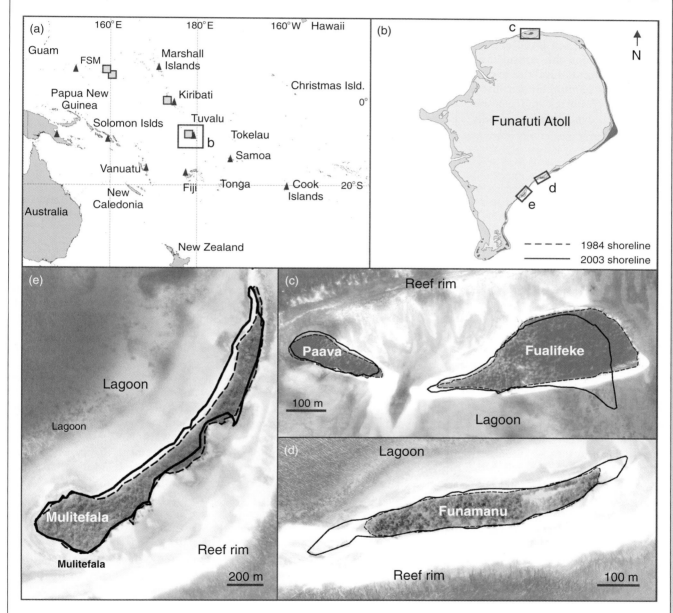

Fig. 16.15 Summary of multi-decadal changes in reef islands in Tuvalu, central Pacific, 1984–2003. (a) Location diagram of the southwest Pacific Ocean showing network of sea-level gauges (triangles) and atolls examined in Webb and Kench (2010); (b) Funafuti Atoll, Tuvalu; and (c-e) examples of changes in island planform characteristics. Dashed lines show 1984 shorelines, solid lines show 2003 shorelines. (Source: Adapted from Webb and Kench 2010. Reproduced with permission of Elsevier.)

between 0.1 and 5.6 hectares. Only 14% of study islands exhibited a net reduction in island area. Despite small net changes in area, islands exhibited larger gross changes (Fig. 16.15). This was expressed as changes in the planform configuration and position of islands on reef platforms. Modes of island change included ocean shoreline displacement toward the lagoon, lagoon shoreline progradation, and extension of the ends of elongate islands. Collectively, these adjustments represent net lagoonward migration of islands in 65% of cases. Results contradict existing paradigms that reef islands will disappear under increased sea level, and have several significant implications for the consideration of island stability under ongoing sea-level rise in the central Pacific. Firstly, islands are geomorphologically persistent features on atoll reef platforms and can increase in island area despite sea-level change. Secondly, islands are dynamic

landforms that undergo a range of physical adjustments in response to changing boundary conditions, of which sea level is just one factor. Thirdly, erosion of island shorelines must be reconsidered in the context of physical adjustments of the entire island shoreline, as erosion may be balanced by progradation on other sectors of shorelines. Results indicate that the style and magnitude of geomorphic change will vary between islands. The findings of the study are consistent with model simulations by Kench and Cowell (2002a) and physical process studies that document mechanisms of reef island morphological change (section 16.4.4). While results indicate islands will remain on reef surfaces, the physical changes observed pose significant management challenges for island communities. Consequently, there is a need to better understand the modes, magnitudes and rates of expected island shoreline change.

The adjustments expected in reef island morphology are readily observed under the current process regime and provide useful analogues for the mode and magnitude of change that may occur in the future (section 16.4.5). However, the pace of change in islands can be expected to increase. Most recently, Webb and Kench (2010) presented the first evidence of morphological change in reef islands in response to sea-level rise in the central Pacific, which are consistent with model projections (Box 16.3).

16.7 Summary

Coral reefs are unique coastal systems that are formed due to interaction between ecological and physical processes. Reef systems are among the most valuable systems on Earth in terms of biodiversity and the ecological and physical goods and services they provide for coastal communities. Coral reefs and reef-associated landforms are considered to be under increasing ecological and physical threat from anthropogenic stress and global climatic change. However, assessments of the ecological condition of reefs alone are not reliable indicators of the geomorphic response of coral reef landforms to future change. Effective management of reefs relies on understanding and maintaining the integrity of ecological and geomorphic processes, which will allow optimal conditions for future geomorphic adjustment. This chapter has examined reefs as geomorphic systems, and explored the controls on evolution and ongoing dynamics of reef landforms at a range of temporal scales.

- Coral reefs possess a variety of geomorphic units, and can be divided into those associated with coral reef structure and reef sedimentary landforms.
- The structure of coral reef platforms represents many thousands, if not millions, of years of deposition of $CaCO_3$ by corals and other reef-dwelling organisms. The distribution of reefs is largely controlled by environmental controls on reef-building organisms, which include water temperature, light, turbidity, salinity, nutrient levels and wave energy. These constraints limit reef-building communities to the shallow waters of tropical and subtropical latitudes.
- Recognizing that reef-building communities are constrained to shallow water depths (<30 m), Darwin proposed a genetic sequence of reef types from fringing reefs to barrier reefs and atolls, controlled by the subsidence of volcanic basement and upward growth of reefs. This model, combined with plate tectonic theory, adequately explains the structural development and distribution of mid-ocean reef systems at millennial timescales.
- Overlying this geophysical template, the morphology of modern reefs reflects sea-level oscillations and reef growth at shorter timescales. Throughout the Quaternary, reefs have experienced alternate periods of exposure, which allowed solutional denudation of reef surfaces, and inundation, allowing the reef to resume vertical growth. Consequently, reef morphology is strongly influenced by the antecedent karst morphology.
- Coral reefs exhibit a range of growth modes in response to sea-level change, which include keep-up, catch-up, give-up and backstepping modes in response to sea-level rise, lateral progradation once reefs attain their vertical

growth limit, and emergence resulting from sea-level fall. The vertical growth rate of coral reefs in the Holocene ranged from 4 to 30 mm/yr.

- Reef islands are a conspicuous sedimentary landform on coral reefs. Reef islands generally formed in the mid- to late Holocene and are low-lying accumulations of sand and gravel, deposited by wave and current processes on reef platforms. Island evolution is controlled by reef platform elevation, accommodation space, process regime and sediment supply.
- Reef islands are dynamic landforms that are in continual readjustment to changes in waves, currents and sediment supply. Such morphological change spans event-to-centennial timescales. The range of morphological responses of islands includes erosion and accretion, entire island migration on reef platforms, and overwash sedimentation on island surfaces that enables vertical island building.
- Consideration of the geomorphic system suggests coral reefs and reef landforms will persist as geomorphic structures in the near future. Contemporary morphodynamics provide useful analogues of the style and magnitudes of future change in reef systems. However, reefs will enter a phase of more rapid morphological adjustment over the coming century, in response to both climatic and anthropogenic forcing. The expected geomorphic response of reefs and islands will vary, depending on the environmental start-up conditions confronting reefs and whether geomorphic processes have been compromised.

Key publications

Hopley, D., Smithers, S.G. and Parnell, K.E., 2007. *The Geomorphology of the Great Barrier Reef: Development, Diversity and Change*. Cambridge University Press, Cambridge, UK.
[*Comprehensive examination of all aspects of the geomorphology of the Great Barrier Reef*]

Kench, P.S., Perry, C.T. and Spencer, T., 2009. Coral reefs. In: O. Slaymaker, T. Spencer and C. Embleton-Haman (Eds), *Geomorphology and Global Environmental Change*. Cambridge University Press, Cambridge, UK, pp. 180–213.
[*Comprehensive and up-to-date synthesis of the geomorphic implications of climate change on coral reefs*]

McLean, R.F. and Woodroffe, C.D., 1994. Coral atolls. In: R.W.G. Carter and C.D. Woodroffe (Eds), *Coastal Evolution*. Cambridge University Press, Cambridge, UK, pp. 267–302.
[*Review of coral atoll development*]

Montaggioni, L.F., 2005. History of Indo-Pacific coral reef systems since the last glaciation: development patterns and controlling factors. *Earth Science Reviews*, **71**, 1–75.
[*Recent and comprehensive review of the development of reefs in the Indo-Pacific*]

References

Aston, J., 1999. Experiences of coastal management in the Pacific Islands. *Ocean & Coastal Management*, **42**, 483–501.

Bayliss-Smith, T.P., 1988. The role of hurricanes in the development of reef islands, Ontong Java Atoll, Solomon Islands. *Geographical Journal*, **154**, 377–391.

Best, B. and Bornbusch, A., 2005. *Global Trade and Consumer Choices: Coral Reefs in Crisis*. American Association for the Advancement of Science, Washington, DC, USA.

Blanchon, P. 2011. Back-stepping. In: D. Hopley (Ed.), *Encyclopedia of Modern Coral Reefs: Structure, Form and Process*. Springer, Dordrecht, the Netherlands, pp. 77–84.

Brander, R.W., Kench, P.S. and Hart, D.E., 2004. Spatial and temporal variations in wave characteristics across a reef platform, Warraber Island, Torres Strait, Australia. *Marine Geology*, **207**, 169–184.

Brown, B.E. and Dunne, R.P., 1988. The impact of coral mining on coral reefs in the Maldives. *Environmental Conservation*, **15**, 159–165.

Burke, L., Selig, E. and Spalding, M., 2002. *Reefs at Risk in Southeast Asia*. World Resources Institute, Washington, DC, USA.

Chappell, J., 1980. Coral morphology, diversity and reef growth. *Nature*, **286**, 249–252.

Cowell, P.J. and Thom, B.G., 1994. Morphodynamics of coastal evolution. In: R.W.G. Carter and C.D. Woodroffe (Eds), *Coastal Evolution*. Cambridge University Press, Cambridge, UK, pp. 33–86.

Darwin, C., 1842. *The Structure and Distribution of Coral Reefs*. Smith, Elder & Co., London, UK.

Flood, P.G., 1986. Sensitivity of coral cays to climatic variations, southern Great Barrier Reef, Australia. *Coral Reefs*, **5**, 13–18.

Gourlay, M.R., 1988. Coral cays: products of wave action and geological processes in a biogenic environment. *Proceedings of the Sixth International Coral Reef Symposium, Townsville*, **2**, 491–496.

Gourlay, M.R. and Colleter, G., 2005. Wave-generated flow on coral reefs – an analysis for two-dimensional horizontal reeftops with steep faces. *Coastal Engineering*, **52**, 353–387.

Hardy, T.A. and Young, I.R., 1996. Field study of wave attenuation on an offshore coral reef. *Journal of Geophysical Research*, **101**(C6), 14311–14326.

Hayne, M. and Chappell, J., 2001. Cyclone frequency during the last 5000 years at Curacao Island, north Queensland, Australia. *Palaeogeography, Palaeoclimatology, Palaeoecology*, **168**, 207–219.

Hopley, D., 1984. The Holocene 'high energy' window on the central Great Barrier Reef. In: B.G. Thom (Ed.), *Coastal Geomorphology in Australia*. Academic Press, Sydney, Australia, pp. 135–150.

Hopley, D., Smithers, S.G. and Parnell, K.E., 2007. *The Geomorphology of the Great Barrier Reef: Development, Diversity and Change*. Cambridge University Press, Cambridge, UK.

Hubbard, D.K., Burke, R.B. and Gill, I.P., 1998. Where's the reef: the role of framework in the Holocene. *Carbonates and Evaporites*, **13**, 3–9.

Hubbard, D.K., Gill, I.P. and Burke, R.B., 2001. The role of framework in modern reefs and its application to ancient systems. In: G.D. Stanley Jr. (Ed.), *The History and Sedimentology of*

Ancient reef Systems. Kluwer Academic/Plenum, New York, USA, pp. 351–386.

Hughes, T.P., Baird, A.H. and Bellwood, D.R. et al., 2003. Climate change, human impacts, and the resilience of coral reefs. *Science*, **301**, 929–933.

IPCC, 2007. Summary for policymakers. In: D. Solomon, D. Qin, M. Manning et al. (Eds), *Climate Change 2007: The Physical Science Basis.* Contribution of Working Group I to the Fourth Assessment Report of the Intergovernmental Panel on Climate Change. Cambridge University Press, Cambridge, UK, pp. 2–18.

Kench, P.S. and Brander, R.W., 2006a. Morphological sensitivity of reef islands to seasonal climate oscillations: South Maalhosmadulu atoll, Maldives. *Journal of Geophysical Research*, **111**, F01001, doi: 10.1029/2005JF000323.

Kench, P.S. and Brander, R.W., 2006b. Wave processes on coral reef flats: implications for reef geomorphology using Australian case studies. *Journal of Coastal Research*, **2**, 209–223.

Kench, P.S. and Cowell, P.J., 2002a. The morphological response of atoll islands to sea-level rise. Part 2: application of the modified shoreface translation model (STM). *Journal of Coastal Research*, **SI34**, 645–656.

Kench, P.S. and Cowell, P.J., 2002b. Variations in sediment production and implications for atoll island stability under rising sea level. *Proceedings of the 9th International Coral Reef Symposium*, Bali, 23–27 October 2000, **2**, 1181–1186.

Kench, P.S. and McLean, R.F., 2004. Hydrodynamics and sediment flux of hoa in an Indian Ocean atoll. *Earth Surface Processes and Landforms*, **29**, 933–953.

Kench, P.S., Parnell, K.E. and Brander, R.W., 2003. A process based assessment of engineered structures on reef islands in the Maldives. *Coasts & Ports Australasian Conference 2003*, paper 75, 10 pp.

Kench, P.S., McLean, R.F. and Nichol, S.L., 2005. New model of reef-island evolution: Maldives, Indian Ocean. *Geology*, **33**, 145–148.

Kench, P.S., McLean, R.F., Brander, R.W. et al., 2006. Geological effects of tsunami on mid-ocean atoll islands: The Maldives before and after the Sumatran tsunami. *Geology*, **34**, 177–180.

Kench, P.S., Perry, C.T. and Spencer, T., 2009a. Coral reefs. In: O. Slaymaker, T. Spencer and C. Embleton (Eds), *Landscape Change in the 21st Century.* Cambridge University Press, Cambridge, UK, pp. 180–213.

Kench, P.S., Smithers, S.L., McLean, R.F. and Nichol, S.L., 2009b. Holocene reef growth in the Maldives: evidence of a mid-Holocene sea level highstand in the central Indian Ocean. *Geology*, **37**, 455–458.

Kench, P.S., Parnell, K.E. and Brander, R.W., 2009c. Monsoonally influenced circulation around coral reef islands and seasonal dynamics of reef island shorelines. *Marine Geology*, **266**, 91–108.

Kinsey, D.W., 1985. Metabolism, calcification and carbon production: 1. Systems level studies. *Proceedings 5th International Coral Reef Congress*, Tahiti, 27 May–1 June 1985, pp. 505–526.

Kinzie III, R.A. and Buddemeier, R.W., 1996. Reefs happen. *Global Change Biology*, **2**, 479–494.

Kleypas, J., McManus, J. and Menez, L., 1999. Environmental limits to coral reef development: where do we draw the line? *American Zoologist*, **39**, 146–159.

Ladd, H.S., Ingerson, E., Townsend, R.C., Russell, M. and Stephenson, H.K., 1953. Drilling on Eniwetok Atoll, Marshall Islands. *US Geological Survey Professional Paper*, **260-Y**, 863–903.

Lough, J.M., 2011. Climate change and coral reefs. In: D. Hopley (Ed.), *Encyclopedia of Modern Coral Reefs: Structure, Form and Process.* Springer, Dordrecht, the Netherlands, pp. 198–210.

Maragos, J.E., 1993. Impact of coastal construction on coral reefs in the US-affiliated Pacific Islands. *Coastal Management*, **21**, 235–269.

Maragos, J.E., Baines, G.B.K. and Beveridge, P.J., 1973. Tropical cyclone creates a new land formation on Funafuti atoll. *Science*, **181**, 1161–1164.

McLean, R.F. and Hosking, P.L., 1991. Geomorphology of reef islands and atoll motu in Tuvalu. *South Pacific Journal of Natural Science*, **11**, 167–189.

McLean, R.F. and Woodroffe, C.D., 1994. Coral atolls. In: R.W.G. Carter and C.D. Woodroffe (Eds), *Coastal Evoluiton: Late Quaternary Shoreline Morphodynamics.* Cambridge University Press, Cambridge, UK, pp. 267–302.

Montaggioni, L.F., 2005. History of Indo-Pacific coral reef systems since the last glaciation: development patterns and controlling factors. *Earth Science Reviews*, **71**, 1–75.

Neumann, A.C. and Macintyre, I., 1985. Reef response to sea level rise: keep-up, catch-up or give-up. *Proceedings 5th International Coral Reef Congress*, Tahiti, 27 May–1 June 1985, pp. 105–110.

Perry, C.T., Spencer, T. and Kench, P.S., 2008. Carbonate budgets and reef production states: a geomorphic perspective on the ecological phase-shift concept. *Coral Reefs*, **27**, 853–866, doi: 10.1007/s00338-008-0418-z.

Roberts, H.H. and Suhayda, M., 1983. Wave current interactions on a shallow reef (Nicaragua). *Coral Reefs*, **1**, 209–260.

Roberts, H.H., Wilson, P.A. and Lugo-Fernandez, A., 1992. Biologic and geologic responses to physical processes: examples from modern reef systems of the Caribbean–Atlantic region. *Continental Shelf Research*, **12**, 809–834.

Scoffin, T.P., 1992. Taphonomy of coral reefs: a review. *Coral Reefs*, **11**, 57–77.

Scott, G.A.J. and Rotondo, G.M., 1983. A model to explain the differences between Pacific plate island atoll types. *Coral Reefs*, **1**, 139–150.

Smithers, S.G., Harvey, N., Hopley, D. and Woodroffe, C.D., 2007. Vulnerability of geomorphological features in the Great Barrier Reef to climate change. In: J.E. Johnson and P.A. Marshall (Eds), *Climate Change and the Great Barrier Reef.* Great Barrier Reef Marine Park Authority and Australian Greenhouse Office, Townsville, Australia, pp. 667–716.

Stanford, J.D., Eelcon, J.R., Hunter, S.E. et al., 2006. Timing of meltwater pulse 1a and climate responses to meltwater injections. *Paleoceanography*, **21**, PA4103, doi: 10.1029/2006PA001340.

Stoddart, D.R., 1963. Effects of Hurricane Hattie on the British Honduras reefs and cays, October 30–31, 1961. *Atoll Research Bulletin*, **95**, 1–142.

Stoddart, D.R. and Steers, J.A., 1977. The nature and origin of coral reef islands. In: O.A. Jones and R. Endean (Eds), *Biology and Geology of Coral Reefs. Vol. IV: Geology 2.* Academic Press, New York, USA, pp. 59–105.

Toscano, M.A. and Macintyre, I.G., 2003. Corrected western Atlantic sea-level curve for the last 11,000 years based on calibrated ^{14}C dates from *Acropora palmata* and mangrove intertidal peat. *Coral Reefs*, **22**, 257–270.

Webb, A.P. and Kench, P.S., 2010. The dynamic response of reef islands to sea-level rise: evidence from multi-decadal analysis of island change in the Central Pacific. *Global and Planetary Change*, **72**, 234–246.

Wilkinson, C., 2004. *Status of Coral Reefs of the World: 2004.* Australian Institute of Marine Science, Townsville, Australia.

Woodroffe, C.D., 2002. *Coasts: Form, Process and Evolution.* Cambridge University Press, Cambridge, UK.

Woodroffe, C.D. and McLean, R.F., 1990. Microatolls and recent sea level change on coral atolls. *Nature*, **344**, 531–534.

Woodroffe, C.D. and Morrison, R.J., 2001. Reef-island accretion and soil development, Makin Island, Kiribati, central Pacific. *Catena*, **44**, 245–261.

Woodroffe, C.D., McLean, R.F., Smithers, S.G. and Lawson, E.M., 1999. Atoll reef-island formation and response to sea-level change: West Island, Cocos (Keeling) Islands. *Marine Geology*, **160**, 85–104.

Woodroffe, C.D., Samosorn, B., Hua, Q. and Hart, D.E., 2007. Incremental accretion of a sandy reef island over the past 3000 years indicated by component-specific radiocarbon dating. *Geophysical Research Letters*, **34**, L03602, doi: 10.1029/2006GL028875.

Vermeer, M. and Rahmstorf, S., 2009. Global sea level linked to global temperatures. *Proceedings of the National Academy of Sciences*, **106**, 21527–21532, doi: 10.1073pnas.0907765106.

17 Coping with Coastal Change

ROBERT J. NICHOLLS[1], MARCEL J.F. STIVE[2]
AND RICHARD S.J. TOL[3]

[1]*Faculty of Engineering and the Environment and Tyndall for Climate Change Research, University of Southampton, Southampton, UK*
[2]*Faculty of Civil Engineering and Geosciences, Delft University of Technology, Delft, The Netherlands*
[3]*School of Economics, University of Sussex, Brighton, UK*

17.1 Introduction

As Chapter 5 shows, coasts are 'risky places', exposed to multiple meteorological and geophysical hazards, including storms and tsunamis, and they represent regions of concentrated population and assets, including many major cities (Kron, 2013). Globally, it is estimated that about 600 million people live within 10 m of sea level (McGranahan et al., 2007). As many as 20 million of these people live below normal high-tide levels, while over 200 million people are exposed to flooding during temporary extreme sea-level events produced by storms (Nicholls, 2010). These threatened low-lying areas already depend on flood risk management strategies of some type, be it natural and/or artificial flood defences, drainage systems, and/or construction methods. Recent storms such as Hurricane Katrina in 2005, Cyclone Nagris in 2008, Typhoon Yasi in 2011, and Hurricane Sandy in 2012 remind us of what can happen in low-lying coastal areas if those management systems fail or are exceeded (Fig. 17.1). Long-term human-induced change, such as failing sediment supplies and artificial subsidence, and processes driven by climate change, such as global sea-level rise, are changing coasts and increasing the likelihood of disaster. Coastal populations and associated assets are also increasing rapidly, reflecting both population and economic growth, and a widespread migration of people to coastal areas. Hence, while coasts are only a small part of the world's surface, they comprise some of the most densely populated and economically active land areas on Earth. Only a major improvement of our ability to cope with these adverse trends can avoid an increase in coastal disasters (see Box 17.1).

This chapter focusses on how to cope with coastal change and its implications. There are two major types of response:

(1) **Mitigation** represents source control of drivers, such as greenhouse gas emissions and groundwater withdrawal (leading to artificial subsidence).

(2) **Adaptation** refers to behavioural changes that range from individual actions through to collective coastal management policy, such as upgraded defence systems, warning systems and land management approaches.

Thus, mitigation addresses the causes of coastal change, whereas adaptation deals with the consequences.

This chapter is structured as follows. First, the different drivers of coastal change are briefly reviewed, drawing on

Coastal Environments and Global Change, First Edition. Edited by Gerd Masselink and Roland Gehrels.
© 2014 John Wiley & Sons, Ltd. Published 2014 by John Wiley & Sons, Ltd. Companion Website: www.wiley.com/go/masselink/coastal

Fig. 17.1 Oblique aerial photographs of Pelican Island and Fire Island, New York, USA, before (upper photo) and after (lower photo) the landfall of Hurricane Sandy. It shows that a new inlet was created. (Source: U.S. Geological Survey 2013.)

Chapters 2 to 6. Then the impacts of coastal change are briefly considered from a physical and a socio-economic perspective, including drawing on experience from locations of rapid change, such as deltas and subsiding cities, and considering direct and indirect impacts. This is followed by a discussion of the responses to the challenges of coastal change, with an emphasis on adaptation, including its implications for coastal zone management. This includes consideration of valuation methods and decision analysis approaches.

17.2 Drivers of coastal change and variability

17.2.1 Historical variability and change

Coasts have been responding to external forcings continually throughout human and geological history. Drivers of coastal change operate at multiple time and space scales, from short-term or episodic events, such as storms or tsunamis, to long-term changes, such as modified sediment budgets in deltas due to upstream dam construction, or even globally, such as global sea-level rise. However, morphological changes are continuous and cumulative, and subject to the magnitude-frequency characteristics of the drivers, both leading to forced and free behaviour. Thus, any systematic understanding of coastal change must be organized in relation to scale (see Fig. 1.7).

The coastal-tract concept (Cowell et al., 2003a) distinguishes a hierarchy of morphodynamic forms and processes to define coastal systems across the complete range of scales. At each scale, morphological system components may be identified that interact dynamically by sharing a common pool of sediment. Long-term prediction (decades

---

CONCEPTS BOX 17.1 Integrated assessment of coastal change

Coastal areas are subject to multiple drivers of change, which operate at multiple scales. Climate change and sea-level rise are often the drivers that first spring to mind, but even these can be split into a range of causes, including global sea-level rise, rising sea-surface temperature, falling ocean pH, and changing wave and storm characteristics and terrestrial run-off. These changes will also interact with other non-climate changes. Relative sea-level rise (RSLR) due to subsidence and changing, and usually falling, sediment supplies are important physical changes, while land-cover changes influence the natural value of the coast and the socio-economic exposure. Hence, the coast can be seen as an interacting coupled system, with a series of interacting and co-evolving sub-systems. At its simplest, we might consider the evolution

of a natural and a socio-economic sub-system as discussed by Klein and Nicholls (1999) (Fig. 17.2).

For any problem, it is important to identify the key drivers and then consider how to analyse them. This includes considering the feedbacks between the sub-systems, with adaptation and local mitigation acting as important feedbacks from the socio-economic system to the physical system. All societal responses need to be consistent with the wider societal and development objectives, and hence require implementation within an integrated coastal management philosophy. This type of analysis constitutes an integrated assessment (or systems analysis) approach, which considers the coastal system, including the natural and human components and their interaction.

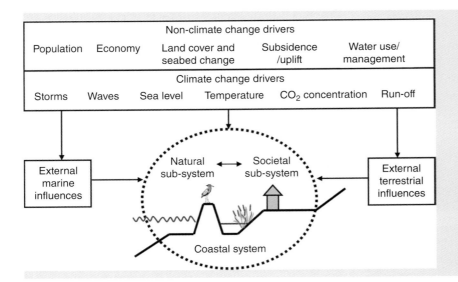

Fig. 17.2 The coastal system comprises interacting natural and human sub-systems, and external terrestrial and marine influences. A range of non-climate and climate drivers can act directly or indirectly on this system. (Source: Adapted from Nicholls et al. 2007. Reproduced with permission of IPCC.)

or longer) of shoreline changes involves movements of the upper shoreface due to its sediment-sharing interaction with the lower shoreface and backshore environments. Interaction with the backshore depends on whether this environment comprises a barrier beach/dune and lagoon, mainland beach/cliff or fluvial delta. On this basis, coastal responses of the morphological system components can be investigated and further resolved in terms of mean-trend and fluctuating components (Stive et al., 1991). These components, which relate to chronic and acute erosion hazards, respectively, have different implications for coastal impacts and the potential management responses.

Chronic erosion hazards are typically associated with imbalances in sediment budget that have ongoing effects and may become progressively worse over time. Spatial gradients in littoral sediment transport and relative (or local) sea-level rise are the most common cause of chronic coastal instability. Acute erosion hazards arise due to episodic erosion events, such as storms, with subsequent slower recovery of morphology to previous conditions. Such fluctuating changes do not cause coastal management problems unless coastal development has been undertaken imprudently in the active zone, or if mean-trend changes displace the region occupied by fluctuating morphology to coincide with pre-existing coastal development. Often, the event changes are seen as the problem under these circumstances because property damage usually occurs episodically, associated with storms. However, the actual problem under these circumstances is the underlying mean-trend change. The only circumstances under which the fluctuating changes become a new hazard in their own right is when the magnitude of maximum fluctuations increases due to changes in environmental conditions, such as an increase in storm intensity and

corresponding water level and wave energy. There is little evidence of such changes to date, but this possibility must be considered in the future.

Thus, the question of scale is fundamental to defining the nature and mode of coastal change. For example, the time series in Fig. 17.3 represents the progradation distance during the postglacial sea-level highstand (i.e. the last 6000 years) with superimposed shoreline fluctuations due to storm erosion and recovery cycles. The time series is synthetic, rather than entailing measured data, but the parameters are based on generalized data from southeast Australia (Cowell et al., 1995). The mean trend rate of progradation is 0.2 m/yr (or 200 m per millennium), consistent with rates evident in late Holocene barrier progradation, for example, in the Netherlands, southeast Australia and northwest Pacific, USA (Cowell et al., 2003b). The synthetic fluctuations involve two sets of sinusoids with periodicities representing a Pacific Decadal Oscillation ($T = 23$ years) and an El Niño Southern Oscillation, ENSO ($T = 7$ years). The top two graphs plot synthetic annual ('averaged') shoreline position and the bottom graph plots monthly 'averages'. For these parameters, the maximum theoretical shoreline fluctuation during the 6000-year highstand is 120 m. This value is comparable to maxima in the Narrabeen and Moruya beach-survey data sets from southeast Australia (Cowell et al., 1995). Since these data sets span only 30 years, the modelled maxima are probably on the low side. Centennial and millennial fluctuations associated with shoreline rotation and curvature changes due to shifts in dominance of directional components of wave climates in southeast Australia are even larger (Goodwin et al., 2006).

The time series in Fig. 17.3 show mean-trend and fluctuating components of large-scale coastal behaviour. The

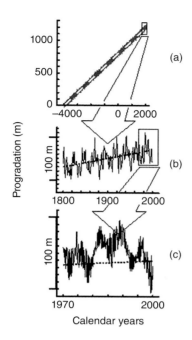

Fig. 17.3 Scales of coastal change and typical time series associated with the three lower-order (longer) morphological scales (see Fig. 1.7). The underlying progradation rate for the geological scale (a) can be resolved using radiometric dating of strand plains, and is of key significance to the engineering timescale (b), but may not be resolvable from measured fluctuations in recurrent surveys spanning several decades (c). (Source: Adapted from Thom and Cowell 2005. Reproduced with permission of Springer Science + Business Media.)

little more than 30 years. Although longer low-resolution data sets of up to 150 years derived from old maps, charts and related data exist in a few locations, such as the USA (e.g. Crowell et al., 1991), these fluctuations are poorly resolved in such records. (The annual Dutch beach lines data from 1843 to 1980 may be the best such measurements in the world (van Straaten, 1961).) For these reasons, mean-trend shoreline changes are more reliably estimated from radiometric dating of progradational sequences (Beets et al., 1992). For coasts on which progradation is absent, mean-trend estimates are less accessible and can only be derived from historic maps or other records, as discussed above, or photogrammetric measurement from quality survey aerial photographs (e.g. Sutherland, 2012).

On any particular scale, net morphological stability reduces to fundamental morphodynamic questions of marine and fluvial sediment availability, the accommodation space available for deposition, and their spatial organization. Relative sea-level rise (RSLR), for instance, increases accommodation space, creating an additional demand for sediments, which are redistributed accordingly. Changes in direction and intensity of wave climates, and magnitude-frequency characteristics of storms, also affect the spatial distribution of accommodation space. These changes would also influence morphological response rates through effects on net sediment fluxes.

Collectively, the balance between the effects of accommodation space, and hence sea-level rise, and sediment supply govern modes of coastal change to the extent responsible for the differentiation between coastal types (Valentin, 1952; Stive et al., 2009). The principal types include mainland beach/cliff and barrier coasts, where the latter in turn exist as a continuum of estuarine-deltaic morphologies. Position in the continuum depends largely on sediment supply relative to the accommodation space. On this basis, for example, deltaic lowlands in which estuaries have been filled by sediments during the Holocene sea-level highstand may revert to estuarine environments due to accelerated sea-level rise, depending on sediment supply. More generally, the available accommodation space also depends on the energy regime, including the relative influence of tides and waves, and the predominant size fraction of the sediment supply (Thorne and Swift, 1991). Changing wave climates are likely to play only a marginal role in transforming coasts in terms of where they lie in the typological continuum.

The gross kinematics of the coastal tract are constrained and steered by sediment-mass continuity. The rate of coastal advance or retreat is determined quantitatively by the balance between the change in sediment accommodation space, caused by sea-level movements, and sediment availability. If the lower shoreface is shallower than required for equilibrium (negative shoreface accommodation), then sand is transferred to the upper shoreface, so that the

trend line shown in each of the plots is for the average progradation rate (0.2 m/yr). Over the full 6000 years of the sea-level highstand, the mean-trend component is so large that the fluctuations become insignificant (Fig. 17.3a). For the 200-year and 30-year time series, the fluctuating component dominates the signal (Fig. 17.3b, c). For the 30-year record, the mean-trend behaviour could not be resolved statistically from the data (Fig. 17.3c). Indeed, for the 30-year data, a linear least-squares best fit (not plotted) gives an average progradation rate of 0.04 m/yr compared to the actual trend of 0.2 m/yr. Depending upon the span of data analysed, it is possible to obtain negative-trend estimates (i.e. shoreline retreat), even though the underlying geological trend is progradational.

Thus, as Cowell et al. (1995) observe, historical data sets generally do not permit resolution of the underlying tendencies in long-term shoreline change because the fluctuating component of large-scale coastal behaviour overwhelms the signal on the decadal, or engineering, timescale. They showed that, on the coast of southeast Australia, mean-trend shoreline change becomes comparable to the decadal fluctuations only after the passage of about 180 years. However, the longest high-resolution data set now spans shoreline tends to advance seaward. This tendency also occurs when relative sea level is falling (coastal emergence). Coastal retreat occurs when the lower shoreface is too deep for equilibrium (positive shoreface accommodation). This sediment-sharing between the upper and lower shoreface is an internal coupling that governs first-order coastal change. The upper shoreface and backbarrier (lagoon, estuary or mainland) are also coupled in first-order coastal change. Sediment accommodation space is generated in the back-barrier by sea-level rise (and reduced by sea-level fall), but the amount of space is also moderated by the influx of mud from coastal sources (e.g. by cliff erosion), or sand and mud from fluvial sources. Any remaining accommodation space is filled by sand transferred from the upper shoreface, causing a retreat of the latter; where this occurs, this will represent a chronic erosion problem (e.g. the US East Coast barriers, which are influenced by tidal inlets).

As discussed above, coastal advance or retreat is determined quantitatively by the balance between sediment availability and sediment accommodation space (sea-level rise). We will therefore discuss existing views on global and local changes in both drivers.

17.2.2   The prospect of future climate change

Climate change is expected to cause a profound series of changes, including rising global sea levels and changing storm, wave and run-off characteristics (see Box 17.1). While all aspects of coastal climate influence coastal change, here we focus on sea-level rise, as this is the most fundamental driver. Sea-level rise is mainly produced by thermal expansion of seawater as it warms, and the melting of land-based ice, comprising components from small glaciers, the Greenland ice sheet, and the West Antarctic ice sheet (Meehl et al., 2007; see also Chapter 2). A rising global sea level of 17 cm was observed through the 20th century, which is faster than the rise during the 19th century (Church et al., 2010). This observed rise is almost certain to continue and is very likely to accelerate through the 21st century. From 1990 to the last decade of the 21st century, a total rise in the range of 18–59 cm has been forecast by the Intergovernmental Panel on Climate Change (IPCC) Fourth Assessment Report (AR4) (Meehl et al., 2007). However, the quantitative AR4 scenarios do *not* provide an upper bound on sea-level rise during the 21st century due to uncertainties concerning the large ice sheets. For this reason, the IPCC estimates have been debated extensively in the literature since 2007, and many higher estimates of future rise have been published based on different types of evidence (see Nicholls et al., 2011). It is concluded that a global rise of sea level exceeding 1 m is a plausible scenario for the 21st century only if Greenland and/or Antarctica ice sheets are significant sources of sea-level rise. The probability of the high-end scenarios is unknown, but, even if relatively

unlikely, their large potential impacts make them highly significant in terms of coastal risk and policy. It is noteworthy that high-end scenarios of sea-level rise have been developed to support long-term flood planning for both London and the Netherlands (e.g. Katsman et al., 2011).

When analysing sea-level rise impacts and responses, it is fundamental that impacts are a product of relative (or local) sea-level rise, rather than global changes alone, as already noted when considering controls on coastal advance and retreat. Relative sea-level change takes into account the sum of global, regional and local components of sea-level change. The underlying drivers of these components are: (1) climate change, changing ocean dynamics, and changes in gravity as ice masses are lost; and (2) non-climate uplift/subsidence processes, such as tectonics, glacial isostatic adjustment (GIA), and natural and human-induced subsidence (Church et al., 2010). Hence, RSLR is only partly a response to climate change, and varies from place to place, as illustrated by the measurements in Fig. 17.4. Despite a global sea-level rise, relative sea level is presently falling, due to ongoing GIA-induced rebound, in some high-latitude locations that were sites of large (kilometre-thick) glaciers in the last glacial maximum (18,000 years ago), such as the northern Baltic and Hudson Bay (see Helsinki; Fig. 17.4). However, most of the world's coasts are experiencing a RSLR. This rise increases from relatively stable coasts (e.g. Sydney; Fig. 17.4), through naturally subsiding coasts, such as New York City (Fig. 17.4), which is experiencing GIA-induced subsidence, to subsiding deltas (see Grand Isle in the Mississippi delta; Fig. 17.4). In subsiding cases, RSLR exceeds global rise. Most dramatically, subsidence in susceptible coastal areas comprising thick sequences of Holocene sediment, such as deltas (Chapter 13), can be enhanced by human activity, especially by drainage and withdrawal of groundwater/fluids. Dramatic RSLR has resulted in many coastal cities built on deltas as a consequence of this, as is illustrated in Fig. 17.4 for Bangkok. Over the 20th century, the parts of Tokyo and Osaka built on deltaic areas subsided up to 5 m and 3 m, respectively, a large part of Shanghai subsided up to 3 m, and Bangkok subsided up to 2 m (Fig. 17.5). As discussed later, this human-induced subsidence can be mitigated by stopping shallow sub-surface fluid withdrawals and managing water levels, but natural 'background' rates of subsidence will continue and RSLR will still exceed global trends in these areas. The four cities cited above have all seen a combination of such policies, combined with the provision of improved flood defence and pumped drainage systems to avoid submergence and/or frequent flooding. In contrast, other coastal cities such as Jakarta and Metro Manila are subsiding rapidly today and hence experiencing rapid growth in flood risk. The World Bank (2010a) concluded that subsidence in many Asian cities is as big an issue as climate-induced sea-level rise; however, little systematic

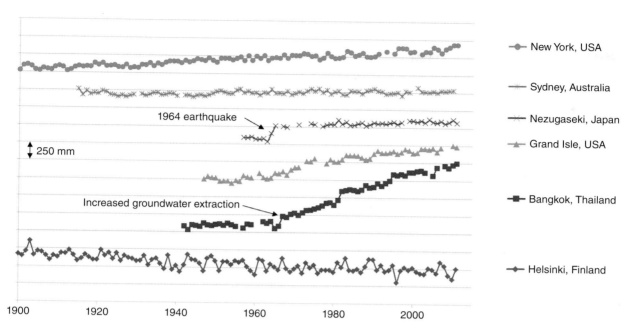

Fig. 17.4 Selected relative sea-level observations from 1900 to 2011, illustrating different trends (offset for display purposes). Helsinki shows a falling trend (–2.0 mm/yr), Sydney shows a gradual rise (0.9 mm/yr), New York is subsiding slowly (3.1 mm/yr), Grand Isle is on a subsiding delta (9.1 mm/yr), Bangkok includes the effects of human-induced subsidence (19.2 mm/yr from 1962 to 2011), and Nezugaseki shows an abrupt rise due to an earthquake in 1964. (Source: Holgate et al. 2013.)

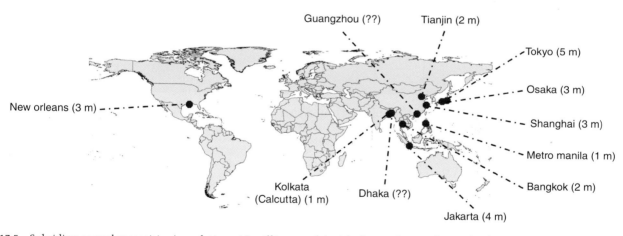

Fig. 17.5 Subsiding coastal megacities (population >10 million people) with the maximum observed subsidence during the 20th century (in metres). Dhaka and Guangzhou are subsiding, but data are limited. New Orleans is also shown, although it is not a megacity. (Source: Adapted from Nicholls 2010. Reproduced with permission of John Wiley & Sons.)

policy response is in place and future flooding problems are not being anticipated. Without a wider approach, including the sharing of relevant experience, the problems of enhanced subsidence are likely to be widely repeated in susceptible areas of expanding coastal cities over the 21st century because there are many candidates in deltaic settings, especially in South, Southeast or East Asia (Hanson et al., 2011). More widely, most populated deltaic areas are threatened by enhanced subsidence to varying degrees (see Chapter 13). Lastly, abrupt changes in relative sea level can occur at tectonically active sites due to earthquakes (see Nezugaseki; Fig. 17.4); for example, significant coastal subsidence of up to 1 m was observed in the Tōhoku earthquake (and subsequent tsunami) in Japan in 2011 (Fig. 17.6).

Fig. 17.6 A photograph of the Japanese port at Ishinomaki, Miyagi Prefecture, which is submerged at high tide due to subsidence during the 2011 earthquake (top two photos). These port areas are being raised to maintain their operational function (bottom two photos). (Source: Photographs by Dr Miguel Esteban, The University of Tokyo.)

Sea-level rise does not happen in isolation, and coasts are changing significantly due to the other coastal drivers, collectively addressed as sediment availability earlier in this chapter. Potential interactions across the impacts of sea-level rise are indicated in Table 17.1 and need to be considered when assessing sea-level rise impacts and responses. For instance, a coast with sufficient sediment availability may not erode given sea-level rise, and, vice versa, a coast with insufficient sediment availability may erode without sea-level rise (Valentin, 1952; Stive et al., 2009). Deltas with 'natural' sediment budgets and deltas with significantly reduced sediment budgets due to dam construction are potential examples of sediment-rich and sediment-starved coasts, respectively (Syvitski et al., 2009). This reinforces the earlier point that coastal change ideally requires an integrated assessment approach in order to analyse the full range of interacting drivers (see Box 17.1).

17.3   Coastal change and resulting impacts

Coastal change is much more than a simple translation of the shoreline. For instance, RSLR causes more effects than just simple submergence (the 'bath-tub' effect): the five main effects are summarized in Table 17.1. Flooding/submergence, ecosystem change and erosion have received significantly more attention than salinization and rising water tables. Along with rising sea level, there are changes to all the processes that operate around the coast. The immediate effect is submergence and increased flooding of coastal land, as well as saltwater intrusion into surface

Table 17.1 The main natural system effects of relative sea-level rise (RSLR) and example adaptation approaches for each. Other interacting factors that could offset or exacerbate these impacts due to other climate change and non-climate factor effects are also shown. Some interacting factors (e.g. sediment supply) appear twice as they can be influenced both by climate and non-climate factors. Adaptation approaches are coded: P, protection (hard or soft); A, accommodation; R, retreat. (Source: Adapted from Nicholls 2010. Reproduced with permission of John Wiley & Sons.)

Natural system effect		Possible interacting factors		Possible adaptation approaches
		Climate	Non–climate	
1 Inundation/ flooding	(a) Surge (flooding from the sea)	Wave/storm climate Erosion Sediment supply	Sediment supply Flood management Erosion Land reclamation	Dykes/surge barriers/closure dams [P – hard] Dune construction [P – soft] Building codes/flood-proof buildings [A] Land-use planning/hazard mapping/flood warnings [A/R]
	(b) Backwater effect (flooding from rivers)	Run-off	Catchment management and land use	
2 Wetland loss (and change)		Carbon dioxide fertilization Sediment supply Migration space	Sediment supply Migration space Land reclamation (i.e. direct destruction)	Nourishment/sediment management [P – soft] Land-use planning [A/R] Managed realignment/forbid hard defences [R]
3 Erosion (of 'soft' morphology)		Sediment supply Wave/storm climate	Sediment supply	Coast defences/seawalls/land claim [P – hard] Nourishment [P – soft] Building setbacks [R]
4 Saltwater Intrusion	(a) Surface waters	Run-off	Catchment management (over-extraction) Land use	Saltwater intrusion barriers [P] Change water abstraction [A/R]
	(b) Groundwater	Rainfall	Land use Aquifer use (over-pumping)	Freshwater injection [A] Change water abstraction [A/R]
5 Impeded drainage/higher water tables		Rainfall Run-off	Land use Aquifer use Catchment management	Drainage systems/polders [P – hard] Change land use [A] Land-use planning/hazard delineation [A/R]

waters. Longer-term effects also occur as the coast adjusts to the new environmental conditions, including wetland loss and change in response to higher water tables and increasing salinity, erosion of beaches and soft cliffs, and saltwater intrusion into groundwater (see Chapter 6). These lagged changes interact with the immediate effects of sea-level rise and generally exacerbate them. For instance, erosion of sedimentary features (e.g. salt marshes, mangroves, sand dunes and coral reefs) will tend to degrade or remove natural protection and hence increase the likelihood of coastal flooding. These effects have been discussed in more detail in Chapters 7 to 16.

A rise in mean sea level also raises extreme water levels. Changes in storm characteristics could also influence extreme water levels. For example, the widely debated potential for an increase in the intensity of tropical cyclones (Chapters 4 and 5) would generally raise extreme water levels in the areas influenced by tropical cyclones (see Fig. 5.2), including the US East and Gulf Coasts (Meehl et al., 2007). Extratropical storms may also intensify in some regions. Changes in mean sea level will also change the propagation of tides and surges, and this could have additional significant effects (positive or negative) on

extreme sea levels. Understanding these changes is an important research topic required to support impact and adaptation assessment.

Changes in natural systems as a result of sea-level rise have many important direct socio-economic impacts on a range of sectors, with these impacts being overwhelmingly negative (Table 17.2). For instance, flooding can damage coastal infrastructure, ports and industry, the built environment, and agricultural areas, and in the worst case lead to significant mortality, as shown recently in Cyclone Nargis (Myanmar) in 2008, and in the developed world in Hurricane Katrina (USA), Hurricane Sandy (USA) and Cyclone Xynthia (France) in 2010 (e.g. Kates et al., 2006; Fritz et al., 2009; Kolen et al., 2012). Erosion can lead to loss of beachfront/cliff-top buildings and related infrastructure, and have adverse consequences for sectors such as tourism and recreation. In addition to these direct impacts, there are also indirect impacts, but these are less well understood (Box 17.2). Indirect effects will have important economic consequences, however, and, via markets, these effects can be felt far from the coast. Thus, sea-level rise and coastal change in general have the potential to trigger a cascade of direct and indirect human impacts.

Table 17.2   Summary of sea-level rise impacts on socio-economic sectors in coastal zones. These impacts are overwhelmingly negative. (Source: Adapted from Nicholls 2010. Reproduced with permission of John Wiley & Sons.)

Coastal socio–economic sector	Sea-level rise natural system effect (see Table 17.1)				
	Inundation/flooding	Wetland loss	Erosion	Saltwater intrusion	Impeded drainage
Freshwater resources	X	x	–	X	X
Agriculture and forestry	X	x	–	X	X
Fisheries and aquaculture	X	X	x	X	–
Health	X	X	–	X	x
Recreation and tourism	X	X	X	–	–
Biodiversity	X	X	X	X	X
Settlements/infrastructure	X	X	X	X	X

X, strong; x, weak; –, negligible or not established.

METHODS BOX 17.2   Direct and indirect impacts of sea-level rise

The direct impacts of sea-level rise are obvious: increased flood risk, greater erosion, wetland change, saltwater intrusion, and, if implemented, the costs of coastal protection. These direct impacts trigger a range of indirect impacts, which are often less obvious but may still be significant (Nicholls and Kebede, 2012). Indirect impacts can be physical, biological and social. Groynes enhance beaches and hence protect a coast against wave action, but also disturb sediment transport, thus enhancing erosion or accretion elsewhere. Wetland loss would affect fish populations and migratory birds. Flooding and erosion can release polluted materials from contaminated ground and landfills, and can also impact on mental as well as physical health.

Land loss and frequent flooding may lead to migration (Tacoli, 2009), and migration has a number of knock-on effects, both positive and negative. Forced migration in particular has negative effects and ideally is to be avoided. There are also indirect economic effects. The value of land and the price of crops will increase as land is lost; this is to the benefit of land owners and farmers everywhere (not just in the coastal zone), but to the detriment of everyone else. Hence, large coastal countries tend to benefit relative to small coastal countries, and landlocked countries benefit even more. Furthermore, as the price of food rises, people will have less money to spend on other things. Protecting against impacts will benefit the building sector and civil engineers. However, investment in coastal protection will crowd-out investment elsewhere in the economy. This slows economic growth, as investment is diverted from productive to defensive capital.

17.4   Impacts of coastal change since 1900

Over the 20th century and early 21st century, coastal change has been widespread, with human agency being responsible for all the major direct and indirect drivers of change. At the same time, the number of people, as well as economic activity and assets per person, have grown rapidly, and disproportionally so in the coastal zone. The largest coastal disasters, based on absolute financial losses, have all occurred in the last 20 to 30 years, reflecting the rapid growth in exposed assets compared to earlier decades (Kron, 2013; see also Table 5.1).

In the popular mind, sea-level rise is often linked to observed coastal change. However, global mean sea-level rise since 1900 is only about 20 cm, and linking this quantitatively to impacts is difficult due to the multiple drivers of change discussed earlier. Reliable long-term data on rising sea levels are only available for a few locations, and growing coastal populations and infrastructure have continued to increase the exposure available to damage. Further flood defences have often been upgraded substantially through the 20th century, especially in those (wealthy) places where there are sea-level measurements. Most of this defence upgrade reflects the expanding populations and wealth in the coastal floodplain and changing attitudes to risk, and RSLR may not have even been considered in the design (e.g. Nicholls and de la Vega-Leinert, 2008). Hence, while global sea-level rise was a pervasive process, other processes obscure its link to impacts, except in some special cases. Most coastal change in the 20th century was therefore a response to multiple drivers of change.

There have certainly been impacts resulting from subsidence, with sites such as Venice being particularly well known (Fletcher and Spencer, 2005). In the Mississippi delta, 1565 km^2 of intertidal coastal marshes and adjacent lands were converted to open water between 1978 and 2000, with RSLR at rates of 5–10 mm/yr being an

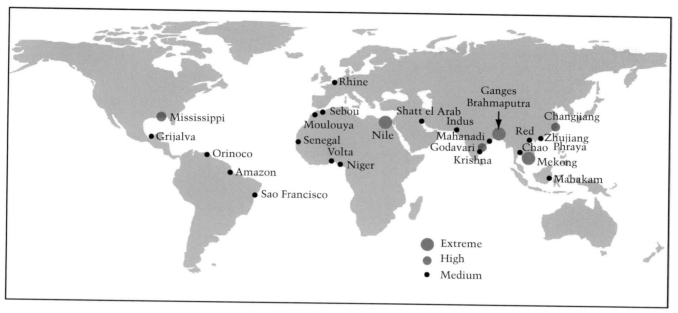

Fig. 17.7 Relative vulnerability of deltas to present estimates of relative sea-level rise to 2050, including deltaic subsidence. The metric used is displaced people, assuming no adaptation, and is defined as follows: extreme = >1 million people displaced; high = 50,000 to 1 million people displaced; medium = 5000 to 50,000 people displaced. (Source: Nicholls et al. 2007. Reproduced with permission of Cambridge University Press. Based on data from Ericson et al. 2006.)

important driver of these changes (Barras et al., 2003). There are significant actual and potential impacts of RSLR rise across all populated deltas (Fig. 17.7), and in and around subsiding coastal cities (e.g. Fig. 17.5), in terms of increased waterlogging, flooding and submergence, and the resulting need for management responses.

In terms of lessons for mitigation and adaptation, nearly all the major developed areas that were impacted by RSLR have been defended, and continue to grow economically and in population, even where the change in RSLR was up to several metres over several decades. New Orleans, however, may be an exception (Kates et al., 2006). Unusually, compared to other low-lying coastal cities, its population peaked in 1965, at more than 625,000, immediately before Hurricane Betsy flooded part of the city. Before Hurricane Katrina in 2005, its population was about 500,000; subsequently, the population has yet to recover to pre-Katrina levels, although $US15 billion have been invested to significantly upgrade defences (which were completed in 2011). Hence, the future of New Orleans will be instructive: will it prosper behind the new defences, will it return to the pre-Katrina rate of economic and demographic decline, or has Katrina accelerated its decline? It would be useful to know whether there are other coastal areas with similar declining population and economic trends and why? In less-developed areas, where the benefits of protection are smaller, coastal retreat due to subsidence has been allowed, such as south of Bangkok,

Thailand, in the Mississippi delta (see earlier) and around Galveston Bay in Texas. Furthermore, in many cases, mitigation of subsidence has been implemented via regulation of groundwater withdrawal, although this is by no means universal.

Hence, observations through the 20th century and the early 21st century reinforce the importance of understanding the impacts of coastal change in the context of multiple drivers of change. Human-induced subsidence is of particular interest, but this remains relatively unstudied in a systematic sense. Observations also emphasize the ability to protect against coastal hazards such as RSLR, especially for the most densely populated areas such as subsiding Asian megacities or around the southern North Sea, including London, the Netherlands and Hamburg.

17.5 Future impacts of coastal change

In the forthcoming decades, all the drivers that operated in the past will continue, with a pervasive trend of urbanization and more intense touristic development, especially in the developing world. These changes will be compounded by human-induced climate change and sea-level rise. Hence, the scale of change is likely to exceed our historic observations, and will require large investments to mitigate and adapt to these changes.

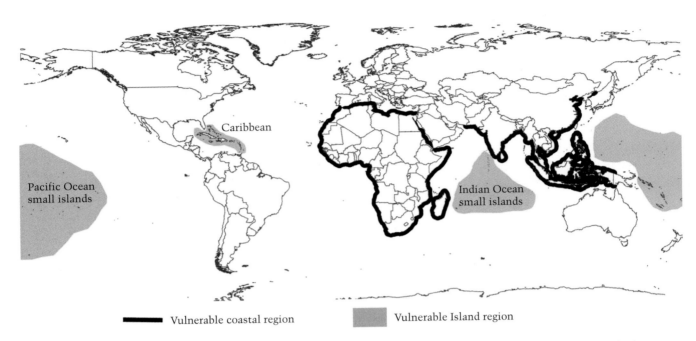

Fig. 17.8 Several regions are vulnerable to coastal flooding caused by future relative or climate-induced sea-level rise. At highest risk are coastal zones with dense populations, low elevations, appreciable rates of subsidence, and/or inadequate adaptive capacity. (Source: Nicholls 2010. Reproduced with permission of John Wiley & Sons.)

The future impacts of coastal change will depend on a range of factors:
- magnitude of forcings,
- coastal physiography,
- level and manner of coastal development, and
- success (or failure) of mitigation and/or adaptation measures.

Taking into consideration the above factors, assessments of the future impacts of sea-level rise have taken place on a range of scales, from local to global. They all suggest potentially large impacts, especially increases in inundation, flooding and storm damage. In absolute numbers, East, Southeast and East Asia, and Africa appear to be most threatened by sea-level rise (Fig. 17.8). Vietnam and Bangladesh appear especially threatened due to their large populations in low-lying deltaic plains (Nicholls et al., 2007). In Africa, Egypt (the Nile Delta) and Mozambique appear to be two potential hotspots for impacts due to sea-level rise. Hotspots also exist outside these regions, such as Guyana, Suriname and French Guiana in South America. There will be significant residual risk in other coastal areas of the world, such as around the southern North Sea, and major flood disasters are possible in many coastal regions. Small island regions in the Pacific, Indian Ocean and Caribbean stand out as being especially vulnerable to sea-level rise impacts (Mimura et al., 2007). The populations of low-lying island nations such as the Maldives or Tuvalu face the real prospect of saltwater intrusion, increased flooding, submergence and, ultimately, forced abandonment. Hence, deltas, small islands and coastal cities are important candidates for developing responses to coastal change, especially adaptation.

17.6 Responding to coastal change

The two potential responses to coastal change are mitigation and adaptation. The former addresses the causes of coastal change, whereas the latter deals with the consequences.

17.6.1 Mitigation

Depending on the driver concerned, mitigation can be local to global in scale. Climate mitigation is tackled on a global scale, and could slow the rise in sea level and reduce its impacts. However, given its strong inertia, mitigation has the effect of stabilizing the *rate* of sea-level rise, rather than stabilizing sea level itself (e.g. Meehl et al., 2012). Hence, sea-level rise will likely continue for centuries and therefore remain an ongoing challenge for the foreseeable future. In essence, the commitment to sea-level rise leads to a commitment to adapt to sea-level rise, with fundamental implications for long-term human use of the coastal zone (Nicholls et al., 2007). Given that the rate of sea-level rise

CASE STUDY BOX 17.3 Adaptation options in coastal areas

Adaptation can be classified in a variety of ways. Here we compare the Intergovernmental Panel on Climate Change (IPCC) adaptation typology with Shoreline Management Planning as practised in the UK. The IPCC typology was first proposed by IPCC CZMS (1990) and has been widely applied since (e.g. Klein et al., 2000). It defines in broad terms how the risk is managed, as defined in Fig. 17.9 and below:

• **(Planned) Retreat**: all natural system effects are allowed to occur, and human impacts are minimized by pulling back from the coast via land-use planning, development controls, etc.
• **Accommodation**: all natural system effects are allowed to occur, and human impacts are minimized by adjusting human use of the coastal zone via flood resilience measures, warning systems, insurance, etc.
• **Protection**: natural system effects are controlled by soft or hard engineering (e.g. nourished beaches and dunes or seawalls), reducing human impacts in the zone that would be impacted without protection.

Examples of each strategy are given in the context of sea-level rise in Table 17.1. Information measures, such as disaster preparedness, hazard mapping and flood warning/evacuation, are also important and in many ways are cross-cutting and complementary of the three approaches.

In some classifications of adaptation responses, the concept of 'attack' (or 'advance the line') as opposed to 'hold the line' has been suggested as a strategy against sea-level rise, which in effect translates into building seaward (or land claim). This approach has a long history in many coastal areas, such as northwest Europe and East Asia, and has been practised to some degree in most coastal cities due to space constraints. Land claim is an active strategy in many coastal countries, such as Singapore, Hong Kong, Dubai and even in the Maldives near the capital Malé, to expand land area as opposed to adapt to coastal change/hazards. However, sea-level rise is increasingly considered in the design of land claim. Another reason to advance seaward, rather than hold the line, is that building seaward does not require the expropriation of land and the demolition of buildings and infrastructure. Raising low-lying coastal areas before development or during redevelopment is a response to flooding and sea-level rise that may be applied more widely.

Shoreline Management Planning is a hierarchical approach, where a set of strategic goals are established at the highest level, based on the natural and socio-economic characteristics of the coast (Nicholls et al., 2013). Management units are normally defined by socio-economic characteristics (e.g. urban areas, agriculture, etc.) and one of four measures are selected: (1) advance the line (i.e. land claim); (2) hold the line (i.e. protect); (3) retreat the line (termed 'managed realignment', and representing a (planned) retreat); and (4) limited intervention (where monitoring is the main measure and a less-managed retreat is expected, often being applied to eroding cliffs). Collectively, (1) and (2) map onto protection, while (3) and (4) map onto retreat. This approach has been applied in England and Wales and is attracting interest elsewhere.

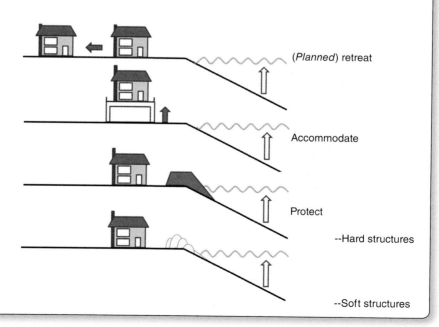

Fig. 17.9 Generic adaptation approaches for coping with coastal hazards (Source: Adapted from IPCC 1990).

controls some impacts, such as wetland loss or coral reef submergence, mitigation will reduce these impacts. In the case of flooding, the absolute rise is of more concern, and many impacts may be delayed rather than avoided due to the commitment to sea-level rise. Importantly, this gives more time to adapt, which is an important benefit of such mitigation, and includes lowering annual adaptation (and damage) costs. Hence, adaptation and mitigation are complimentary policies for climate change in coastal areas (Nicholls et al., 2007). The fundamental goal of mitigation in the context of coastal areas is to reduce the risk of passing irreversible thresholds concerning the breakdown of the two major ice sheets, and constrain the commitment to sea-level rise to a rate, and ultimate rise, that can be adapted to at a reasonable economic and social cost. Quantification of the most appropriate mixtures of mitigation and adaptation has been done in principle (Tol, 2007; de Bruin et al., 2009), but solid policy advice awaits improved process understanding and better data.

Mitigation of human-induced subsidence has also already been discussed, and needs to be considered in areas where humans are contributing to subsidence. This translates into measures to control/reduce groundwater extraction and manage water levels, which have been successfully implemented in a number of cities and delta areas to date, including in parts of Tokyo and all of Shanghai. In Bangkok, subsidence has been controlled in the central districts, but continues in the expanding suburbs at a significant rate of 10–30 mm/yr (Phien-Wej et al., 2006). Hence, the problem is evolving, but not yet solved – this situation is probably more widespread than appreciated. In some subsiding cities, such as Jakarta, comprehensive mitigation measures remain to be implemented, even though subsidence is recognized as an important process.

Human actions have also often reduced sediment supply to the coast, due to upstream dam construction and cliff protection. This is well illustrated in deltaic areas, as discussed in Chapter 13, and on open coasts, as discussed in Chapter 15. As recognition of these issues has increased, there has been growing interest in reversing these trends (a mitigation approach as defined here), such as removal of cliff defences in the UK and removal of small dams in California. However, the scope of reactivating natural sediment supplies may be limited, and the alternative of artificial shore nourishment and sediment management in general as an adaptation option is discussed later.

17.6.2 Adaptation

Adaptation to coastal change involves responding to both extreme events and long-term change. Two high-level classifications of coastal adaptation options are summarized in Box 17.3, namely the IPCC approach comprising retreat, accommodate and protect, and the Shoreline Management Approach (Fig. 17.9). Given the large and rapidly growing concentration of people and activity in the coastal zone, autonomous (or spontaneous) adaptation processes alone will not be able to cope. Furthermore, adaptation in the coastal context is widely seen as a public rather than a private responsibility (Klein et al., 2000). Therefore, people generally expect all levels of government to play a role in developing and facilitating appropriate adaptation measures.

It is not simply a political or societal choice between these adaptation strategies. The shape and nature of the coastal zone dictate that certain interventions are simple and cheap, while other interventions are difficult and expensive, if not impossible. The choice of an appropriate adaptation strategy is closely linked to both the level of vulnerability and the land use (infrastructure, living, recreation, agriculture, etc.), and thus the social, economic and cultural value of the coast and the amount of available funding.

Of the numerous coastal archetypes around the globe, we select three important ones and link these to existing adaptation responses that lead to different adaptation strategies:

(1) Linear coastal reaches with a high potential for recreational use. The coast of Spain is a good example of this type of coast. Most coastal reaches are narrow with limited vulnerabilty to flooding, so erosion is the major concern. These stretches are managed through zonation, in which the use and management responsibilities are clearly indicated.

(2) Linear coastal reaches or coastal environments that have limited potential for multiple use. This coastal type is commonplace in the UK, where large parts of the coast are eroding soft cliffs with low population density, low tourism values and limited vulnerability to flooding (e.g. Holderness, Norfolk, southern Isle of Wight, Isle of Purbeck, Dorset). It is now recognized that managed retreat is the most economic option, and this also maximizes sediment supply to adjacent low-lying coasts where beaches contribute to flood defence (Dawson et al., 2009). However, legacy issues will often need to be addressed.

(3) Coastal environments, often containing a delta, an estuary or a tidal river, that have a high potential for multiple uses. The Thames, Elbe and the Netherlands, and the numerous deltas that contain many of the world's coastal megacities are examples of this type of coast. Deltas are particularly vulnerable to flooding, and flood management is crucial because of the high social and economic values present. A combination of adaptation strategies is necessary to reconcile the many conflicting interests along this coastal type.

The first two coastal archetypes are most prevalent, based on coastal length, but the last archetype, while less frequent based on coastal length, is most important as it supports the largest populations and the bulk of economic activity threatened by coastal hazards, and these areas are often of great environmental significance.

Through human history, improved technology has increased the range of adaptation options in the face of coastal hazards, and there has been a move from retreat and accommodation to hard protection and active seaward advance via land claim, as described for the Netherlands by van Koningsveld et al. (2008). Rising sea level is one factor calling automatic reliance on protection into question, and the appropriate mixture of protection, accommodation and retreat is now being seriously debated. In practice, many real-world responses may be hybrids, and combine elements of more than one approach. For example, flood protection needs to consider the residual risk that remains for all protected areas, and this suggests that it might be combined with flood forecast and warning systems, evacuation plans, and some form of insurance. Adaptation for one sector may also exacerbate impacts elsewhere: a good example is coastal squeeze of intertidal and shallow coastal habitats, where onshore migration due to rising sea levels is prevented by fixed hard protection. Note that soft protection, accommodation and retreat reduce this problem and allow coastal habitats to persist in place or migrate landward as sea levels rise. Hence, coastal management needs to consider the balance between protecting socio-economic activity/human safety on the one hand, and the habitats and ecological services of the coastal zone on the other hand. While the 20th century saw large losses of coastal habitats due to direct and indirect destruction, most coastal countries now aspire to protect these areas and their ecosystem services. Unfortunately, coastal change and especially sea-level rise threaten such initiatives.

The most appropriate timing for an adaptation response needs to be considered in terms of anticipatory versus reactive planned adaptation (or in practical terms, what should we do today, versus wait and see until tomorrow?) (Fig. 17.10). **Autonomous adaptation** represents the spontaneous adaptive response (e.g. increased vertical accretion of coastal wetlands given sea-level rise within the natural system, or market-price adjustments as our understanding of coastal risk improves within the socio-economic system). **Planned adaptation** (by the socio-economic system) can serve to reduce vulnerability by a range of (anticipatory or reactive) measures. Adaptation normally reduces the magnitude of the impacts that would occur in their absence. Hence, impact assessments that do not take autonomous adaptation and/or (proactive or reactive) planned adaptation into account will generally estimate larger impacts (determining worst-case or potential impacts, rather than actual or residual impacts) (Fig. 17.10). The coastal zone is an area where there is significant scope for anticipatory adaptation, as many decisions have long-term implications (e.g. Hallegatte, 2009). Examples of anticipatory adaptation in coastal zones for sea-level rise

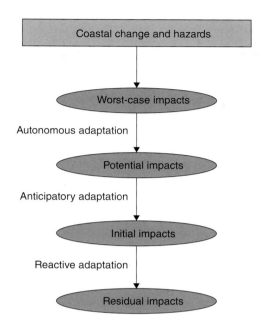

Fig. 17.10　The different stages of adaptation and their influence on impacts. (Source: Adapted from IPCC 1990. Reproduced with permission of IPCC.)

include upgraded flood defences and drainage systems, higher-elevation designs for new land claim and coastal bridges, building standards/regulations to promote flood proofing and resilience, and building setbacks to prevent development in areas threatened by erosion and flooding.

In general, new issues such as sea-level rise will exacerbate existing pressures and problems, so there are important synergies in considering adaptation to climate change in the context of existing problems. In some cases, the focus of future conditions may help identify 'win-win' situations, where such adaptation measures are worthy of implementation just based on solving today's problems (e.g. Dawson et al., 2009; Hallegatte, 2009). Adaptation measures are more likely to be implemented if they offer immediate benefits in terms of reducing impacts of current hazards as well as addressing long-term change.

There is considerable experience of adapting to coastal change, including climate variability, and we can draw on this experience to inform decision-making. Importantly, adaptation to coastal problems is a multi-stage process, as shown in Fig. 17.11. Each of these stages operates within multiple policy cycles (e.g. Klein et al., 2000; Hay, 2009). The constraints on approaches to adaptation due to broader policy and development goals should also be carefully considered. The growth of modelling and information technology is increasing the capability of all these stages: as models become more sophisticated, so a range of virtual worlds can be explored to better understand the choices that society faces (e.g. Fig. 17.12 and Fig. 17.13). However, the

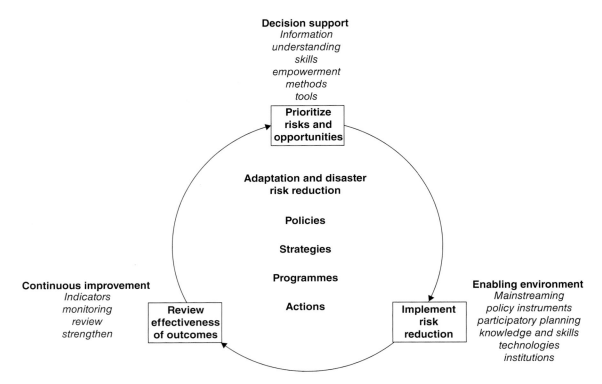

Fig. 17.11 The adaptation process. (Source: Adapted from Hay 2009. Reproduced with permission of United Nations International System for Disaster Reduction.)

reactive nature of most adaptation to date should be noted, and changing to a more proactive approach as advocated here represents a significant paradigm shift in management. Monitoring and evaluation are critical and yet easily ignored, and are essential to a 'learning by doing approach', which is appropriate to the large uncertainties associated with adaptation and coastal management. The design and construction of the Delta Works in the Netherlands following the 1953 storm surge disaster recognized and followed such an approach. Shortening of the southern Netherlands coastline by closing several estuarine branches was undertaken by first closing the minor branches, building experience, and taking on the larger closures subsequently. Supporting such an approach, technological developments are also greatly improving our ability to monitor, store and distribute relevant data, but these efforts need to be fostered and disseminated more widely. In many countries, there is limited capacity to address today's coastal problems, let alone consider tomorrow's problems. Therefore, promoting coastal adaptation should include developing coastal management capacity, as has already been widely recommended (e.g. Adger et al., 2007; USAID, 2009).

In terms of selecting adaptation options, there are widespread views that retreat will be the dominant response to sea-level rise, and, by implication, to other adverse coastal change. This reflects a variety of factors, including: (1) a

limited and poor understanding of the possible protection options; (2) a perception that these protection options are unaffordable anyway; and (3) a perception of the sea as a relentless and unforgiving enemy. However, many protection options are available, and our skill at implementing them is growing. Furthermore, benefit-cost models that compare protection with retreat generally suggest that it is worth investing in widespread protection, as populated coastal areas have high economic value compared to the cost of protection (see Box 17.4). These results suggest that significant resources are available for adapting to coastal change, and further that protection can be expected to be a significant component of the portfolio of responses. With or without protection, small-island and deltaic areas stand out as relatively more vulnerable in most of these analyses, and the impacts fall disproportionately on poorer countries. Even though optimal in a benefit-cost sense, protection costs could also overwhelm the capacity of local economies, especially when they are small, such as islands (Fankhauser and Tol, 2005; Nicholls and Tol, 2006). While adaptation is essentially a local activity, these funding challenges should be an issue of international concern.

There have been some global cost estimates for adaptation to sea-level rise. These normally focus on the costs of upgrading existing defence infrastructure, as this is consistent with the United Nations Framework Convention on Climate

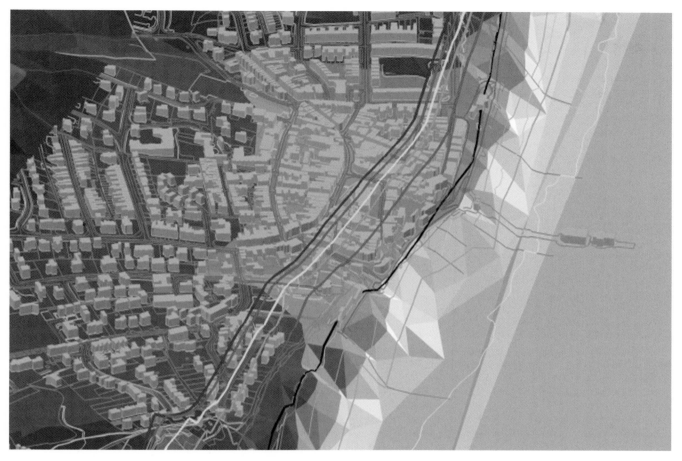

Fig. 17.12 Improving information for coastal planning and adaptation: a visualization of simulated cliff retreat if coastal defences were removed at Cromer, Norfolk, UK. The lines show change every 10 years. Note that the recession decreases. (Source: Dawson et al. 2009. Reproduced with permission of Springer-Verlag Dordrecht.) For colour details, please see Plate 40.

Change. (In other words, they are the additional costs of responding to sea-level rise rather than the cost of building defences from scratch.) IPCC CZMS (1990) estimated the total costs of defending against a 1-m sea-level rise at $US500 billion. Hoozemans et al. (1993) doubled these costs to $US1000 billion. Tol (2002a,b) estimated, using an economic model that balanced dryland and wetland loss, forced migration and the incremental defence investment, that the optimal annual protection costs for 1 m of global sea-level rise were only $US13 billion/yr, in line with the earlier Hoozemans et al. results. More recently, the World Bank (2010b) and Nicholls et al. (2010) have estimated coastal adaptation costs for the developing world alone at $US26–89 billion/yr by the 2040s, with the cost depending on the magnitude of sea-level rise. Maintenance costs of defences are also included, and were found to be significant as the stock of defences grows. However, none of these studies address the adaptation deficit, which is the gap between what adaptation measures exist and what would be optimal today. For

example, Hurricane Sandy illustrated an adaptation deficit for New York City, and significant adaptation deficits are apparent in other assessments (e.g. Hinkel et al., 2012). At the least, this increases the costs of adaptation in general, and protection in particular, by a substantial and presently unknown amount; it has the potential to radically change the adaptation pathway we select and is a priority for further assessment.

While there is widespread awareness of the need to adapt to coastal change, only a few countries or locations are comprehensively preparing for this challenge, with climate change generally being the trigger for action. Examples of locations that are preparing include London and the Netherlands (Box 17.5). Both these locations have considered all the relevant drivers of change, including a wide range of sea-level rise scenarios of up to 5 m and 4 m, respectively, implicitly thinking far into the future (well beyond 2100). Importantly, adaptation pathways are identified that are a logical sequence of

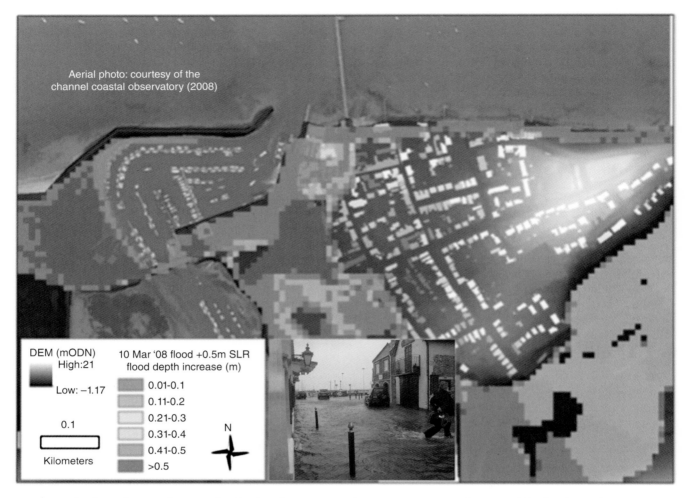

Aerial photo: courtesy of the
channel coastal observatory (2008)

DEM (mODN)
High:21

Low: −1.17

0.1

Kilometers

10 Mar '08 flood +0.5m SLR
flood depth increase (m)

0.01-0.1
0.11-0.2
0.21-0.3
0.31-0.4
0.41-0.5
>0.5

N

Fig. 17.13   Flood simulations of Yarmouth, Isle of Wight, UK, using the LISFLOOD-FP inundation model. This shows the increase in flood depth due to a 0.5-m rise in sea-level, assuming a repeat of the flood of 10 March 2008 (see inset photograph) (Source: Yarmouth Harbour Commissioners 2008), see Wadey et al. 2012. For colour details, please see Plate 41.

METHODS BOX 17.4   Choosing protection versus retreat for coastal change

In adaptation terms, the simplest choice is between protection and retreat; i.e. the investment costs of protection on the one hand, and the loss of the land and its assets, plus the displacement of its resident population, on the other hand. This choice is considered here. Three-way trade-offs between protection, accommodation and retreat are more complex, and for simplicity, only the two-way protection-retreat trade-off is considered here.

While coping with coastal change will be more challenging in a more crowded and warmer world, coastal protection is not new and it is likely to follow the same principles as earlier public investments in safety measures (e.g. Penning-Rowsell et al., 2013). There are two key decisions about coastal protection; namely, the

type of protection and the *level* of protection. Academics are fond of studying what decisions should be made, and benefit-cost analysis is a classic tool for supporting such decisions. The costs of coastal protection are weighted against the benefits of reduced erosion and improved flood safety. Discounting is important in this trade-off, as current expenditure is compared to future benefits. Discounting establishes an equivalence between the value of money now and money in the future. If the interest rate is 5%, then €105 next year is worth €100 today, because €100 put in the bank for a year would grow to €105. Uncertainty needs to be considered as well, particularly as some decisions and consequences are irreversible.

Standard benefit-cost analysis is done in monetary terms, and anything that is not priced is ignored. There are two ways to improve on this. Firstly, one can use monetary valuation methods to attach a price to goods and services that are not traded on markets (Champ et al., 2003). Such methods have been regularly applied, for instance to determine the worth of ecosystem services such as coastal wetlands, measured through the services provided for recreation, fisheries, coastal protection, water purification, and so on. Alternatively, one can employ multi-criteria analysis (Ishizaka and Nemery, 2013). Trade-offs are then made in multiple dimensions at once (rather than in a single, monetary dimension). Multi-criteria analysis reveals whether certain decisions dominate other choices for all aspects we would care about; and highlights how, say, lower costs come at the expense of higher risks.

Besides studying what should be done, one can also empirically research what decisions have been made about coastal protection and why. This reveals few, if any, examples where policy-makers took the recommendations of a benefit-cost analysis and implemented them with no other consideration. In reality, decisions about coastal protection are idiosyncratic, depending on the social construction of the problem (rather than its objective nature), and the particular balance of interests and powers of those that happen to make the decision. Although there are common empirical tendencies – richer societies have higher levels of protection, democratic societies find it difficult to implement disruptive infrastructure projects, etc. – there is a substantial amount of noise, with many seemingly random decisions (although in a democratic society, there will always be a supporting consultant's report).

Hence, benefit-cost analysis provides a good theoretical underpinning of coastal decision-making, which can be applied in modelled assessments (e.g. Tol, 2007). However, real-world decisions can be expected to deviate from this ideal.

CASE STUDY BOX 17.5   Present and future economic damage due to flooding: estimates for the Netherlands

The Central Agency for Statistics in the Netherlands estimates the national wealth as five times the national income, without taking account of ecological, landscape and cultural values (van Tongeren and van de Veen, 1997). Based on this definition, the national wealth was about €2750 billion in 2007. Since an estimated 65% of this wealth lies in flood-prone areas, the total wealth that is *potentially* under threat due to flooding is about €1800 billion. The actual, or real, economic damage due to flooding should (part of) a flood protection system fail, has been estimated at between €10 and €50 billion for each individual area protected by dykes; these areas are referred to as dyke rings. There are 53 dyke rings in the Netherlands, with protection levels varying between a 90% probability of withstanding an overtopping event that is predicted to occur every 10,000 or 1250 years (following national laws per dyke ring), implying that the flooding probability varies between 1 in 100,000 to 1 in 12,500 per year.

With the identification of failure mechanisms for dykes other than wave overtopping and/or overflow, notably piping (cf. New Orleans during Hurricane Katrina, 2005), there now exists a strong debate on how realistic these flooding probability estimates are (Cunningham et al., 2011). For this reason, the Second Delta Committee (Deltacommissie, 2008) suggests interpreting the design water-level probabilities as flooding probabilities for the time being. This interpretation is motivated by recent insights from flooding scenarios (Jonkman, 2007), indicating that it is most unlikely that the (major) dyked areas will be inundated completely. The location where, and the physical circumstances under which, a dyke is breached will have a marked effect on the resulting economic damage. Also, the damage caused by a flood depends on the size of the area inundated, the water depth in that area, and the duration of the episode.

Aerts et al. (2008) estimated the economic damage from flooding of all the dyke rings to be approximately €190 billion, including consideration of water depth per dyked area. This includes both direct and indirect damage (see Box 17.2). The estimated future potential damage would increase to €400–800 billion in 2040 and €3700 billion in 2100, in the absence of any additional protection measures and given a sea-level rise of 24–60 cm in 2040 and 150 cm in 2100. The factors governing calculations of estimated future potential damage are economic growth combined with indirect damage. Prior to Hurricane Katrina, potential damage in New Orleans was estimated at $US16.8 billion or €12.3 billion. However, after the disaster, it appeared that the direct damage to dwellings, government buildings and public infrastructure alone was US$27 billion or €19.7 billion (IPET, 2008), illustrating that it is essential to update economic growth and indirect damage figures regularly.

To manage future flood risk in the Netherlands, the Second Delta Committee (Deltacommissie, 2008) led to the establishment of the Delta Commission, which is now taking a strategic national perspective on flood risk management.

protection and related measures as a function of sea-level rise rather than time. This confirms that there are options available for large rises in sea level, and in these cases protection seems feasible for the long term, and adaptation is explicitly being considered as a process. This is an effective way to deal with the uncertainty of future sea-level rise and other dimensions of coastal change. In general, urban areas are expected to be a major focus for these adaptation efforts, given the concentration of people and assets, and their ability to fund large adaptation investments.

17.7   Concluding thoughts

This chapter illustrates that planning for coastal change is a multi-dimensional problem that crosses many disciplines. Coastal change has important implications, but the actual outcome will depend on our responses, both in terms of mitigation and adaptation, and their success or failure. For adaptation in general, and protection in particular, the likely success or failure is an important uncertainty, which deserves more attention as there are widely divergent views on this with regard to sea-level rise, and this influences how this issue is considered (e.g. Nicholls and Tol, 2006). 'Pessimists' tend to view our ability to adapt to sea-level rise as being limited, resulting in alarmingly high impacts, including widespread human displacement from coastal areas. In contrast, 'optimists' stress a high technical ability to protect, and the high benefit-cost ratios in developed areas leading to widespread protection. From this perspective, a major consequence of sea-level rise is the diversion of investment to coastal adaptation, including widespread protection, with indirect effects on economic growth, but with limited direct damage.

Optimists have empirical evidence to support their views that sea-level rise is not a big problem in terms of the subsiding megacities that have been protected and are also thriving (e.g. Bangkok). Importantly, these analyses suggest that improved protection under rising sea levels is more likely and rational than is widely assumed. This can be generalized to all coastal change, and the common assumption of a widespread retreat from the shore is not inevitable, and coastal societies will have more choice in their response to changing coasts. However, the pessimists also have evidence to support their view. Firstly, the published protection costs are incremental costs of adapting to sea-level rise, assuming the existence of well-adapted protection infrastructure. This is not the case in much of the world, and the adaptation deficit needs to be assessed and quantified. Secondly, assumptions of substantial future population and especially economic growth in coastal areas reinforce the conclusion that protection is worthwhile: lower growth and greater inequalities of wealth may mean less damage in monetary terms, but it will also lower the ability to protect. Thirdly, the benefit-cost approach implies a proactive attitude to protection, while historical experience shows most protection has been a reaction to actual or near-disaster. Therefore, high rates of sea-level rise (and possibly more intense storms and other rapid changes) may lead to more frequent coastal disasters and damage, even if the ultimate reactive response is upgraded protection. Fourthly, disasters (or adaptation failures) could trigger a loss of confidence, decline and abandonment of coastal areas. This could have a profound influence on society's future choices concerning coastal protection. A cycle of local economic decline is not inconceivable, potentially undermining the economics of adaptation. This is reinforced if capital is increasingly mobile and the will for collective action declines. Taking the issue of sea-level rise as an example, disinvestment from coastal areas may be triggered, even without disasters actually occurring: for example, the economies of small islands may be highly vulnerable if investors avoid these areas due to concern about future risks (Barnett and Adger, 2003). Lastly, the retreat and accommodation responses have long lead times, and benefits are greatest if planning and implementation occurs soon, which it may not. Hence, adaptation may not be as successful as some assume, especially for larger rises in sea level. There is much work remaining in order to understand these diverse issues, which will involve engineers, natural scientists and social scientists working together to understand the coastal system.

Coastal change is clearly a threat to the growing and urbanizing coastal populations, and as such demands a response. The commitment to sea-level rise means that an ongoing adaptation response is essential through the 21st century and beyond. However, mitigation can reduce the commitment to sea-level rise, specifically the large potential Greenland and West Antarctic ice sheet contributions. Scientists need to better understand these threats, including the implications of different mixtures of mitigation and adaptation, and different mixtures of adaptation, as well as the need to engage with the coastal and climate policy process, so that these scientific perspectives are heard. While research must continue at all scales, from local to global, much will be learnt about adaptation in practice, and this engagement is critical in order to promote more appropriate adaptation options, as well as the opportunity to learn from the experience of these projects.

17.8   Summary

Coastal environments are dynamic at many timescales, and direct and indirect human influences, such as climate change, are increasingly influencing these dynamics, as discussed in earlier chapters of this book. At the same time, coasts contain a large and growing population and economy, including metropolises such as London, New York, Tokyo, Shanghai, Mumbai and Lagos. They also contain productive coastal habitats that are both declining and increasingly being valued for their ecosystem services. Short-term variability can produce a range of hazards, such as flooding from storms and tsunamis, while longer-term change often degrades resources and may exacerbate the damages during episodic events. If these tendencies continue, as seems almost inevitable, there will be significant impacts unless we can better cope with coastal change and variability. Fortunately, there is widespread experience of a range of strategies for responding to these challenges, and the sophistication of these options continues to grow. Appropriate responses include mitigation (or source control) and/or adaptation (responding to the change). The adaptive responses can be characterized as: (1) protect; (2) accommodate; or (3) retreat approaches. Improving information measures, such as disaster preparedness, hazard mapping and flood warning/evacuation, are also important, and in many ways are complementary to the three adaptive approaches.

- Coping with coastal change involves analysis of *all* the drivers of change, as they operate across a range of scales, from individual storms to long-term issues such as sea-level rise and climate change.
- Developing such an understanding involves an integrated assessment (or systems analysis) approach, which considers the complete coastal system, including the natural and human components, as well as their interaction.
- All coping responses need to be consistent with wider societal and development objectives, and hence require implementation within an integrated coastal management philosophy.
- Adaptation is expected to be widespread, with coastal urban areas being a major focus for these efforts with their concentrations of people and assets; proactive adaptation plans are already being formulated for urban areas such as London, the Netherlands and Ho Chi Min City.
- Some of the major adaptation challenges will be in many developing countries, reflecting a large adaptation deficit; deltaic areas and small islands are the most vulnerable settings.

Key publications

Aerts, J., Botzen, W., Bowman, M.J., Ward, P.J. and Dircke, P. (Eds), 2012. *Climate Adaptation and Flood Risk in Coastal Cities.* Earthscan, London, UK.
[*A comprehensive consideration of adaptation in coastal cities to flooding, including case studies*]

French, P.W., 2002. *Coastal Processes and Protection: Processes*, Problems and Solutions. Routledge, London, UK.
[*An integrated perspective concerning coastal geomorphic problems and management solutions*]

Kamphuis, J.W., 2010. *Introduction to Coastal Engineering and Management*(2nd edn). World Scientific, Singapore.
[*A broad and comprehensive discussion of the principle and application of coastal engineering and management*]

Kay, R. and Alder, J., 2005. *Coastal Planning and Management* (2nd edn). CRC Press, Boca Raton, FL, USA.
[*A comprehensive discussion of coastal planning and management, including implementation*]

McFadden, L., Nicholls, R.J. and Penning-Rowsell, E. (Eds), 2007. *Managing Coastal Vulnerability*. Elsevier, Oxford, UK.
[*Multi-disciplinary perspectives and synthesis of managing coastal vulnerability*]

Moser, S.C., Jeffress Williams, S. and Boesch, D.F., 2012. Wicked challenges at Land's End: managing coastal vulnerability under climate change. *Annual Review of Environment and Resources*, 37, 51–78.
[*A comprehensive and up-to-date review concerning managing coastal vulnerability to climate change*]

Nicholls, R.J., Townend, I.H., Bradbury, A.P., Ramsbottom, D. and Day, S.A., 2013. Planning for long-term coastal change: experiences from England and Wales. *Ocean Engineering*, 71, 3–16, doi: http://dx.doi.org/10.1016/j.oceaneng.2013.01.025.
[*Considers the history of shoreline management planning in England and Wales and its fitness for purpose under climate change*]

References

Adger, W.N., Agrawala, S., Mirza, M.M.Q. et al., 2007. Assessment of adaptation practices, options, constraints and capacity. In: M.L. Parry, O.F. Canziani, J.P. Palutikof, P.J. van der Linden and C.E. Hanson (Eds), *Climate Change 2007: Impacts, Adaptation and Vulnerability*. Contribution of Working Group II to the Fourth Assessment Report of the Intergovernmental Panel on Climate Change. Cambridge University Press, Cambridge, UK, pp. 717–743.

Aerts, J., Sprong, T. and Bannink, B. (Eds.), 2008. *Aandacht voor veiligheid*. BSIK Klimaat voor Ruimte, DG Water, The Hague, the Netherlands, http://www.klimaatvoorruimte.nl and http://www.deltacommissie.com/doc/Aandacht%20voor%20veiligheid%20.pdf.

Barnett, J. and Adger, W., 2003. Climate dangers and atoll countries. *Climatic Change*, 61, 321–337, doi: 10.1023/b:clim.0000004559.08755.88.

Barras, J., Beville, S., Britsch, D. et al., 2003. *Historical and Projected Coastal Louisiana Land Changes: 1978–2050*. Open-File Report 03–334. US Geological Survey, Reston, VA, USA.

Beets, D.J., Vandervalk, L. and Stive, M.J.F., 1992. Holocene evolution of the coast of Holland. *Marine Geology*, 103, 423–443, doi: 10.1016/0025-3227(92)90030-1 .

Champ, P.A., Boyle, K.J. and Brown, T.C., 2003. *A Primer on Non-Market Valuation*. Kluwer, Dordrecht, the Netherlands.

Church, J., Woodworth, P.L., Aarup, T. and Stanley, W.S. (Eds), 2010. *Understanding Sea-Level Rise and Variability*. Wiley-Blackwell, Chichester, UK.

Cowell, P.J., Roy, P.S. and Jones, R., 1995. Simulation of large-scale coastal change using a morphological behavior model. *Marine Geology*, **126**, 45–61.

Cowell, P.J., Stive, M.J.F., Niedoroda, A.W. et al., 2003a. The coastal-tract (part a): a conceptual approach to aggregated modeling of low-order coastal change. *Journal of Coastal Research*, **19**, 812–827.

Cowell, P.J., Stive, M.J.F., Niedoroda, A.W. et al., 2003b. The coastal-tract (part 2): applications of aggregated modeling of lower-order coastal change. *Journal of Coastal Research*, **19**, 828–848.

Crowell, M., Leatherman, S.P. and Buckley, M.K., 1991. Historical shoreline change – error analysis and mapping accuracy. *Journal of Coastal Research*, **7**, 839–852.

Cunningham, A.C., Bakker, M.A.J., van Heteren, S. et al., 2011. Extracting storm-surge data from coastal dunes for improved assessment of flood risk. *Geology*, **39**, 1063–1066, doi: 10.1130/g32244.1.

Dawson, R.J., Dickson, M.E., Nicholls, R.J. et al., 2009. Integrated analysis of risks of coastal flooding and cliff erosion under scenarios of long term change. *Climatic Change*, **95**, 249–288, doi: 10.1007/s10584-008-9532-8.

de Bruin, K.C., Dellink, R.B. and Tol, R S.J., 2009. AD-DICE: an implementation of adaptation in the DICE model. *Climatic Change*, **95**, 63–81, doi: 10.1007/s10584-008-9535-5.

Deltacommissie, 2008. *Working Together with Water: A Living Land Builds for its Future*. Findings of the Deltacommissie 2008. Deltacommissie, The Hague, the Netherlands, http://www.deltacommissie.com/doc/deltareport_full.pdf.

Ericson, J.P., Vorosmarty, C.J., Dingman, S.L., Ward, L.G. and Meybeck, M., 2006. Effective sea-level rise and deltas: causes of change and human dimension implications. *Global and Planetary Change*, **50**, 63–82, doi: 10.1016/j.gloplacha.2005.07.004.

Fankhauser, S. and Tol, R.S.J., 2005. On climate change and economic growth. *Resource and Energy Economics*, **27**, 1–17, doi: 10.1016/j.reseneeco.2004.03.003.

Fletcher, C.A. and Spencer, T., 2005. *Flooding and Environmental Challenges for Venice and its Lagoon: State of Knowledge*. Cambridge University Press, Cambridge, UK.

Fritz, H.M., Blount, C.D., Thwin, S., Thu, M.K. and Chan, N., 2009. Cyclone Nargis storm surge in Myanmar. *Nature Geoscience*, **2**, 448–449, doi: 10.1038/ngeo558.

Goodwin, I.D., Stables, M.A. and Olley, J.M., 2006. Wave climate, sand budget and shoreline alignment evolution of the Iluka-Woody Bay sand barrier, northern New South Wales, Australia, since 3000 yr BP. *Marine Geology*, **226**, 127–144 , doi: 10.1016/j.margeo.2005.09.013.

Hallegatte, S., 2009. Strategies to adapt to an uncertain climate change. *Global Environmental Change: Human and Policy Dimensions*, **19**, 240–247, doi: 10.1016/j.gloenvcha.2008.12.003.

Hanson, S., Nicholls, R., Ranger, N. et al., 2011. A global ranking of port cities with high exposure to climate extremes. *Climatic Change*, **104**, 89–111, doi: 10.1007/s10584-010-9977-4.

Hay, J.E., 2009. *Institutional and Policy Analysis of Disaster Risk Reduction and Climate Change Adaptation in Pacific Island Countries*. United Nations International System for Disaster Reduction (UNISDR) and the United Nations Development Programme (UNDP), Suva, Fiji.

Hinkel, J., Brown, S., Exner, L., Nicholls, R.J., Vafeidis, A.T. and Kebede, A.S., 2012. Sea-level rise impacts on Africa and the effects of mitigation and adaptation: an application of DIVA. *Regional Environmental Change*, **12**, 207–224, doi: 10.1007/s10113-011-0249-2.

Hoozemans, F.M.J., Marchand, M. and Pennekamp, H.A., 1993. *Sea Level Rise. A Global Vulnerability Analysis. Vulnerability Assessments for Population, Coastal Wetlands and Rice Production on a Global Scale*. Delft Hydraulics and Rijkswaterstaat, Delft and The Hague, the Netherlands.

IPCC CZMS, 1990. *Sea-Level Rise: A World-Wide Cost Estimate of Coastal Defence Measures*. Appendix D in Report of the Coastal Zone Management Subgroup, Response Strategies Working Group of the Intergovernmental Panel on Climate Change. Ministry of Transport, Public Works and Water Management, The Hague, the Netherlands.

IPET, 2008. *Performance Evaluation of the New Orleans and Southeast Louisiana Hurricane Protection System. Vol. 1: Executive Summary and Overview*. Interagency Performance Evaluation Taskforce (IPET), Army Corps of Engineers, Maryland/West Virginia, USA.

Ishizaka, A. and Nemery, P., 2013. *Multi-Criteria Decision Analysis*. Wiley-Blackwell, Oxford, UK.

Jonkman, S.N., 2007. Loss of Life Estimation in Flood Risk Assessment – Theory and Applications. PhD Thesis. Civil Engineering and Geosciences, Technical University of Delft, the Netherlands, http://repository.tudelft.nl/view/ir/uuid%3Abc4fb945-55ef-4079-a606-ac4fa8009426/.

Kates, R.W., Colten, C.E., Laska, S. and Leatherman, S.P., 2006. Reconstruction of New Orleans after Hurricane Katrina: a research perspective. *Proceedings of the National Academy of Sciences of the United States of America*, **103**, 14653–14660, doi: 10.1073/pnas.0605726103.

Katsman, C.A., Sterl, A., Beersma, J.J. et al., 2011. Exploring high-end scenarios for local sea level rise to develop flood protection strategies for a low-lying delta – the Netherlands as an example. *Climatic Change*, **109**, 617–645, doi: 10.1007/s10584-011-0037-5.

Klein, R.J.T. and Nicholls, R.J., 1999. Assessment of coastal vulnerability to climate change. *Ambio*, **28**, 182–187.

Klein, R.J.T., Aston, J., Buckley, E.N. et al., 2000. Coastal adaptation technologies. In: B. Metz, O. Davidson, J. Martens, S. van Rooijen and L. van Wie Mcgrory (Eds), *Methodological and Technological Issues in Technology Transfer*. A Special Report of the Intergovernmental Panel on Climate Change. Cambridge University Press, Cambridge, UK, pp. 349–372.

Kolen, B., Slomp, R. and Jonkman, S.N., 2012. The impacts of storm Xynthia February 27/28, 2010 in France: lessons for flood risk management. *Journal of Flood Risk Management*, **6**(3), 261–278, doi: 10.1007/s10113-011-0249-2.

Kron, W., 2013. Coasts: the high-risk areas of the world. *Natural Hazards*, **66**, 1363–1382, doi: 10.1007/s11069-012-0215-4.

McGranahan, G., Balk, D. and Anderson, B., 2007. The rising tide: assessing the risks of climate change and human settlements in low elevation coastal zones. *Environment and Urbanization*, **19**, 17–37, doi: 10.1177/0956247807076960.

Meehl, G.A., Stocker, T.F., Collins, W.D. et al., 2007. *Global Climate Projections. Climate Change 2007: The Physical*

Science Basis. Contribution of Working Group I to the Fourth Assessment Report of the Intergovernmental Panel on Climate Change. Cambridge University Press, Cambridge, UK.

Meehl, G.A., Hu, A.X., Tebaldi, C. et al., 2012. Relative outcomes of climate change mitigation related to global temperature versus sea-level rise. *Nature Climate Change*, **2**, 576–580, doi: 10.1038/nclimate1529.

Mimura, N., Nurse, L., McLean, R.F. et al., 2007. Small islands. In: M.L. Parry, O.F. Canziani, J.P. Palutikof, P.J. van der Linden and C.E. Hanson (Eds), *Climate Change 2007: Impacts, Adaptation and Vulnerability*. Contribution of Working Group II to the Fourth Assessment Report of the Intergovernmental Panel on Climate Change. Cambridge University Press, Cambridge, UK, pp. 687–716.

Nicholls, R.J., 2010. Impacts of and responses to sea-level rise. In: J. Church, P.L. Woodworth, T. Aarup and S. Wilson (Eds.), *Understanding Sea-Level Rise and Variability*. Wiley-Blackwell, Chichester, UK, pp. 17–51.

Nicholls, R.J. and de la Vega-Leinert, A.C., 2008. Implications of sea-level rise for Europe's coasts: an introduction. *Journal of Coastal Research*, **242**, 285–287, doi: 10.2112/07A-0002.1.

Nicholls, R.J. and Kebede, A.S., 2012. Indirect impacts of coastal climate change and sea-level rise: the UK example. *Climate Policy*, **12**, S28–S52, doi: 10.1080/14693062.2012.728792.

Nicholls, R.J. and Tol, R.S.J., 2006. Impacts and responses to sea-level rise: a global analysis of the SRES scenarios over the twenty-first century. *Philosophical Transactions of the Royal Society A*, **364**, 1073–1095, doi: 10.1098/rsta-2006.1754.

Nicholls, R.J., Wong, P.P., Burkett, V.R. et al., 2007. Coastal systems and low-lying areas. In: M.L. Parry, O.F. Canziani, J.P. Palutikof, P.J. van der Linden and C.E. Hanson (Eds), *Climate Change 2007: Impacts, Adaptation and Vulnerability*. Contribution of Working Group II to the Fourth Assessment Report of the Intergovernmental Panel on Climate Change. Cambridge University Press, Cambridge, UK, 315–356.

Nicholls, R.J., Brown, S., Hanson, S. and Hinkel, J., 2010. *Economics of Coastal Zone Adaptation to Climate Change*. The International Bank for Reconstruction and Development/The World Bank Washington, DC, USA.

Nicholls, R.J., Marinova, N., Lowe, J.A. et al., 2011. Sea-level rise and its possible impacts given a 'beyond 4 degrees C world' in the twenty-first century. *Philosophical Transactions of the Royal Society A*, **369**, 161–181, doi: 10.1098/rsta.2010.0291.

Nicholls, R.J., Townend, I.H., Bradbury, A.P., Ramsbottom, D. and Day, S.A., 2013. Planning for long-term coastal change: experiences from England and Wales. *Ocean Engineering*, **71**, 3–16, doi: 10.1016/j.oceaneng.2013.01.025.

Penning-Rowsell, E.C., Priest, S., Parker, D. et al., 2013. *Flood and Coastal Erosion Risk Management: A Manual for Economic Appraisal*. Routledge, London, UK.

Phien-Wej, N., Giao, P.H. and Nutalaya, P., 2006. Land subsidence in Bangkok, Thailand. *Engineering Geology*, **82**, 187–201.

Stive, M.J.F., Nicholls, R.J. and De Vriend, H.J., 1991. Sea-level rise and shore nourishment – a discussion. *Coastal Engineering*, **16**, 147–163, doi: 10.1016/0378-3839(91)90057-n.

Stive, M.J.F., Cowell, P. and Nicholls, R.J., 2009. Beaches, cliffs and deltas. In: O. Slaymaker, T. Spencer and C. Embleton-Hamann (Eds), *Geomorphology and Global Environmental Change*. Cambridge University Press, Cambridge, UK, pp. 158–179.

Sutherland, J., 2012. Error analysis of Ordnance Survey map tidelines, UK. *Proceedings of the Institution of Civil Engineers: Maritime Engineering*, **165**, 189–197, doi: 10.1680/maen.2011.10.

Syvitski, J.P.M., Kettner, A.J., Overeem, I. et al., 2009. Sinking deltas due to human activities. *Nature Geoscience*, **2**, 681–686, doi: 10.1038/ngeo629.

Tacoli, C., 2009. Crisis or adaptation? Migration and climate change in a context of high mobility. *Environment and Urbanization*, **21**, 513–525, doi: 10.1177/0956247809342182.

Thorne, J.A. and Swift, D.J.P., 1991. Sedimentation on continental margins. II: Application of the regime concept. In: D.J.P. Swift, G.F. Oertel, R.W. Tillman and J.A. Thorne (Eds), *Shelf Sand and Sandstone Bodies – Geometry, Facies and Sequence Stratigraphy*. International Association of Sedimentologists, Special Publication 14. Blackwell Scientific Publications, Oxford, UK.

Thom, B.G. and Cowell, P.J., 2005. Coastal changes. In: M.L. Schwartz (Ed.), *Encyclopedia of Coastal Science*. Springer, Dordrecht, the Netherlands, pp. 251–253.

Tol, R.S.J., 2002a. Estimates of the damage costs of climate change. Part II: Dynamic estimates. *Environmental and Resource Economics*, **21**, 135–160, doi: 10.1023/a:1014539414591.

Tol, R.S.J., 2002b. Estimates of the damage costs of climate change. Part I: Benchmark estimates. *Environmental and Resource Economics*, **21**, 47–73, doi: 10.1023/a:1014500930521.

Tol, R.S.J., 2007. The double trade-off between adaptation and mitigation for sea level rise: an application of FUND. *Mitigation and Adaptation Strategies for Global Change*, **12**, 741–753.

USAID, 2009. *Adapting to Coastal Climate Change: A Guidebook for Development Planners*. US Agency for International Development, Washington, DC, USA.

Valentin, H., 1952. Die Küsten der Erde. *Petarmanns Geographisches Mitteilungen Ergänzsungheft*, **246**, 118.

van Koningsveld, M., Mulder, J.P.M., Stive, M.J.F., VanDervalk, L. and VanDerWeck, A.W., 2008. Living with sea-level rise and climate change: a case study of the Netherlands. *Journal of Coastal Research*, **24**, 367– 379, doi: 10.2112/07a-0010.1.

van Straaten, L.M.J.U., 1961. Directional effects of winds, waves and currents along the Dutch North Sea coast. *Geologie en Mijnbouw*, **40**, 363–391.

van Tongeren, D. and van de Veen, P., 1997. *De Nationale Balans en de Overheidsbalans*. [In Dutch] (The National Balance and the Government Balance). M&O.006. Centraal Bureau voor de Statistiek, Voorburg/Heerlen, the Netherlands, http://www.cbs.nl/NR/rdonlyres/F7AEA517-5B8D-4EA5-8946-C48426F3F9E7/0/mo006.pdf.

Wadey, M., Nicholls, R.J. and Hutton, C., 2012. Coastal flooding in the Solent: an integrated analysis of defences and inundation. *Water*, **4**, 430–459, doi: 10.3390/w4020430.

World Bank, 2010a. *Climate Risks and Adaptation in Asian Coastal Megacities. A Synthesis Report*. World Bank, Washington, DC, USA, http://siteresources.worldbank.org/EASTASIAPACIFICEXT/Resources/226300-1287600424406/coastal_megacities_fullreport.pdf.

World Bank, 2010b. *Economics of Adaptation to Climate Change – Synthesis Report*. World Bank, Washington, DC, USA, http://documents.worldbank.org/curated/en/2010/01/16436675/economics-adaptation-climate-change-synthesis-report.

Geographical Index

Subject Index

Coastal Environments and Global Change, First Edition. Edited by Gerd Masselink and Roland Gehrels.
© 2014 John Wiley & Sons, Ltd. Published 2014 by John Wiley & Sons, Ltd. Companion Website: www.wiley.com/go/masselink/coastal